The Best of

Ciarcia's Circuit Cellar

by Steve Ciarcia

McGRAW-HILL, INC.

New York St. Louis San Francisco Auckland Bogotá
Caracas Lisbon London Madrid Mexico Milan
Montreal New Delhi Paris San Juan São Paulo
Singapore Sydney Tokyo Toronto

This book is dedicated to my wife, Jeannette. After learning to reprogram our Home Control System by herself, there was no other fitting reward. Of course, sitting in the dark may have been incentive enough.

Library of Congress Cataloging-in-Publication Data

Ciarcia, Steve
[Ciarcia's Circuit cellar. Selections]
The best of Ciarcia's Circuit cellar / by Steve Ciarcia.
p. cm.
Selected articles which have been reprinted from v. 1-7 of Ciarcia's Circuit cellar.
Includes index.
ISBN 0-07-011019-0 (hard): —ISBN 0-07-011025-5 (pbk.)
1. Microcomputers. I. Title. II. Title: Ciarcia's Circuit cellar.
TK7888.3.C5825 1992
621.39 '16—dc20 91-41843
CIP

1 2 3 4 5 6 7 8 9 0 MAL/MAL 9 7 6 5 4 3 2 1

ISBN 0-07-011019-0 [HC]
ISBN 0-07-011025-5 [PBK]

The sponsoring editor for this book was Daniel Gonneau, the editing supervisor was Jane Palmieri, and the production supervisor was Pamela Pelton.

Printed and bound by Malloy Lithographers, Inc.

The articles included in *The Best of Ciarcia's Circuit Cellar* have been reprinted from the following publications:

Articles 1-5 from *Ciarcia's Circuit Cellar*, Volume 1
Articles 6-7 from *Ciarcia's Circuit Cellar*, Volume 2
Articles 8-13 from *Ciarcia's Circuit Cellar*, Volume 3
Articles 14-17 from *Ciarcia's Circuit Cellar*, Volume 4
Articles 18-22 from *Ciarcia's Circuit Cellar*, Volume 5
Articles 23-28 from *Ciarcia's Circuit Cellar*, Volume 6
Articles 29-37 from *Ciarcia's Circuit Cellar*, Volume 7

The author of the programs and hardware schematics provided with this book has carefully reviewed them to ensure their performance in accordance with the specifications described in the book. Neither the author nor BYTE Publications Inc., however, make any warranties whatever concerning the programs or hardware schematics. They assume no responsibility or liability of any kind for errors in the programs or hardware schematics or for the consequences of any such errors. The programs and hardware schematics are the sole property of the author and have been registered with the United States Copyright Office.

Contents

Preface vii

1 No Power for Your Interfaces? (October 1978) 1

2 Try an 8 Channel DVM Cocktail (December 1977) 7

3 Add More Zing to the Cocktail (January 1978) 18

4 Let Your Fingers Do the Talking
Part 1: Add a Noncontact Touch Scanner to Your Video Display (August 1978) 28
Part 2: Scanner Applications (September 1978) 37

5 I've Got You in My Scanner (November 1978) 41

6 Communicate on a Light Beam (May 1979) 49

7 Joystick Interfaces (September 1979) 60

8 Handheld Remote Control for Your Computerized Home (July 1980) 66

9 Home In on the Range!: An Ultrasonic Ranging System (November 1980) 78

10 Build a Low-Cost Logic Analyzer (April 1981) 90

11 Build a Z8-Based Control Computer with BASIC
Part 1 (July 1981) 100
Part 2 (August 1981) 110

12 Build an Unlimited-Vocabulary Speech Synthesizer (September 1981) 120

13 Switching Power Supplies: An Introduction (November 1981) 132

14 Build a Computerized Weather Station (February 1982) 142

15 Use Voiceprints to Analyze Speech (March 1982) 156

16 Build an RS-232C Breakout Box (April 1983) 166

17 Build an RS-232C Code-Activated Switch (May 1983) 176

18 Build the RTC-4 Real-Time Controller (July 1983) 184

19 Build a Power-Line Carrier-Current Modem (August 1983) 195

20 Keep Power-Line Pollution Out of Your Computer (December 1983) 202

21 A Musical Telephone Bell (July 1984) 211

22 An Ultrasonic Ranging System (October 1984) 218

23 Understanding Linear Power Supplies (January 1985) 229

24 Build the BASIC-52 Computer/Controller (August 1985) 239

25 Build the SB180 Single-Board Computer
Part 1: The Hardware (September 1985) 252
Part 2: The Software (October 1985) 268

26 Build an Analog-to-Digital Converter (January 1986) 280

27 Real-Time Clocks: A View toward the Future (March 1986) 292

28 Computer on Guard! (April 1986) 305

29 Build a Gray-Scale Video Digitizer
Part 1: Display/Receiver (May 1987) 315
Part 2: Digitizer/Transmitter (June 1987) 327

30 Using the ImageWise Video Digitizer: Image Processing (July 1987) 337

31 Build an Infrared Remote Controller (February 1987) 345

32 Build a Trainable Infrared Master Controller (March 1987) 355

33 Computers on the Brain
Part 1 (June 1988) 367
Part 2 (July 1988) 376

34 Build the Circuit Cellar IC Tester
Part 1: Hardware (November 1987) 383
Part 2: Software and Operation (December 1987) 391

35 Why Microcontrollers?
Part 1 (August 1988) 397
Part 2 (September 1988) 405

36 The SmartSpooler
Part 1: The Spooler Hardware (April 1988) 415
Part 2: Software and Operation (May 1988) 423

37 Build an Intelligent Serial EPROM Programmer (October 1986) 429

Products Available 447

Index 449

Preface

Whenever I run into fans at conventions or computer club meetings, invariably they ask how I started writing the projects contained in this book. If the publishing world was apolitical someone might have published an unbiased account by now, but because I was not an independent author, no other magazine could interview me and relate those events without appearing to promote BYTE at the same time. I guess it is left to me to answer the question.

Did you know that my first BYTE project was published in a competing publication? What is even more unbelievable is that this particular project resulted from my pursuit of a computer game. Anyone who knows me now would have a hard time believing that I don't play computer games or own a Nintendo, and I have never spent a quarter on a video game.

When I was a high school student (mid-sixties), I made a college prospecting trip to MIT and visited one of their computer labs. One of the more intriguing "toys" was a computer graphics "space-war" game. I can't say whether there was a whole building of mainframe computers or what running the game, but I was instantly hooked on the technology. The player sat in front of a 21-inch black-and-white screen that depicted a certain sector of the known universe. The player navigated his way across the galaxy using joystick-controlled thrusters, phasers, and a lot of luck. Of course, he had to compensate for gravitational effects, thrust vectors, asteroid encounters, alien intruders, and what not. Needless to say, the thrill of dealing with a dynamically changing artificial intelligence, albeit a game, was intriguing. I had to have one of these.

Even though I later attended an engineering school that had the latest mainframes, they neglected to dedicate one to a space-war game. Perhaps if they had, I wouldn't have visualized building the game a few years later when a little California company named Intel began making some interesting "microprocessor" chips.

While most people waited until the 8080 or 6502 to discover computing, I had a system running on the 8008 processor shortly after it appeared. I paid $120 for that 8008 chip and to this day I remember the sweaty feeling as I powered it up the first time. Would it be puff, pop, or hummm? Well, it worked.

I had an LED and switch entry interface panel, 4K bytes of RAM, 4K of EPROM, a tape-cassette interface, serial and parallel I/O ports, a CRT terminal based on Don Lancaster's TV Typewriter, and a 2K-byte hand-assembled operating system. (Comparable technology today would be an 80486 with 160-Mbyte hard disk and super VGA. The cost would be comparable too.) I even built the computer using 5400 series military-grade logic chips to ensure its reliability, and a power supply that could melt iron rebars.

The only elements missing for my game were a graphics display, a joystick, and a lot of programming. At the time, the typical graphics terminal cost at least $15,000. The controllers for them were the size of a refrigerator and the local power generating plant had to be notified before they were turned on.

Just as I was about to give up, I ran into an ad from a small company in New Hampshire that sold surplus equipment from Sanders Associates and DEC. While there were the assorted $30,000 bargain PDP-11s and $10,000 1-Mbyte hard drives, there was a closeout listing for a truckload of Sanders Associates 12-inch graphics displays with vector driver boards but no outboard controller units. The price was an unbelievable $250 each! After all, who could use it without a controller?

Within 2½ hours of reading the ad, I was on the dock of a dusty old warehouse in Nashua, New Hampshire, bargaining with the dealer for a Sanders unit. He seemed like a nice fellow who really wanted to unload this junk to as many happy hobbyists as there existed. He even threw in an extra set of driver boards to "help spread the word." Grinning profusely, I drove home with the terminal in the back seat of my Volvo.

Back then the difference between big computers and little computers was that big computers needed big interfaces and little computers often needed only little ones. Rather than a refrigerator-sized unit, I built an 8-bit, parallel-port driven vector generator in less than a ½ cubic foot. The Sanders display itself was essentially equivalent to an oscilloscope driven in the XY-mode. The controller I built was technically nothing more than two 8-bit D/A converters (one X and one Y) tuned for the deflection range of the Sanders unit, some added slope timing controls so you could see the screen trace between end points, and a blanking signal generator. Needless to say the display worked wonderfully. I could draw complete pictures on the screen using a simple technique of describing end points and drawing lines (called vectors) between them.

Around the same time I started reading a new magazine dedicated to computerists called BYTE and I began corresponding with the then editor in chief Carl Helmers as well. When I told him I had a vector graphics display suitable for microcomputers, he asked if I would write an article about it. While I had never written anything more than technical manuals, I agreed to try. The resulting article described the physical assembly of the vector graphics display and the program for generating a line drawing of the Starship Enterprise on the screen. A neat first try if I do say so myself.

Because I was a complete neophyte, trusting beyond belief back then, I also decided to do a favor for the one person who had helped solve my problem. I sent a copy of the unpublished manuscript to the surplus dealer and explained that his "product" would be featured in an upcoming issue of BYTE. Perhaps seeing how easily an interface could be built would help him sell his terminals.

About five weeks later, as I was browsing through a computer magazine considered to be a BYTE competitor, I came across the vector graphics article I had written. It had the surplus guy's name on it along with mine as coauthor! In addition, there was an ad for a complete graphics controller kit based on the article as well as the Sanders Terminals. Say what?

After a few frantic phone calls, I found out how politics between magazines is something to watch out for. This surplus guy turned out to be fast friends with a competing publisher, who had a personal feud going with BYTE. I suppose publishing my article first was a scoop intended to embarrass BYTE. When I confronted them with this discovery, they made profuse apologies and immediately sent half of the article fee (remember: coauthors). They regretted that they hadn't had time to notify me properly, but they had "held the presses" to publish "my" project and I should feel gratified to see it in print. As further pacification, the publisher guaranteed to publish the next article I wrote too. But what about BYTE?

I called Carl Helmers to apologize for being a jerk with this surplus guy. Through this and conversation with others, I learned this stuff was old hat and these two organizations had been playing "hide the manuscript" and "who's really publishing it anyway" for quite a while. Bad attitude and subterfuge were standard operating procedure. In fact, Carl was the one who apologized to me over the whole situation. He promised to publish the vector graphics article if I rewrote it.

Needless to say, I did the rewrite and the new article turned out to be a smashing hit among the readership. Of course, the other publication would never talk to me again because I had joined the enemy. Carl then asked me to write a series of three projects, which were equally well-received. Shortly thereafter we agreed upon a regular project presentation each month. Here we are, 100 or so projects later.

The projects I presented in BYTE have taken up seven volumes of *Ciarcia's Circuit Cellar*. When McGraw-Hill asked me to pick the "best" projects for *The*

Best of Ciarcia's Circuit Cellar, I had a real dilemma. I could have picked 50 or 60 right off the top that I considered "best," but that surely wouldn't work. Instead, I chose the projects that had the most reader response, as measured by correspondence on the Circuit Cellar BBS, and those that enjoy continued technical support should questions still arise with new readers.

While technology has surely changed over the years and my 8008 has indeed been replaced with the latest processor, I'd like to say that Circuit Cellar projects are meant to be ageless. The majority of my designs are intended as presentations of real technology and real applications. They are not easily superseded just because higher speed processors or denser memory chips are now available.

Instead, Circuit Cellar projects are valuable because they work and are cost-effective. To this day I am told there isn't an electroencephalograph on the market for under $4000 that compares to the HAL-4 you can build here for under $200. Similarly, my presentation of an 8052-based embedded controller is attributed with spawning the entire market for such devices. I could go on about all these projects, but I think you'll recognize their merit as you build them yourselves.

The Best of Ciarcia's Circuit Cellar is based on the hundred or so projects that were originally presented in BYTE. Leaving BYTE has not stopped the support of these past projects or the introduction of new ones. We now have our own magazine, *The Computer Applications Journal*, that continues to present mine and other similar application-oriented projects. In addition, we operate a multiline BBS to support both readers of the magazine and Circuit Cellar projects builders. If you want to know more contact:

The Computer Application Journal
4 Park Street
Vernon, CT 06066

Magazine: (203) 875-2199
BBS: (203) 871-1988
Fax: (203) 872-2204

—*Steve Ciarcia*

1

No Power For Your Interfaces?

Build a 5 W DC to DC Converter

Recently I attended a local computer club meeting where we discussed the question of power supplies. Many people were remarking that, while they enjoyed building the projects in my articles, often their power supplies were not compatible with the multiple voltages I required. Many of the newer single board computers that some members owned contained only a hefty +5 V supply and a note that the user should add additional supplies if the basic board is expanded.

This is not an industry copout by any means. The newest digital designs from companies like Intel are made to run on +5 V and this is considered an advance in technology. The 8080A processor requires +12, +5 and -5 V for operation, while the new 8085 uses only a single +5 V supply. As long as all other components such as universal asynchronous receiver-transmitters (UARTs), programmable memories, erasable read only memories (EROMs) and read only memories (ROMs) in the computer are all +5 V, we can eliminate additional power supplies and save money. Computer manufacturers have done just that.

This situation does not cause any problems as long as the user stays with the basic unit, or expands it using single +5 V supply devices. Erasable read only memories such as the Intel 2716 and programmable peripheral interfaces such as the 8255 are designed specifically for this application.

The problem arises when the single supply computer tries to be communications compatible with the rest of the world, or when a bipolar analog interface is added. The RS-232C interface generally requires + and -12 V potentials, and digital to analog converters such as the Motorola 1408L8, which run on +5 and -12 to -15 V.

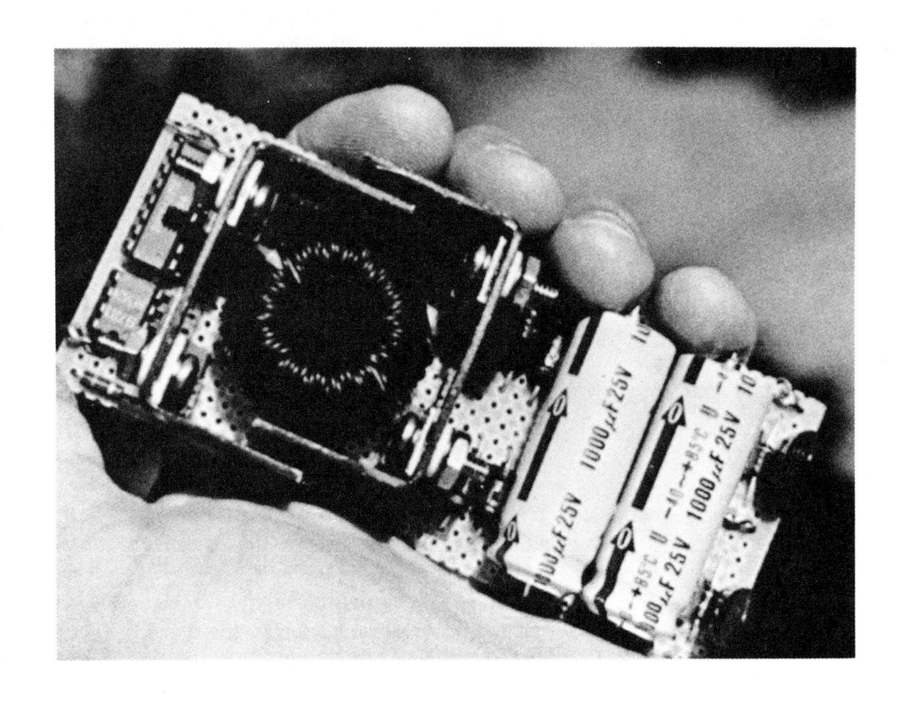

Photo 1: 5 W DC to DC converter, which produces 0.2 A at +12 and -12 VDC from a 5 VDC source. The circuit uses a special custom wound toroidal transformer (see figures 5a and 5b). Note: the prototype shown uses 1000 µF 25 V capacitors, which were later replaced with 100 µF 25 V versions.

The Whole World Isn't TTL Compatible

What is the experimenter to do when a -15 V supply is needed and the computer has only +5 V, or when one wishes to tie an RS-232 terminal into a system? Obviously the answer is to add an additional power supply or two—but, what kind?

Power supply requirements should be based on load requirements. If 0.5 A at +15 V is needed to power a particular interface, then perhaps a 1 A traditional transformer-rectifier-filter-regulator design is in order. More often than not, though, the interface might use one or two dual supply

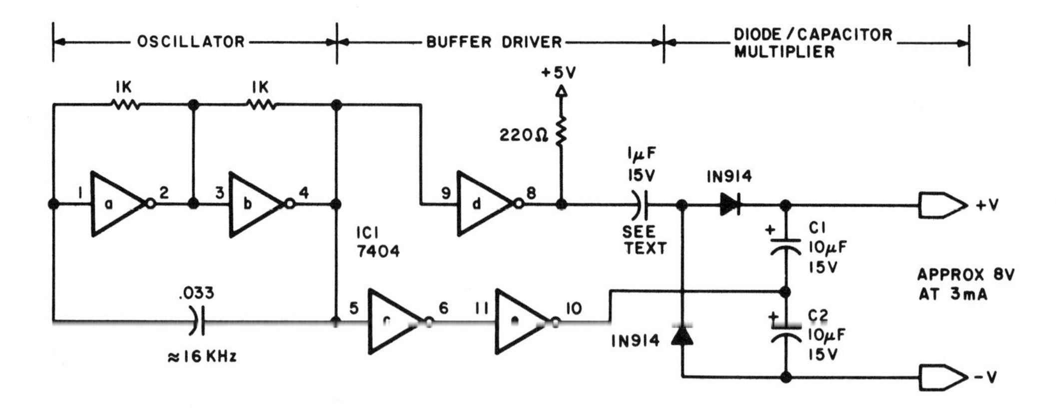

Figure 1: Typical DC to DC converter, a device used to convert one DC voltage into another. The oscillator section supplies a train of square waves to the buffer drivers. On the first half cycle, capacitor C1 is charged to approximately 4 V, and on the second half cycle, C2 is charged to -4 V. The voltage across the two capacitors is twice the input voltage, or approximately 8 V (open circuit). The 1 mF capacitor between IC1d and the two diodes isolates the circuit so that the 8 V can be referenced to ground.

integrated circuits and require only 50 mA, or if the interface is designed with CMOS circuitry, the current requirement could be 5 mA or less. While the 60 Hz transformer design may be more than adequate, the volume and weight of the low frequency magnetics is bulky and may not fit easily within the present enclosure.

The DC to DC Converter

In an application that requires higher voltage at low current, the DC to DC converter is the natural choice for the designer. As its name implies, it converts one DC voltage to another, usually a higher one. All DC to DC converters incorporate oscillator sections to provide AC either to drive transformers or to drive diode-capacitor voltage multipliers. The converters operate at high frequencies to reduce transformer weight. We'll explore the particulars later.

A DC to DC converter need not be low power, but the designs and applications presented here are specifically for low current and limited space applications. The majority of the circuits occupy less than 2 square inches (12.9 square cm).

A DC to DC converter draws its power from some major power bus, such as a +5 V or +12 V computer supply, and converts this source voltage to a higher level of either the same or reversed polarity. The simplest configuration is shown in figure 1. IC1a and IC1b form the oscillator which is common to all DC to DC converters. IC1c, IC1d and IC1e are buffers with the outputs of IC1d and IC1e 180 degrees out of phase,

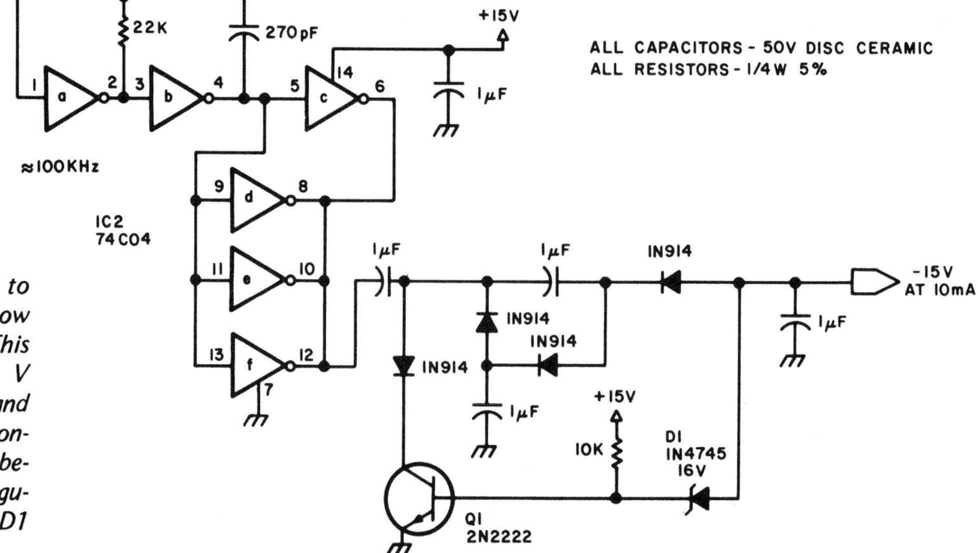

Figure 2: A CMOS DC to DC converter used for low current applications. This circuit produces -15 V from a +15 V source and provides a relatively constant output voltage because of the shunt regulator formed by diodes D1 and Q1.

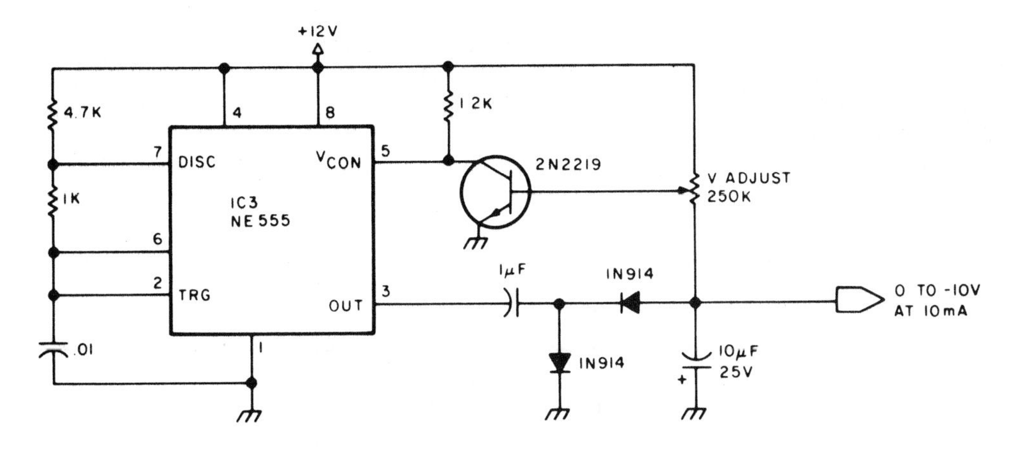

Figure 3: A variable output DC to DC converter capable of producing 0 to −10 V.

simulating a pseudo AC signal to the voltage multiplier. During the first half cycle, the capacitor, C1, is charged to approximately 4 V, and during the second half cycle, C2 is oppositely charged. The voltage across the two capacitors is twice the input voltage, or approximately 8 V (open circuit). If this circuit were not isolated from the drivers (IC1d and IC1e), neither +V nor −V line can be grounded or the multiplier section will be shorted out. The 1 mF 15 V capacitor between pin 8 and the junction of the two IN914 diodes provides isolation and allows the −V lead to be grounded. The output is then approximately 8 V, referenced to ground.

Inverting Supplies

Most often DC to DC supplies are used where a negative voltage is required to power a bipolar linear interface or a dual supply large scale integrated circuit such as a keyboard encoder.

Figures 2 and 3 are examples of converters which would be suitable for these low current applications. Figure 2 produces −15 V from a +15 V source and provides a relatively constant output voltage because of the shunt regulator formed by the diode, D1, and the transistor, Q1. Changing the zener diode, D1, to 13 V makes the output −12 V instead of −15 V. The circuit outlined in figure 3 uses the voltage control input of an NE555 timer circuit to produce a variable output of 0 to −10 V.

Dual Voltage Converters

In most cases single voltage converters use diode steering and charged capacitor voltage multiplication. Transformers or other inductive devices must be incorporated if dual outputs are a requirement. Figure 4 is a very simple ±15 V converter which is powered from a +5 V supply.

Type	Function	+5 V	−5 V	−12 V	+12 V
Ay-5-1013A	UART	20 mA		18 mA	
2708	1 K x 8 EROM	10 mA	45 mA		65 mA
2716 (Intel)	2 K x 8 EROM	100 mA			
2716 (TI)	2 K x 8 EROM	22 mA		12 mA	45 mA
MC1408L8	8 bit digital to analog converter	8 mA		20 mA	
LM301	op amp			3 mA	3 mA
LM741	op amp			2.8 mA	2.8 mA
MM5559	33 bit serial to parallel converter	10 mA		20 mA	

Table 1: Worst case current requirements for a variety of integrated circuits.

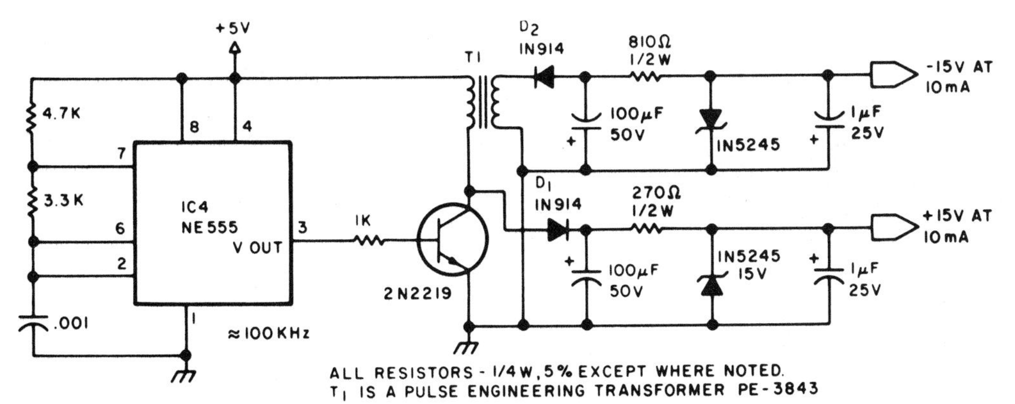

Figure 4: Low current dual voltage output DC to DC converter which supplies −15 and −15 V from a +5 V input.

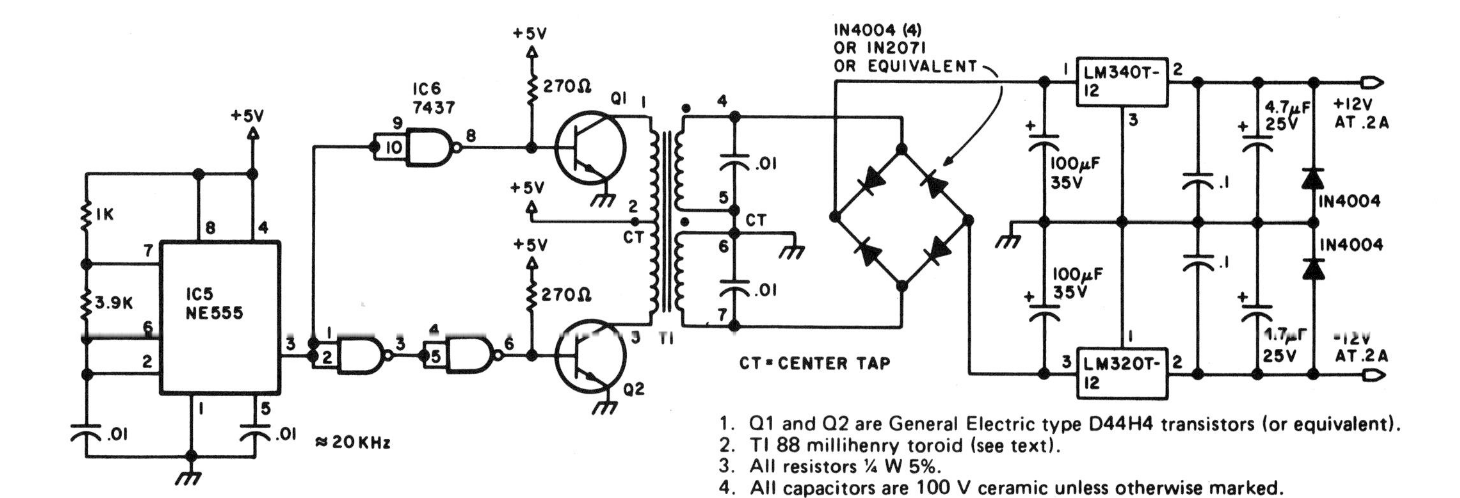

Figure 5a: 5 W DC to DC converter pictured in photo 1, which produces 0.2 A at +12 and -12 V from a 5 V source. See figure 5b for details of winding a toroidal transformer for this circuit.

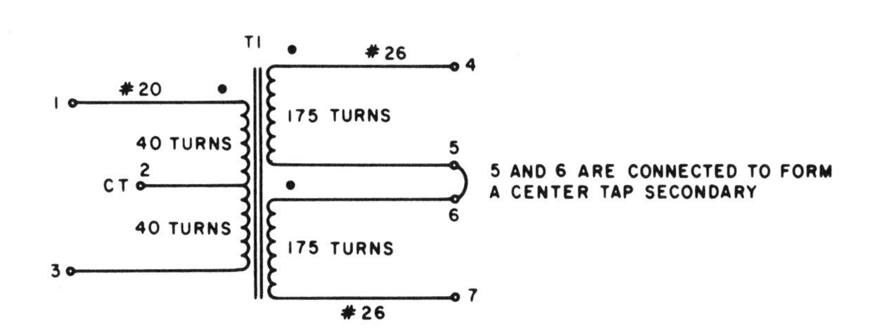

1. Use enamel or Fomvar coated wire for each winding.
2. Be careful when winding not to scratch protective insulation.
3. Primary consists of 80 turns of #20 wire with center tap.
4. Secondaries can be wound as two #26 wire, 175 turn windings or as a single 350 turn winding with center tap.
5. For toroid source see text.
6. Use sandpaper or similar material to remove insulation from terminal wires before soldering.

Figure 5b: Toroid winding details for the custom transformer used in the circuit of figure 5a (see photos 2 thru 5).

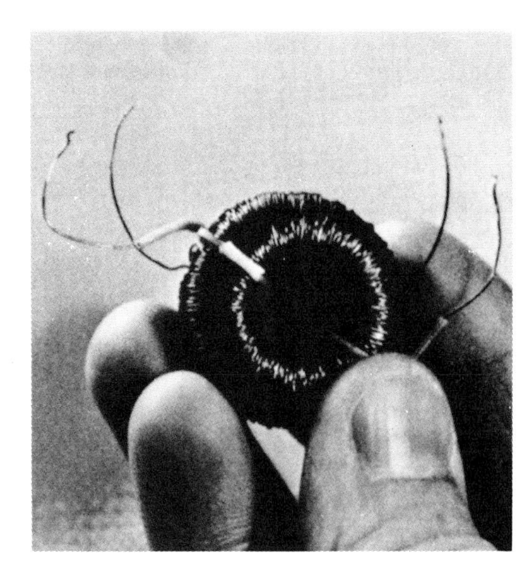

Photo 2: Surplus 88 millihenry toroidal transformer rewound with two secondaries of 175 turns of #26 wire each (after first unwinding the existing two windings of approximately 350 turns each). The unit is used in the circuit of figure 5a.

Number	Type	+5 V	Gnd
IC1	7404	14	7
IC2	74C04	14	7
IC3	555	8	1
IC4	555	8	1
IC5	555	8	1
IC6	7437	14	7

Table 2: Power wiring table for figures 1 thru 5.

A 100 kHz oscillator switches a transistor on and off, inducing a current into the primary of transformer, T1. The voltage produced at the secondary is rectified and regulated to -15 V.

As with all inductive devices which are pulsed, a high voltage spike is reflected back to the collector of the transistor. Rather than shunting this voltage, as would be the case when we put a diode across a coil, D1 routes this spike to a filter and regulator combination to provide a +15 V output.

Building a DC to DC Converter

One of the first things to determine after deciding to use a DC to DC converter in your system is just how much current it must provide. Table 1 lists the typical voltages and operating current requirements (worst case) of a sampling of devices.

It should be apparent from this listing that EROMs are power-hungry devices and will use more than the 10 mA that the converters discussed thus far can supply. For this reason the unit described in figure 5 is designed to produce a full 200 mA at ±12 V.

This design uses a push/pull inverter technique to create AC which drives transformer, T1. T1 is a toroid transformer and its doughnut shape is quite unlike the more

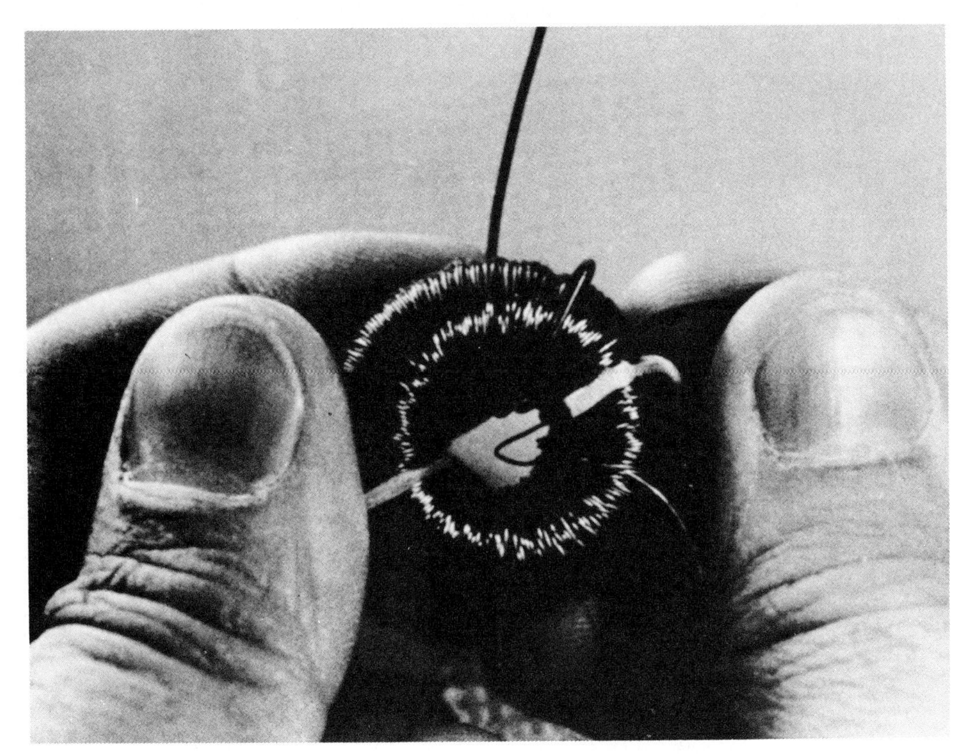

Photo 3: Adding the primary winding, step 1: wind 40 turns of # 20 wire evenly around the toroid.

common rectangular filament transformers. The shape and style of the toroid are specifically designed for high frequency operation, which is the main attribute of this inverter design. Heavy magnetic cores are necessary only for low frequencies such as 60 Hz. Since this converter's switching speed is 20 kHz, relatively little magnetic material is necessary, and high power output can be obtained.

The toroid in this design is a surplus 88 millihenry toroid, frequently advertised in the amateur radio magazines. A source I have found is: M Weinschenker, POB 353, Irwin PA 15642. Order 88 millihenry unpotted toroids.

There are two ways to wind this toroid. Since it presently contains two windings of approximately 350 turns each, adding a primary sounds most logical. In reality though, 180 turns of #20 wire couldn't possibly fit in the remaining space, and the number of windings seems to vary from source to source. To obtain a properly wound toroid, it is best to first completely unwind the toroid and then rewind two 175 turn secondaries. The rewound toroid looks like photo 2. Since inductors exhibit an output polarity that is important when tying two secondaries in series, it is advisable to mark the starting lead on each coil and wind each in the same direction. It is not catas-

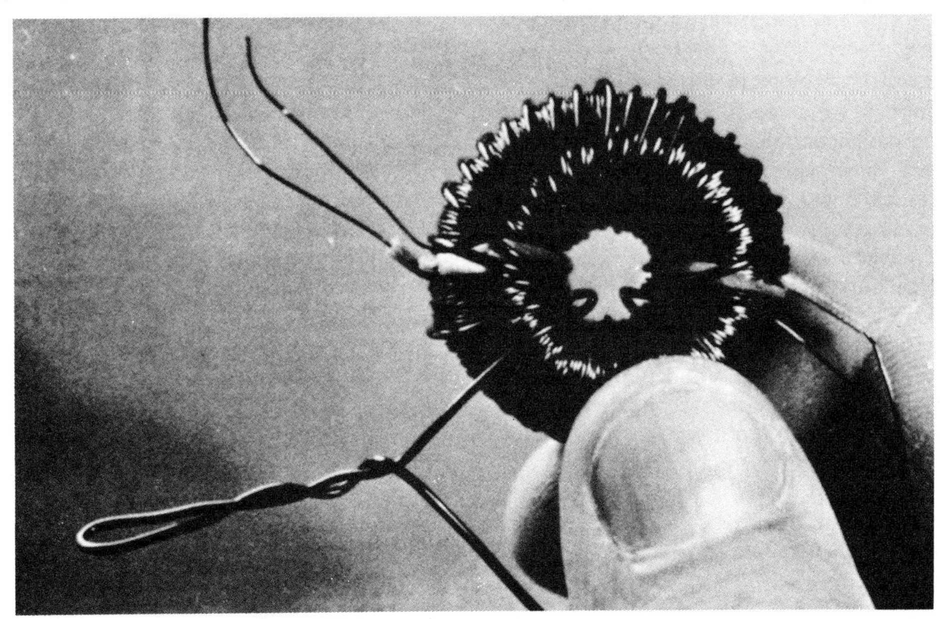

Photo 4: Adding the primary winding, step 2: make a loop for the center tap and continue with 40 additional turns.

Photo 5: The completed transformer. The ends of all enameled wires should be cleaned of insulation before soldering.

trophic if you don't. Polarity can be determined empirically later.

The primary is wound with #20 wire over the two secondaries as in photo 3, and should be distributed evenly around the toroid. When 40 turns have been wound, make a loop in the wire so that it will stick out (as shown in photo 4) and then continue winding the next 40 turns in the same direction. The complete toroid should look like photo 5.

The design outlined in figure 5a is a DC inverter. The NE555 20 kHz oscillator sources the high current 7437 buffers which are necessary to drive the push/pull transistor combination of Q1 and Q2. The continuous on/off action of the transistors produces an alternating current of 20 kHz in the primary winding of the toroid. This in turn induces a voltage proportional to the ratio of the primary to secondary turns, times the primary input voltage into the secondary winding. With approximately 4 V into the primary (taking into account the collector to emitter voltage drop, V_{CE}, transistors Q1 and Q2), 18 to 20 V should be present on each secondary.

The output of the toroid is treated as it would be in a traditional DC regulator design. The two secondaries are connected in series (terminals 4 and 5 connected) to produce 45 V between terminals 3 and 6. If a low voltage is obtained instead of 45 V, then the secondaries are out of phase and the terminals of one of the coils should be reversed. The two terminals which are connected at this point are the center tap and should be grounded.

Four diodes and two capacitors function as the full wave rectifier and filter input to a pair of 3 terminal voltage regulators. The result is a well-regulated + and −12 V supply with output current in excess of 200 mA on each. Overall conversion efficiency is better than 50%.

One note to keep in mind when testing this device: since the output is 5 W with 50% efficiency, the continuous input current to the converter will be approximately 2 A (at 5 V). Peak current will be higher at each clock transition. Use a supply with sufficient current capabilities or it will degrade the performance of the converter and possibly not even work. ■

Photo 1: ICs on ICE. Pictured are all the components necessary to build the 8 channel 3½ digit computer controlled voltmeter described in this article.

2

On a Test Equipment Diet?

Try an 8 Channel DVM Cocktail

About three weeks ago, I was testing a new 8 bit analog to digital converter which I had just built for an upcoming magazine article: this one, in fact. It was a high speed successive approximation analog to digital converter which performed 200,000 conversions a second, and it worked fine. I had intended to use it for some speech digitization experiments. During the testing phase, however, I became exasperated from continually moving my digital voltmeter (DVM) probes around the circuit to take readings and having to stop to make the same calculations repeatedly. To speed the process up, I wrote a BASIC program which would do the number crunching, provided I typed in the voltage values correctly. More often, though, all I wanted was to monitor a few voltage levels simultaneously.

After stringing my two DVMs, an analog volt-ohm meter (VOM), and my oscilloscope all over the bench to aid in my testing, I concluded that there must be a better way. It's old hat to use one channel of a dual trace scope to troubleshoot the other trace, so it was natural to consider using the analog to digital converter to monitor itself. While the thought was momentarily gratifying, the low resolution inherent with eight bits and clumsy binary conversion made me reconsider.

While thinking over this dilemma I was leaning back in my reclining desk chair with one elbow on my computer and my feet up on my printer. I realized that I should move some of the junk so that I'd have more room in the basement. I concluded that what I needed were eight DVMs. This insane desire was quickly eradicated and replaced by a more economically sound idea. I had designed a 4 channel 8 bit digital to analog converter to run with BASIC. It was only natural to design a multichannel analog to digital converter which also interfaced to BASIC.

12 bit analog to digital converters and 3½ digit DVM chips come in a variety of configurations. Converters which specifically state that they are 12 bit converter modules can have either binary or binary coded decimal (BCD) outputs, but are almost universally parallel binary output devices. The end of conversion signal results in immediate data output. The computer just has to scan the data lines and translate

Motorola MC14433 3½ Digit Analog to Digital Converter Specifications:

Accuracy: ± 0.05% of reading ± 1 count
Two voltage ranges: 1.999 V and 199.9 mV
Up to 25 conversions per second
Input impedance > 1000 megohms
Auto zero
Single positive voltage reference
Auto polarity
Drives CMOS or low power Schottky loads
On chip system clock
Over, under, and auto ranging signals available

them into meaningful notation. Chips which are specifically referred to as 3½ or 4½ digit DVM large scale integrated circuit (LSI) chips do not have this luxury. In general, their output is a combination of serial and parallel, one digit at a time. Interfacing to a parallel output analog to digital converter would be far easier with regard to the computer software, but as is generally the case, one never gets something for nothing. 12 bit parallel analog to digital converters are expensive. Most are designed to cover high speed data acquisition applications. Speed (1000 to 100 K conversions per second) costs money.

This leaves us with the 3½ digit DVM LSI chips. They run very slowly by comparison (1 to 50 conversions a second), but cost an order of magnitude less. Software to perform the serial to parallel conversions is a bit more involved, but once it's written, who cares?

One of the latest chips to hit the market is the Motorola MC14433, a 3½ digit low power complementary MOS analog to digital converter. Its specifications (relative to computer applications) are listed in the box on the previous page.

The MC14433 is a modified dual ramp integrating analog to digital converter. This is outlined in figure 1.

The conversion sequence is divided into two integration periods: unknown and reference. During the V_{in} or unknown input integration sequence, the unknown voltage is applied to an integrator with a defined integration time constant for a predetermined time limit. The result is that the voltage level at the output of the integrator will be a function of the unknown voltage input. More positive input voltages will result in higher levels at the integrator output.

During the second cycle of the integration sequence, V_{in} is replaced at the input of the integrator with a negative 2.000 V reference. The output of the integrator starts to move toward zero while the digital circuitry in the chip keeps track of the time it takes to make it to zero again. The time difference between the two integration sequences is then a function of their voltage difference. Since the integration time constants are the same for both periods, if 2.000 V were the unknown applied voltage, t_2 would be equal to t_1. The unknown voltage is equivalent to the ratio of the periods, times the voltage reference, V_{ref}. This is also known as a ratiometric converter. Quite a mouthful. The full scale range of the converter is determined by the level of V_{ref}. Changing V_{ref} to .200 V will make the same 1999 count represent a 199.9 mV full scale. (Obviously, V_{ref} could be set to any value within the voltage limitations of the chip. But, remember, full scale will still be 1999 counts even if it represents 2.463 V, if for example that were V_{ref}.)

Figure 1: A simplified functional representation of the Motorola MC14433 3½ digit analog to digital converter. 1a shows a block diagram of the device; 1b shows the two integration periods used to convert the input voltage to a 3½ digit decimal number. During time t_1, the unknown voltage (V_{in}) is applied to an integrator having a predefined integration time constant (τ) for a preset time. During t_2 a known negative voltage is presented to the integrator. The time needed for the integrator to return to the 0 level is therefore a function of the unknown voltage. A digital counter keeps track of this time, from which V_{in} can be calculated.

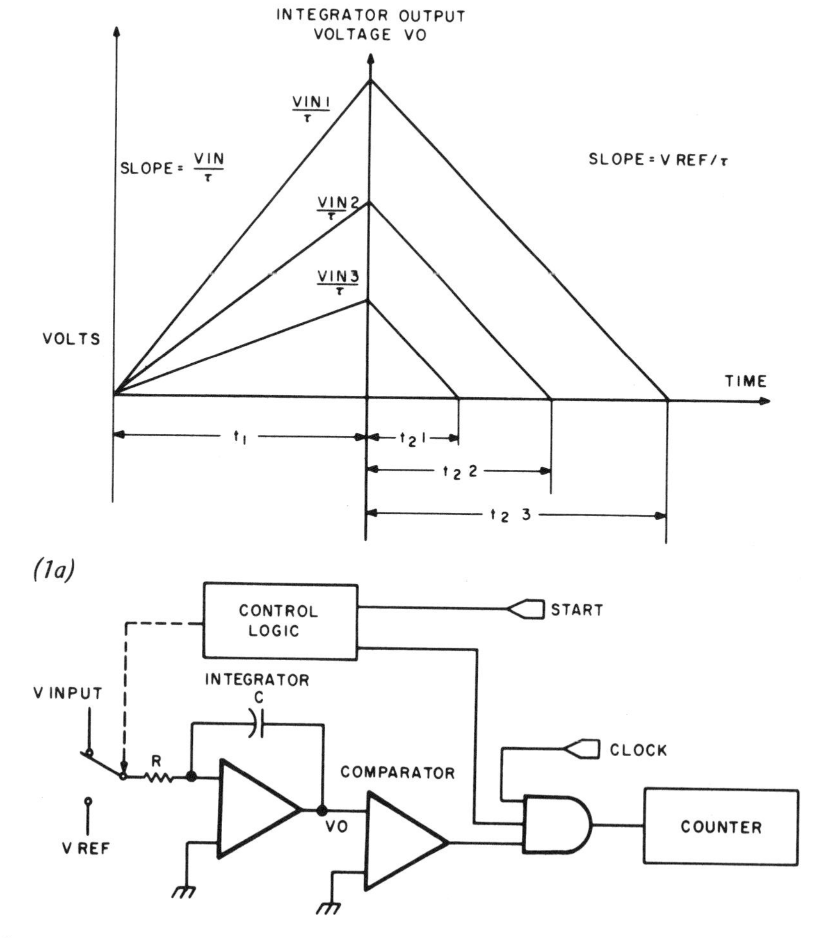

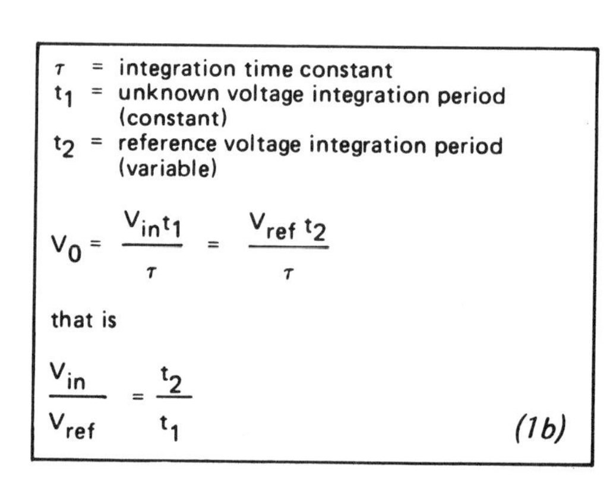

τ = integration time constant
t_1 = unknown voltage integration period (constant)
t_2 = reference voltage integration period (variable)

$$V_O = \frac{V_{in} t_1}{\tau} = \frac{V_{ref} t_2}{\tau}$$

that is

$$\frac{V_{in}}{V_{ref}} = \frac{t_2}{t_1}$$

(1b)

Making a DVM Chip Computer Compatible

There are more bus configurations than I know what to do with lately, so I set up this interface to run from decoded input and output ports. Whether they be memory mapped IO or not, we do not care, as long as the outputs are latched and the inputs can be driven by low power Schottky TTL devices.

To fully utilize this eight channel 3½ digit DVM, we must design the correct hardware interface and write a universal software driver.

Hardware and Data Format

Figure 2 details the schematic of the 8 channel interface board. IC1 is the MC14433 DVM chip. With the values chosen, it will perform approximately 25 conversions a second. Reducing the 68 K resistor between pins 10 and 11 to about 27 K will increase this to about 50 conversions per second. This is an out of specification condition and, though probably successful, is dependent on individual parts.

Each output pin of IC1 has the power to drive one LS TTL load. Since all input ports are not necessarily low power, we provide IC3 and IC4 as buffers. They are 74LS04s and while they are capable of driving regular TTL, they do invert the output data of the DVM. Any driver program must complement the BCD and digit data it receives from this interface before using it.

IC2 is a MC1403 precision voltage reference chip and supplies the V_{ref} input. This IC will vary only 7 mV over a range of 0° to 70°C from its nominal 2.5 V output. While a zener diode might also supply an adequate reference voltage, the temperature drift characteristics of the average zener would negate the value of a 3½ digit converter if used beyond a 5 or 10°C temperature variation. A precision voltage integrated circuit is an absolute must if this circuit is to be used for practical applications.

IC5 is a 7474 which is used here as a set-reset flip flop. The end conversion signal from IC1 sets it, and an output bit from the computer resets it after reading the output data.

IC6 is an 8 input CMOS multiplexer. Its address lines are tied directly to a latched

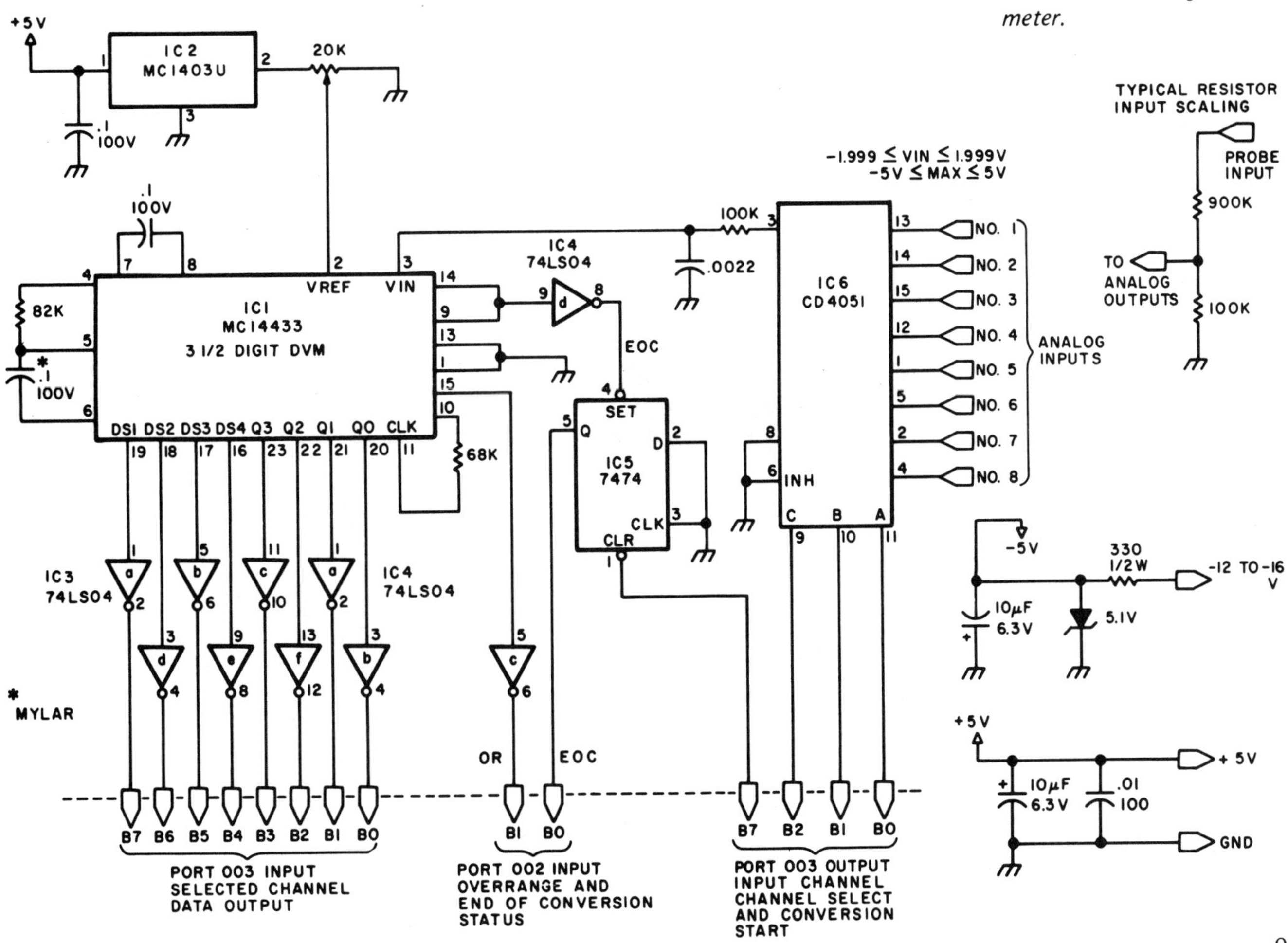

Figure 2: Circuit for the 8 channel 3½ digit voltmeter.

output port. The usual conversion sequence is to set the channel information to the multiplexer, clear the EOC flip flop and wait for an end of conversion signal. More on this later.

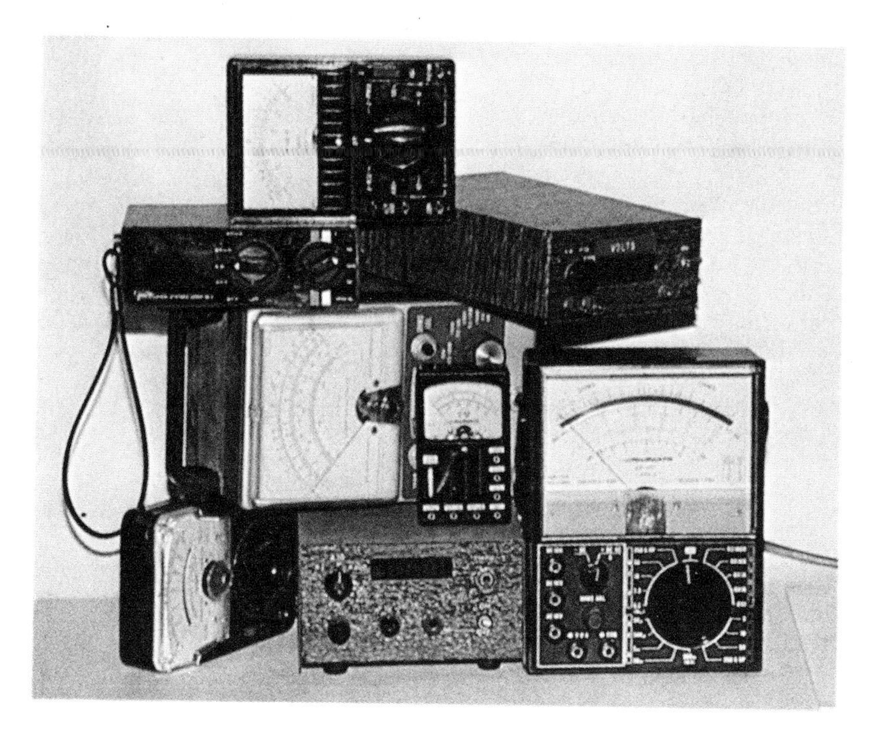

Photo 2: Eight meters (some are multimeters, others are voltmeters) which could be replaced (at least for DC voltage measurements) by the computerized 8 channel voltmeter described here.

Data Format

As I stated earlier, the data from the DVM to the computer is both serial and parallel. There are four digit select lines and four BCD data lines (see table 1).

With respect to what the computer sees through the 74LS04 buffers, the digit select output is low when the respective digit is selected. The most significant digit (½ digit DS1) goes low immediately after an EOC pulse, followed by the remaining digits sequencing from most significant to least significant digit (MSD to LSD). An interdigit blanking time of two clock periods is included to ensure that the BCD data has settled. The multiplex clock rate is equal to the system clock frequency divided by 80.

During the ½ digit (DS1), the polarity and certain status bits are available. It would be confusing to list the status bits, since they are not being used in this application for autoranging. The polarity will be Q_2 and a "1" will indicate negative. The ½ digit value will appear on Q_3 and a "1" will indicate high.

The interface is summarized by port allocations in table 1. (Note: I have assigned particular port numbers to each byte. These designations will run directly with the software driver provided. If the reader wishes to assign different port numbers, that is fine, but remember to modify the driver software to reflect the changes.)

Photo 3: Prototype board for the 8 channel 3½ digit voltmeter.

Designing an Analog to Digital Converter Software Driver

For a hardware personality like me, software is a tedious task. I don't like writing any more than I have to and if it is possible to write a universal piece of code which is compatible with any operating system, all the better. Units such as the digital to analog converter I presented in "Control the World" do not need software drivers because the hardware is explicitly designed to be independent of computer timing. *Timing* is the key word. A "software driver" is the same as its hardware counterpart. Both serve to couple the computer to external devices and synchronize the timing. The most obvious driver already existing in a computer system like my Digital Group system is the asynchronous data link to the tape cassette, video display and printer. The computer is instructed through this program to perform explicitly timed operations which result in the correct serial input and output.

The 3½ digit DVM interface is not unlike a communications driver. To effectively obtain data from the interface, the computer must synchronize itself to the integrated circuit and perform a set instruction repertoire to demultiplex the input data stream. There is a certain trade-off between hardware and software. Another ten or 15 chips could be added to the interface board so that it requires no more software

Photo 4: An illustration of the accuracy of the computerized voltmeter. A Data Precision 4½ digit digital multimeter and the author's system simultaneously measure a C cell battery. The computer value is 1.540 V compared with the Data Precision reading of 1.5402 V.

Table 1: IO port data formats.

Command Output Byte (Port 003 OUT) (Enable = 1 Disable = 0)

- B7 = EOC/Interrupt Enable/disable
- B6, B5, B4, B3 } Future Expansion
- B2, B1, B0 } Channel Select, 0-7

Status Input Byte (Port 002 IN)

- B7, B6, B5, B4, B3, B2 } Not Used
- B1 = Out of Range ($-1.999 < V_{in} > 1.999$)
- B0 = End of Conversion

Data Input Byte (Port 003 IN)

Bit	When true =	Symbol	IC1 Pin Number
B7 = 1st digit (MSD):	B7→0	DS1	19
B6 = 2nd digit	B6 } N/A	DS2	18
B5 = 3rd digit	B5 } N/A	DS3	17
B4 = 4th digit	B4 } N/A	DS4	16
B3 } BCD Digit Value	B3 = 1/2 digit value	Q_3	23
B2 } BCD Digit Value	B2 = Polarity	Q_2	22
B1 } BCD Digit Value	B1 = N/A	Q_1	21
B0 } BCD Digit Value	B0 = Status Bit	Q_0	20

Listing 1: An assembly program for driving the 8 channel 3½ digit voltmeter in figure 3. It is designed to run on the Z-80 and is assembled to occupy memory page octal 140.

```
ASSM 140000 140000

140000                       0100 *
140000                       0110 *** MC14433 3 1/2 DIGIT A/D CONVERTER DRIVER
140000                       0120 *
140000                       0125 * REV 1.6
140000                       0130 *
140000                       0140 DIP    EQU  3      DATA INPUT PORT NUMBER
140000                       0150 SIP    EQU  2      STATUS INPUT PORT NUMBER
140000                       0160 COP    EQU  3      COMMAND OUTPUT PORT NUMBER
140000                       0170 EEOC   EQU  200    ENABLE EOC INPUT
140000                       0180 DEOC   EQU  000    DISABLE EOC INPUT
140000                       0190 *
140000                       0200 *
140000                       0210 * CONVERTED CHANNEL DATA BUFFERS
140000                       0220 *
140000 000 000               0230 CHAN0  DW   000000
140002 000 000               0240        DW   000000
140004 000 000               0250 CHAN1  DW   000000
140006 000 000               0260        DW   000000
140010 000 000               0270 CHAN2  DW   000000
140012 000 000               0280        DW   000000
140014 000 000               0290 CHAN3  DW   000000
140016 000 000               0300        DW   000000
140020 000 000               0310 CHAN4  DW   000000
140022 000 000               0320        DW   000000
140024 000 000               0330 CHAN5  DW   000000
140026 000 000               0340        DW   000000
140030 000 000               0350 CHAN6  DW   000000
140032 000 000               0360        DW   000000
140034 000 000               0370 CHAN7  DW   000000
140036 000 000               0380        DW   000000
140040                       0390 *
140040                       0400 * INTERMEDIATE DATA BUFFERS
140040                       0410 *
140040 000                   0420 POLVAL DB   000    LAST POLARITY VALUE (0=POSITIVE)
140041 000                   0430 CHAN   DB   000    CURRENT CHANNEL NUMBER
140042 000 000               0440 CCP    DW   000000 COMMAND CHANNEL PARAMETER
140044 000 000               0450 STATUS DW   000000 RETURN STATUS PARAMETER
140046                       0460 *
140046                       0470 *
140046                       0480 *** START A/D CONVERTER
140046                       0490 *
140046                       0500 *     INPUT PARAMETER=DE REGISTER WITH CHANNEL SELECT BITS
140046                       0510 *                     SET FOR DESIRED CHANNEL (BIT 0=1
140046                       0520 *                     FOR CHANNEL 0, ETC.)
140046                       0530 *     OUTPUT PARAMETER=HL REGISTER(BIT 0 FOR CHANNEL 0
140046                       0540 *                      WHERE 0=GOOD VALUE;1=OUT OF RANGE)
140046                       0550 *
140046 353                   0560 START  EX   DE,HL  SAVE INPUT PARAMETER
140047 042 042 140           0570        LD   (CCP),HL
140052 257                   0580        XOR  A      INITIALIZE CHANNEL NUMBER
140053 375 041 040 140       0590        LD   IY,POLVAL INITIALIZE INTERMEDIATE DATA POINTER
140057 375 167 001           0600        LD   (IY+1),A ZERO CHANNEL NUMBER
140062                       0610 *
140062                       0620 * START A/D CONVERTER AND ESTABLISH POLARITY
140062                       0630 *
140062 006 002               0640        LD   B,2    CYCLE TWO TIMES
140064 076 007               0650        LD   A,7    SELECT CHANNEL 8
140066 323 003               0660 AGAIN  OUT  COP    SELECT CHANNEL
140070 366 200               0670        OR   EEOC
140072 323 003               0680        OUT  COP    ENABLE EOC INPUT
140074 333 002               0690 WAIT   IN   SIP    READ STATUS
140076 313 107               0700        BIT  0,A    TEST FOR EOC
140100 050 372               0710        JR   Z,WAIT JUMP IF NOT TRUE
140102 020 362               0720        DJNZ AGAIN  JUMP IF NOT DONE
140104 006 200               0730        LD   B,200  SELECT DGIT 1
140106 315 361 140           0740        CALL RDIG   READ DIGIT
140111 016 000               0750        LD   C,0    POLARITY=POSITIVE
140113 313 122               0760        BIT  2,D    TEST POLARITY BIT
140115 040 001               0770        JR   NZ,POS JUMP IF POSITIVE
140117 014                   0780        INC  C      POLARITY=NEGATIVE
140120 375 161 000           0790 POS    LD   (IY+0),C SAVE CURRENT POLARITY
140123                       0800 *
140123                       0810 * SELECT NEXT CHANNEL FOR CONVERSION
140123                       0820 *
140123 072 042 140           0830 SELNXT LD   A,(CCP) LOAD CHANNEL COMMAND PARAMETER
140126 313 077               0840        SRL  A        TEST NEXT CHANNEL BIT
140130 062 042 140           0850        LD   (CCP),A RESTORE
140133 070 010               0860        JR   C,SEL001 JUMP IF CHANNEL SELECTED
140135 312 355 140           0870        JP   Z,RAPUP
140140 375 064 001           0880 INCCN  INC  (IY+1) INCREMENT CHANNEL NUMBER
140143 030 356               0890        JR   SELNXT
140145 335 041 000 140       0900 SEL001 LD   IX,CHAN0 LOAD DATA BUFFER BASE ADDRESS
140151 026 000               0910        LD   D,0
140153 375 136 001           0920        LD   E,(IY+1) LOAD CURRENT CHANNEL NUMBER
140156 313 043               0930        SLA  E       CALCULATE BUFFER OFFSET
140160 313 043               0940        SLA  E
140162 335 031               0950        ADD  IX,DE
140164                       0960 *
140164                       0970 * SELECT CHANNEL AND START CONVERSION
140164                       0980 *
140164 072 041 140           0990 SCSC   LD   A,(CHAN) LOAD CHANNEL NUMBER
140167 323 003               1000        OUT  COP    SELECT CHANNEL
140171 366 200               1010        OR   EEOC   ENABLE EOC OUTPUT
140173 323 003               1020        OUT  COP    COMMAND A/D CONVERTER
140175                       1030 *
140175                       1040 * WAIT FOR EOC
140175                       1050 *
140175 333 002               1060 WEOC   IN   SIP    READ CONVERTER STATUS
140177 313 107               1070        BIT  0,A    TEST FOR EOC
140201 050 372               1080        JR   Z,WEOC JUMP IF NOT READY
140203 313 117               1090        BIT  1,A    TEST FOR OVERANGE
140205 040 124               1100        JR   NZ,OVER JUMP IF TRUE
140207                       1110 *
140207                       1120 * CONVERSION DONE;PROCESS FIRST (MSD) DIGIT
140207                       1130 *
140207 006 200               1140 MSD0   LD   B,200  SELECT DIGIT 1
140211 315 361 140           1150        CALL RDIG   WAIT AND READ DIGIT 1
140214 057                   1160        CPL
140215 017                   1170        RRCA RIGHT  JUSTIFY DIGIT VALUE
140216 017                   1180        RRCA
140217 017                   1190        RRCA
140220 346 001               1200        AND  1      ISOLATE
140222 036 000               1210        LD   E,0    INITIALIZE STATUS BYTE
140224 113                   1220        LD   C,E
140225                       1230 *
```

than the digital to analog converter board, but the cost justification is not there.

Driver programs can be triggered by either a poll from another program or an interrupt which initiates execution. While both can be equally effective in certain applications, using interrupt initiated drivers which give the appearance of simultaneous computer operation can be hazardous. By now, most experimenters have mastered BASIC and are trying to find more challenging applications. But consider for a moment the BASIC interpreters most systems are provided with. They may execute divinely, but they have no source listing and therefore cannot be modified very easily. If a program utilizes information provided through interrupt driven peripherals, but has no way of knowing when the information will arrive, it is of no use. Attempting to add interrupt analog data acquisition to unsourced sequentially interpreted BASIC is more than I intend to explain in this article.

Adding this DVM interface to BASIC requires a polled driver. A machine language program is written which can be inserted anywhere in the computer's memory (assuming it's assembled to execute there, of course) and called as a subroutine when the peripheral is to be exercised. The Digital Group Maxi BASIC, like many others, has instructions which allow memory and IO port manipulation as well as calling machine language subroutines. It is this latter call instruction which initiates the analog to digital conversion cycles and communicates with the interface driver program. When it executes this call instruction, it passes a channel convert code in the DE register pair. The driver program returns control to the BASIC interpreter at the conclusion

Constructing the Interface

1. Use IC sockets and solder in all passive components.
2. Turn on the power and ensure that the correct supply voltages are presented to ICs 1, 2 and 6. Turn off power.
3. Insert IC2 and apply power. The output at pin 2 should be 2.5 V and should not drift. Adjust the pot so that there is exactly 2.000 V on IC1 pin 2. Turn off power.
4. Insert the rest of the ICs including the MC14433. Be careful when inserting the 4051 and MC14433. You are now ready to wire the board to some convenient input and output ports and see if it flies.
5. Turn on power. A driver program obviously is necessary to see if the circuit actually works and I have included one. If you are really anxious, you can try a couple of quickies: an oscilloscope attached to digit select or data lines will tell you immediately if the circuit is running. You should see square waves of various duty cycles. Another method is to write a short program which scans the end of conversion bit (remember to reset it first) and halt. If it halts, there must be an EOC.

of the analog to digital conversion. This provides a convenient method of synchronization. BASIC waits for the driver to finish storing the converted input data before trying to use it. Perhaps the next level is to write an interrupt driver which continually updates a value in the interpreter's tables of variables; but this would require a source listing and further documentation of the interpreter in order to accomplish the goal.

The Driver Is a Relocatable Subroutine

The actual program which interfaces to and stores the values to the DVM chip is written in the form of a single callable subroutine. To maintain the relocatability of the subroutine to any page in memory, all information necessary for the proper execution of the driver is provided at the time of the call. The additional information about which channels are to be converted is loaded into the DE registers at the time of the call. One bit of the E register is allocated for each analog to digital channel. Channel 1 is the least significant bit and channel 8 is the most significant. Setting a "1" value for the channel bit will tell the driver to convert that channel and a "0" means to ignore it. Loading E with binary 10 110 011 will indicate to the driver that channels 1, 2, 5, 6 and 8 are to be converted. Setting all bits to "1" will cause all channels to be read and converted. Indicating to the driver which, if any, channels are to be read rather than scanning all of them is a method of saving time. By computer standards, this analog to digital interface is slow; it is better not to waste any more time than is necessary.

The driver starts the conversion process by selecting a channel address to convert. This is accomplished by looking at the least significant bit of the E register. If it is a "1" it will convert on that channel. If it is a "0" it shifts and inspects the next bit, and so on until it finds one that is set. When a bit set condition is found, the channel address of that particular channel is sent out via port 003 to the analog input multiplexer IC6 and the end of conversion flip flop IC5 is reset. The DVM then starts the process of converting the analog input signal.

Demultiplexing the output of the DVM is fairly straightforward. The processor hangs in a loop waiting for an end of conversion signal. When this happens, the program knows that the next four digits of data are what is wanted. The DVM integrated circuit sets each of the digit select lines successively, and the program records the values of the four data lines each time. It strips the status and polarity bits from the most significant digit (the 3½ digit) and reformats the value into four bytes of memory. The three whole digits will be stored in BCD notation and occupy three of the bytes. The ½ digit, polarity and out of range will be located in the remaining data byte. Polarity is indicated by setting the most significant bit. A positive reading is a zero condition and negative is a one in that bit. The ½ digit value can only be a one or zero and occupies the least significant bit of the quantity. Out of range is accomplished with a little program manipulation. If the driver detects that the incoming reading is not within range, it sets the equivalent of +2 in the ½ digit byte. Obviously, this is an illegal condition for a DVM capable of only counting to 1999, but it is easy for BASIC to check the authenticity of the data by checking that all incoming values are between −1999 and +1999. The driver program continues to do

Listing 1, continued:

```
140225                  1240 * TEST POLARITY OF CHANNEL
140225                  1250 *
140225 313 122          1260 MSD1   BIT  2,D     TEST POLARITY
140227 040 017          1270        JR   NZ,MSD2 JUMP IF POSITIVE
140231                  1280 *
140231                  1290 * NEGATIVE POLARITY
140231                  1300 *
140231 014              1310        INC  C
140232 036 200          1320        LD   E,200  LOAD NEGATIVE SIGN
140234 375 313 000 106  1330        BIT  0,(IY+0) TEST PREVIOUS POLARITY
140240 040 022          1340        JR   NZ,MSD3 JUMP IF ALSO NEGATIVE
140242 375 313 000 306  1350        SET  0,(IY+0) MAKE PREVIOUS VALUE NEGATIVE
140246 030 314          1360        JR   SCSC   CONVERT AGAIN
140250                  1370 *
140250                  1380 * POSITIVE POLARITY
140250                  1390 *
140250 375 313 000 106  1400 MSD2   BIT  0,(IY+0) TEST PREVIOUS POLARITY
140254 050 006          1410        JR   Z,MSD3 JUMP IF ALSO POSITIVE
140256 375 313 000 206  1420        RES  0,(IY+0) MAKE PREVIOUS VALUE POSITIVE
140262 030 300          1430        JR   SCSC   CONVERT AGAIN
140264                  1440 *
140264                  1450 * SAVE MSD AND CURRENT POLARITY
140264                  1460 *
140264 263              1470 MSD3   OR   E      ADD POLARITY SIGN TO MSD
140265 335 167 000      1480        LD   (IX+0),A SAVE IN DATA BUFFER
140270 375 161 000      1490        LD   (IY+0),C SAVE CURRENT POLARITY
140273                  1500 *
140273                  1510 * PROCESS 2ND DIGIT
140273                  1520 *
140273 313 010          1530        RRC  B      SELECT DIGIT 2
140275 315 361 140      1540        CALL RDIG   WAIT AND READ DIGIT
140300 346 017          1550        AND  017    ISOLATE
140302 335 167 001      1560        LD   (IX+1),A STORE SECOND DIGIT
140305                  1570 *
140305                  1580 * PROCESS 3RD DIGIT
140305                  1590 *
140305 313 010          1600        RRC  B      SELECT 3RD DIGIT
140307 315 361 140      1610        CALL RDIG   WAIT AND READ DIGIT
140312 346 017          1620        AND  017    ISOLATE
140314 335 167 002      1630        LD   (IX+2),A STORE
140317                  1640 *
140317                  1650 * PROCESS 4TH DIGIT
140317                  1660 *
140317 313 010          1670        RRC  B      SELECT 4TH DIGIT
140321 315 361 140      1680        CALL RDIG   WAIT AND READ DIGIT
140324 346 017          1690        AND  017    ISOLATE
140326 335 167 003      1700        LD   (IX+3),A STORE
140331 030 205          1710        JR   INCCN
140333                  1720 *
140333                  1730 * LOAD 2.000 OVERRANGE VALUE INTO DATA BUFFER
140333                  1740 *
140333 076 002          1750 OVER   LD   A,2    LOAD MSD VALUE
140335 335 167 000      1760        LD   (IX+0),A
140340 257              1770        XOR  A
140341 335 167 001      1780        LD   (IX+1),A LOAD LSD VALUES
140344 335 167 002      1790        LD   (IX+2),A
140347 335 167 003      1800        LD   (IX+3),A
140352 303 140 140      1810        JP   INCCN
140355                  1820 *
140355                  1830 * END OF CHANNEL CONVERSIONS
140355                  1840 *
140355 052 044 140      1850 RAPUP  LD   HL,(STATUS)
140360 311              1860        RET         RETURN TO CALLER
140361                  1870 *
140361                  1880 *
140361                  1890 * READ DIGIT ROUTINE
140361                  1900 *
140361 333 003          1910 RDIG   IN   DIP    READ DATA BYTE
140363 057              1920        CPL         CONVERT TO HIGH TRUE LOGIC
140364 127              1930        LD   D,A    SAVE COPY
140365 240              1940        AND  B      TEST FOR GIVEN DIGIT READY
140366 050 371          1950        JR   Z,RDIG JUMP IF NOT
140370 172              1960        LD   A,D    RESTORE A REGISTER
140371 311              1970        RET         RETURN TO CALLER
```

Table 2: Power wiring table for figure 3.

IC Number	Type	+5V Pin	−5V Pin	GND Pin
IC1	MC14433	24	12	1&13
IC2	MC1403	1		3
IC3,4	74LS04	14		7
IC5	7474	14		7
IC6	CD4051	16	7	8

Note: All resistors ¼ W 5% unless otherwise noted.
All capacitors are 100 V ceramics unless otherwise noted.

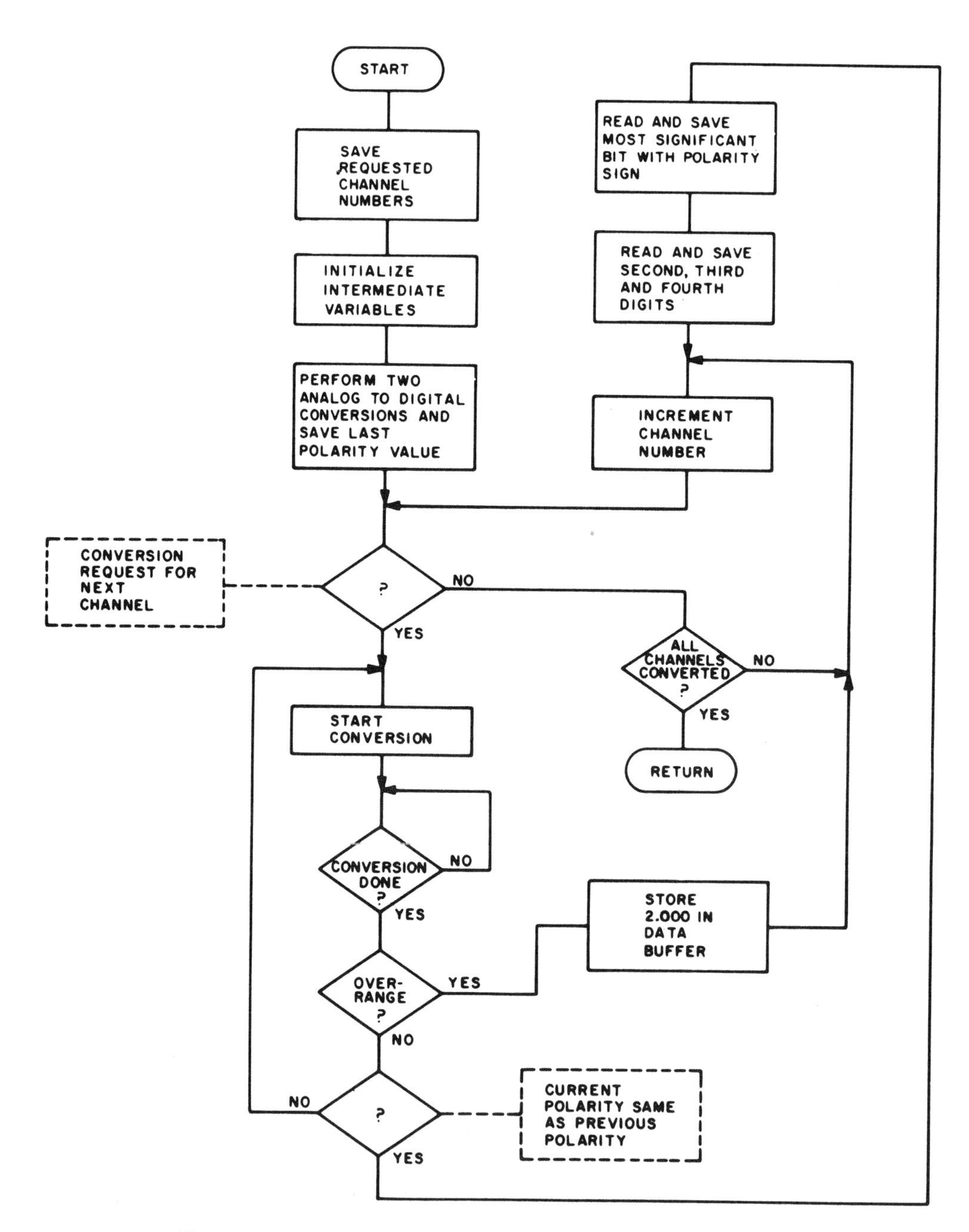

Figure 3: Flowchart of the digital voltmeter driver program of listing 1.

this same sequence until all designated channels have been converted.

There is a slight peculiarity with DVM chips: they don't like changes in polarity. The first conversion after a change in polarity will be 0.000 and will have to be discarded. In a single channel DVM this wouldn't present a problem, but when reading eight channels, some will be negative inputs and others will be positive.

The initial conversion also has the same problem to contend with, since the conversion history when the driver is not active is unknown. The solution is to write a smarter driver. Following a call, the driver program initializes the interface and determines the polarity. After that, any time the polarity changes between successive readings on designated channels, another conversion is initiated and stored. Figure 3 is a simplified flow diagram showing the logical design of the driver.

The end product of the driver is a 32 byte memory resident table which contains the eight 4 byte values corresponding to the eight channels. The values are sequentially arranged in the table. A simple formula locates a particular channel location at L + (4(N-1)) where L is the starting address of the table and N is the channel number. A complete assembly listing of the DVM driver is outlined in listing 1. It is made to run on a Z-80 and is assembled to occupy page 140 (octal).

The driver can be assembled for practically any portion of memory, but take care not to overlap into operating system or source files. If you own Digital Group software, there are some alternatives depending on what version you have. For people with straight (non-universal) 32 character Z-80 Maxi BASIC Version 1.0, page 012 is empty and has been left for future expansion. If you have the 64 character Maxi BASIC Version 1.1, it's better not to try to bury the driver within the interpreter unless you're an experienced programmer. Owners of 8080 systems have only to reassemble the code using 8080 instructions and locate it in a similar manner. The logic behind the driver is not so involved that it necessitates using the Z-80. Any microprocessor should be able to work with the interface.

Using the Interface with BASIC

This DVM interface is specifically designed to run with a BASIC interpreter such as Maxi BASIC or the equivalent. Listing 2 illustrates a BASIC program which does data acquisition and computes results from this input data. Often, the best method of explanation is to illustrate the actual use

Listing 2: A BASIC program (written in Maxi BASIC) which performs data acquisition and computes results from the output of the 8 channel digital voltmeter.

```
LIST

100 REM
110 REM
120 REM 8 CHANNEL 3 1/2 DIGIT SCANNING PROGRAM -S.CIARCIA
130 REM REV 1.5
140 REM SPECIAL ANALYSIS SECTION ---
150 REM TTL TO MOS VOLTAGE LEVEL CONVERTER
160 REM
170 REM
180 REM
190 LET M1=24576
192 REM PAGE 140(OCTAL)
200 REM M1 IS SET TO BE THE DECIMAL STARTING LOCATION OF
210 REM THE VALUE TABLE
220 LET M2=24614
230 REM M2 IS THE MACHINE LANGUAGE CALL ADDRESS LOCATION FOR THE A/D
240 LET M3=10
250 REM M3 IS THE GAIN. IN THIS APPLICATION, THE RANGE OF THE CONVERTER
260 REM IS +19.99 TO -19.99 VOLTS
270 REM TO USE THE CONVERTER FOR -1.999 TO +1.999, LET M3=1
280 GOTO 300
290 PRINT'TO REPEAT THE SAME SELECTION, TYPE AN X'
291 PRINT'TO SELECT A NEW OPTION,TYPE AN O' :INPUT B$
292 IF B$='X' THEN GOTO 420
294 PRINT : PRINT : PRINT
300 PRINT'                    OPTION LIST'
310 REM WE START THE PROGRAM WITH AN OPTION LIST
320 PRINT'--------------------------------------------------'
330 PRINT'0 ----SELECT CHANNELS'
340 PRINT'1 ----SCAN AND DISPLAY ALL CHANNELS'
350 PRINT'2 ----SCAN AND DISPLAY SELECTED CHANNELS ONCE'
360 PRINT'3 ----SCAN AND DISPLAY SELECTED CHANNELS CONTINUOUSLY'
370 PRINT'4 ----SCAN CHANNEL 1 CONTINUOUSLY 100 TIMES'
380 PRINT'5 ----GO TO SPECIAL ANALYSIS SUBROUTINES'
390 REM THESE ROUTINES ARE DEPENDENT UPON THE PARTICULAR A/D APPLICATION
400 PRINT'6 ----EXIT'
410 PRINT'WHICH OPTION ':INPUT S
420 IF S=0 THEN 520
430 IF S=1 THEN 930
440 IF S=2 THEN 1060
450 IF S=3 THEN 1280
460 IF S=4 THEN 1380
470 IF S=5 THEN 1470
480 IF S=6 THEN PRINT'THANKYOU' :END
490 GOTO 410
500 REM
510 REM FIRST WE DETERMINE WHICH ANALOG CHANNELS TO READ
520 PRINT 'INDICATE YOUR CHOICES WITH A Y OR N AFTER THE CHANNEL NUMBER
530 FOR C=1 TO 8
540 PRINT'CHANNEL ';C,
550 INPUT A$
560 REM ACCEPT ONLY TRUE INPUTS
570 IF A$='Y' THEN LET A(C)=1 :GOTO 610
580 IF A$='N' THEN LET A(C)=0 :GOTO 610
590 PRINT'INPUT A Y FOR YES OR A N FOR NO'
600 GOTO 540
610 NEXT C
620 GOTO 290
630 REM
640 REM
650 REM
660 REM SET D EQUAL TO THE DECIMAL MEMORY ADDRESS OF THE
670 REM BEGINNING OF VALUE TABLE
680 REM THIS SUBROUTINE DETERMINES THE 3 1/2 DIGIT VALUE
690 REM FROM THE TABLE IN MEMORY
700 LET Q1=EXAM(D)
710 LET Q=Q1
720 IF Q1>=128 THEN LET Q=Q1-128
730 D=D+1
740 LET W=EXAM(D)
750 D=D+1
760 LET E=EXAM(D)
770 D=D+1
780 LET R=EXAM(D)
790 LET D=D+1
800 LET Y=Q+(.1*W)+(.01*E)+(.001*R)
810 LET Y1=M3*Y
820 RETURN
830 REM
840 REM THIS SUBROUTINE PRINTS OUT THE VOLTAGE VALUES
850 PRINT'CHANNEL ';X;' IS ';
860 IF Q1<128 THEN PRINT' '; :GOTO 880
870 IF Q1>=128 THEN PRINT '-';
880 IF M3=10 THEN PRINT %5F2;Y1;' VOLTS' :GOTO 900
890 PRINT %6F3;Y1;' VOLTS'
900 RETURN
910 REM
920 REM
930 LET B=CALL(M2,255)
940 REM THE CALL INSTRUCTION TELLS THE A/D INTERFACE TO START CONVERTING
950 REM 255 IS ALL BITS SET
960 REM THIS WILL CAUSE THE A/D TO CONVERT AND STORE ALL EIGHT CHANNELS
970 LET D=M1
980 REM D IS THE START ADDRESS OF THE VALUE TABLE
990 FOR X=1 TO 8
1000 GOSUB 700
1010 REM GET 3 1/2 DIGIT VALUE FROM MEMORY
1020 IF Y>=2 THEN PRINT'CHANNEL ';X;' IS OUT OF RANGE' :GOTO 1040
1030 GOSUB 850
1040 NEXT X
1050 GOTO 290
1060 LET D=M1
1070 GOSUB 1130
1080 GOTO 290
1090 REM
1100 REM
1110 REM
```

Listing 2, continued:

```
1120 REM THIS SUBROUTINE PRINTS ONLY THE SELECTED CHANNELS
1130 LET L=A(1)*1+A(2)*2+A(3)*4+A(4)*8+A(5)*16+A(6)*32+A(7)*64+A(8)*128
1140 REM THIS EQUATION SETS THE BIT PATTERN FOR THE CALL TO THE A/D
1150 LET H=CALL(M2,L)
1160 REM H WILL RETURN FROM THE CALL WITH THE HL REG. VALUE BUT
1170 REM IS NOT BEING USED PRESENTLY IN THIS PROGRAM
1180 FOR X=1 TO 8
1190 GOSUB 700
1200 LET Z=A(1)+A(2)+A(3)+A(4)+A(5)+A(6)+A(7)+A(8)
1210 IF Z=0 THEN PRINT"NO CHANNELS HAVE BEEN SELECTED" :EXIT 290
1220 IF A(X)=0 THEN 1250
1230 IF Y>=2 THEN PRINT"CHANNEL ";X;" IS OUT OF RANGE" : GOTO 1250
1240 GOSUB 850
1250 NEXT X
1260 RETURN
1270 GOSUB 1130
1280 LET D=M1
1290 REM THIS SUBROUTINE IS A CONTINUOUS LOOP --- EXIT WITH RESET SWITCH
1300 FOR J=1 TO 1000
1310 LET D=M1
1320 GOSUB 1130
1330 PRINT : PRINT : PRINT
1340 NEXT J
1350 GOTO 290
1360 REM
1370 REM THIS SUBROUTINE CONTINUOUSLY SCANS AND PRINTS CHANNEL 1
1380 FOR R=1 TO 100
1390 LET H=CALL(M2,1)
1400 REM GO AND CONVERT CHANNEL 1 ONLY
1410 LET D=M1
1420 GOSUB 700
1430 LET X=1
1440 GOSUB 850
1450 NEXT R
1460 GOTO 290
1470 LET B=CALL(M2,255)
1480 REM SCAN AND STORE ALL CHANNELS
1490 LET D=M1
1500 FOR X=1 TO 8
1510 GOSUB 700
1520 LET V(X)=Y1
1530 REM SET VALUES INTO AN 8 VALUE ARRAY
1540 NEXT X
1550 REM CHECK CALIBRATION
1560 IF V(8)>=2.006 THEN PRINT"OUT OF CALIBRATION" :GOTO 290
1570 IF V(8)<=1.994 THEN PRINT"OUT OF CALIBRATION" :GOTO 290
1580 PRINT"TTL TO MOS LEVEL CONVERTER ---TTL LOW INPUT STATE":PRINT:PRINT
1590 PRINT"DIODE D1"
1600 PRINT"VOLTAGE DROP =",V(2)-V(1);" VOLTS"
1610 PRINT :PRINT"R1"
1620 LET T1=(V(2)-V(3))/2200
1630 PRINT"CURRENT = ";T1;" AMPS"
1640 PRINT"POWER = ";T1*T1*2200
1650 PRINT :PRINT"Q1"
1660 PRINT"VCE Q1 = ";V(4)-V(3);" VOLTS"
1670 PRINT"VBE Q1 = ";V(3);" VOLTS"
1680 PRINT"VCB Q1 = ";V(4);" VOLTS"
1690 PRINT :PRINT"R2"
1700 LET T2=(V(5)-V(4))/4700
1710 PRINT"R2 DROP = ";V(5)-V(4);" VOLTS"
1720 PRINT"CURRENT = ";T2;" AMPS"
1730 PRINT"POWER = ";T2*T2*4700;" WATTS"
1740 PRINT :PRINT"R3"
1750 LET T3=(V(6)-V(7))/4700
1760 PRINT"R3 DROP = ";V(6)-V(7);" VOLTS"
1770 PRINT"CURRENT = ";T3;" AMPS"
1780 PRINT"POWER = ";T3*T3*4700;" WATTS"
1790 PRINT:PRINT"Q2"
1800 PRINT"VBE Q2 = ";V(5)-V(4);" VOLTS"
1810 PRINT"VCE Q2 = ";V(7)-V(5);" VOLTS"
1820 PRINT"VCB Q2 = ";V(7)-V(4);" VOLTS"
1830 PRINT"IC OF Q2 = ";T3;" AMPS"
1840 PRINT"POWER DISSIPATION = ";(V(7)-V(5))*T3;" WATTS"
1850 PRINT:PRINT:PRINT"SUPPLY VOLTAGES"
1860 PRINT V(6)
1870 PRINT"-";V(5)
1880 GOTO 290
READY


RUN

                    OPTION LIST
----------------------------------------------------
0 ----SELECT CHANNELS
1 ----SCAN AND DISPLAY ALL CHANNELS
2 ----SCAN AND DISPLAY SELECTED CHANNELS ONCE
3 ----SCAN AND DISPLAY SELECTED CHANNELS CONTINUOUSLY
4 ----SCAN CHANNEL 1 CONTINUOUSLY 100 TIMES
5 ----GO TO SPECIAL ANALYSIS SUBROUTINES
6 ----EXIT
WHICH OPTION
?1
CHANNEL  1 IS   3.54 VOLTS
CHANNEL  2 IS   2.93 VOLTS
CHANNEL  3 IS    .63 VOLTS
CHANNEL  4 IS -10.75 VOLTS
CHANNEL  5 IS -11.02 VOLTS
CHANNEL  6 IS   5.04 VOLTS
CHANNEL  7 IS -11.45 VOLTS
CHANNEL  8 IS   2.00 VOLTS
TO REPEAT THE SAME SELECTION, TYPE AN X
TO SELECT A NEW OPTION,TYPE AN 0
?0

                    OPTION LIST
----------------------------------------------------
0 ----SELECT CHANNELS
1 ----SCAN AND DISPLAY ALL CHANNELS
2 ----SCAN AND DISPLAY SELECTED CHANNELS ONCE
3 ----SCAN AND DISPLAY SELECTED CHANNELS CONTINUOUSLY
4 ----SCAN CHANNEL 1 CONTINUOUSLY 100 TIMES
5 ----GO TO SPECIAL ANALYSIS SUBROUTINES
6 ----EXIT
WHICH OPTION
?0
INDICATE YOUR CHOICES WITH A Y OR N AFTER THE CHANNEL NUMBER
CHANNEL  1     ?Y
CHANNEL  2     ?N
CHANNEL  3     ?N
CHANNEL  4     ?Y
CHANNEL  5     ?Y
CHANNEL  6     ?N
CHANNEL  7     ?N
CHANNEL  8     ?N
TO REPEAT THE SAME SELECTION, TYPE AN X
TO SELECT A NEW OPTION,TYPE AN 0
?0

                    OPTION LIST
----------------------------------------------------
0 ----SELECT CHANNELS
1 ----SCAN AND DISPLAY ALL CHANNELS
2 ----SCAN AND DISPLAY SELECTED CHANNELS ONCE
3 ----SCAN AND DISPLAY SELECTED CHANNELS CONTINUOUSLY
4 ----SCAN CHANNEL 1 CONTINUOUSLY 100 TIMES
5 ----GO TO SPECIAL ANALYSIS SUBROUTINES
6 ----EXIT
WHICH OPTION
?2
CHANNEL  1 IS   3.54 VOLTS
CHANNEL  4 IS -10.75 VOLTS
CHANNEL  5 IS -11.02 VOLTS
TO REPEAT THE SAME SELECTION, TYPE AN X
TO SELECT A NEW OPTION,TYPE AN 0
?0

                    OPTION LIST
----------------------------------------------------
0 ----SELECT CHANNELS
1 ----SCAN AND DISPLAY ALL CHANNELS
2 ----SCAN AND DISPLAY SELECTED CHANNELS ONCE
3 ----SCAN AND DISPLAY SELECTED CHANNELS CONTINUOUSLY
4 ----SCAN CHANNEL 1 CONTINUOUSLY 100 TIMES
5 ----GO TO SPECIAL ANALYSIS SUBROUTINES
6 ----EXIT
WHICH OPTION
?5
TTL TO MOS LEVEL CONVERTER ---TTL LOW INPUT STATE

DIODE D1
VOLTAGE DROP =        -.61 VOLTS
R1
CURRENT =  1.0454545E-03 AMPS
POWER =  2.4045452E-03
Q1
VCE Q1 =  10.12 VOLTS
VBE Q1 =  .63 VOLTS
VCB Q1 =  10.75 VOLTS
R2
R2 DROP =  .27 VOLTS
CURRENT =  5.7446809E-05 AMPS
POWER =  1.5510639E-05 WATTS
R3
R3 DROP =  -6.41 VOLTS
CURRENT =  -1.3638298E-03 AMPS
POWER =  8.742149E-03 WATTS
Q2
VBE Q2 =  .27 VOLTS
VCE Q2 =  .43 VOLTS
VCB Q2 =  .7 VOLTS
IC OF Q2 =  -1.3638298E-03 AMPS
POWER DISSIPATION =  -5.8644681E-04 WATTS
SUPPLY VOLTAGES
 5.04
- 11.02
TO REPEAT THE SAME SELECTION, TYPE AN X
TO SELECT A NEW OPTION,TYPE AN 0
?0

                    OPTION LIST
----------------------------------------------------
0 ----SELECT CHANNELS
1 ----SCAN AND DISPLAY ALL CHANNELS
2 ----SCAN AND DISPLAY SELECTED CHANNELS ONCE
3 ----SCAN AND DISPLAY SELECTED CHANNELS CONTINUOUSLY
4 ----SCAN CHANNEL 1 CONTINUOUSLY 100 TIMES
5 ----GO TO SPECIAL ANALYSIS SUBROUTINES
6 ----EXIT
WHICH OPTION
?6
THANKYOU
READY
```

of a device. This program, while being general in nature, provides specific reference to the value of mating BASIC and analog acquisition.

Figure 4 is a circuit of a TTL to MOS voltage level converter. Its use is to convert 0 and 5 V TTL levels to +5 V and −12 V MOS logic levels. It is a relatively simple circuit, but it shows how BASIC can work for you.

Up to this point I have said that the input

range of the DVM is ±1.999 V. By putting resistor voltage dividers in series with the multiplexer channel inputs, other ranges can be accommodated. A 900 K-100 K resistor divider network will change the input range to ±19.999 V. Some channels can be set for 20 V ranges. With the present CD4051, though, separate resistor dividers are needed on the inputs because the maximum voltage handling capability of the 4051 is the range of its power supply. Relays, which could pass the high voltages, could be configured to allow use of only one selectable divider network, but for now we are limited. If you put resistor dividers on the inputs, the only necessity is to instruct the program to multiply the particular channel reading by an appropriate ranging factor. In this particular case, all input channels have been set for ±19.99 V ranges, and the multiplier is ten.

The program presents an option list. It allows general application as an acquisition and data logging tool. With it, one can select to read and print all eight channels, particular channels, or log a single channel continuously. Option 5 is what it's all about. It automatically records the input voltages and computes the circuit parameters such as power dissipation and voltage drops. A very complicated circuit example would probably have been more impressive, but that is merely a case of applying programming talents to the same set of input data.

One further note of explanation: the call instruction in Maxi BASIC has been misinterpreted by some people. It is not a directly executable instruction, but is rather used in a statement like LET X = CALL (2560,9). The BASIC interpreter will go to memory location decimal 2560 and start executing a machine language subroutine. The number in parentheses after the comma is the value which is put in the D and E registers at the same time. This is a 16 bit value with a range of 0 to 65,535. When the machine language subroutine is finished, it returns to the interpreter. X will then have a value equal to whatever was in the H and L registers when the subroutine ended.

Conclusion

Having eight channels is better than having one, especially if it doesn't cost any more. I've attempted to present a low cost solution to a usually expensive data acquisition problem. As is always the case with computers, the maximum utilization of the device is dependent upon the programmer, and as my college textbooks used to say, this is an exercise left to the reader.■

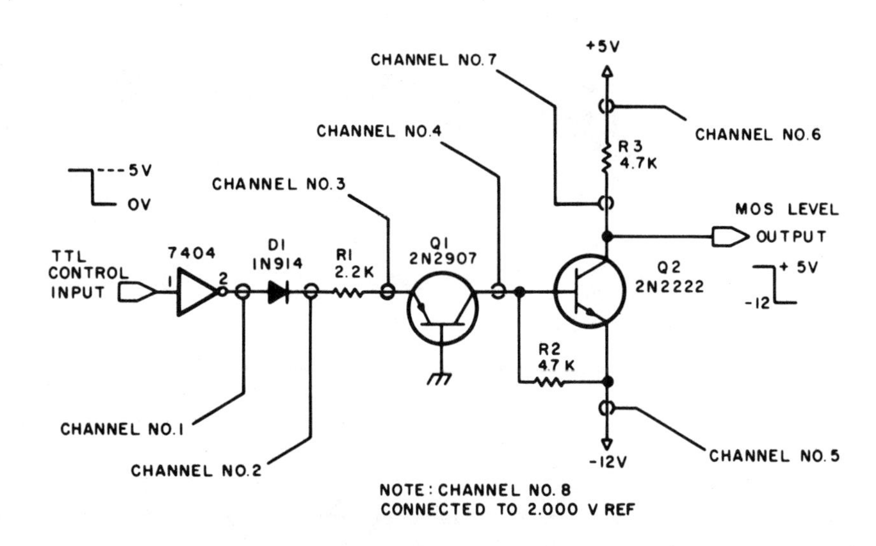

Figure 4: A sample circuit illustrating the use of the 8 channel 3½ digit voltmeter. The circuit is a TTL to MOS voltage level converter.

The author would like to extend special thanks to Dave Hardenbrook for his help in writing the DVM driver program.

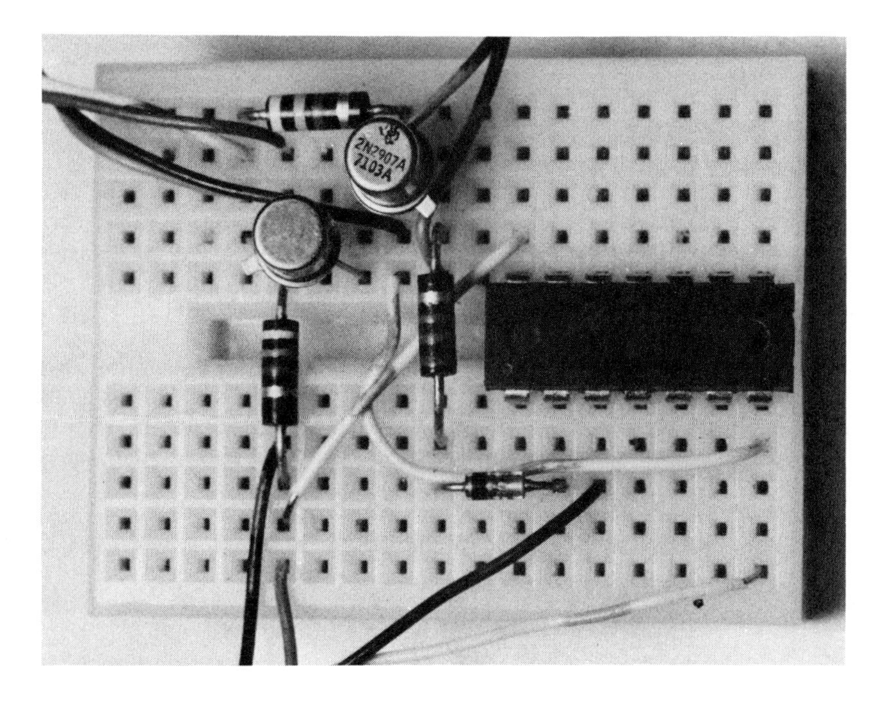

Photo 5: The breadboard circuit of the schematic in figure 4 used to test the 8 channel voltmeter.

Photo 1: The prototype board for the expanded digital voltmeter.

3

Add More Zing to the Cocktail

"Try an 8 Channel DVM Cocktail" brought you a design for an 8 channel 3 1/2 digit 0 to 2 V digital voltmeter (DVM) interface. The article introduced multiplexed analog data acquisition by means of a construction project.

I'm sure that the majority of the readers who have built the DVM will be satisfied with the results. There is of course that small group of problem makers who don't believe the whole world exists in the range from 0 to 2 VDC: a point well taken.

Actually, I planned to expand the capabilities of the basic DVM all along. I'll elaborate in detail; the end result will be a DVM interface with these additional specifications:

Super Cocktail DVM

- 8 programmable input channels
- AC or DC capability
- programmable gain of 1, 10, or 100
- ranges of 0 to 200 mV, 0 to 2 V, 0 to 20 V, or 0 to 200 V
- input overvoltage protection

I had hoped that by presenting the basic 0 to 2 V interface first, more readers would attempt to build it due to its low cost. The extra capabilities presented in this article can be added directly to the previously described hardware interface.

A Quick Review of the Interface Hardware

This DVM is designed around the Motorola MC14433 3 1/2 digit low power complimentary MOS analog to digital converter. The MC14433 is a modified dual ramp integrating analog to digital converter with multiplexed binary coded decimal (BCD) output. With the resistor values chosen it will perform approximately 25 conversions per second.

The full scale voltage value (ie: the value represented by the 3-1/2 digits after any input voltage division) is set by an MC1403 voltage reference integrated circuit. With 2.000 V applied to the V_{Ref} input of the MC14433, full scale is ±1.999 V. If 0.200 V is applied, full scale would be ±0.1999 V or ±199.9 mV.

The MC14433 can directly drive one LS TTL load. Since not all parallel input ports are LS TTL compatible, 74LS04s act as buffers and drivers on all digital voltmeter integrated circuit output pins. Data output is of course inverted and must be complemented before use.

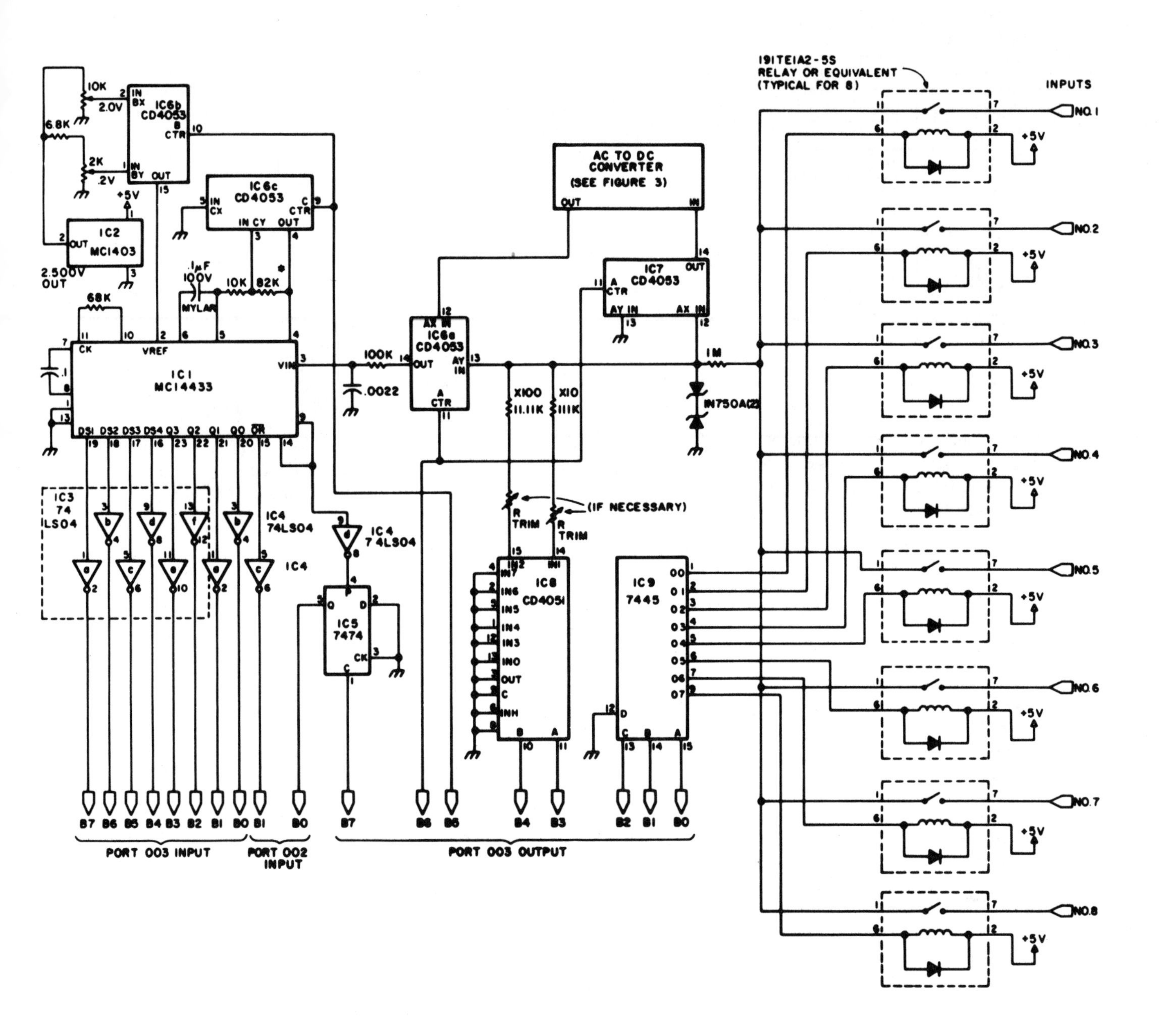

Figure 1a: The modified digital voltmeter, an expansion of the basic design presented in "Try an 8 Channel DVM Cocktail!" (page 7). Changes include the addition of an input multiplexer, made up of eight dual in line package relays, and IC9, a 1 of 10 decoder. The modification allows the voltmeter to handle a wider variety of input voltages, both AC and DC. All resistors are 5% 1/4 W, and all capacitors are 100 V ceramic, unless otherwise indicated.

Number	Type	+5 V	GND	−5 V
IC1	MC14433	24	13	12
IC2	1403	1	3	
IC3	74LS04	14	7	
IC4	74LS04	14	7	
IC5	7474	14	7	
IC6	CD4053	16	8	7
IC7	CD4053	16	8	7
IC8	CD4051	16	8	7
IC9	7445	16	8	

Table 1: Power wiring table for figure 1.

The data from the digital voltmeter to the computer is serial and parallel. There are four digit select lines and four binary coded decimal (BCD) data lines:

pin 23	Q_3	(Most significant bit)
pin 22	Q_2	
pin 21	Q_1	BCD, digit value outputs
pin 20	Q_0	
pin 19	DS1	(Most significant digit)
pin 18	DS2	
pin 17	DS3	Digit select outputs
pin 16	DS4	

With respect to what the computer sees through the 74LS04 buffers, the digit select output is low when the respective digit is selected. The most significant digit (1/2 digit DS1) goes low immediately after an end of conversion pulse followed by the remaining digits sequencing from the most significant to the least significant digit. An interdigit blanking time of two clock periods is included internally to ensure that the BCD data has settled.

During the 1/2 digit (DS1), the polarity and certain status bits are available. Polarity is on Q_2 and a 1 will indicate negative. The 1/2 digit will appear on Q_3 and a 1 will indicate high.

Enhancements to the Basic DVM Interface

Photo 1 and figure 1 illustrate the fully modified DVM interface. It retains the basic interface structure outlined in the original article, but with some additional goodies. The interface is designed for attachment to decoded 8 bit parallel input and output ports and can be polled by a machine language subroutine. More on this later.

The following is a summary of the interface by port allocations. (Note: I have assigned particular octal port numbers to each byte. These designations will run directly with the software driver provided. If the reader wishes to assign some other port numbers, this is fine, but remember to modify the driver software to reflect the changes.)

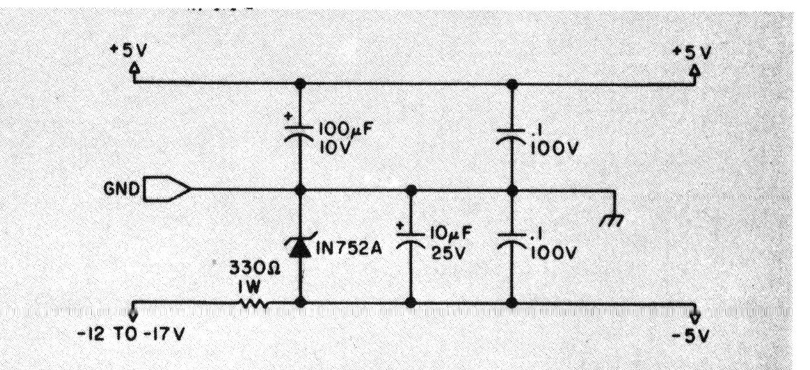

Figure 1b: A circuit enabling the experimenter to derive −5 VDC from an existing power supply having any output from −12 to −17 VDC. −5 VDC is needed to power the various CMOS switches used in this design (see figure 1c).

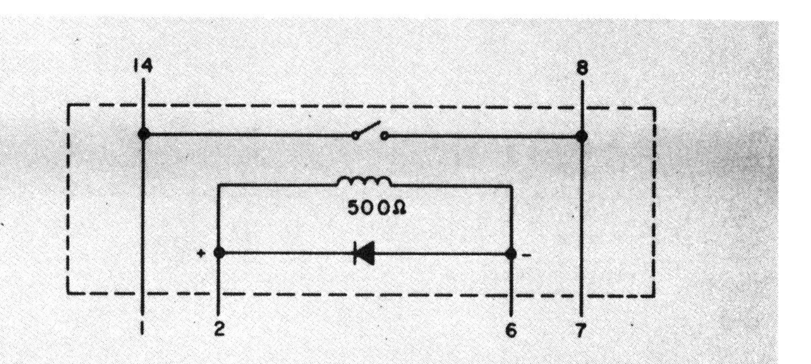

Figure 1c: Pin diagram of a Sigma relay, type 191TE1A2-5S 14 pin dual in line package. For further details and prices, contact SIGMA Instruments, Braintree MA 02184.

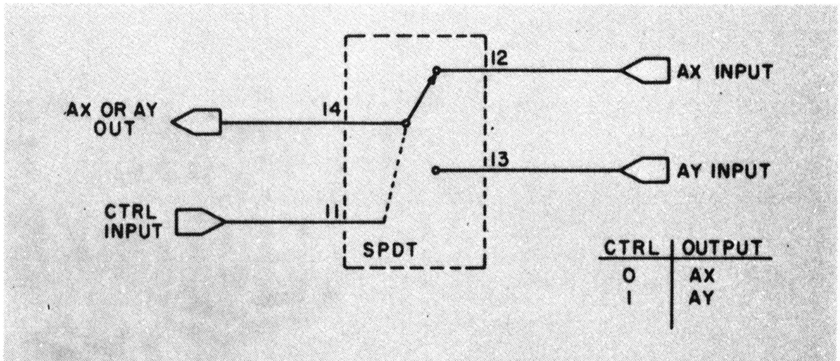

Figure 1d: Functional description of one switching section of a CD 4053 CMOS switch. The device acts like a remote controlled single pole double throw switch.

Command Output Byte (Port 003 Out)

B7 = EOC enable or disable (Disable = 0 Enable = 1)
B6 = AC or DC select (AC = 0; DC = 1)
B5 = 2.0 V or 0.2 V V_{Ref} select (2.0 V = 0; 0.2 V = 1)
B4, B3 = Gain Code
B2, B1, B0 = Channel select, 0 to 7

B_4	B_3	Gain
0	0	×1
0	1	×10
0	0	×100
1	1	N/A (will result in ×1)

Status Input Byte (Port 002 In)

B7, B6, B5, B4, B3, B2 = Not used
B1 = Out of range {−1.999 V > V_{IN} > 1.999 V}
B0 = End of conversion {−199.9 mV > V_{IN} > 199.9 mV}

Data Input Byte (Port 003 In)

B7 = 1st digit
B6 = 2nd digit
B5 = 3rd digit
B4 = 4th digit
B3, B2, B1, B0 = BCD digit value

Most significant digit: when low true = B7 = 0
B6, B5, B4 = N/A
B3 = 1/2 digit value
B2 = Polarity
B1 N/A
B0 = Ranging status bit

The most obvious change to the newly modified DVM board is the input multiplexer. Up to this point all the inputs were multiplexed through a CMOS CD4051 integrated circuit. This device performs quite satisfactorily for inputs in the range of 0 to 2 V. The maximum input voltage range it can handle is limited by its supply voltage (in this case ±5 V). Even if the supplies were increased to ±9 V (18 V absolute, which is the maximum supply for a CD4051B) a separate voltage divider would still be required at each input channel to keep the applied voltages within safe limits. To have a 0 to 200 V range selectable unit incorporating a CD4051 input maximum would require having eight separate programmable dividers and an overvoltage protection circuit on each channel in case the wrong divider values are chosen.

The preferred approach is to have only one divider network and one overvoltage circuit, but such an alternative requires that the input multiplexer be capable of handling all input voltage levels from 0 to 200 V! The answer is to use relays. Not the big 10 A clunkers you see in surplus catalogs, but the new generation of (dual in line package) reed relays such as the Series 191 by SIGMA (see photo 2). These particular relays can be driven directly by TTL logic, exhibit maximum 1 ms bounce, and have a rated life of 100 million operations.

The new input multiplexer section consists of dual in line package relays RL1 thru RL8 and a 1 of 10 decoder, IC9. When a latched output port 3 bit channel address is impressed on the input lines of IC9, it puts a 0 voltage level on the output pin corresponding to that address and pulls in the proper channel select relay. The outputs of all eight relays are wired together and are next directed to the overvoltage and gain divider circuitry.

The input impedance of the DVM chip is very high (on the order of 1000 MΩ). Placing a 1 MΩ resistor in series with the relay outputs facilitates the addition of protection circuitry without compromising the interface's capabilities. This current limiting 1 MΩ resistor and two back-to-back zener diodes limit the absolute voltage seen by the MC14433 to about ± V. (The MC14433's absolute limit is its power supply range, even though its usable input range is ±2 V.) If the correct gains are chosen and programmed to the interface, this protection should never be required. But, no one is perfect.
perfect.

In addition to this function, the 1 MΩ resistor is one leg of a programmable divider network. Figure 2 shows the input subsystem in simplified terms. An AC to DC converter is also included and will be explained later. SW1 and SW2, parts of IC8, represent the gain selection section. The switches are illustrated in a unit gain DC input mode. When an input relay is closed, its applied voltage is sent directly to the DVM integrated circuit input through the 1 MΩ resistor. The AC to DC converter is switched out of the system and with both SW1 and SW2 open, no dividers are in the circuit. If 1.400 V is applied through a closed relay, the DVM will read 1.400 V for that channel. If, on the other hand, 150 V is suddenly switched in on another relay with this SW1 and SW2 setting, the chips would be fried were it not for Z1 and Z2. At

Photo 2: The Sigma dual in line package relay, type 191TE1C2-5S, similiar to that used in the design of the expanded digital voltmeter's input multiplexer.

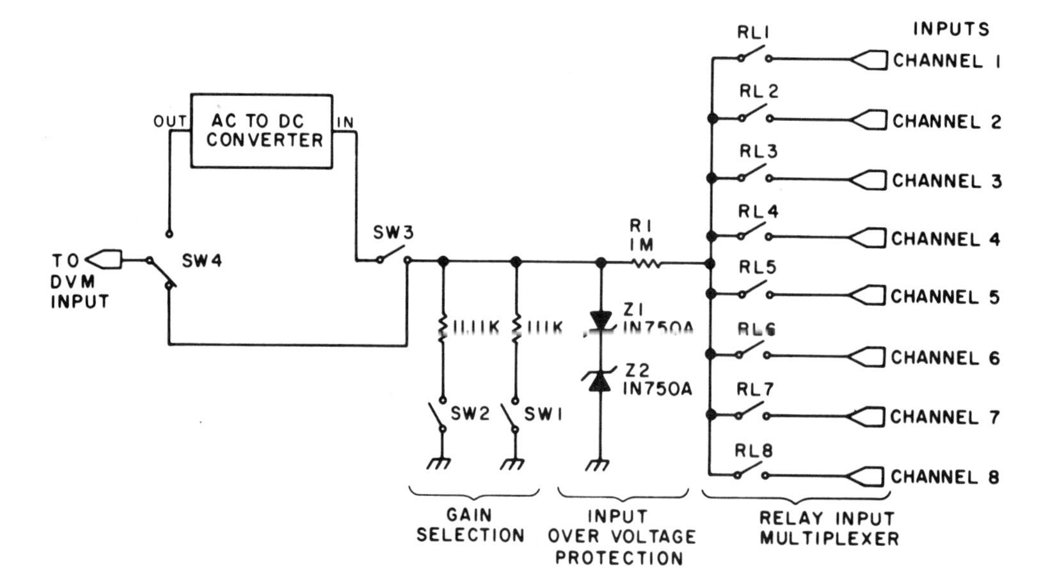

Figure 2: A simplified representation of the input section of the expanded digital voltmeter. The amount of gain and the AC to DC conversion option are selectable by means of CMOS switches.

voltages of less than 4 absolute (±4 V) the diodes do nothing. When inputs exceed this absolute value, Z1 and Z2 clamp them to 4 V. The data acquired by the computer will indicate an out of range condition, since it is over 2 V, but at least it will not have evaporated.

How Do We Read 0 to 200 V Inputs?

Closing switch 2 forms a 10:1 divider network. If 8 V is applied and switch 1 is closed, the result is:

$$VA = \left(\frac{8}{R_1 + R_2}\right) \times R_2$$

$$= \left(\frac{8}{1.111\ M\Omega}\right) \times 111\ K$$

$$= 0.799\ V$$

$$\approx 0.800\ V$$

As you can see, the result of closing SW1 and applying the 111 K resistor is to divide the 8 V by 10 to get 0.799 V. Proper trimming of this 111 K resistor will give an output of 0.800 V. This value is compatible with the DVM integrated circuit input range, and when read by the computer, will be equal to 0.800 V. The programmer should keep in mind that a divider is used on this channel and should multiply the result by 10 to obtain 8.00 V.

Closing switch 2 forms a 100:1 divider. The mathematics is the same except that the divider resistor is 11.11 K instead of 111 K. An 8 V input appears at the DVM input as 0.080 V, while 150 V becomes 1.500 V.

AC to DC Converter

An additional bonus of this interface is AC to DC conversion on any input channel. Figure 3 shows the schematic of the AC to DC converter section of the interface. Bit 6 of output port 003 controls the application of this function. When it is high, SW3 and SW4 are in the positions shown in figure 2. In this state the AC to DC converter is switched out of the circuit and the DVM gets its input directly from the divider section. When bit 6 is programmed to be a low level, switches 3 and 4 switch to their alternate positions and route the input signal from the divider network through the AC to DC converter. The resulting signal is equal to the average RMS value of the applied input signal. This is basically the same type of circuit as the kind included in many single channel digital meters.

The AC to DC converter consists of three sections of an LM324, IC10. IC10A is a high impedance input buffer with variable offset adjustment. When the converter is switched into play, it must have a high input impedance to avoid loading down the divider network. IC10B is the actual AC to DC conversion section. Its output is a current proportional to the AC input voltage. IC10C converts this current to a voltage and provides ripple filtering. One consideration to keep in mind is that, since this is a multiplexing analog to digital converter, the usual DC blocking capacitor at the input of the AC to DC converter has been removed. Given the particular circuit impedances and desired frequency range the converter should cover, the input capacitor had to be removed because it couldn't respond quickly enough. The result is an AC to DC converter that will pass both AC and DC

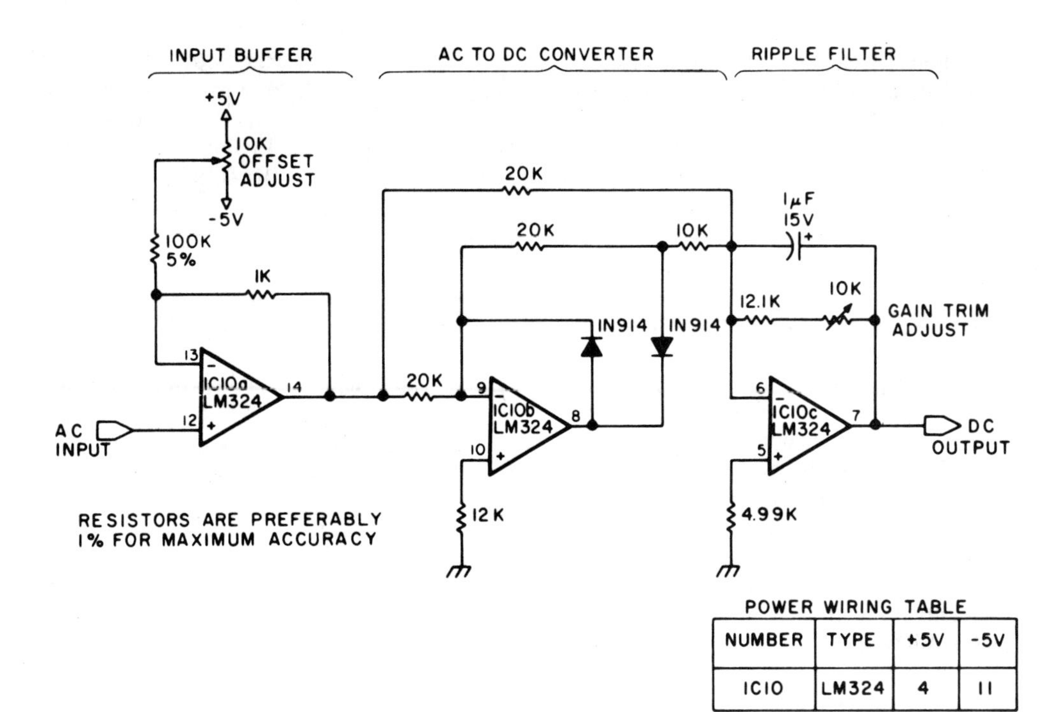

NUMBER	TYPE	+5V	-5V
IC10	LM324	4	11

Figure 3: Circuitry of the AC to DC converter used by the digital voltmeter. All resistors are 1% except where shown.

signals; only the AC converted signals should be used, however.

When a 1.0 V peak AC signal (60 Hz) is applied to the converter, the output should be +0.707 VDC. If by accident the AC converter is switched into a DC signal, the output of the converter will be 1.414 times the true DC input. Keep a close watch on your program command byte to the interface.

Adding a 199.9 mV Range

Up to this point we have discussed additions to the basic circuit that allow range selections of 0 to 2 V, 0 to 20 V, and 0 to 200 V. Circuit changes can be incorporated to extend the DVM range in the opposite direction. Figure 4 illustrates the voltage reference and range selection setup of this interface. The MC14433 can also be configured to cover a range of 0 to ±199.9 mV. When bit 5 of port 003 is low, switches 5 and 6 are in the positions shown. A ratiometric converter has a range determined by the applied V_{Ref}, which would be 2.000 V in this instance. With SW5 open, the integrating time constant is set by using a 82 K resistor (this can be formed by a series combination of resistors). With bit 5 set to a 1, the converter changes its V_{Ref} level to 0.200 V and its integration resistor to 10 K. These changes are the only ones necessary for 0 to 0.2 V range selection.

A Less Complicated Driver

The driver was explained in "Try an 8 Channel DVM Cocktail" in detail. It was designed to be as fast as possible. The relay multiplexer added this month unfortunately cannot operate at that speed without modification. I have included a new driver subroutine written especially for this application (see listing 1).

The interface driver is a relocatable subroutine which is polled by a call instruction. The driver is written for page 140 (octal) but is easily relocatable. It occupies less than one 256 byte page of memory and is written for the Z-80 processor. It is especially designed to run with an Extended BASIC which has instructions to access memory and IO ports, and can call a machine language program.

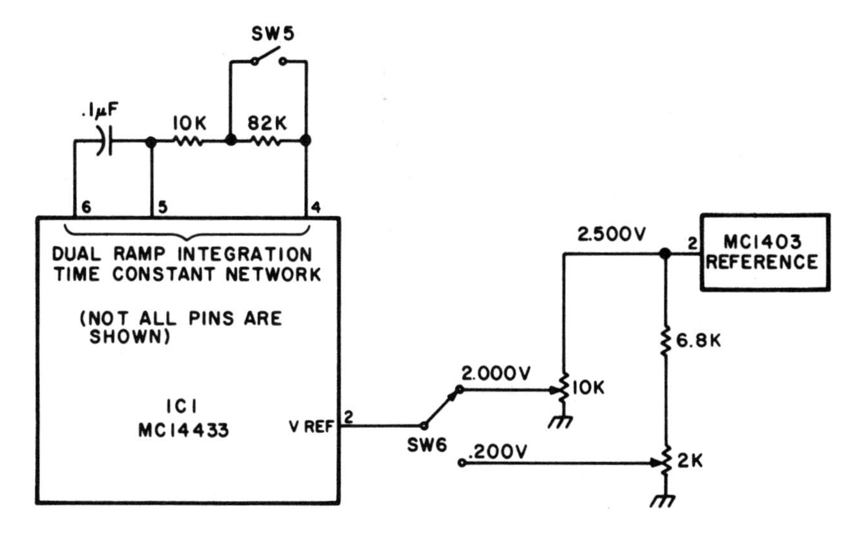

Figure 4: A simplified detail of figure 1, showing the V_{ref} and integration time constant circuitry.

The driver is exercised by a call instruction. In Digital Group Maxi BASIC, the call instruction looks like this: LET X=CALL (24611,64). The BASIC interpreter goes to decimal location 24611 to execute the call and decimal 64 is put in the DE register pair. To the driver, this call is a signal to perform an analog to digital conversion. The contents of the DE register tell it which channel to convert, whether it should be AC or DC, and which V_{Ref} and gain to use. One channel is converted every time the driver is called. The information sent in the DE register at the time of the call is the command output byte (port 003), and each bit has the designations previously listed. The only difference is that bit 7 (the enable disable bit to the analog to digital converter) is sent out as a 0 when doing a call. The driver will set it to an enable condition after it has pulled in the proper relay and allowed a 1.3 ms bounce delay.

When the driver concludes its operation, it has acquired a 3 1/2 digit voltage reading from the DVM which is represented by four bytes. These four bytes are placed in a table in memory. The eight channels of data constitute a 32 byte table. The location of a particular channel's data can be found by a simple expression:

4 byte data location
starts at $L + [4(N-1)]$

where L = starting address of table
N = channel number (1 to 8)

To use the converter with BASIC, the program merely calls for a particular channel conversion and then extracts the appropriate data from the table. Listing 2 is a BASIC program which details the entire procedure.

Conclusion

I often see construction projects which are beyond the means of some experimenters. With these two articles I've attempted to reverse the trend by giving the complete design of a low cost DVM interface. By adding more components, such as relays, this interface can become a full fledged data acquisition system. ■

CAUTION:
One caution should be kept in mind when using this interface to measure AC signals: the ground on the interface board is the same ground as the computer. If you use the interface board to read 115 VAC line voltage, a potential short circuit exists unless either the computer or the measured voltage is isolated. Since isolating the computer equipment would constitute a violation of many electrical codes, only isolated AC signals should be read. A common measurement case which meets this criterion is the AC secondary section of a low voltage power supply such as the unit which runs your computer.

Listing 1: An assembly language program for driving the MC14433 3 1/2 digit analog to digital converter, written for the Z-80.

```
140000                     0140 DIF    EQU  3      DATA INPUT PORT NUMBER
140000                     0150 SIP    EQU  2      STATUS INPUT PORT NUMBER
140000                     0160 COP    EQU  3      COMMAND OUTPUT PORT NUMBER
140000                     0170 EEOC   EQU  200    ENABLE EOC INPUT
140000                     0180 DEOC   EQU  000    DISABLE EOC INPUT
140000                     0190 *
140000                     0200 *
140000                     0210 * CONVERTED CHANNEL DATA BUFFERS
140000                     0220 *
140000 000 000             0230 CHAN0  DW   000000
140002 000 000             0240        DW   000000
140004 000 000             0250 CHAN1  DW   000000
140006 000 000             0260        DW   000000
140010 000 000             0270 CHAN2  DW   000000
140012 000 000             0280        DW   000000
140014 000 000             0290 CHAN3  DW   000000
140016 000 000             0300        DW   000000
140020 000 000             0310 CHAN4  DW   000000
140022 000 000             0320        DW   000000
140024 000 000             0330 CHAN5  DW   000000
140026 000 000             0340        DW   000000
140030 000 000             0350 CHAN6  DW   000000
140032 000 000             0360        DW   000000
140034 000 000             0370 CHAN7  DW   000000
140036 000 000             0380        DW   000000
140040                     0390 *
140040                     0400 * INTERMEDIATE DATA BUFFERS
140040                     0410 *
140040 000                 0430 CHAN   DB   000    CURRENT CHANNEL NUMBER
140041 000 000             0440 CCP    DW   000000 COMMAND CHANNEL PARAMETER
140043                     0460 *
140043                     0470 *
140043                     0480 *** START A/D CONVERTER
140043                     0490 *
140043                     0550 *
140043 173                 0560 START  LD   A,E
```

See page 447 for a listing of parts and products available from Circuit Cellar Inc.

Listing 1, continued:

```
140044 062 041 140          0570          LD    (CCP),A
140047 346 007              0580          AND   007
140051 062 040 140          0590          LD    (CHAN),A
140054 335 041 000 140      0600          LD    IX,CHANO
140060 026 000              0910          LD    D,0
140062 137                  0920          LD    E,A
140063 313 043              0930          SLA   E       CALCULATE BUFFER OFFSET
140065 313 043              0940          SLA   E
140067 335 031              0950          ADD   IX,DE
140071                      0960 *
140071                      0970 * SELECT CHANNEL AND START CONVERSION
140071                      0980 *
140071 006 003              0985          LD    B,3     SET CYCLE COUNT
140073 072 041 140          0990 SCSC     LD    A,(CCP)
140076 323 003              1000          OUT   COP     SELECT CHANNEL
140100 315 243 140          1005          CALL  DELAY
140103 366 200              1010          OR    EEOC    ENABLE EOC OUTPUT
140105 323 003              1020          OUT   COP     COMMAND A/D CONVERTER
140107                      1030 *
140107                      1040 * WAIT FOR EOC
140107                      1050 *
140107 333 002              1060 WEOC     IN    SIP     READ CONVERTER STATUS
140111 313 107              1070          BIT   0,A     TEST FOR EOC
140113 050 372              1080          JR    Z,WEOC  JUMP IF NOT READY
140115 020 354              1085          DJNZ  SCSC
140117 313 117              1090          BIT   1,A     TEST FOR OVERANGE
140121 040 066              1100          JR    NZ,OVER JUMP IF TRUE
140123                      1110 *
140123                      1120 * CONVERSION DONE;PROCESS FIRST (MSD) DIGIT
140123                      1130 *
140123 006 200              1140 MSDO     LD    B,200   SELECT DIGIT 1
140125 315 232 140          1150          CALL  RDIG    WAIT AND READ DIGIT 1
140130 057                  1160          CPL
140131 017                  1170          RRCA RIGHT    JUSTIFY DIGIT VALUE
140132 017                  1180          RRCA
140133 017                  1190          RRCA
140134 346 001              1200          AND   1       ISOLATE
140136 036 000              1210          LD    E,0     INITIALIZE STATUS BYTE
140140 313 122              1220          BIT   2,D     TEST POLARITY
140142 040 002              1230          JR    NZ,MSD3 JUMP IF POSITIVE
140144 036 200              1240          LD    E,200   LOAD POLARITY SIGN
140146                      1440 *
140146                      1450 * SAVE MSD AND CURRENT POLARITY
140146                      1460 *
140146 263                  1470 MSD3     OR    E       ADD POLARITY SIGN TO MSD
140147 335 167 000          1480          LD    (IX+0),A SAVE IN DATA BUFFER
140152                      1500 *
140152                      1510 * PROCESS 2ND DIGIT
140152                      1520 *
140152 313 010              1530          RRC   B       SELECT DIGIT 2
140154 315 232 140          1540          CALL  RDIG    WAIT AND READ DIGIT
140157 346 017              1550          AND   017     ISOLATE
140161 335 167 001          1560          LD    (IX+1),A STORE SECOND DIGIT
140164                      1570 *
140164                      1580 * PROCESS 3RD DIGIT
140164                      1590 *
140164 313 010              1600          RRC   B       SELECT 3RD DIGIT
140166 315 232 140          1610          CALL  RDIG    WAIT AND READ DIGIT
140171 346 017              1620          AND   017     ISOLATE
140173 335 167 002          1630          LD    (IX+2),A STORE
140176                      1640 *
140176                      1650 * PROCESS 4TH DIGIT
140176                      1660 *
140176 313 010              1670          RRC   B       SELECT 4TH DIGIT
140200 315 232 140          1680          CALL  RDIG    WAIT AND READ DIGIT
140203 346 017              1690          AND   017     ISOLATE
140205 335 167 003          1700          LD    (IX+3),A STORE
140210 311                  1710 RAPUP    RET
140211                      1720 *
140211                      1730 * LOAD 2.000 OVERRANGE VALUE INTO DATA BUFFER
140211                      1740 *
140211 076 002              1750 OVER     LD    A,2     LOAD MSD VALUE
140213 335 167 000          1760          LD    (IX+0),A
140216 257                  1770          XOR   A
140217 335 167 001          1780          LD    (IX+1),A LOAD LSD VALUES
140222 335 167 002          1790          LD    (IX+2),A
140225 335 167 003          1800          LD    (IX+3),A
140230 030 356              1810          JR    RAPUP
140232                      1870 *
140232                      1880 *
140232                      1890 * READ DIGIT ROUTINE
140232                      1900 *
140232 333 003              1910 RDIG     IN    DIP     READ DATA BYTE
140234 057                  1920          CPL           CONVERT TO HIGH TRUE LOGIC
140235 127                  1930          LD    D,A     SAVE COPY
140236 240                  1940          AND   B       TEST FOR GIVEN DIGIT READY
140237 050 371              1950          JR    Z,RDIG  JUMP IF NOT
140241 172                  1960          LD    A,D     RESTORE A REGISTER
140242 311                  1970          RET           RETURN TO CALLER
140243 016 377              1980 DELAY    LD    C,377
140245 015                  1990 DEL1     DEC   C
140246 310                  2000          RET   Z
140247 030 374              2010          JR    DEL1
```

Listing 2: A supervisory program for controlling the expanded digital voltmeter written in extended BASIC.

```
120 REM 8 CHANNEL 3 1/2 DIGIT AC/DC PROGRAMABLE RANGE DVM -S.CIARCIA
130 REM REV. 1.9
140 REM BOARD CHECK OUT PROGRAM
150 REM
160 REM
170 LET M1=24576
180 REM THIS IS PAGE 140(OCTAL)
190 LET M2=24611
200 REM THIS IS THE CALL ADDRESS
210 REM
220 PRINT
230 PRINT
240 PRINT"DO YOU WANT TO SCAN PREVIOUSLY CHOSEN CHANNELS OR"
250 PRINT"SELECT NEW ONES ?   SCAN OR SELECT OR STOP"
260 INPUT S$
265 IF S$="STOP" THEN GOTO 2000
270 IF S$="SCAN" THEN GOTO 830
280 PRINT"SELECT ALL VALUES OR CHANGE ONE CHANNEL"
290 PRINT"ALL OR ONE"; :INPUT S$
300 IF S$<>"ONE" THEN GOTO 420
310 PRINT
320 PRINT"WHICH CHANNEL DO YOU WISH TO CHANGE "; :INPUT C
330 PRINT"PRESENTLY CHOSEN VALUES ARE "
340 IF D(C)=1 THEN R1=.2 ELSE R1=2.0
350 PRINT"VREF.=";R1;" VOLTS   DIVIDER GAIN IS X";F(C);"   CONDITIONING IS FOR ";
360 IF C(C)=1 THEN PRINT"DC" ELSE PRINT"AC"
370 LET A(C)= 1 :GOSUB 590
380 GOSUB 750
390 PRINT"ANOTHER CHANNEL TO CHANGE ?  Y OR N"; :INPUT R$
400 IF R$<>"N" THEN GOTO 320
410 GOTO 830
420 PRINT
425 PRINT"INPUT CHANNEL PARAMETERS"
430 PRINT"GAIN MULTIPLIER IS  1,10 OR,100"
440 PRINT"ENTER CHANNEL PARAMETERS AS REQUIRED"
450 PRINT :PRINT: PRINT
460 FOR C=1 TO 8
470 PRINT"DO YOU WANT TO READ CHANNEL ";C;"        Y OR N OR EXIT";
480 INPUT A$
490 IF A$="EXIT" THEN GOTO 240
500 LET A(C)=0
510 IF A$="N" THEN GOTO 710
520 IF A$="Y" THEN LET A(C)=1
530 IF A$<>"Y" THEN GOTO 470
540 GOSUB 590
550 GOTO 700
560 REM
570 REM
580 REM THIS IS THE PARAMETER SETTING SUBROUTINE
590 PRINT"GAIN ",
600 INPUT B(C)
610 LET F(C)=B(C)
620 LET E(C)=0
630 IF B(C)=10 THEN LET E(C)=8 :GOTO 650
640 IF B(C)=100 THEN LET E(C)=16 :GOTO 650
650 PRINT"ENTER 1 FOR DC OR 0 FOR AC",
660 INPUT C(C)
670 PRINT"ENTER 1 FOR .2 VOLT, OR 0 FOR 2.0 VOLT'DVM VREF.";
680 INPUT D(C)
690 RETURN
700 PRINT
710 NEXT C
720 REM X1 TO X8 ARE THE CALL SETPOINTS
730 GOSUB 750
740 GOTO 810
750 FOR J=1 TO 8
760 LET X(J)=64*C(J)+32*D(J)+E(J)+J-1
770 REM X(J) IS LOADED WITH THE BIT PATTERN WHICH IS
780 REM PUT IN THE DE REG. PAIR DURING THE CALL INSTRUCTION
790 NEXT J
800 RETURN
810 PRINT
820 PRINT
830 REM THIS ROUTINE DETERMINES WHICH CHANNELS ARE TO BE CONVERTED"
840 FOR C=1 TO 8
850 IF A(C)=0 THEN GOTO 870
860 LET H=CALL(M2,X(C))
870 NEXT C
880 REM THIS ROUTINE PRINTS THE VALUES IN THE MEMORY TABLE
890 LET Z=A(1)+A(2)+A(3)+A(4)+A(5)+A(6)+A(7)+A(8)
900 IF Z=0 THEN PRINT"NO CHANNEL PARAMETERS HAVE BEEN CHOSEN" :GOTO 450
910 LET D=M1
920 FOR L=1 TO 8
930 GOSUB 1030
940 IF A(L)=0 THEN 990
950 IF D(L)=0 THEN GOTO 970
960 IF Y1>=.2 THEN PRINT"CHANNEL ";L;" IS OUT OF RANGE" : GOTO 990
970 IF Y1>=2 THEN PRINT"CHANNEL ";L;" IS OUT OF RANGE" :GOTO 990
980 GOSUB 1230
990 NEXT L
1000 GOTO 170
1010 REM
```

Listing 2, continued:

```
1020 REM
1030 REM THIS ROUTINE EXAMINES THE MEMORY TABLE
1040 REM AND CONVERTS THE 4 BYTES TO A 3 1/2 DIGIT VOLTAGE
1050 LET Q1=EXAM(D)
1060 LET Q=Q1
1070 IF Q1>=128 THEN LET Q=Q1-128
1080 D=D+1
1090 LET W=EXAM(D)
1100 D=D+1
1110 LET E=EXAM(D)
1120 D=D+1
1130 LET R=EXAM(D)
1140 LET D=D+1
1150 LET Y=Q+(.1*W)+(.01*E)+(.001*R)
1160 LET Y1=Y
1170 LET Y=B(L)*Y
1180 IF D(L)=1 THEN LET Y1=Y/10
1190 RETURN
1200 REM
1210 REM
1220 REM
1230 REM THIS SUBROUTINE PRINTS OUT THE VOLTAGE VALUES
1240 PRINT"CHANNEL ";L;" IS ";
1250 IF Q1<128 THEN PRINT" "; : GOTO 1270
1260 IF Q1>=128 THEN PRINT"-";
1270 IF D(L)=1 THEN PRINT %7F4;Y1;" VOLTS "; :GOTO 1310
1280 IF B(L)=100 THEN PRINT %5F1;Y;" VOLTS ";
1290 IF B(L)=10 THEN PRINT %5F2;Y;" VOLTS ";
1300 IF B(L)=1 THEN PRINT %6F3;Y;" VOLTS ";
1310 IF C(L)=0 THEN PRINT"AC";
1320 PRINT
1330 RETURN
2000 END
READY
```

Listing 3: A sample of the program in listing 2 being used to read five different inputs.

```
DO YOU WANT TO SCAN PREVIOUSLY CHOSEN CHANNELS OR
SELECT NEW ONES ?   SCAN OR SELECT OR STOP
?SELECT
SELECT ALL VALUES OR CHANGE ONE CHANNEL
ALL OR ONE?ALL

INPUT CHANNEL PARAMETERS
GAIN MULTIPLIER IS  1,10 OR,100
ENTER CHANNEL PARAMETERS AS REQUIRED

DO YOU WANT TO READ CHANNEL  1        Y OR N OR EXIT?Y
GAIN        ?1
ENTER 1 FOR DC OR 0 FOR AC      ?1
ENTER 1 FOR .2 VOLT, OR 0 FOR 2.0 VOLT DVM VREF.?0

DO YOU WANT TO READ CHANNEL  2        Y OR N OR EXIT?Y
GAIN        ?1
ENTER 1 FOR DC OR 0 FOR AC      ?1
ENTER 1 FOR .2 VOLT, OR 0 FOR 2.0 VOLT DVM VREF.?0

DO YOU WANT TO READ CHANNEL  3        Y OR N OR EXIT?N
DO YOU WANT TO READ CHANNEL  4        Y OR N OR EXIT?N
DO YOU WANT TO READ CHANNEL  5        Y OR N OR EXIT?N
DO YOU WANT TO READ CHANNEL  6        Y OR N OR EXIT?Y
GAIN        ?10
ENTER 1 FOR DC OR 0 FOR AC      ?1
ENTER 1 FOR .2 VOLT, OR 0 FOR 2.0 VOLT DVM VREF.?0

DO YOU WANT TO READ CHANNEL  7        Y OR N OR EXIT?Y
GAIN        ?10
ENTER 1 FOR DC OR 0 FOR AC      ?0
ENTER 1 FOR .2 VOLT, OR 0 FOR 2.0 VOLT DVM VREF.?0

DO YOU WANT TO READ CHANNEL  8        Y OR N OR EXIT?Y
GAIN        ?100
ENTER 1 FOR DC OR 0 FOR AC      ?1
ENTER 1 FOR .2 VOLT, OR 0 FOR 2.0 VOLT DVM VREF.?0


CHANNEL  1 IS   1.515 VOLTS
CHANNEL  2 IS    .114 VOLTS
CHANNEL  6 IS - 9.48 VOLTS
CHANNEL  7 IS   9.40 VOLTS AC
CHANNEL  8 IS  118.2 VOLTS
```

4

Let Your Fingers Do the Talking

Part 1: Add a Noncontact Touch Scanner to Your Video Display

"Thanks for coming, Steve. I'm glad we were finally able to schedule this meeting. This problem we have is driving us crazy." Fred scurried over to me in the waiting room and shook my hand.

"Let's get you signed in at guard headquarters and then I'll introduce you to Ted."

This was the first time Fred and I had ever met. But his look of relief told me he thought I was some kind of engineering whiz kid. I picked up my briefcase and we walked to the guard's desk. The place was the prototype for a blue chip company waiting room. Decked out with numerous perfectly blended chairs and sofas, it gave the impression of slick tastefulness, and above all *money*. Current issues of various news and business magazines were arranged neatly on the highly polished end tables. I imagined somewhere within the inner depths of the company walls a heavy walnut grained office door with a brushed brass plate reading "Customer Coordinator for Waiting Room Impressions."

My "life in a big company" fantasies were interrupted as I signed my name to the guest card. Signing my name and title was the least significant thing they had me do. There were questions of citizenship, social security number and sex, statements that I represented an equal opportunity employer, and a list of subversive organizations to which I might belong. The urge to check them all and watch the bells and whistles go off was curtailed by my basic marketing instincts.

I passed the card, which ultimately revealed more information than even my wife knew about me, to the guard. He frowned and scrutinized the card carefully. The delay was agonizing as he examined every detailed answer.

"I'll have to inspect that briefcase, buddy," he said.

Surely I'm no buddy of yours, sir, I thought to myself. I fully expected the frown I usually receive when my briefcase is inspected at airports. The inspectors thumb through the piles of paperwork and, upon discovery of an issue of BYTE, quickly cover up this unusually titled magazine and gulp an embarrassed "Next!" Much to my surprise the guard seemed uninterested.

"OK, here's your visitor's badge. Remember, you have to be escorted at all times," he said, and whisked me away with a sweep of his hand.

Fred appeared relieved. I was now on the inside and hoped I could help alleviate his urgent problem.

"OK, Fred, what's your problem and how can my company help you?" This was a basic marketing question for our type of business which specializes in technical solutions through custom electronics – which really means providing engineering consulting to companies who have become embroiled in political debate over the latest in-house technical fiasco.

"We'll get to Ted's office in a few seconds and I'll let him explain. Basically we need a black box."

Before I could get the functional requirements from Fred, we arrived at Ted's office. Being introduced to Ted as "director of marketing" elicited a certain degree of respect, because in his company this was a vice-presidential position. Ted motioned me to a seat at his mahogany conference table near the window overlooking the company golf course. After asking how we wanted our coffee, he stated in a very businesslike manner, "I presume Fred has filled you in on the problem?"

Fred jumped in before I could answer, "I'm sorry, Ted, I haven't had a chance to."

Ted stood up, rotated his body 90 degrees and pointed to the video display terminal in the corner. "That's my problem! Or rather the computer types downstairs who program it!"

I looked at the display. It was a standard graphics terminal similar to those available from several manufacturers.

Ted continued, "Programmers program computers for other programmers! They never think of the user. I drag that terminal to board meetings so we can review marketing figures, and I spend half my time entering 8 digit passwords, hitting escape and control keys to select options, and answering endless quantities of mindless interrogation." Ted was getting a little hot under the collar. "Time is money in those meetings and here in my office. I don't want to spend all day playing true confessor with a computer! Its function is to display information and that's all the interaction I want."

Ted's problem was not unusual. Where a program requires that the next entry be a control R, one had better type a control R. In higher level systems operators need all kinds of cross reference manuals to communicate in the different languages.

"Look," Ted turned on the display and typed the log-on password and terminal identification. Various options were displayed. "This is what I mean. If I want one of these options, I have to type a 5 digit code, wait to give a particular file number and then some other code."

As displays flashed on the screen I couldn't help but offer the obvious question, "Ted, why can't your programmers just change the software to allow single or 2 digit entry?"

"That would be fine if the software weren't already written. We're talking about millions of dollars worth of software and I'm using only a small portion within a large operating system. I want to be able to choose what I want simply."

Ted needed a "black box" and he knew exactly what he wanted.

"I want something to replace this keyboard for the limited specific application of menu selection and display. Put a log-on button on it. When I press log-on it will send whatever information is necessary. The user should know only that he or she has to log into the system—that's all. Next, give me a key that will send the necessary message to get into the menu programs I use and then I can select the options by number. You send any other messages that are necessary."

Ted was not discussing the usual black box. He was promoting the idea of intelligent rather than dogmatic communication with the computer. A person at this level in the corporate structure could not be expected to maintain the code word and syntax library of the average programmer downstairs. What he wanted was only logical. I left the meeting with the feeling that here was a man who also realized it was time to fight rather than conform.

Perhaps if computers were programmed less for interaction with computer peripherals and more with the human operator in mind, people would be less afraid of them. Ted's application was specific and repetitive but he was still burdened with the general system protocol. In a company that probably had a thousand programmers generating software, his cry to change everything to allow simple input and output (IO) for his application would be fighting an uphill battle. He knew this and also realized that it was easier to change it at his end.

We would make Ted's black box for him and it would solve his immediate problem, but what of the future?

Do your computer input devices limit you? Many personal computer systems have this problem.

Consider a simple program to teach your child mathematics. Such a program in its least complicated form might involve a multiple choice and printout something like this:

```
4 x 8 = 28, 30, 31, 32, or 35
   The right answer is?
```

Most BASICs would require typing 32 and a carriage return. Don't forget the carriage return! Remember, you have to conform to the input protocol of the BASIC.

Now, before I explain what I'm driving at, let me give another example. Say you want to use your system for a home management application, such as putting together a shopping list. You could list out the following on the screen:

```
1. Milk            6. Peanut butter
2. Butter          7. Dog biscuits
3. Margarine       8. Cheese
4. Eggs            9. Coffee
5. Rice           10. Tomatoes
      control P for next page
```

Obviously, the number and a carriage return could be entered to choose the items that would be ultimately listed out as a shopping list. A few pages along in the listings, though, the entry data will get more complex strictly from the sheer volume of possible choices. Most homemakers would tire of the complexity of such a system even though the concept of just choosing items from a list sounds simple.

The solution is to watch the way our young mathematics student might react when we display the expression 4 X 8 on the screen. The natural response is to *point* to the answer!

The homemaker would appreciate using a system that communicates in straightforward terms. Display a list of groceries and let the user point to the desired items.

A New Data Input Device

How do you point to a particular selection on a video display generated menu? The computer needs to know how to interpret your response regardless of the input device. The ASCII keyboard is strictly an input code to the computer. There are unique codes for each switch on the keyboard. The computer doesn't know the location of the particular key that prints an R or a Q. It recognizes only a 7 bit code for these letters. If you don't have a keyboard on your computer, but want to check out some software that needs very little typed entry, you could use seven toggle switches. It would be very slow, but the computer wouldn't care. All it's concerned about is that you present the code it wants.

The same goes for any device attached to a computer. The most obvious way to point to a video display screen and have the computer understand it is to use a light pen. Such units have been described before in BYTE so I won't go into too much detail here (see the references at the end of the article). All a light pen interface does is present to the computer, usually in the same manner as a keyboard input, a code representing a position on the video display screen. This code has to be translated by the program from a position into an action. More on this later.

But, why use a light pen? This again makes the operator conform more than necessary.

Fingers Came Before Light Pens!

Though not capable of the same positional resolution as the light pen, it is possible to design an interface that allows a noncontact data input. Photo 1 is a picture of the prototype designed to illustrate such a technique. It is an infrared scanning system that serves as a low resolution noncontact digitizer. In this particular case it is mounted on the front of a video display to approximate the function of a light pen, but it could just as easily be laid over a typed sheet of

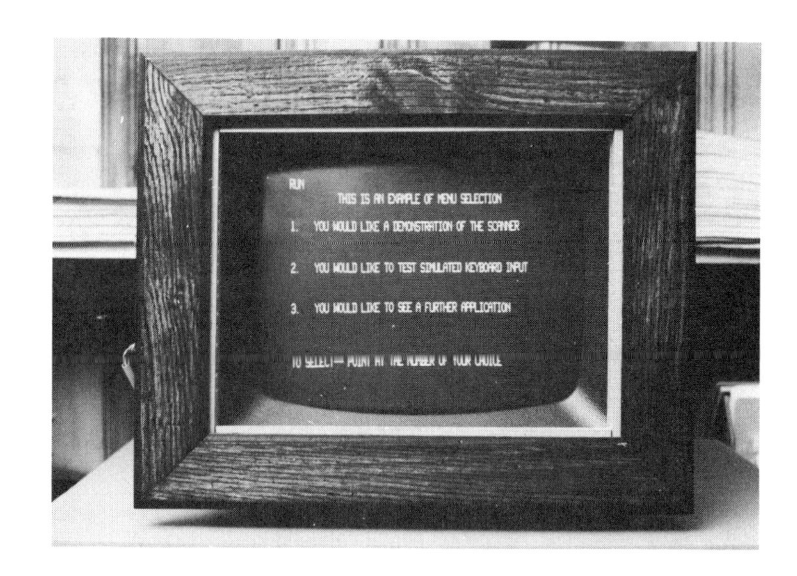

Photo 1a: Noncontact scanning digitizer in action with a BASIC program.

Photo 1b: Side view of the video monitor showing circuitry mounted on two printed circuit boards on either side of the picture frame.

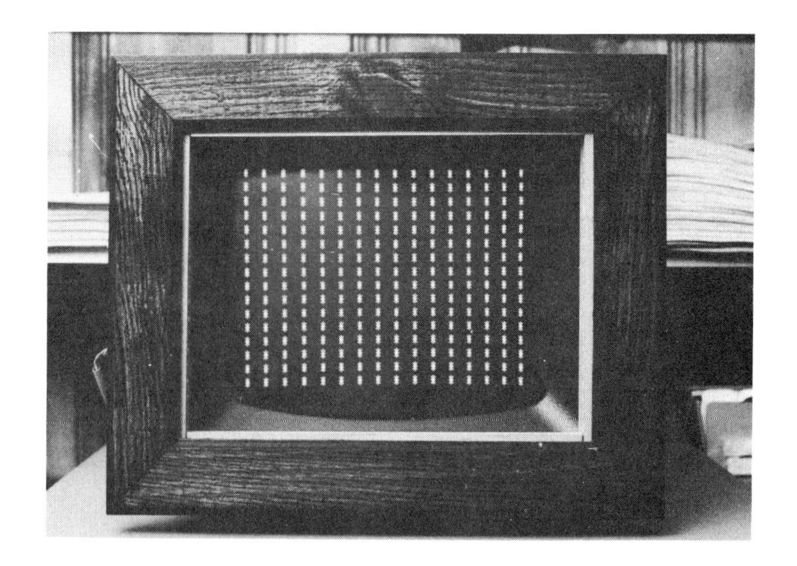

Photo 2: Display showing locations of the 256 points of the array.

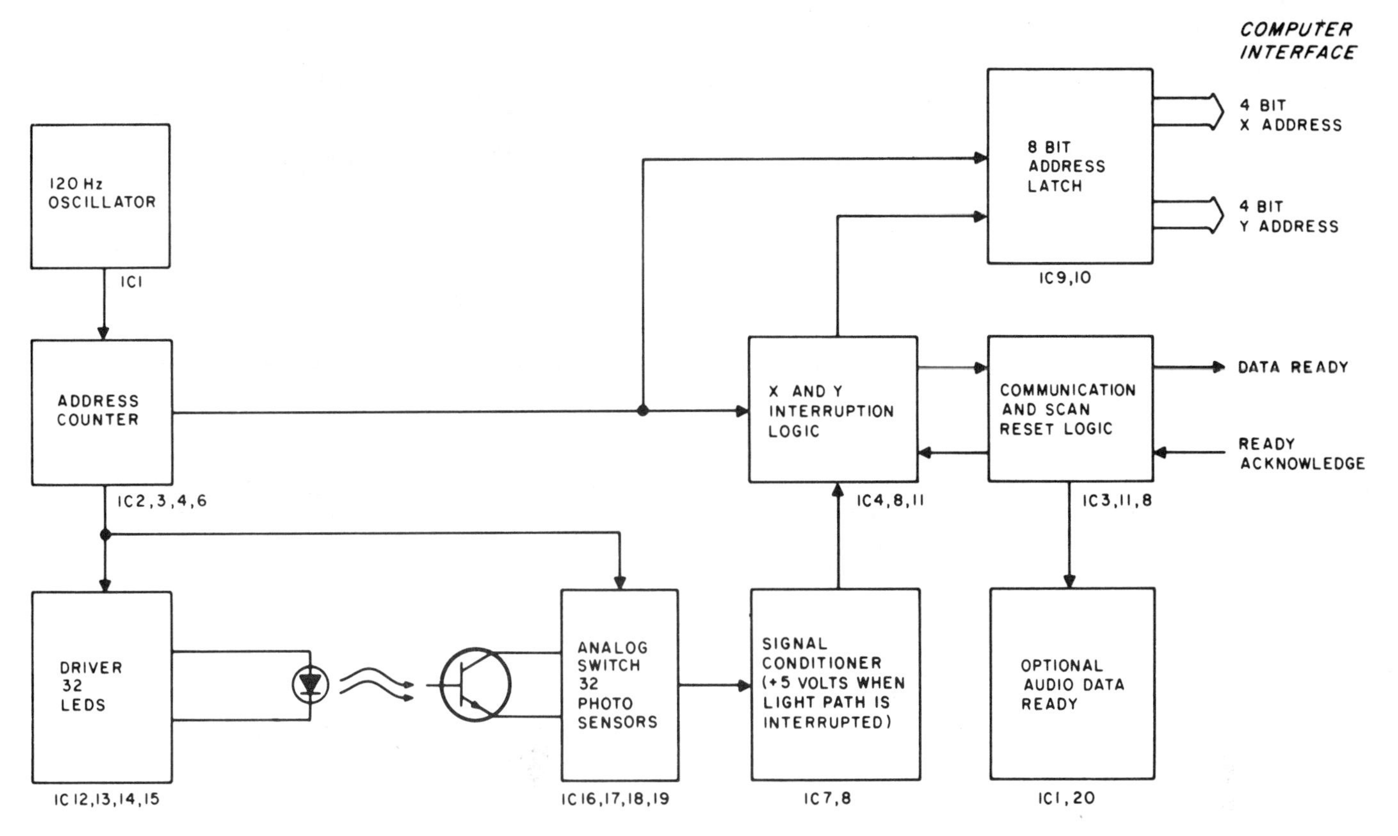

Figure 1: Block diagram of the noncontact scanning digitizer. Two rows of 16 pairs of LEDs and phototransistors are placed opposite each other in front of a video display. When the user breaks the infrared light beams with a finger or other object, a signal is sent to the computer giving the coordinates of the point in question.

paper in which position coordinates could be translated into usable relationships. I refer to it as a touch panel or touch scanner for lack of a better word.

Build Your Own Touch Panel

The touch panel is an elaborate infrared scanner. There are 32 pairs of infrared light emitting diode (LED) transmitters and receivers mounted around the perimeter of the screen. There are 16 on the X (or horizontal) axis and 16 on the Y (or vertical) axis. The resolution of such a device is therefore 16 by 16, and there are 256 individual points. Photo 2 shows this grid system.

Figure 1 is the block diagram of the system, and figure 2 shows the detailed schematics of the system. The noncontact digitizer is basically a hardware stepping circuit that turns on each transmitter/receiver pair sequentially and checks to see if anything (like a finger or a pencil) is blocking the beam. The transmitters and receivers are on opposite sides of the board, as illustrated in figure 3. The lower left corner is position (0,0) in a Cartesian coordinate system. The upper right is location (15,15).

The hardware first turns on the pair D_0 and Q_0 and then sequences down the line along the horizontal (X) axis to D_{15} and Q_{15}. Only one pair is energized at any one time. If any of the beams within these 16 pairs is obstructed, the 4 bit binary code for that location is loaded into IC9. The scan continues in the Y direction in a similar manner and the 4 bit Y position is loaded into IC10. If the hardware senses that something is obstructing an X and Y beam within one scan around the perimeter, it sets a data ready flag and stops the scanner.

The data presented to the computer is an 8 bit word representing a 4 bit X coordinate and a 4 bit Y coordinate. These lines are simply tied to a parallel input port, in the same manner as all the other devices I design. The data ready bit can be read either as a single bit input on another port, or as a control line on a more intelligent interface. When the program senses that the data ready is high, it reads the scanner data and momentarily pulses the ready reset line low to start the scan cycle again.

Use a Picture Frame

The heart of the system is the LEDs and phototransistors shown in photo 3. The device on the left is a General Electric LED 56 and the photodarlington detector used with

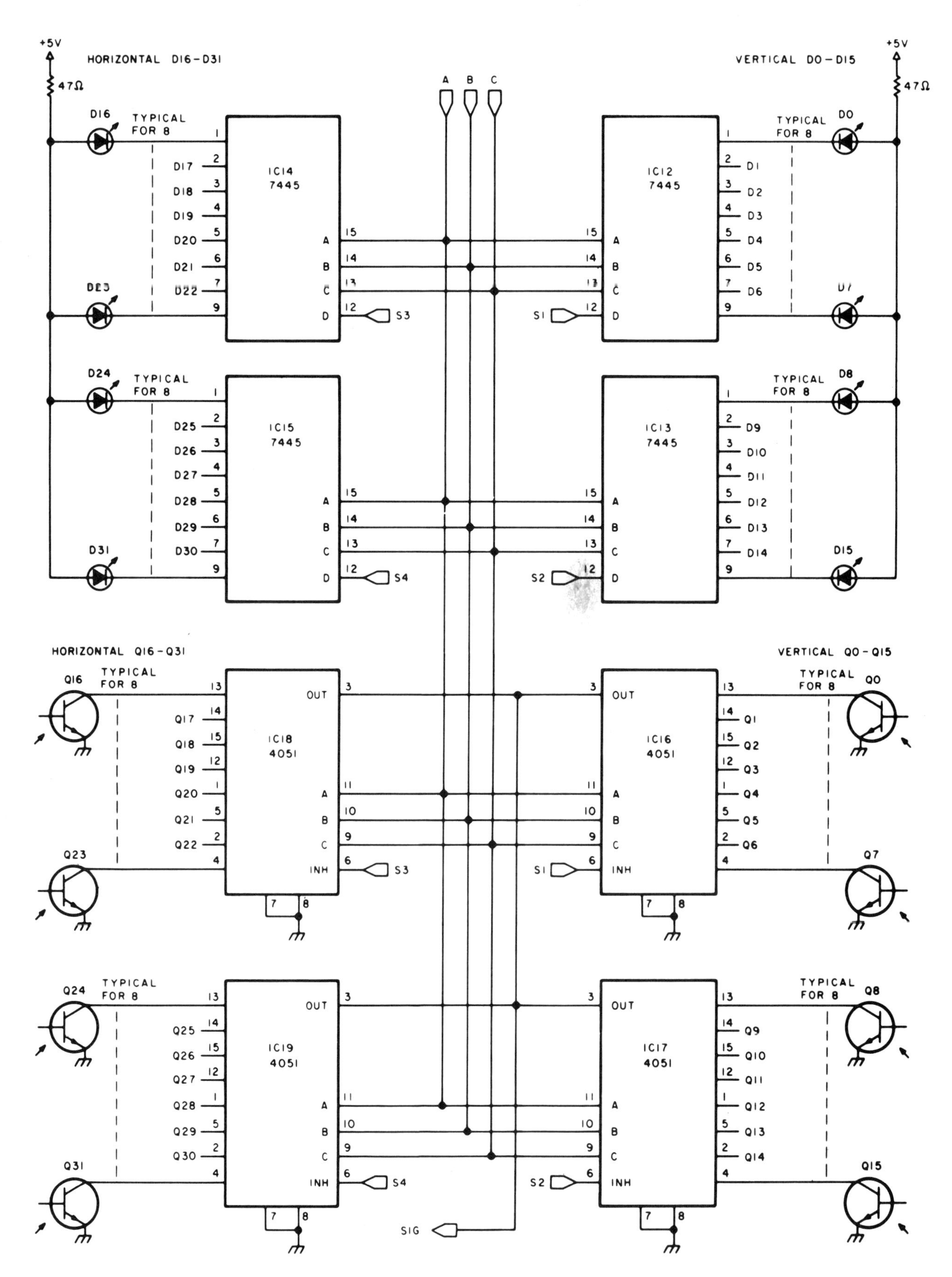

Figure 2a: LED driver and optical receiver circuitry for the noncontact digitizer. Each transmitter/receiver pair (consisting of an LED and phototransistor) is activated sequentially via lines A, B and C. D0 and Q0 are turned on first, and the sequence continues down the horizontal axis to D15 and Q15. If any of the beams is broken, the 4 bit binary code for that location is loaded into IC9 (see figure 2b). The scan continues in the Y direction and the 4 bit Y position is loaded into IC10. Any obstruction causes the data ready flag to be set and the scanner to be halted.

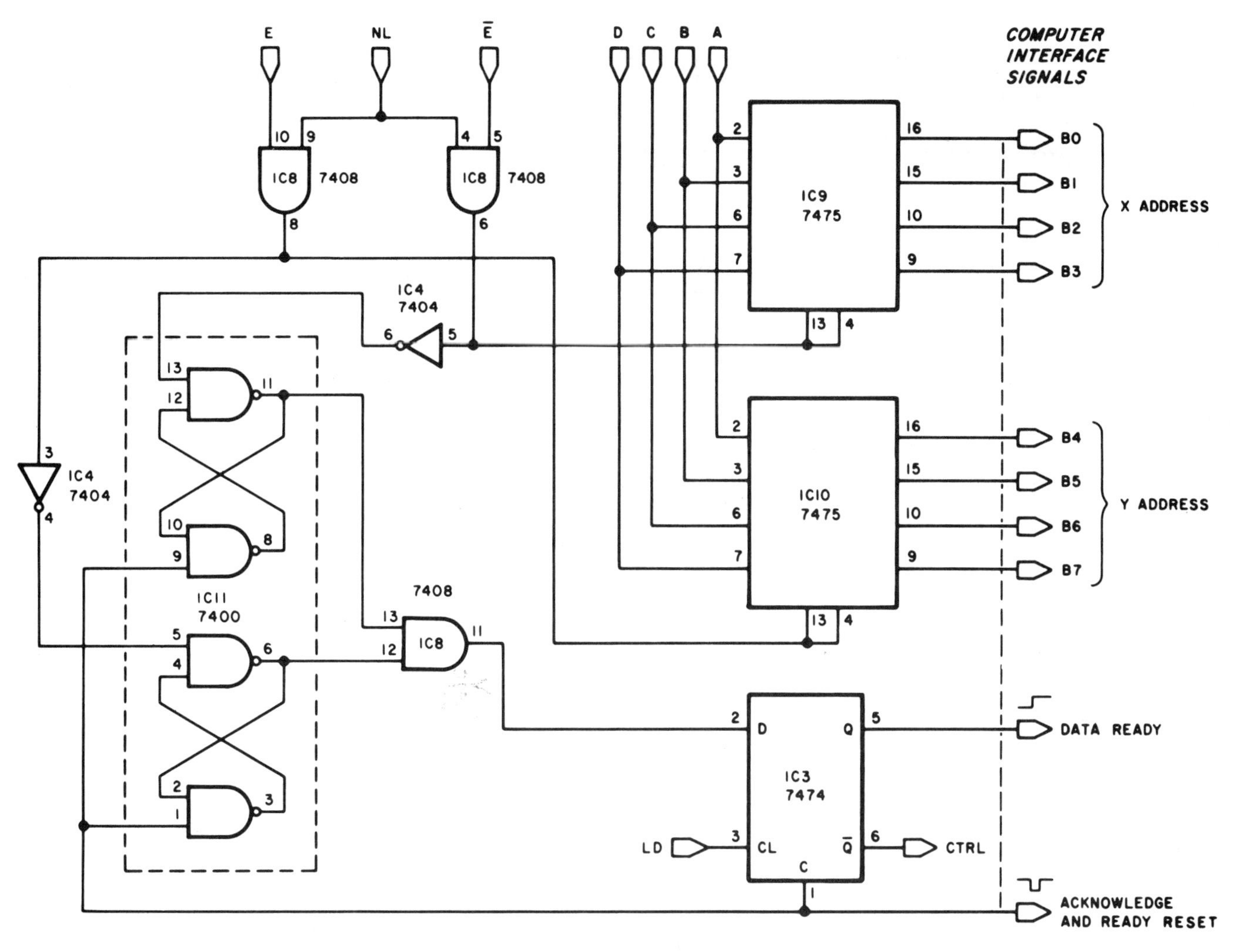

Figure 2b: Interface circuitry for the noncontact digitizer. Data presented to the computer is in the form of an 8 bit word representing a 4 bit X coordinate and a 4 bit Y coordinate. These lines are tied to the parallel input port of the computer.

Notes on figure 2

1. All capacitors are 25 V ceramics unless otherwise specified.
2. All resistors are ¼ W 5 percent unless otherwise specified.
3. ⏚ denotes signal ground.
4. ICs 16 thru 19 are CMOS devices and should be handled carefully.
5. Additional LEDs on prototype unit are for testing purposes only.
6. Q0 thru Q31: GE LED56 infrared emitter.
   D0 thru D31: GE L14F2 photodarlington infrared detector.

NOTE: Any one building a unit from these designs should be advised that they are covered by a number of patents by the University of Illinois and may not be sold for profit.

IC	Type	+5 V	Gnd
1	7400	14	7
2	7493	5	10
3	7474	14	7
4	7404	14	7
5	74155	16	8
6	74123	16	8
7	LM311	8	1
8	7408	14	7
9	7475	5	12
10	7475	5	12
11	7400	14	7
12	7445	16	8
13	7445	16	8
14	7445	16	8
15	7445	16	8
16	CD4051	16	8
17	CD4051	16	8
18	CD4051	16	8
19	CD4051	16	8
20	74121	14	7

Table 1: Power wiring table for the noncontact digitizer.

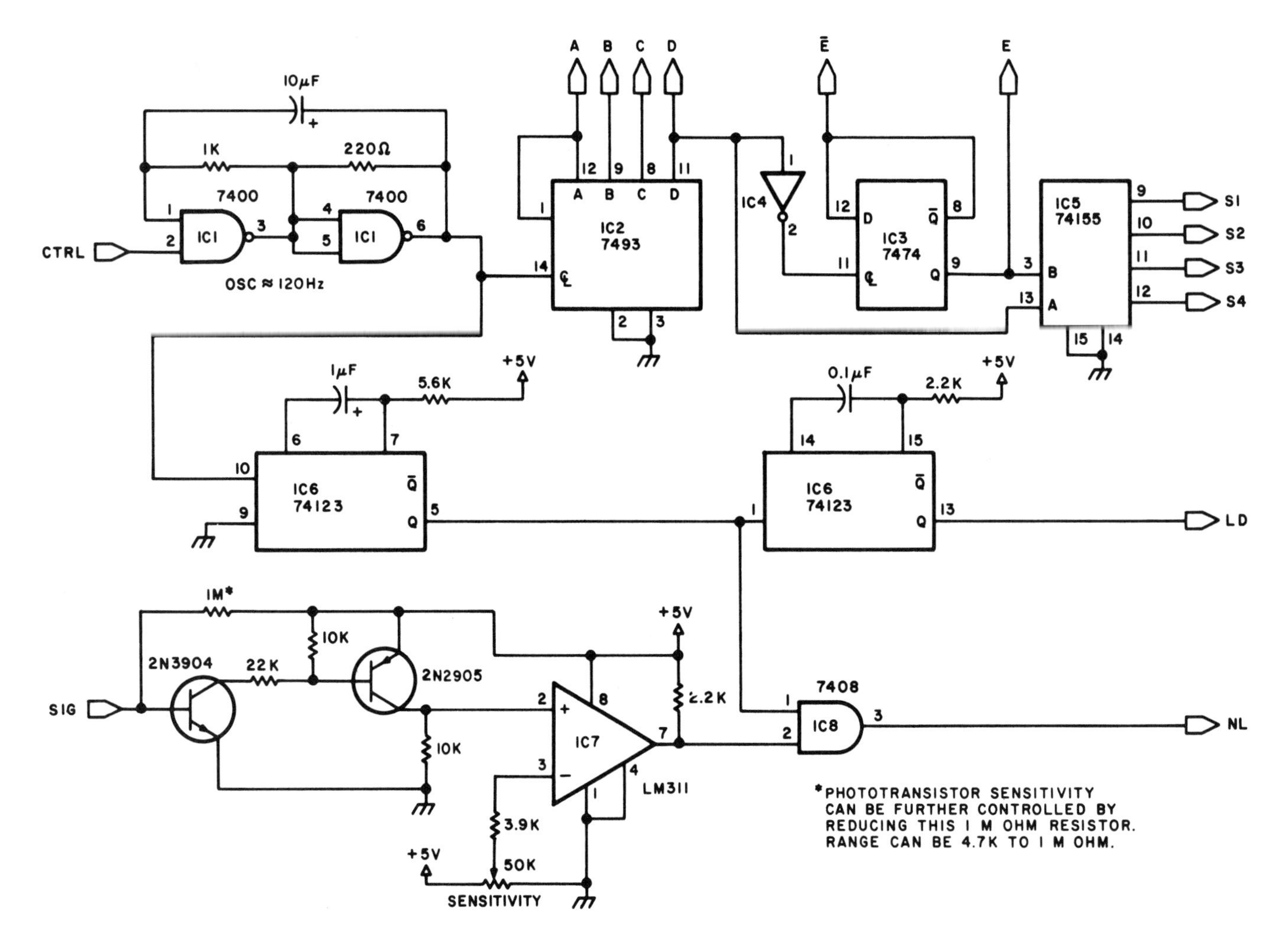

Figure 2c: Address decoder and phototransistor signal conditioning circuitry for the noncontact digitizer. IC2 is a counter driven by the oscillator at upper left. When a phototransistor is activated, the SIG line goes high, activating line NL, which stores the 4 bit address of the interrupted beam (see figure 2b). The scanner is finally halted via the CTRL line. The computer then reads the coordinates and reactivates the scanner.

See page 447 for a listing of parts and products available from Circuit Cellar Inc.

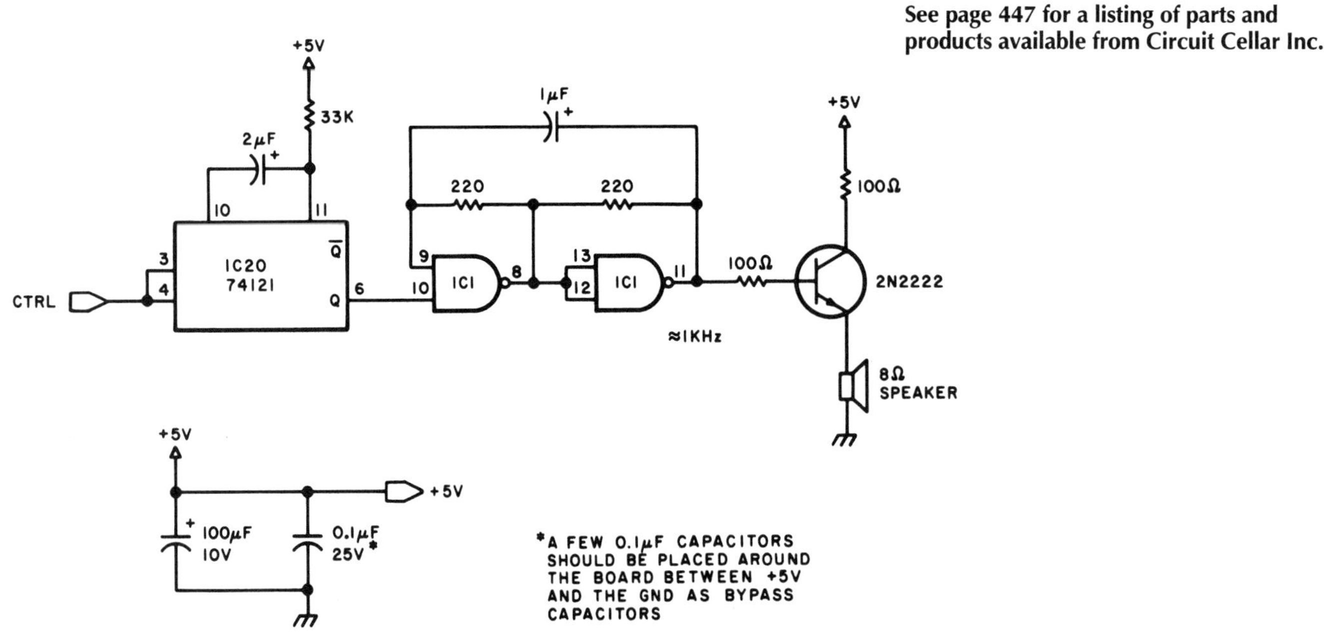

Figure 2d: Optional audio data ready signal circuit, which causes an audible beep on a speaker whenever a pair of beams is obstructed and sets the data ready signal.

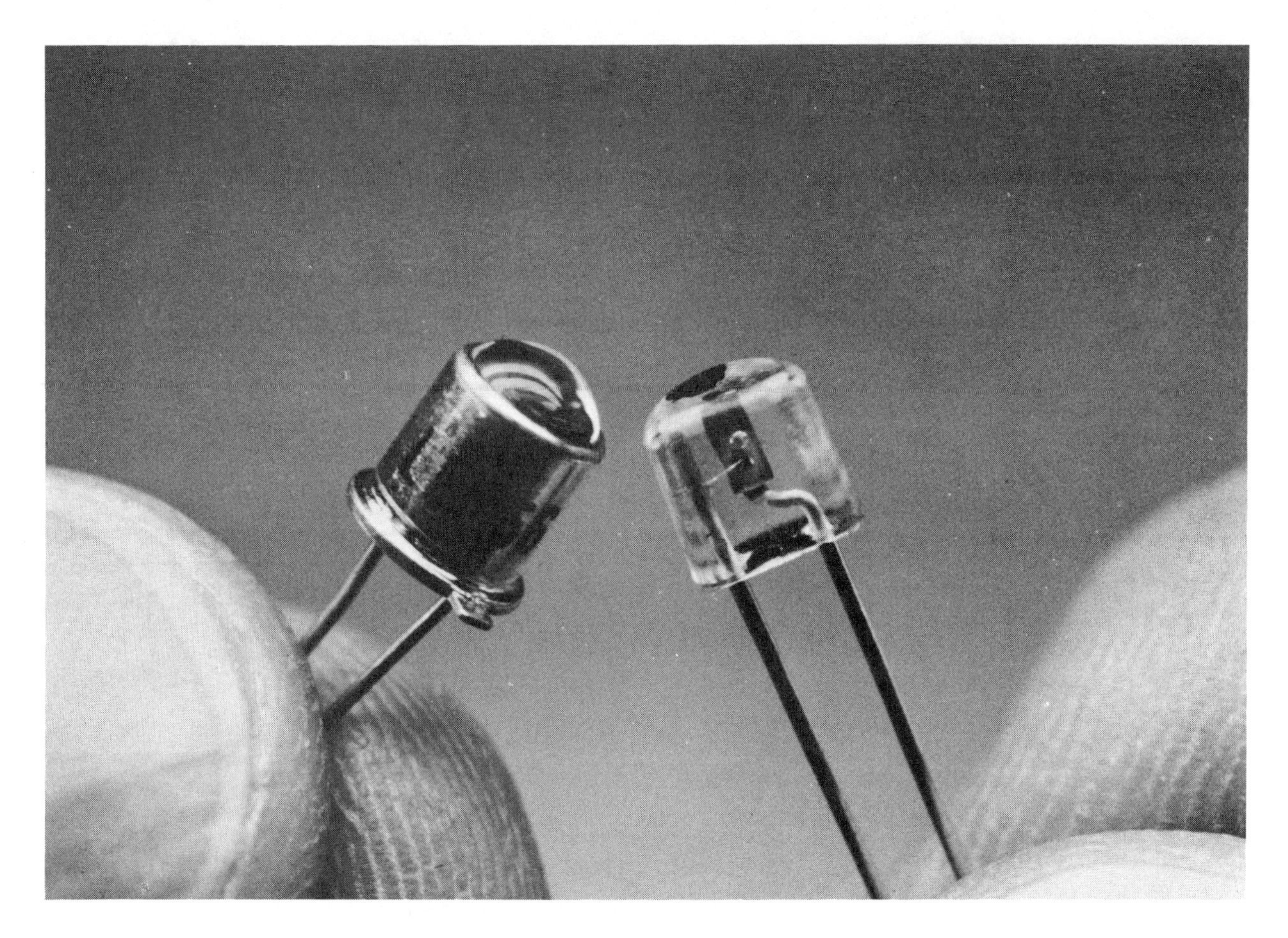

Photo 3: Lensed type GE LED56 light emitting diode (left) and nonlensed H17B1 photodarlington infrared detector. Pairs of either type can be used to transmit and receive infrared light, respectively, for use in the noncontact scanning digitizer.

it is the L14FZ. These units have built-in glass lenses and are very sensitive. A much less expensive though equally capable opto-electric pair is the H17B1 shown on the right in photo 3. Because it has no lens, it requires considerably more shielding from ambient light, but it will work if properly aligned. I have checked the operation of both devices and recommend the lensed type if you intend to use the touch scanner in high ambient light environments. The prototype described here used LED56s and L14F2s.

The frame that holds all the electronics is a $4 discount store wooden picture frame. Half inch (1.27 cm) wooden strips glued around the edges hold the phototransistors and LEDs in evenly spaced, recessed, ¼ inch (0.63 cm) holes. This technique is shown in photos 4a and 4b.

The entire assembly is attached to the picture frame and can be secured to the front of a video display. The display in these photos is a 12 inch (30.76 cm) surplus Phase 4 monitor.

One further addition to the hardware to aid users of the scanner is audio feedback to confirm that a position coordinate has been selected. The data ready strobe triggers a 0.1 second beep on a small speaker.

Photo 4a: Mounting the photodarlington detectors.

Calibration and Testing

There is virtually nothing to calibrate or test on this unit. The only adjustment is the

Notes for figure 3

1. Scan is sequential from D0 thru D31.
2. Only one LED is on at any one time.
3. Scan rate is approximately four samples per second.
4. Total detectable points = 256.

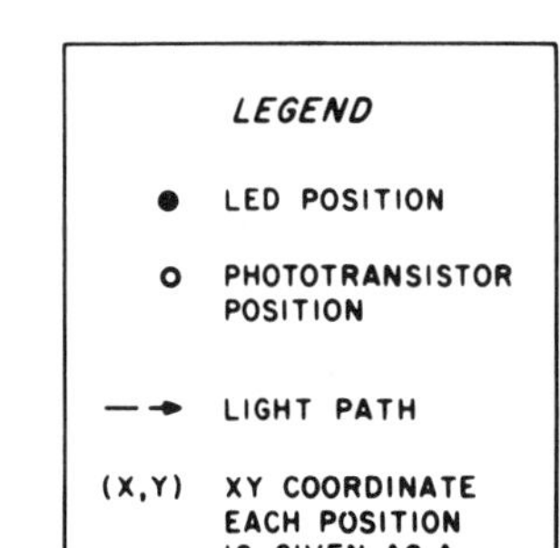

Figure 3: LED and phototransistor placement for the 16 by 16 Cartesian coordinate noncontact digitizer.

sensitivity control on the phototransistor amplifier. Direct sunlight or incandescent lights will cause saturation of the input and disable the scanner. The only other important consideration is mechanical alignment: the LED and phototransistor constituting each pair must be exactly opposite and in direct alignment.

The program in listing 1 is a simple BASIC program to exercise the scanner and provide the operator with an indication of its operational integrity. It is written in Micro Com 8 K Zapple BASIC. The decimal coordinates of X and Y will be output as your finger is moved across the scanned area. This is the only routine that has to be added to any BASIC program to exercise the scanner. If set up as a subroutine by changing line 210 to a RETURN statement, the routine will turn the scanner on when called and return to the main program with a value in variable D representing the coordinates to which you pointed. The main program then responds appropriately.

Obviously the scanner would be more efficiently driven by a machine language program, but I feel most users will be interested in utilizing this device with a high level language. The relatively slow scan rate allows considerable leeway.

In Part 2 I'll pursue the software (in BASIC) necessary to drive this scanner effectively. The major emphasis will be the use of menus and keyboard substitutions.■

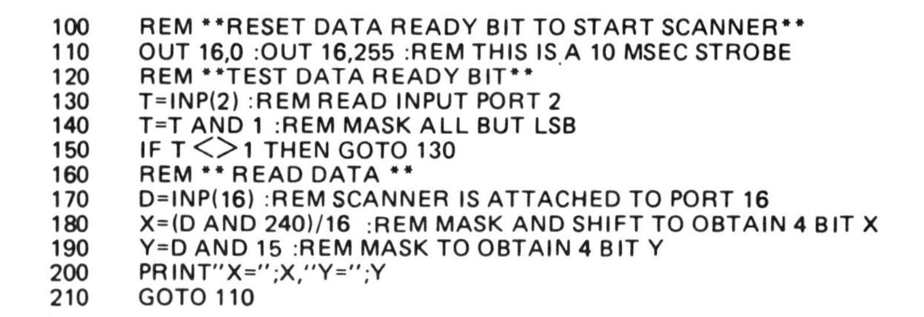

```
100   REM **RESET DATA READY BIT TO START SCANNER**
110   OUT 16,0 :OUT 16,255 :REM THIS IS A 10 MSEC STROBE
120   REM **TEST DATA READY BIT**
130   T=INP(2) :REM READ INPUT PORT 2
140   T=T AND 1 :REM MASK ALL BUT LSB
150   IF T<>1 THEN GOTO 130
160   REM ** READ DATA **
170   D=INP(16) :REM SCANNER IS ATTACHED TO PORT 16
180   X=(D AND 240)/16  :REM MASK AND SHIFT TO OBTAIN 4 BIT X
190   Y=D AND 15 :REM MASK TO OBTAIN 4 BIT Y
200   PRINT"X=";X,"Y=";Y
210   GOTO 110
```

Listing 1: Program written in 8 K Zapple BASIC to exercise the scanner.

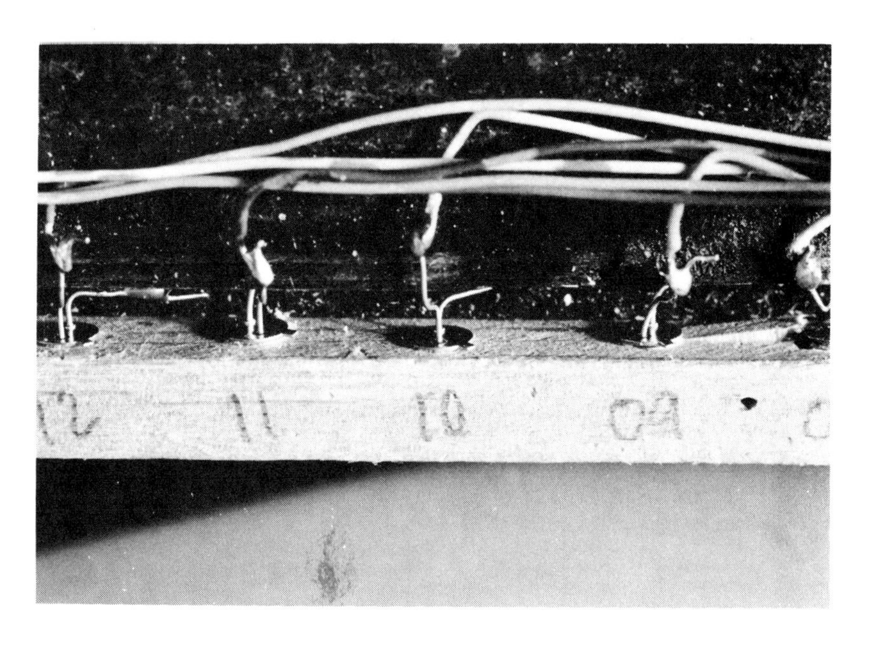

Photo 4b: Mounting the LEDs.

REFERENCES

1. Loomis, Sumner S, "Let There Be Light Pens," January 1976 BYTE, page 26.
2. Webster and Young, "Add a $3 Light Pen to Your Video Display," February 1978 BYTE, page 52.

4

Let Your Fingers Do the Talking

Part 2: Scanner Applications

In "Let Your Fingers Do the Talking, Part 1: Add a Noncontact Touch Scanner to Your Video Display" page 28, I detailed the hardware design of a noncontact touch scanner which sits over a conventional video screen. This system, though lower in resolution, allows a fingertip to simulate the function of a light pen and with proper programming can become as important a peripheral as the common ASCII keyboard.

Quick Hardware Review

The scanner consists of 32 pairs of infrared light emitting diode transmitters and photo transistor receivers arranged around the perimeter of a picture frame. There are 16 pairs on the X axis and 16 pairs on the Y axis. The hardware logic sequentially activates the 32 pairs, first in the X direction (horizontal) then in the Y direction (vertical). If a physical obstruction is placed in the plane of the scan, one X and one Y beam are interrupted. The corresponding X and Y beam addresses are stored when this happens. Since there are 16 pairs per axis, each coordinate can be represented by a 4 bit code and both the X and Y addresses can be packed into one data byte.

The end result of the hardware logic is a very simple scanner to computer interface. The scanner output is one 8 bit byte containing the 4 bit X and 4 bit Y addresses. The only other signals are a little something often referred to as hand shaking. A data ready line is set to a high level output when the scanner has sensed an obstruction.

This data ready signal can be tied to a parallel input port and scanned as I have done, used as a control line on a peripheral interface circuit, or used directly to generate a processor interrupt. If the touch panel is to be exercised in BASIC, the first method will prove to be easiest. The latter method, normally used with a machine language program rather than BASIC, will be the most efficient from a memory utilization standpoint.

I continue to use BASIC wherever the interface data processing speed allows it. In this way I can write illustrative program examples which are not tied to a particular processor. Of course, the speed advantages of machine language may be useful if your programs using the touch panel have a lot to do; so feel free to strike out on your own using these BASIC programs as a model.

Whatever the software method utilized to recognize the data ready bit, the program action must be the same. After the data ready bit goes high, the data byte is stored and the data ready is reset by momentarily pulsing the ready reset line low. In BASIC, the easiest way to do this is to tie the ready reset line to one bit on a parallel output port (it need only be a strobe rather than a latched output) and then sequentially execute two OUT instructions. The 10 ms pulsewidth I get on my machine is the result of the time it takes for BASIC to respond. The program examples presented in the listings use the following port allocations (in decimal):

Photo 1: The basic information returned from the touch panel is a coordinate pair for one of 256 possible finger sized locations on the video display's face. Here, using the program in listing 3, the displayed coordinates 10 and 9 correspond to the point just touched on the screen.

Data Ready	–	Input Port 2 (least significant bit)
Ready Reset	–	Output Port 16 (least significant bit)
X, Y Coordinate	–	Input Port 16 (b_7-b_4 is X address) (b_3-b_0 is Y address).

```
100 REM THIS IS THE ONLY SOFTWARE NECESSARY TO EXERCISE THE
    SCANNER
110 REM *** RESET SCANNER ***
120 OUT 16,0 : OUT 16,255 : REM THIS WILL GIVE A SHORT RESET PULSE TO
    PORT 16
130 REM *** TEST DATA READY ***
140 T=INP(2) : REM THE DATA READY SIGNAL IS BIT 0 OF PORT 2
150 T=T AND 1 :REM MASK ALL BUT BIT 0
160 IF T<>1 THEN GOTO 140 :REM TEST TO SEE IF DATA READY IS SET
170 REM *** READ DATA ***
180 D=INP (16) :REM SCANNER IS ATTACHED TO PORT 16
190 D1=(D AND 240)/16 :REM MASK AND SHIFT RIGHT 4 BITS
195 REM D1 IS THE X COORDINATE
200 D2=D AND 15
205 REM D2 IS THE Y COORDINATE
210 RETURN :REM RETURN IS ONLY NECESSARY IF CALLED AS A
    SUBROUTINE
```

Listing 1: Subroutine used to determine activated coordinates on the scanner.

```
10  PRINT "MY SCREEN ITCHES!! PLEASE SCRATCH IT!"
20  GOSUB 100 :REM ACTIVATE SCANNER
30  PRINT "OH!! THAT FEELS SO GOOOOOOOD!!!"
40  END
```

Listing 2: Example of using the entire video screen as a push button.

```
100 S=USR(255) :REM THIS IS A SCREEN CLEAR FOR DG Z-80
110 PRINT'THIS IS A TEST OF TOUCH INPUT'
120 PRINT'THE SCREEN IS CURRENTLY BEING SCANNED BY AN ARRAY'
130 PRINT'    INFRARED LEDS AND OPTICAL SENSORS'
140 PRINT
150 PRINT'POINT AT THE SCREEN SOMEPLACE '
160 GOSUB 1000 :REM GOTO THE SCANNER SUBROUTINE AND RETURN WITH COORDINATES
170 PRINT'                   THANKYOU'
180 PRINT
190 PRINT
200 PRINT'THE SCANNER HARDWARE SAYS THAT YOU TOUCHED LOCATION'
210 PRINT'          X-';D1,'AND  Y-';D2,'      ON A 16X16 GRID'
220 GOSUB 2500 :REM CALL SLIGHT DELAY TIMER
250 S=USR(255) :REM CLEAR SCREEN
260 PRINT'LET ME DEMONSTRATE THE COORDINATE SYSTEM'
270 PRINT'POINT YOUR FINGER AT THE SCREEN AND I'II PRINT OUT (X,Y)'
280 PRINT'TO EXIT JUST POINT TO LOCATION (15,15) ---UPPER RIGHT'
290 GOSUB 1000 :REM CALL SCANNER
300 S=USR(255) :REM CLEAR SCREEN
310 IF D1=15 THEN 320 ELSE 330
320 IF D2=15 THEN END
330 PRINT
340 PRINT D1,D2; :REM PRINT COORDINATES
350 GOTO 290
1000 REM *** RESET SCANNER ***
1010 OUT 16,0 :OUT 16,255
1050 REM *** TEST DATA READY ***
1060 T=INP(2)
1070 T=T AND 1
1080 IF T<>1 THEN GOTO 1060
1090 REM *** READ DATA ***
1100 D=INP(16)
1110 D1=(D AND 240)/16 :REM THIS IS THE X VALUE
1120 D2=D AND 15 :REM THIS IS THE Y VALUE
1130 RETURN
2500 FOR W=1 TO 2000
2510 NEXT W
2520 RETURN
```

Listing 3: This program outputs the coordinates of the point you are touching on the screen. The output of the program can be used at a higher level to indicate some object that is printed on the screen.

Using the Touch Panel

Using the touch panel in any BASIC program, whether it be game or instructional, will necessitate having a subroutine to read and reset the scanner placed somewhere within the BASIC program. The total software necessary to exercise the touch panel is shown in listing 1.

If a GOSUB 100 command is encountered, BASIC vectors to this subroutine and begins execution. This subroutine will not return until someone touches the screen. Variable D1 would contain the X coordinate and D2 would contain the Y value. Each call to this subroutine results in returning to the main program with the X, Y address of a single touched point. To obtain ten touch inputs would require calling this routine ten times.

The simplest program utilizing the scanner would be one which sensitizes the entire screen to act as one giant push button. Such a program is similar to a press any key option on a keyboard.

The program in listing 2 prints "MY SCREEN ITCHES!! PLEASE SCRATCH IT!" on the video screen, waits for someone to touch any place on the screen and then responds with the message in line 30. Notice that we did not use the coordinate information from the scanner because we only needed to take advantage of the fact that the subroutine returns only if data is *ready*.

Test the Coordinate System

If one builds the touch panel, the first program written should be one that illustrates the coordinate system dynamically, such as the program in listing 3. (All BASIC programs in this article are written in Micro Com 8 K Zapple BASIC.)

After printing an opening comment on the video screen, the program calls the scanner subroutine as before. This time when it returns, it prints out the X and Y coordinate which was touched as shown in photo 1. The rest of the program is a repeat of this basic cycle with one exception. The values of D1 and D2 are both compared to 15 after each scan. Should you point at coordinate position (15,15) the program ends.

Converting Position to Function

So far we have displayed only the raw output of the scanner and have not used it in its true application. Telling you that you are pointing to location (4,2) illustrates that the touch panel functions, but does no use-

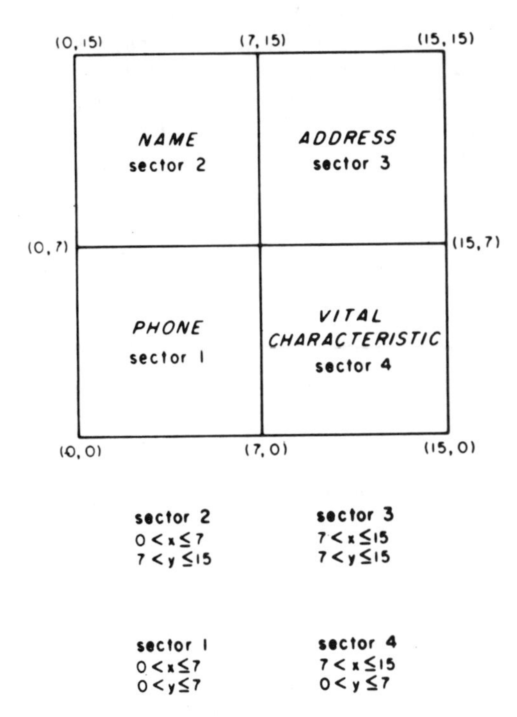

Figure 1: Physical arrangement of sectors on the screen as used by program in listing 4.

ful work. If instead some letter or word were at (4,2) and the program used this higher function output rather than just the numerical coordinate, we'd have something.

Fortunately it isn't all that difficult. By dividing the scanner system into fields and having each field represent a function, we can do useful work. A 2 level program must be written. First, it should have the capability of formatting the screen so that the printing is beneath the proper touch coordinate. Then, after returning from the scanner subroutine, it must translate this position value into the function designated by the printing on the screen.

A simple program which divides the screen into four fields or sectors and performs a function dependent on which sector is touched is shown in listing 4. Figure 1 describes the mathematical relationship between the coordinate system and the BASIC program of listing 4.

After printing the opening lines on the screen the program calls for the data from the scanner. The X coordinate (D1) is first tested to see if it is greater than 7. If it is, then either sector 3 or 4 must have been chosen. If D1 is less than 7 then it must be sector 1 or 2. After choosing whether it is the right or left half of the screen the test is repeated with the Y coordinate. In theory, this binary search method would require no more than eight such tests if all 256 points were designated as separate fields.

A further extension of this binary search

See page 447 for a listing of parts and products available from Circuit Cellar Inc.

Photo 2: Here is a picture of an experiment which was backed up by a fairly long BASIC program: using the screen as the input device for a simulation of an ordinary 4 function calculator. The imagination of the user, to use a well-worn cliche, is the only limitation upon trying experiments with special purpose keyboards and interactive sequences on the screen. Use of the video display behind the touch panel area makes easily altered software the determining factor — rather than physical tools in the workshop.

```
100 S=USR(255)
110 PRINT'NAME                                   ADDRESS'
120 FOR L=1 TO 12
130 PRINT
140 NEXT L
150 PRINT'PHONE                            VITAL CHARACTERISTIC'
160 GOSUB 1000
170 IF D1>7 THEN GOTO 300 ELSE GOTO 200
200 REM THIS ROUTINE DECIDES IF YOU ARE POINTING TO
202 REM SECTOR 1 OR 2
210 IF D2<7 THEN PRINT'UNLISTED NUMBER' :GOTO 2000
220 PRINT'BRENDA (THE LITTLE WOOFER) CIARCIA'
230 GOTO 2000
300 REM THIS ROUTINE DECIDES IF YOU ARE POINTING TO
302 REM SECTOR 3 OR 4
310 IF D2<7 THEN PRINT'SCOTTISH TERRIER -- FOUR LEGGED BURGLAR ALARM' :GOTO 2000
320 PRINT'BOX 582   GLASTONBURY,CONN. 06033'
330 GOTO 2000
990 REM
992 REM SCANNER SUBROUTINE
1000 OUT 16,0 :OUT 16,255
1010 T=INP(2)
1020 T=T AND 1
1030 IF T<>1 THEN GOTO 1010
1040 D=INP(16)
1050 D1=(D AND 240)/16
1060 D2=D AND 15
1070 RETURN
2000 FOR N=1 TO 2000
2010 NEXT N
2020 GOTO 100
```

Listing 4: Illustration of a BASIC program which simulates a 4 function calculator menu. The program inputs the screen coordinate position of a specific menu function.

Photo 3: Touch panel input using the program of listing 5. The line of text at the bottom of the display was entered by touching the index finger to each letter in turn. The photo is shown with the letter P about to be pressed.

```
100 REM THIS PROGRAM DISPLAYS A KEYBOARD ON THE CRT SCREEN
110 REM AND ILLUSTRATES DATA ENTRY WITHOUT A PHYSICAL KEYBOARD
120 REM JUST POINT AT THE LETTERS AND IT WILL 'TYPE' YOUR MESSAGE
200 PRINT"A          B          C          D          E          F          G          H"
210 PRINT
215 PRINT
220 PRINT"I          J          K          L          M          N          O          P"
230 PRINT
235 PRINT
240 PRINT"Q          R          S          T          U          V          W          X"
250 PRINT
255 PRINT
260 PRINT"Y          Z                     SPACE"
261 PRINT
262 PRINT
263 PRINT
264 PRINT
265 GOSUB 2500
268 GOSUB 1000
270 IF D2>=12 THEN PRINT CHR$(D1/2+65);:GOTO 265
280 IF D2>=10 THEN PRINT CHR$(D1/2+73);:GOTO 265
290 IF D2>=8  THEN PRINT CHR$(D1/2+81);:GOTO 265
300 IF D2>=5  THEN 302 ELSE 310
302 IF (D1/2+89)>91 THEN PRINT CHR$(32);:GOTO 265
303 PRINT CHR$(D1/2+89);:GOTO 265
310 IF D2=0 THEN GOTO 320 ELSE 330
320 IF D1=15 THEN 330 ELSE GOTO 265
330 S=USR(255) :REM CLEAR SCREEN
340 PRINT"TO RETRY EXERCISE----TOUCH SCREEN"
350 GOSUB 1000
360 GOTO 200
1000 OUT 16,0 :OUT 16,255 :REM LINES 1000-1070 READ THE SCANNER DATA
1010 T=INP(2)
1020 T=T AND 1
1030 IF T<>1 THEN GOTO 1010
1040 D=INP(16)
1050 D1=(D AND 240)/16
1060 D2=D AND 15
1070 RETURN
2500 FOR A=0 TO 500 :REM THIS IS A SHORT DELAY
2510 NEXT A
2520 RETURN
```

Listing 5: Keyboard simulation program.

```
100 REM THIS IS A SIMPLE PROGRAM TO ILLUSTRATE SIMULTANEOUS
110 REM DATA INPUT FROM EITHER THE TOUCH PANEL OR THE KEYBOARD
120 Q=INP(0) :REM KEYBOARD IS ATTACHED TO PORT 0
130 REM MSB IS KEYBOARD STROBE  --- BITS 0 TO 6 ARE 7 BIT ASCII
140 IF Q>0 THEN GOTO 220 :REM CHECK KEYBOARD STROBE
150 T=INP(2) :REM SCANNER DATA READY IS PORT 2 LSB
160 T=T AND 1
170 IF T<>1 THEN GOTO 120
180 D=INP(16) :D1=(D AND 240)/16 :D2=D AND 15 :REM READ SCANNER COORDINATE
190 PRINT"PANEL TOUCHED AT LOCATION ("D1;D2")"
200 GOSUB 240
210 GOTO 120
220 PRINT"KEYBOARD KEY ";CHR$(INP(0));" PRESSED"
230 GOTO 120
240 OUT 16,0 :OUT 16,255 :REM RESET SCANNER HARDWARE
250 RETURN
```

Listing 6: Method for scanning two input devices simultaneously on a Digital Group Z-80 system.

concept is used in the calculator of photo 2. While never meant to replace the hand held calculator it uses a routine similar to the previous example to determine the action of each of the 16 possible entries. The picture is included to present the reader with one of the many possible applications of the scanner. The program, however, is quite long and difficult to explain in an introductory article such as this.

Simulated Keyboard

One use of the touch panel would be the simulation of direct keyboard entry. Obviously this technique is valuable only where limited data entry is required. Large menu selection programs with numerous choices displayed may not always have the particular item of interest. By having one of the available selections be a keyboard display and entry routine such as photo 3 and listing 5, the miscellaneous entry could be accommodated. The program of listing 5 displays a keyboard on the video screen and allows one to *type* by pointing to the individual characters. The example does not include punctuation and a carriage return, but they could be easily accommodated.

One final note. Using the touch panel need not eliminate the standard ASCII keyboard as an input device. By using the BASIC INPUT command, keyboard entry is still available to the user as is the scanner through a callable subroutine. A program could be written where some entries come from the touch panel and others from the keyboard. A more versatile program would allow input from either device at any time.

Listing 6 is a simple program which demonstrates how BASIC can scan two input devices simultaneously and provide appropriate response.

I hope that this touch panel design will spark the creative interests of other computer enthusiasts. In a field where technology advances by leaps and bounds and product obsolescence can be described in months, innovative ideas are necessary to extend the concept of creative home computing. By adding advanced peripherals and high level languages, system obsolescence is delayed considerably. ■

Photo 1: *A stepper motor controlled scanning sensor capable of detecting both infrared and visible light. A photo detector is mounted at the focal point of an inexpensive parabolic solar cigarette lighter. The computer controlled unit is capable of following a moving flashlight, detecting headlights in a driveway, and many other applications.*

I've Got You in My Scanner

A Computer Controlled Stepper Motor Light Scanner

"Boy, sitting here is really relaxing, isn't it, Lloyd?" I leaned back in the recliner and looked out of my living room window at the dense forest no further than 30 feet from where we sat. The sliding glass doors were open and occasionally some furry little animals could be seen darting in and out through the underbrush feverishly searching for dinner. The setting midsummer sun created an orange and yellow background for the beautiful scene. I wondered why I had ever waited for five years before moving out of the city. All that noise and congestion. This was so peaceful.

"It's very nice up here, Steve. I especially like the big driveway. You have lots of room to park cars. When are you going to chop down all these trees and put in a lawn like everybody else?"

Although I was not actually far enough removed from such suburban beatitudes to scoff with impunity, I piped back, "Bah! We moved out here to get away from the rest of the world. The last thing I want to do is be reminded of civilization, whether that means people or grass." The alternative I much preferred was to turn into a leaf shrouded computer hermit. "Lloyd, if I could figure a way to put in a moat with alligators, I'd do it," I said, tongue in cheek.

Such was the tone of our conversation for the next few hours. Whenever Lloyd ventured into the Connecticut wilderness (as he called it) he would stop by and visit me. Because of Lloyd's practical knowledge of computer related subjects I often used him as a sounding board for article ideas. His diplomatic responses sometimes disguised his opinions so tactfully, though, that I wasn't always sure what he really thought.

Before we knew it the sun had set and we were enveloped in darkness. The moonlight cast a silvery glow across the tops of the trees but hardly penetrated to the underbrush. But the moonlight was of small consequence to us. Even with the additional dim light escaping from the next room there was barely enough illumination to discriminate facial expressions, but there was sufficient backlighting for the little night creatures to observe us. Having lived in my new house for two months I had finally become accustomed to the nocturnal sounds and no longer experienced heart failure whenever I detected a pair of eyeballs peering at me from between the tree limbs. It was, after all, the domain of the owl and deer and I was the intruder.

I was less sure of the effect on Lloyd. Far away from the accustomed roar of jets at JFK and the traffic jams on the Long Island Expressway, he was suddenly very quiet, almost subdued, as he stared out the window into the darkness. Suddenly his eyes became focused on something in the distance and, gripping the arms of the chair tightly enough to leave an impression, he craned his neck to get a closer look. Something had obviously attracted his interest.

"I saw something!" he said.

"It's probably some possum checking us out or some other small animal after the dinner scraps I put outside."

"No, it's no animal. At least no small one. I thought I saw a light too. How many possums glow in the dark?"

"Don't be an alarmist. There's nothing to worry about."

"Look, there it is again, Steve! I think someone's out there."

I, too, saw a form way off in the darkness. It was definitely an erect biped moving between the trees and making considerable noise as it went.

We jumped from our chairs and crouched together looking out through the screen. The same thought came to both of us: "Is it Bigfoot?!"

"Wait a minute," I said in a hushed tone. "This is Connecticut. That's absurd! How can it be Bigfoot? Besides, since when does Bigfoot carry a flashlight?"

The bright beam of a flashlight shot from the stranger's hand. The dim light revealed a large man in coveralls dragging a heavy sack and carrying something over his shoulder.

"He must have a gun!" Lloyd gulped, and we both dropped to floor level. "Quick! Call the police or something! Better yet turn on the outside flood lights. Maybe it will scare him away."

"Look, Lloyd, if you want to become a moving target walking across the room to the light switch go right ahead."

"How come your burglar alarm hasn't turned the lights on?"

I thought about the alarm system for a moment and then answered, "I've got sensors all over the driveway and the road leading to the house. I didn't put them out in the woods because it's more likely that someone would come down the road rather than hike through the woods."

"How come nobody told *him* that?"

"Look, it'll pick him up anyway if he comes within 50 feet of the house."

Before Lloyd could reply, the man in the woods stopped in a clear area. The object slung over his shoulder wasn't a gun, but a shovel. He started to dig.

"Steve, do you think he's burying a body?"

I gathered up what courage I could and decided to go out and confront the perpetrator before my front yard looked like the aftermath of Dunkirk. "Come on, let's find out what he's doing."

As we approached, the man ignored our presence and kept digging. Occasionally he pointed his flashlight into the hole, then dumped the contents of the shovel into the sack. Was there buried treasure on my property?

"Excuse me, sir? Excuse me?" I said softly but with resolve. When I did not receive a response I stepped closer and repeated a little louder, "Excuse me, sir?!"

"Shhhh, Sonny! Da ya wanna scare all these critters away? It's hard enough making a living these days without everyone getting into the act. This here is my mound, Sonny!"

Mound? Sonny? I listened to his voice closely now and examined his features as best as I could in the moonlight. His accent was definitely Maine — deep woods Maine, and I put his age conservatively at 70. He seemed harmless enough, but I still had some unanswered questions.

"Sir, do you mind telling me what you're digging?"

He swung the shovel up over his shoulder and turned toward me. His face was weather-beaten and aged, yet there was a youthful glint in his eyes. The gravity of the situation evaporated as he answered, "Worms."

"You're digging *what?*" Lloyd chimed in.

"Worms," he answered again. "This hea mound," he pointed at the area where he was digging," is one of the best night crawler mounds in the county. Youst to be a farm around here, few yease back. This was the compost heap. Worms love it, ya know." He chuckled as he explained the worm breeding business to us city slickers. "I been diggin' around here off an on for 30 yease. Then someone came along and put a house on it." He pointed a boney finger at my place.

"I had no idea . . . ," I said, somewhat embarrassed.

"No matter . . . them's still my worms! I got ten spots just like this one and I'm transplantin' my worms. You know what a night crawler like this is worth around here, Sonny?" He reached into the hole and suddenly I had a handful of worms held in my face. I took great care to take shallow breaths lest I accidentally gasp and inhale one of the squirmers.

"No sir. I don't fish."

"Well, they're worth plenty. And I got to

dig a lot of them durin summa cause there ain't no worms in winta, Sonny."

His logic was irrefutable. He obviously earned his living at this. I felt a bit sad for the old codger. His digging really wasn't an inconvenience as long as he only took the worms and left the dirt. I didn't know enough about night crawlers to know the best time of night for harvest but I was sure we could work something out. I held my hand out to his and said, "You don't have to transplant your worms. What's a few worms between friends?"

A Modification to the Alarm System

At the conclusion of this episode I couldn't help but be concerned about the detection logic of the sophisticated alarm I had installed. There were sensors across critical points in the driveway and the road leading to the house that could detect the presence of a car or person. But, because of the likelihood of false triggering by wild animals, I hesitated to place similar detectors in the woods surrounding the house. I had thought the woods were impassable, but I guess I was wrong. The common denominator for anyone trying to make it through those woods at night is the necessity of a light. It should seem easy in principle to just place a light activated switch out there and activate the sequence when it detects some light source. Unfortunately, since the sensitivity would have to be relatively high, it would no doubt be accidentally triggered from lightning bolts and wayward fireflies. Complex integration and delay logic could

Circuit Cellar experimental setup showing the parabolic scanner as it detects a light bulb and a candle.

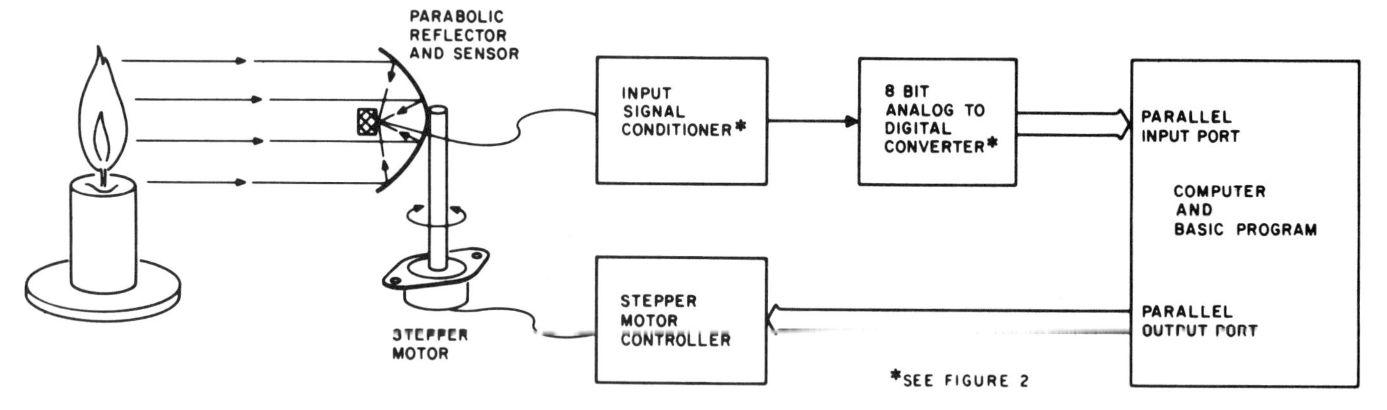

Figure 1: Block diagram of closed loop optical scanning system. The three main sections are the optics, consisting of a parabolic reflector and visible and infrared detector, the signal conditioner and input interface, and the stepper motor.

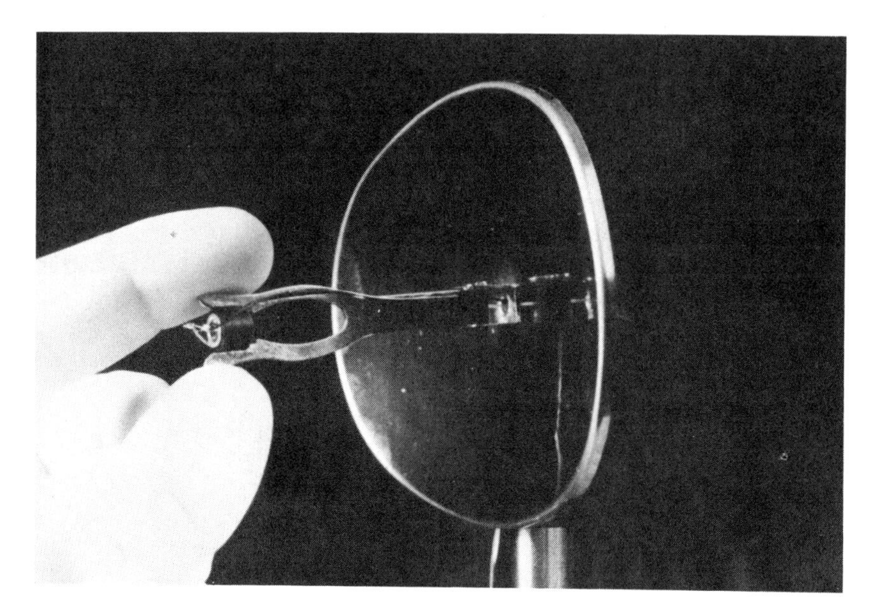

Photo 2: The parabolic reflector, used to gather light for detection by the photo detector.

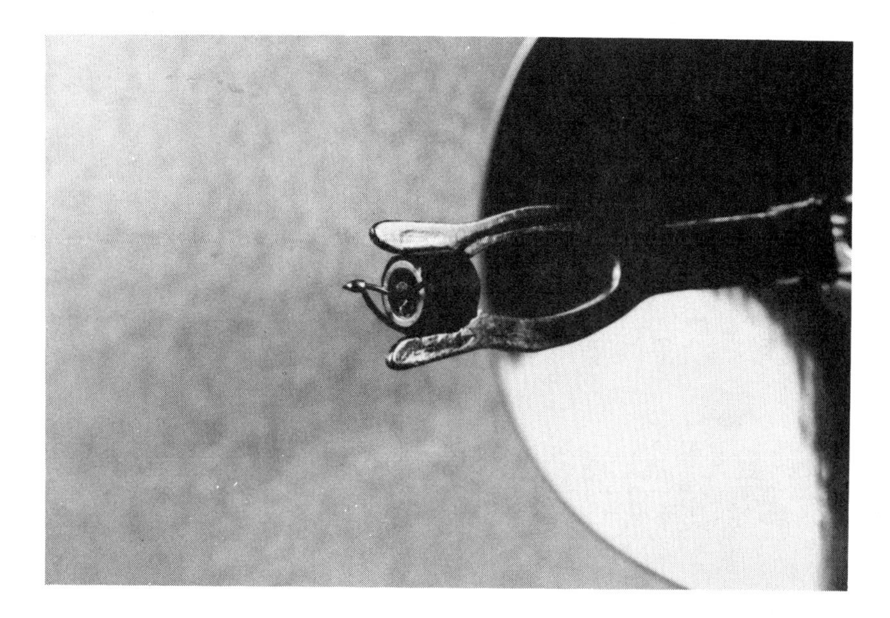

Photo 3: Closeup of the photo detector glued to the pronged holder.

be incorporated which would eliminate many false alarms but light level is still the only detection quantity.

Since a light source such as a flashlight or motor vehicle would have to move to approach the house, motion is another necessary parameter to consider. Most motion detection systems are passive beams whose sequential interruption triggers an appropriate response to a time and distance algorithm. Such a system of infrared or visible light beams, sufficient to protect four or five acres of property, would be prohibitively expensive if it incorporated a laser light source, and probably couldn't work reliably without one.

If we accept the premise that anyone coming through the woods on my property would need a flashlight or lantern, etc, then to detect the presence of an intruder requires a combination of light level and motion.

There are a number of methods that achieve the desired result. The most straightforward is to use a television camera, digitize the image, and after adjusting for ambient light changes, compare it to a previous digitized image. Many of the most sophisticated alarm systems incorporate this feature. While it is not beyond the capability of the more than modest home computer, it would be expensive in this application.

If You're Trying to Detect Motion — Move the Detector

Detecting motion with a light level sensor requires that a quantity of them be placed throughout the detection area. As the source moves, the relative light levels reaching all the sensors can be plugged into an equation and the location of the source computed. Tracking an object is simply a case of repeating this snapshot technique a number of times. Unfortunately, the concept is about

Figure 2: Optical signal conditioner and analog to digital converter that convert light input into a proportional digital output. The analog to digital converter shown was described in "Talk to Me: Add a Voice to Your Computer for $35".

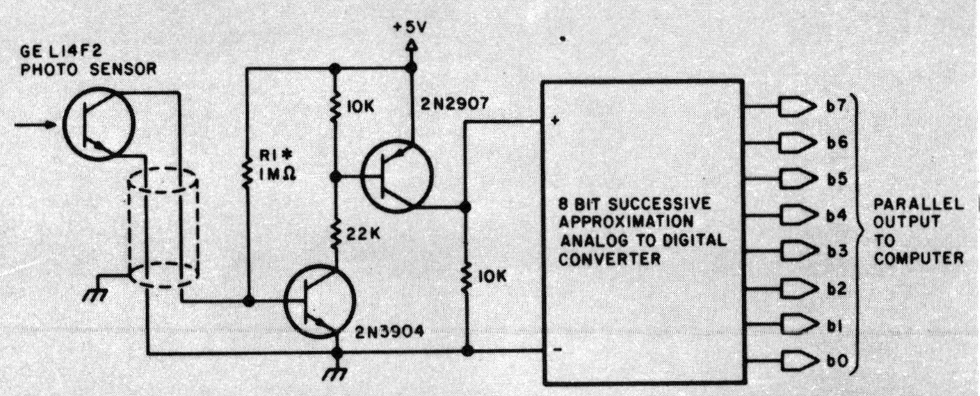

*Raising the value of this resistor will increase sensitivity. Lowering it will reduce sensitivity (range of resistor values is 4.7 k to 3.3 MΩ).

the only part that's simple.

An alternative approach is to point a light sensor at a source and then move the sensor to a new location. If eventually the source is again detected in this new position the source must have motion! This of course presumes that there aren't so many sources that placement of the sensor inadvertently coincides with a stationary source. Small but discrete steps of sensor displacement will increase the resolution of this method.

Build a Light Sensitive Scanning System

I wish to back off a bit at this point and explain that this design is not merely a motion detection system. That is one of its numerous applications and, as previously stated, it is the idea that prompted its development. The design is a simple, yet effective, light sensitive scanning system. A sort of passive radar (radio detecting and ranging system) if you will. It incorporates a sensitive visible and infrared light detector that is highly directional. In addition, it has the ability to accurately position itself on a rotational axis and sweep a wide area, much like a radar.

Figure 1 is a block diagram of the device. The scanning system consists of three prime components: optics, including sensor and reflector; signal conditioner and input interface; and finally, for closed loop control, a rotational positioning mechanism consisting of a 7.5° resolution stepper motor. The completed unit allows the computer to position its sensor in a known direction, read an analog value of the light level in that direction, and move to another point or track a moving source (more on this later).

The prime consideration in any light detection system is the optics. To take full advantage of any positioning mechanics, the light sensor must be highly directional. This is usually done with a series of lenses, the whole affair resembling a telescope. This technique is quite expensive and heavy. Instead of lenses, a highly polished parabolic reflector can be used to concentrate the light. One such device ideal for this application is an inexpensive parabolic mirror sold by Radio Shack for under $2 as a solar cigarette lighter. The unit, shown in photo 1, has a fork tipped hinged prong which extends from the center to hold the cigarette. Already designed to be at the focal point of the mirror, it serves as the perfect mounting bracket for the photo sensor. A GE L14F2

```
100 REM INFRARED SENSOR TEST PROGRAM
110 REM
120 REM
130 REM THIS PROGRAM CAUSES A SOUND SOURCE ATTACHED
140 REM TO LSB OF PORT 16 TO -BEEP- WITH   A PERIOD PROPORTIONAL
150 REM TO THE AMOUNT OF LIGHT SEEN BY THE LIGHT SENSOR
160 REM REV. 1.1      S.CIARCIA
170 REM
180 OUT 16,0
190 X=INP(16)
200 IF X<230 THEN GOSUB 220
210 GOTO 190
220 OUT 16,255
230 FOR T=0 TO X+5
240 NEXT T
250 OUT 16,0
260 RETURN
```

Listing 1: Program written in Micro Com 8 K Zapple BASIC that reads the light level from the analog to digital converter and converts it to a proportional pulse width on output port 16 (in my particular system configuration).

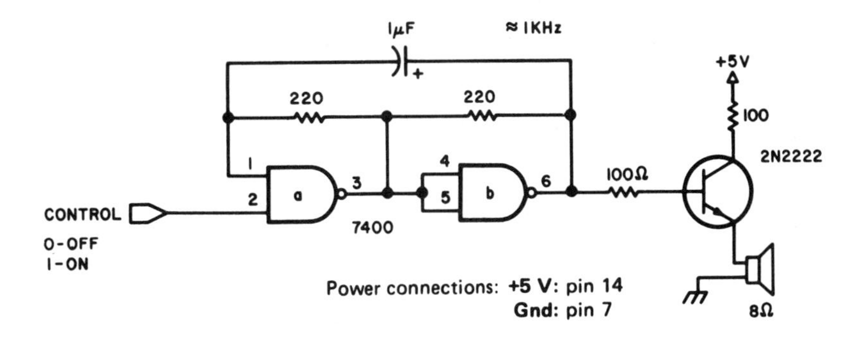

Figure 3: Sound source circuit (use with optical scanner test program above).

Photo 4: North American Phillips stepper motor, Model K82701-P2, and the SAA1027 controller circuit used to drive it.

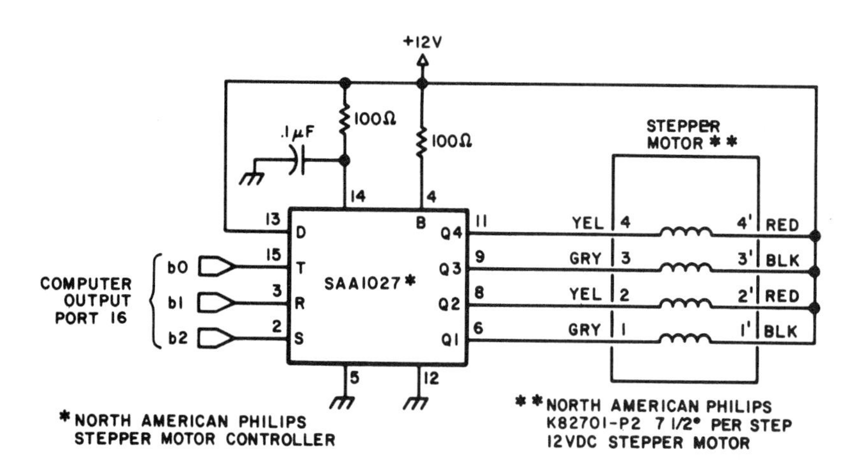

Figure 4: Connecting the stepper motor to the integrated circuit controller.

```
100 REM THIS PROGRAM DRIVES THE STEPPER MOTOR IN A BACK AND FORTH MOTION
110 OUT 16,1 :OUT 16,255 :REM PRESET STEPPER CONTROLLER
120 REM
130 REM
140 REM GO 25 STEPS CLOCKWISE
150 FOR D=0 TO 24
160 REM BIT 2 IS SET HIGH AND BIT 0 IS TOGGLED TO GO CLOCKWISE
170 OUT 16,5
180 GOSUB 390
190 OUT 16,4
200 NEXT D
210 REM
220 REM RETURN SCAN DELAY
230 FOR S1=0 TO 10
240 GOSUB 390
250 NEXT S1
260 OUT 16,1 :OUT 16,255
270 REM GO 25 STEPS COUNTERCLOCKWISE
280 FOR D=0 TO 24
290 REM BITS 1 AND 2 ARE HELD HIGH AND BIT 0 IS TOGGLED TO
300 REM GO COUNTERCLOCKWISE
310 OUT 16,7
320 GOSUB 390
330 OUT 16,6
340 NEXT D
350 GOTO 110
360 REM
370 REM IN BETWEEN STEP DELAY TIMER
380 REM DELAY TIME SET BY VALUE OF T1
390 FOR T=0 TO 5
400 NEXT T
410 RETURN
```

Listing 2: A BASIC program that drives the stepper motor and demonstrates the sweep action. It initializes the stepper motor, drives 25 steps clockwise, waits a short period, and then returns to its initial position.

infrared photo Darlington inserted into a phenolic sleeve is glued to the cigarette holder at the focal point as in photos 2 and 3. The lens of the photo sensor should face the reflector.

While the photo detector is infrared by design, it is highly sensitive to visible light as well. By choosing the infrared unit, a detection system can be designed that utilizes the best of both spectra.

The output of the photo sensor is essentially a current proportional to the light hitting it. The signal conditioner section of figure 2 converts this to an analog voltage level. The sensitivity of the photo detector is governed by resistor R1; changing this resistance value will affect both sensitivity and dynamic range. For the computer to read this voltage it must be converted to a digital quantity. While in theory any method, such as voltage to frequency, or voltage to pulse width, etc, could have been used, I'm a purist. The output of the signal conditioner is fed to an 8 bit successive approximation analog to digital converter. The details of this design were outlined in the June 1978 Ciarcia's Circuit Cellar in an article entitled "Talk to Me!"

Two slight modifications were made to the circuit for this application. The sample rate was reduced by placing a 0.01 μF capacitor in parallel with the 150 pF component already between pins 1 and 6 of IC1, and the offset potentiometer was readjusted to allow full scale unipolar operation (ie: 0 V input would give hexadecimal 00 output and +5 V input would give hexadecimal FF output).

The parallel output of the analog to digital converter is attached to an 8 bit parallel input port. Either an assembly language or a BASIC program can be used to read and display this quantity by querying the input port (input port 16 in my examples).

Exercising the device with a BASIC program is relatively straightforward. Listing 1 is a program written in Micro Com 8 K Zapple BASIC which reads the light level from the analog to digital converter and converts it to a proportional pulse width on output port 16. If a Sonalert or the circuit of figure 3 is attached to the least significant bit of port 16, it will beep. The beep rate will change as the reflector is pointed toward various light intensities. Printing out the analog to digital conversion value will give an accurate account of the sensitivity and dynamic range.

Add a Stepper Motor for Positioning

Now that we have an effective light sensor, we must add rotational mechanics to provide *sweep*. The simplest method for rotating this relatively lightweight reflector is to mount it directly on the shaft of a stepper motor.

An inexpensive stepper motor is available from North American Philips. This unit (shown in photo 4) is relatively small, and a single integrated circuit controller is all that is needed to interface it to a computer. The particular unit in this article is a 12 VDC 7.5° stepper motor. This means that there are 48 steps per revolution, and, if one were trying to scan a 180° field of view, the stepper should oscillate between 24 clockwise and counterclockwise steps. This would give the impression of "sweep."

The electronics of the stepper are outlined in figure 4. Three bits of a parallel output port are necessary to control the direction and speed of the motor. The three signals are S, R and T, for set, rotation and trigger. When first engaging the motor it should be set to a known condition by pulsing the set input low while keeping the trigger input high. Once initialized, the direction of rotation must be chosen. This is done by setting the R input low if clockwise rotation is desired and high for counterclockwise rotation. An actual step is initiated by simultaneously making a 0 to 1 logic transition on trigger input T. By repeatedly toggling this bit, continuous motion will result. The stepper motor in this article is capable of 200 steps per second.

A BASIC program which drives the stepper motor and demonstrates the sweep action is outlined in listing 2. It initializes the stepper, drives 25 steps clockwise, waits a short period, and then returns to its initial position.

Making a Scanning System

To produce a closed loop controlled scanning system, the reflector and photo sensor are attached to the stepper motor shaft by any convenient means. I glued the reflector to a sleeve which attached to the shaft of the motor. The concept of closed loop control comes from the ability of this unit to position itself, take a light reading, and perform some further action as a result. This could be to step to a new location or to stop and remain stationary on any source above a certain light level.

Listing 3 is the BASIC program of such an exerciser which seeks out and points at a light source. As the parabolic reflector steps through its sweep, it checks the reading of the analog to digital converter and compares it to a set point. If the set point is exceeded, the program will stop stepping and point at this source. Should the light be extinguished or obstructed, the sweep resumes until it finds another source of sufficient intensity.

```
100 REM THIS PROGRAM SIMULATES A CLOSED LOOP -RADAR-
110 REM IT SCANS BACK AND FORTH 25 STEPS IN EACH DIRECTION
120 REM LOOKING FOR A PRESET LIGHT LEVEL EITHER PRODUCED BY
130 REM OR REFLECTED FROM SOME OBJECT IN ITS SCAN PATH
140 REM IT WILL STOP SCANNING AND REMAING POINTING AT ANY SUCH
150 REM OBJECT IT FINDS. IF THE OBJECT MOVES, THE SCANNER WILL FOLLOW
160 REM
170 REM
180 OUT 16,1 :OUT 16,255 :REM PRESET STEPPER CONTROLLER
190 FOR D=0 TO 24
200 OUT 16,5
210 REM TAKE ONE CLOCKWISE STEP
220 GOSUB 550
230 OUT 16,4
240 GOSUB 480 :REM READ SENSOR
250 NEXT D
260 REM
270 REM RETURN SCAN DELAY
280 FOR S1=0 TO 10
290 GOSUB 550
300 NEXT S1
310 REM
320 REM
330 OUT 16,1 :OUT 16,255 :REM PRESET STEPPER CONTROLLER
340 FOR D=0 TO 24
350 OUT 16,7
360 REM TAKE ONE COUNTERCLOCKWISE STEP
370 GOSUB 550
380 OUT 16,6
390 GOSUB 480 :REM READ SENSOR
400 NEXT D
410 FOR D=0 TO 25
420 NEXT D
430 GOTO 180
440 REM -RADAR- SENSOR READ ROUTINE
450 REM A/D INPUT IS ATTACHED TO INPUT PORT 16
460 REM LOW LIGHT LEVEL IS A VALUE OF 255 AND HIGH INTENSITY
470 REM IS AN INPUT VALUE OF 0
480 X=INP(16) :REM READ A/D CONVERTER
490 L=10 :REM PRESET LEVEL SET ....THIS WOULD BE A BRIGHT LIGHT
500 IF X<L THEN GOTO 480
510 RETURN
520 REM
530 REM
540 REM DELAY TIMER TO COMPLETE MECHANICAL MOTION BEFORE READING SENSOR
550 FOR T=0 TO 25
560 NEXT T
570 RETURN
```

Listing 3: A BASIC program that causes the scanner system to seek out and point to a light source. The scanner tracks the light source as it moves. If the light source is extinguished or obstructed, the sweep resumes until another source of sufficient intensity is found.

This is a rather rudimentary program but it incorporates all the basic structure to which enhancements such as motion detection and tracking can be added. It will, as now written, follow a flashlight as someone walks across a room. It is left as an exercise for the reader to drop a net over the perpetrator.

There are a few other little things you can try after you've built this gadget. The sketch on page 43 shows the portion of my basement (the "Circuit Cellar") immediately adjacent to the computer system. After modifying the BASIC program of listing 3 to print out a number on a scale of 1 to 9 (a period is 0) indicating relative intensity, and turning on a light and lighting a candle, I initiated a single scan across the room. Listing 4 is a printout of that scan. The sensitivity of the device had to be set very high to pick up the candle, and the result was rather interesting. The scan allowed the computer to "see" around the room in front of it.

```
. . . . . . . . . . 1 . . . 1 6 9 7 3 1 . . . .
```

Listing 4: A single scan of the room containing the light bulb and candle. A modification of the program in listing 3 to print numbers on a scale of 1 to 9 (a period is 0) indicating relative intensity.

```
. . . . . . . . . . . . . . . . . . . . . . . .
. . . . . . . . . . . . . . . . . . . . . . . .
. . . . . . . . . . . . . . . . . . . . . . . .
. . . . . . . . . . . . . . . . . . * . . . . .
. . . . . . . . . . . . . . . . . * * . . . . .
. . . . . . . . . . . . . . . . . * * * . . . .
. . . . . . . . . . . . . . . . * * * . . . . .
. . . . . . . . . . * . . . . . * . . . . . . .
. . . . . . . . . . . . . . . . . . . . . . . .
. . . . . . . . . . . . . . . . . . . . . . . .
```

Listing 5: Ten sweeps of the room. The relatively large size of the light used in the experiment accounts for the large number of asterisks at the right.

See page 447 for a listing of parts and products available from Circuit Cellar Inc.

There is an intense light source to the left and a rather low level one to the right. By incorporating gain selection (changing the 1 MΩ resistor in the signal conditioner) under program control the computer could reduce the gain selectively to determine the origin of each light.

One further experiment entailed taking numerous sweeps and combining them to form a digitized computer picture. First the program was changed back to a threshold detector again. As it scanned the 25 steps it would print out an asterisk (*) for anything that exceeded this threshold and a period (.) otherwise. A protractor was attached to the arm of the tripod so that the angle of the reflector could be adjusted by a known increment each time the computer stopped between scans. The result was as illustrated in listing 5. The ten scans form a computer's eye picture of the wall. Again, because of the dynamic range differences between the candle and the light, the incandescent bulb appears much larger than it actually is.

Conclusion

Here's a simple device that can detect and track infrared and visible light sources. See what you can do with it. I don't want to leave anyone with the impression that I'm waiting for a burglar with a million candlepower flashlight to come tripping through the woods. This is but one sensor in a larger system, and the infrared capabilities, which I neglected to discuss in detail, are its primary application.

There have been numerous articles on light seeking robots. With this detector it is quite possible that the mechanics and software could be reduced considerably. I've often thought about building a robot, but my mechanical talent is nonexistent. When I can build one with a screwdriver and a soldering iron only, I'll write about it. (My thanks to Lloyd Kishinsky for graphics ideas used in this article.)■

Photo 1: Example of an optical fiber transmitting a very bright light. The conductor is a single 40 mil plastic fiber. The light is generated by a helium-neon laser.

6
Communicate on a Light Beam

Coming up out of the Circuit Cellar is a rare occurrence, to the point where some of my friends have accused me of being a mushroom. I prefer to be likened to a mole—a more dignified species. We share a common bond of subterranean existence and fear of bright sunlight, but the mole's predicament is dictated by nature, and mine by choice.

The Circuit Cellar is by no means a hole in the ground. It's heated, well-lit and looks more like a living room than a cellar. Even though it affords all the comforts of home, there are those occasions when a change of environment is required. It's not enough to walk out in the driveway, take a deep breath and run back into the cellar. Sometimes a complete change of surroundings is needed to shock the mind out of the doldrums and spark creativity (eg: a vacation). Since I usually don't have time for vacations, I take "business excursions for purposes of cerebral detoxification" or "ECDs" for short.

For two months I had been wrestling with the details of an article on fiber optics and laser communications (this one). The hardware was completed very quickly, as with most of my projects, but the text dragged on for weeks. Lighting the wood stove in the Circuit Cellar became an all too easy chore using the piles of scrap paper I was generating. My graphospasms (ie: writer's cramps) were not bearing fruit. One time I even found myself sitting at my desk pushing pencils through the electric pencil sharpener until it started smoking.

During times like this there was only one place to go — New Hampshire — to see the Colonel. My father-in-law, Colonel Foster, was the one person who could break me out of this slump. Between stories about old army buddies and spending the war in the Aleutians waiting for an invasion I would surely find some inspiration.

"Colonel? Are you there?" After anxiously dialing his telephone number and saying hello, I was left with silence at the other end of the line. . .

"Colonel?"

"Be right with you, Steve." As the receiver was picked up again he apologized, "Sorry Steve, my man was at bat and I had to see the hit. You're a Red Sox fan, aren't you?"

It would be in bad taste for me to suggest that my subterranean hideaway provided all the spiritual stimulation I needed and that chasing a little ball around in the grass was not in my spectrum of pursuits.

"I quite understand your enjoyment of the game, Colonel. I hope your team wins," I replied, evading his question. During my statement I heard him roar again in response to the activities on the television. When I sensed a lull, possibly precipitated by a commercial, I continued, "Colonel, I need to get away. How would you like some company tonight?"

"Sure, you know you're always welcome. I haven't had anyone to tell a good army story to in a long time."

I told him I'd pack all the gear in the car and be there in three hours. Possibly I would feel better about writing once I arrived.

The Colonel, sensing the termination of the commercial, quickly responded, "Three hours is great. The game is still in the first inning. If you hurry you may get here before it's over. . .gotta go now."

One of the good things about living in New England is that everything is close. It was a scant 3 hour drive between Connecticut and New Hampshire, but I dragged it out an extra half hour so I wouldn't be competing with the Red Sox for the Colonel's attention. As I pulled into the garage he came out to greet me.

"Howdy," he said, slapping me on the back. From his exuberance I could tell that the Red Sox had just won the game.

"Come on in and get settled. I'm expecting a telephone call. . .oops, there it is now."

Leaving the electronics junk in the car I followed him into the house. He was still wearing his lucky Red Sox baseball cap as he spoke.

"Chester, wasn't the game great? I thought they were going to blow it in the 6th. . .You bet, I'm ready for tomorrow's game. If they can play like that again, the pennant is in the bag. . ."

Suddenly Colonel Foster's expression changed to amazement, then anger. He grabbed his cap, slung it into the chair he was standing near and complained, "Darn woman again!. . .What do you mean lucky! The Red Sox won through skill, not luck!. . .Go play with your WATS lines and let Chester and me talk." It was obvious that suddenly there was a third party to their conversation.

"Beatrice, I don't care if you think it was an error. It was ruled as a single!. . . Yes, I know the 6th looked bad but that still doesn't mean they're just lucky. . ."

It was becoming an argument between the Colonel and Beatrice. A hint as to her identity was provided when he responded, "Beatrice, would you keep your opinions to yourself and let me talk to Chester? Chester, come on over for a private talk!"

He slammed the reciever down on the phone, put his baseball cap back on, and slumped into the easy chair. "I just can't carry on a baseball conversation with that woman around."

"Who's Beatrice?"

"The switchboard operator for the town. We don't have all that new computer telephone stuff you city slickers have. We have Beatrice. When it's business or personal she's good and keeps her nose out. But, when it's baseball, Beatrice has to get her two cents in!"

(Obviously what the Colonel and Chester needed was an alternate means of communication, such as CB.)

"I've got a great idea, Colonel. Why don't you and Chester use CB radios instead of the telephone?" The Colonel led me to the bookcase in the study. I found myself staring directly at a CB radio. He flipped it on and said, "Tune in channel 19 and listen." The radio came to life. "Breaker one nine. . .breaker one nine. . . this is your Big Mama on this one niner. . . all you 18 wheelers just put the hammer to the floor and let Big Mama be your guide. . .I'll have a Smokey report in five, but first, the weather. . ."

My eyes opened wide. "Is that Bea. . ."

"Beatrice? You're darn tootin' it is. She's got an antenna tower on her house and radio gear that would put an FCC test laboratory to shame. I swear she's running a full gallon."

"We tried CB a while back and it was useless." This time the conversation came from behind. Chester had let himself in and joined us in the study. He continued, "It all started when we telephoned the games to the tower."

"Tower?"

"I'm sorry, I guess the Colonel didn't tell you." Walking over to the window of the study and pointing to the mountain top roughly two miles away, he said, "You see that structure on top of that hill? That's my tower. Well, not exactly *my* tower. I just work there. It's a combination fire tower and radio relay station. Occasionally I have to sit up there and monitor equipment during important transmissions."

"What's that got to do with Beatrice?"

"With all the interference from the equipment up there I can't use a radio or television to watch the Red Sox."

(This was beginning to take on the aspects of a good mystery.)

"The Colonel would tune in the game on his television set here, telephone me in the tower and then lay the receiver near the television so I could listen to the game. When Beatrice found out, she'd bust in and add her commentary to the game. Do you know what it's like having a nosey Howard Cosell-type beating on your ear for three hours at a time?"

I could only offer my sympathy. If there were a solution short of stringing two miles of wire I didn't see it yet. But I would continue to think about it.

"Tomorrow is a very important Red Sox game. The pennant may hinge on it. Unfortunately, tomorrow is also a day I

Warning: *due to the nature of lasers, any prolonged skin contact or viewing of the laser beam is hazardous.*

have to spend in the tower. I really want to listen to the game, but Beatrice is tough to listen to."

I ran over to the window, looked at the tower in the distance, and noted the glass windows circling the observation deck. "What's the weather report for tomorrow?"

"Cloudy and cool, I think," Chester answered.

"Good! Clear weather. . .Colonel, could the television set be moved in this room for the game tomorrow?"

"I suppose so. Why?"

I scanned the study looking for a convenient AC power outlet and spied one by the window.

"Perfect," I said.

Both the Colonel and Chester were a little perplexed at my behavior.

"What if I told you there were a way for Chester to listen to tomorrow's game undisturbed by Beatrice?"

"We've tried everything. What are you planning?"

"Wait here and I'll show you." I dashed off to my car and took a tripod, a long white rectangular instrument, a small black box with a lens at one end and a few patch cords out of the trunk. Dragging all the equipment into the study, I proceeded to assemble it, much to their amazement.

"What's all this, Steve?" the Colonel asked.

With as straight a face as I could muster I replied, "It's a laser."

Both men, army veterans of two wars and thirty years' service, took two steps back and exclaimed, "A laser?" It was instantly apparent that the words "laser" and "death ray" were synonymous for them. Before I let them think I planned to rub out Beatrice, I quickly continued my explanation.

"There are big lasers and little lasers. This is a little one. It won't burn anything or hurt anyone if used properly. Eye protection is the only consideration necessary on this particular laser."

"Do you always carry this stuff around with you?" the Colonel asked.

"No. It just happens to be the topic of this month's article for *BYTE*."

"What has this got to do with tomorrow's game?" Chester asked.

"We're going to transmit the game to you in the tower on a beam of light."

Their eyes opened wider but they remained receptive.

"Let me demonstrate."

I took the transistor radio, tuned it to a station and placed it on the coffee table. Taking a long patch cord, I plugged one end in the radio earphone jack, automatically silencing the radio speaker, and plugged the other into the rear of the laser. Aiming the laser, I turned it on. A red spot, about 1/8 inch diameter, shone brightly on the wall 15 feet away.

"You're sure that won't burn the wall?"

"Trust me."

Next, I picked up the black box with the lens on it and turned it on. I walked over to the illuminated spot on the wall and interrupted the laser beam path with the box. When the beam intersected with the lens, music was heard!

"That's the radio station you tuned in, all right," Chester said.

"Colonel, take that poker from the fireplace and wave it back and forth in front of the laser so it interrupts the beam."

"Why. . .the radio goes on and off," he exclaimed a minute later.

"Correction, Colonel. The radio doesn't go off, only the receiver, when it no longer 'sees' the modulated laser light beam. Notice in addition that the beam barely spreads out at all over the 15 feet to the wall."

"I think I get what you're driving at, Steve."

"You've got it. Chester takes the receiver up to the tower tomorrow, aims it at this window using the gun sight scope on top. Then we turn on the laser which, instead of being connected to the radio, comes from the television. Voila! Instant uninterrupted Red Sox baseball. And, no Beatrice!"

"Will it really work, Steve?" Chester asked.

"Sure, and tomorrow we'll prove it."

Before the next comment from anyone, the telephone rang and Colonel Foster answered it. Chester and I listened and smiled.

"Look, Beatrice, your team doesn't have a chance for the pennant. . .Are you still claiming that that was an error?. . . It wasn't just luck in the 6th I tell you. . ."

Chester and I laughed. Beatrice was really giving the Colonel a run for his money, but there was a twinkle in his eye as he spoke. The Colonel was living what he enjoyed most – baseball. First on television and then blow by blow with Beatrice.

Communicate on a Light Beam

Most experimenters have never considered using a modulated light beam for data communication. I'm not suggesting that everyone throw out their twisted pair RS-232 lines and replace them with laser beams, but I do ask you to consider the commercial advantages of such a concept and try a few experiments.

When discussing modulated light communications, a definition of terms is in order. The two most often heard are lasers and fiber optics. It is important to recognize that one is a light source and the other is a light conductor. It is not necessary for them to be used together but this is often the case. I'll explain more about each later.

A full duplex optical communication link is shown schematically in figure 1. It consists of two pairs of optical transmitters and receivers which allow data to flow in two directions simultaneously. Data from the base to the remote travels on one line, while data from the remote to the base is on the other. This is a dedicated duplex hookup. Unlike the ones you've probably used, this one uses fiber optic cable rather than wire. In its commercial applications it can offer the following advantages:

- Immunity to strong electrical or magnetic noise. Fiber optic material is usually glass or plastic and since there is no electrical conduction there can be no induced electrical noise.
- High electrical isolation. Since the data conductor is a dielectric material, the isolation between the transmitter and receiver is a function of distance.
- Higher bandwidth and lighter cable. Optical modulation systems have inherently higher data rate capabilities and glass and plastic weighs less than copper. Bandwidth is typically 100 megabits.
- Lower loss than coaxial systems. New low loss fibers extend transmission distance.
- Negligible crosstalk. If each fiber optic channel is optically sheathed there is no crosstalk. Even adjacent unsheathed fibers rarely interfere with each other.
- Ultimately lower cost than either coaxial or twisted-wire systems. The raw material (sand) used in making fiber optics is abundant, while copper gets increasingly more expensive. Cost for a data transmission system is ultimately based on dollars per megabit times distance. Since fiber optic systems have higher bandwidths, the cost factor is slowly moving in their favor.

Key ingredients in any optical communications system are the transmitters and receivers. The ultimate data rate is a function of how fast the transmitter can turn on and off, sending one bit of information, and whether the light sensitive receiver can track this transition. If the data rate is very low, say, 110 bps in your experimental setup, a simple incandescent light and cadmium sulfide photocell will suffice. Higher data rates require much faster response and dictate use of LEDs (light emitting diodes) and phototransistors or photodiodes. Common red LEDs will easily handle 100 K bits per second and most common phototransistors, if properly biased, will also suffice. Higher frequencies require specially fabricated LEDs or, if the transmission line is especially long, then laser diodes might be in order.

It is important to know what each of the components in the system is and the way its selection affects the other components. The designs illustrated in this article are included to demonstrate a workable low frequency system which the personal computer enthusiast may wish to build. The physical elec-

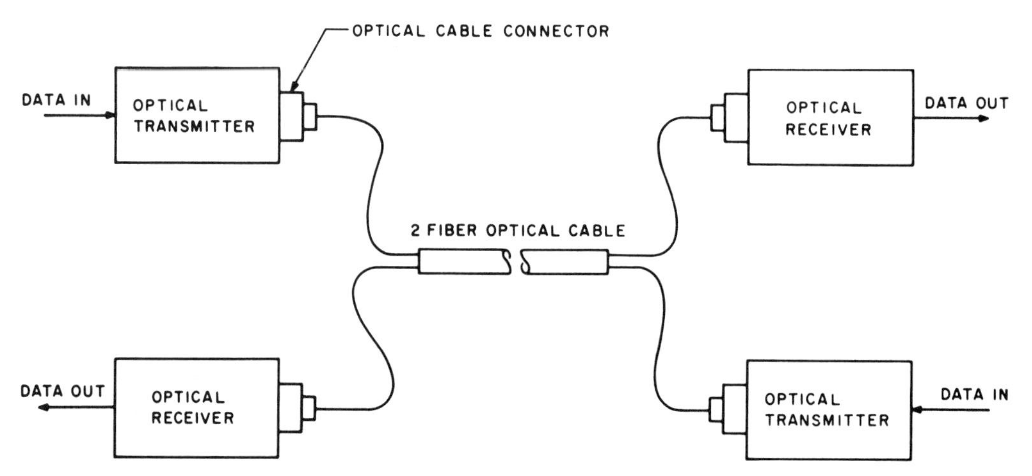

Figure 1: Block diagram of full duplex optical communications link.

tronics of high frequency commercial systems differ considerably, but the physical laws and general concepts are the same.

Fiber Optics

Fiber optics are just what they sound like – glass fibers which conduct light rather than electricity. To understand optical fibers we must look at a few definitions. An example of reflection and refraction is illustrated in figure 2. When a light ray strikes a boundary, partial reflection and partial transmission take place. The materials on either side of the boundary have particular constants n_1 and n_2 respectively (called *indices of refraction*) associated with them. These constants are dependent upon wavelength of the light transmission and the speed of light through the material. Reflection and refraction are related as follows:

$$\text{Reflection } \theta_1 = \theta_1'$$
$$\text{Refraction } n_1 \sin\theta_1 = n_2 \sin\theta_2$$

The fiber has a *core*, a light transmitting material of higher index of refraction surrounded by a *cladding* or optical insulating material of a lower index of refraction. Figure 3a is a pictorial representation of a single fiber. Light enters the fiber at an infinite number of angles but only those rays entering the fiber at an angle less than the *critical acceptance angle* are transmitted. Light is propagated within the core of a multimode fiber at specific angles of internal reflection. When a propagating ray strikes the core/cladding interface, it is reflected and zigzags down the core. This is further illustrated in figure 3b.

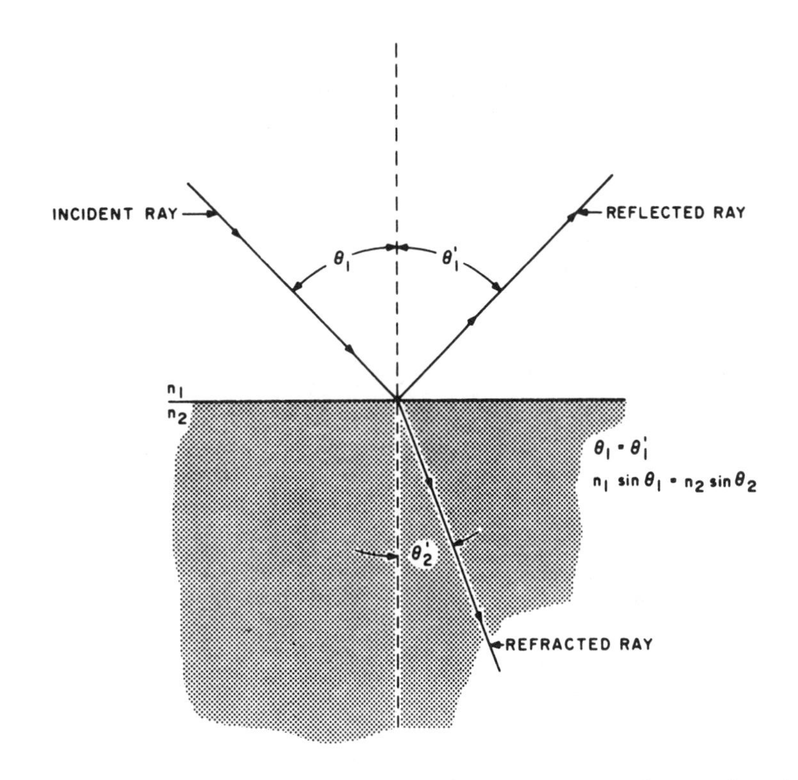

Figure 2: An example of reflection and refraction at an interface, such as the side of the optical cable.

Figure 3: Pictorial diagram of a single fiber illustrating the cladding and core boundary. Only light entering within the "acceptance cone" will be guided down the optical fiber as in figure 3b. Any rays outside this cone are not transmitted.

(3a)

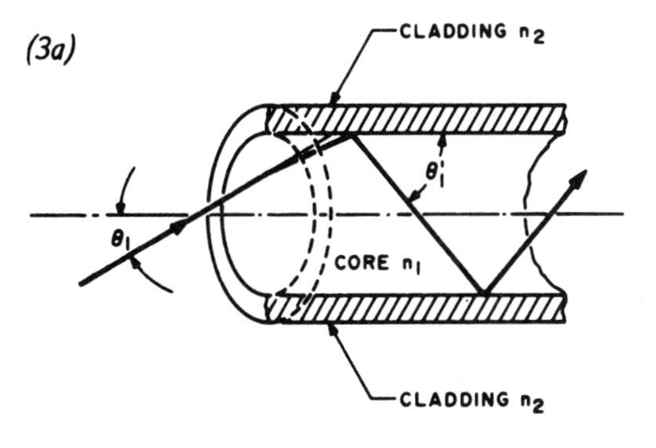

(3b)

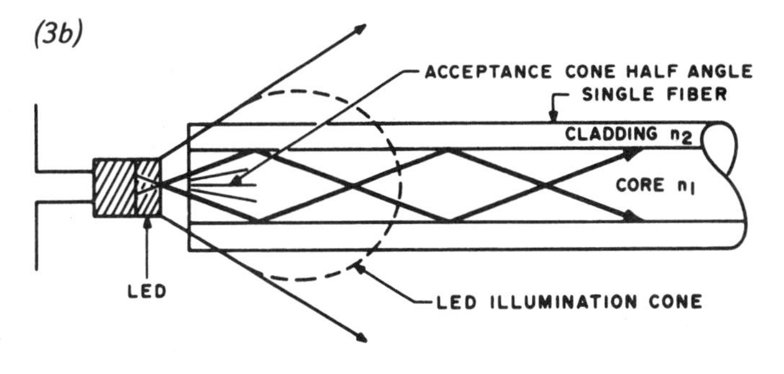

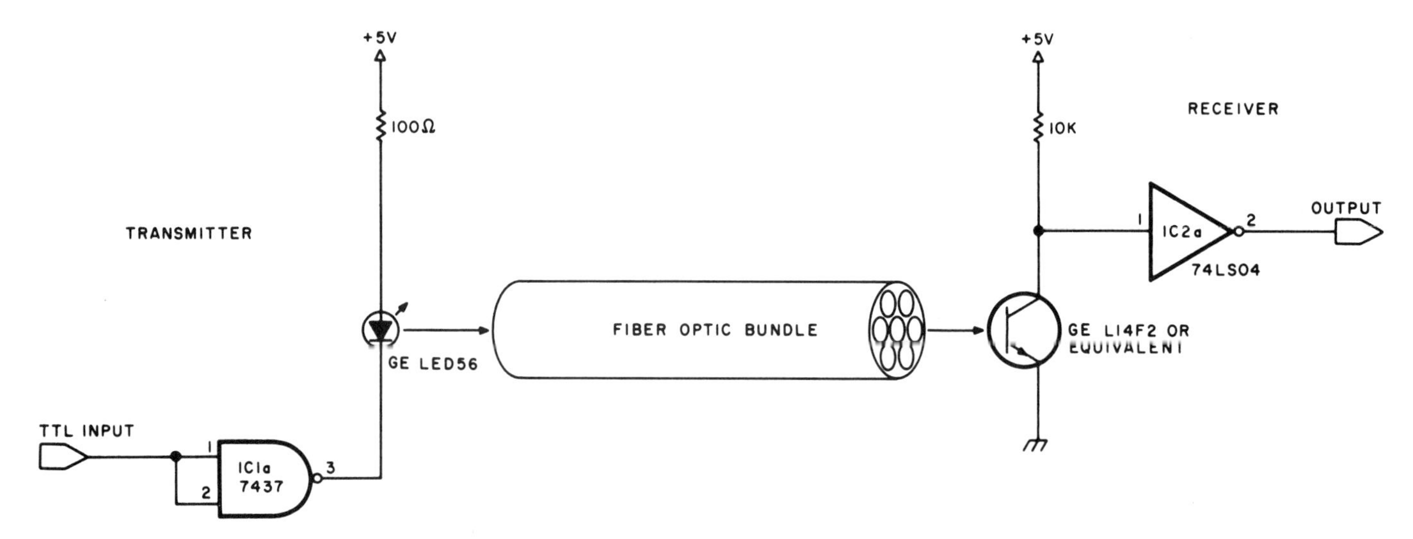

Figure 4: Schematic of a simple low speed and short distance fiber optics communications interface.

Photo 1 demonstrates that a very bright light can be transmitted through a single fiber. In this example the conductor is a single 40 mil plastic fiber with a helium-neon laser as an illumination source.

A fiber optic transmission system using readily available components can be constructed by any interested experimenter. A simple interface is shown in figure 4. An LED driven by a 7437 NAND buffer is focused into the end of a fiber optic bundle. The light emitted at the other end is focused on a phototransistor. When the light strikes the phototransistor it effectively grounds the input of the 74LS04, producing a high output. The connection between the LED, fiber optics, and phototransistor is facilitated through use of special optical connectors. Photo 2 shows an assortment of the type which should be used to build the interface in figure 4.

Lasers

The circuit of figure 4 is useful for only a short distance. This is due primarily to the low intensity of a standard LED. For greater distances a more intense light source is needed. This calls for a device such as a laser, an acronym that stands for *light amplification*

Photo 2: Special connectors necessary to use fiber optics properly. Shown here (starting in the upper right corner and continuing clockwise) are a fiber optic cable with an end connector, a phototransistor in a T0-18 package, an extension coupling which allows two cables to be connected, and a bulkhead receptacle containing either an LED (light emitting diode) or phototransistor.

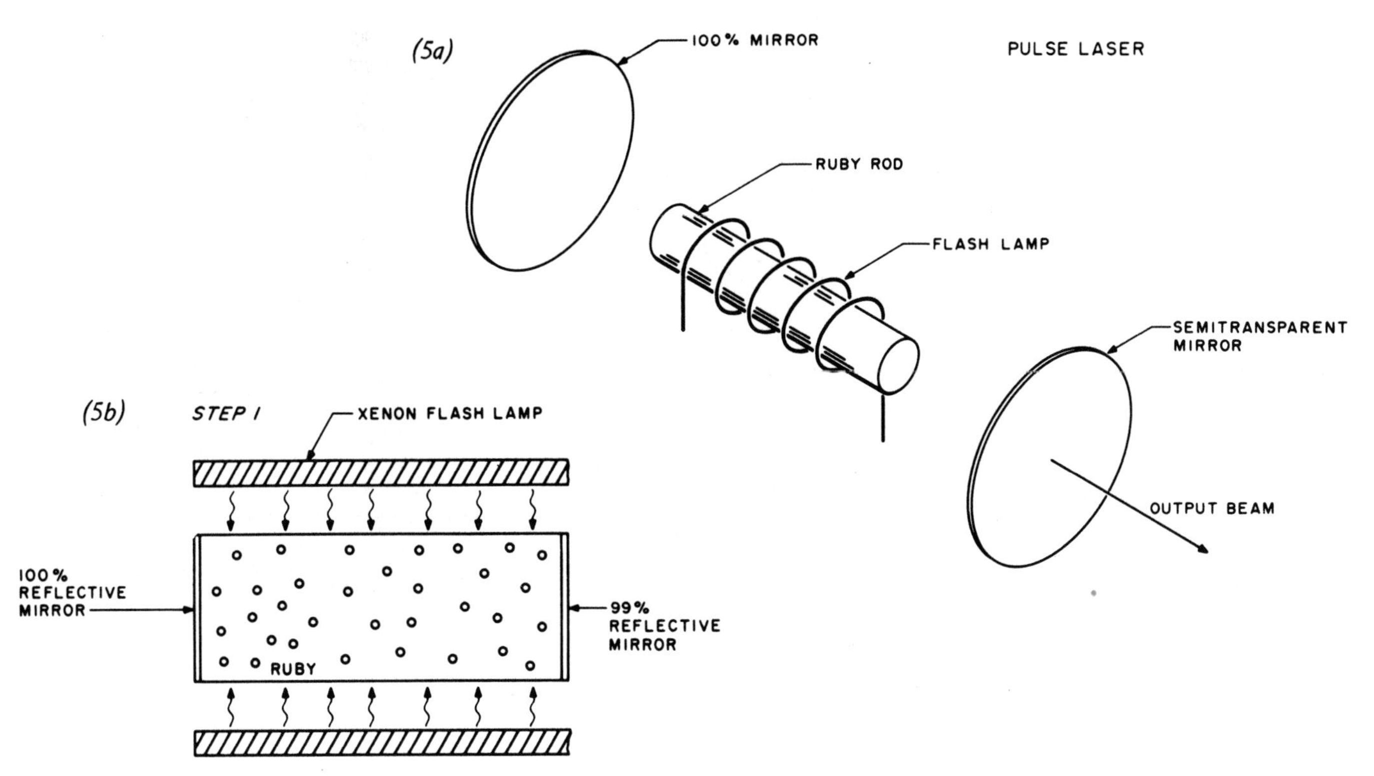

by *stimulated emission* of *radiation.* Light from a laser is all the same frequency, unlike the output of an incandescent bulb. Laser light is referred to as *coherent,* and has a high energy density. It can travel great distances without diverging from a tight beam.

The basic requirements for the creation of a laser are quite simple. We need a material that can absorb and release energy. Next, we need an energy source for exciting this material and a container to hold and control the lasing action, such as a glass tube or solid crystal.

In the actual lasing process, the laser material is placed inside the container, and then stimulated by means of an energy source into the emission of light waves. The laser beam is created by channelling the energy of these light waves into a particular and controlled direction. The result is a highly concentrated, brilliant beam of tremendous power. Figure 5 is a schematic of the first laser invented by Dr Theodore Maiman and a pictorial description of the lasing process.

The ruby laser is a pulse type laser which only produces a light output when the xenon lamp flashes. The best flash lamp can be fired only a few hundred times a second without extensive cooling apparatus. In a ruby laser this pulse mode operation is suitable for cutting stone and welding steel, but not for data communications, because the duty cycle is too short and the energy density too high for low cost fiber optics. The solution is to use a laser that operates continuously, such as a helium-neon gas laser

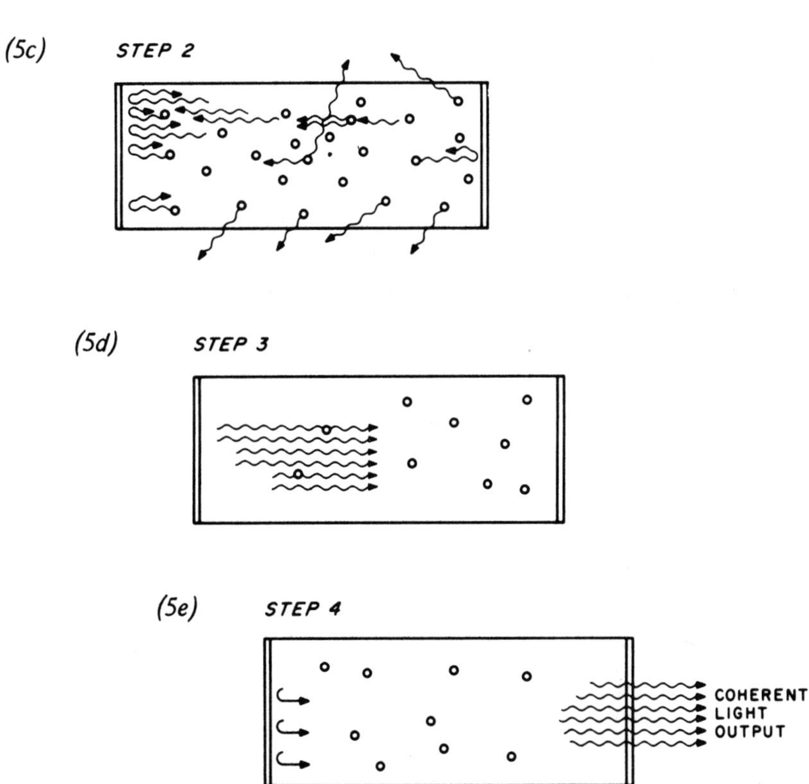

Figure 5: The first laser, invented by Dr Theodore Maiman, was made from a ruby rod excited by a xenon flash lamp. A schematic representation is shown in figure 5a. The laser builds up energy by the following process. In figure 5b the flash lamp is fired thereby exciting the electrons in the ruby rod. As the electrons drop back to their original energy level (step 2, figure 5c) they emit photons in random directions. In-step collisions of photons with other excited electrons start a wave front between mirrors (figure 5d). After many reflections back and forth between the mirrors, a wave front is built up until it contains sufficient energy to pass through the slightly less reflective of the two mirrors. This light output consists of coherent light.

(figure 6) or a laser diode which can be pulsed often enough to carry useful data.

The He-Ne laser uses mirrors and electrical excitation in a manner similar to the solid crystal type except that the lasing action is continuous. Photo 3 shows a He-Ne laser in operation. The particular unit has a power output of 2.2 mW and is made by Metrologic Inc. This type of laser can be modulated (the power supply high voltage is modulated) and used to drive a fiber optic bundle, but it is not normally used in that application. The light output of a He-Ne laser is usually red.

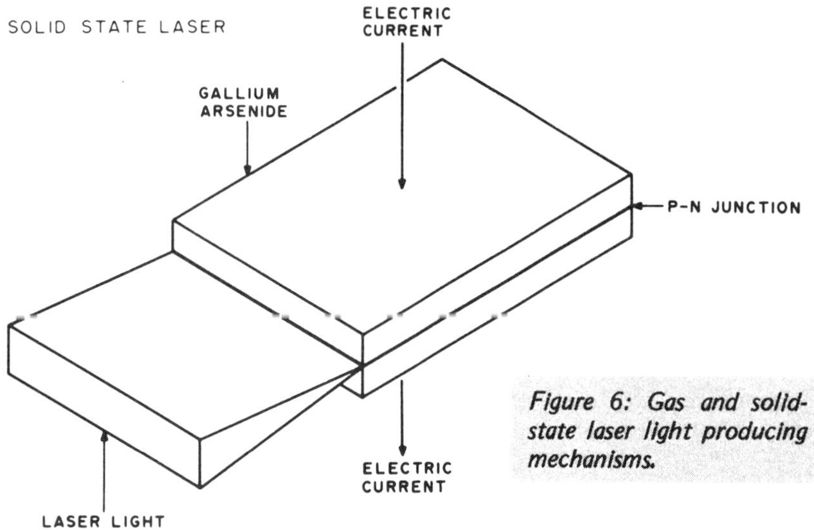

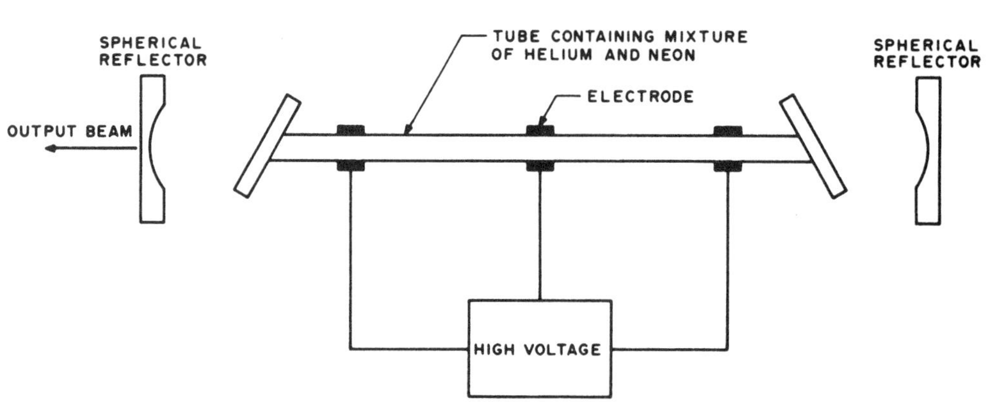

Figure 6: Gas and solid-state laser light producing mechanisms.

Photo 3: A laser on a tripod shooting across my living room. The laser is a 2.2 mW unit built by Metrologic Instruments of Bellmawr NJ 08031 (this particular model is the ML-969). This picture was taken at night; the trees outside are illuminated by outside flood lamps.

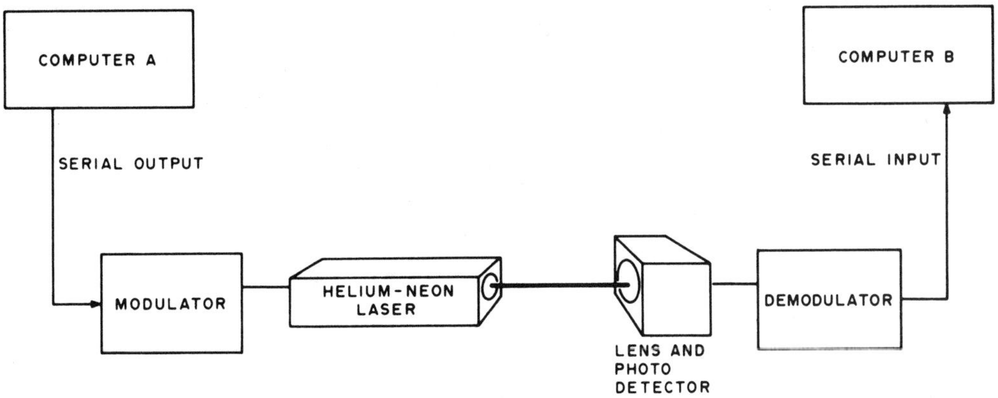

COMPUTER A
SERIAL OUTPUT
MODULATOR
HELIUM-NEON LASER
LENS AND PHOTO DETECTOR
DEMODULATOR
SERIAL INPUT
COMPUTER B

Figure 7: System configuration necessary for one computer to transmit data to another via a helium-neon laser beam. The schematic for the modulator and demodulator are shown in figures 8 and 9, respectively.

The most economical high intensity light source for long runs of fiber optics is the laser diode. Don't be so whimsical as to run out and buy one thinking you are going to make a ray gun – it should be just as easy to use as an LED. Laser diodes get very hot in operation and are generally operated only in pulse mode. An 8 W laser diode sold through the surplus dealer can have an average power of only a few hundred microwatts when used in pulse mode operation. Using laser diodes in continuous operation is beyond the talents and resources of most hobbyists and must be left to the commercial ranks for the moment. The light output from a laser diode is infrared and invisible to the human eye.

Communicating on a Laser Beam

While it is possible to demonstrate communication with a laser diode, it is much more dramatic with a He-Ne laser since you can see the beam. A He-Ne laser can be modulated, but it cannot be turned on and off rapidly like an LED or diode. Instead the light intensity is modulated by the data signal. The Metrologic laser I used is a type ML-969 "modulatable" laser. It has a BNC connector on the rear and accepts a 0 thru 1 V input for 0 to 15 per cent intensity modulation. Any greater degree of modulation shuts off the lasing action.

Figure 7 illustrates the system configuration necessary to transmit data from one computer to another. Figure 8 is the schematic of a FSK (frequency shift keyed) modulation interface which can be used as the input to the laser. A 4800 Hz frequency reference produced by IC1 is divided by IC2 to give either 2400 Hz or 1200 Hz for a 1 or 0 logic input respectively. The modulation input to the laser can be any 1 V input up to 500 kHz bandwidth. A transistor radio is a good test source for experiments.

The receiver is shown in figure 9. The laser beam is directed at the phototransistor. With no modulation, the sensitivity is adjusted to set the phototransistor in the middle of its linear range. With the modulation turned on, the trigger adjust control is turned until the modulation data is seen at test point 1. If using a transistor radio as the source, the analog output can be obtained at this point and the rest of the circuit is unnecessary.

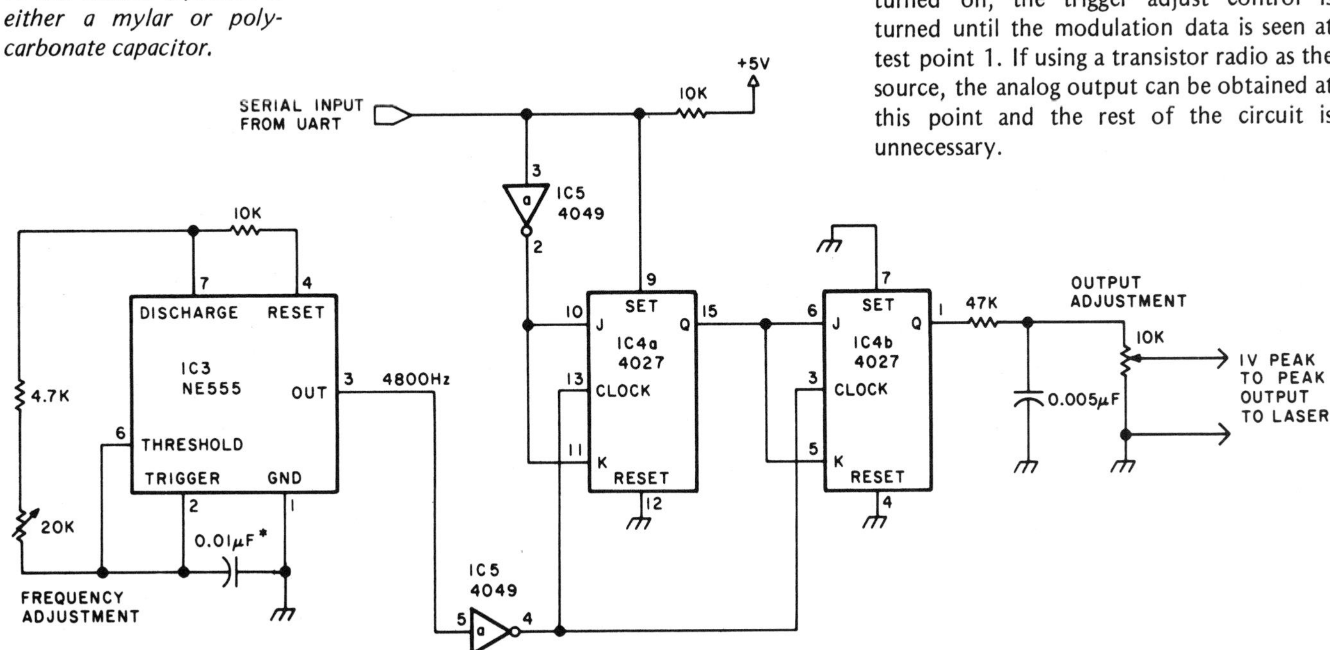

Figure 8: A frequency shift keyed laser modulation interface. This circuit accepts input from the computer's UART (universal asynchronous receiver and transmitter). A logic 1 input produces a 2400 Hz output. An input of logic 0 produces a 1200 Hz output. The power connections for the integrated circuits are shown in table 1. The starred capacitor is either a mylar or polycarbonate capacitor.

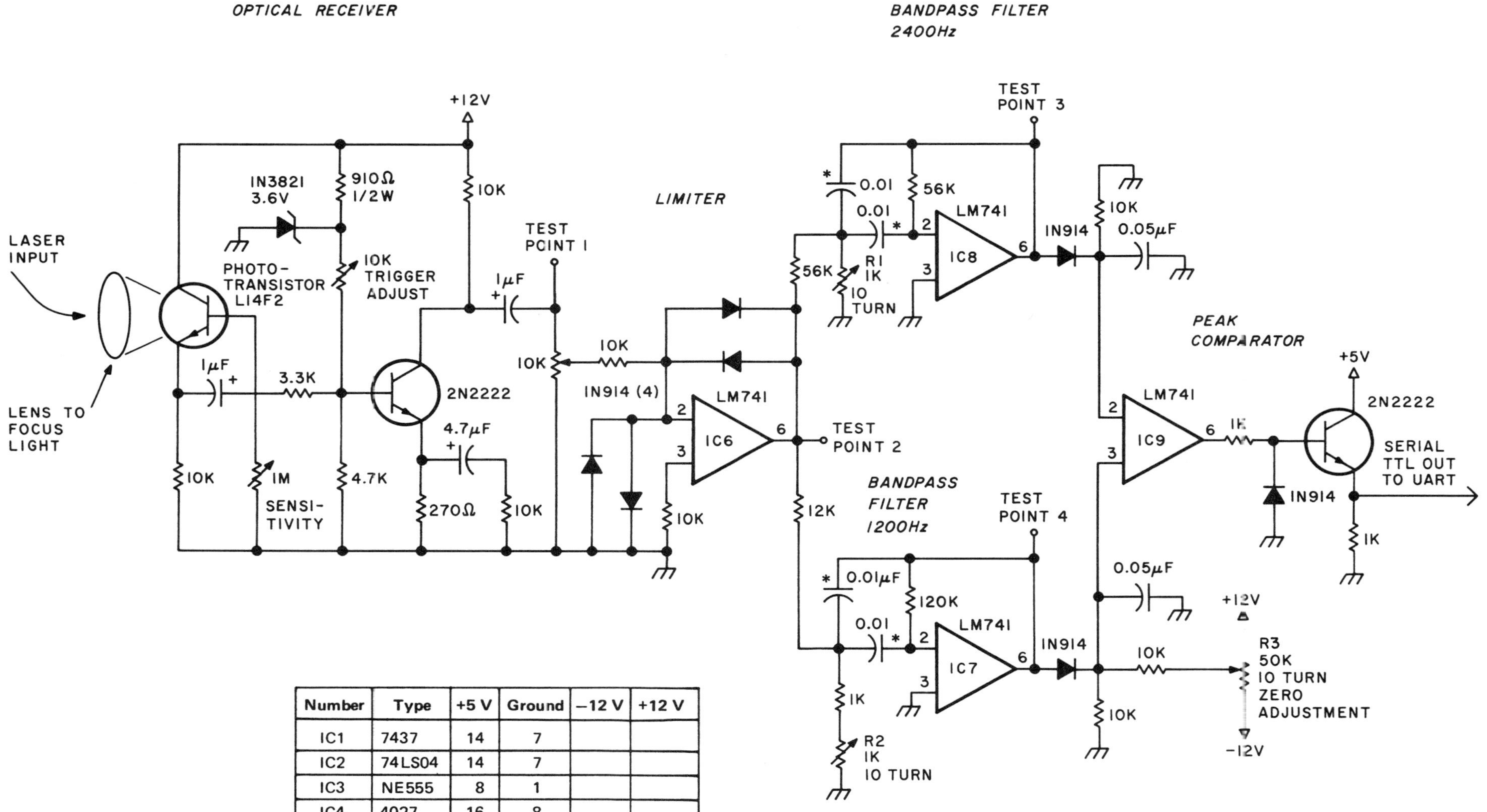

Number	Type	+5 V	Ground	−12 V	+12 V
IC1	7437	14	7		
IC2	74LS04	14	7		
IC3	NE555	8	1		
IC4	4027	16	8		
IC5	4049	1	8		
IC6	LM741			4	7
IC7	LM741			4	7
IC8	LM741			4	7
IC9	LM741			4	7

Table 1: Power pin connections for the integrated circuits used in constructing the laser communicator.

Figure 9: Modulated laser beam serial data receiver. The demodulator consists of two bandpass filters, one for 2400 Hz and the other for 1200 Hz. The power connections are given in table 1. The starred capacitors are mylar or polycarbonate capacitors. All resistors are 1/4 W unless otherwise specified. All diodes are type 1N914.

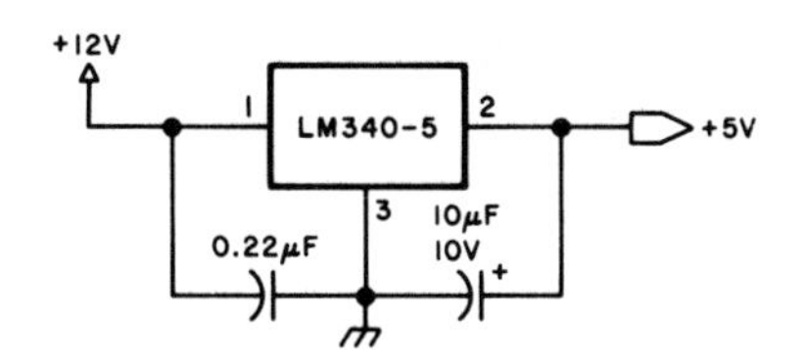

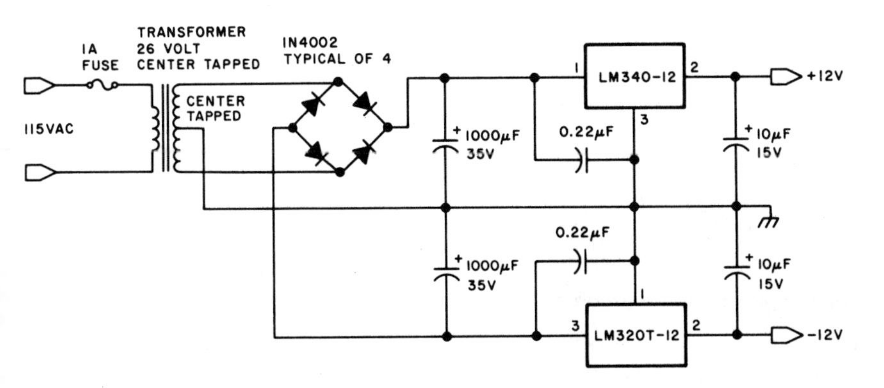

Figure 10: A triple voltage power supply for the laser modulator.

Integrated circuits 1 thru 4 form a frequency shift keyed demodulator with a TTL (transistor-transistor logic) output which is sent to a UART (universal asynchronous receiver-transmitter). To tune this section, first connect a 1200 Hz signal source to test point 1. Turn potentiometer R2 until the output amplitude of IC3 test point 4 peaks. Then applv 2400 Hz to test point 1 and adjust R1 until the amplitude at test point 3 also peaks. R3 adjusts the point at which circuit's output switches between logic levels. It should be set to follow the input at test point 1 with the shortest response time.

While the 15 per cent modulation could be detected directly and converted to NRZ (nonreturn to zero) formatted data, the receiver circuitry would be far more complicated. The combination of amplitude and frequency modulation techniques is intended to add significantly to the chances that an experimenter will have success building it. The critical parameters (as with any optical system) are alignment and light level. And, while you may never have to transmit a Red Sox baseball game across two miles of New Hampshire woods, it's nice to know how if you ever have to do it.■

LONG DISTANCE COMMUNICATION

Dear Steve,

I saw your article "Communicate on a Light Beam," and became very interested. I have an application which requires sending data up to a kilometer at speeds from 2000 to 9600 characters per second (cps). Your descriptions of the fiber optic cable and the light-emitting diode (LED) transmission circuits seem to be ideal, if they are cost-effective.

Could you give more details of the distances which the circuits can drive and the addresses of the suppliers of the fiber optic components?

R H Fields

Realize, of course, that the circuits presented, while possibly usable in commercial applications, are presented more to introduce the reader to the concept of fiber optic communications than solve any particular application problem. Their usability in a 1 kilometer data link depends upon more than just the electronic parameters of the circuit. The laser probably can drive such a length, but cable losses and mechanical/optical connections are going to be an important factor in any success.

When you speak of 9600 cps, that is approximately 100 k bits per second (bps) and is a reasonable transmission rate. However, response time of the receiver electronics is going to be much more critical than a 10 k rate. Given the length of cable as 1 kilometer, I would caution you that a certain intensity must be maintained at the output to achieve this response.

Rather than try to reinvent the wheel or try to second-guess the technical people who really know the field, I think you would be better off purchasing a commercial system. The following is a list of American companies that deal in fiber optics. I am sure they will have a cost-effective solution for you:

Corning Glass Works
Telecommunications Dept
Corning, NY 14830
(607) 974-8812

Dupont Co
Plastic Products and Resins Dept
Wilmington, DE 19898
(302) 774-7850

Fiberoptic Cable Corp
POB 1492
Framingham, MA 01701
(617) 875-5530

Galileo Electro-Optics Corp
Galileo Park
Sturbridge, MA 01618
(617) 347-9191

General Cable Corp
500 W Putnam Ave
Greenwich, CT 06830
(203) 661-0100

ITT
Electro-Optical Products Div
Roanoke, VA 24019
(703) 563-0371

Quartz Products Corp
688 Somerset St
Plainfield, NJ 07061
(201) 757-4545

Times Fiber Communications Inc
358 Hall Ave
Wallingford, CT 06492
(203) 265-2361

Valtec Corp
Electro Fiberoptics Div
West Boylston, MA 01583
(617) 835-6083

For further descriptive information on the use of fiber optics, I suggest you refer to the January 5, 1978 issue of EDN magazine and an article entitled "Designer's Guide to Fiber Optics."...Steve

7
Joystick Interfaces

Photo 1: A typical joystick with 4 potentiometers.

Photo 2: Note how moving the stick moves the gimbal arrangement, which in turn changes the settings of the potentiometers.

The thought that often comes to mind when the word joystick is mentioned to a computer enthusiast is of a spacewar-type game. A photon torpedo is fired from an opponent's starship, and the thruster joystick is deftly moved to reposition the craft out of its path. All of this occurs without having to take your eyes off the screen. Eye/hand coordination is almost "instinctive." With a glance to the upper right of the video screen, the joystick is tilted to the upper-right corner of its 360° range. This moves the spacecraft toward that coordinate. Reverse thrust is accomplished by moving the joystick in the opposite direction, as though you are pulling back on the throttle of a real craft. Such is the general experience with joysticks. However, the potential use of these devices greatly exceeds that of game playing.

A joystick, for those people who are unfamiliar with one, is shown in photo 1. It is an electromechanical device with resistance outputs proportional to the X,Y displacement of a central ball and lever. Photo 2 illustrates the mechanical connections to the potentiometers.

When the stick is positioned in the center of its axes, the X and Y potentiometers show resistances in the center of their ranges. When the stick is tilted to the upper right, both potentiometers are at their full-resistance limit, while the opposite (lowest resistance) is true when in the lower-left position. The outputs of the 2 potentiometers accurately track, as if on an X,Y coordinate axis, the position of the joystick. It should be noted that while it takes only 2 potentiometers to define 2-dimensional travel, most joysticks are manufactured with 4 potentiometers. This is a remnant of the days when joysticks were connected directly to the 4 deflection-plates of a cathode ray tube (video screen).

It is one thing to *consider* interfacing a joystick to a computer, and quite another to *do* it. A joystick is a mechanical X,Y positioning device. Even with proportional output resistances, an input interface must be designed to convert position from an analog to a digital representation which can be used by the computer. A further consideration is the resolution, or percent, of full-scale travel per bit sensitivity. Is the application so gross that center and full-scale are the only points of interest, as in a

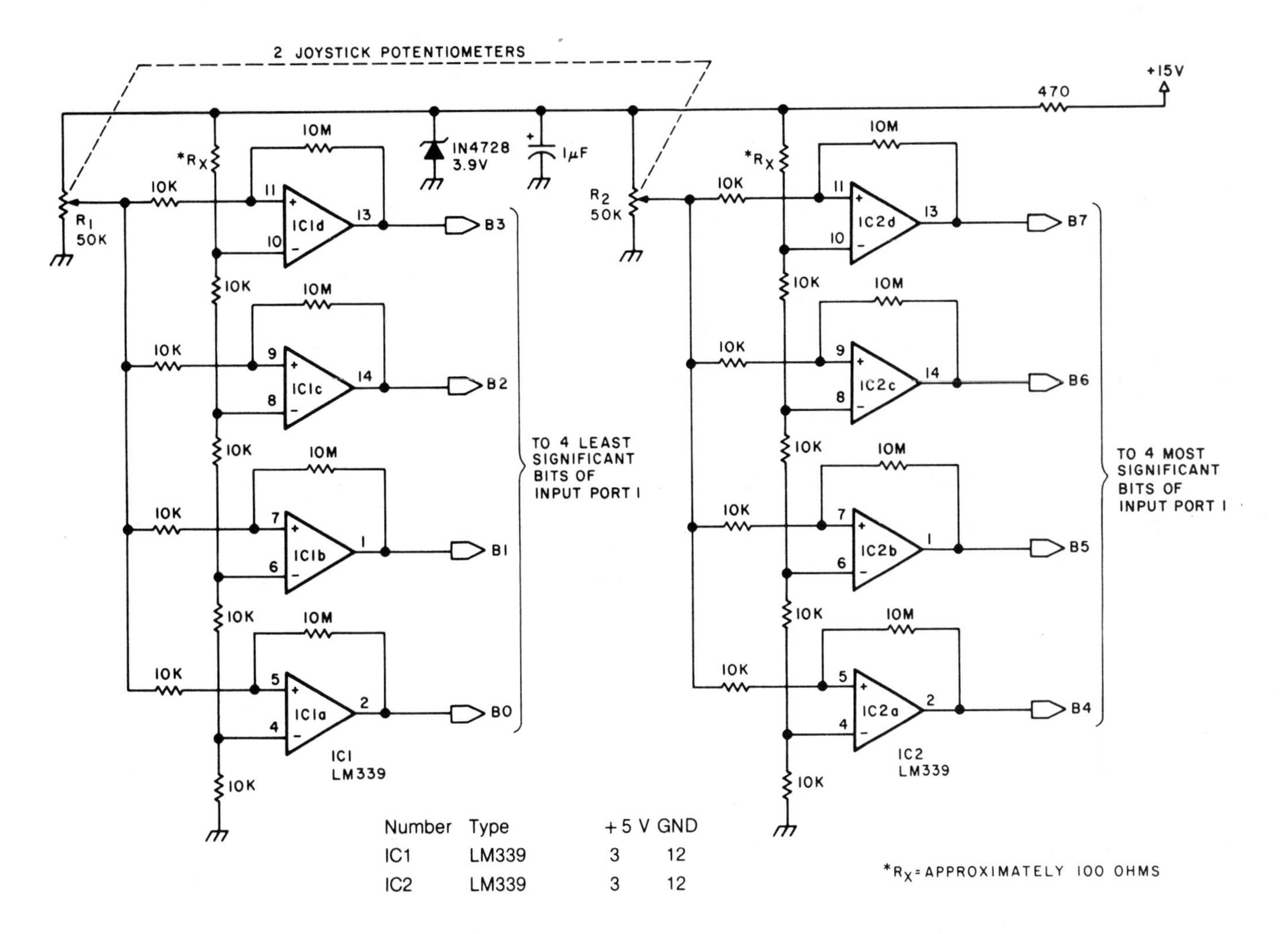

Number	Type	+5 V	GND
IC1	LM339	3	12
IC2	LM339	3	12

Figure 1: *Low-resolution static interface. This interface is for a 2-potentiometer joystick. For 4-potentiometer joysticks, build a second circuit like this one, and interface it to another input port. Note that if the comparator does not trigger at full-scale setting, a small resistor may have to be added at Rx (marked with asterisk).*

game control, or is the application one which requires fine control, such as a cursor-positioning device in a high-resolution graphics system?

All joystick interfaces are not created equal. There is a trade-off between hardware and software. The lower the resolution, the fewer the parts. The higher the resolution, the greater the electrical complexity or the software interaction with the interface. It is also important to recognize that computer systems which operate *only* in a high-level language like BASIC cannot use an interface design that requires an assembly-language subroutine as an integral component. In such instances only a static interface can be used.

Included in this presentation are 4 interface designs which should cover most requirements, as well as demonstrate the considerable differences between them. The 4 types are:

- low-resolution static
- high-resolution fully static hardware
- software-driven pulse-width modulated
- high-resolution analog-to-digital

Low-Resolution Static Interface

First of all, *static* simply means that the interface hardware determines the potentiometer position value and presents it in constant, parallel digital form to the computer. When the interface is attached to any parallel input port, this joystick value can be read with a single INPUT command in BASIC. As far as the computer is concerned, the value is fully static, and the computer reads whatever data is there when the INPUT is executed. The interface hardware has the responsibility of asynchronously updating the digital value as the stick is moved.

Often the joystick is simply used to indicate relative direction and magnitude. In a wheelchair, for instance, full linear control of speed and direction would require rather expensive drive electronics. Most chairs use simple relay contacts and provide 2 or 3 selectable speeds. A joystick control built for this application would not have to have a resolution of 8 bits, but could, in fact, suffice with 2. Figure 1 shows a low-resolution static output joystick interface suitable for use in this application.

Each potentiometer is connected as a voltage divider between a reference voltage source of 3.9 V and ground. The voltage output of each potentiometer is, in turn, fed to a 2-bit, parallel analog-to-digital converter. This type of converter uses 4 comparators set for 25%, 50%, 75%, and 100% of full scale. If a voltage, when applied, is less than 0.975 V, all comparator outputs will be at 0 V. At 1.0

```
        MVI    B      clear B
        OUT    FF,0   trigger one-shots
AGAIN   INR    B      increment B register
        IN     FF     read potentiometers
        ANA    01     isolate bit 0
        JNZ    AGAIN  continue as long as one-shot is high
        HLT           value is in B register
```

Listing 1: A typical assembly-language program for using the joystick interface of figure 4. After the one-shots are triggered, the program loops and checks the status of bit 0. When this bit is set, the conversion value is in register B. This program assumes that there is only 1 value being checked, and it is being input through bit 0.

V, corresponding to the joystick being moved 25% of full scale, the least-significant bit (LSB) of the converter will be a logic 1, while the other bits are low. Similarly, at full input all comparators will be triggered, and bits 0 thru 3 will be logical 1s.

Additional encoding logic can be added to produce a true 2-bit representation from the 4 comparators, but it is just as easy for a computer to interpret it directly. With a 4-bit connection as shown, used in a BASIC program, 25% of full scale would be 1 decimal, 50% of full scale would be 3 decimal, 75% of full scale would be 7 decimal, and full scale would be 15 decimal. It should be easy to trigger any action by a coincidence with these values. The real significance of this method is that the potentiometer position is presented statically to the computer and requires no other interaction. This makes it ideal for direct use with BASIC.

High-Resolution Static Interface

It is quite possible that 2 bits of resolution is not enough for your application, but direct compatibility with a slow, high-level language is still a requirement. Expanding the parallel comparator method will work in theory, but you must realize that a 4-bit analog-to-digital converter uses 15 comparators, and an 8-bit, parallel analog-to-digital converter needs 255 comparators! So much for that method.

Realizing that the output of the joystick is a variable resistance, we can use this to advantage. This resistance can set the *time constant* of a function which has a pulse width proportional to joystick position. Figure 2 illustrates an interface design which uses this technique.

The 2 joystick potentiometers R1 and R2 control the pulse width of a one-shot (monostable multivibrator). The one-shot has a pulse width of 35 ms when the potentiometer is at 50 k ohm full scale and something less than 100 μs at 0% of full scale. A 7.5 kHz clock signal asynchronously triggers the one-shots. When the one-shot fires, its duration is proportional to the joystick position and will vary from approximately 0 to 35 ms. Using midscale pulse width of 17 ms as an example, the circuit timing is as in figure 3.

On the leading edge of the one-shot signal, a *clear* pulse is generated through an edge detector configured 7486 device. The clear pulse resets the **two 7493s which form an 8-bit counter. Once cleared, the counters start counting clock pulses for the duration of the one-shot's period. On its trailing edge, a *load* pulse is generated which loads this 8-bit counter into an 8-bit storage register. The computer is connected to read this 8-bit value through a parallel input port. Successive clearing and counting operations update the register every 35 ms or so (worst case). The clock rate is 7.5 kHz which has a period of 133 μs. If the one-shot has a pulse width of 17 ms, the 127 clock pulses would be gated to the counter. Of a total possible 255 counts, 127 would represent 50% of full scale.**

Software-Driven Interfaces

So far I have discussed only static interfaces. If the computer used with the joystick has sufficient speed and excess computing time available, then it is reasonable to use the computer to directly determine the one-shot period.

Figure 4 shows a circuit which directly connects to the computer bus and demonstrates this technique. The circuit as shown is wired for I/O (input/output) port decimal 255 or hexadecimal FF. The 4 joystick potentiometers are used as the timing resistors on 4 NE555-type one-shots. When an OUT 0, FF is executed in assembly language, it triggers all 4 one-shots. To keep track of the pulse widths, a 74125 3-state driver gates the one-shot outputs onto the data bus during an IN FF instruction. By looping through this program a number of times and keeping track of the logic levels of the 4 one-shots, the computer can accurately determine joystick position in terms of loop counts of instruction times. Listing 1 is a program which does this for 1 potentiometer.

High-Resolution Analog-to-Digital

While all methods are in *some* way analog-to-digital converters, the last **method is in fact an 8-bit absolute-analog-to-digital converter, typical of the type used in computerized measurement applications. IC1 is an 8-bit** digital-to-analog converter that produces an output voltage proportional to a digital input applied to pins 5 thru 12. For a complete explanation of this device, I refer you to a previous " Ciarcia's Circuit Cellar" article, "Control the World" (September 1977 BYTE, page 30). This article also outlines calibration and test procedures.

The 3 basic sections are a computer-controlled voltage source (ICs 1 and 2), an analog-input multiplexer (IC3) which selects an individual joystick potentiometer by a 2-bit address code, and a comparator (IC4) which compares these voltages. In operation, the digital-to-analog converter is first set to 0 V out (hexadecimal 00 digital input to it) and 1 potentiometer is selected through the multiplexer. If V0 from the digital-to-analog converter is less than V_{in} from the potentiometer, the output will be logic 0. Next, the digital-to-analog converter input setting is incremented, and the comparator output is checked again.

Eventually an input count will be reached which will exceed V_{in}. The comparator output will then be a logic 1. The digital-to-analog converter input count is now the value of the voltage V_{in}. The worst case requires 256 iterations using this

Figure 2: High-resolution, static interface. Each potentiometer in the joystick controls the pulse width of a one-shot. The pulse width can vary from 35 ms at full-scale to 100 μs at 0. If a joystick with 4 potentiometers is used, a duplicate circuit may be constructed for the 3rd and 4th potentiometers.

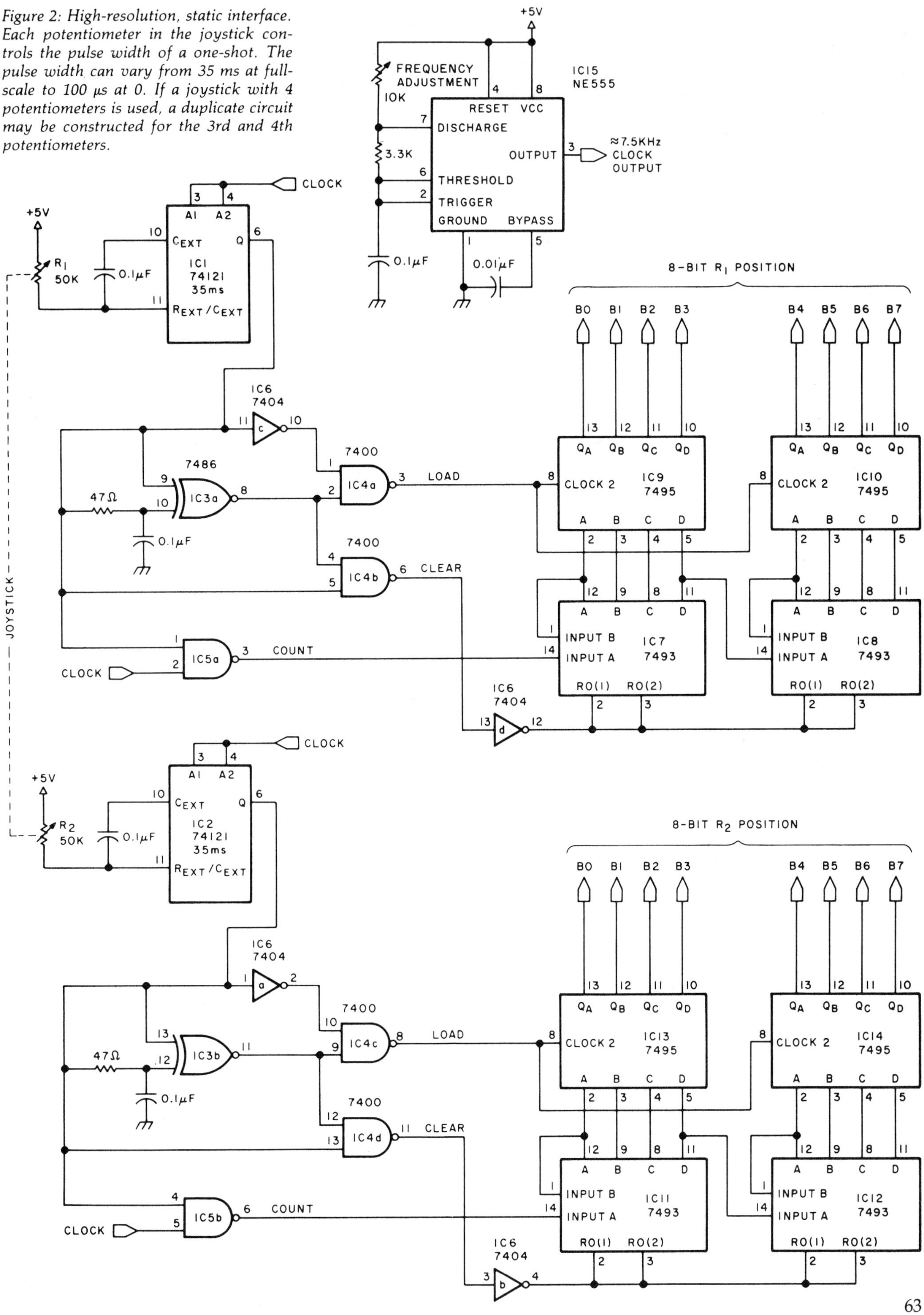

Number	Type	+5 V	GND
IC1	74121	14	7
IC2	74121	14	7
IC3	7486	14	7
IC4	7400	14	7
IC5	7400	14	7
IC6	7404	14	7
IC7	7493	5	10
IC8	7493	5	10
IC9	7495	14	7
IC10	7495	14	7
IC11	7493	5	10
IC12	7493	5	10
IC13	7495	14	7
IC14	7495	14	7
IC15	NE555	8	1

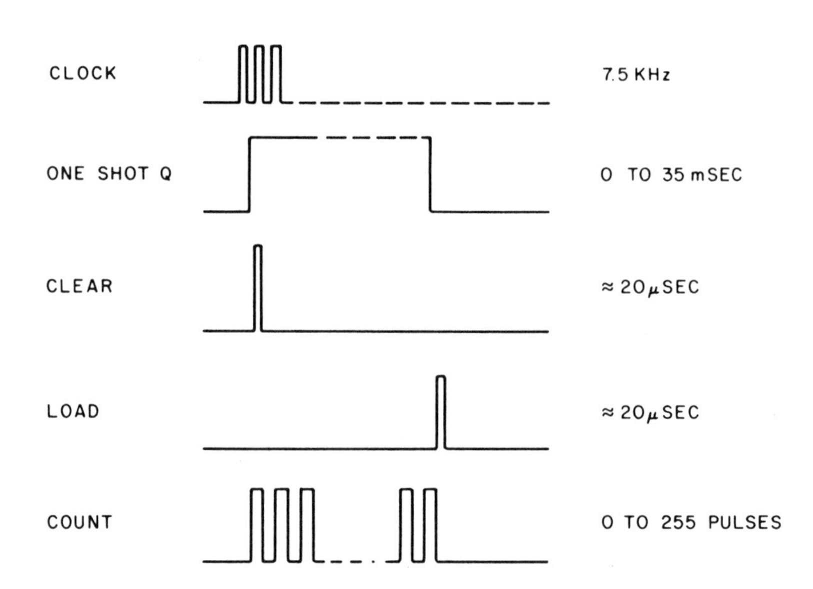

Figure 3: Timing diagram for interface of figure 2. The driving clock signal is 7.5 kHz. The one-shot can be triggered for periods of 0 to 35 ms, depending upon the position of the joystick. When a reading is to be taken, the counters are cleared. Counts are made until the one-shot signal drops, and then a load signal is sent to the interface. At this point the counter is read to determine the position of the joystick.

Number	Type	+5 V	GND
IC1	NE556	14	7
IC2	NE556	14	7
IC3	7430	14	7
IC4	7400	14	7
IC5	74125	14	7

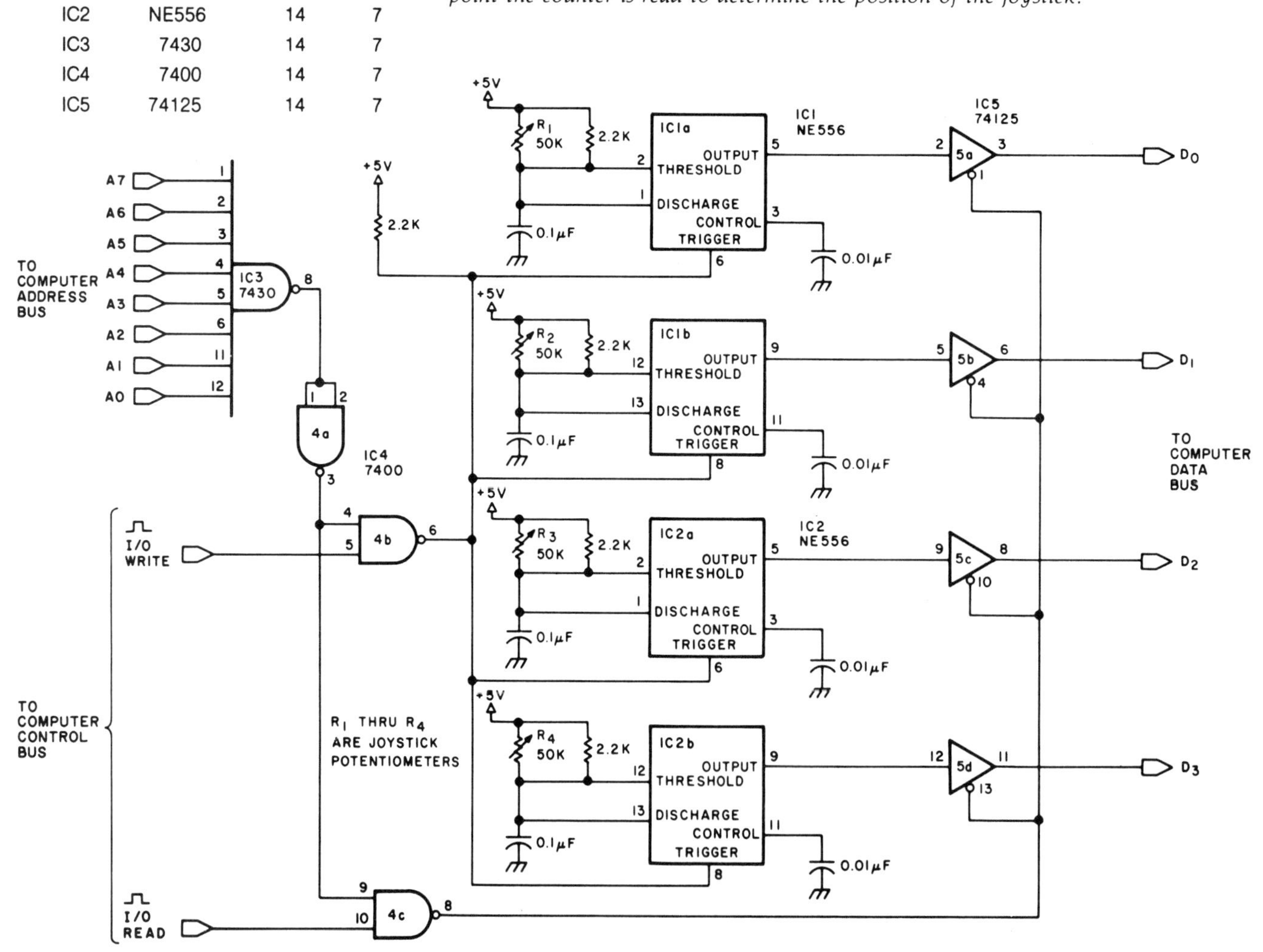

Figure 4: Software-driven interface. If the computer can directly read the input from the joystick interface, the hardware required can be greatly simplified. When hexadecimal FF is output to port 0, all 4 one-shots are triggered. The pulse width is then determined by a program running through a short loop looking at the logic levels of the 4 one-shots. Listing 1 shows a typical program for this application.

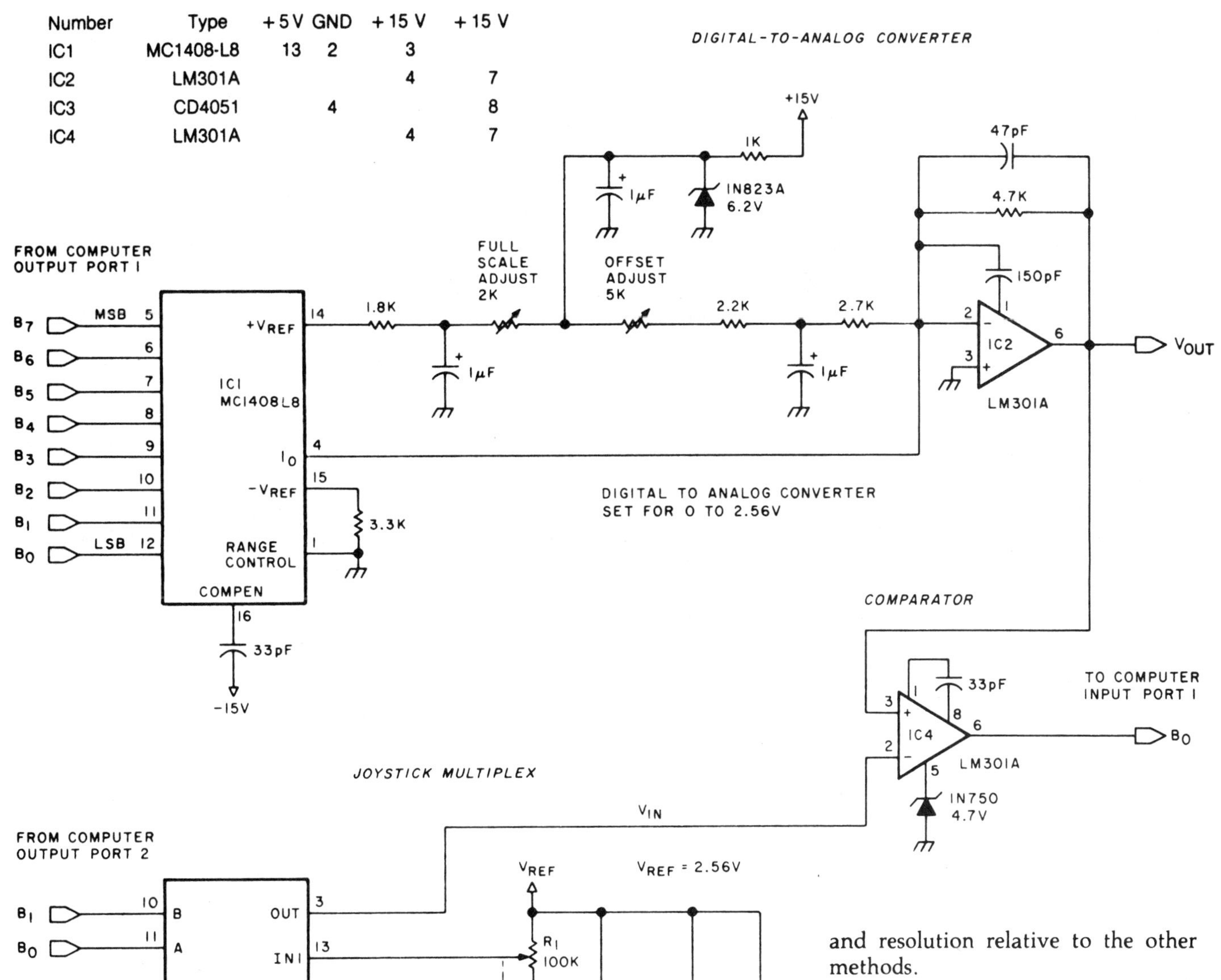

Number	Type	+5 V	GND	+15 V	+15 V
IC1	MC1408-L8	13	2	3	
IC2	LM301A			4	7
IC3	CD4051		4		8
IC4	LM301A			4	7

Figure 5: High-resolution analog-to-digital conversion. This hardware-oriented device multiplexes 4 voltage inputs (from the joystick potentiometers) and has the capability of handling 4 more voltages.

method. A better technique is successive approximation where the computer progresses through a binary search to "zero in" on the final value. A full explanation of successive approximation is delineated in my article entitled "Talk to Me: Add a Voice to Your Computer for $35" (June 1978 BYTE, page 142).

With the digital-to-analog converter set for a full-scale value of 2.56 V, each count is equivalent to 10 mV. Only 4 channels of the CD4051 are used for the joysticks, leaving another 4 channels as auxiliary inputs from external sources. Thus it is possible for this interface to serve a dual role because of its high accuracy and resolution relative to the other methods.

You should now realize that both the design and construction of a joystick interface are influenced by many factors. It is not unusual to find one manufacturer charging $50 for a joystick, while another charges $200. Resolution, accuracy, and software interaction are the prime considerations. Where static inputs are required, the hardware will necessarily be more complicated. Resolution and accuracy ultimately determine the complexity of the interface.

For simple spacewar-type games, the circuit of figure 1 should suffice. For more demanding applications such as cursor control in a high-resolution graphics system, figure 5 may be the optimum choice. Be careful when buying joystick interfaces. Make sure that they mate with your program requirements and your system's abilities. ■

8

Handheld Remote Control for Your Computerized Home

Remote control is on the minds of many people these days. The Busy Box AC remote-control interface for household appliances has been received with great interest, whetting the appetites of most experimenters.

The Busy Box interface, which connects the BSR X-10 Home Control System (as sold by Sears) to a personal computer, is intended to facilitate inexpensive AC remote control. When attached to a computer such as a Radio Shack TRS-80, it can easily turn on the television set precisely at 7 o'clock as you flop in your easy chair after a hard day at the office. (Delivering slippers is still the dog's job.) By using a sufficient quantity of the remote output modules and coordinating software, the appearance of a completely computer-controlled home can be obtained.

This control is limited, however. Without physically typing on the computer keyboard, there is no direct method for the operator to command the computer to turn on remote-control channel 2 of the X-10, or for the computer to verify that this activation has in fact occurred. It is left to the operator to either indicate the status of each output or clear everything at the start. While this might at first seem to be unimportant in most domestic control applications, it is a major annoyance. In critical control applications, it is a definite liability.

Control systems incorporating no feedback of status are called *open-loop* systems. When feedback on the effect of control outputs is provided, the control loop is completed, and the system is then referred to as a *closed-loop* control system.

While various methods may be employed to directly drive the BSR system with control information, any change in output status is not relayed back to the computer. The control computer does not know that you have overridden the system and manually turned on the TV. Nor does it know that you have just changed your mind about staying up to watch the late, late show. It is easy for you to get out of synchronization with preprogrammed timing, and you may find yourself suddenly sitting in the dark with the TV off.

For example: I thought I'd like to automatically shut off the TV set after the late show in case I fell asleep. I set the real-time software to shut off the X-10's remote-control channel 4 at 2:30 AM. It turned out, however, that the only one who could keep time was my computer. Even allowing an extra hour to make up for changes in schedules and interruptions, invariably everything would go black just as Charlie Chan gathered everyone together in the living room to disclose the identity of the murderer.

True, I could have kept changing the shut-off time in software, but that would have cured only one symptom and would not have attacked the real problem. I needed a way to tell the computer, "Tonight, I will retain control of the TV set," and to tell the control system when it can take over again. It is difficult to have effective automatic activation of household appliances unless both controllers, you and the computer, communicate directly. It sounds easier to do than it is.

Closing the Loop

Two immediate alternatives come to mind. One is to pull the plug on the computer when you walk in the house. Then you are the only one in command. Or, if you prefer to retain automated control, you can always command the control system directly through the computer. For a truly computer-controlled environment, this is the only possible solution.

The limiting factor in constructing remote-control devices lies more in the complexity of checking out the hardware than in the communication techniques employed.

The next question is "How do we indicate to the control system what actions it is to perform when we are not there sitting at the console?" There has to be some facility for remote communication. In its least complicated form, this facility might be nothing more than a single switch at the end of a long cable. This switch can be used to initiate execution of particular control programs on the computer, or to let the computer know that a specific, controlled event has occurred. While the idea of a 100-foot cable might sound rather questionable to most computer users, it is inexpensive and it will work.

However, it is a rather cumbersome approach. Since I have an aversion to being attached to my computer by an umbilical cord, and since the actual feedback mechanism does not have to be especially complicated, I propose a less conspicuous connection using wireless communication. More on this later.

Handheld Remote Control

The most convenient communication mechanism is a handheld transmitter or controller. On it can be a button, or buttons, which you press to initiate various computerized activities, which can range from running a Star Trek game program automatically to activating and deactivating the house security system. Entire chains of events can be triggered by a single output command. And, by utilizing the simple radio-frequency (RF) transmitter I have designated, remote operation of AC-powered appliances can be carried out from much greater distances than presently accommodated with the standard BSR X-10 system.

The purpose of this article is to present a circuit for the construction of a handheld, transmitting communication device. With a receiver attached appropriately to an input port on the computer, and using software that coordinates its activities, we can effectively have a "handheld remote controller."

The limiting factor in constructing remote-control devices lies more in the complexity of checking out the hardware than in the communication techniques employed. I shall describe three different systems which can function as control communicators. They vary considerably in complexity of construction. Your choice of which to build should be tempered somewhat by a frank assessment of your engineering abilities. Use of the latest large-scale integrated circuits in these systems does not necessarily make them easier to check out. The three designs to be discussed are:

- biphase frequency-shift-keyed (FSK) transmitter/receiver
- complementary metal-oxide semiconductor (CMOS) large-scale integration (LSI) remote-control transmitter/receiver integrated circuits
- single-channel transmitter/receiver using inexpensive walkie-talkies

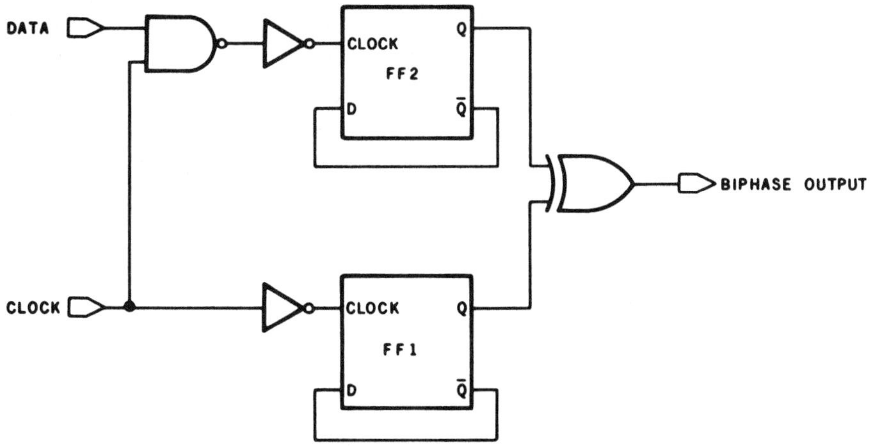

Biphase Frequency-Shift-Keyed Communication

If we wanted to communicate an off/on signal through wires to a computer we would simply use two voltage levels. "On" could be a +5 V potential and "off" could be a ground (0 V) potential. Does it sound familiar? In a wireless communication link, we cannot use DC voltage levels. The simplest alternative is to use two bursts of tones at different frequencies instead. Communicating more than a single control function over this same link is accomplished by serializing the data and time-multiplexing its transmission.

Usually when we hear the word "serial" we think of the standard serial-communication system protocol employing universal asynchronous receiver/transmitters (UARTs), etc. This form of serialization is but one of the many possible techniques and is not necessarily the most convenient for our purposes.

Figures 1 and 2 demonstrate *biphase encoding* and *decoding* of clock and data signals. This method uses relatively few components and allows recovery of the clock signal as well as the data.

To encode data we use the circuit of figure 1a. It consists of two type-D divide-by-two flip-flops sending their outputs into an exclusive-OR gate. The resulting output is demonstrated in the timing diagram of figure 1b. Close comparison of the input data and the biphase-encoded output shows how the process works. You will notice that if the data input is at a 0 level during one clock period, the output level changes *once*, but if the input is at a logic-1 level during the clock period, the output changes *twice*. These changes are called *transitions* and can be either 1-to-0 or 0-to-1 logic changes.

Figure 1a: *Diagram of a circuit that performs biphase encoding of data for transmission to the BSR X-10 Home Control System. Two type-D divide-by-two flip-flops are gated into an exclusive-OR, producing the output shown in the timing diagram of figure 1b.*

Figure 1b: *Timing diagram of the biphase encoder of figure 1a. The clock and data pulses are combined into the biphase output.*

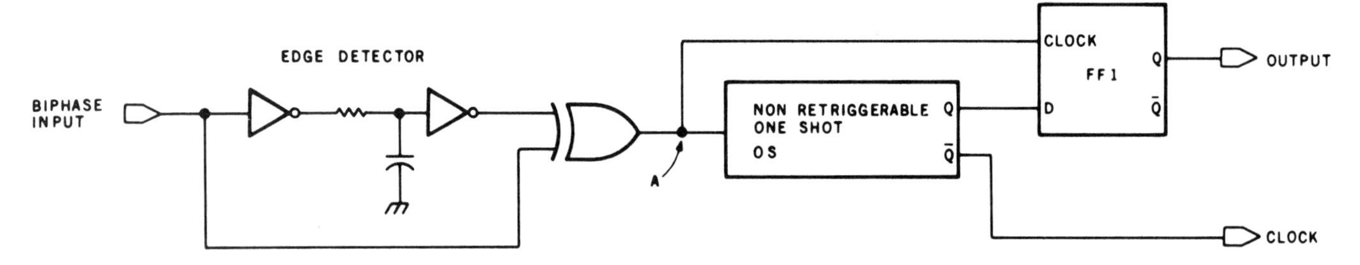

Figure 2a: *Diagram of a possible circuit for a biphase data decoder. The cycle time of the nonretriggerable one-shot (a monostable multivibrator) is set equal to about three-quarters of the duration of a bit in the incoming data stream. The corresponding timing diagram is presented in figure 2b.*

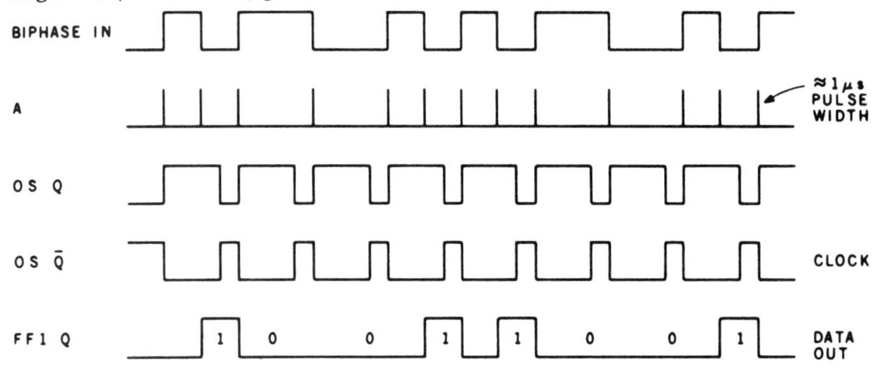

Figure 2b: *Timing diagram of the biphase data decoder of figure 2a.*

Simply stated, during a clock period there are two transitions for a logic-1 data input and only one transition for a logic-0 input. Biphase encoding relies upon timing between transitions rather than absolute voltage level. This makes the method relatively immune to power-line transients, power-supply fluctuations, and filter phase shifts.

Recovering the biphase-encoded data is done with the decoder circuit of figure 2a. It consists of an edge detector which produces a 1 μs pulse upon detecting any transition in the input voltage, a nonretriggerable one-shot (that is, a monostable multivibrator) set for a period equal to approximately three-fourths of the clock period of the originating transmission, and a flip-flop. As biphase data is presented to the circuit, it produces the edge-detector pulses shown as line A on the timing diagram of figure 2b.

When the first input pulse comes along, it fires the one-shot. (For a clock rate of 35 Hz, the one-shot period is set for 21 ms). If another edge is detected before the one-shot times out, it clocks a logic 1 out of the flip-flop. If a second pulse does not occur during the one-shot period, the output stays at a logic-0 level. Also, since voltage transitions in the received data coincide exactly with the transmitter clock, the receiver's one-shot becomes synchronized to the transmitter. Neglecting duty cycle, this clock rate is exactly the same as the transmitter's clock rate.

This technique has a few advantages that experimenters will appreciate. Since the one-shot is set at 75% of the optimum clock period, it essentially allows up to a 25% variation in timing between the transmitter and receiver while still maintaining synchronization. Compare this to about a 5% tolerance for the usual methods of serial data transmission!

Photo 1: *The handheld biphase FSK remote-control transmitter, which uses the circuit shown in figure 3. The transmit (XMIT) switch is a momentary power switch; the other four switches are used to set a 4-bit control-channel code.*

Number	Type	+5 V	GND	−12 V	+12 V
IC1	74LS151	16	8		
IC2	74LS93	5	10		
IC3	74LS00	14	7		
IC4	74LS04	14	7		
IC5	74LS74	14	7		
IC6	74LS86	14	7		
IC7	CO4049	14	7		
IC8	CO4027	14	7		
IC9	NE555	8	1		
IC10	74LS04	14	7		
IC11	74LS86	14	7		
IC12	74121	14	7		
IC13	74LS74	14	7		
IC14	74LS95	14	7		
IC15	74LS95	14	7		
IC16	74LS20	14	7		
IC17	74LS75	5	12		
IC18	LM741			4	7
IC19	LM741			4	7
IC20	LM741			4	7
IC21	LM741			4	7

Table 1: *Power connections for integrated circuits of figure 3.*

No crystals or elaborate clock generators are necessary for low-speed transmissions, either.

A Functional Biphase Remote-Control Unit

Figure 3 and figure 4 are the complete schematic diagrams of a functional sixteen-channel biphase FSK remote-control unit. The transmitter circuit of figure 3, including an RF transmitter, is packaged in the unit shown in photo 1. The circuit includes a switch- and sync-word scanner, an FSK modulator, and a biphase encoder. The four switches are used to set a 4-bit control code.

Because this unit is handheld, I have chosen to use a very low clock rate to reduce possible data errors. At 35 Hz, it takes approximately one quarter of a second to send the 4-bit switch status. Functionally, IC2 (a 74LS93) counter is attached to an 8-input multiplexer (IC1, a 74LS151 device). The first 4 bits of the multiplexer input are hardwired to a binary code of 1001 and the last 4 bits are connected to the data-input switches. As the counter (IC2) increments, IC1 steps through the binary sync word 1001 and then the four switch settings. This process keeps repeating for as long as power is applied to the circuit. This data is in turn encoded and modulated so that

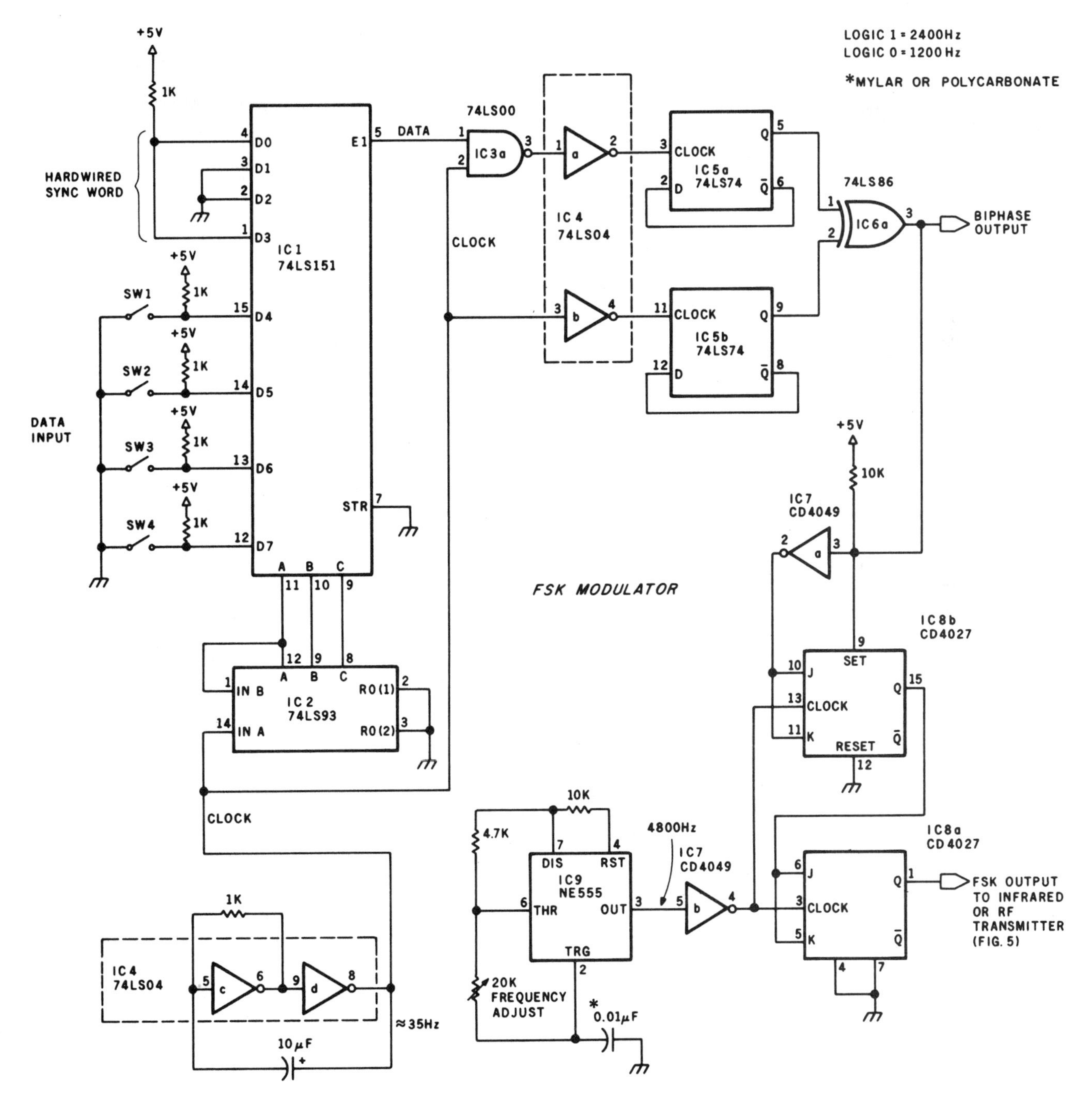

Figure 3: *Schematic diagram of a functional sixteen-channel biphase FSK remote-control transmitter. This circuit, with the addition of an RF modulator, is packaged in the handheld unit illustrated in photo 1.*

Power for the handheld device is provided by a rechargeable nickel-cadmium (ni-cad) battery of 4.8 V potential. If the transmitter is to be used for extended periods without recharging, CMOS parts may be substituted for the transistor-transistor logic (TTL) parts listed here.

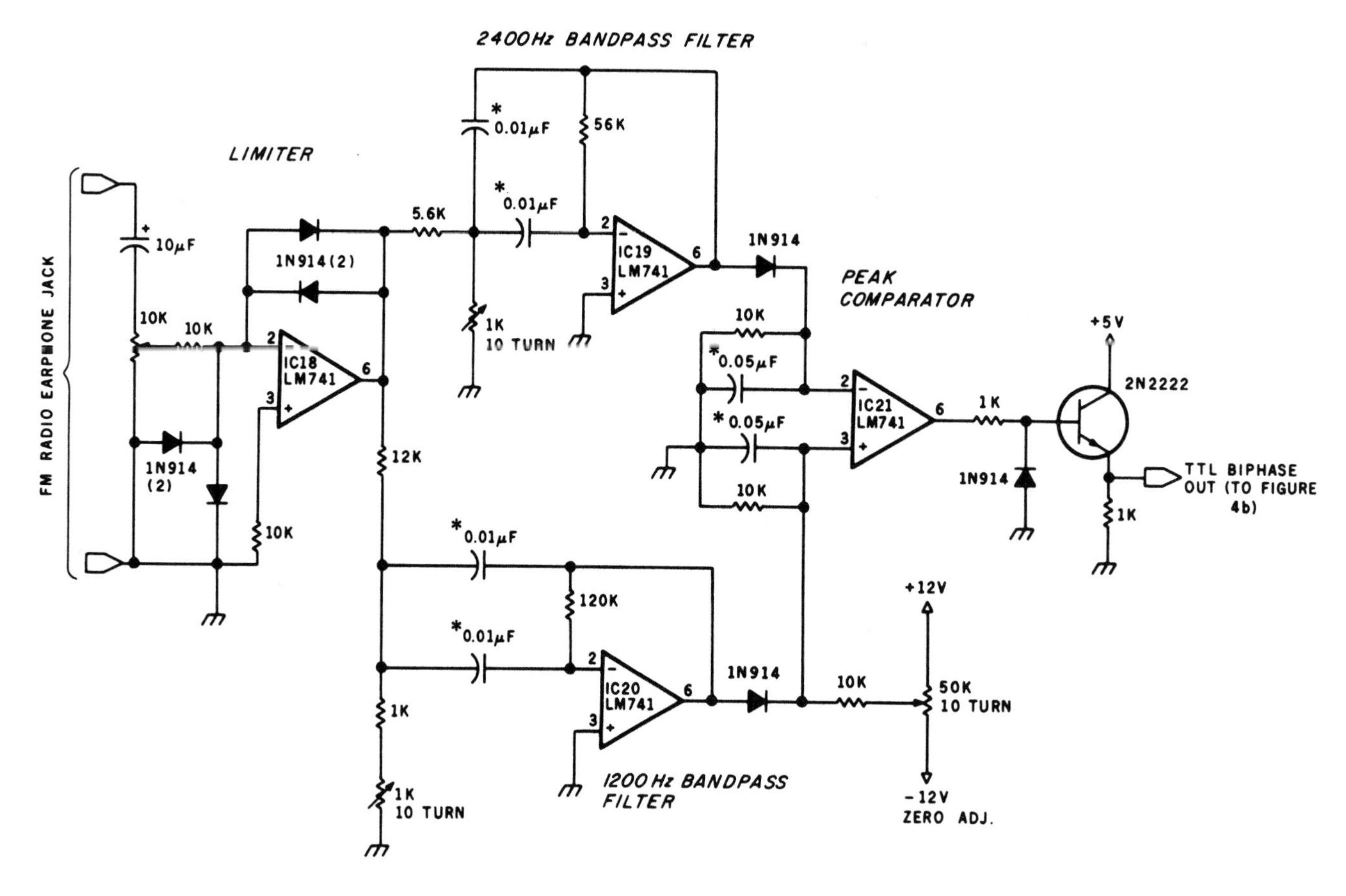

Figure 4a: *Section of schematic diagram of the biphase FSK remote-control receiver. Here is the FSK demodulator, which connects to the audio output of an FM radio receiver and produces a TTL-level biphase output, which is further processed by the circuit in figure 4b. The capacitors marked with asterisks (*) should be mylar or polycarbonate types.*

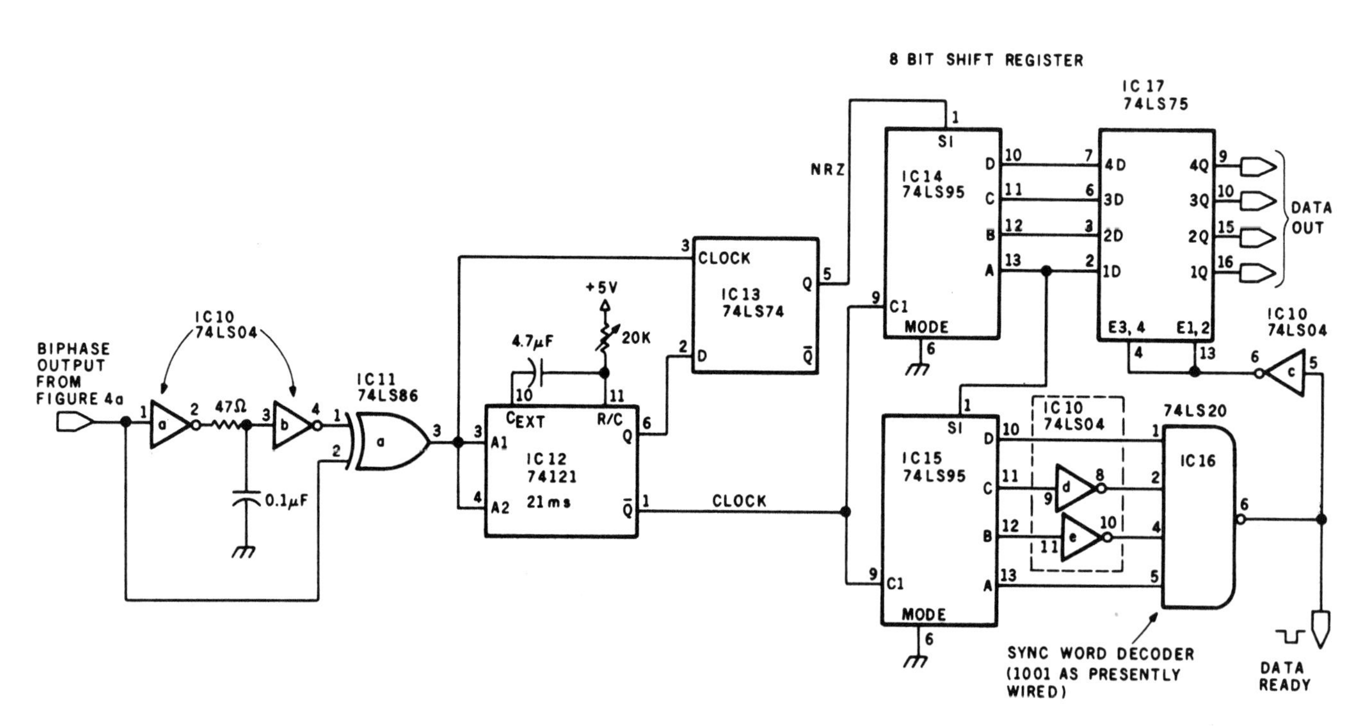

Figure 4b: *Section of schematic diagram of the biphase FSK remote-control receiver; here is shown the biphase decoder and data discriminator, which decodes 4 bits of data and sends them to an input port on the computer. The period of the one-shot is set approximately equal to three-quarters of the transmitter clock period.*

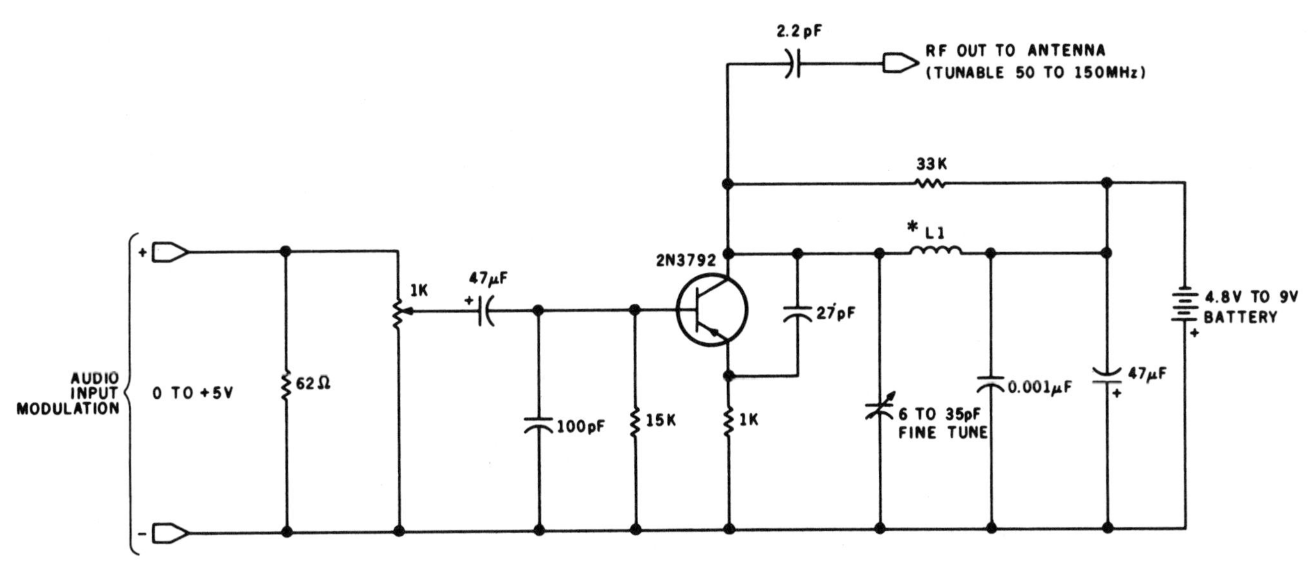

Figure 5: *Schematic diagram of a VHF (very high frequency) transmitter, tunable 50 to 150 MHz, for use in the wireless remote-control system. This circuit is also suitable for use in wireless microphone systems. The coil L1 consists of eight turns of number-26 wire wrapped around a coil form with one-quarter-inch inside diameter.*

it can be conveniently transmitted on a frequency-modulated (FM) RF carrier using the transmitter circuit of figure 5 or sent as an infrared-light pulse train using the circuit of figure 6. In either case, a logic 1 is encoded as a burst of 2400 Hz signal and a logic 0 is 1200 Hz. These tones can be easily received on an ordinary FM radio if the transmitter is set to a frequency between 88 and 108 MHz.

At the receiving end, the FM-radio audio output is first processed through an FSK demodulator. (See figure 4a.) This consists of an amplitude limiter, two bandpass filters, and a peak comparator. The resultant output should be the same as the biphase signal being transmitted. IC10, IC11, IC12, and IC13 separate this output into data and clock signal as I previously outlined and as shown in figure 4b. This data is in turn shifted into an 8-bit shift register. When the sync word gets to the end of the shift register (IC10 and IC16 decode a binary 1001) the contents of the last 4 bits shifted in will be clocked into holding register IC17. These 4 bits reflect the switch settings on the transmitter. They will be updated every one-quarter of a second if the transmitter remains on. The data-out lines in figure 4b are connected to the computer.

The transmitter circuit as I have shown it uses low-power Schottky (LS) transistor-transistor logic (TTL) devices. If powered by a 4.8 V rechargeable nickel-cadmium bat-

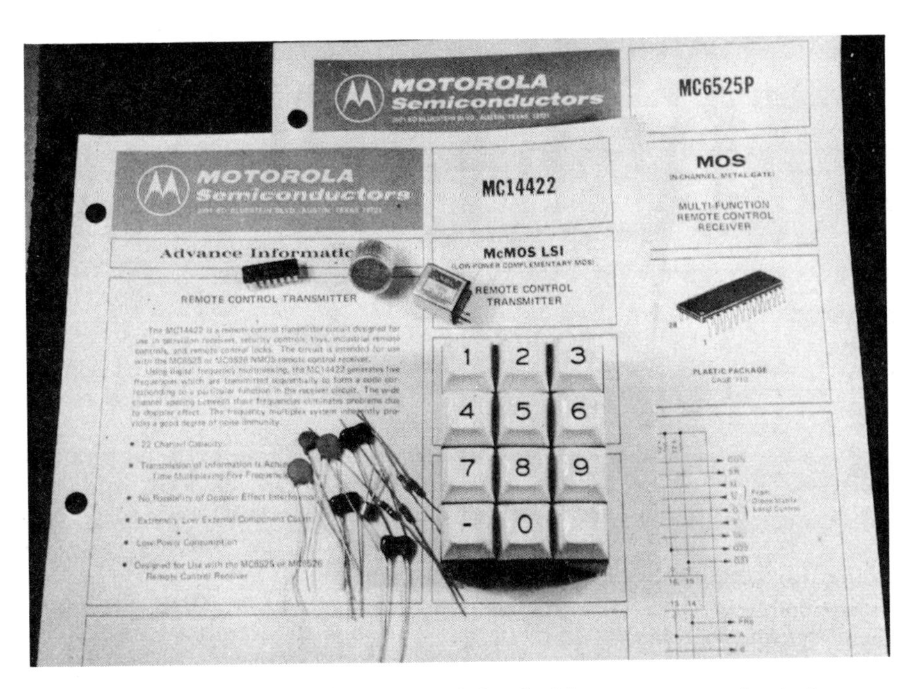

Photo 2: *Assembled here are the parts needed to build a remote-control unit that uses the Motorola MC14422 CMOS LSI remote-control transmitter and MC6525 receiver.*

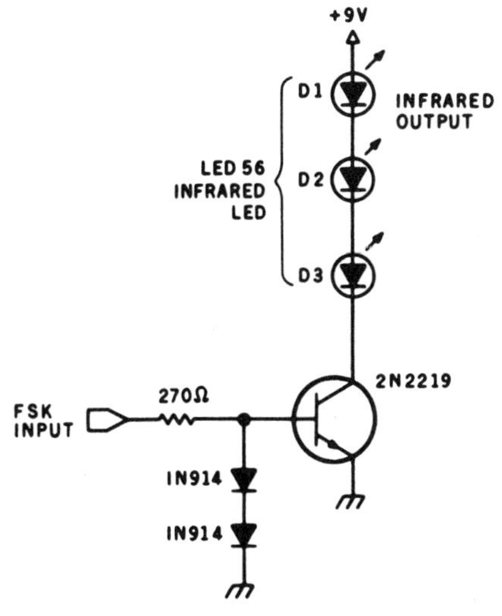

Figure 6: *Schematic diagram of an infrared-light transmitter for use in the wireless control application, where use of the radio-frequency transmitter is impractical or undesirable. A series of LEDs emits bursts of infrared light that carry the control information to the phototransistor receiver illustrated in figure 9.*

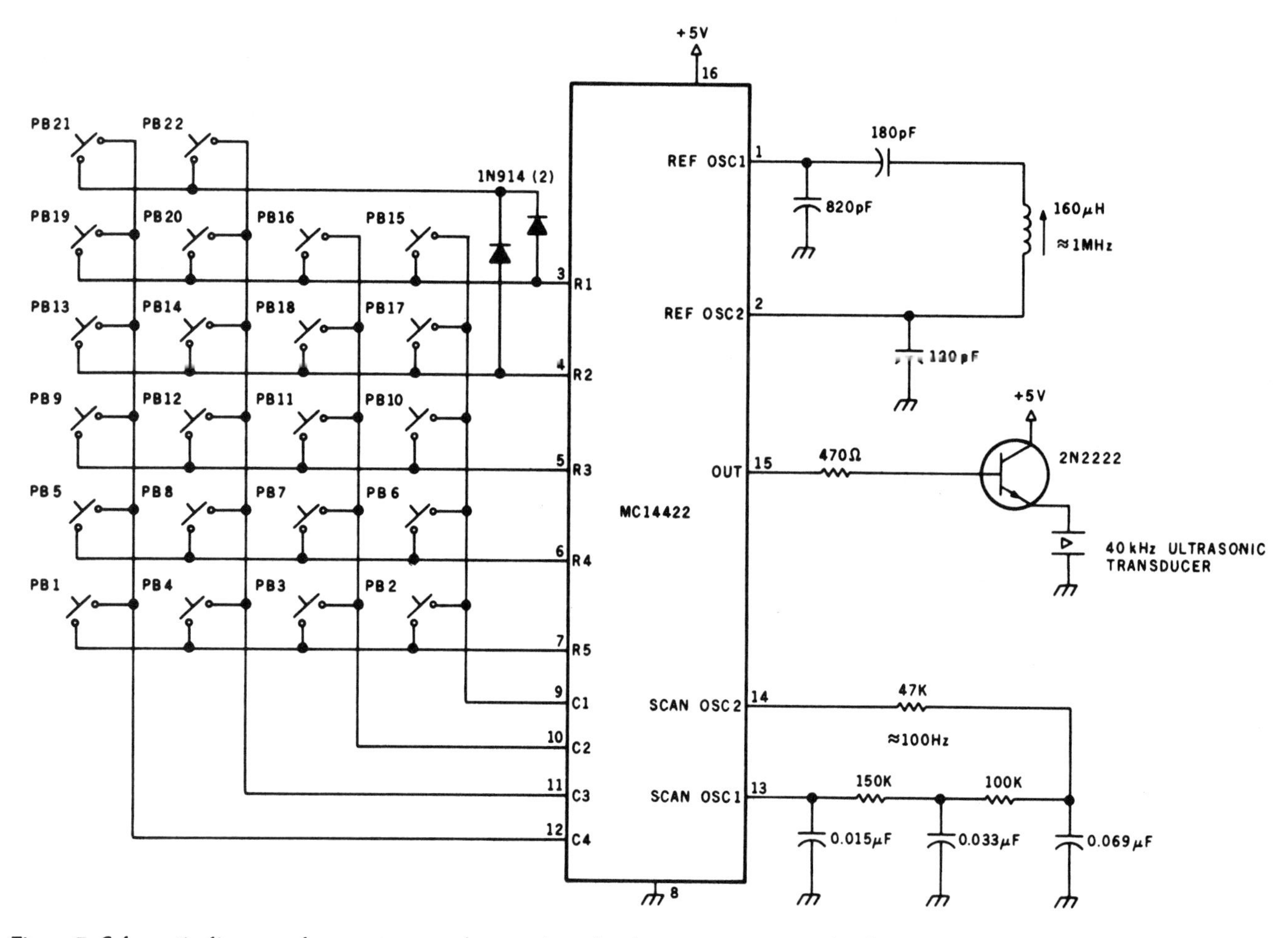

Figure 7: *Schematic diagram of a remote-control transmitter that incorporates a specialized integrated circuit, the Motorola MC14422 (a device mostly used in remote controls for television sets). Signals are transmitted as ultrasonic sound. The frequency-encoding scheme is presented in tables 2 and 3.*

Channel	Matrix Connections Pin to Pin		Transmitted Frequencies t1	t2	t3	t4
1	7	12	fe			
2	7	9	fe	fa		
3	7	10	fe		fb	
4	7	11	fe	fa	fb	
5	6	12	fe			fc
6	6	9	fe	fa		fc
7	6	10	fe		fb	fc
8	6	11	fe	fa	fb	fc
9	5	12	fe			fd
10	5	9	fe	fa		fd
11	5	10	fe		fb	fd
12	5	11	fe	fa	fb	fd
13	4	12			fb	
14	4	11		fa	fb	
15	3	9				fc
16	3	10		fa		fc
17	4	9			fb	fc
18	4	10		fa	fb	fc
19	3	12				fd
20	3	11		fa		fd
21	3,4	12			fb	fd
22	3,4	11		fa	fb	fd

Table 2: *Control channels, key closures, and corresponding transmitted tone frequencies used by the Motorola MC14422 remote-control transmitter used in the circuit of figure 7. See table 2 for the exact frequencies used. The column headers "t1", "t2", "t3", and "t4" refer to the sequential time segments during the total transmission; this is part of the transmission protocol.*

Frequencies	Output Frequency	Division Ratio
fa	34.688 kHz	f2/26.5
fb	36.048 kHz	f2/25.5
fc	37.519 kHz	f2/24.5
fd	39.116 kHz	f2/23.5
fe	42.755 kHz	f2/21.5

Table 3: *Frequencies of control tones used in the Motorola MC14422 remote-control transmitter in figure 7.*

tery, it consumes less than 50 mA. In this unit, the transmit switch is in fact a power switch. Sending a remote command requires only 1 or 2 seconds of transmitter operation. While CMOS devices *could* be substituted, it isn't necessary, considering the low duty cycle of the unit.

CMOS LSI Remote-Control Integrated Circuits

The idea of handheld remote control isn't new. The television industry

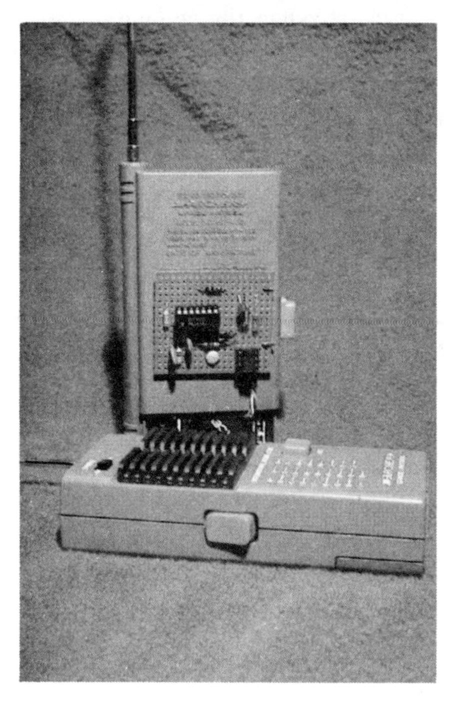

Photo 3: *Two walkie-talkies can be used in a simple remote-control scheme. One walkie-talkie, used as the receiver, is modified by connection to the circuit illustrated in figure 10a. The new circuit is attached to the rear of the walkie-talkie's case, as shown. These walkie-talkies are sold by Radio Shack as catalog number 60-4001; they incorporate a Morse-code tone oscillator necessary to the control scheme. Transmission is on a frequency of 49.86 MHz.*

has had it for years. To meet the demand for remote-controlled TV sets, special CMOS remote-control integrated circuits were designed, employing LSI. It is possible to use a pair of these for computerized remote control as well. Figure 7 shows the Motorola MC14422 remote-control transmitter part and a typical application circuit. Photo 2 illustrates the number of components necessary to implement this circuit. Figure 8 outlines the receiver circuit which uses the Motorola MC6525 receiver.

These specialized devices are very powerful. They use digital frequency multiplexing and transmit any or all of the five different frequencies sequentially to form a code corresponding to the particular selected function. These frequencies range from 34 to 43 kHz and accommodate twenty-two control channels including the ability to remotely adjust three analog-output signals in the receiver.

My experience in building these circuits warrants some mention. While the transmitter section (figure 7) went together easily, the receiver portion (figure 8) is a bear, and entailed difficulty in alignment. Even though it finally worked quite well, I don't recommend it for the novice builder. I'm presenting it in this article because it is a high-level controller and may be "just what the doctor ordered" for some individuals. Anyone interested in further information on these devices should contact a Motorola distributor directly.

Photo 4: *Three wires and an optional resistor are needed to attach the remote-control modification to the receiving walkie-talkie.*

Single-Channel Walkie-Talkie Remote-Control Interface

It isn't necessary for you to run out and spend much money on expensive integrated circuits or to spend 3 weeks building an elaborate circuit, if you can be satisfied with a less sophisticated control system. My final offering consists of two Radio Shack walkie-talkies (catalog number 60-4001) and a two-chip circuit. When I bought them, they were $10 each. Other walkie-talkies can be used, but the Morse-code tone generator built into these units is necessary for the operation of my circuit design.

This interface, shown in figure 10, is terribly simple. To transmit a command, turn on the transmitter and press the Morse-code button. This transmits a tone. At the receiving end, the circuit just signals the computer that it has received a tone.

The circuit of figure 10 is an AC amplifier with some bandpass characteristics. The tone frequencies on these walkie-talkies vary all over the lot. A narrow bandpass filter is useless. Instead, the circuit detects a minimum threshold of a midfrequency AC signal. More often than not this is from the tone generator and is the signal we want.

The circuit can be assembled on a piece of perforated circuit-layout board and screwed to the back of the walkie-talkie. (See photo 3.) Connections to the board are made from inside the walkie-talkie case and consist of only three wires. Wire A picks up +9 V power while wire B is walkie-talkie ground. Wire C is either attached to the speaker or disconnected and soldered to a 6.8 to 10-ohm resistor connected to ground. Wire D in my circuit is tied to the wire-C-and-resistor junction. Finally, the filter/amplifier (IC1) output goes to an opto-isolator. This device shields the walkie-talkie from the high-frequency electrical noise present on the computer input connections. Without it, walkie-talkie reception is so poor as to be useless.

The receive walkie-talkie is connected to 1 bit of a parallel input port on the computer, as shown, and is switched on. The Data-Received light-emitting diode (LED) will be off;

the input to the computer will be a logic 1. The LED will remain off as the transmitter is switched on. When the Morse-code tone button on the transmitter is pressed, the LED will light and the computer input will go to a logic 0. That is the simple principle by which the whole interface operates.

You are not limited to a single on/off remote-control operation (as with our switch on the cable) if you want to consider adding a little software such as that in listing 1. This simple BASIC routine monitors the walkie-talkie interface, waiting for an input signal (here, a logic 0 on bit 0 of port 3). When the computer detects an input signal, the program simply starts a 10-second sampling routine and counts how many times the tone button is pressed during the sample period. Three pulses could mean "go to application program 3," and six

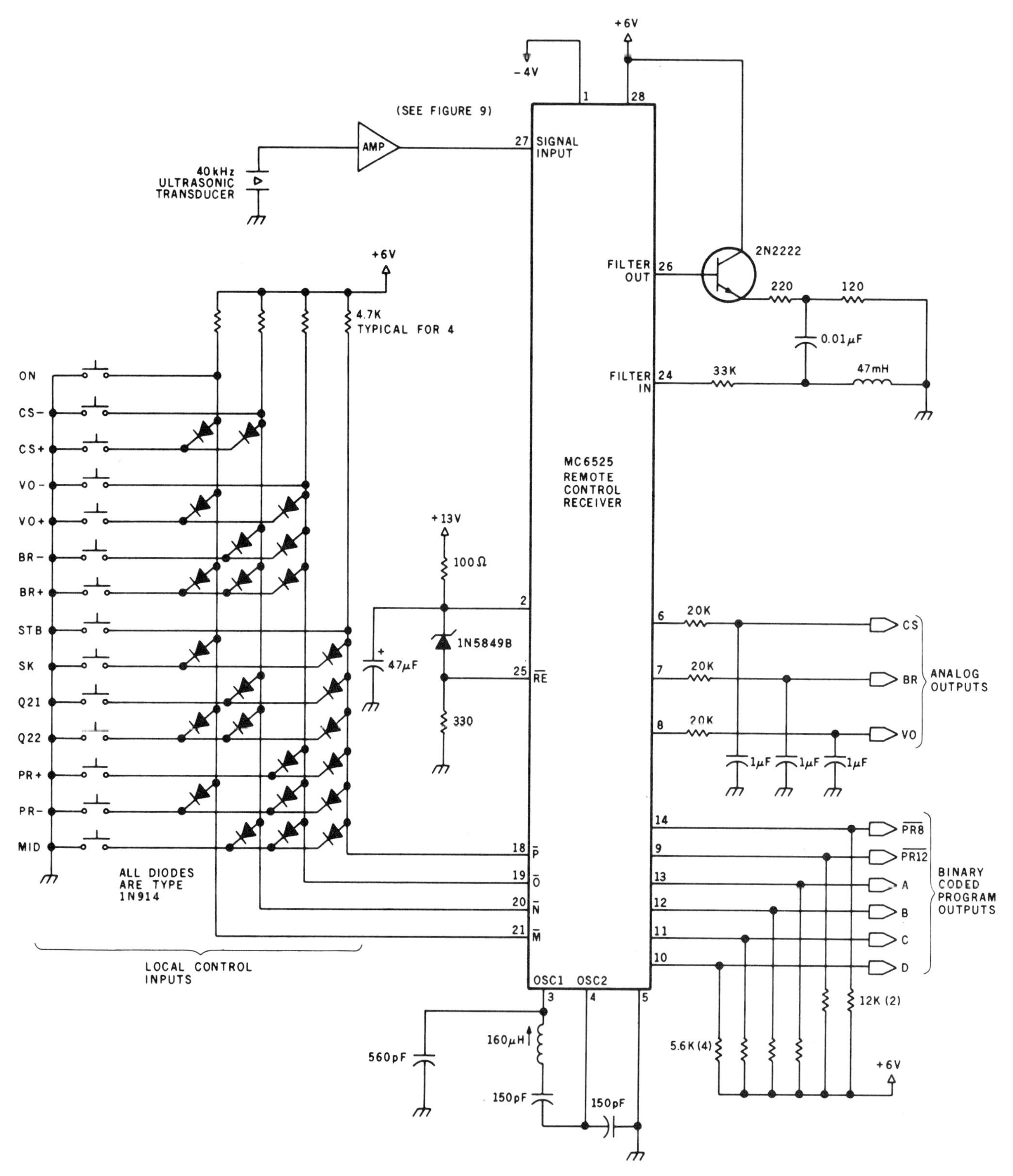

Figure 8: *Schematic diagram of a remote-control receiver that employs the Motorola MC6525, a part chiefly used for television remote control.*

pulses could mean "go to program 6." Using this sampling technique, it would not be difficult to actually send remote-control instructions in Morse code. (You may consult the code table imprinted on the front of the walkie-talkie.) That would provide twenty-six or so control functions with a maximum number of only four pulses on the key. [*See the October 1976 BYTE for several discussions of how to decode Morse code using computer software....* **RSS**]

In Conclusion

There are many other ways to remotely activate control programs through a computer. I have outlined only three. Other special LSI integrated circuits exist that are equally as powerful as the two discussed here. But it is impossible to cover them all. Perhaps one of the three designs I have presented is suitable for use on your system.

In the meantime, I have a few other applications in mind for these remote-control devices. As soon as I get the interfaces designed and tested, I'll be back with an article on a remote-controlled whatchamacallit.

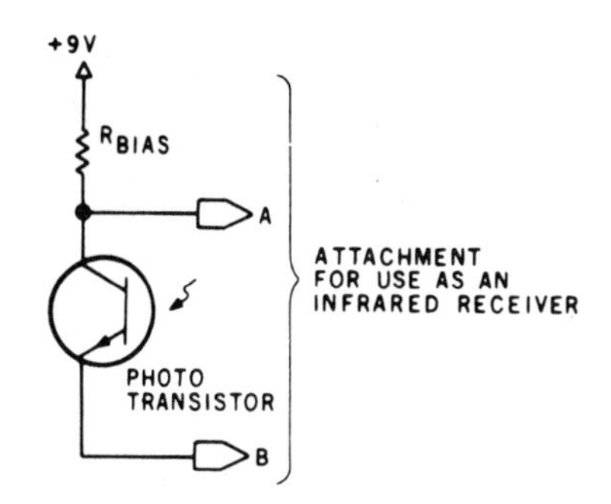

Figure 9a: *This phototransistor circuit can be substituted for the ultrasonic transducer in figure 9b to change the preamplifier circuit into a receiver for the infrared transmitter shown in figure 6.*

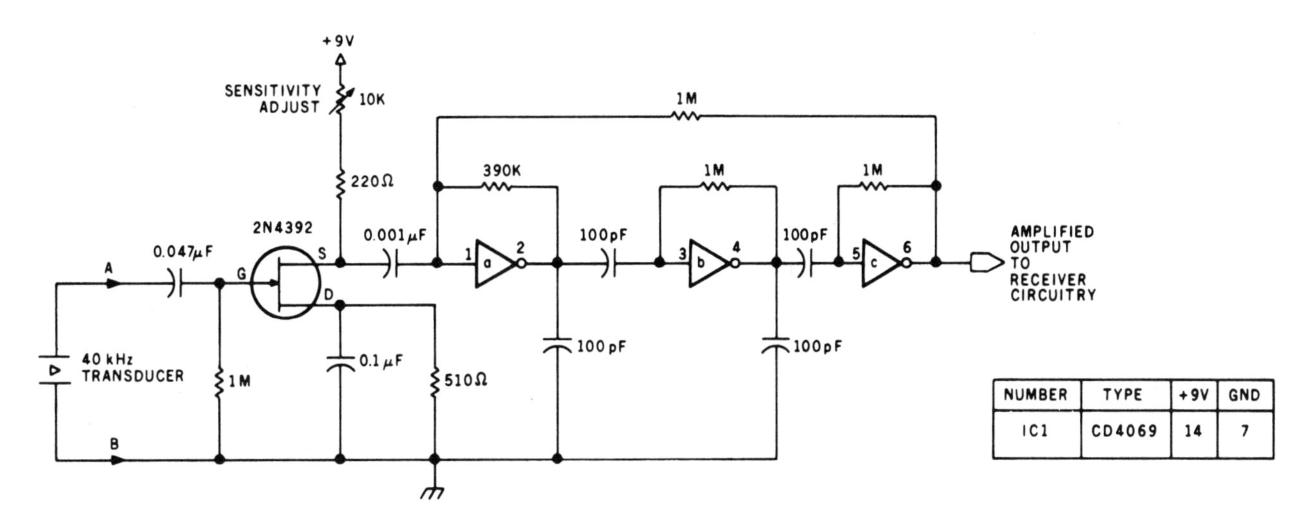

NUMBER	TYPE	+9V	GND
IC1	CD4069	14	7

Figure 9b: *Schematic diagram of a preamplifier circuit to be used with the ultrasonic remote-control receiver.*

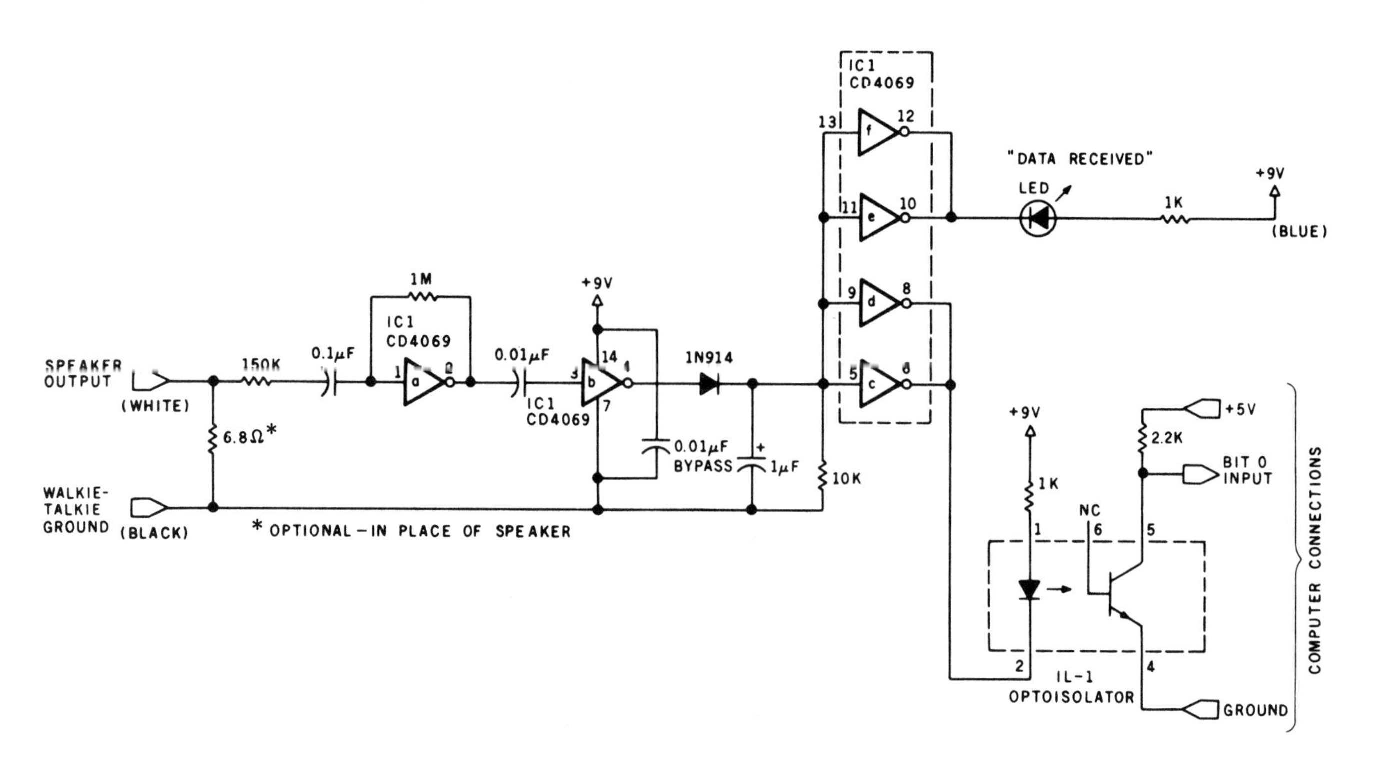

Figure 10a: *A receiver circuit that allows use of inexpensive walkie-talkies for the transmission of control data. The two walkie-talkies used in this application were Radio Shack catalog number 60-4001 types, which have an integral Morse-code audio oscillator. One walkie-talkie is used as the handheld remote transmitter, the other as the receiver.*

This arrangement allows transmission of only a single bit of control information, but this 1 bit can be modulated in various ways (even in Morse code) to control various functions. Decoding of this modulation must be carried out by a program running on the host computer.

A 0.01 μF bypass capacitor must be soldered directly to pins 7 and 14 on the CD4069.

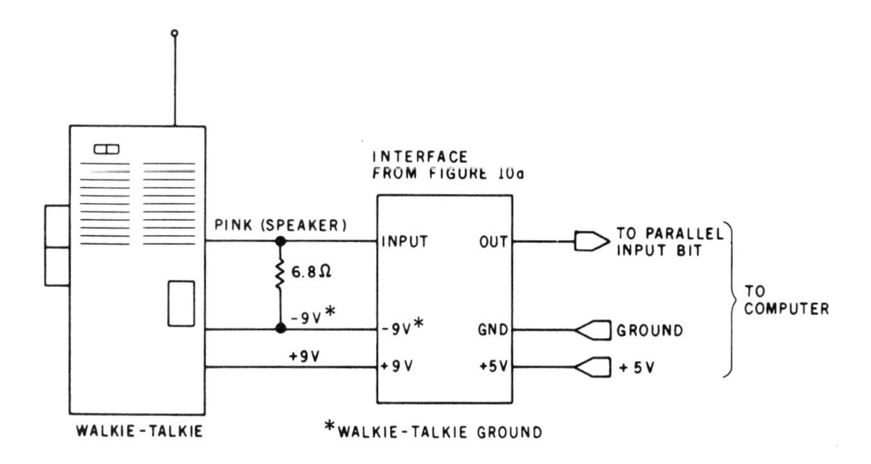

Figure 10b: *In the walkie-talkie remote-control scheme, one walkie-talkie is used as the receiver for the RF transmission of the other. The circuit of figure 10a is attached to the receiving walkie-talkie.*

Listing 1: *A BASIC program that monitors the walkie-talkie interface of figure 10a, waiting for a logic 0 on bit 0 of input port 3. When the logic 0 appears, the program starts a 10-second sampling period. During the period, the program counts how many times the tone button on the transmitter is pressed (that is, how many times a logic 0 is detected at the input port). The number of pulses received can be used to indicate which of the various control routines is to be activated. A more sophisticated program could decode more complex information from the transmitter, even information encoded in Morse code.*

```
100 REM  TIME SEQUENCED REMOTE CONTROL INTERFACE
110 REM
120 REM  THIS PROGRAM MONITORS THE STATUS OUTPUT OF A
130 REM  WALKIE-TALKIE (W/T) CONNECTED TO BIT 0 OF PORT 3
140 REM  AND DEMONSTRATES HOW IT CAN BE USED TO VECTOR
150 REM  TO VARIOUS CONTROL PROGRAMS BY REMOTE ACTIVATION
160 REM
170 REM
180 REM READ PORT 3 AND CHECK FOR AN OUTPUT FROM THE W/T
190 X=INP(3) :Y=X-1
200 IF INP(3)=X THEN GOTO 200
210 IF INP(3)=Y THEN GOTO 220 ELSE 200
220 PRINT"START":FOR D=25000 TO 25100
230 REM SET MEMORY LOCATIONS 25000 TO 25100 FOR BIT MAP
240 REM ANY 100 BYTE SEGMENT CAN BE DESIGNATED
250 GOTO 440
260 POKE D,M :REM STORE A BIT MAP OF W/T OUTPUT FOR GATE TIME
270 GOSUB 520
280 NEXT D
290 PRINT"END":REM SIGNIFIES END OF GATE TIME
300 Z=0:T=0 :A=0
310 FOR D=25000 TO 25100 :REM EXAMINE BIT MAP
320 IF PEEK(D)=0 THEN GOTO 560
330 A=0
340 NEXT D
350 REM T= TOTAL PULSES DURING GATE TIME
360 REM AT THIS POINT BRANCH TO OTHER PROGRAMS BASED UPON
370 REM THE VALUE OF T.  IF T=2 THEN GOTO APPLICATION PROGRAM #2
380 PRINT"BRANCH TO APPLICATION PROGRAM ";T
390 GOTO 190 :REM RETURN TO BEGINNING
400 REM
410 REM
420 REM READ INPUT TWICE AND VERIFY THAT IT IS TRUE
430 REM IF TRUE THEN MEMORY BIT IS A 0. IF NOT TRUE THEN M=1
440 H=INP(3)
450 GOSUB 520
460 H1=INP(3) :IF H<>H1 THEN GOTO 270
470 IF H=Y THEN M=0 ELSE M=1
480 GOTO 260
490 REM
500 REM
510 REM READ SAMPLE DELAY --- 10 SEC. GATE TIME
520 FOR W=1 TO 20: NEXT W :RETURN
530 REM
540 REM
550 REM INCREMENT PULSE TOTAL AT 1 TO 0 TRANSITIONS OF BIT MAP
560 IF A=0 THEN A=A+1 :T=T+1
570 GOTO 340

READY
```

9

Home In on the Range!

An Ultrasonic Ranging System

Each month I try to present a hardware project that is both interesting and relatively easy to build. Unfortunately, it's not as simple as picking a topic and quickly whipping up some circuit. More often than not, I have a number of potential topics and projects on the fire at the same time. Some are in limbo and just waiting for the right parts. Others are postponed when it turns out that the necessary hardware is something that could be better built by NASA (National Aeronautics and Space Administration) than by a computer hobbyist.

One topic that has always interested me is the concept of automatic ranging. I became involved with this idea when I wrote an article entitled "I've Got You In My Scanner," November 1978 BYTE, page 76. The original article was about an infrared sensor and parabolic reflector mounted to rotate on a stepper-motor shaft. With computer-controlled stepping, the result was something like the sweep of a radar antenna. The project was sensitive to infrared and visible light.

The scanner, parabolic-reflector, and stepper-motor combination could easily tell the direction of a light source to an angular resolution of 7.5°. It could make a 180° sweep, stop, and then follow the brightest object in its field of view. By recognizing the absence of known light sources (when the light path is blocked), it could even function as part of an intrusion alarm.

Diagrams and schematics of the Ultrasonic Ranging System Designer's Kit were provided through the courtesy of Polaroid Corporation.

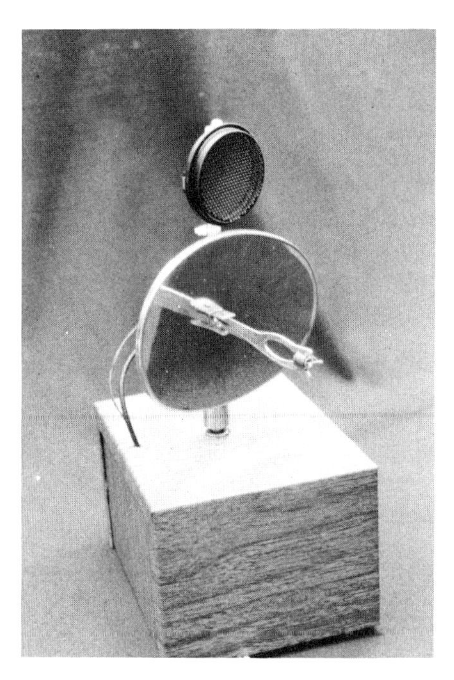

Photo 1: *A computer-controlled, stepper-motor-driven infrared and ultrasonic ranging scanner. An infrared-sensitive photo Darlington transistor (GE L14F2) is mounted at the focus of a parabolic reflector, which is attached to the shaft of a stepper motor; the ultrasonic transducer is mounted above it.*

The infrared sensor and drive mechanism were described in a previous Circuit Cellar article, "I've Got You in My Scanner! A Computer Controlled Stepper Motor Light Scanner."

However, even though it could "see," the infrared scanner could not tell how far an object was in front of it, or detect the presence of a nonluminous body crossing its path. What I really wanted was a device that could provide the computer with range as well as direction. That's when I started hanging around the camera shop.

Polaroid to the Rescue

The automatic focusing system on the Polaroid SX-70 Sonar OneStep Land camera intrigued me. I had considered tearing a camera apart just to use the ranging unit for my scanner, but sanity prevailed and I went back to designing my own circuit. Somewhere between thoughts of "Who'd really build this thing anyway?" and "I hope everyone can find all these components," I started seeing ads from Polaroid offering just what I wanted, without the camera.

The solution came in the form of an Ultrasonic Ranging System Designer's Kit sold by Polaroid for $150. The kit contains a technical manual, two instrument-grade electrostatic ultrasonic transducers, a modified SX-70 ultrasonic circuit board, an experimental demonstrator display board, and two Polapulse 6 V batteries. With this unit I was able to enhance my original infrared-scanner design to include automatic range detection. The new scanner system incorporating the Polaroid unit is shown in photo 1. More on this later.

Polaroid Ultrasonic Ranging System

The Polaroid Ultrasonic Ranging System Designer's Kit (shown in photo 2) costs $150 and is available from:

Polaroid Corporation
Battery Division
784 Memorial Drive
Cambridge MA 02139
telephone (617) 577-4681

Two primary components compose the ranging unit. They are the electrostatic transducer (see photo 3) and the ultrasonic transceiver board (see photo 4). Together these components are capable of detecting the presence and distance of objects within a range of approximately 0.9 feet (0.3 meters) to 35 feet (10.6 meters) with a resolution of ± 1.2 inches (± 30 mm, or 0.29% of range).

In operation, a pulse is transmitted toward a target, and the resulting echo is detected. The elapsed time between initial transmission and echo detection can be used to find the distance by taking this round-trip time and multiplying it by the speed of sound. For a transmitted pulse to leave the transducer, strike a target 2 feet (0.61 meters) away, and return to the transducer, it requires 3.55 ms (1.78 ms per foot, or 5.84 ms per meter, during the round trip).

Essential to system operation is the transducer (shown disassembled in photo 5). It acts as a speaker in the transmit mode and as an electrostatic microphone in the receive mode. The transducer is 1.5 inches (38.1 mm) in diameter and consists of a 0.003 inch (0.07 mm)-thick gold-plated foil stretched over a concentrically

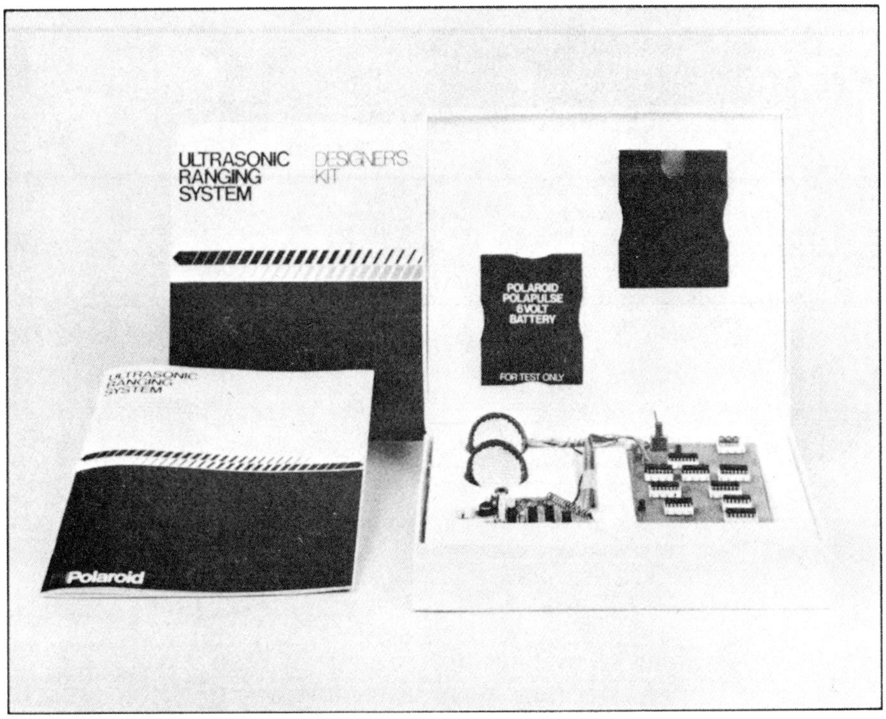

Photo 2: *Polaroid Ultrasonic Ranging System Designer's Kit, which includes ultrasonic sonar transducers, electronic circuitry, and a detailed specifications booklet.*

Photo 3: *Close-up view of the Polaroid Ultrasonic Transducer.*

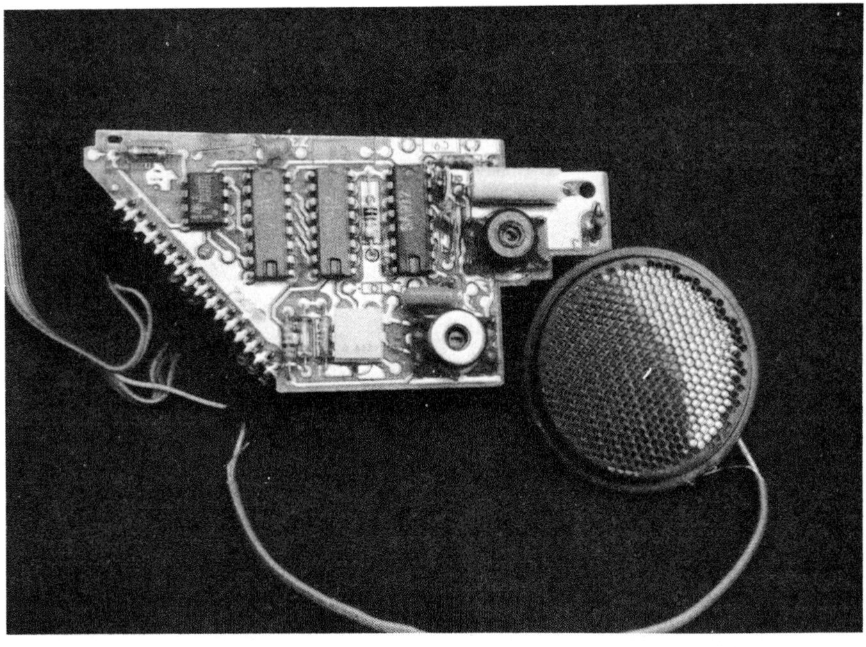

Photo 4: *Close-up of the ultrasonic circuit board, which contains custom analog and digital integrated circuits.*

grooved aluminum plate. When the metallic backplate is in proximity to the foil, it forms a capacitor. The foil is the moving element which converts electrical energy into sound and the returning echo into electrical energy.

The diameter of the transducer determines the directionality of the transducer. The acoustical signal-strength lobe pattern, or acceptance angle, during operation is shown in figure 1. The graph indicates that the transducer is fairly directional.

When the unit is activated, the transducer emits a sound pulse. The crystal-controlled electrical pulse

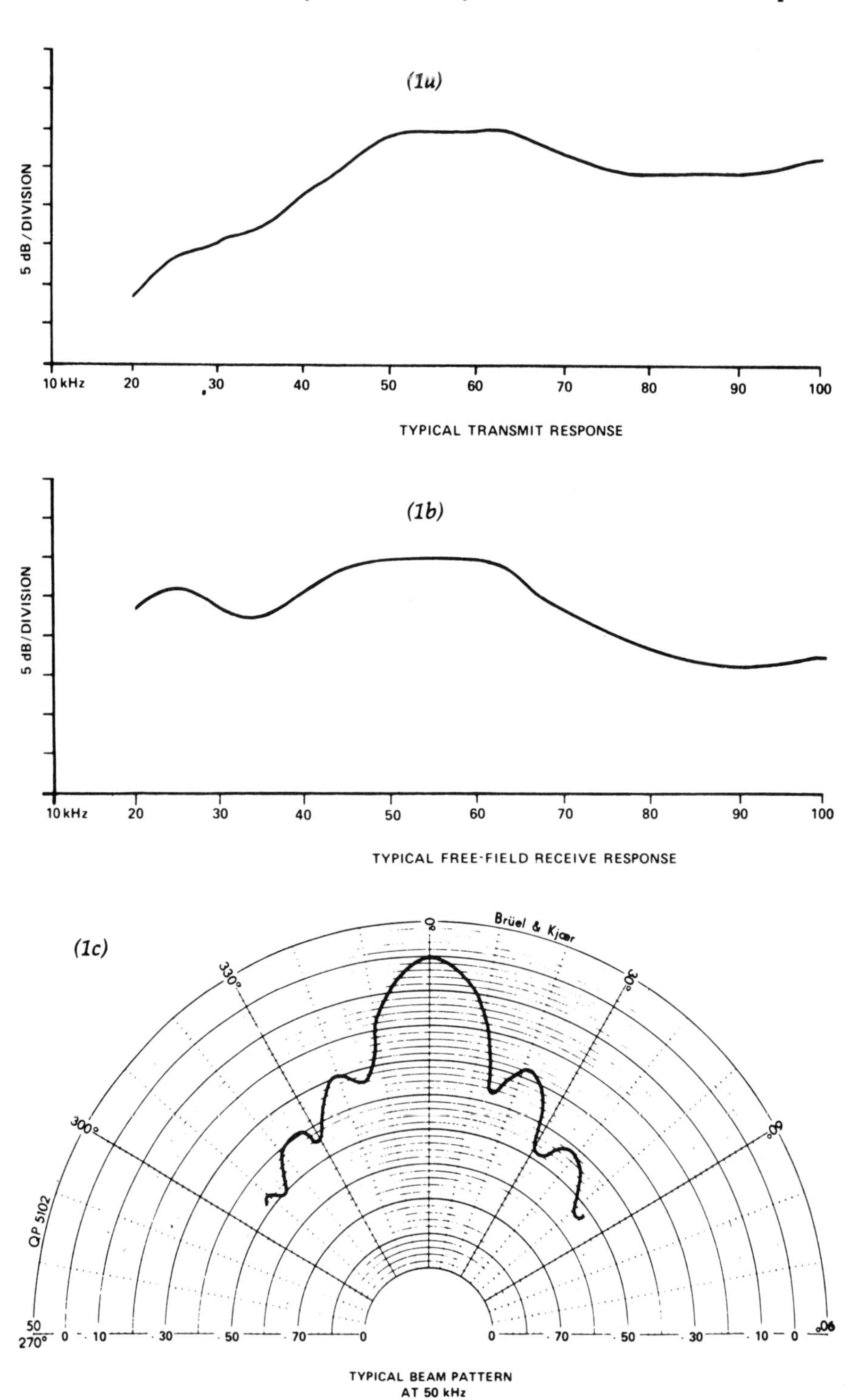

Figure 1: *Typical transmission frequency-response curve (1a), reception frequency-response curve (1b), and radial-beam pattern (1c) of the Polaroid ultrasonic transducer. The beam pattern was measured at 50 kHz, with dB values normalized to on-axis response.*

Photo 5: *Expanded view of the Polaroid ultrasonic sonar transducer. Behind a honeycomb grill, a 0.003-inch (0.07 mm)-thick gold-coated foil stretches over a concentrically grooved aluminum plate. The retainer at left holds the parts in place.*

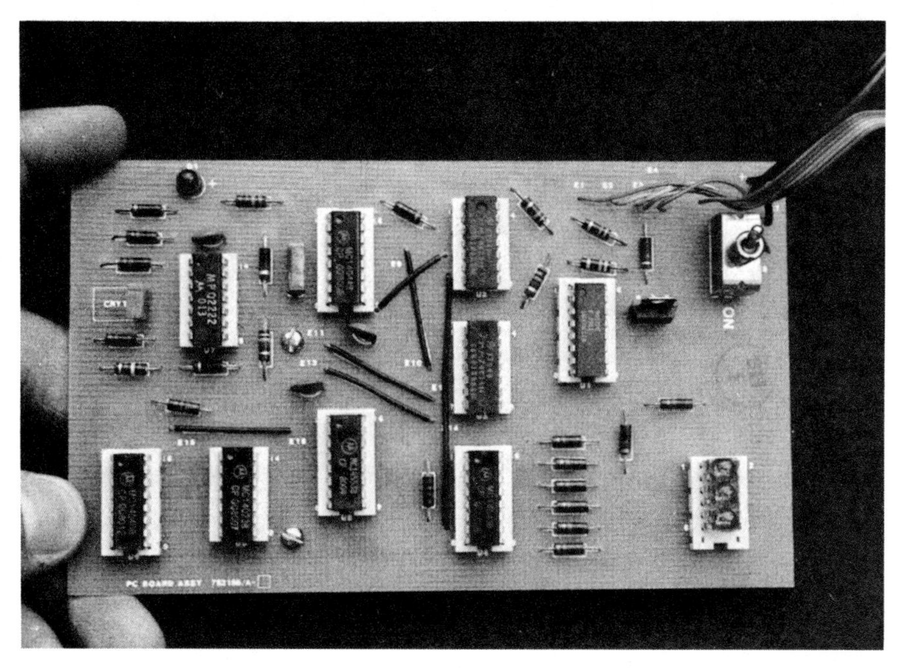

Photo 6: *The EDB, which contains the electronic circuitry shown in figure 4. The three-digit LED display is at the upper right.*

generated by the driver circuit is a 300 V high-frequency 1 ms "chirp" consisting of fifty-six pulses at four carefully chosen frequencies: eight cycles at 60 kHz, eight cycles at 57 kHz, sixteen cycles at 53 kHz, and twenty-four cycles at 50 kHz. This combination is used to overcome certain topographical characteristics of the area into which the signal is being transmitted, where a single frequency might be cancelled and no echo would be received.

The ultrasonic circuit board controls both the transmit and receive operating modes. It contains both digital and analog circuitry. In addition to transmitting the chirp and processing the echo, this circuit also tailors the amplifier sensitivity depending upon the object distance. Lower amplification is needed for close echoes, while higher amplification is needed for distant echoes. This is accomplished by increasing the amplifier gain and Q (ratio of reactance to resistance) in steps. Figure 2 is a block diagram of the ultrasonic circuit board.

Experimental Demonstration Board

The ultrasonic circuit board previously described is a modified camera assembly. The EDB (Experimental Demonstration Board, shown in photo 6) is not a camera component; it was designed specifically as a user interface to the ultrasonic board.

Photo 7: *The prototype of the interface circuit of figure 5 has been attached to the EDB. The interface allows a computer to read the three-digit distance value.*

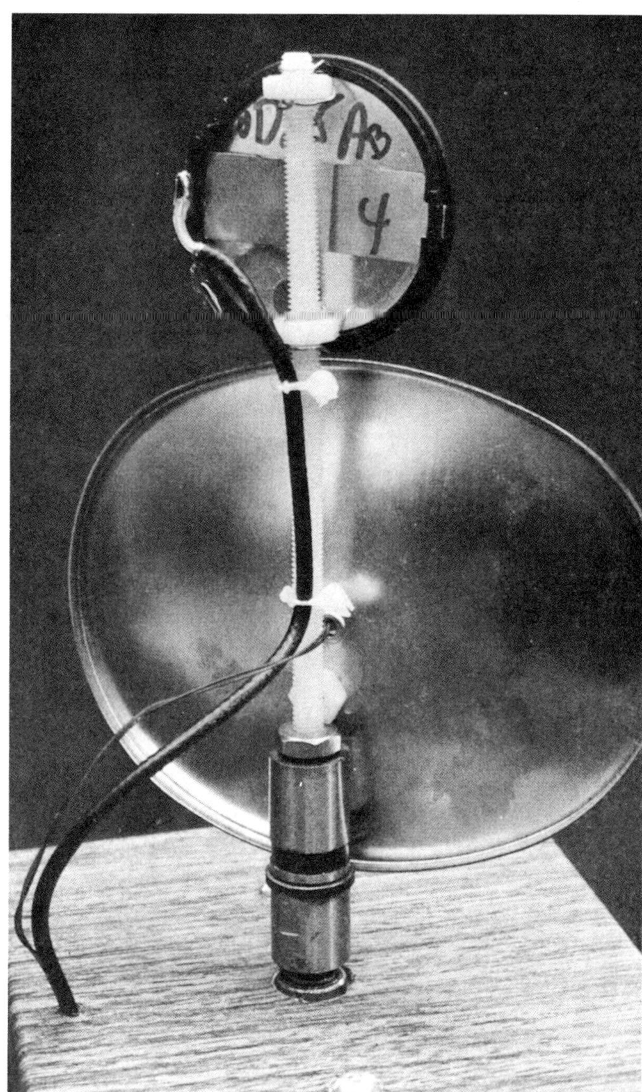

Photo 8: *Close-up of the back side of the reflector and transducer of the scanner, showing the mounting apparatus.*

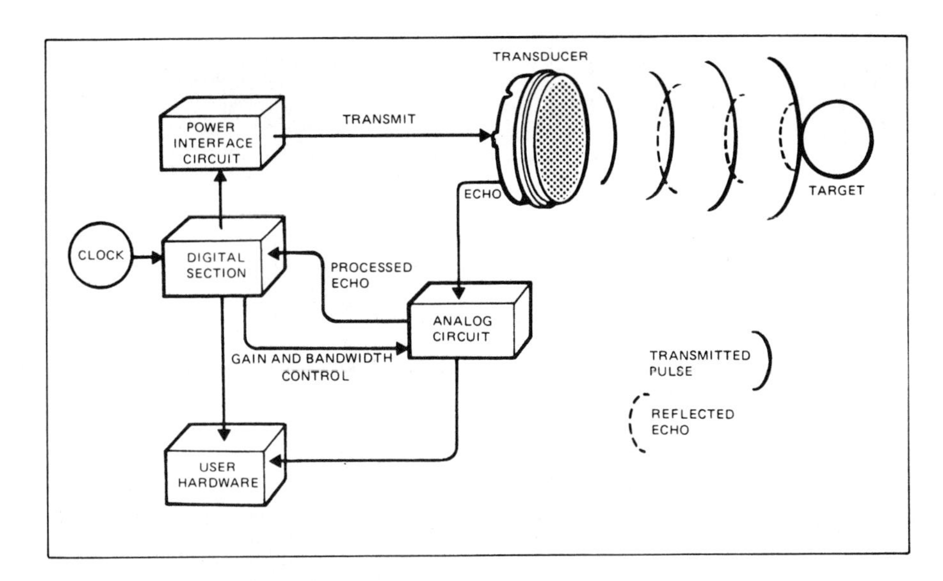

Figure 2: *Block diagram of the ultrasonic circuit. The circuit board contains a variety of custom components and is slightly modified from the unit used in SX-70 Land cameras. This circuit, as well as the EDB, is powered by a 6 V Polapulse battery. It seemed to work acceptably with a 5 VDC power supply.*

The block labelled "User Hardware" can be the EDB or any interface that can convert the ultrasonic circuit board's time-gated output into useful form.

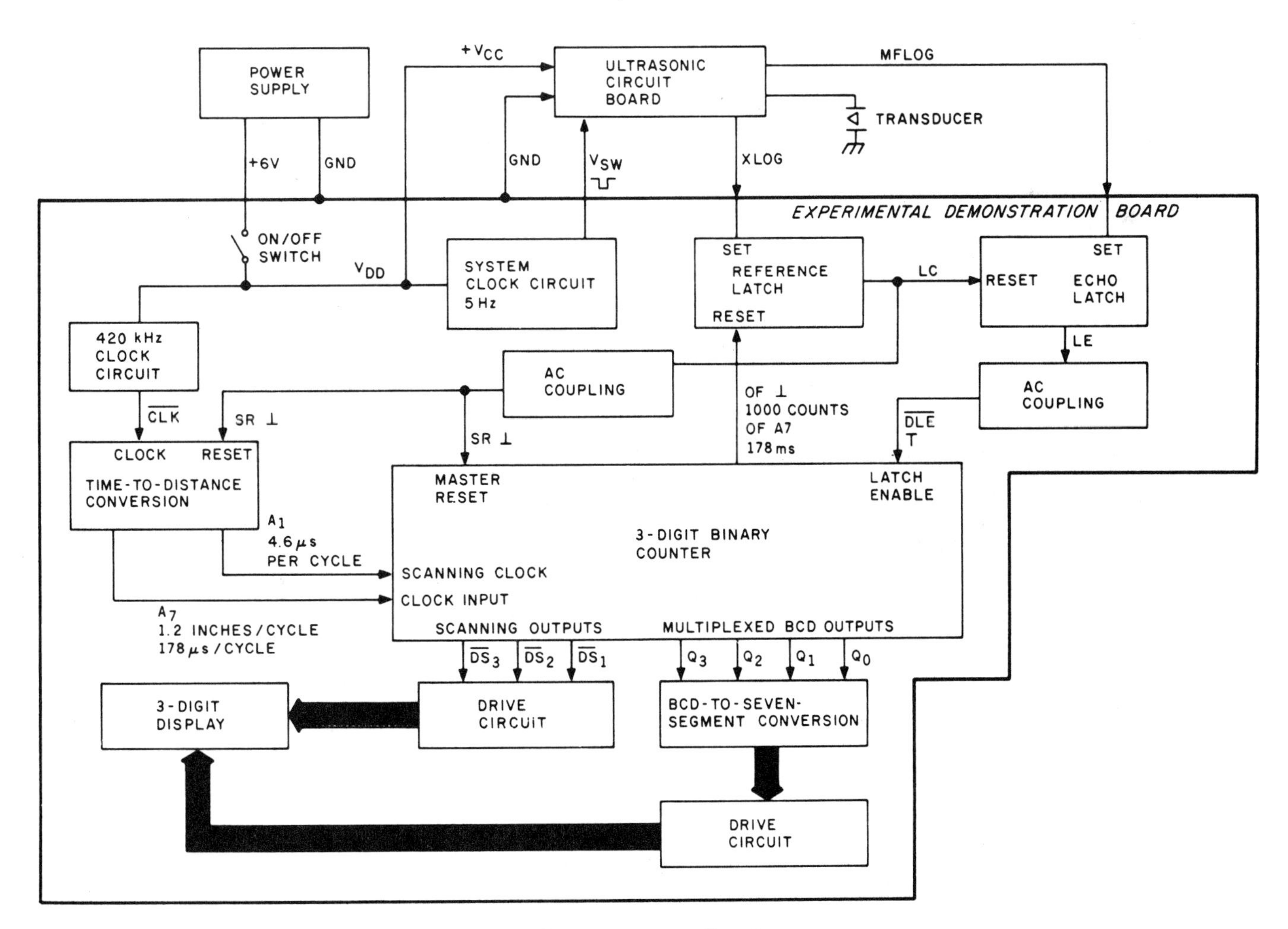

Figure 3: *Block diagram of the Polaroid Experimental Demonstration Board.*

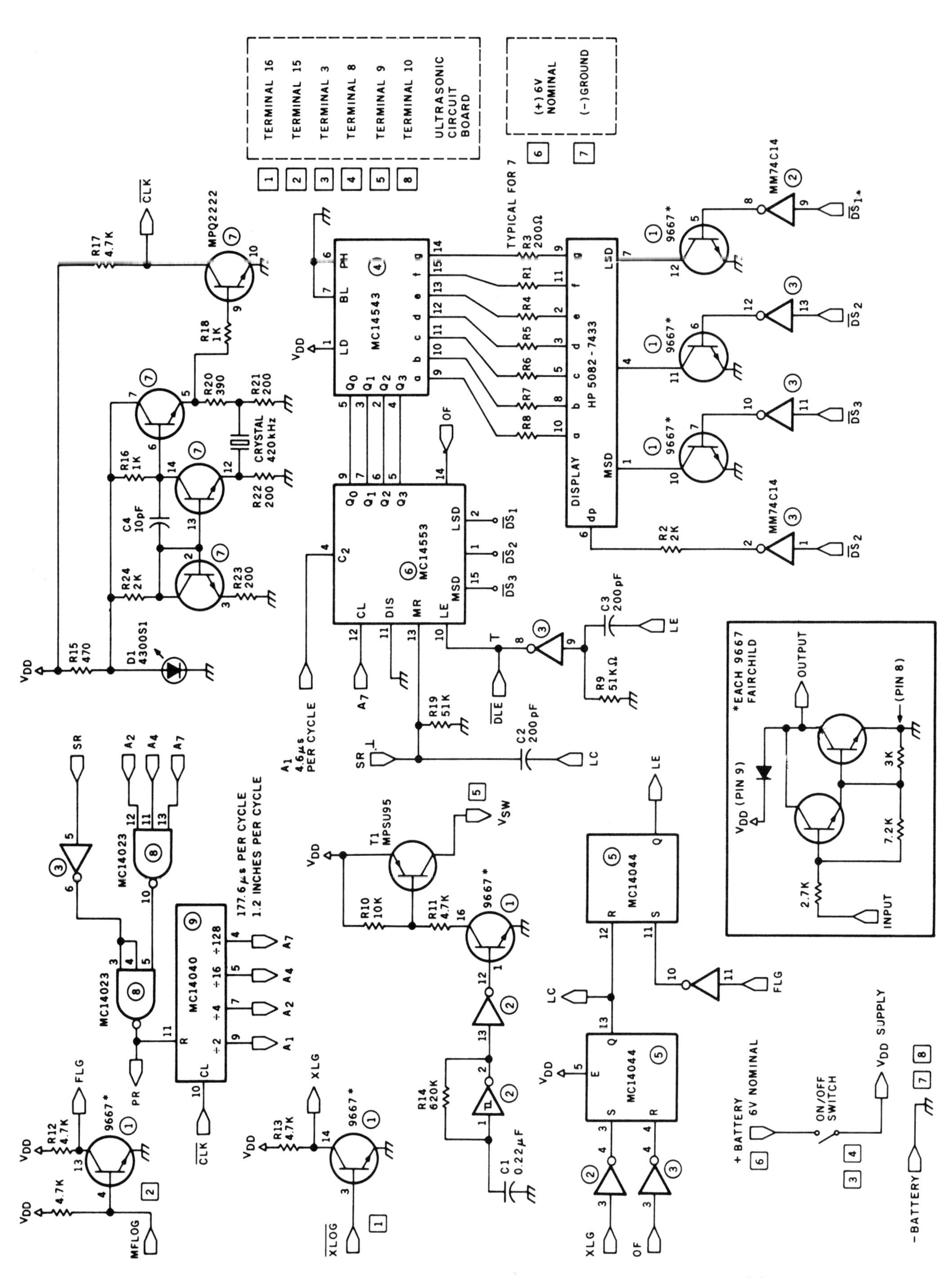

Figure 4: *Schematic diagram of the EDB. This board contains all the necessary circuitry to convert the raw data of the sonar transmit/receive time interval into a numeric distance value and display it on a three-digit LED display.*

Bit 1	Bit 0	Output Digit to Computer
0	0	$\overline{DS_1}$ (LSD)
0	1	$\overline{DS_2}$
1	0	$\overline{DS_3}$ (MSD)
1	1	n/a

Table 1: *Correspondence of the 2-bit digit-select codes with the EDB output data sent to the computer.*

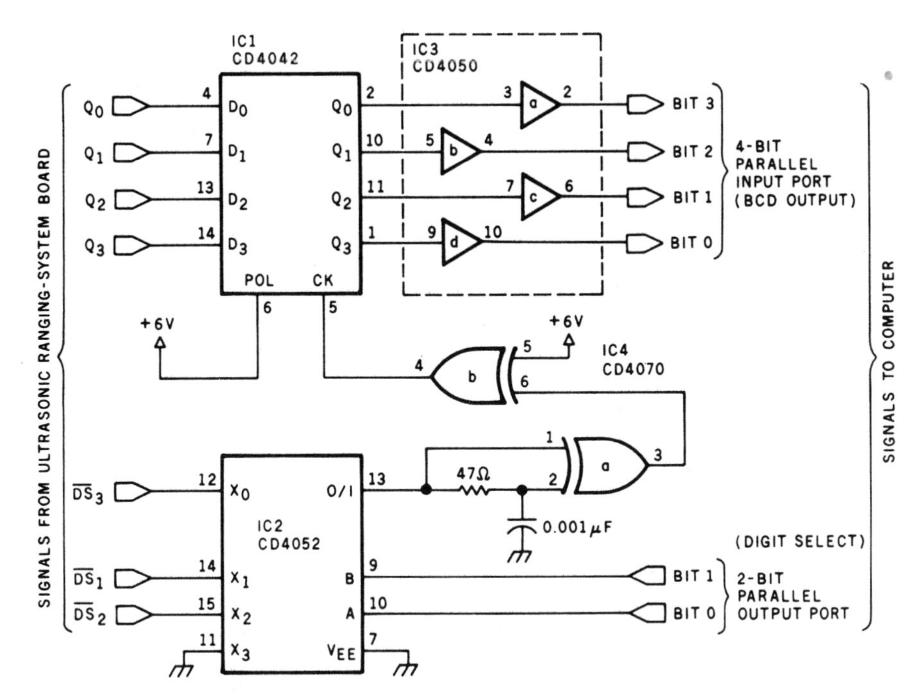

Figure 5: *Schematic diagram of an interface that allows a computer to directly read the three-digit LED display of the EDB, using four integrated circuits. Through 2 bits of a parallel output port, the computer sends a digit-select code and then reads the corresponding BCD value of the selected digit through 4 bits of a parallel input port.*

Number	Type	+6 V	GND
IC1	CD4042	16	8
IC2	CD4052	16	8
IC3	CD4050	1	8
IC4	CD4070	14	7

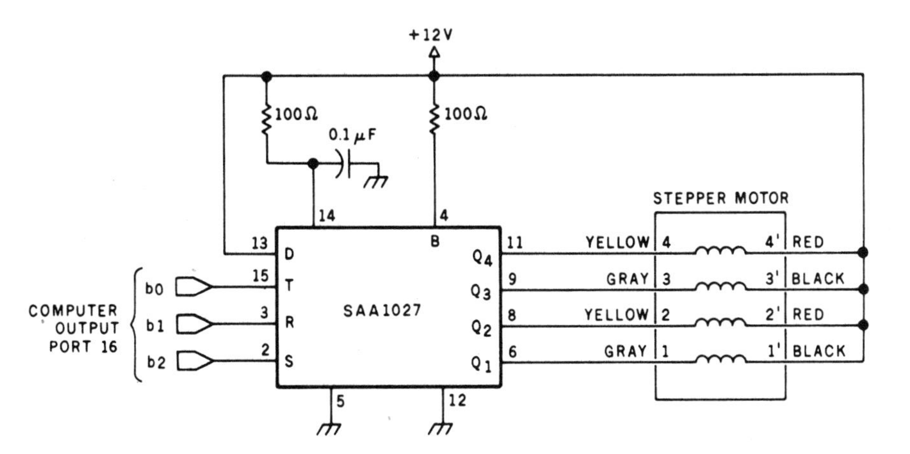

Figure 6: *Stepper motor and controller used in the infrared and ultrasonic scanner. The motor is a North American Philips K82701-P2 type, which turns 7.5° per step. It operates on 12 VDC.*

The SAA1027 integrated circuit is available from Signetics or from North American Philips, Cheshire, Connecticut, (203) 272-0301.

The EDB contains all the necessary electronic circuitry to convert the transmit/receive time interval into a figure indicating distance (in feet) and present it on a three-digit LED (light-emitting diode) display. Figure 3 is a block diagram of the EDB, while figure 4 shows the schematic diagram.

Connecting the EDB to the computer requires some thought. The output of the EDB is a three-digit display with a numeric output range of 00.9 to 35.0 in increments of 0.1 feet. The multiplexed display is controlled by a three-digit binary counter with strobed digit-select lines. It uses a single BCD (binary-coded decimal)-to-7-segment decoder/driver. At any instant, only one digit is energized, but because of the persistence of human vision, they all appear to be illuminated. Unfortunately, this multiplexed display output is not very computer-compatible and requires additional interface circuitry.

Decoding the EDB Output

Figure 5 is the schematic diagram of a four-integrated-circuit interface that decodes the counter output on the EDB and latches the digits while the computer reads them. Essentially the circuit consists of a three-input demultiplexer (IC2), an edge detector (IC4), a 4-bit latch (IC1), and an output buffer (IC3). The four-chip circuit is conveniently mounted on a piece of perforated circuit board and attached to the rear of the EDB, as illustrated in photo 7.

When the MSD (most-significant digit) of the LED display is energized, the $\overline{DS_3}$ line is low. The data on Q_0 thru Q_3 at this time form the BCD value of that number. Similarly, when $\overline{DS_2}$ goes low, the data lines will hold the second digit value. IC2 is a 4-to-1-line demultiplexer with the three digit strobes as inputs. A 2-bit TTL (transistor-transistor logic)-compatible parallel output from the computer determines which of these channels is routed through the multiplexer. To get $\overline{DS_1}$, the LSD (least-significant digit), the input code to the EDB interface would be 00. A binary code of 10 would set channel 3, allowing $\overline{DS_3}$ to go through. A summary of the codes is given in table 1.

The inputs to IC2 are offset by one channel due to the peculiar timing of the EDB. While the $\overline{DS_3}$ line is physically tied to channel 0 and would appear to be addressed with a 00 input code, the edge-detector timing of the circuit is such that we are not latching the current digit's value, but the *next* digit's value, when we address the channel. However illogical it may seem, the codes that work are stated in table 1.

When we have selected which digit we want to read by setting the proper multiplexer-input code, that digit value will be latched into IC1 and available as a BCD value to the computer. IC3 buffers the CMOS (complementary metal-oxide semiconductor) voltage levels of the EDB to the TTL level required by most computers. To read a three-digit range, we simply set the three multiplexer codes in succession. To obtain the distance indication, just add the three values as follows:

Distance = (MSD) × 10 + (2nd digit) × 1 + (LSD) × 0.1

This interface design is essentially speed-independent and can be driven equally well by an assembly-language or BASIC program. Listing 1 is a BASIC program that reads and displays the three-digit range determined by the ultrasonic ranging system.

A More Sophisticated Scanner

The original article, "I've Got You in My Scanner!," previously mentioned, has been reprinted in the book *Ciarcia's Circuit Cellar,* volume 1, available from BYTE Books. Photo 8 is a close-up of the updated version of the scanner, which now includes the ultrasonic ranging detector. The basic scanner consists of a North American Philips stepper motor (12 V type K82701-P2) and integrated-circuit controller (SAA1027) with an infrared-sensitive photo Darlington transistor (General Electric type L14F2) fixed at the focus of a parabolic reflector mounted on the shaft. I used a Radio Shack solar cigarette lighter, catalog number 61-2797, as the parabolic reflector. The driver circuit for the stepper motor is outlined in figure 6. The original article explained the infrared sensing system in detail.

The new scanner has the ranging detector mounted on the stepper-motor shaft, above the parabolic reflector. Both point in the same direction. The stepper motor is driven through the SAA1027 with 3 bits of a parallel output port. To drive the motor clockwise, bit 1 is set low, bit 2

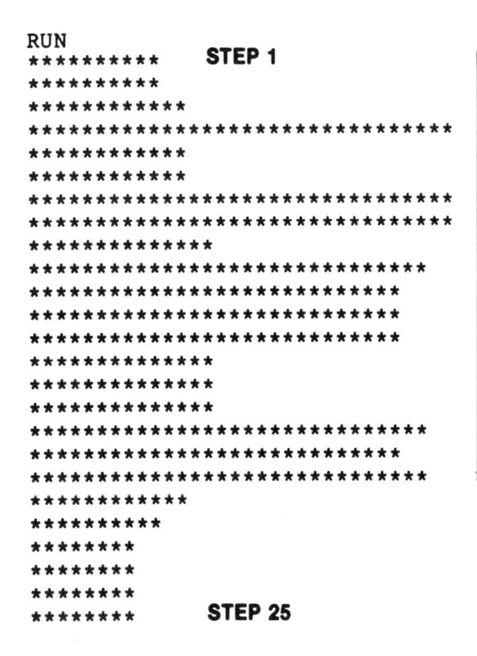
```
RUN
**********          STEP 1
**********
************
*********************************
************
************
*********************************
*********************************
**************
*******************************
*****************************
*****************************
*****************************
**************
**************
**************
*******************************
*****************************
*******************************
************
**********
********
********
********
********            STEP 25
```

Figure 7a: *Bar graph of distance measurements taken by the scanning system as the ultrasonic transducer was pivoted in twenty-five steps through a 180° sweep around the Circuit Cellar (each asterisk represents approximately one-half foot). Note correspondence with floor plan in figure 7b.*

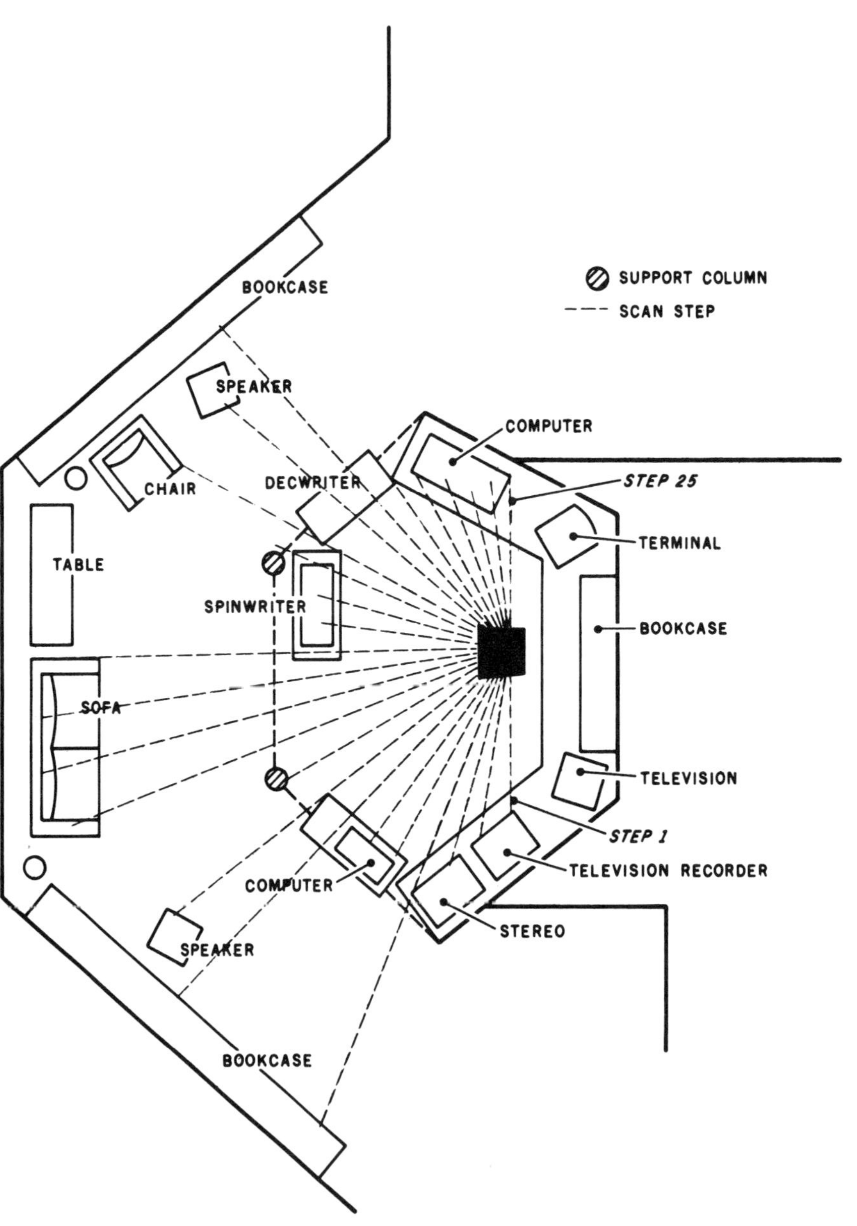

Figure 7b: *Floor plan of Circuit Cellar showing location of scanner and beam paths to room objects during the twenty-five steps in the scanning sweep. Bar graph of figure 7a shows relative distance to the nearest obstruction in the beam path at each step.*

Listing 1: *A BASIC program that uses the interface circuit shown in figure 5 to read the three-digit distance value from the EDB and display the distance on the computer printer. A sample execution follows the BASIC-language statements.*

```
100 REM   THIS PROGRAM ALLOWS A COMPUTER TO READ AND DISPLAY
110 REM   DISTANCE AS MEASURED BY THE POLAROID ULTRASONIC
120 REM   RANGING SYSTEM DEMONSTRATOR BOARD. RANGE .9 TO 35 FT.
130 REM
140 REM
150 GOSUB 250
160 PRINT"DISTANCE TO TARGET IS ";S;" FEET"
170 GOTO 150
180 REM
190 REM
200 REM   THIS ROUTINE SETS AND READS THE 3 DIGITS ON THE
210 REM   RANGING BOARD.
220 REM   IT IS A THREE STEP PROCESS: SET THE DIGIT; READ THE
230 REM   DIGIT VALUE; AND MASK OFF EVERYTHING EXCEPT THE 4 BIT
240 REM   CHARACTER.
250 FOR T=0 TO 2
260 OUT 16,T
270 S(T)=INP(16)
280 S(T)=S(T) AND 15
285 S=(S(2)*10)+(S(1)*1)+(S(0)*.1)
290 NEXT T
300 RETURN


RUN

DISTANCE TO TARGET IS  3.3  FEET
DISTANCE TO TARGET IS  3.4  FEET
DISTANCE TO TARGET IS  3.5  FEET
DISTANCE TO TARGET IS  3.4  FEET
DISTANCE TO TARGET IS  3.3  FEET
DISTANCE TO TARGET IS  3.4  FEET
DISTANCE TO TARGET IS  3.3  FEET
DISTANCE TO TARGET IS  3.4  FEET
DISTANCE TO TARGET IS  3.4  FEET
DISTANCE TO TARGET IS  3.5  FEET
DISTANCE TO TARGET IS  3.3  FEET
```

Listing 2: *A BASIC program that causes the scanner to make a 180° scanning sweep in twenty-five steps and prints the distance measurements in the form of a bar graph. Figure 7a shows the output from the execution of this program on the system set up in the Circuit Cellar.*

```
100 REM THIS PROGRAM MAKES A 180 DEGREE SCAN AND RECORDS THE
110 REM DISTANCE TO SOLID OBJECTS EVERY 7.5 DEGREES.
120 REM
130 REM STEPPER MOTOR CONTROLLER ATTACHED TO PORT 18
140 REM ULTRA SONIC RANGING UNIT ATTACHED TO PORT 16
150 REM
160 REM
170 DIM Z(25)
180 OUT 18,1 :OUT 18,255 :REM PRESET STEPPER CONTROLLER
190 REM
200 REM CLOCKWISE SCAN
210 REM BIT 2 IS SET HIGH AND BIT 0 IS TOGGLED
220 FOR D=0 TO 24
230 OUT 18,5
240 GOSUB 470
250 OUT 18,4
260 NEXT D
270 REM
280 REM COUNTERCLOCKWISE SCAN
290 REM BITS 1 AND 2 ARE HELD HIGH AND BIT ZERO IS TOGGLED
300 FOR D=0 TO 24
310 OUT 18,7
320 GOSUB 570
330 OUT 18,6
340 NEXT D
350 REM
```

is held high, and bit 0 is toggled to produce each step. To drive the motor counterclockwise, bits 1 and 2 are held high, and bit 0 is toggled for each step. The new scanner can read the distance at each step.

Listing 2 is a program that causes the scanner to make a 180° scan and prints out the distance measurements in the form of a bar graph, demonstrated here in figure 7a.

To help you understand the mode of operation and value of the ranging device, I have also sketched the area of the Circuit Cellar where the measurements were taken. (See figure 7b.)

The scanner (the dark object in figure 7b) was placed on a tripod at a height of 5 feet (1.5 meters), about 2 feet (0.6 meters) in front of my desk area. The parabolic reflector was pointed 90° to the left of center so that a 180° scan resulted in it ending up pointing 90° right of center. At each of the twenty-five steps it took to reach this point, it measured the distance to the nearest obstruction to its line of detection. For comparison, the dotted lines in figure 7b show where each step should have been and what should have been in the way of the sonar "beam."

The program of listing 2 printed the graph bar corresponding to each step, starting with step 1. At the position reached after step 1, the system recorded a distance of about 5 feet (1.5 meters) to the VTR (videotape recorder) on the counter top. The same result was obtained for the next two steps. At the position reached after step 4 (about 30° around), the scanner was pointing between the stereo system and the TRS-80 computer on the desk to the right. This was indicated by a reading of about 15 feet (4.6 meters), measuring the distance to the bookcase on the far wall.

The next couple of steps had the TRS-80 directly in the path of the scanner beam, and then the path of the beam was open to the far wall again for a couple of steps. The rest of the scan was similarly significant in that the range detector accurately described the perimeter from its viewpoint. Most important, however, was the demonstration of the sensitivity of the ranging device. At steps 9 and 16, the only object in the path between the scanner and the wall was a 4-inch (10 cm) ceiling-support column

Listing 2 continued:

```
360 REM
370 REM PLOT RANGES AS BAR GRAPH
380 FOR D=0 TO 24
390 FOR W=1 TO INT(Z(D))
400 PRINT"**";
410 NEXT W
420 PRINT" "
430 NEXT D
440 GOTO 220
450 REM
460 REM
470 REM STEP DELAY AND RANGE SAMPLE ROUTINE
480 FOR T=0 TO 2
490 OUT 16,T
500 S(T)=INP(16) :S(T)= S(T) AND 15
510 NEXT T
520 Z(D)=(S(2)*10)+(S(1)*1)+(S(0)*.1)
530 FOR Q=0 TO 10 :NEXT Q
540 RETURN
550 REM
560 REM
570 FOR Q1=0 TO 100 :NEXT Q1
580 RETURN
```

Listing 3: *A short BASIC program that demonstrates one method for using the ultrasonic scanning device in a security system.*

```
100 REM THIS PROGRAM DEMONSTRATES HOW THE ULTRASONIC RANGING
110 REM BOARD CAN BE USED AS AN INTRUSION DETECTOR.
120 REM
130 REM
140 A=1 :GOSUB 220 :REM  TAKE FIRST DISTANCE READING
150 GOSUB 330
160 A=2 :GOSUB 220 :REM  TAKE SECOND DISTANCE READING
170 IF ABS(X(1))-ABS(X(2))>=.3 THEN GOTO 280
180 IF ABS(X(2))-ABS(X(1))>=.3 THEN GOTO 280
190 GOTO 140 :REM CONTINUE SCAN
200 REM
210 REM
220 FOR T=0 TO 2
230 OUT 16,T
240 S(T)=INP(16) :S(T)=S(T) AND 15
250 NEXT T
260 X(A)=(S(2)*10)+(S(1)*1)+(S(0)*.1)
270 RETURN
280 PRINT" I GOT YOU IN MY SCANNER AT ";X(2);" FEET."
290 REM AN ALARM ROUTINE WOULD BE PLACED HERE
300 GOTO 140
310 REM
320 REM
330 REM SAMPLE RATE DELAY TIMER
340 FOR Y=0 TO 200 :NEXT Y
350 RETURN


RUN

I GOT YOU IN MY SCANNER AT 11.4 FEET.
```

about 7 feet (2.1 meters) away. In both cases the obstruction was accurately identified.

We now have a device that can rotate to a particular position and accurately measure the distance to any object it "sees." A practical use of the range detector is as a security device. When the wall is known to be 16 feet (4.8 meters) away from the scanner, a sudden reading of 9 feet (2.7 meters) indicates that someone or something just moved in front of the range detector. The program of listing 3 allows the range detector to be used as a motion detector.

In Conclusion

I have demonstrated only two uses for the Polaroid Ultrasonic Ranging System Demonstrator Kit. The majority of applications I've heard about thus far have been independent projects that utilize the ranging system *without* the additional capabilities of a computer. They include a walking cane (with audio feedback) for the visually handicapped, a 0 to 35 foot (0 to 11 meter) altimeter for the *Gossamer Albatross* aircraft (for its English Channel crossing), and as an electronic "dip stick" for measuring liquid levels in storage tanks.

I hope that once you realize how easy it is to attach this automatic ranging system to a computer, you'll have as much fun experimenting with it as I have. Unfortunately, a new problem has arisen. Until now, one of the major reasons I haven't attempted to build a robot was the amount of expense and technical effort required to make it "see." Now I'll have to find a new excuse.■

Auto Warning

Dear Steve,

In a book on microcomputers that I read, the author predicted that an automotive warning device that would tell drivers they were too close to another vehicle would be devised.

It occurred to me that such a gadget might be realized right now using the Polaroid development kit and a simple single-board computer. Software, it seems to me, might be the biggest hurdle. What

do you think?
Bob Crafts
Edgartown MA

The Polaroid ranging sensor is definitely usable for a driver-warning device like the kind you mentioned. However, I don't see this sensor being used as a crash-avoidance device because its response time is a little slow. I have seen one company using the device on each side of a car's fenders, with a dash-mounted display for the driver. In my mind, while this may work, its feasibility and production is another matter. It would seem to be rather expensive unless produced in large quantities.

When using the Polaroid development kit in an automobile, you must try to isolate the ignition noise from any power being drawn from the car's electrical system. From my experience, the Polaroid ranging kit is also electromagnetic interference and static sensitive. If used in a car, it should be in a shielded enclosure. . . . **Steve**

Smart Wheelchair Project

Steve Ciarcia's article "Home in on the Range! An Ultrasonic Ranging System" (November 1980 BYTE, page 32) was excellent. I would, however, like to make BYTE readers aware of another project that has incorporated the Polaroid Ultrasonic Ranging technology. The project was funded by the Veterans Administration Rehabilitative Engineering Research and Development Center of Palo Alto, California. The participants, Karen Altman, Rick Epstein, Leslie Gerding, Wayne Ledger, and Dave Parker, were graduate students last year at Stanford Mechanical Engineering.

The objective was to design, develop, and successfully fabricate a "smart" electronic wheelchair. Its construction included ten ultrasonic sensors, eight of which were used to detect approaching obstacles or the presence of a wall on either side of the chair. The remaining sensors were focused on the user's head from two angles.

The chair has many modes of operation: the most important is the head-control mode. Here, the user directs the movements of the chair by head motions. To move the chair forward, the user positions his or her head toward the front of the chair. Similar operations control the three remaining directions. In effect, the user's head is a proportional-control joystick. One can readily see that this type of noncontacting control would be helpful for people who have no usable arm function.

In operation, the front-facing ultrasound sensors detect the presence of obstacles in the chair's path. When such an obstacle comes within a predetermined distance, the chair automatically slows and stops before running into it. If the "obstacle" moves away, the chair will follow at a fixed distance.

Side sensors serve to detect walls. A mode to "follow that wall" enables a chair to travel parallel to the chosen wall at a fixed distance. Open doorways are detected and passed over, but a discontinuity of more than a few feet disables the wall-following mode and waits for further commands from the user.

A "cruise control" mode does not use any additional sensors, but instead relies on wheel-speed data obtained from two optical shaft encoders. Once in this mode, the chair proceeds at a constant speed and heading despite changes in terrain.

A final mode allows the head to be moved without affecting the chair.

The user initializes the system to the range of his or her head motion by means of a "training" program that instructs the user to center the head, to move it to the left or right, and forward or backward. The program uses this information to calibrate the position/speed algorithm as well as set up a dead band around the user's rest position.

The hardware presently consists of a Z80 microprocessor, 64 K bytes of memory, and an external disk-drive system. Once the program is loaded, the disk is disconnected and the user drives off. The software executive is written in BASIC, with a majority of the actual real-time program coded in machine language and as arithmetic function calls. The listing consumes 40 pages.

The current construction phase will shrink the initial hardware and software configuration by one-third. A final design will capture the features on a single printed-circuit board.

The approach taken in pursuit of the interface between the ultrasound sensors and the microprocessor is considerably different from the method described in Steve's article. Since the Polaroid kits were not available at the time of construction, several new cameras were sacrificed to acquire the parts required. In addition, the computer interface was done not at the EDB level, but at the custom ultrasound board level. To perform a ranging, the computer generates a transmit request pulse via a convenient parallel output bit. The output from the board is then interrogated to start a software timing loop that is terminated by the received echo signal. The number of times the loop is performed gives a fairly precise measure of the range. Dividing this value by an appropriate factor will yield the range in whatever units are required. In the course of the project, a resolution of about a quarter of an inch was obtained over distances ranging from 9 inches to 20 feet (depending on surface characteristics).

Additional information about this ongoing project can be obtained by writing me at the address below.

David L Jaffe
Palo Alto VA Medical Center
Rehabilitative Engineering Research and Development Center
3801 Miranda (153)
Palo Alto CA 94304■

10

Build a Low-Cost Logic Analyzer

The Digital Age has spawned a variety of electronic troubleshooting aids, including logic probes, integrated-circuit test clips, multi-trace oscilloscopes, and logic analyzers. All are useful, up to a point, but it is important to know when to use a particular test instrument and how much you can depend on it.

If the logic states of signal lines were the only information needed, a simple voltage measurement would suffice in digital troubleshooting. But *timing,* rather than absolute voltage level, is the more important consideration in digital systems. Most digital systems operate by setting discrete logic conditions on bus lines and then *strobing* that data through the system at the occurrence of edges of specific clock pulses. A system operates correctly only if all the parallel states are set correctly at a specific instant in time. The system fails if any single logic state is in error at any clock time during program execution.

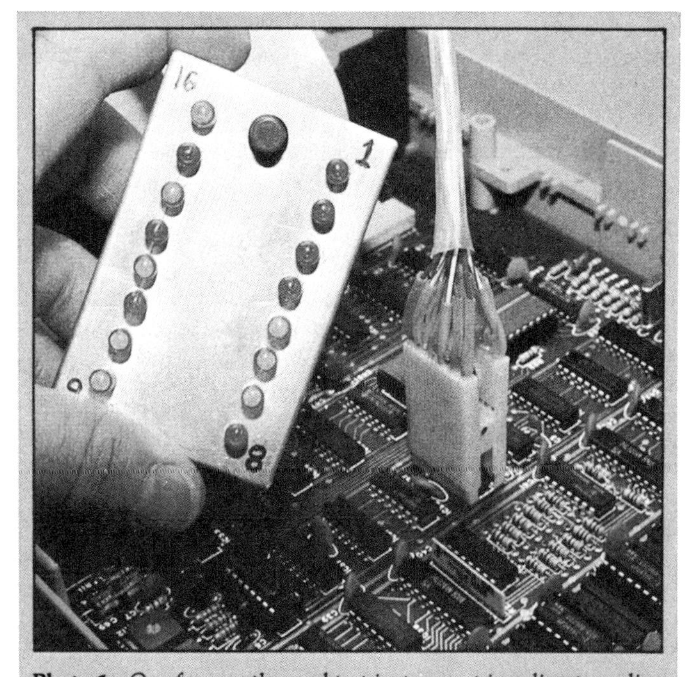

Photo 1: *One frequently used test instrument is a direct-reading state indicator. The sixteen indicators are transistor-driven incandescent lamps or LEDs (light-emitting diodes). The indicator panel is attached to a "chip-clip" connector so that the logic states on any TTL (transistor-transistor logic) or LS (low-power Schottky-diode-clamped) TTL dual in-line package can be read while the circuit is energized. The display is most valid for static conditions.*

The first special digital instrument was the logic probe. A schematic diagram of a typical logic probe is shown in figure 1. This device accurately indicates the logic state on LED (light-emitting diode) indicators at any selected point in a circuit. However, it is a static device and will not follow rapidly clocked digital logic other than to indicate general activity. Even when the concept is expanded to include fourteen or sixteen separate indicators on the probe (as shown in photo 1), effective use still depends on stopping the system clock (or slowing it substantially) to examine static logic states. Unfortunately, stopping the clock changes the dynamics of circuit operation and may, in many instances, mask the true cause of problems.

More frequently, digital-logic errors are dynamic and occur during clock-state transitions. The errors are often due to timing problems associated with the propagation of signals through the circuit or with miscuing of multiplexed components. Because the logic state at clock transitions often determines either proper operation or failure, a more suitable test instrument would be one that provides the operator with a view of all logic activity coincident with the transition of the clock.

To most people this sounds like a job for a multi-trace oscilloscope with its sweep triggered from the system clock. An oscilloscope can in many instances be of value, but unless it is an expensive storage-tube scope, fast system-clock rates can make viewing difficult. Also, viewing two signals with respect to each other in real time is of little help when the error occurs intermittently and involves more signals than can be viewed simultaneously.

What Is a Logic Analyzer?

One solution to the digital-troubleshooting dilemma is called a logic analyzer. This is an instrument that displays a "truth table" of the activity of the digital circuit being tested under actual operating conditions. After you have selected a key combination of input signals, called a *trigger* or *sync word,* and activated the analyzer, it stores all signal-input logic states for a specific number of system-clock transitions. Depending upon the sophistication of the particular unit, many commercial logic analyzers can accommodate 32 or more inputs and store up to 256 clock cycles before and after the trigger event.

A logic analyzer acts like an electronic time machine.

In effect, a logic analyzer acts like an electronic time machine. When sequentially displayed in the order it was acquired, the stored data can be used to form state tables or timing diagrams of the circuit's operation.

For example, a logic analyzer might be used to troubleshoot a malfunctioning microcomputer I/O (input/output) port that keeps receiving consistent but wrong data. You don't know whether the error is caused by the wrong data being sent to the output register or by an incorrect address signal strobing the register at the wrong time (try troubleshooting this kind of problem with just an oscilloscope). You can find out by connecting the logic analyzer to the address and data buses of the microcomputer.

Set the trigger-word switches to produce a trigger pulse when the address bus contains the I/O port address. When the trigger pulse occurs, you can examine the logic states on the data bus with the analyzer to see what value was being loaded into the port register at the occurrence of the trigger pulse, as well as those states following the pulse. It is like having an 8- to 32-channel oscilloscope with the display frozen in time on a specific clock cycle.

Commercial logic analyzers are generally stand-alone instruments with integral video-monitor or oscilloscope displays. They can present stored data in a variety of ways. A *data-domain* analyzer ordinarily displays logic states as lists of 1s and 0s. The listings are sequential and in either binary, octal, or hexadecimal format. This display method is particularly helpful when you are debugging address-bus problems. In such cases, data is most easily read as

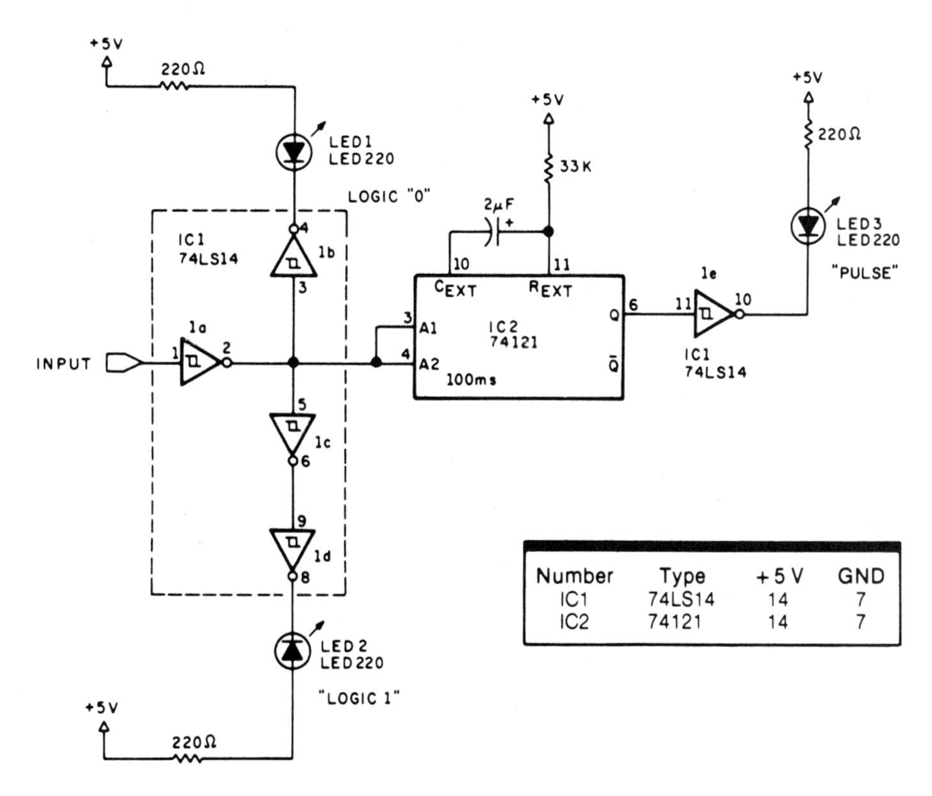

Figure 1: *A simple logic probe that uses two integrated circuits. When a logic-0 signal voltage is applied to the input, the "logic 0" LED will light. When a logic-1 signal voltage is applied to the input, the "logic 1" LED indicator will light. If the input oscillates between the 0 and 1 states, the "Pulse" LED indicator will also light.*

4-digit hexadecimal values.

For hardware troubleshooting, a *time-domain* analyzer is preferred. This unit presents the stored data in timing-diagram format. The result appears like the display of an 8- or 16-channel oscilloscope. The vertical scale has a high-voltage value that represents a logic 1 and a low-voltage value that represents a logic 0. The data signals are plotted with respect to each other and can be displayed as a function of actual time.

A third data format is the *mapped* mode. Essentially, the display screen is divided into an x, y coordinate system, and data points are plotted as dots on the screen. In some units, vectors between dots connect successive data points so that it is easier for an operator to trace sequential activity in the device under test. The process of interpreting this kind of display is essentially one of recognizing a "good" pattern and identifying wild vectors. Presumably, a properly operating program will have a repeatable pattern. Any discrepancies will show up as an extra dot or "wild vector."

The various types of logic-analyzer display formats are shown in figure 2 on page 94.

Regardless of the display format, all logic analyzers share a common internal structure. Generally, they incorporate the subsystems outlined in the block diagram of figure 3. All logic analyzers have some form of input conditioning, trigger-word selection and comparison, memory, and display (LEDs, oscilloscope, or raster-display tube, etc). The combination of capabilities is usually a function of price, which can range from $2500 to $10,000.

Photo 2: *The prototype logic analyzer described in this article. The switches on the left are for setting the trigger (sync) word.*

Photo 3: *Inside the box of photo 2 is the circuit of the analyzer as shown schematically in figure 4. Seventeen integrated circuits are used.*

A Low-Cost Logic Analyzer

Obviously, we cannot hope to construct a logic analyzer that is equivalent to an $8500 Hewlett-Packard unit. However, we can design a special logic analyzer as a peripheral device of a personal computer. By utilizing the display and processing power of the computer, we can greatly enhance the capabilities of a relatively simple hardware interface. Also, for those readers interested in the concept but not quite ready to grab their soldering irons, I will outline a method that demonstrates how to use your present computer to perform logic-analyzer functions totally in software. First, the hardware approach.

Figure 4 is the schematic diagram of a low-cost eight-input logic-analyzer interface that requires only one and a half parallel I/O ports (9 output and 6 input bits) for complete operation. It is easily expandable to 16 or even 32 inputs.

All probe inputs and clock signals are conditioned through Schmitt triggers to reduce noise and false triggering. When the sync word, set on external switches (SW1 through SW8), appears on the input lines, the analyzer automatically collects and stores 16 sequential words representing input status at the instant of either an internal or external clock signal (usually the system clock). It can operate on either edge of the clock pulse and store data at frequencies as fast as 5 MHz. The prototype interface is shown in photo 2.

Unlike commercial logic analyzers, this unit has no integral CRT (cathode-ray tube) display: it has eight externally controlled LEDs. It depends instead upon the computer to display the list of stored data. After the interface has taken sixteen samples, it sends a Scan Complete signal to the computer. A computer program sets the Read/Write line to the Read mode and sets a 4-bit address to access the contents of the 16-word scratch-pad memory. As the 4-bit address is incremented, the appropriate 8-bit output is placed on the analyzer's data-output lines from the scratch-pad memory and is stored by the computer. In addition, as the computer reads the scratch-pad memory, the contents of each location are displayed on eight LEDs. If the addresses are changed slowly, or are otherwise physically set, the 16 stored words can be viewed directly without a special display program.

Once the data has been acquired by the computer, a format-and-display program lists the values on the computer's display in binary, octal, or hexadecimal format, simulating a commercial analyzer display. To gather an additional 16 words, the computer program merely sets the Read/Write line to the Write mode and toggles the Sample Enable line. The BASIC program in listing 1 on page 118 exercises the interface and displays the output shown in listing 2.

Inside the Interface

The analyzer hardware (shown in photo 3) has an interface consisting of seventeen integrated circuits. Input signals are fed through IC1 and IC2, which are hex Schmitt-trigger inverters. Photo 4 shows typical test connections. These conditioned outputs are in turn buffered and gated through to the memory section by IC3, a type-74LS240 8-input bus driver. The output of this driver is compared to eight preset switches through two 74L85 4-bit comparators (IC7 and IC8). (Trigger-word initiation is disabled by setting all switches to the logic-1 state. Storage will commence on the first clock pulse after Sample Enable.) If the switch settings and data input are equal, a pulse is generated which stores the current input data. The first word stored is usually the sync word (assuming that the trigger word and external clock-pulse edge are synchronous).

On the trailing edge of the WE (memory-write-enable) pulse, the 4-bit memory-address counter IC9 is incremented. Data will be stored again at the occurrence of the next edge (positive or negative as selected) of the clock pulse.

Photo 4: *The analyzer is intended for use while a circuit is in dynamic operation. Connection to the circuit can be done with the "chip-clip" method shown in photo 1, or by using separate test probes. The latter is more versatile. The circuit shown under test is the Disk-80 expansion interface from last month's Circuit Cellar.*

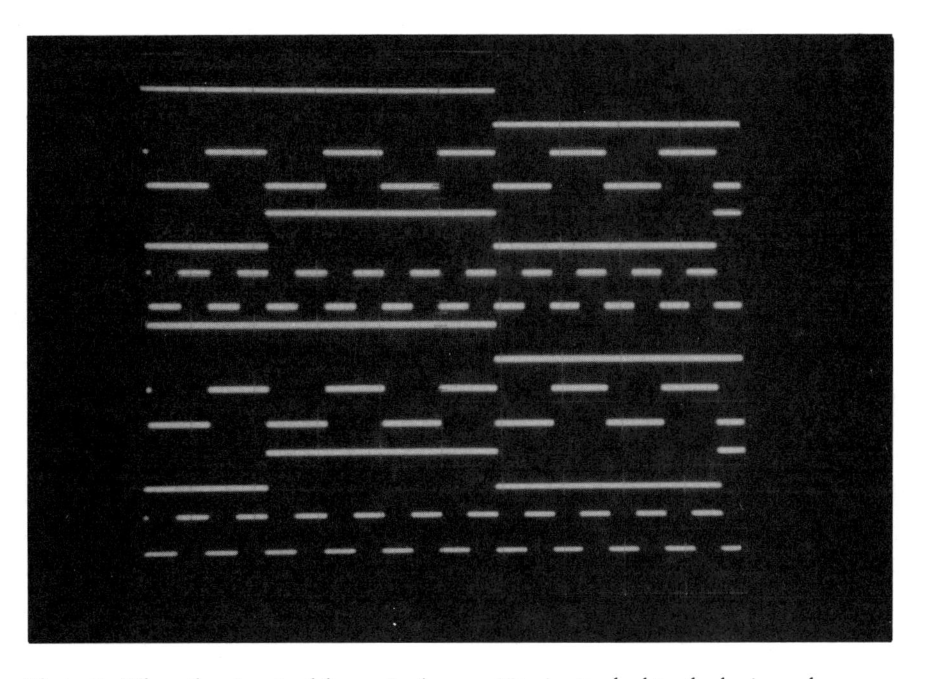

Photo 5: *When the circuit of figure 5 (on page 96) is attached to the logic analyzer, a data-domain display can be converted to a time-domain display. Essentially nothing more than an eight-channel scope multiplexer, this circuit greatly expands the display potential of the average oscilloscope, as the photo demonstrates.*

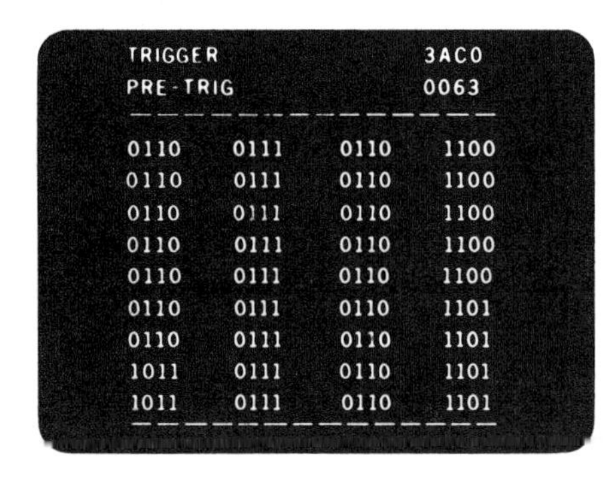

(2a)

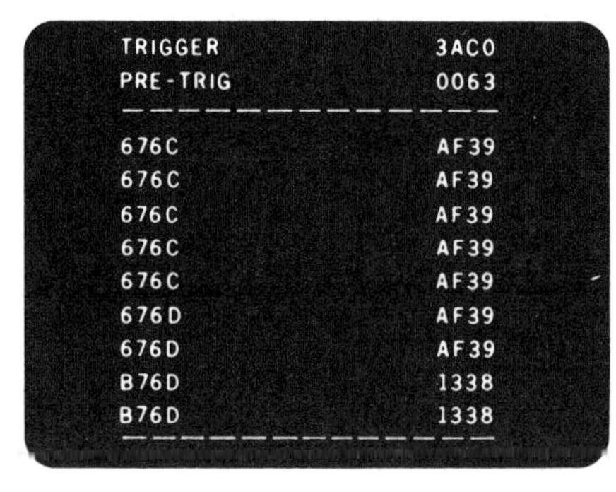

(2b)

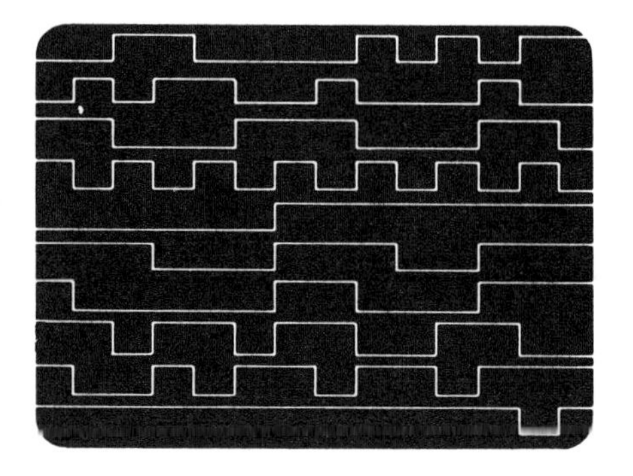

(2c)

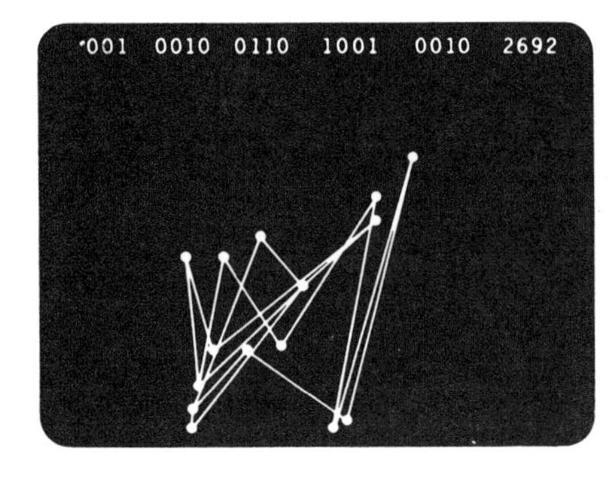

(2d)

Figure 2: *The data acquired by a logic analyzer can be displayed in various formats. The different types are:*

(2a) The ones and zeros logic-state display. In this format, binary words are plotted against clock pulses in a matrix m bits wide by n clock pulses deep. This format is used most often where word flow or data sequence is of prime concern.

(2b) Same as 2a except that the data is listed in hexadecimal notation. Hexadecimal listings are most frequently used in logic analyzers specifically designed for microprocessor troubleshooting, where thirty-two to forty inputs are not uncommon.

(2c) The timing-diagram display. In the timing format, data words are plotted against time. This format is used most often for hardware troubleshooting to detect incorrect timing between signals.

(2d) Vector-display analyzer. In the vector-display format, data words define points on an x, y coordinate system. Usually, the data word is divided in half with a separate D/A converter attached to each segment. One output goes to the display's x input and the other goes to the y input.

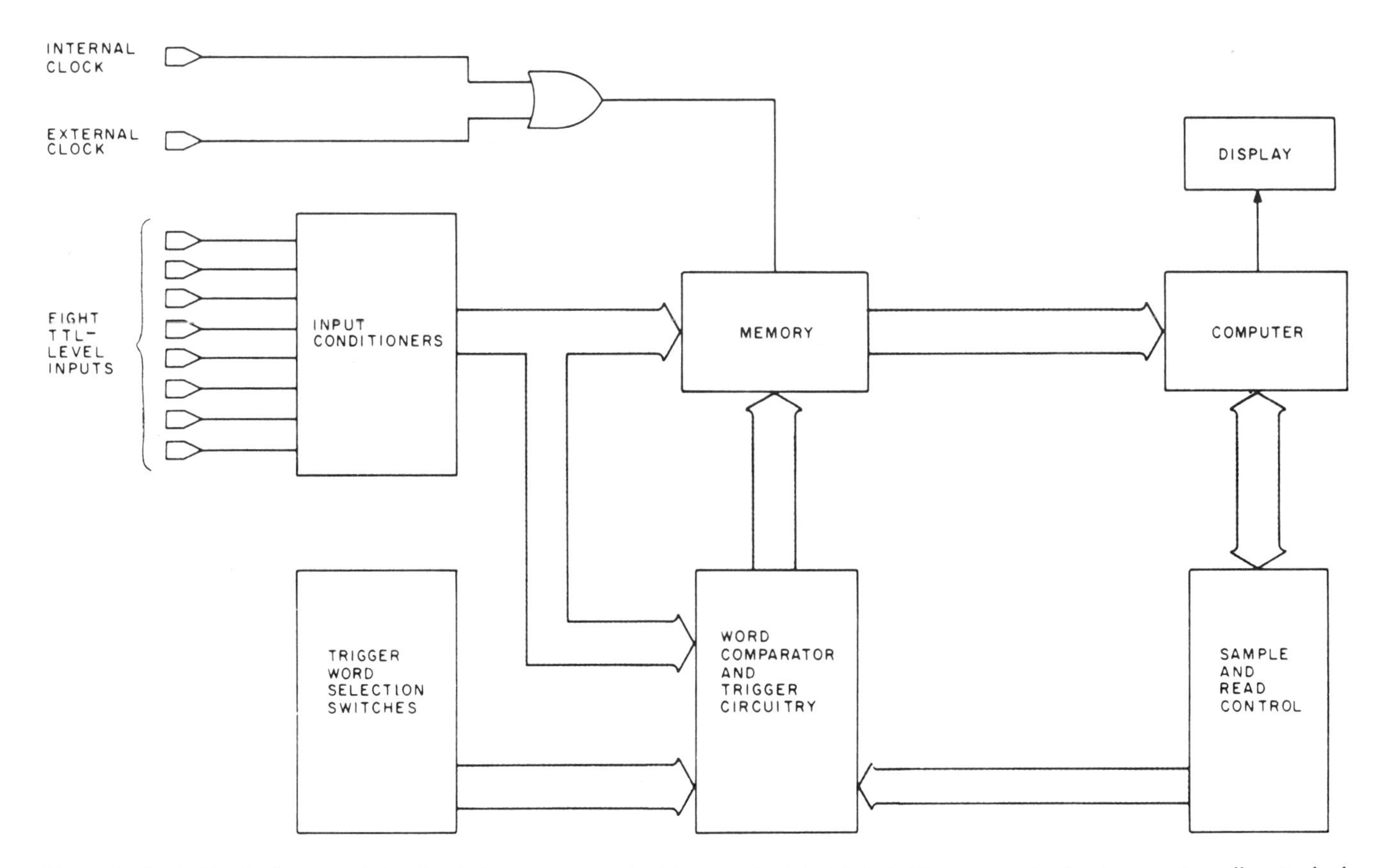

Figure 3: *Basic block diagram of the simple logic analyzer. In this case, the block labeled "computer" refers to an externally attached personal computer. In commercial units, the computer and display are integral components of the logic analyzer.*

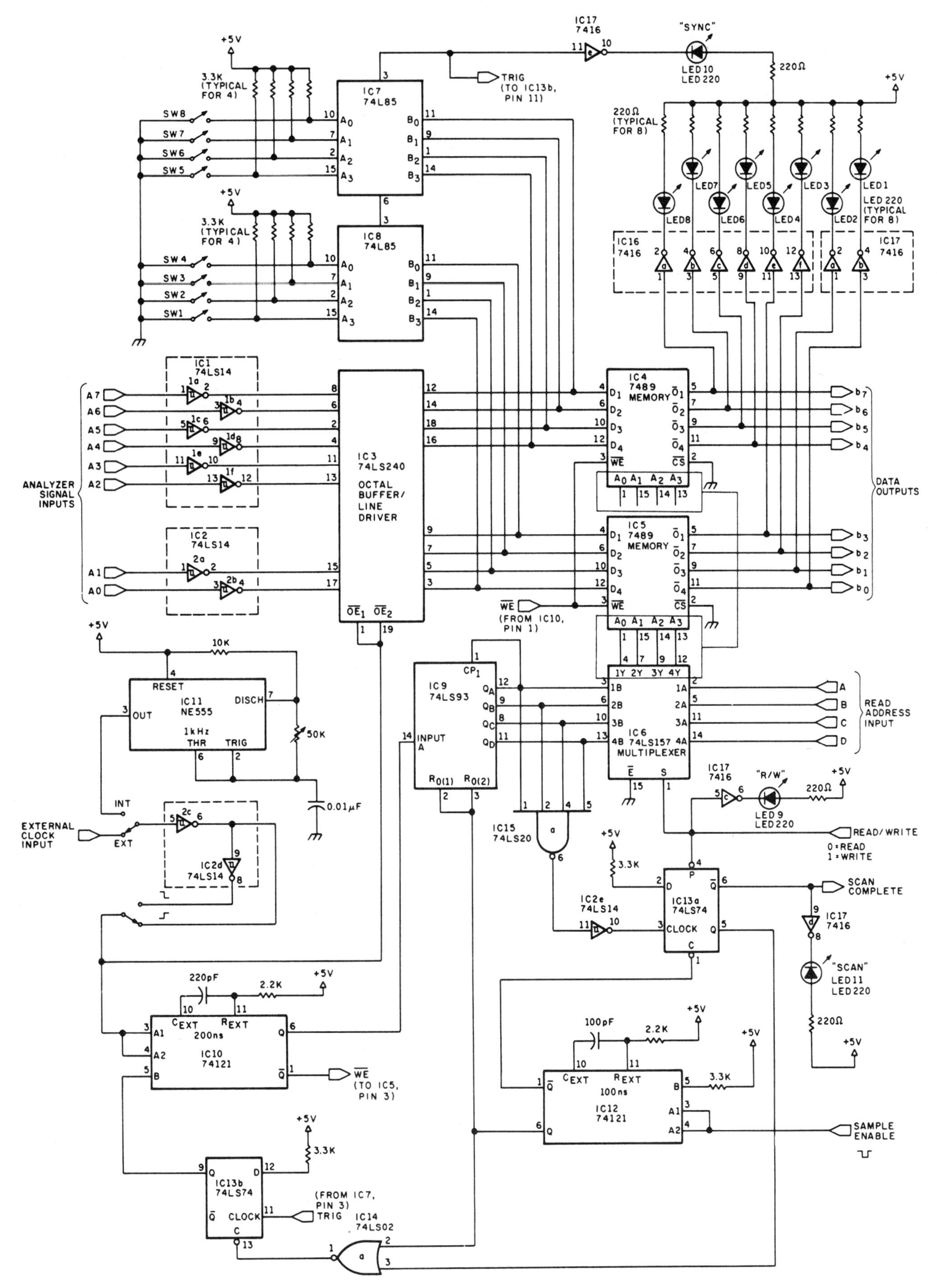

Figure 4: *Schematic diagram of an eight-input logic analyzer. One and a half parallel I/O ports are required for operation. Note that the 74L85 integrated circuits used here have a different pinout specification from the 74LS85. User connections are on the left; computer connections are on the right.*

Number	Type	+5 V	GND
IC1	74LS14	14	7
IC2	74LS14	14	7
IC3	74LS240	20	10
IC4	7489	16	8
IC5	7489	16	8
IC6	74LS157	16	8
IC7	74L85	16	8
IC8	74L85	16	8
IC9	74LS93	5	10
IC10	74121	14	7
IC11	NE555	8	1
IC12	74121	14	7
IC13	74LS74	14	7
IC14	74LS02	14	7
IC15	74LS20	14	7
IC16	7416	14	7
IC17	7416	14	7

Table 1: *Power connections for integrated circuits of figure 4, on page 116.*

When sixteen samples have been taken, the 4-bit memory address is binary 1111. IC13 and IC14 detect this condition and set the Scan Complete line to a logic 0. This also disables further storage until the interface is reset with a Sample Enable pulse to IC2.

Reading the contents is simply a matter of setting the Read/Write line to a logic 0 and placing an appropriate 4-bit address on the Read Address input lines. When an address is set on these lines, the data-output lines of the analyzer will contain the contents of that memory location. The eight LEDs will also display that value.

Creating a Time-Domain Display

As previously mentioned, the display format available from this interface is generally a listing of 1s and 0s. This is quite useful under most circumstances but not as appealing to hardware buffs as a timing-diagram-type output. Even if your computer has graphics capability, writing a program to simulate a multi-trace oscilloscope display requires considerable software expertise.

The logic-analyzer interface can be converted to a time-domain display with relatively little extra hardware and only a single-line BASIC program. Figure 5 is the schematic diagram of the additional circuitry. Essentially, it consists of a dual 4-input digital multiplexer and 2-bit D/A (digital-to-analog) converter, which offsets each of the four channels when displayed. In effect, it

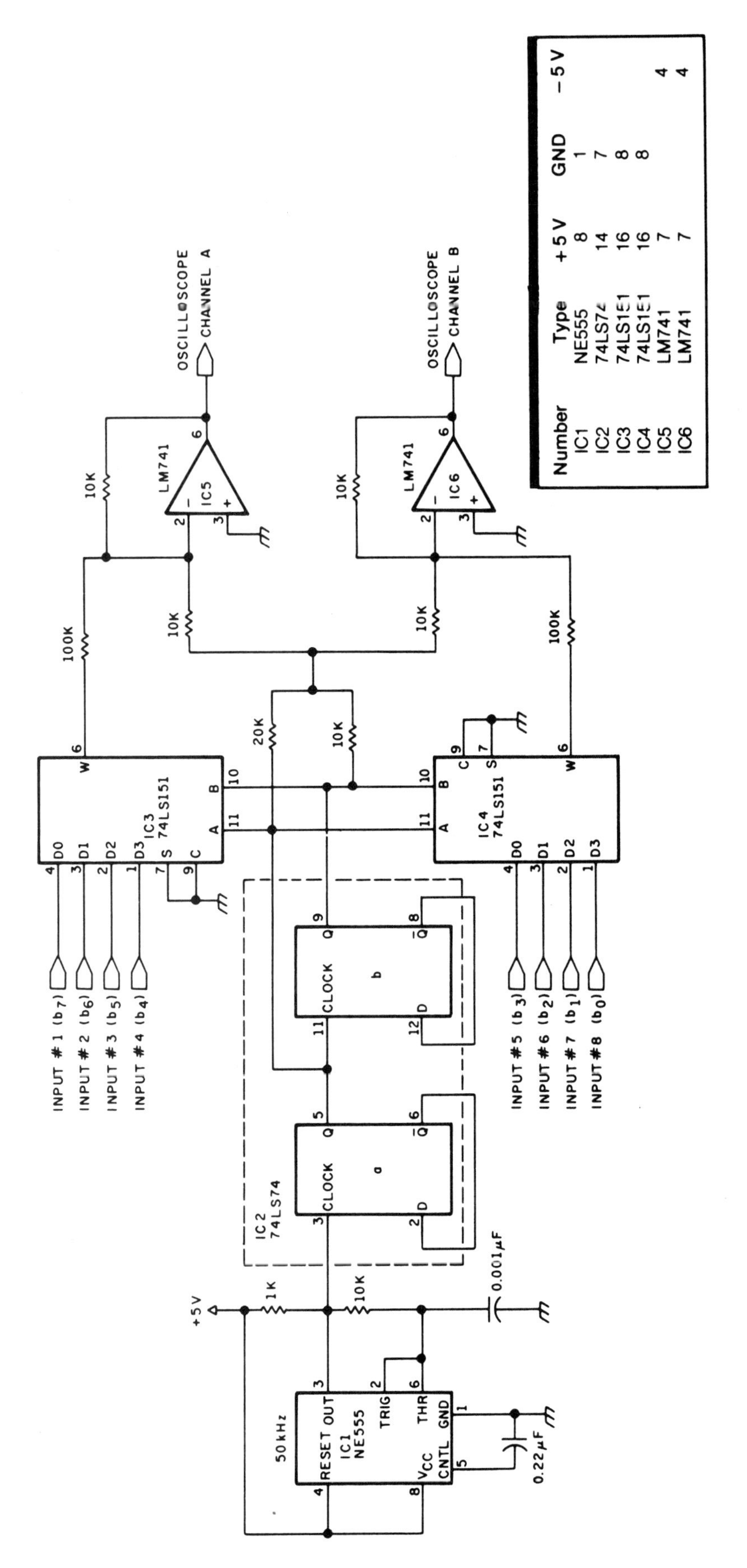

Figure 5: *Eight-channel display multiplexer, which facilitates display of eight TTL inputs on a standard dual-trace oscilloscope. Its intended use is to convert the data-domain output from the circuit of figure 4 into a time-domain display on an oscilloscope.*

allows a dual-trace oscilloscope to display eight channels simultaneously. Such a display appears in photo 5.

Conversion from data-domain to time-domain operation is not as difficult as it might seem. Consider the operation of the analyzer for a moment. Once the 16-word buffer is full, the data can be read out at any rate. If we cycle the read addresses very quickly, the outputs will form a repetitive pattern which can be easily viewed on an oscilloscope. The fast cycling can be accomplished using a 4-bit counter and oscillator source attached to the address-input lines or by using a simple program statement like:

```
100 FOR X = 0 TO 255:OUT 16,X:
    NEXT X:GOTO 100
```

Using a dual-trace oscilloscope, you can view two signals, or, with the circuit of figure 5, you can view all eight data channels simultaneously. Since there is no system clock to contend with and the pattern repeats every sixteen steps, triggering problems are reduced and the display is stationary. All other interface operations remain the same.

Adding a Vector-Display Capability

If you are determined to hunt "wild vectors," the same technique employed to provide a timing plot lends itself to vector display. Using the same methods to cycle the buffer data on the output lines of the analyzer, substitute D/A converters for the multiplexer in figure 5. Typically, two 4-bit D/A converters are needed. One would be attached to the 4 high-order bits and the other to the 4 low-order bits. One D/A converter is attached to the x-axis scope input and the other to the y-axis input. When the buffer is cycled, a unique vector pattern will appear on the screen, describing the 16 data words stored in the analyzer's buffer. (A more informative discussion on this approach to troubleshooting was one of my previous articles, "A Penny Pinching Address State Analyzer," February 1978 BYTE, page 6. It has been reprinted in *Ciarcia's Circuit Cellar*, Volume I, available from BYTE Books.)

Listing 1: *A BASIC program that exercises the computer/logic analyzer interface, displaying output through the computer's normal output devices.*

```
100 REM Logic Analyzer Program
110 REM
120 REM data in on port 16, scan complete on bit 0 of port 17
130 REM read enable and sample enable are bits 6 and 7
140 REM of port 16
150 REM read address is bits 0 thru 3 of port 16
160 REM memory locations 25000 to 25015 is set aside as the data
170 REM buffer
180 PRINT"LOGIC ANALYZER"
190 PRINT:PRINT"Enable New Sample or List Analyzer Buffer";
200 PRINT" (E or L)";
210 INPUT A$
220 IF A$ ="E" THEN 250
230 IF A$ ="L" THEN 380
240 GOTO 190
250 REM Enable Logic Analyzer and take 16 readings
260 REM pulse sample enable line and set read/write line=0
270 OUT 16,255:OUT 16,0: OUT 16,255
280 REM
290 REM test scan complete line
300 IF INP(17) =255 THEN GOTO 300
310 REM when scan is completed store readings in table
320 FOR S=25000 TO 25015
330 N=S-25000
340 REM set read address and store analyzer output
350 OUT 16,N :A=INP(16):POKE S,A
360 NEXT S
370 GOSUB 380
380 REM Ones and Zeros data-domain display routine
390 PRINT:PRINT
400 PRINT"D7 D6 D5 D4     D3 D2 D1 D0"
410 FOR S=25000 TO 25015 :X=PEEK(S)
420 FOR N=7 TO 0 STEP -1
430 W=X AND 2^N
440 IF W>0 THEN PRINT"1  "; ELSE PRINT"0  ";
450 IF N=4 THEN PRINT"    ";
460 NEXT N
470 PRINT"   SAMPLE #";S-24999
480 NEXT S
490 GOTO 190

READY
```

Listing 2: *Sample output produced by the program of listing 1.*

```
RUN
LOGIC ANALYZER

Enable New Sample or List Analyzer Buffer (E or L)? E


D7 D6 D5 D4     D3 D2 D1 D0
0  0  0  0      1  1  1  0     SAMPLE # 1
1  1  0  1      0  1  0  1     SAMPLE # 2
0  0  1  1      1  0  1  0     SAMPLE # 3
1  0  0  0      1  0  0  0     SAMPLE # 4
0  0  0  0      0  0  0  1     SAMPLE # 5
1  1  1  1      0  1  0  1     SAMPLE # 6
1  1  0  0      1  1  0  1     SAMPLE # 7
0  0  0  1      0  0  0  0     SAMPLE # 8
0  0  0  1      0  1  1  0     SAMPLE # 9
1  1  1  1      0  0  0  1     SAMPLE # 10
1  1  1  0      0  0  1  1     SAMPLE # 11
0  0  1  0      0  0  1  0     SAMPLE # 12
1  0  1  0      0  1  1  0     SAMPLE # 13
0  0  0  0      0  0  0  1     SAMPLE # 14
0  0  0  1      1  1  1  1     SAMPLE # 15
1  1  0  0      1  1  0  1     SAMPLE # 16

Enable New Sample or List Analyzer Buffer (E or L)?
```

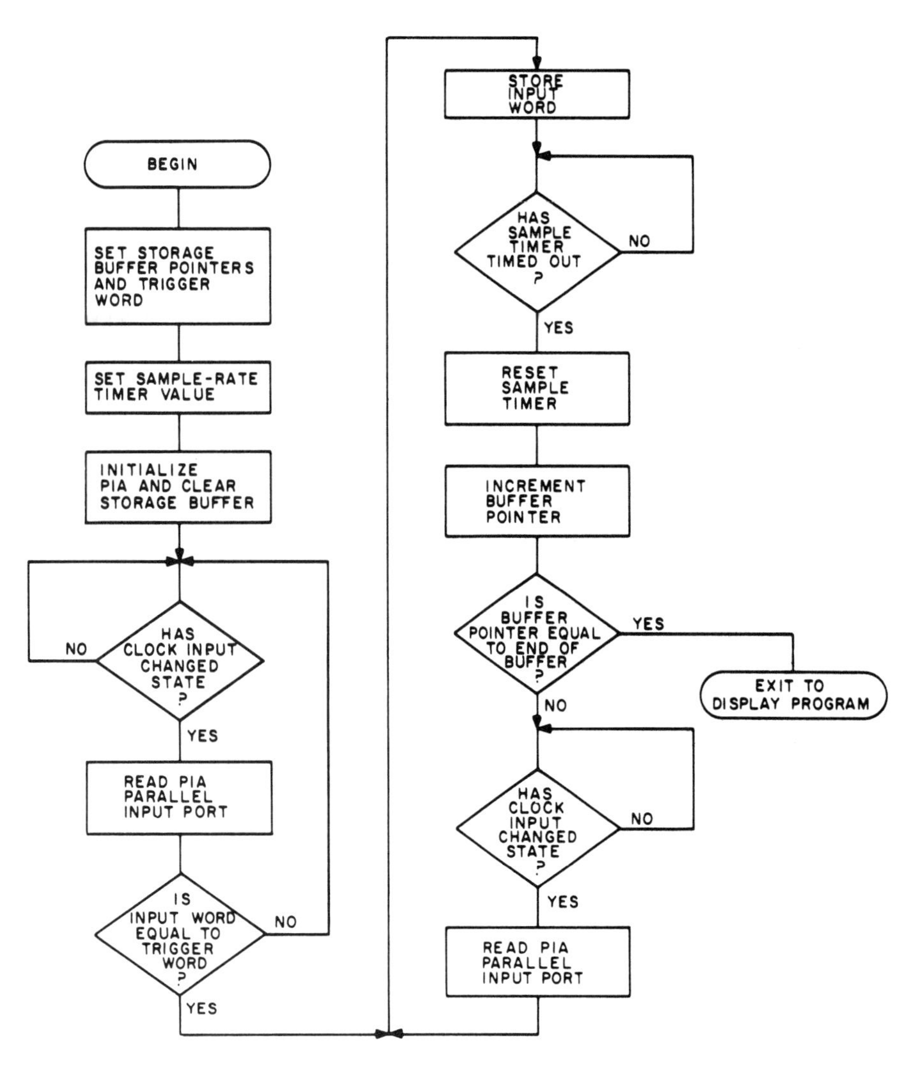

Figure 6a: *Flowchart of a software logic analyzer. Using a Motorola 6820 PIA (Peripheral Interface Adapter), this sequence of operations is all that is required to demonstrate logic-analyzer functions in software. This method is limited in speed of operation by the execution time of the program.*

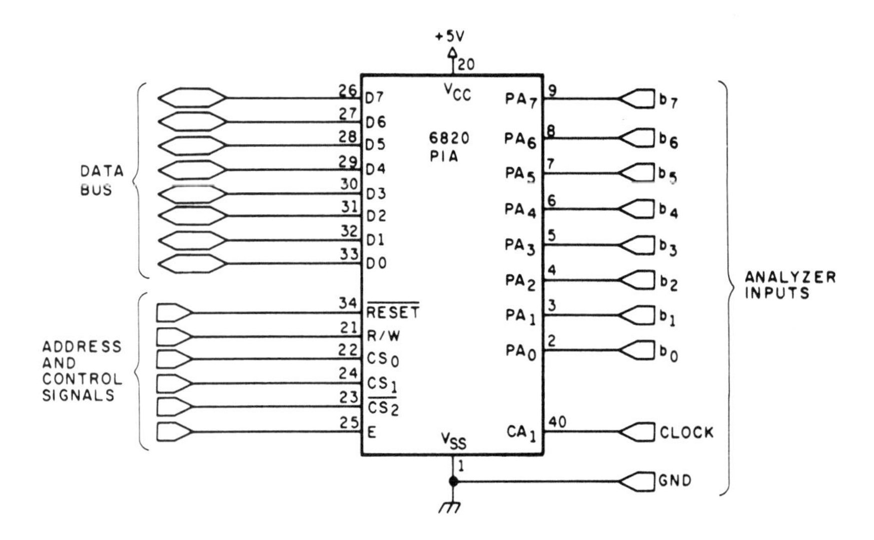

Figure 6b: *Pinout chart of the Motorola 6820 PIA used by the algorithm of figure 6a.*

Logic-Analyzer Functions Created Through Software

While I generally prefer to demonstrate hardware interfaces in my articles, the functions of a logic analyzer can easily be simulated in software if data-acquisition speed (under 20 kHz) is not critical. While it may not be appropriate for testing microcomputer bus signals, it should work for slower applications.

Figure 6 is a flow diagram outlining the specific steps involved in accomplishing this function. While any existing parallel input port will suffice, the Motorola 6820 PIA (Peripheral Interface Adapter) shown has a separate clock input, which greatly facilitates proper timing.

In Conclusion

As digital hardware becomes more complex, the instruments used in troubleshooting and debugging these circuits must themselves become more sophisticated. This sophistication, however, need not always be provided in the form of a commercially produced test instrument. Often the solution can be intelligent application of existing equipment with limited modifications.

The logic analyzer I have described can be used for all types of troubleshooting and testing of digital circuits. However, its true flexibility is revealed when the instrument captures the extremely fast data flowing in a microcomputer and generates a stationary timing diagram with the results. Built from scratch, combined with an oscilloscope, and exercised by a computer, this interface costs only a fraction of the price of commercial analyzers, yet approximates many of their features.

Tools of the Course

Dear Steve,

I am beginning implementation of a hands-on microcomputer experimentation and interfacing course here at the University of Dubuque Theological Seminary. I want to establish a digital-microprocessor laboratory. What would you consider to be the *minimum* test equipment necessary? Our financial resources are somewhat limited, so your advice would be most helpful.

Terry A Ward
Dubuque IA

At the very minimum, I would recommend that you get an oscilloscope. If you can afford it, it should be dual-trace and have at least a 15 MHz bandwidth. With it, you can troubleshoot many pieces of equipment and perform some logic-analyzer functions. If you can afford it, of course, a logic analyzer is always a good piece of equipment to have around. However, you can spend so much time teaching people how to use a logic analyzer that you don't have any time left in the course.

Other than an oscilloscope, the only other piece of equipment that you probably need would be a simple digital voltmeter (DVM) or digital multimeter.

Often the things that are needed when teaching students are not the things that you can buy off the shelf as test equipment. Frequently, simpler equipment, such as a buffered LED (light-emitting diode) that functions as a logic probe, is what's necessary.

A logic probe, 'scope, and a DVM should take care of practically anything that would arise. . . . **Steve**

'Scope Trials

Dear Steve,

I'm faced with the decision to buy an oscilloscope or to continue using a homemade logic probe. What bandwidth 'scope would you recommend: 30 MHz or 50 MHz? (The 16-bit microprocessors are getting into the 10 MHz range, and I want my investment to last.) The problem is that the 50 MHz 'scope is twice the price of the 30 MHz one.

I'd prefer a logic analyzer, but most are designed for specific microprocessors and are just too expensive.

Mel K Schmuldt
San Jose CA

The choice of a 'scope must be a trade-off between required operating needs and price. Rarely will you have to deal with the 20 MHz clock frequencies of the new microprocessors. Most likely you will just check to see if the clock is present. A frequency counter is the better instrument to measure period.

In general, most of the signals you will be trying to observe will be at far lower frequencies. You would find very little difference between a 30 and 50 MHz 'scope when displaying a 1 MHz signal.

More important factors to be concerned about when buying a 'scope that will be used primarily on digital circuitry are the precision of the trigger *and* sweep *electronics and a dual-trace (*not *dual-beam) display. Frequently, 'scopes are used to compare two signals while being triggered by a third. If the trigger circuitry is not particularly stable, the comparison of the signals is invalid and misleading (unfortunately, detecting these errors is very difficult). Also, it is often desirable to view the actual trigger signal or wait a specific time interval before starting the sweep.* Trigger view *and* delayed sweep *are expensive options.*

In my opinion, the most economical choice for a computer hobbyist is a 15 to 25 MHz dual-trace 'scope that has a time-base range between 200 ns and 0.5 s (without the time-base magnifier). Vertical sensitivity should be at least 10 mV per division. Delayed sweep and trigger view are not necessary. This type of 'scope probably costs about $1200.

If you are planning to do digital design, then you must be more particular about your needs. The market is wide open, and it is not unusual to pay $5000 to $15,000 for some 'scopes. My biggest complaint about top-end 'scopes is that they have so many bells and whistles that you need a road map to find the on/off switch.

Finally, if you are determined to buy a 50 MHz 'scope, I suggest the Tektronix Model 455 (about $2200). A comprehensive list of the 'scopes on the market is available in the September 1980 Electronic Products *magazine. . . .* **Steve**

11

Build a Z8-Based Control Computer with BASIC, Part 1

I hope you believe me when I say that I have been waiting years to present this project. For what has seemed an eternity, I have wanted a microcomputer with a specific combination of capabilities. Ideally, it should be inexpensive enough to dedicate to a specific application, intelligent enough to be programmed directly in a high-level language, and efficient enough to be battery operated.

My reason for wanting this is purely selfish. The interfaces I present each month are the result of an overzealous desire to control the world. In lieu of that goal, and more in line with BYTE policy, I satisfy this urge by stringing wires all over my house and computerizing things like my wood stove.

There are many more places I'd like to apply computer monitoring and control. I want to modify my home-security system to use low-cost *distributed* control rather than central control. I want to try my hand at a little energy management, and, of course, I am still trying to find some reason to install a microcomputer in a car. (How about a talking dashboard?)

Generally, the projects I present each month are designed to be attached to many different commercially available microcomputers through existing I/O (input/output) ports. Most of my projects are applicable for use on the small (by IBM standards) computers owned by many readers, but, unfortunately, a typical home-computer system cannot be stuffed under a car seat.

The Z8-BASIC Microcomputer is a milestone in low-cost microcomputer capability.

The time has come to present a versatile "Circuit Cellar Controller" board for some of these more ambitious control projects. I decided not to adapt an existing single-board computer, which would be larger, more expensive, and generally limited to machine-language programming. Instead, I started from scratch and built exactly what I wanted.

The microcomputer/controller I developed is called the Z8-BASIC Microcomputer. Its design and application will be presented in a two-part article beginning this month. In my opinion, it is a milestone in low-cost microcomputer capability. It can be utilized as an inexpensive tiny-BASIC computer for a variety of changing applications, or it can be dedicated to specialized tasks, such as security control, energy management, solar-heating-system monitoring, or intelligent-peripheral control. [**Editor's Note:** *We are using the term "tiny BASIC" generically to denote a small, limited BASIC interpreter. The term has been used to refer to some specific commercially available products based on the Tiny BASIC concept promulgated by the People's Computer Company in 1975....***RSS**]

The entire computer is slightly larger than a 3 by 5 file card, yet it includes a tiny-BASIC interpreter, 4 K bytes of program memory, one RS-232C serial port and two parallel I/O ports, plus a variety of other features. (A condensed functional specification is shown in the "At a Glance" text box.) Using a Zilog Z8 microcomputer integrated circuit and Z6132 4 K by 8-bit read/write memory device, the Z8-BASIC Microcomputer circuit board is completely self-contained and optimized for use as a dedicated controller.

To program it for a dedicated application, you merely attach a user terminal to the DB-25 RS-232C connector, turn the system on, and type in a BASIC program using keywords such as GOTO, IF, GOSUB, and LET. Execution of the program is started by typing RUN. If you need higher speed than BASIC provides, or if you just want to experiment with the Z8 instruction set, you can use the

GO@ and USR keywords to call machine-language subroutines.

Once the application program has been written and tested with the aid of the terminal, the finished program can be transferred to an EPROM (erasable programmable read-only memory) via a memory-dump program and the terminal disconnected. Next, the 28-pin Z6132 memory component is removed from its socket and either a type-2716 (2 K by 8-bit) or type-2732 (4 K by 8-bit) EPROM is plugged into the lower 24 pins. (The choice of EPROM depends upon the length of the program.) When the Z8 board is powered up, the stored program is immediately executed. *The EPROM devices and the Z6132 read/write memory device are pin-compatible.* Permanent program storage is simply a matter of plugging an EPROM into the Z6132's socket.

There is much more power on this board than is alluded to in this simple description. That is why I decided to use a two-part article to explain it. This month, I'll discuss the design of the system and the attributes of the Z8 and Z6132. Next month, I'll describe external interfacing techniques, a few applications, and the steps involved in transferring a program into an EPROM.

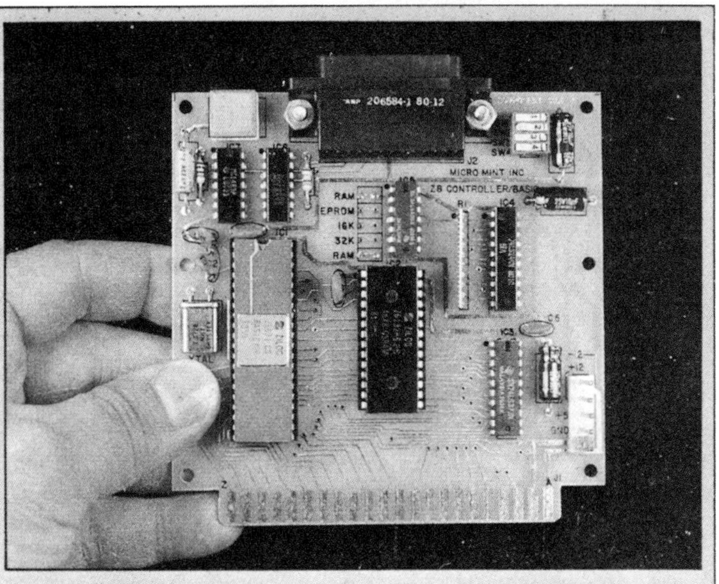

Photo 1: *A prototype of the versatile "Circuit Cellar Controller," formally called the Z8-BASIC Microcomputer. The printed-circuit board measures 4 by 4½ inches and has a 44-pin (two-sided 22-pin) edge connector with contacts on 0.156-inch centers. A 2716 or 2732 EPROM can be substituted for the Z6132 Quasi-Static memory, plugging into the same socket.*

Single-Chip Microcomputers

The central component in the Z8-BASIC Microcomputer is a member of the Zilog Z8 family of devices. The specific component used, the Z8671, is just one of them. Unlike a micro*processor*, such as the well-known Zilog Z80, the Z8 is a single-chip micro*computer*. It contains programmable (read/write) memory, read-only memory, and I/O-control circuits, as well as circuits to perform standard processor functions. Microprocessors such as the Z80 or the Intel 8080 require support circuitry to make a functional computer system. A single-chip microcomputer, on the other hand, can function solely on its own.

The concept is not new. Single-chip microcomputers have been around for quite a while, and millions of them are used in electronic games. The designers of the Z8, however, raised the capabilities of single-chip microcomputers to new heights and provided many powerful features usually found only in general-application microprocessors.

Typically, single-chip microcomputers have been designed for microcontroller applications and optimized for I/O processing. On a 40-pin dual-inline package, as many as 32 of the pins can be I/O related. A ROM-programmed single-chip microcomputer used in an electronic chess game might offer a thousand variations in game tactics, but it could not be reprogrammed as a word processor. The ability to reorient processing functions and reallocate memory has generally been the province of microprocessors, with their memory-intensive architecture.

The Z8 architecture (shown in figure 1a on page 148) allows it to serve in either memory- or I/O-intensive applications. Under program control, the Z8 can be configured as a stand-alone microcomputer using 2 K to 4 K bytes of internal ROM, as a traditional microprocessor with as much as 120 K to 124 K bytes of external memory, or as a parallel-processing unit working with other computers. The Z8 could be used as a controller in a microwave oven or as the processor in a stand-alone data-entry terminal complete with floppy-disk drives.

Getting Specific: The Z8671

The member of the Z8 family used in this project is the Z8671. This component differs from the garden-variety Z8601 chiefly in the contents of the ROM set at the factory. The pinout specification of the Z8671 is shown in figure 1b, and the package is shown in photo 2 on page 149. The Z8671 package contains the processor circuitry, 2 K bytes of ROM (preprogrammed with a tiny-BASIC interpreter and a debugging monitor), 32 I/O lines, and 144 bytes of programmable (read/write) memory.

The operational arrangement of memory-address space is shown in figure 1c. The internal read/write memory is actually a register file (illustrated in figure 2) composed of 124 general-purpose registers (R4 thru R127), 16 status-control registers (R240 thru R255), and 4 I/O-port registers (R0 thru R3). Any general-purpose register can be used as an accumulator, address pointer, index register, or as part of the internal stack area. The significance of these registers will be explained when I describe the tiny-BASIC/Debug interpreter/monitor.

The 32 I/O lines are grouped into four separate ports and treated internally as 4 registers. They can be configured by software for either input or output and are compatible with

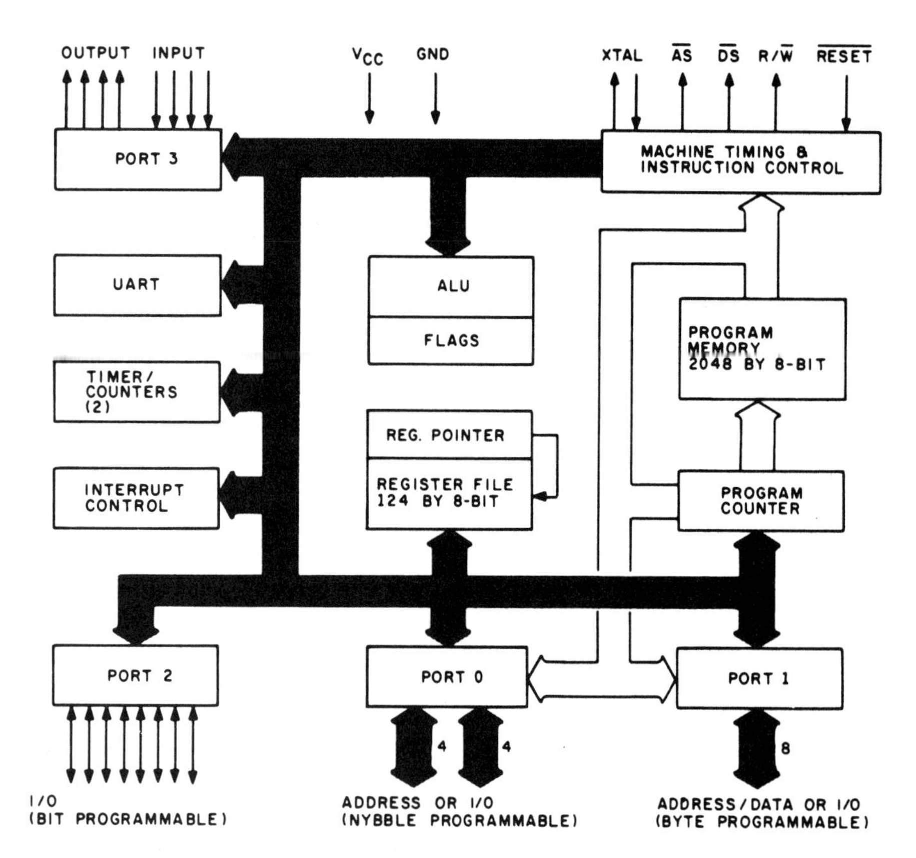

Figure 1a: *Block diagram of the Zilog Z8-family single-chip microcomputers. Their architecture allows these devices to serve in either memory- or I/O-intensive applications. This figure and figures 1b, 1c, 2, 3, and 4 were provided through the courtesy of Zilog Inc.*

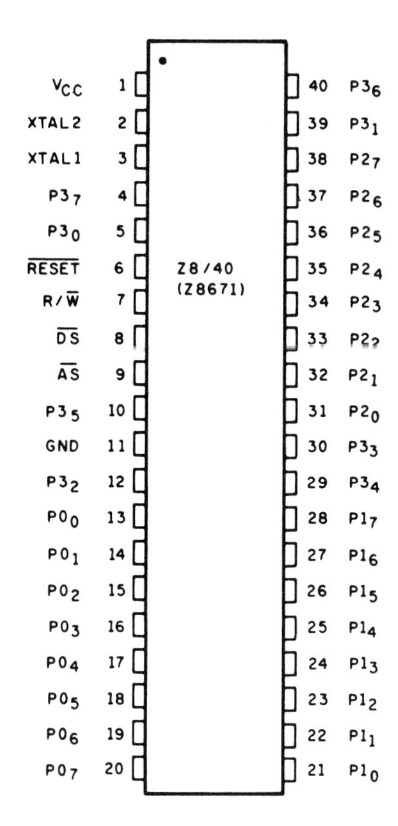

Figure 1b: *Pinout specification of the Zilog Z8671 microcomputer. The Z8671 is a variant of the basic Z8601 component of the Z8 family. The Z8671 is used in this project because it contains the BASIC/Debug interpreter/monitor in read-only memory. Other members of the Z8 family are supplied in different packages, chiefly to support system-development work.*

LSTTL (low-power Schottky transistor-transistor logic). In addition, port 1 and port 0 can serve as a multiplexed address/data bus for connection of external memory and peripheral devices.

In traditional nomenclature, port 1 transceives the data-bus lines D0 thru D7 and transmits the low-order address-bus signals A0 thru A7. Port 0 supplies the remaining high-order address lines A8 thru A15, for a total of 16 address bits. This allows 62 K bytes of *program* memory (plus 2 K bytes of ROM) to be directly addressed. If more memory is required, one bit in port 3 can be set to select another memory bank of 62 K bytes, which is referred to as *data* memory. In the Z8-BASIC Microcomputer presented here, a separate data-memory bank is not implemented, and program and data memory are considered to be the same.

The Z8 has forty-seven instructions, nine addressing modes, and six interrupts. Using a 7.3728 MHz crystal (producing a system clock rate of 3.6864 MHz) most instructions take about 1.5 to 2.5 μs to execute. Ordinarily, you would not be concerned about single-chip-microcomputer instruction sets and interrupt handling because the programs are mask-programmed into the ROM at the factory. In the Z8671, however, only the BASIC/Debug interpreter is preprogrammed. Using this interpreter, you can write machine-language programs that can be executed through subroutine calls written in BASIC. This feature greatly enhances the capabilities of this tiny computer and potentially allows the software to control high-speed peripheral devices. (A complete discussion of the Z8 instruction set and interrupt structure is beyond the scope of this article. The documentation accompanying the Z8-BASIC Microcomputer Board describes the instruction set in detail.)

The final area of concern is communication. The Z8 contains a full-duplex UART (universal asynchronous receiver/transmitter) and two counter/timers with prescalers. One of the counters divides the 7.3728 MHz crystal frequency to one of eight standard data rates. With the Z8671, these rates range between 110 and 9600 bps (bits per second) and are switch- or software-selectable.

A block diagram of the serial-I/O section is shown in figure 3. Serial data is received through bit 0 of port 3 and transmitted from bit 7 of port 3. While the Z8 can be set to transmit odd parity, the Z8671 is preset for 1 start bit, 8 data bits, no parity, and 2 stop bits. Received data must have 1 start bit, 8 data bits, at least 1 stop bit, and no parity (in this configuration).

Quasi-Static Memory

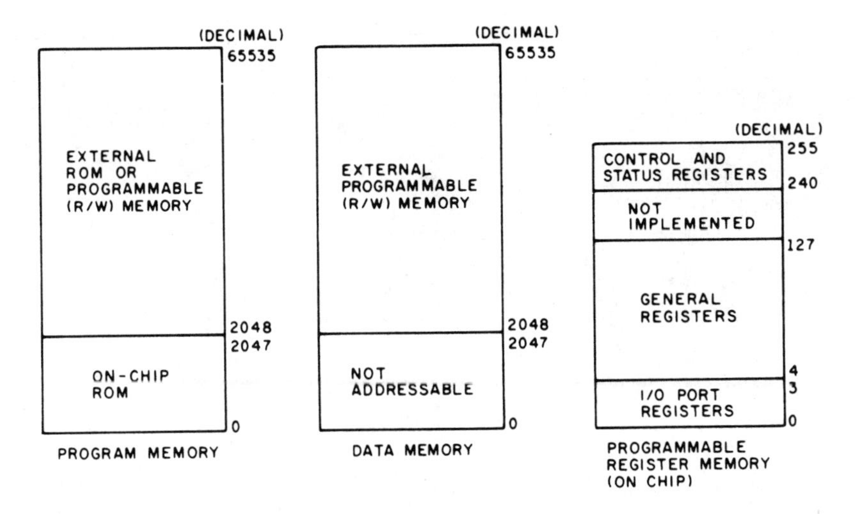

Figure 1c: *The operational arrangement of memory-address space in the Z8 family. The regions labeled "program memory" and "data memory" may map to the same physical memory, or two separate banks may be used, selected through one bit of I/O port 3. The internal programmable (read/write) memory is a register file containing 124 general-purpose registers, 16 status-control registers, and 4 I/O-port registers.*

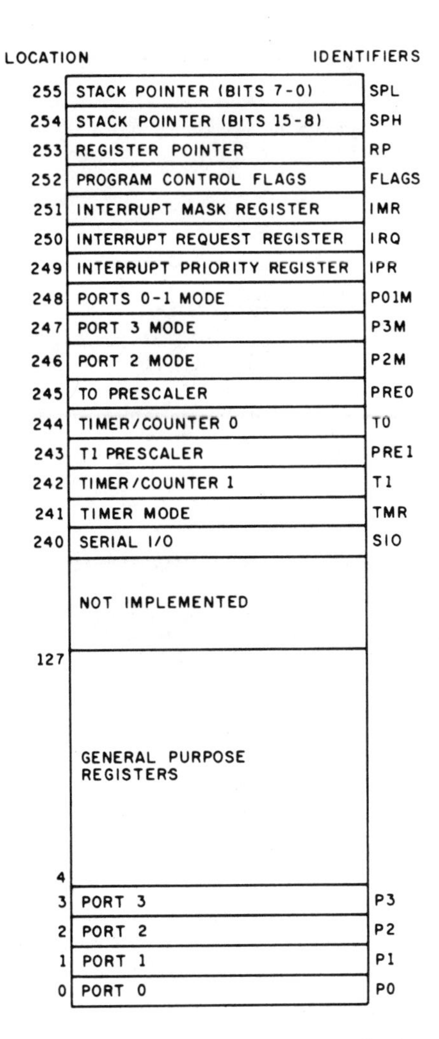

Figure 2: *An expanded view of the register-memory section of figure 1c, showing the organization of the register file. Any general-purpose register can be used as an accumulator, address pointer, index register, or as part of the internal stack area.*

A limiting factor in small controller designs has always been the trade-off between memory size and power consumption. To keep the number of components down and simplify construction, a designer generally selects a limited quantity of static memory. Frequently, the choice is to use two type-2114 1 K by 4 NMOS (negative-channel metal-oxide semiconductor) static-memory devices. In practice, however, the 1 K-byte memory size thereby provided is rather limited. It would be much better to expand this to at least 4 K bytes. Unfortunately, eight 2114 chips require considerably more circuit-board space and consume about 0.7 amps at +5 V. Not only would this make the design ill suited for battery power, it could never fit on my 4- by 4½-inch circuit board.

Another approach is to use dynamic memory, as in larger computers. Dynamic memory costs less, bit for bit, than static memory and consumes little power. Unfortunately, most dynamic-memory components require three separate operating voltages and special refresh circuitry. Adding 4 K bytes of dynamic memory would probably take about twelve chips. The advantages gained in reduced power consumption hardly justify the expense and effort.

The solution to this problem, surprisingly enough, also comes from Zilog, in the form of the Z6132 Quasi-Static Memory. The Z6132, shown in photo 4 on page 151, is a 32 K-bit dynamic-memory device, organized into 4 K 8-bit (byte-size) words. It uses single-transistor dynamic bit-storage cells, but the device performs and controls its own data-refresh operations in a manner that is completely invisible to the user and the rest of the system. This eliminates the need for external refresh circuitry. Also, the Z6132 requires only a +5 V power supply. The result is a combination of the design convenience of static memory and the low power consumption of dynamic memory. All 4 K bytes of memory fit in a single 28-pin dual-inline package, which typically draws about 30 milliamps.

An additional benefit in using the Z6132 is that it is pin-compatible with standard type-2716 (2 K by 8-bit) and type-2732 (4 K by 8-bit) EPROMs. This feature is extremely beneficial when you are configuring this Z8 board for use as a dedicated controller. As previously mentioned, the Z6132 can be removed and an EPROM inserted in the low-order 24 pins of the same socket. Thus, any program written and operating in the Z6132 memory can be placed in a

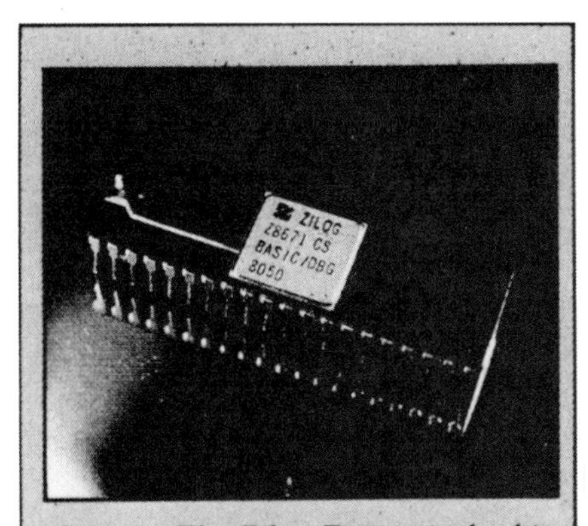

Photo 2: *The Zilog Z8671 single-chip microcomputer, a member of the Z8 family of devices. This dual-inline package contains the processor circuitry, 2 K bytes of ROM, 32 I/O lines, and 144 bytes of programmable memory.*

Photo 3: *A photomicrograph of the silicon chip containing the working parts of a Z8 microcomputer.*

See page 447 for a listing of parts and products available from Circuit Cellar Inc.

Photo 4: *The Zilog Z6132 Quasi-Static Memory device, shown with the hood up. This component stores 32 K bits in the form of 4 K bytes in invisibly refreshed dynamic-memory cells.*

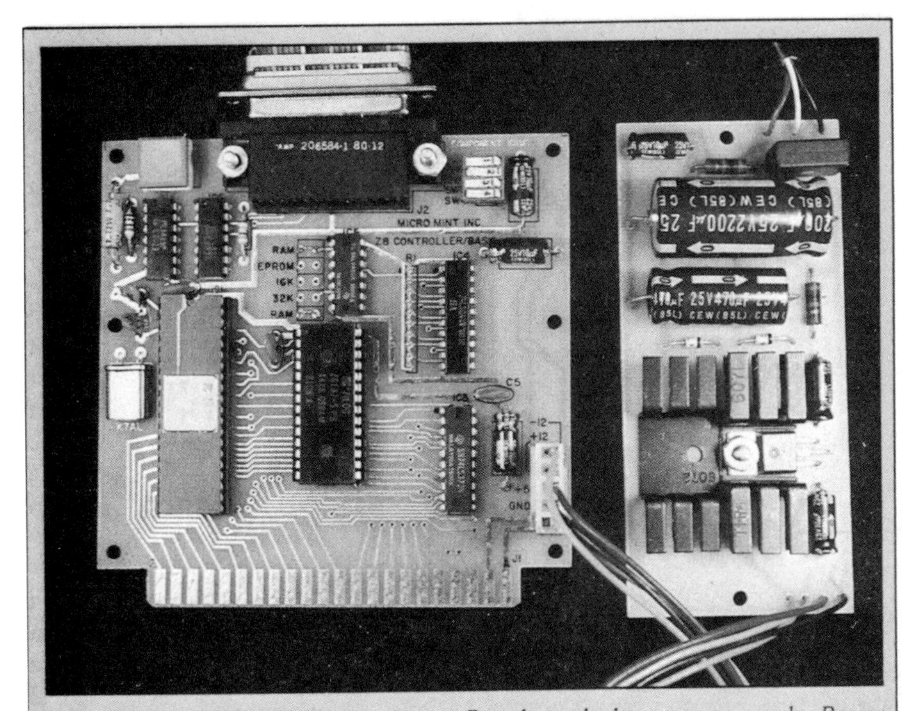

Photo 5: *The Z8-BASIC Microcomputer Board attached to a power supply. Power can be supplied either through the separate power connector, as shown, or through the edge connector.*

At a Glance

Name
Z8-BASIC Microcomputer

Processor
Zilog Z8-family Z8671 8-bit microcomputer with programmable (read/write) memory, read-only memory, and I/O in a single package. The Z8671 includes a 2 K-byte tiny-BASIC/Debug resident interpreter in ROM, 144 bytes of scratch-pad memory, and 32 I/O lines. System uses 7.3728 MHz crystal to establish clock rate. Two internal and four external interrupts.

Memory
Uses Z6132 4 K-byte Quasi-Static Memory (pin-compatible with 2716 and 2732 EPROMs); 2 K-byte ROM in Z8671. Memory externally expandable to 62 K bytes of program memory and 62 K bytes of data memory.

Input/Output
Serial port: RS-232C-compatible and switch-selectable to 110, 150, 300, 1200, 2400, 4800, and 9600 bps.
Parallel I/O: two parallel ports; one dedicated to input, the other bit-programmable as input or output; programmable interrupt and handshaking lines; LSTTL-compatible.
External I/O: 16-bit address and 8-bit bidirectional data bus brought out to expansion connector.

BASIC Keywords
GOTO, GO@, USR, GOSUB, IF...THEN, INPUT, LET, LIST, NEW, REM, RETURN, RUN, STOP, IN, PRINT, PRINT HEX. Integer arithmetic/logic/operators: +, −, /, *, and AND; BASIC can call machine-language subroutines for increased execution speed; allows complete memory and register interrogation and modification.

Power-Supply Requirements
+5 V ±5% at 250 mA
+12 V ±10% at 30 mA
−12 V ±10% at 30 mA
(The 12 V supplies are required only for RS-232C operation.)

Dimensions and Connections
4- by 4½-inch board; dual 22-pin (0.156-inch) edge connector. 25-pin RS-232C female D-subminiature (DB-25S) connector; 4-pole DIP-switch data-rate selector.

Operating Conditions
Temperature: 0 to 50°C (32 to 122°F)
Humidity: 10 to 90% relative humidity (noncondensing)

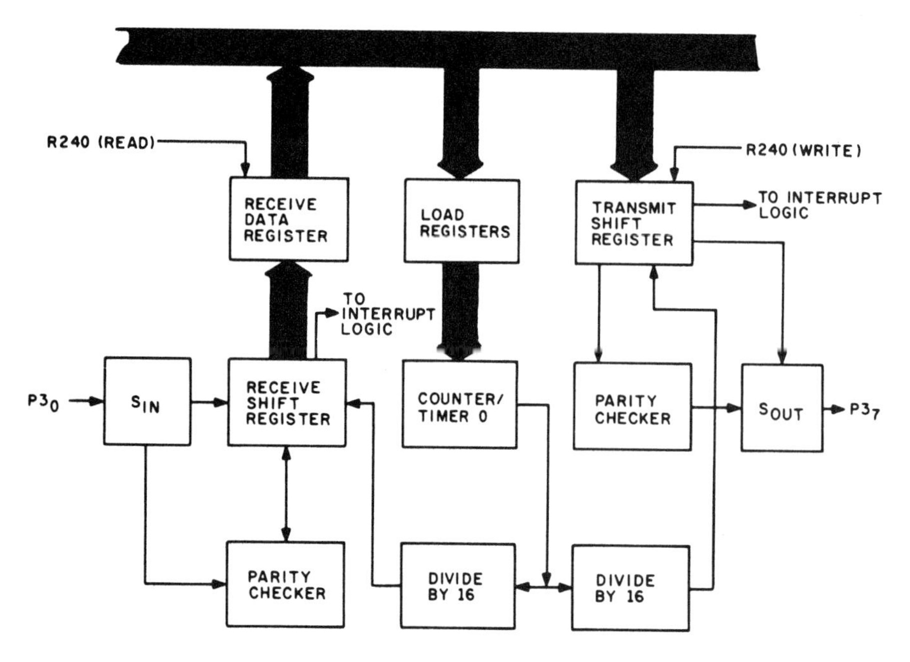

Figure 3: *Block diagram of the serial-I/O section of the Z8-family microcomputers. The Z8 contains a full-duplex UART (universal asynchronous receiver/transmitter). The data rates are derived from the clock-rate crystal frequency. Serial data is received through bit 0 of port 3 and is transmitted from bit 7 of port 3. An interrupt is generated within the Z8 whenever transmission or reception of a character has been completed.*

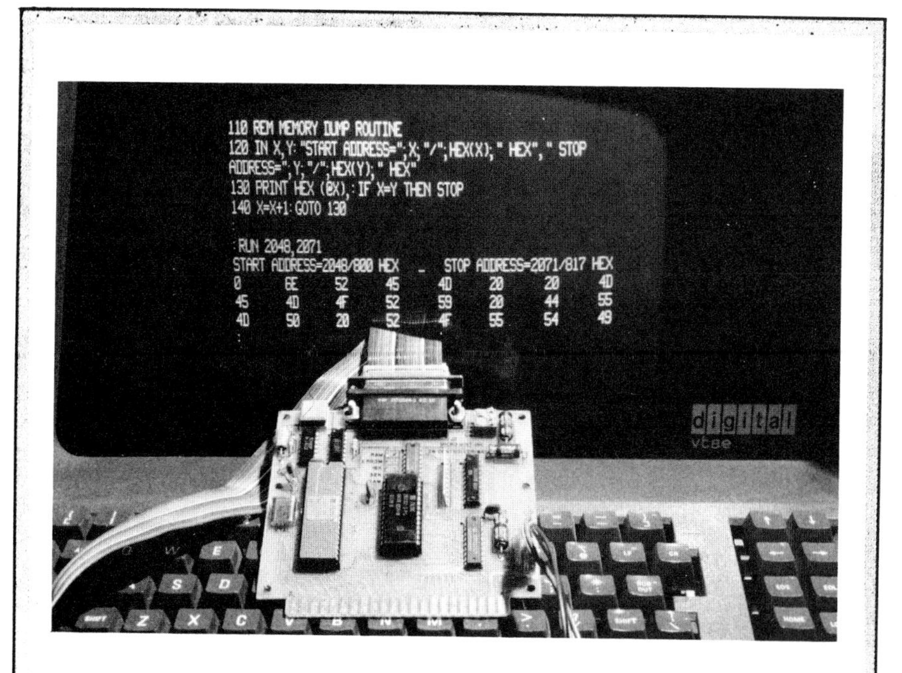

Photo 6: *The Z8-BASIC Microcomputer in operation, communicating with a video terminal (here, a Digital Equipment Corporation VT8E). A memory-dump routine, written using the BASIC/Debug interpreter, is shown on the display screen. The starting address of the dump is the beginning of the user-memory area; the hexadecimal values displayed are the ASCII (American Standard Code for Information Interchange) values of the characters that make up the first line of the memory-dump program.*

nonvolatile EPROM. (There are some limitations placed on the number of subroutine calls and variables allowed by this substitution because variable data and return addresses must be stored in the Z8's register area instead of in external read/write memory.)

Z8-BASIC Microcomputer

Figure 5 on pages 154 and 155 is the schematic diagram of the seven-integrated-circuit Z8-BASIC Microcomputer Board, shown in prototype form, with a power supply, in photo 5. IC1 is the Z8671 microcomputer, the member of the Z8 family that contains Zilog's 2 K-byte BASIC/Debug software in read-only memory. IC2 is the Z6132 Quasi-Static Memory, and IC3 is an 8-bit address latch. Under ordinary circumstances, the Z6132 is capable of latching its address internally, but IC3 is included to allow EPROM operation. IC4 and IC5 form a hard-wired memory-mapped input port used to read the data-rate-selection switches. IC6 and IC7 provide proper voltage-level conversion for RS-232C serial communication.

The seven-integrated-circuit computer typically takes about 200 milliamps at +5 V. The +12 V and −12 V supplies are required only for operating the RS-232C interface. Power required is typically about 25 milliamps on each.

The easiest way to check out the Z8-BASIC Microcomputer after assembly is to attach a user terminal to the RS-232C connector (J2) and set the data-rate-selector switches to a convenient rate. I generally select 1200 bps, with SW2 closed and SW1, SW3, and SW4 open. After applying power, simply press the RESET push button.

Pressing RESET starts the Z8's initialization procedure. The program reads location hexadecimal FFFD in memory-address space, to which the data-rate-selector switches are wired to respond. When it has acquired this information, it sets the appropriate data rate and transmits a colon to the terminal. At this point, the Z8 board is completely operational and programs can be entered in tiny BASIC.

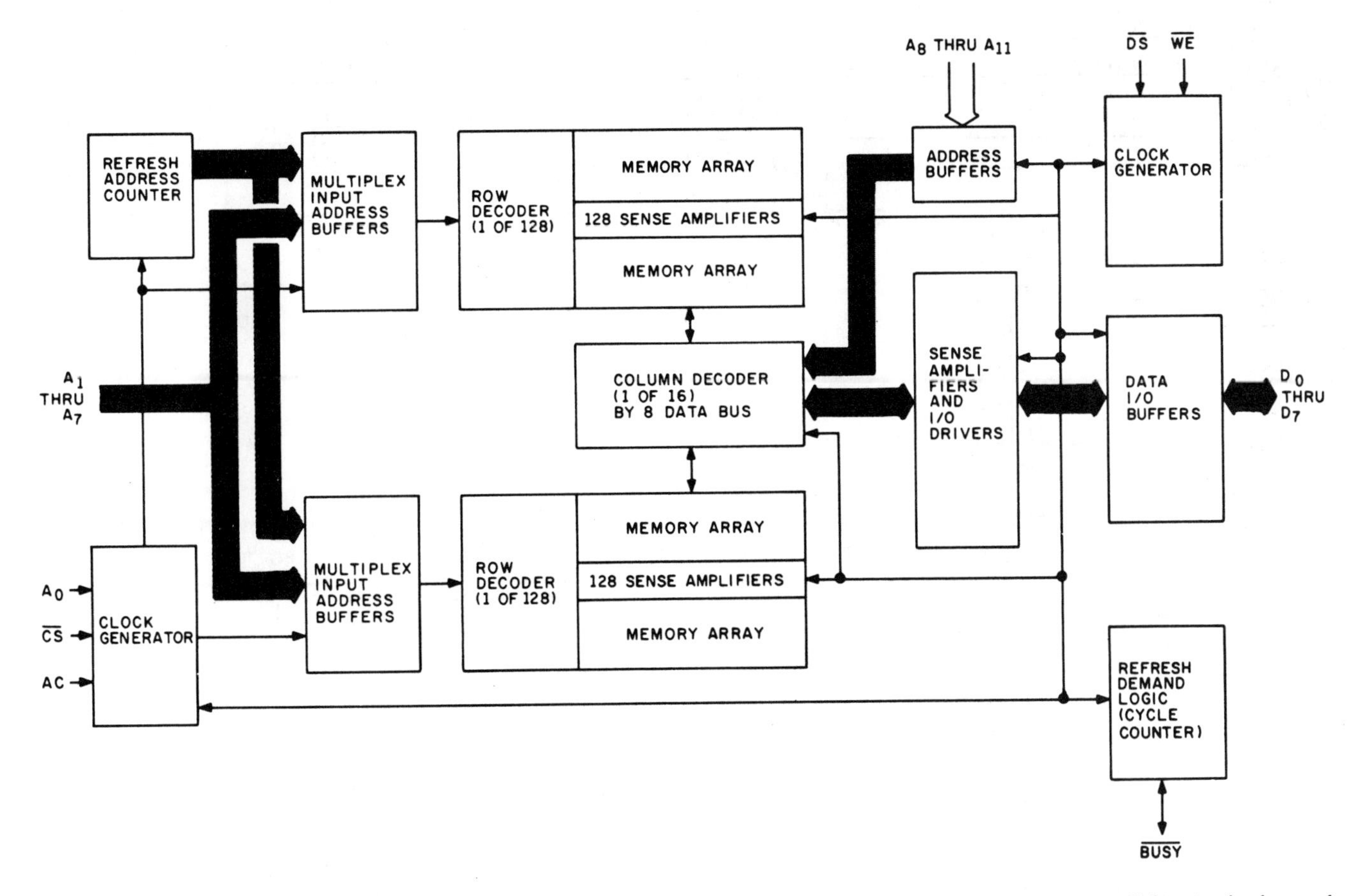

Figure 4: *Block diagram of the Zilog Z6132 Quasi-Static Memory component. This innovative part stores 32 K bits in the form of 4 K bytes, using single-transistor dynamic random-access bit-storage cells, but all refresh operations are controlled internally. The memory-refresh operation is completely invisible to the user and the other components in the system. The Z6132 draws about 30 milliamps from a single +5 V power supply.*

(With the simple address selection employed in this circuit, the data-rate switches will be read by an access to any location in the range hexadecimal C000 thru FFFF. This should not unduly restrict the versatility of the system in the type of application for which it was designed.)

BASIC/Debug Monitor

I'll go into the features of the tiny-BASIC interpreter in greater detail next month, but I'm sure you are curious about the capabilities present in a 2 K-byte BASIC system.

Essentially an integer-math dialect of BASIC, Zilog's BASIC/Debug software is specifically designed for process control. It allows examination and modification of any memory location, I/O port, or register. The interpreter processes data in both decimal and hexadecimal radices and accesses machine-language code as either a subroutine or a user-defined function.

BASIC/Debug recognizes sixteen keywords: GOTO, GO@, USR, GOSUB, IF...THEN, INPUT, IN, LET, LIST, NEW, REM, RUN, RETURN, STOP, PRINT, and PRINT HEX. Standard syntax and mathematical operators are used.

The Z8 board is not my idea of what should be available; it is available now.

Twenty-six numeric variables, designated by the letters A thru Z, are supported. Variables can be used to designate program line numbers. For example, GOSUB B*100 and GOTO A*B*C are valid expressions.

In my opinion, the 2 K-byte interpreter is extremely powerful. Because it operates easily on register and memory locations, arrays and blocks of data can be easily manipulated. (Full appreciation of the Z8-BASIC Microcomputer comes after a complete review of the operating manuals and a little experience. Documentation approximately 200 pages long is supplied with the unit; the documentation is also available separately.)

In Conclusion

It's easy to get spoiled using a large computer as a simple control device. I have heard of many inexpensive interfaces that, when attached to any computer, supposedly perform control and monitoring miracles. Frequently overlooked, however, is the fact that implementation of these interfaces often requires the software-development tools and hardware-interfacing facilities of relatively large systems. The Z8-BASIC Microcomputer, with its interpretive language, virtually eliminates the need for costly development systems with memory-consuming text editors, assemblers, and debugging programs.

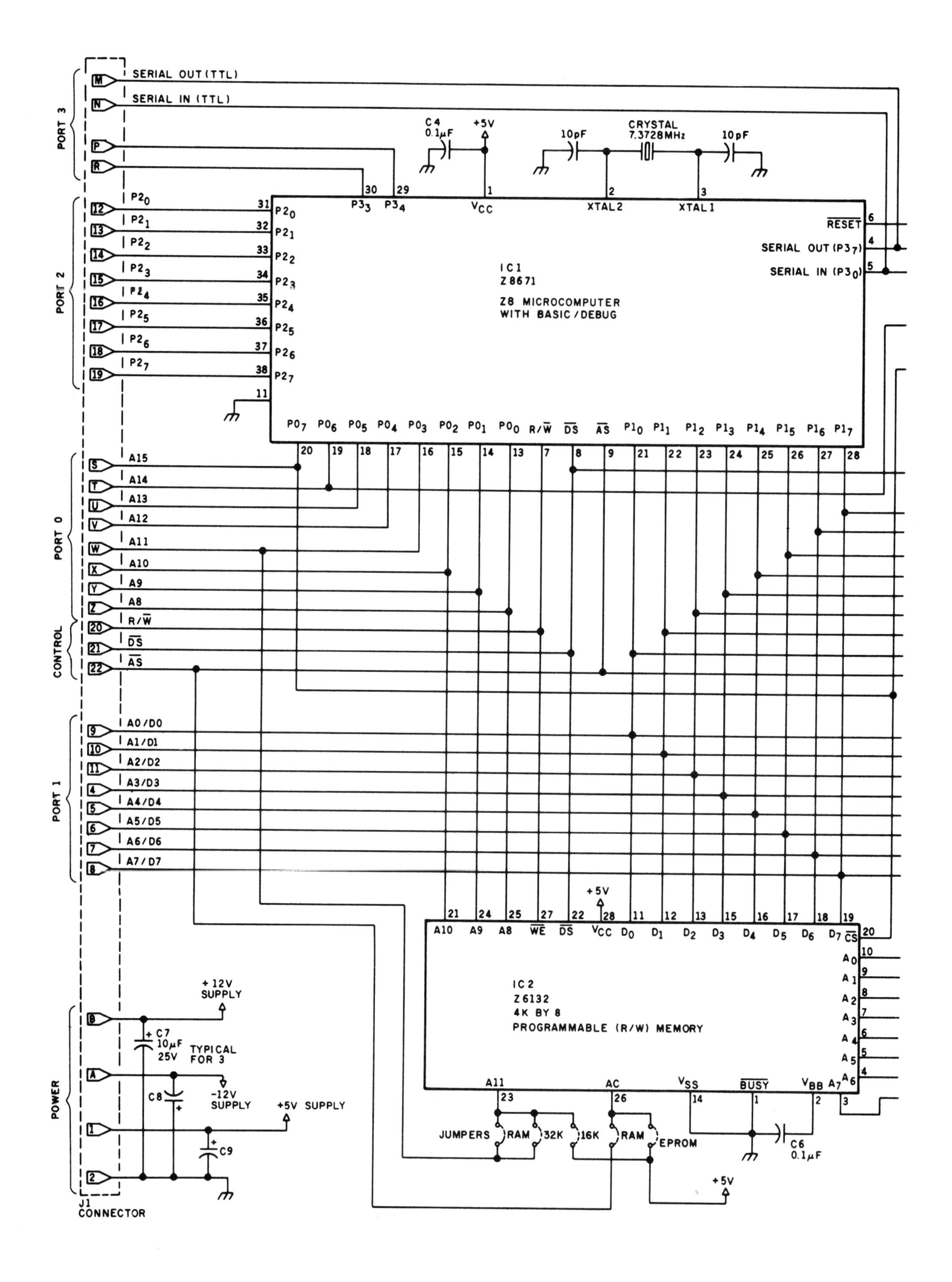

SERIAL OUT (TTL)
SERIAL IN (TTL)
PORT 3
PORT 2
PORT 0
CONTROL
PORT 1
POWER
J1 CONNECTOR
C4 0.1μF
+5V
10pF
CRYSTAL 7.3728MHz
10pF
IC1
Z8671
Z8 MICROCOMPUTER WITH BASIC/DEBUG
RESET
SERIAL OUT (P37)
SERIAL IN (P30)
XTAL2
XTAL1
VCC
A15
A14
A13
A12
A11
A10
A9
A8
R/W
DS
AS
A0/D0
A1/D1
A2/D2
A3/D3
A4/D4
A5/D5
A6/D6
A7/D7
IC2
Z6132
4K BY 8
PROGRAMMABLE (R/W) MEMORY
+12V SUPPLY
C7 10μF 25V
TYPICAL FOR 3
C8
-12V SUPPLY
+5V SUPPLY
C9
JUMPERS
RAM
32K
16K
RAM
EPROM
C6 0.1μF
A11
AC
VSS
BUSY
VBB

If you need a proportional motor-speed control for your solar-heating system, you don't have to dedicate your Apple II or shut off your heating system when you balance your checkbook. From now on, there is a small, cost-effective microcomputer specifically designed for such applications. The Z8 board described in this article is not my idea of what *should be* available; it *is* available now.

Acknowledgment

Special thanks to Steve Walters and Peter Brown of Zilog Inc for help in production of this article.

See page 447 for a listing of parts and products available from Circuit Cellar Inc.

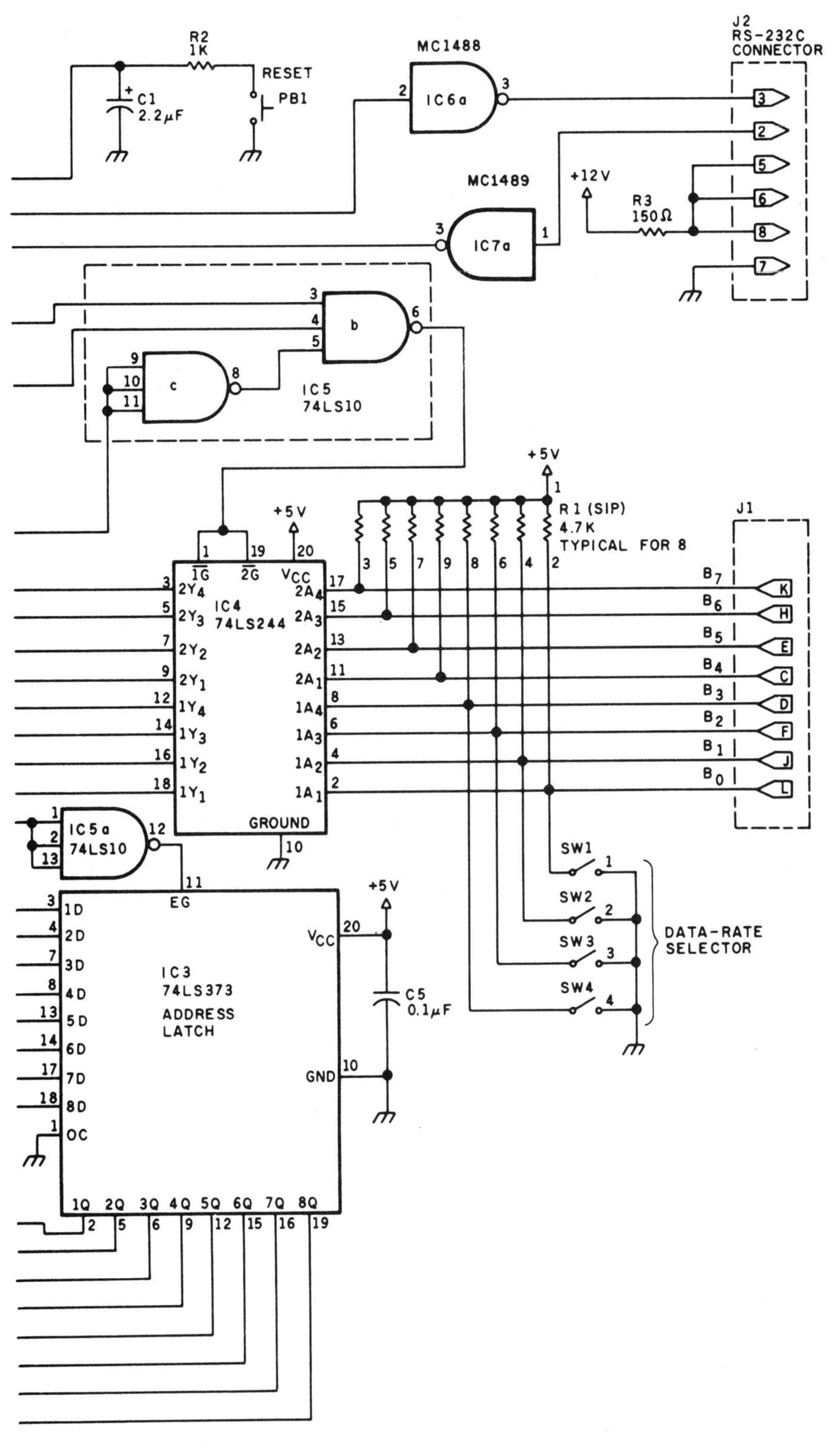

Figure 5: *Schematic diagram of the Circuit Cellar Z8-BASIC Microcomputer. Five jumper connections are provided so different memory devices can be used. For general-purpose use and program development, the 4 K-byte Z6132 read/write memory device will be used; for dedicated applications, two kinds of EPROMs can be substituted in the same integrated-circuit socket. Standard 450 ns type-2716 or type-2732 EPROM chips can be used. The connection labeled "32 K" should be closed if a type-2732 EPROM is installed; the connection labeled "16 K" should be closed for use of a type-2716 EPROM.*

The pull-up resistors adjacent to IC4 (the 74LS244 buffer) are contained in a SIP (single-inline package).

Number	Type	+5 V	GND	−12 V	+12 V
IC1	Z8671	1	11		
IC2	Z6132	28	14		
IC3	74LS373	20	10		
IC4	74LS244	20	10		
IC5	74LS10	14	7		
IC6	MC1488		7	14	1
IC7	MC1489	14	7		

11

Build a Z8-Based Control Computer with BASIC, Part 2

The Z8-BASIC Microcomputer system described in this two-part article is unlike any computer presently available for dedicated control applications. Based on a single-chip Zilog Z8 microcomputer with an onboard tiny-BASIC interpreter, this unit offers an extraordinary amount of power in a very small package. It is no longer necessary to use expensive program-development systems. Computer control can now be applied to many areas where it was not previously cost-effective.

The Z8-BASIC Microcomputer is intended for use as an intelligent controller, easy to program and inexpensive enough to dedicate to specific control tasks. It can also serve as a low-cost tiny-BASIC computer for general interest. Technical specifications for the unit are shown in the "At a Glance" box.

Last month I described the design of the Z8-BASIC Microcomputer hardware and the architectures of the Z8671 microcomputer component and Z6132 32 K-bit Quasi-Static Memory. This month I'd like to continue the description of the tiny-BASIC interpreter, discuss how the BASIC program is stored in memory, and demonstrate a few simple applications.

Process-Control BASIC

The BASIC interpreter contained in ROM (read-only memory) within the Z8671 is officially called the Zilog BASIC/Debug monitor. It is essentially a 2 K-byte integer BASIC which has been optimized for speed and flexibility in process-control applications.

There are 15 keywords: GOTO, GO@, USR, GOSUB, IF...THEN, INPUT, IN, LET, LIST, NEW, REM, RUN, RETURN, STOP, PRINT (and PRINT HEX). Twenty-six numeric variables (A through Z) are supported; and numbers can be expressed in either decimal or hexadecimal format. BASIC/Debug can directly address the Z8's internal registers and all external memory.

Photo 1: *Z8-BASIC Microcomputer. With the two "RAM" jumpers installed, it is configured to operate programs residing in the Z6132 Quasi-Static Memory. A four-position DIP (dual-inline pin) switch (at upper right) sets the serial data rate for communication with a user terminal connected to the DB-25S RS-232C connector on the top center. The reset button is on the top left.*

Byte references, which use the "@" character followed by an address, may be used to modify a single register in the processor, an I/O port, or a memory location. For example, @4096 specifies decimal memory location 4096, and @%F6 specifies the port-2 mode-control register at decimal location 246. (The percent symbol indicates that the characters following it are to be interpreted as a hexadecimal numeral.) To place the value 45 in memory location 4096, the command is simply, @4096=45 (or @%1000=%2D).

Command abbreviations are standard with most tiny-BASIC interpreters, but this interpreter allows some extremes if you want to limit program space. For example:

```
IF 1>X THEN GOTO 1000
        can be abbreviated
IF 1>X 1000

PRINT"THE VALUE IS ";S

        can be abbreviated
"THE VALUE IS ";S
```

```
IF X=Y THEN IF Y=Z
THEN PRINT "X=Z"
        can be abbreviated
IF X=Y IF Y=Z "X=Z"
```

One important difference between most versions of BASIC and Zilog's BASIC/Debug is that the latter allows variables to contain statement numbers for branching, and variable storage is not cleared before a program is run. Statements such as GOSUB X or GOTO A*E−Z are valid. It is also possible to pass values from one program to another. These variations serve to extend the capabilities of BASIC/Debug.

In my opinion, the main feature that separates this BASIC from others is the extent of documentation supplied with the Z8671. Frequently, a computer user will ask me how he can obtain the source-code listing for the BASIC interpreter he is using. Most often, I have to reply that it is not available. Software manufacturers that have invested many man-years

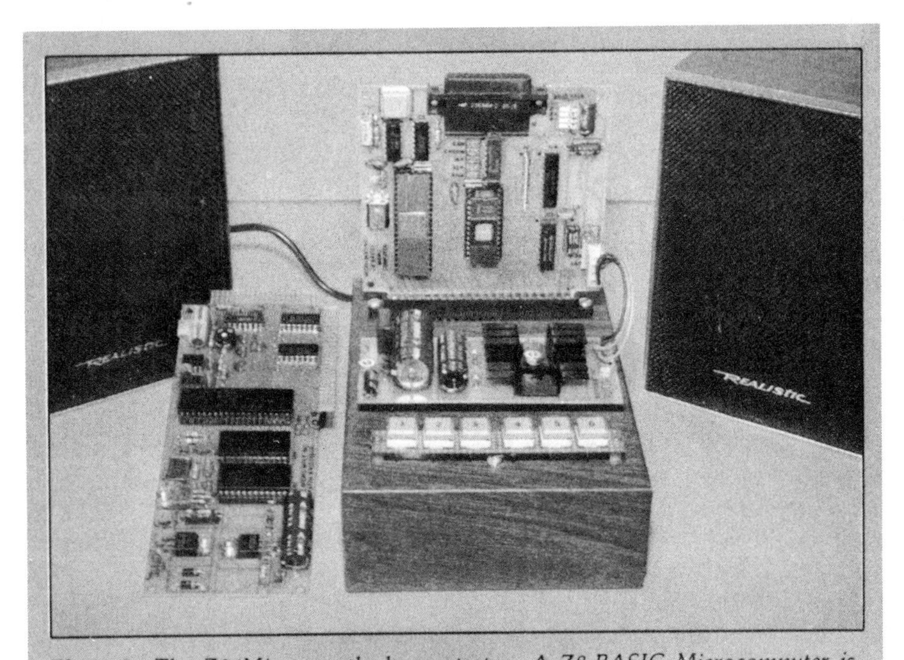

Photo 2: *The Z8/Micromouth demonstrator. A Z8-BASIC Microcomputer is configured to run a ROM-resident program that exercises the Micromouth speech synthesizer presented in the June Circuit Cellar article. A Micromouth board similar to that shown on the left is mounted inside the enclosure. Six pushbutton switches, connected to a parallel input port on the Z8 board, select various speech-demonstration sequences. The Micromouth board is driven from a second parallel port on the Z8 board.*

At a Glance

Name
Z8-BASIC Microcomputer

Processor
Zilog Z8-family Z8671 8-bit microcomputer with programmable (read/write) memory, read-only memory, and I/O in a single package. The Z8671 includes a 2 K-byte tiny-BASIC/Debug resident interpreter in ROM, 144 internal 8-bit registers, and 32 I/O lines. System uses 7.3728 MHz crystal to establish clock rate. Two internal and four external interrupts.

Memory
Uses Z6132 4 K-byte Quasi-Static Memory (pin-compatible with 2716 and 2732 EPROMs); 2 K-byte ROM in Z8671. Memory externally expandable to 62 K bytes of program memory and 62 K bytes of data memory.

Input/Output
Serial port: RS-232C-compatible and switch-selectable to 110, 150, 300, 1200, 2400, 4800, and 9600 bps.
Parallel I/O: two parallel ports; one dedicated to input, the other bit-programmable as input or output; programmable interrupt and handshaking lines; LSTTL-compatible.
External I/O: 16-bit address and 8-bit bidirectional data bus brought out to expansion connector.

BASIC Keywords
GOTO, GO@, USR, GOSUB, IF...THEN, INPUT, LET, LIST, NEW, REM, RETURN, RUN, STOP, IN, PRINT, PRINT HEX. Integer arithmetic/logic operators: +, −, /, *, and AND; BASIC can call machine-language subroutines for increased execution speed; allows complete memory and register interrogation and modification.

Power-Supply Requirements
+5 V ±5% at 250 mA
+12 V ±10% at 30 mA
−12 V ±10% at 30 mA
(The 12 V supplies are required only for RS-232C operation.)

Dimensions and Connections
4- by 4½-inch board; dual 22-pin (0.156-inch) edge connector. 25-pin RS-232C female D-subminiature (DB-25S) connector; 4-pole DIP-switch data-rate selector.

Operating Conditions
Temperature: 0 to 50°C (32 to 122°F)
Humidity: 10 to 90% relative humidity (noncondensing)

in a BASIC interpreter are not easily persuaded to give away its secrets.

In most cases, however, a user merely wants to know the location of the GOSUB...RETURN address stack or the format and location of stored program variables. While the source code for BASIC/Debug is also not available (because the object code is mask-programmed into the ROM, you couldn't change it anyway), the locations of all variables, pointers, stacks, etc, are fixed, and their storage formats are defined and described in detail. The 60-page BASIC/Debug user's manual contains this information and is included in the 200 pages of documentation supplied with the Z8-BASIC Microcomputer board. (The documentation is also available separately.)

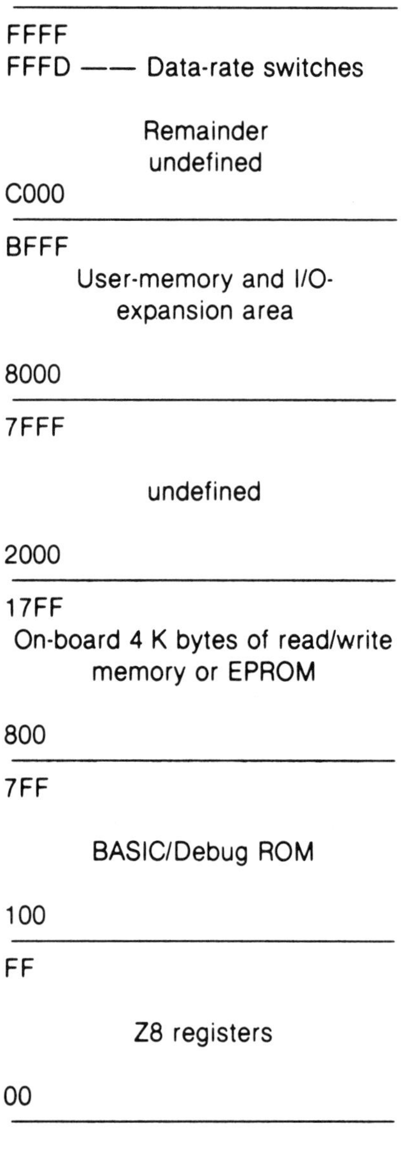

FFFF
FFFD —— Data-rate switches
Remainder undefined
C000

BFFF
User-memory and I/O-expansion area
8000

7FFF
undefined
2000

17FF
On-board 4 K bytes of read/write memory or EPROM
800

7FF
BASIC/Debug ROM
100

FF
Z8 registers
00

Figure 1: *A simplified hexadecimal memory map of the Z8-BASIC Microcomputer.*

Memory Allocation

Z8-family microcomputers distinguish between four kinds of memory: internal registers, internal ROM, external ROM, and external read/write memory. (A slightly different distinction can also be made between program memory and data memory, but in this project this distinction is unnecessary.) The register file resides in memory-address space in hexadecimal locations 0 through FF (decimal 0 through 255). The 144 registers include four I/O- (input/output) port registers, 124 general-purpose registers, and 16 status and control registers. (No registers are implemented in hexadecimal addresses 80 through EF [decimal addresses 128 through 239]).

The 2 K-byte ROM on the Z8671 chip contains the BASIC/Debug interpreter, residing in address space from address 0 to hexadecimal 7FF (decimal 0 to 2047). External memory starts at hexadecimal address 800 (decimal 2048). A memory map of the Z8-BASIC Microcomputer system is shown in figure 1.

When the system is first turned on, BASIC/Debug determines how much external read/write memory is available, initializes memory pointers, and checks for the existence of an auto-start-up program. In a system with external read/write memory, the top page is used for the line buffer, program-variable storage, and the GOSUB...RETURN address stack. Program execution begins at hexadecimal location 800 (decimal 2048).

When BASIC/Debug finds no external read/write memory, the internal registers are used to store the variables, line buffer, and GOSUB...RETURN stack. This limits the depth of the stack and the number of variables that can be used simultaneously, but the restriction is not too severe in most control applications. In a system without external memory, automatic program execution begins at hexadecimal location 1020 (decimal 4128).

In a system that uses an external 2 K-byte EPROM (type 2716), wraparound addressing occurs, because the state of the twelfth address line on the address bus (A11) is ignored. (A 4 K-byte type-2732 EPROM device does use A11.) A 2716 EPROM device inserted in the Z6132's memory socket will read from the same memory cells in response to accesses to both logical hexadecimal addresses 800 and 1000. Similarly, hexadecimal addresses 820 and 1020 will be treated as equivalent by the 2716 EPROM. Therefore, when a 2 K-byte 2716 EPROM is being used, the auto-start address, normally operating at hexadecimal 1020, will begin execution of any program beginning at hexadecimal location 820. For the purposes of this discussion, you may assume that programs stored in EPROM use type-2716 devices and that references to hexadecimal address 820 also apply to hexadecimal address 1020.

Program Storage

The program-storage format for BASIC/Debug programs is the same in both types of memory. Each BASIC statement begins with a line number and ends with a delimiter. If you were to connect a video terminal or teletypewriter to the RS-232C serial port and type the following line:

```
100 PRINT "TEST"
```

it would be stored in memory beginning at hexadecimal location 800 as shown in listing 1.

The first 2 bytes of any BASIC statement contain the binary equivalent of the line number (100 decimal equals 64 hexadecimal). Next are bytes containing the ASCII (American Standard Code for Information Interchange) values of characters in the statement, followed by a delimiter byte (containing 00) which indicates the end of the line. The last statement in the program (in this case the only one) is followed by 2 bytes containing the hexadecimal value FFFF, which designates line number 65535.

The multiple-line program in listing 2 further illustrates this storage format.

Listing 1: *Simple illustration of BASIC program storage in the Z8-BASIC Microcomputer.*

```
        100      P    R    I    N    T         "    T
800   00   64   50   52   49   4E   54   20   22   54
      E    S    T    "
80A   45   53   54   22   00   FF   FF
```

Listing 2: *A multiple-line illustration of BASIC program storage.*

```
        100 A=5
        200 B=6
        3005 "A*B=";A*B

        100      A    =    5         200       B    =
800   00   64   41   3D   35   00   00   C8   42   3D
      6            3005    "    A    *    B    =    "
80A   36   00   0B   BD   22   41   2A   42   3D   22
      ;    A    *    B
814   3B   41   2A   42   00   FF   FF
```

One final example of this is illustrated in listing 3 on page 160. Here is a program written to examine itself. Essentially, it is a memory-dump routine which lists the contents of memory in hexadecimal. As shown, the 15-line program takes 355 bytes and occupies hexadecimal locations 800 through 963 (decimal 2048 through 2499). I have dumped the first and last lines of the program to further demonstrate the storage technique.

I have a reason for explaining the internal program format. One of the useful features of this computer is its ability to function with programs residing solely in EPROM. However, the EPROMs must be programmed externally. While I will explain how to serially transmit the contents of the program memory to an EPROM programmer, some of you may have only a manual EPROM programmer or one with no communication facility. But if you are willing to spend the time, it is easy to print out the contents of memory and manually load the program into an EPROM device.

Dedicated-Controller Use

The Z8-BASIC Microcomputer can be easily set up for use in intelligent control applications. After being tested and debugged using a terminal, the control program can be written into an EPROM. When power is applied to the microcomputer, execution of the program will begin automatically.

The first application I had for the unit was as a demonstration driver for the Micromouth speech-processor board I presented two months ago in the June issue of BYTE. (See "Build a Low-Cost Speech-Synthesizer Interface," June 1981 BYTE, for a description of this project, which uses National Semiconductor's Digitalker chip set.) It's hard to discuss a synthesized-speech interface without demonstrating it, and I didn't want to carry around my big computer system to control the Micromouth board during the demonstration. Instead, I quickly programmed a Z8-BASIC Microcomputer to perform that task. While I was at it, I set it up to demonstrate itself as well.

The result (see photo 2 on page 111) has three basic functional components. On top of the box is a Z8-BASIC Microcomputer (hereinafter called the "Z8 board") with a 2716 EPROM installed in the memory integrated-circuit socket, the Z8-board power supply (the wall-plug transformer module is out of view), and six pushbutton switches. Inside the box is a prototype version of the Micromouth speech-processor board (a final-version Micromouth board is shown on the left).

The Micromouth board is jumper-programmed for parallel-port operation (8 parallel bits of data and a data-ready strobe signal) and connected to I/O port 2 on the Z8 board. The Micromouth BUSY line and the six pushbuttons are attached to 7 input bits of the Z8 board's input port mapped into memory-address space at hexadecimal address FFFD (decimal 65533).

The most significant 3 bits of port FFFD are normally reserved for the data-rate-selector switches, but with no serial communication required, the data rate is immaterial and the switches are left in the open position. This makes the 8 bits of port FFFD, which are brought out to the edge connector, available for external inputs. In this case, pressing one of the six pushbuttons selects one of six canned speech sequences.

Coherent sentences are created by properly timing the transmission of word codes to the speech-processor board. This requires nothing more than a single handshaking arrangement and a table-lookup routine (but try it without a computer sometime). The program is shown in listing 4a.

The first thing to do is to configure the port-2 and port-3 mode-control registers (hexadecimal F6 and F7, or decimal 246 and 247). Port 2 is bit-programmable. For instance, to configure it for 4 bits input and 4 bits output, you would load F0 into register F6 (246). In this case, I wanted it configured as 8 output bits, so I typed in the BASIC/Debug command @246=0 (set decimal location 246 to 0).

The data-ready strobe is produced using one of the options on the Z8's port 3. A Z8 microcomputer has data-available and input-ready handshaking on each of its 4 ports. To set the proper handshaking protocol and use port 2 as I have described, a code of hexadecimal 71 (decimal 113) is placed into the port-3 mode-control register. The BASIC/Debug command is @247= 113. The RDY2 and $\overline{\text{DAV2}}$ lines on the Z8671 are connected together to produce the data-available strobe signal.

Lines 1000 through 1030 in listing 4a have nothing to do with demonstrating the Micromouth board. They form a memory-dump routine that illustrates how the program is stored in memory. You notice from the memory dump of listing 4b that the first byte of the program, as stored in the

ROM, begins at hexadecimal location 820 (actually at 1020, you remember) rather than 800 as usual. This is to help automatic start-up. The program could actually begin anyplace, but you would have to change the program-pointer registers (registers 8 and 9) to reflect the new address. The 32 bytes between 800 and 820 are reserved for vectored addresses to optional user-supplied I/O drivers and interrupt routines.

Programming the EPROM

The first EPROM-based program I ran on the Z8-BASIC Microcomputer was manually loaded. I simply printed out the contents of the Z6132 memory using the program of listing 3 and entered the values by hand into the EPROM programmer. This is fine once or twice, but you certainly wouldn't want to make a habit of it. Fortunately, there are better alternatives if you have the equipment.

Listing 3: *A program (listing 3a) that examines itself by dumping the contents of memory in printed hexadecimal form. Listing 3b shows the first and last lines of the program as dumped during execution.*

(3a)

```
100 PRINT"ENTER START ADDRESS FOR HEX DUMP ";:INPUT X
102 PRINT"THE LIST IS HOW MANY BYTES LONG ";:INPUT C
103 PRINT:PRINT
105 B=X+8 :A=X+C
107 PRINT"ADDRESS                    DATA":PRINT
110 PRINT HEX (X);"        ";
120 GOSUB 300
130 X=X+1
140 IF X=B THEN GOTO 180
150 GOTO 120
180 IF X>=A THEN 250
200 PRINT:PRINT:B=X+8:GOTO 110
250 PRINT:STOP
300 PRINT HEX (@X);: PRINT"  ";
310 RETURN
:
```

(3b)

```
:RUN
ENTER START ADDRESS FOR HEX DUMP ? 2048
THE LIST IS HOW MANY BYTES LONG ? 30
```

ADDRESS	DATA							
	100		P	R	I	N	T.	"
800	0	64	50	52	49	4E	54	22
	E	N	T	E	R	sp	S	T
808	45	4E	54	45	52	20	53	54
	A	R	T	sp	A	D	D	R
810	41	52	54	20	41	44	44	52
	E	S	S	sp	F	O	R	sp
818	45	53	53	20	46	4F	52	20
:								
:								

```
:RUN
ENTER START ADDRESS FOR HEX DUMP ? 2360
THE LIST IS HOW MANY BYTES LONG ? 45
```

ADDRESS	DATA							
	O	P		300		P	R	I
938	4F	50	0	1	2C	50	52	49
	N	T	sp	H	E	X	sp	(
940	4E	54	20	48	45	58	20	28
	@	X	)	;	:	sp	P	R
948	40	58	29	3B	3A	20	50	52
	I	N	T	"	sp	sp	"	;
950	49	4E	54	22	20	20	22	3B
		310		R	E	T	U	R
958	0	1	36	52	45	54	55	52
	N		65535					
960	4E	0	FF	FF	0	0	0	0

Many EPROM programmers are peripheral devices on larger computer systems. In such cases, it is possible to take advantage of the systems' capabilities by downloading the Z8 program directly to the programmer.

The programmer shown in photo 3 is a revised version of the unit I described in a previous article, "Program Your Next EROM in BASIC" (March 1978 BYTE, page 84). It was designed for type-2708 EPROMs, but I have since modified it to program 2716s instead. All I had to do was lengthen the programming pulse to 50 ms and redefine the connections to four pins on the EPROM socket. It still is controlled by a BASIC program and takes less than 2½ minutes to program a type-2716 EPROM device. Refer to the original article for the basic design.

Normally, the LIST function or memory-dump routine cannot be used to transmit data to the EPROM programmer because the listing is filled with extraneous spaces and carriage returns. It is necessary to write a program that transmits the contents of memory without the extra characters required for display formatting. The only data received by the EPROM programmer should be the object code to load into the EPROM.

In writing this program we can take advantage of the Z8's capability of executing machine-language programs directly through the USR and GO@ commands. The serial-input and serial-output subroutines in the BASIC/Debug ROM can be executed independently using these commands. The serial-input driver starts at hexadecimal location 54, and the serial-output driver starts at hexadecimal location 61. Transmitting a single character is simply done by the BASIC statement

Listing 4: *A program (listing 4a) that demonstrates the functions of the Micromouth speech synthesizer, operating from a type-2716 EPROM. The simple I/O-address decoding of the Z8 board allows use of the round-figure address of 65000. The program uses a table of vocabulary pointers that has been previously stored in the EPROM by hand. Listing 4b shows a dump of the memory region occupied by the program, proving that storage of the BASIC source code starts at hexadecimal location 820.*

(4a)

```
100 @246=0:@247=113
110 X=@65000 :A=%1400
120 IF X=254 THEN @2=0
130 IF X=253 THEN GOTO 500
140 IF X=251 THEN A=A+32 :GOTO 500
150 IF X=247 THEN A=A+64 :GOTO 500
160 IF X=239 THEN A=A+96 :GOTO 500
170 IF X=223 THEN A=A+128 :GOTO 500
180 IF X=222 THEN N=0 :GOTO  300
200 GOTO 110
300 @2=N :N=N+1 :IF N=143 THEN 110
310 IF @65000<129 THEN 310
320 GOTO 300
500 @2=@A :A=A+1
510 IF @65000<129 THEN 510
520 IF @A=255 THEN GOTO 110
530 GOTO 500
1000 Q=2048
1005 W=0
1010 PRINT HEX(@Q),:Q=Q+1
1015 W=W+1 :IF W=8 THEN PRINT" ":GOTO 1005
1020 IF Q=4095 THEN STOP
1030 GOTO 1010
:
```

(4b)

```
:goto 1000
FF      FF      FF      FF      FF      FF      FF      FF
FF      FF      FF      FF      FF      FF      FF      FF
FF      FF      FF      FF      FF      FF      FF      FF
FF      FF      FF      FF      FF      FF      FF      FF
0       64      40      32      34      36      3D      30
3A      40      32      34      37      3D      31      31
33      0       0       6E      58      3D      40      36
35      30      30      30      20      3A      41      3D
25      31      34      30      30      0       0       78
49      46      20      58      3D      32      35      34
20      54      48      45      4E      20
0! AT 1015
:
```

GO@ %61,C

where C contains the value to be transmitted. A serial character can be received by

C=USR (%54)

where the variable C returns the value of the received data.

To dump the entire contents of the Z6132 memory to the programmer, the statements in listing 5 should be included at the end of your program.

Execution begins when you type GOTO 1000 as an immediate-mode command and ends when all 4 K bytes have been dumped. The transmission rate (110 to 9600 bps) is that selected on the data-rate-selector switches.

Conceivably, this technique could also be used to create a cassette-storage capability for the Z8 board. In theory, a 3- or 4-line BASIC program can be entered in high memory (you can set the pointer to put the program there) to read in serial data and load it in lower memory. Changing the program pointer back to hexadecimal 800 allows the newly loaded program to be executed. Since the Z8-BASIC Microcomputer already has a serial I/O port, any FSK (frequency-shift keyed) modem and cassette-tape recorder can be used for cassette data storage.

I/O for Data Acquisition

Data acquisition for process control is the most likely application for the Z8-BASIC Microcomputer. Low-

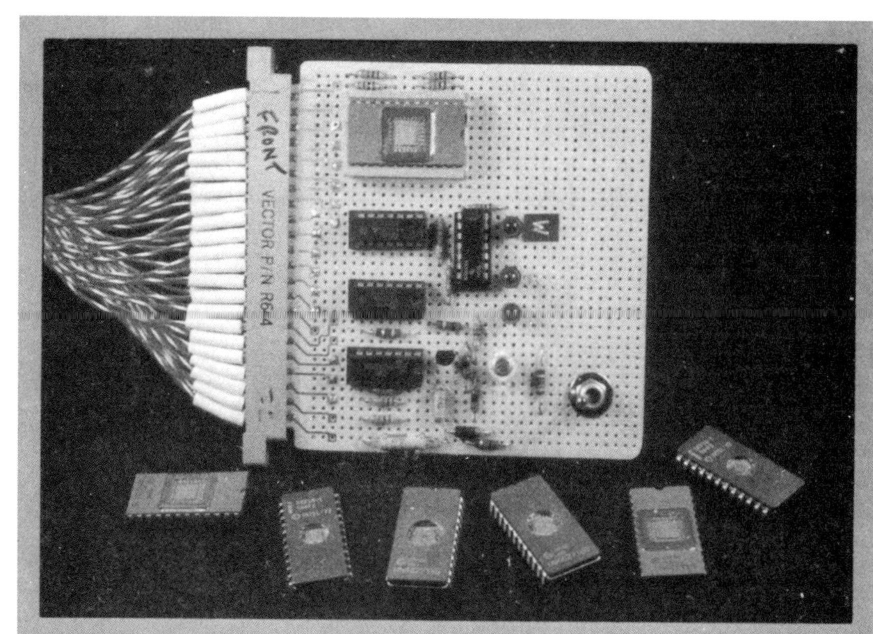

Photo 3: *Type-2716 EPROM programmer, adapted from "Program Your Next EROM in BASIC" (March 1978 BYTE, page 84). The circuit, which is driven through parallel ports, programs a 2716 in about 2½ minutes and is controlled by a BASIC program.*

Listing 5: *BASIC statements that print out the entire contents of the 4 K bytes of user memory, for use with a communicating EPROM programmer.*

```
1000 X=%800 :REM BEGINNING OF
     USER MEMORY
1010 GO@ %61,@X :REM TRANSMIT
     CONTENTS OF LOCATION X
1020 X=X+1 :IF X=%1801 THEN
     STOP
1030 GOTO 1010
```

Listing 6: *A simple BASIC program segment to demonstrate the concept of the "black box" method of modifying data being transmitted through the Z8-BASIC Microcomputer.*

```
100 @246=0:@247=113 :REM SET PORT
    2 TO BE OUTPUT
110 @2=X :REM X EQUALS THE DATA
    TO BE TRANSMITTED
```

cost distributed control is practical, substituting for central control performed by a large computer system. Analog and digital sensors can be read by a Z8-BASIC Microcomputer, which then can digest the data and reduce the amount of information (experiment results or control parameters) stored or transmitted to a central point. Control decisions can be made by the Z8-BASIC Microcomputer at the process locality.

The Z8 board can be used for analog data acquisition, perhaps using an A/D (analog-to-digital) converter such as that shown in figure 2. This 8-bit, eight-channel A/D converter has a unipolar input range of 0 to +5 V (although the A/D integrated circuit can be wired for bipolar operation), with the eight output channels addressed as I/O ports mapped into memory-address space at hexadecimal addresses BF00 through BF07 (decimal 48896 through 48903). When the Z8671 performs an output operation to the channel address, the channel is initialized for acquiring data, while data is read from the channel when the Z8671 performs

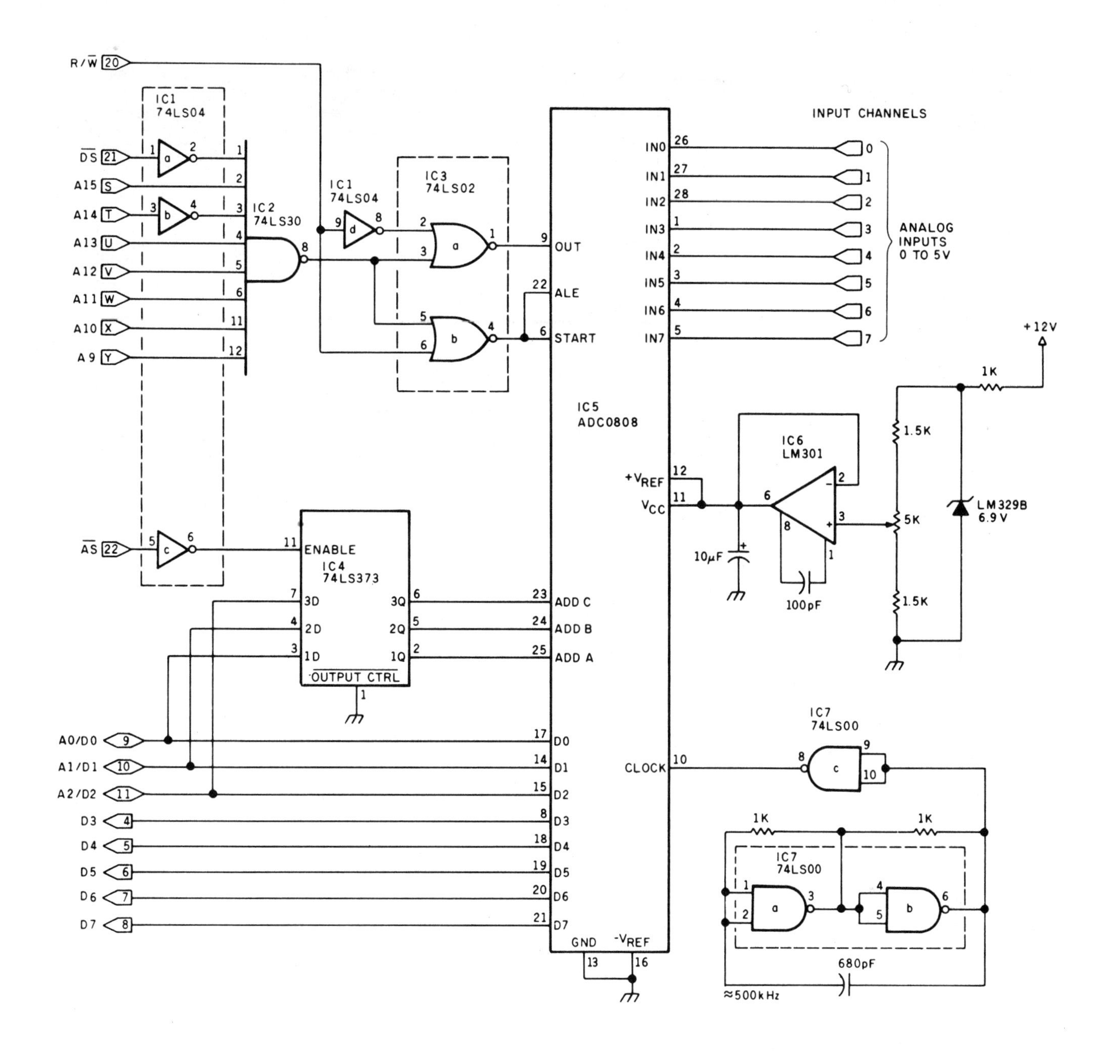

Number	Type	+5V	GND	+12V
IC1	74LS04	14	7	
IC2	74LS30	14	7	
IC3	74LS02	14	7	
IC4	74LS373	20	10	
IC5	ADC0808	see schematic diagram		
IC6	LM301		4	7
IC7	74LS00	14	7	

Figure 2: *Schematic diagram of an A/D converter. This 8-bit, eight-channel unit has a unipolar input range of 0 to +5 V, with the eight output channels addressed as I/O ports mapped into memory-address space at hexadecimal addresses BF00 through BF07.*

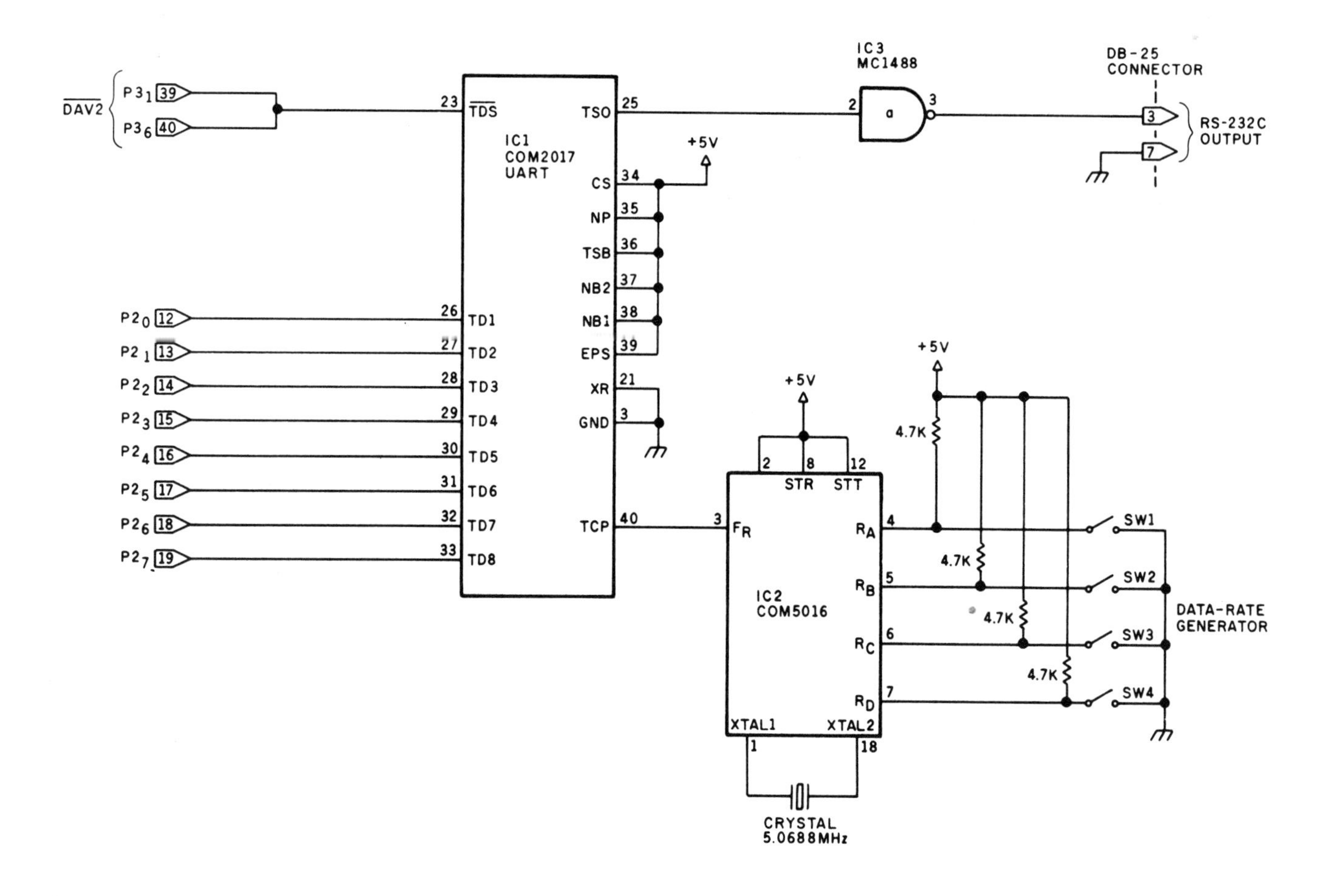

Number	Type	+5V	GND	−12V	+12V
IC1	COM2017	1	3	2	
IC2	COM5016	2	11		9
IC3	COM1488		7	1	14

Figure 3: *Schematic diagram of an RS-232C serial output port for the "black box" communication application of the Z8-BASIC Microcomputer. The Z8671 must be configured by software to provide the proper signals: one such signal, $\overline{DAV2}$, is derived from two bits of I/O port 3 on the Z8671. The pin numbers shown in the schematic diagram for $P3_1$ and $P3_6$ are pins on the Z8671 device itself, not pins or sections on the card-edge connector, as are $P2_0$ through $P2_7$.*

an input operation on the channel's address.

Intelligent Communication

Another possible use for the Z8-BASIC Microcomputer is as an intelligent "black box" for performing predetermined modification on data being transmitted over a serial communication line. The black box has two DB-25 RS-232C connectors, one for receiving data and the other for retransmitting it. The intelligence of the Z8-BASIC Microcomputer, acting as the black box, can perform practically any type of filtering, condensing, or translating of the data going through.

Perhaps you have an application where continuous raw data is transmitted, but you would rather just keep a running average or flag deviations from preset limits at the central monitoring point rather than contend with everything. The Z8 board can be programmed to digest all the raw data coming down the line and pass

See page 447 for a listing of parts and products available from Circuit Cellar Inc.

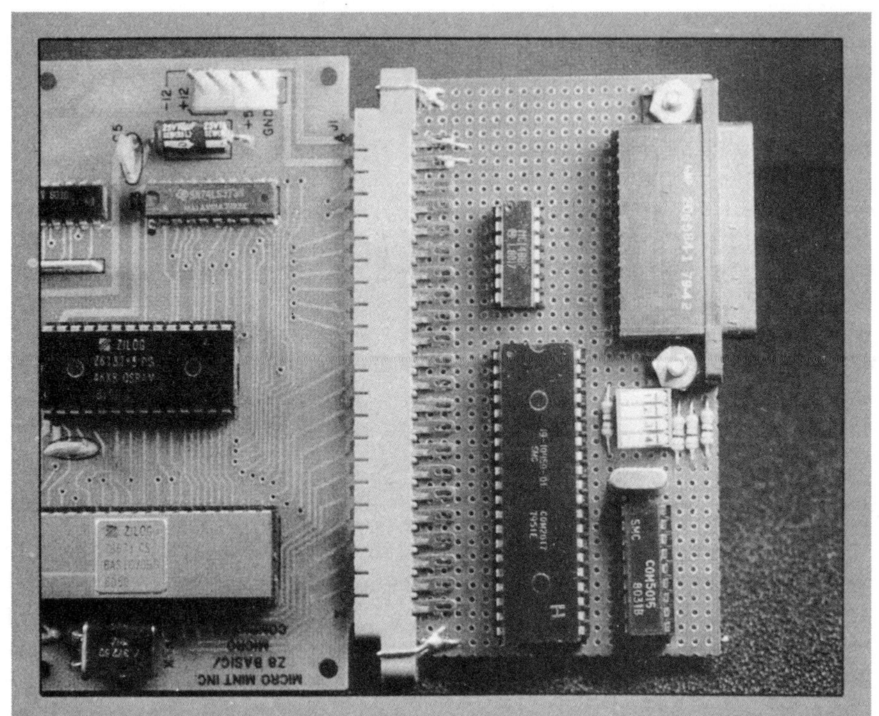

Photo 4: *A three-integrated-circuit hardwired serial output port for the Z8-BASIC Microcomputer. Connected to port 2, any program data sent to register 2 will be transmitted serially at the data rate selected on the four-position DIP switch (between 50 to 19200 bps). The Z8 board, configured with two serial ports, is used to process raw data moving through it. Data is received on one side, digested, and retransmitted in some more meaningful form from the other port. Such a configuration could also be used to connect two peripheral devices that have radically different data rates.*

Photo 5: *When the Z8-BASIC Microcomputer is used with a ROM-resident program, the two jumpers used with the Z6132 are removed, and the EPROM jumper is installed instead. When using a type-2716 16 K-bit (2 K-byte) EPROM device, the "16 K" jumper is installed. If a type-2732 32 K-bit (4 K-byte) EPROM is used instead, the "32 K" jumper is installed. The EPROM is inserted in the lower 24 pins of the 28-pin Z6132 socket (IC2) as shown.*

on only what's pertinent.

Another such black-box application is to use the Z8 board as a printer buffer. Photo 4 shows the interface hardware of one specific application, which I used to attach a high-speed computer to a very slow printer. The host computer transmitted data to the Z8 board at 4800 bps. Since the receiving serial port used had to be bidirectional to handshake with the host computer, I added another serial output to the Z8 board for transmitting characters to the printer. Only three integrated circuits were required to add a serial output port. A schematic diagram is shown in figure 3 on page 118. The UART (universal asynchronous receiver/transmitter, shown as IC1) is driven directly from port 2 on the Z8 board (port 2 could also be used to directly drive a parallel-interface printer), and IC2 supplies the clock signal for the desired data rate. Of course, the UART could have been attached to the data and address buses directly, but this was easier.

Transmitting a character out of this serial port requires setting the port-2 and port-3 mode-control registers as before. After that, any character sent to port 2 will be serially transmitted. The minimum program to perform this is shown in listing 6 on page 116. This circuit can also be used for downloading programs to the EPROM programmer.

In Conclusion

It is impossible to describe the full potential of the Z8-BASIC Microcomputer in so few pages. For this reason, considerable effort has been taken to fully document its characteristics. I have merely tried to given an introduction here.

I intend to use the Z8-BASIC Microcomputer in future projects. I am interested in any applications you might have, so let me know about them, and we can gain experience together.

Special thanks to Steve Walters and Peter Brown of Zilog Inc for their aid in producing these articles.

12

Build an Unlimited-Vocabulary Speech Synthesizer

The alarm clock that jolts you out of sweet dreams with a monotone buzz is a thing of the past. State-of-the-art technology is the clock that prods you out of slumber with a *voice* that speaks your own language: "The time is 6 o'clock."

The artificial voice is becoming an increasingly important and potentially indispensable part of the interface between man and machine. Electronic speech synthesis is a young but rapidly evolving technology. It won't be long before that speaking alarm clock will also announce your entire day's appointment schedule. It will be no less unusual for the computer in your car to recount its mechanical ills as you drive to work. For now, however, electronic speech synthesis is still a relatively new concept.

In a previous Circuit Cellar article ("Build a Low-Cost Speech-Synthesizer Interface," June 1981 BYTE) I described the design of an inexpensive, limited-vocabulary, computer-controlled electronic speech synthesizer called the Micromouth. This speech processor, based on the National Semiconductor Digitalker chip set, was an attempt to introduce personal computer users to artificial speech. Considering the response it received, I believe many of you are now listening to everything your computer has to say.

This month I wish to return to the topic of computer-controlled electronic voice synthesis and introduce you to the Votrax SC-01 speech synthesizer chip.

Instead of waveform digitization or linear-predictive coding, the SC-01 uses phoneme synthesis, which allows the SC-01 to speak an unlimited vocabulary simply by sequentially pronouncing the individual phonemes (basic sound units) that make up words in the English language.

Many other articles have been written that describe in detail the theory of phoneme synthesis and the workings of the Votrax SC-01 integrated circuit. A few appropriate references are given at the end of this article. Instead of discussing many theoretical concepts at length in my limited space here, I prefer to concentrate on the design of a practical, computer-controlled, phonetic speech synthesizer.

This month's construction project, shown in photo 1, is called the Sweet Talker speech synthesizer. It uses the SC-01 integrated circuit to allow synthesis of an almost unlimited vocabulary (limited only by the size and complexity of the controlling program running on the computer). The Sweet Talker circuit contains I/O (input/output) signal buffering, a clock oscillator, an audio filter, and an amplifier. The circuit board provides protection from static electricity for the SC-01 in addition to being a convenient package. (Protection for the SC-01 is important, because it is both expensive—$70—and delicate. It's made using complementary metal-oxide semiconductor technology.) The standard Sweet Talker synthesizer can be connected to any microcomputer through a parallel port, while a special version of the unit can be plugged into the I/O bus of the Apple II computer.

Votrax is a trademark of Federal Screw Works, Inc.

Digitalker is a trademark of National Semiconductor Corporation.

Special thanks to Ray Long of Custom Photo and to Greg Peterson and Phil Walton of Tech Circuits for their expert printed-circuit design and production talents.

Speech-Synthesis Review

Three major techniques are presently used to synthesize the human voice: formant synthesis, linear-predictive coding (LPC), and waveform digitization. They differ in the number of bits of data required to construct a word.

Formant synthesis is essentially an electronic modeling of the natural resonances of the human vocal tract. Bands of resonant frequencies in the vocal spectrum, called formants, are generated by excitation sources and then passed through variable filters.

One variation of the formant technique is called phoneme synthesis, in

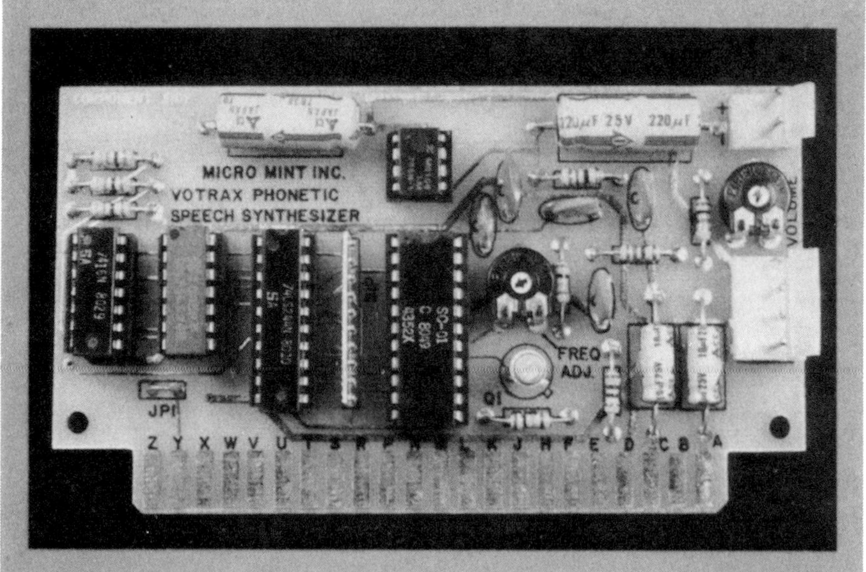

Photo 1: *The assembled Sweet Talker phonetic-speech-synthesizer circuit board. The Votrax SC-01 phoneme-synthesizer integrated circuit supports a vocabulary limited only by the size and complexity of the computer program that controls the Sweet Talker. Any English word may be constructed from phonemes, the basic building blocks of speech. The circuit board shown is a prototype of the parallel-port version of the Sweet Talker; the Apple II plug-compatible version is not shown.*

Photo 2: *The Votrax SC-01 is a 22-pin CMOS integrated circuit which functions as an electronic model of the human voice.*

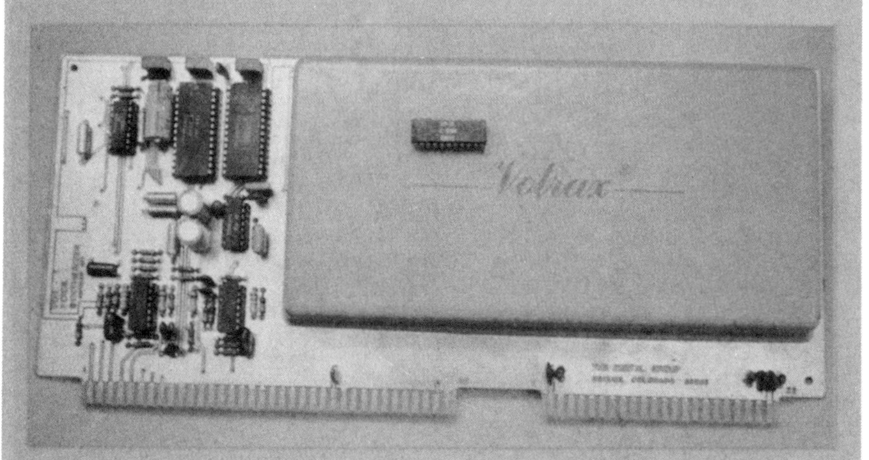

Photo 3: *Before the development of the SC-01, Votrax used many medium-scale-integrated circuits and discrete components to perform the same functions. All these components were mounted in modules such as the VSL-type shown. When this particular synthesizer card was purchased three years ago, the price was $600. Compare this both in size and price to the $70 SC-01 (sitting on top).*

which the spectral parameters are derived from basic sound units that make up words. A phoneme-generator circuit is used to reproduce these sounds. In such a circuit, each phoneme is given a numeric code, and the synthesizer circuit (discrete or integrated) utters phoneme sounds corresponding to codes it receives when it is activated. Words and sentences are assembled by simply stringing the phoneme codes together. The electronic voice so generated is intelligible, but has a slight mechanical quality. Continuous speech using phoneme synthesis can typically be generated with a data rate of less than 100 bps (bits per second).

Linear-predictive coding is similar to formant synthesis in that both techniques are based in the frequency domain and use similar hardware to model the vocal tract. The quality of speech is often better than formant or phoneme synthesis, but a higher data rate (1200 to 2400 bps) is needed for continuous speech.

Waveform digitization is the third method of speech synthesis, in which the amplitude characteristics of a vocal waveform are stored and reproduced. The quality of speech is better than the other two methods, but the data rate for continuous speech is very high, and storing sufficient amounts of data conveniently can be a problem. Various schemes of compressing the data have been devised; one of the more successful is used in the National Semiconductor Digitalker system, which I described in my June Circuit Cellar article.

Votrax SC-01

The 22-pin Votrax SC-01 integrated circuit, shown in photo 2 and in the diagrams in figure 1, contains a digital code translator, or phoneme controller, and an electronic analog of the human vocal tract. The phoneme controller translates a 6-bit phoneme code and a 2-bit pitch code into a matrix of spectral parameters which in turn adjusts the vocal-tract analog to synthesize the phonemes.

In the first part of the vocal-tract section, there are a pair of variable-frequency oscillators for simulating

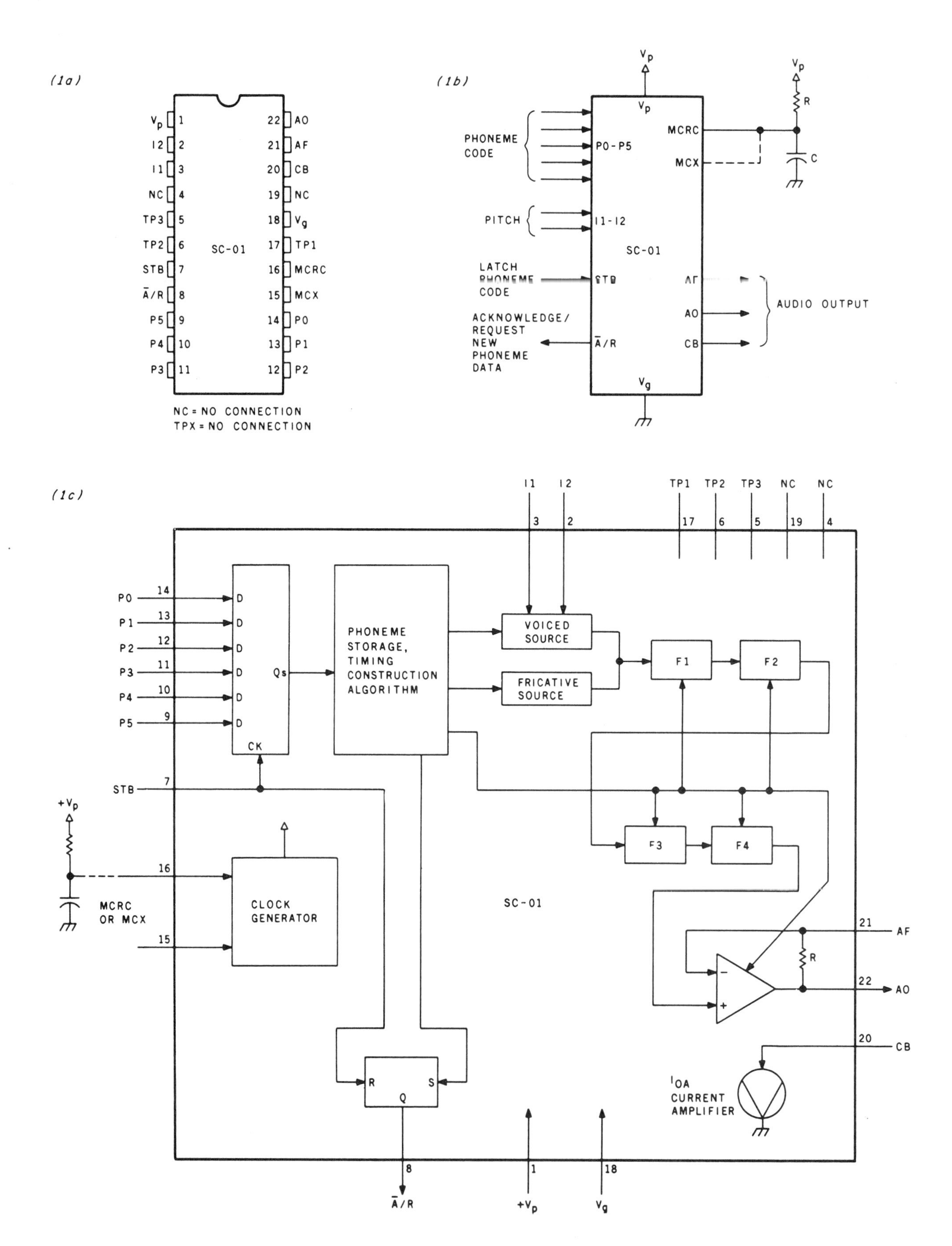

Figure 1: *Technical characteristics of the Votrax SC-01 Speech Synthesizer Chip. Shown are the pinout designations (1a), the scheme of data flow through the circuit (1b), and a block diagram of the internal structure (1c). This figure is reproduced courtesy of the Votrax Division of Federal Screw Works, Inc.*

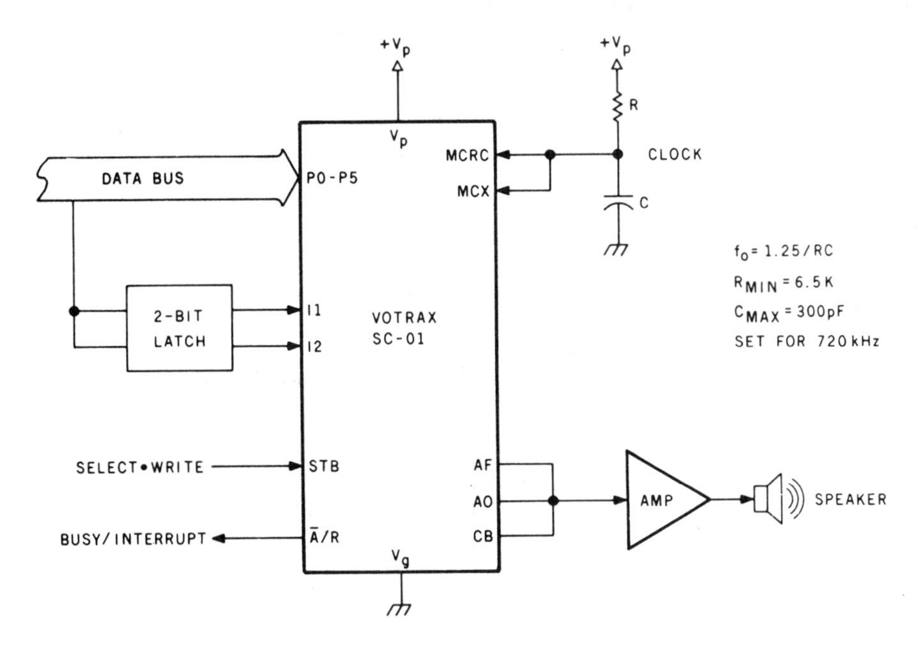

Figure 2: *A diagram of the general connection scheme of the SC-01.*

vocal-cord-produced periodic sounds and a pseudorandom (pink-noise) signal generator that simulates the sound of rushing air. The output signals from these sources are shaped by a bank of four analog band-pass filters that simulate the vocal-tract cavities. The filter outputs, in turn, are directed through a preamplifier to an external amplifier and speaker.

The SC-01 phoneme synthesizer is a CMOS (complementary metal-oxide semiconductor) integrated circuit which should be operated within the range of 7 to 14 V (V_p). The phoneme-input lines (P0 through P5) are 5 V level-compatible and self-latching. (Here "5 V level-compatible" means matching LSTTL [low-power Schottky transistor-transistor logic] levels with an *external* pull-up resistor.) The two pitch-control lines, on the other hand, must have external latches and must be switched at the same input voltage as the SC-01's power supply. Handshaking with external control circuitry is accomplished through two control lines: strobe (STB) and acknowledge/request ($\overline{A}$/R). The STB line can be either CMOS or 5 V level, while the $\overline{A}$/R line is CMOS level only.

The output pitch of the phonemes is controlled by the frequency of the clock signal, which can be applied from an external source or set internally with a resistor and capacitor combination. The clock frequency is nominally 720 kHz, but subtle variations of pitch are induced through "automatic inflection" to prevent the synthesized voice from sounding too monotonous or "robot-like." Two independent pitch-control lines, I1 and I2, are available for gross variations in pitch so that the chip can speak with more than one voice. Referred to as "manual inflection" controls, I1 and I2 operate in addition to the automatic-inflection system already present. I have found that the 6-bit phoneme code alone is sufficient, and the two pitch-control lines can be ignored. A diagram of the general connection scheme is shown in figure 2.

Listed in table 1 on page 124 are the 64 phonemes defined for the English language (two produce silent periods of different lengths; one causes synthesis to stop). A phoneme sound is generated when a 6-bit phoneme code is placed on the control-register input lines (P0 through P5) and latched by pulsing the strobe (STB) input. Each phoneme is internally timed and has a duration of 47 to 250 ms (milliseconds); some phonemes last longer than others, and variations in the clock frequency affect the phoneme durations. The $\overline{A}$/R line goes from a logic 1 to a logic 0 when a phoneme is sounding.

There are two general methods for using the SC-01. One method, shown in figure 3a on page 125, configures the chip in an independently acting, self-timed circuit which asynchronously extracts phoneme codes from a

Photo 4: *The Sweet Talker phonetic-speech-synthesizer board can be driven through any parallel output port. One port which can be used is a Centronics-compatible parallel printer port. When using this connection, phonemes for words to be spoken are transmitted using LPRINT statements in BASIC. Pictured above is the Sweet Talker board combined with its power supply and connected to a TRS-80 Model III computer and a speaker.*

Hexadecimal Phoneme Code	Phoneme Symbol	ASCII Character	Duration (ms)	Example Word
00	EH3	@	59	jacket
01	EH2	A	71	enlist
02	EH1	B	121	heavy
03	PA0	C	47	no sound
04	DT	D	47	butter
05	A2	E	71	make
06	A1	F	103	pail
07	ZH	G	90	pleasure
08	AH2	H	71	honest
09	I3	I	55	inhibit
0A	I2	J	80	inhibit
0B	I1	K	121	inhibit
0C	M	L	103	mat
0D	N	M	80	sun
0E	B	N	71	bag
0F	V	O	71	van
10	CH	P	71	chip
11	SH	Q	121	shop
12	Z	R	71	zoo
13	AW1	S	146	lawful
14	NG	T	121	thing
15	AH1	U	146	father
16	OO1	V	103	looking
17	OO	W	185	book
18	L	X	103	land
19	K	Y	80	trick
1A	J	Z	47	judge
1B	H	[	71	hello
1C	G	\	71	get
1D	F	]	103	fast
1E	D	↑	55	paid
1F	S	←	90	pass
20	A	(space)	185	tame
21	AY	!	65	jade
22	Y1	"	80	yard
23	UH3	#	47	mission
24	AH	$	250	mop
25	P	%	103	past
26	O	&	185	cold
27	I	'	185	pin
28	U	(	185	move
29	Y	)	103	any
2A	T	*	71	tap
2B	R	+	90	red
2C	E	,	185	meet
2D	W	–	80	win
2E	AE	.	185	dad
2F	AE1	/	103	after
30	AW2	0	90	salty
31	UH2	1	71	about
32	UH1	2	103	uncle
33	UH	3	185	cup
34	O2	4	80	bold
35	O1	5	121	aboard
36	IU	6	59	you
37	U1	7	90	June
38	THV	8	80	the
39	TH	9	71	thin
3A	ER	:	146	bird
3B	EH	;	185	ready
3C	E1	<	121	be
3D	AW	=	250	call
3E	PA1	>	185	no sound
3F	STOP	?	47	no sound

Note: T must precede CH to produce "CH" sound.
D must precede J to produce "J" sound.

Table 1: *The sixty-four phonemes defined for the English language. Two of these produce silence; one causes synthesis to cease.*

dedicated memory buffer. Typically a 32- or 64-character FIFO (first-in, first-out) buffer is attached to the computer bus and loaded with the phoneme codes under program control. Once loaded, the codes are shifted out one at a time as the STB and $\overline{A}$/R lines change states. This self-clocking technique can also be used with an EPROM (erasable programmable read-only memory) and a counter when the SC-01 is to speak a canned message without computer control.

While use of a FIFO buffer reduces the main processor's waiting time when exercising relatively slow (typically 70 bps) peripheral devices such as the SC-01, buffers are expensive. Interface-hardware costs can be measurably reduced by a second scheme: using the computer system (running an appropriate program) to time the transmission of phoneme codes to the SC-01, as outlined in figure 3b. This method sends codes to the synthesizer chip through a latched parallel output port and monitors the synthesizer's activities (via the $\overline{A}$/R line) through an input port or interrupt line.

The latter is the technique I chose to use in my design. Interestingly enough, eliminating the extra hardware doesn't really complicate computer/synthesizer interaction nor does it require a sophisticated machine-language driver program like those ordinarily associated with software-controlled peripheral devices. The program code to control the synthesizer can be as simple as an LPRINT statement in BASIC. More on this later.

Sweet Talker

The schematic diagram of the Sweet Talker speech-synthesizer circuit board is shown in figure 4 on page 126. The phoneme-code bits are sent in parallel to the SC-01 (IC3) and buffered through IC1 (a 74LS244 three-state octal buffer). Pull-up resistors assure that a logic-1 input to the SC-01 will be at least 4 V as required. Unless the board is being used with external address circuitry, the $\overline{\text{Enable}}$ input line (on connector J1,

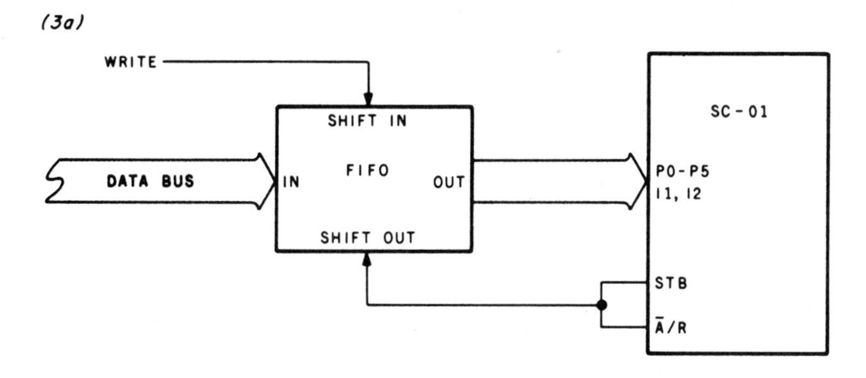

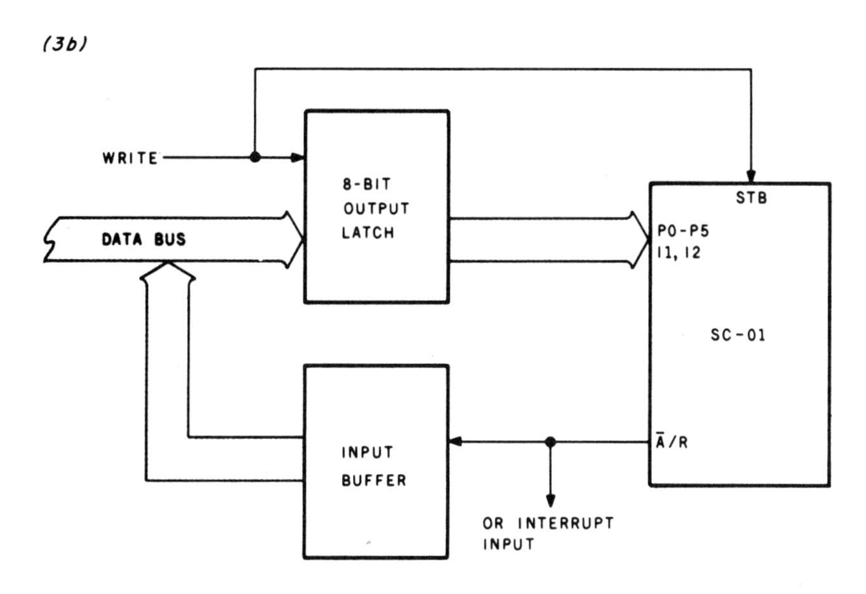

Figure 3: *Two methods of interfacing the SC-01 to a microcomputer data bus. Figure 3a shows an independently acting, self-timed circuit with a FIFO (first-in, first-out) buffer. Figure 3b shows a circuit that allows the computer to time the transmission of data to the SC-01.*

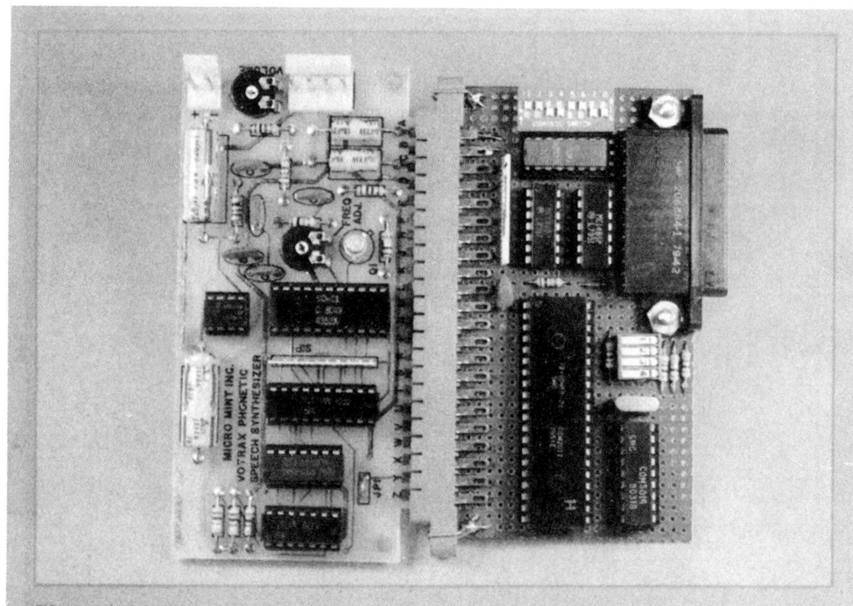

Photo 5: *Many computers have serial, rather than parallel, I/O ports. With a little extra hardware, it is possible to add a serial interface to the Sweet Talker.*

pin 12) should be grounded, thereby continuously enabling the buffer.

The two manual-inflection inputs (I1 and I2) are also buffered through IC1. The SC-01 cannot store these signals, and storage must be provided externally. A 74LS74 type-D flip-flop (IC2) is configured as a 2-bit latch. It is clocked synchronously with the SC-01's strobe input. Unlike the phoneme inputs, however, the inflection lines are not 5 V compatible. Two sections of a 7416 open-collector inverter (IC4) are used with pull-up resistors to level-shift these data inputs to CMOS levels. Since the automatic inflection is generally adequate, the manual-inflection inputs (J1, pins 16 and 20) can be left open or grounded when not in use.

The SC-01 can use either its internal clock or an external clock. External clock signals are applied through pin 15 on the SC-01 while pin 16 is grounded. My design uses the internal clock-signal generator, instead. The clock frequency is determined by an R/C (resistor/capacitor) combination attached to pins 15 and 16. The frequency is adjusted through potentiometer R8 and nominally set for 720 kHz. Slight adjustments to this control will vary the pitch of the speech. The easiest way to set this potentiometer is by ear. Simply output a sequence of phonemes to the SC-01 and set R8 for the most pleasant-sounding voice.

The process of sounding a phoneme begins when the 6-bit phoneme code is latched into the SC-01's control register. Latching occurs when the rising edge of a positive-going strobe pulse is received on pin 7 of the SC-01, the STB input line. The synthesizer will continue to sound the same phoneme until another phoneme code or a stop code is loaded.

The Sweet Talker circuit board can be set up (through jumper connections) to accept either a normally high or a normally low strobe signal from the external computer. Two sections of the inverter IC4 are involved in this flexibility. The normally high strobe signal connects to the Sweet Talker board through pin Y of the edge connector J1, while the normally

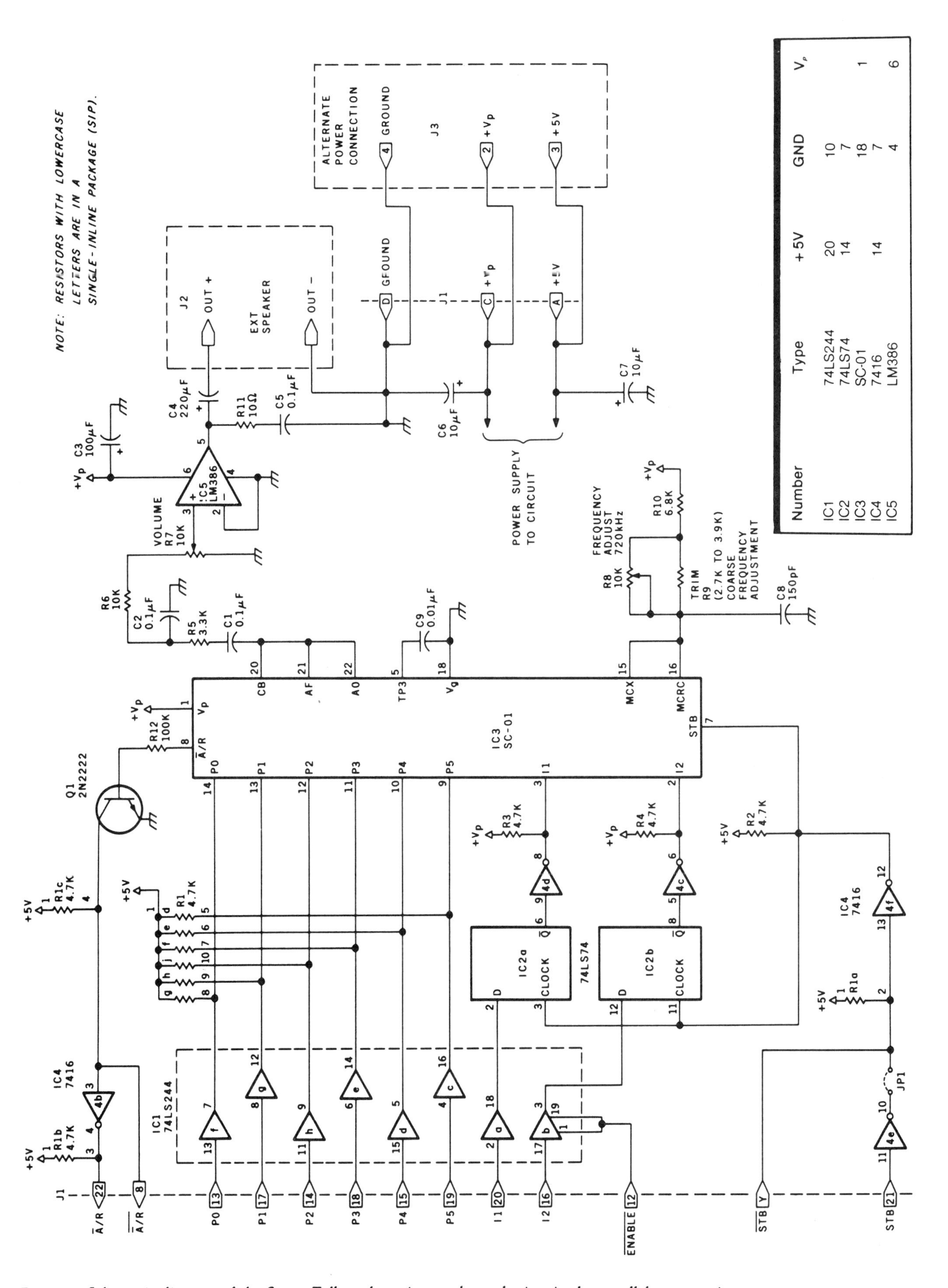

Figure 4: *Schematic diagram of the Sweet Talker phonetic speech synthesizer in the parallel-port version.*

low strobe signal connects through pin 21. The two inputs cannot be used in exactly the same way, however, because of some timing restrictions imposed by the SC-01.

The SC-01 senses the positive-going edge of the strobe pulse arriving on its pin 7, but unlike typical TTL latches (which operate in a few nanoseconds), the SC-01 requires some setup time before it can accept the strobe signal. This setup time must meet two requirements:

- The data on the phoneme-input lines P0 through P5 must have been stable for 450 ns before the rising edge of the strobe pulse arrives.
- The logic level on the STB input of the SC-01 (pin 7) must have been low during at least 72 clock periods (approximately 100 microseconds) before it goes high for the strobe pulse.

The staggered timing of the phoneme data and the strobe pulse makes interfacing the SC-01 directly to a microcomputer data bus difficult without the use of an external data latch (an output port).

Furthermore, in some cases (depending upon the method of connection), when the Sweet Talker board is being driven through a parallel output port that uses a DAV (data available) strobe signal, you may have to add a one-shot (monostable multivibrator) to the circuit to stretch the signal out so that the logic level at pin 7 of the SC-01 stays low long enough. The DAV strobe signal of a typical microprocessor is less than a microsecond in length.

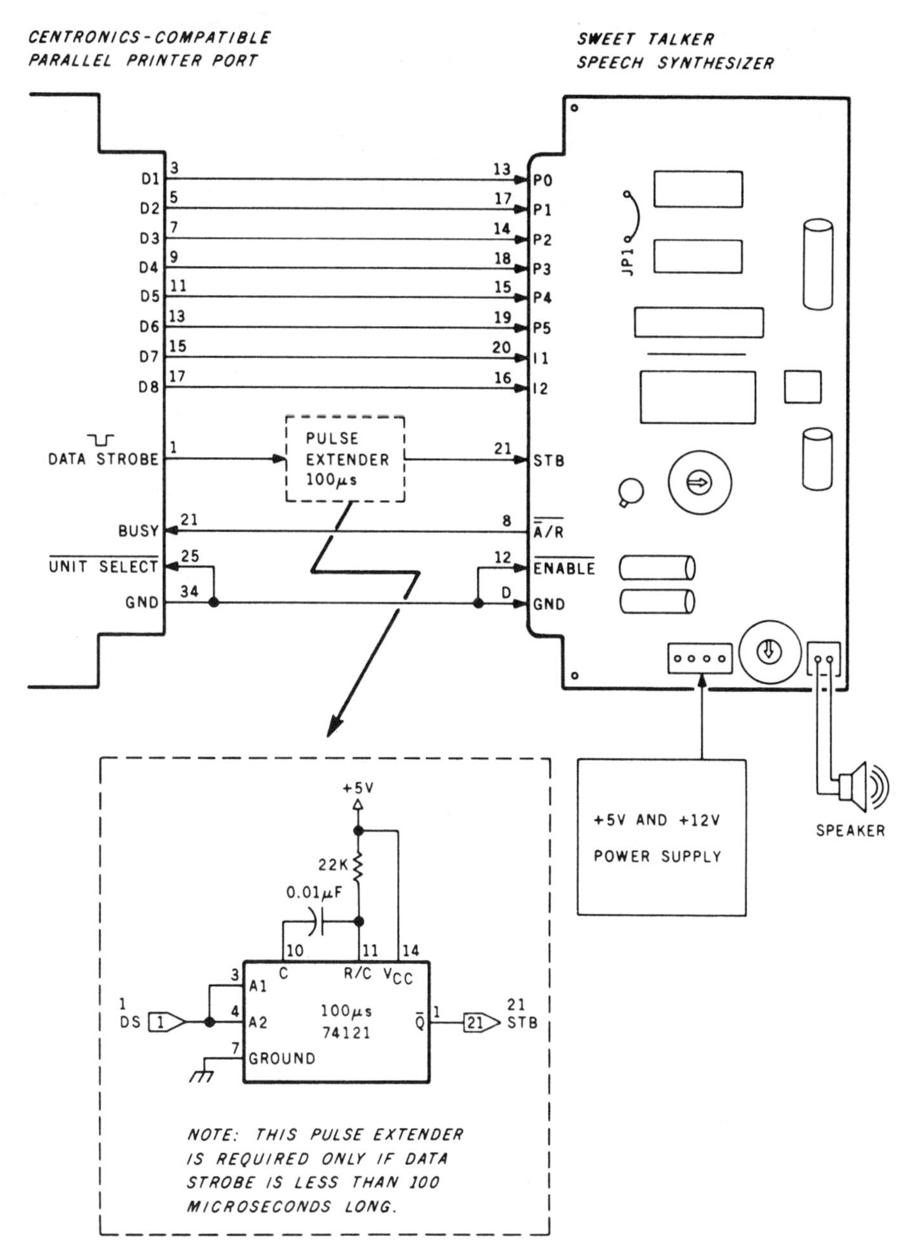

Figure 5: *Diagram of the connections between a Centronics-compatible parallel printer port (as on a TRS-80 Model I or III) and the Sweet Talker circuit board. Note that on some computers the Unit Select line may need to be connected to +5 V.*

Approximately 500 nanoseconds after the rising edge of the strobe pulse, the A/R line (pin 8) of the SC-01 goes to a logic 0, indicating that the synthesizer chip is busy. Transistor Q1 and IC4 convert the CMOS output of pin 8 to LSTTL levels. The A/R output can be monitored by the controlling computer in either of two ways: directly through an input port or connected to an interrupt line. In either case, when the A/R line returns to the logic-1 level, the SC-01 is ready to receive another phoneme code.

The remaining components on the Sweet Talker board make up the amplifier and filter sections. Capacitors C1 and C2 and resistors R5 and R6 form a simple low-pass audio filter. The audio signal is then amplified by an LM386 1-watt amplifier (IC5) to drive an 8-ohm speaker directly. Potentiometer R7 controls the volume, and the speaker connects to the 2-pin connector on one corner of the board.

The board operates on power-supply voltages of +5 V and V_p. V_p can be any voltage between +7 and +14 V. I generally use +12 V. Power can be applied either through the edge connector or the 4-pin power header. Pin assignments on the power header are arranged exactly the same as those on the Z8-BASIC Microcomputer board presented in my last two articles, and the Sweet Talker board can conveniently use the same power supply. (See "Build a Z8-Based Control Computer with BASIC," Part I is found in this volume on page 100, Part II on page 110.)

Table 2: *A list of useful words with their Votrax-notation phonemes, for ease in program coding.*

Word	Phonemes	Word	Phonemes
A	A1, AY, Y	keyboard	K, AY, Y, B, O1, O2, R, D
able	A1, Y, B, UH3, L	kill	K, I1, I3, L
about	UH1, B, UH2, AH2, U1, T	knowledge	N, AH1, UH3, L, I3, D, J
actual	AE1, EH3, K, T, CH, U1, UH3, L		
add	AE1, EH3, D	L	EH1, EH3, UH3, L
adjust	UH1, D, J, UH1, UH3, S, T	language	L, AE1, EH3, NG, G, W, I1, D, J
		large	L, AH1, R, D, J
B	B, E1, Y	left	L, EH1, EH3, F, T
back	B, AE1, AE1, K	length	L, EH1, EH3, NG, TH
basic	B, A1, Y, S, I2, K	listen	L, I1, I3, S, I2, N
been	B, EH1, EH3, N		
before	B, Y, F, O2, O2, R	M	EH1, EH2, M
better	B, EH1, EH3, T, ER	make	M, A1, AY, Y, K
		many	M, EH2, EH2, N, Y
C	S, E1, Y	match	M, AE1, EH3, T, CH
came	K, A1, AY, Y, M	memory	M, EH1, EH3, M, ER, Y
can	K, AE1, EH3, N	message	M, EH1, EH3, S, I2, D, J
car	K, AH2, UH3, R		
catalog	K, AE2, EH3, DT, UH3, L, AW2, AW2, G	N	EH1, EH2, N
change	T, CH, A1, AY, Y, N, D, J	name	N, A1, AY, Y, M
		near	N, AY, I1, R
D	D, E1, Y	need	N, E1, Y, D
data	D, A2, Y, DT, UH1	next	N, EH1, EH3, K, PA0, S, T
date	D, A2, AY, Y, T	none	N, UH1, UH3, N
decide	D, Y, S, AH2, EH3, Y, D		
decision	D, Y, S, I2, ZH, UH3, N	O	O2, O1, U1
deliver	D, Y, L, I2, V, ER	object	UH1, B, D, J, EH1, EH3, K, T
		obsolete	AH1, UH3, B, S, UH3, L, AY, Y, T
E	E1, Y	often	AW2, AW2, F, I3, N
early	ER, R, L, Y	omit	O1, U1, M, I1, I3, T
either	E1, Y, THV, ER	other	UH1, UH3, THV, ER
empty	EH2, EH3, M, P, T, Y		
end	EH2, EH3, N, D	P	P, E1, Y
exact	EH2, EH3, G, PA0, Z, AE2, EH3, K, T	package	P, AE1, EH3, K, I1, D, J
		paper	P, A1, Y, P, ER
F	EH1, EH2, F	part	P, AH1, R, T
fact	F, AE2, EH3, K, T	person	P, ER, S, UH1, N
fault	F, AW, L, T	phone	F, O1, U1, N
final	F, AH2, Y, N, UH3, L		
first	F, ER, R, S, T	Q	K, Y1, IU, U1, U1
follow	F, AH1, AW2, L, O1, U1	qualify	K, W, AW1, L, I1, F, AH1, EH3, Y
		quantity	K, W, AH1, N, T, I3, T, Y
G	D, J, E1, Y	question	K, W, EH1, EH3, S, T, CH, UH3, N
game	G, A2, AY, Y, M	quick	K, W, I1, I3, K
good	G, OO1, OO1, D	quiet	K, W, AH1, EH3, AY, I2, T
great	G, R, A2, Y, T		
ground	G, R, AH1, UH3, W, N D	R	AH1, UH2, ER
grow	G, R, O1, U1	raise	R, A1, AY, Y, Z
		reach	R, E1, Y, T, CH
H	A1, AY, Y, T, CH	ready	R, EH1, EH3, D, Y
hand	H, AE1, EH3, N, D	remain	R, E1, M, A1, AY, Y, N
have	H, AE1, EH3, V	resistor	R, E1, Z, I1, S, T, ER
hear	H, AY, I3, R		
heavy	H, EH1, V, Y	S	EH1, EH2, S
high	H, AH1, EH3, Y	safe	S, A1, AY, Y, F
		sale	S, A1, A2, AY, UH3, L
I	AH1, EH3, I3, Y	schedule	S, K, EH1, EH3, D, J, IU, U1, L
important	I1, I3, M, P, O2, O2, R, T, EH3, N, T	scrap	S, K, R, AE1, EH3, P
include	I1, I3, N, K, L, IU, U1, U1, D	section	S, EH1, EH3, K, SH, UH3, N
inform	I1, I3, N, F, O2, O2, R, M		
insert	I1, N, S, R, R, T	T	T, E1, AY, Y
instead	I1, I3, N, S, T, EH1, EH3, D	talk	T, AW, K
		technical	T, EH1, EH3, K, N, I3, K, UH3, L
J	D, J, EH3, A1, AY, Y	terminal	T, ER, M, EH3, N, UH2, L
job	D, J, AH1, UH3, B	think	TH, I1, I3, NG, K
join	D, J, O1, UH3, I3, AY, N	time	T, AH1, EH3, Y, M
joy	D, J, O1, UH3, I3, AY		
judge	D, J, UH1, UH2, D, J	U	Y1, IU, U1, U1
jump	D, J, UH1, UH2, M, P	under	UH2, UH2, N, D, ER
		uniform	Y1, IU, U1, N, I3, F, O1, R, M
K	K, EH3, A1, AY, Y	until	UH2, UH2, N, T, I1, I3, L
keep	K, E1, Y, P	up	UH1, UH2, P
key	K, E1, Y	urgent	R, R, D, J, I3, N, T

Table 2 continued:

Word	Phonemes
us	UH1, UH2, S
use	Y1, IU, U1, U1, Z
V	V, E1, AY, Y
vacant	V, A1, Y, K, EH3, N, T
valid	V, AE1, UH3, L, I1, D
value	V, AE1, EH3, L, Y1, IU, U1
vendor	V, EH1, EH3, N, D, ER
vent	V, EH1, EH3, N, T
verify	V, EH1, R, I3, F, AH1, EH3, Y
very	V, EH1, R, Y
via	V, E1, AY, UH2, UH3
victor	V, I1, I3, K, T, ER
voice	V, O1, UH3, I3, AY, S
void	V, O1, UH3, I3, AY, D
volt	V, O2, O2, L, T
volume	V, AH1, UH3, L, Y1, IU, U1, M
W	D, UH1, B, UH3, L, Y1, IU, U1
wage	W, A1, AY, Y, D, J
wait	W, A1, AY, Y, T
want	W, AH1, UH3, N, T
was	W, UH1, UH3, Z
wash	W, AW, SH
water	W, AH1, UH3, T, ER
watt	W, AH1, UH3, T
wave	W, A1, AY, Y, V
we	W, E1, Y
weapon	W, EH2, EH2, P, UH1, N
Wednesday	W, EH1, N, Z, D, A1, I3, Y
week	W, E1, Y, K
weigh	W, A2, A2, Y
went	W, EH1, EH3, N, T
west	W, EH1, EH3, S, T
wet	W, EH1, EH3, T
what	W, UH3, UH1, T
wheel	W, E1, Y, L
when	W, EH1, EH3, N
where	W, EH3, A2, EH3, R
which	W, I1, I3, T, CH
while	W, AH1, EH3, I1, UH3, L
whiskey	W, I1, I3, S, K, AY, Y
white	W, UH3, AH2, Y, T
who	H, IU, U1, U1
will	W, I1, I3, L
window	W, I1, N, D, O1, U1
winter	W, I1, I3, N, T, ER
wire	W, AH1, EH3, AY, R
with	W, I1, I3, TH
withdraw	W, I1, I3, TH, D, R, AW
without	W, I1, I3, TH, UH2, AH2, U1, T
word	W, ER, R, D
work	W, ER, R, K
wrong	R, AW, NG
X	EH1, EH2, K, PA0, S
X-ray	EH1, EH2, K, PA0, S, R, A1, I3, Y
Y	W, AH1, EH3, I3, Y
Yankee	Y1, AE1, EH3, NG, K, E1, Y
yard	Y1, AH1, R, D
year	Y1, AY, I3, R
yellow	Y1, EH1, EH3, L, O1, U1
yes	Y1, EH3, EH1, S
yesterday	Y1, EH3, EH1, S, T, ER, D, A1, I3, Y
yet	Y1, EH1, EH3, T
your	Y, O2, O2, R
Z	Z, E1, Y
zap	Z, AE1, EH3, P
zero	Z, AY, I1, R, O1, U1
zone	Z, O1, U1, N
zulu	Z, IU, U1, L, IU, U1

Speaking in Phonemes

Table 1 lists the 64 basic phonemes of the English language. At first glance, it appears complicated, but it is easy to understand and use. Take the word "call," for example. It is made up of three distinct phonemes, as follows (expressed in Votrax notation):

K, AW, L

which correspond to the hexadecimal codes:

19 3D 18

Similarly, the word "disk" is broken into the phonemes:

D, I1, S, K

which correspond to the hexadecimal codes:

1E 0B 1F 19

See reference 3 for more details on this process.

Causing the synthesizer to speak either of these words is done simply by sending the hexadecimal codes sequentially to it. This is most easily done through a parallel I/O port under control of a program written in BASIC. Typically, if the synthesizer were connected to port 0 on your computer, the routine for saying "call" would be coded as follows:

```
100  DATA 25, 61, 24  :REM
     Decimal Phoneme codes for
     "call"
110  FOR A=1 TO 3 : READ P(A)
     :NEXT A
130  FOR A=1 TO 3
140  OUT 0,P(A)  :REM Latch
     Phoneme code into SC-01
150  IF INP(0)=0 THEN GOTO
150  :REM Continue if A/R not busy
160  NEXT A
170  OUT 0,63  :REM Send STOP
     code to SC-01
180  END
```

Essentially any word or series of words can be spoken using this method. It isn't necessary for you to acquire special knowledge about word sounds to use a phonetic speech synthesizer, because many lists of word and phoneme equivalents are available. Table 2 is a list of some common words. A more extensive list is in preparation.

Easy Interfacing

What could be easier than pretending that the Sweet Talker speech synthesizer is a parallel-interfaced printer? It just so happens that many computers already have a parallel output port in the form of a Centronics-compatible parallel printer port. This connection is avail-

Number	Type	+5V	GND	−12V	+12V
IC1	COM5016	2	11		9
IC2	COM2017	1	3	2	
IC3	CD4049	1	8		
IC4	MC1489	14	7		
IC5	MC1488		7	1	14

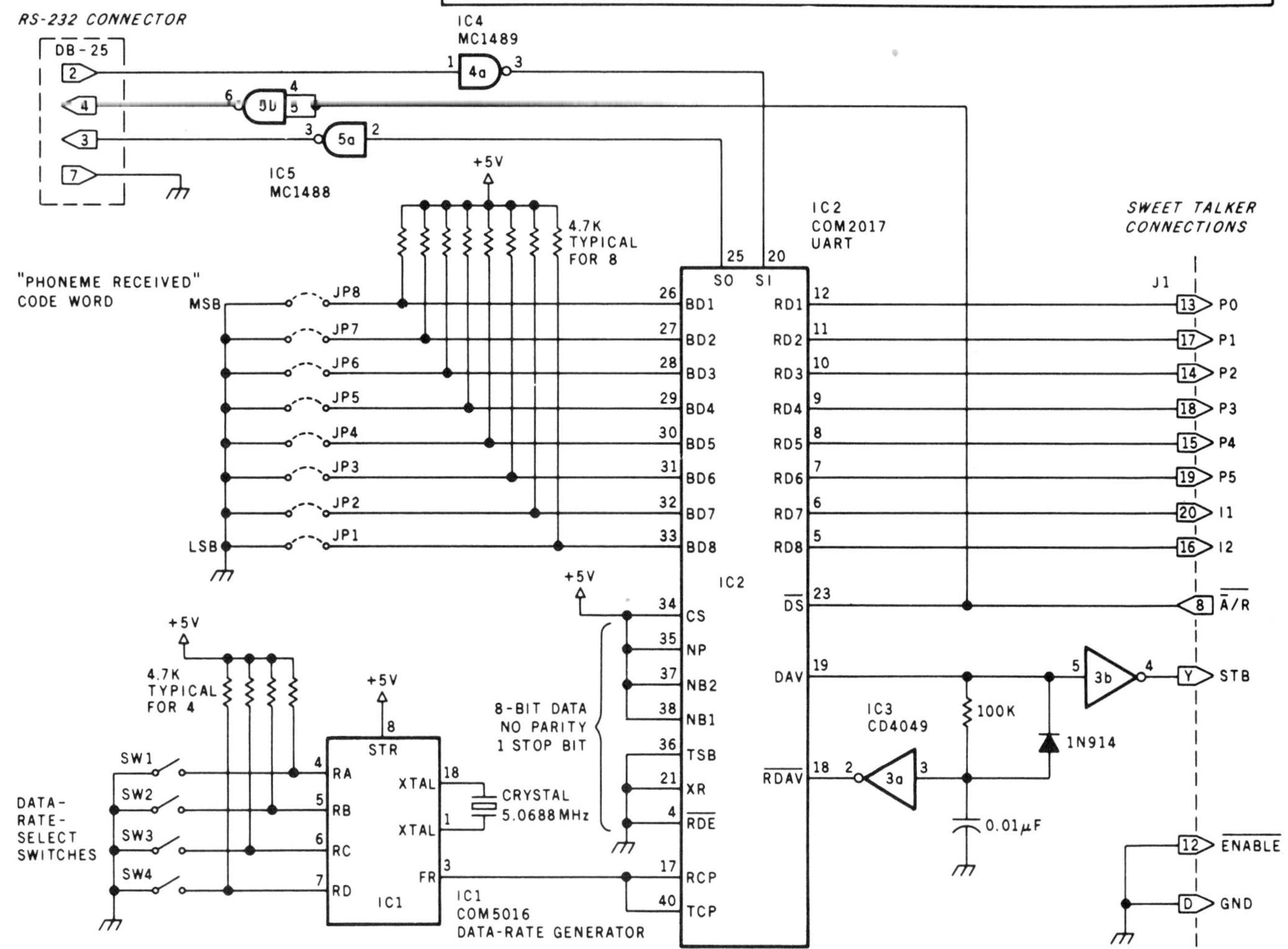

Figure 6: *Schematic diagram of a serial I/O interface for the Sweet Talker board, to be used in place of parallel I/O. The serial communication protocol is RS-232C.*

able on all Radio Shack TRS-80 Model III and expanded Model I computer systems, as well as many others. By connecting the Sweet Talker board as shown in figure 5, it is possible to fool the computer into thinking that the Sweet Talker is a printer, whereupon we can use LPRINT statements to drive it. The same machine-language routine in the BASIC interpreter that normally transfers ASCII (American Standard Code for Information Interchange) character strings to the printer will also work with the speech synthesizer.

A BASIC LPRINT statement will transmit any ASCII characters between the double quotes (except the quotation marks themselves and perhaps a few control codes) whether they spell out something humanly coherent or not. As table 1 illustrates, all of the phonemes correspond to ASCII characters which produce the equivalent 6-bit code (the lowercase letters "a" through "z" correspond to hexadecimal codes 21 through 3A). It is possible, therefore, to type an "@" for the EH3 phoneme (hexadecimal code 00) or a ">" for PA1 (hexadecimal code 3E). Using this technique, the program statement for saying "call" would be:

```
100 LPRINT "Y=X"
```

It's a good idea to add a stop phoneme (corresponding to ASCII "?") after the end of the word to cancel the last phoneme. The line then becomes:

```
100 LPRINT "Y=X";"?";
```

The Sweet Talker speech synthesizer attaches to and handshakes with the computer in the same manner as a printer would. The $\overline{A}$/R output is connected to the Busy input, and the Unit Select line is grounded to simulate printer attachment. The Sweet Talker's Enable input should also be grounded. The computer's LPRINT driver routine sends one character to the "printer" (speech synthesizer) and then checks the Busy line before sending another. When the Busy line is high again, the next character (phoneme) is sent.

The only area for concern is the pulse width of the data strobe (attached to J1 pin 21 with jumper JP1 installed), as I previously mentioned. If it is less than 100 μs, a type-74121 monostable multivibrator should be added as indicated in figure 5. If you are unsure of the duration, add the circuit anyway.

Once the interface is attached, a simple program can be used to test phoneme combinations. For example, sending "S*1L/@*KY" will cause the unit to say "automatic," and "Y2M*KIMB677" will make it say "continue." A simple test program requires only three lines:

```
100  INPUT A$
110  LPRINT A$;"?";
120  GOTO 100
```

Using a Serial Interface

Your computer might not have a parallel printer port, but a serial one instead. While the interface is more complicated, you can also use LPRINT statements to drive the additional circuitry shown in figure 6. This circuit is a full-duplex RS-232C serial interface which is capable of receiving a phoneme transmitted serially from the computer, converting it back to parallel form, and strobing it into the SC-01.

The timing relationships between the interface and the computer become slightly more complex. Whenever a phoneme is loaded into the SC-01, the A/R line drops and the RS-232C Data Terminal Ready signal goes low. After the phoneme has concluded, the UART (universal asynchronous receiver/transmitter) transmits a jumper-selected character (optionally preset on UART pins 26 through 33) and raises the Data Terminal Ready line again. Proper timing from the host computer can be accomplished either by sending successive characters only in response to the "phoneme-concluded" code or by monitoring the state of the Data Terminal Ready line.

The communication rate between the host and the synthesizer is switch-selectable from 50 to 19,200 bps using the COM5016 data-rate generator as shown. Communication is hard-wire selected for 8-bit data words, no parity bit, and 1 stop bit. A more in-depth discussion of the data-rate generator and UART was given in one of my previous Circuit Cellar articles ("I/O Expansion for the TRS-80, Part 2: Serial Ports," June 1980 BYTE, page 42).

In Conclusion

What can you do with a computer-controlled speech synthesizer? I'm sure you have a few ideas. In any case, the benefits of electronic speech synthesis will surely propagate as more people learn how to use it.

References

1. Ciarcia, Steve. "Build a Low-Cost Speech-Synthesizer Interface," BYTE, June 1981, page 46.
2. Ciarcia, Steve. "Talk to Me: Add a Voice to Your Computer for $35," BYTE, June 1978, page 142, reprinted in *Ciarcia's Circuit Cellar*, Volume I. Peterborough NH: BYTE Books, 1979, page 77.
3. Fons, Kathryn and Tim Gargagliano. "Articulate Automata: An Overview of Voice Synthesis," BYTE, February 1981, page 164.
4. Gargagliano, Tim and Kathryn Fons. "Text Translator Builds Vocabulary for Speech Chip," *Electronics*, February 10, 1981, page 118.
5. Gargagliano, Tim and Kathryn Fons. "The TRS-80 Speaks: Using BASIC to Drive a Speech Synthesizer," BYTE, October 1979, page 113.
6. Gargagliano, Tim and Kathryn Fons. "A Votrax Vocabulary," BYTE, June 1981, page 384.
7. Lin, Kun-Shan, Gene Frantz, and Kathy Goudie. "Software Rules Give Personal Computer Real Word Power," *Electronics*, February 10, 1981, page 122.
8. Weinrich, David W. "Speech-Synthesis Chip Borrows Human Intonation," *Electronics*, April 10, 1980, page 113.
9. Wiggins, Richard and Larry Brantingham. "Three-Chip System Synthesizes Human Speech," *Electronics*, August 31, 1978, page 109.

13

Switching Power Supplies

An Introduction

Since the advent of the three-terminal integrated-circuit voltage regulator, it seems that everyone has become a power-supply expert. No longer are ten pages of calculations required to produce a design for even a modest power supply, thanks to the wide tolerances and relatively sturdy architecture of these devices. After you have purchased a few readily available parts, you can have your completed supply running in a few hours.

Three-terminal regulators have become so easy to use that few experimenters stop to consider how inefficient they are. For example, if we were to design a 5-volt 1.5-amp power supply from commonly available parts, we would probably use a 12.6 VAC transformer, a bridge rectifier, a filter capacitor, a three-terminal regulator (such as an LM317), and a few discrete components. (A 6.3 VAC transformer is only marginally usable.)

Photo 1: *Ferrite toroids and pot cores are available in various sizes and compositions. These units are from the Ferroxcube Division of Amperex Electronics Corporation, 5083 Kings Hwy, Saugerties NY 12477, (914) 246-2811.*

Set up normally, the transformer, rectifier, and filter produce about 16 volts. With the LM317's output adjusted for 5 volts, 11 volts would be dropped across the regulator. ("Dropped" really means "consumed" in this case.) Power is dissipated by the regulator in an amount equivalent to the difference between the regulator's input and output voltages multiplied by the current through it. In this instance,

$$(16V - 5V) \times 1.5\ A = 16.5\ W$$

The conversion efficiency is

$$V_{OUT}/V_{IN} = 5/16 = 31\%$$

The LM317 and similar linear regulators (shown in figure 1) such as the 7805 and LM340 are all called *series dissipative regulators*. They function in a linear mode, simulating a variable resistance between the input-voltage source and the load. There are other factors involved, but basically the linear series regulator simulates a varying resistance. Within a specified range of variation in the input voltage and load current, the regulator maintains a constant output voltage by dissipating the excess power as heat. Unfortunately, as we see in this example, it consumes 16.5 watts producing the desired 7.5-watt output.

In most applications, however, the ease of use and relative low cost of linear series regulators far outweigh the inherent lack of efficiency. The linear series regulator is well suited for medium-current applications with a small voltage differential, where the power dissipation can be handled with heat sinks and cooling fans. When electricity costs only five cents per kilowatt-hour (1000 watts for one hour), it's hard to get concerned about losing 16.5 watts.

Why Use a Switching Regulator?

Power-supply efficiency usually isn't important unless size, heat dissipation, or total power consumption is limited. If the power source for our regulator were a battery, we would have to be more careful about how much energy is converted for useful work and how much is thrown away as heat.

Efficiency is really the name of the game. In a series dissipative regulator, conversion efficiency is directly related to the input/output voltage differential. As the difference between the two voltages increases, efficiency decreases. The power radiated by the regulator represents a loss to the system and limits the amount of power deliverable to the load.

It would be far better if the regulator consumed no power and if all the power were channeled to the load. While perfect conversion efficiency is impossible, the inherent fault in using series dissipative regulators is the linear operating mode of the series-pass transistor.

If, however, the transistor is used as a *switch* (in saturated operation) rather than as a variable resistor (in linear mode), the series-pass transistor consumes very little power. (This is not a new discovery. Designers have always been aware of the technique, but it required some time to develop cheap, fast switching transistors and inexpensive, low-loss ferrite materials from which cost-effective switching supplies could be built.)

A regulator constructed to operate in this manner is called a *series switching regulator*. The same series-pass transistor switches between cutoff and saturation at a high frequency, producing a pulse-width-modulated square wave of amplitude V_{IN}. This waveform is then filtered through a low-pass LC (inductance/capacitance) filter, producing an average DC output potential (V_{OUT}) proportional to the pulse width and frequency. The efficiency of such a regulator is generally independent of the voltage differential and can approach 95% in good designs.

Switching regulators come in various circuit configurations, a few of which are the flyback, feed-forward, and push-pull types. I am limiting this discussion primarily to another type, nonisolated single-ended (single-polarity) switching regulators, because they are the easiest to understand and build. Although simple, they are nonetheless quite useful. Unlike the typical three-terminal dissipative regulator, the switching regulator can be directly configured to operate in any of three modes: step-down, step-up, or polarity-inverting.

Photo 2: *A commercially sold switching power supply that combines both linear-regulator and switching-regulator technology to provide outputs of ±12 volts and +5 volts. Note the use of the toroidal inductor and the relatively small heat sinks. This unit was made by Conver Corporation, 10631 Bandley Dr, Cupertino CA 95014, (408) 255-0151.*

Switching-Regulator Basics

Figures 2a through 2c on page 199 outline the three common modes of switching-regulator operation. Basically, the switching regulator consists of a power source which supplies a voltage V_{IN}, a "switch" Q1, and an LC filter. (It is assumed in all cases that a load is connected between the output and ground and that Q1 is actually a transistor controlled through external circuitry.) The way the components are connected determines the output mode.

In the *step-down* regulator of figure 2a, the basic circuit operation is to close switch (transistor) Q1 for a time T_{ON}, and then open it for time T_{OFF}. The total, $T_{ON} + T_{OFF}$, is called the switching period T. Neglecting the saturation voltage of Q1 (V_{SAT}) and the diode (V_{DIODE}), the voltage at the input to the inductor is $+V_{IN}$ during the time T_{ON} and zero during T_{OFF}. (But remember, these other voltage drops must be included in our calculations when we are choosing actual components.)

When Q1 is closed, a step increase in voltage is applied to the inductor coil, which has the value L. However, current flowing through an inductor cannot change instantaneously; instead it increases linearly according to the factor $L(di/dt)$, building a magnetic field. This reduces any instantaneous current change seen by the load. When Q1 opens, the magnetic field in the inductor decays linearly, supplying power to the load. The current path is completed through the forward-biased flyback diode D1.

In this type of switching regulator, the inductor and capacitor form a low-pass filter. High-frequency pulses are applied to the input, and an averaged DC level comes out. The peak-to-peak ripple voltage is a function of the switching period T and the values of the inductance L and capacitance C. As the frequency of operation is increased, the voltage ripple is reduced, but the supply becomes less efficient.

Figure 2b illustrates the circuit configuration of the basic *step-up* switching-type voltage regulator. In this type of regulator, closing Q1 during T_{ON} charges the inductor. When Q1 is opened, the inductor discharges through D1 into the load. The output voltage is determined by the rate of discharge according to the equation

$$V = L(di/dt)$$

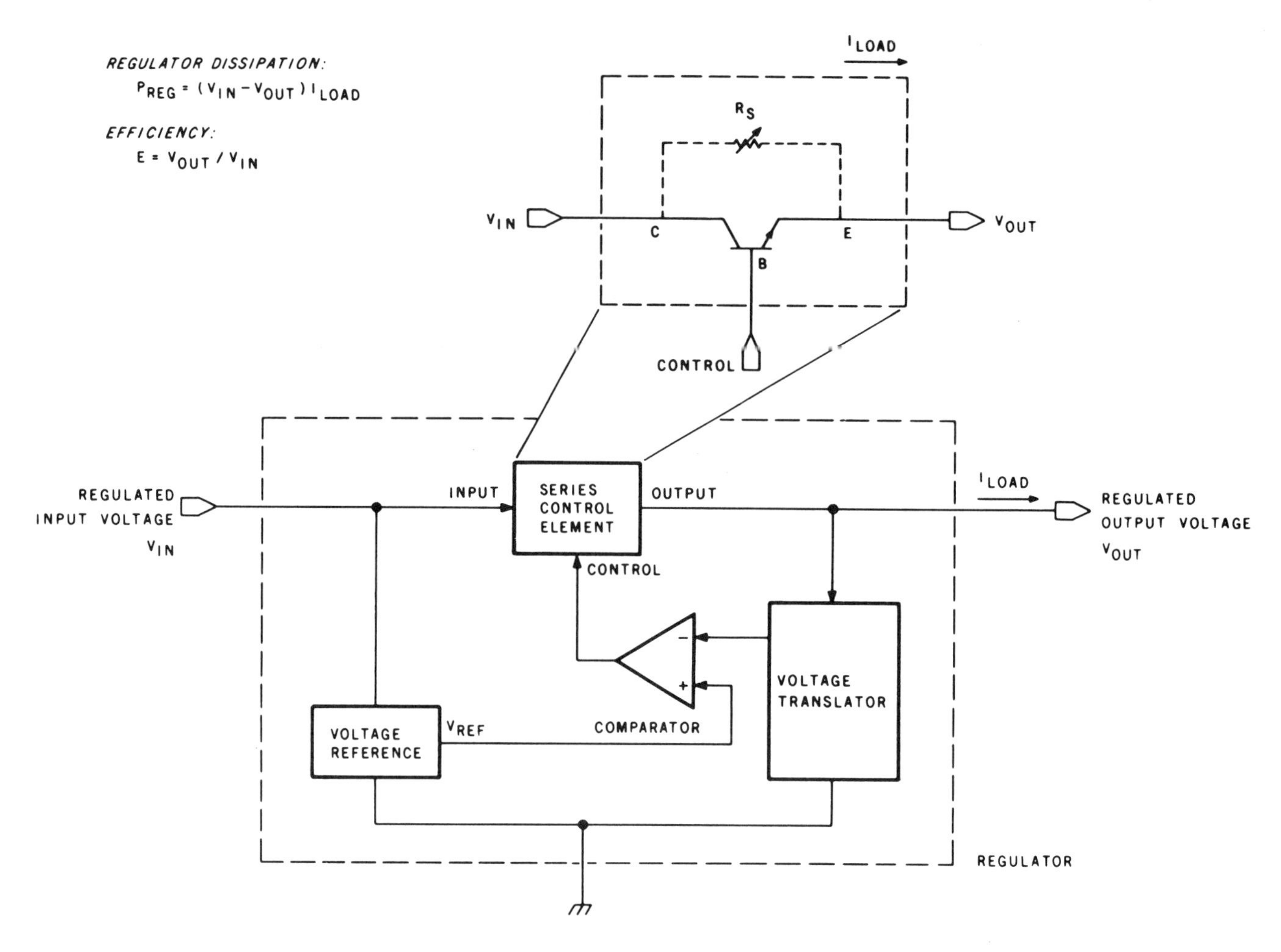

Figure 1: *Block diagram of a typical linear series voltage regulator, such as the LM317, LM340 or 7805. The series regulator acts as a varying resistance, dissipating excess power as heat to maintain a constant output voltage.*

A fast discharge delivers an output voltage higher than input V_{IN}.

Figure 2c shows a *polarity-inverting* switching regulator. As in the other cases, closing Q1 charges the inductor during T_{ON}. When Q1 is opened during T_{OFF}, there is a "kickback" voltage produced by the inductor as it discharges. This effect occurs elsewhere, too. For years, many of you have probably been putting reverse-biased diodes across relay coils, perhaps without thinking about it. The purpose of the diode is to dampen the high-voltage spike produced when a pulse is applied to the inductive relay coil. In the power-supply situation, rather than short out the voltage, diode D1 directs this opposite-polarity voltage to the load.

Generally speaking, in all three cases, the output voltage V_{OUT} is regulated by controlling the ratio of T_{ON}/T. This *duty cycle* can be altered in a number of ways depending upon the control method. Two of the more common approaches are variable pulse width (pulse-width modulation) and variable frequency. In a pulse-width-modulated switching regulator, the switching period T is fixed and the "on" time T_{ON} varied. Conversely, in a variable-frequency regulator, T_{ON} is fixed and the "off" time T_{OFF} varied.

You can experiment with a simple design for a nonisolated single-ended switching voltage regulator.

The variable-frequency switching regulator is generally easier to design and build, since the magnetic flux developed in the inductor coil during the fixed on-time determines the amount of power deliverable to the load. This eases the design of the inductor because the inductor's operating region within its characteristic curve is precisely defined. Operating frequency, which increases proportionally with the load, is primarily a function of the inductance L, capacitance C, and voltages V_{IN} and V_{OUT}.

The fixed-frequency pulse-width-modulated switching regulator varies the duty cycle to change the average power delivered to the load. This method is particularly advantageous for systems employing transformer-coupled output stages and is most

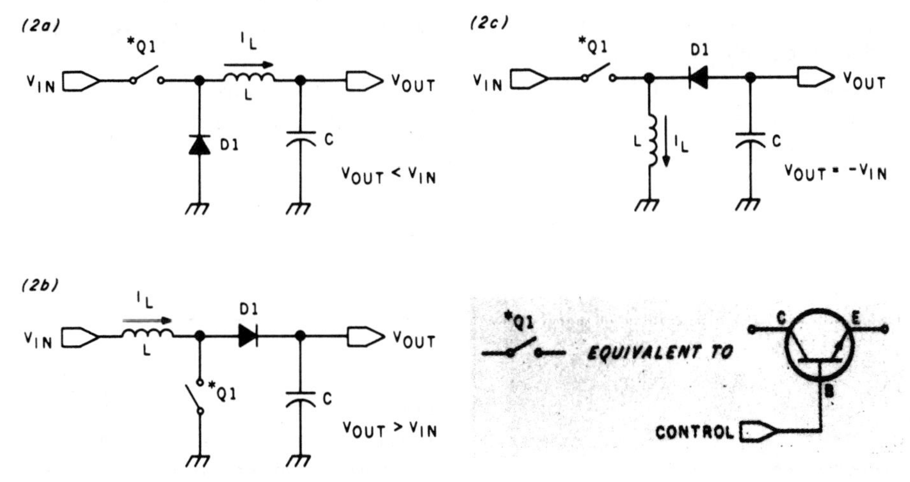

Figure 2: *The configurations of switching voltage regulators: the voltage-step-down (2a), voltage-step-up (2b), and polarity-inverting (2c) arrangements are depicted. The component shown as a switch is assumed to be a bipolar transistor driven in saturated switching mode. Some commercial designs may use field-effect transistors.*

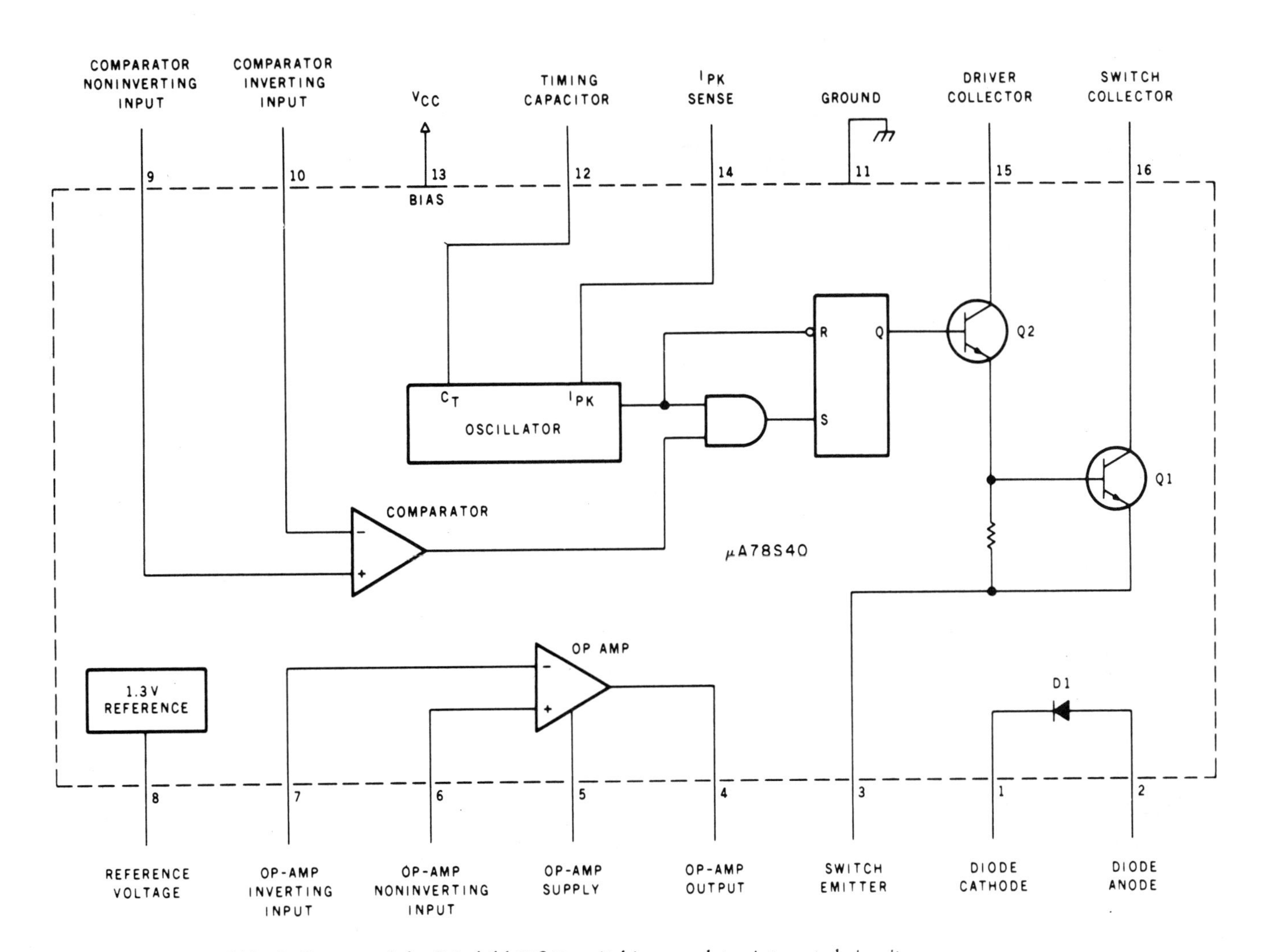

Figure 3: *Functional block diagram of the Fairchild 78S40 switching-regulator integrated circuit.*

often used in commercial switching supplies with multiple outputs. This method is more complex and uses more components than variable-frequency supplies, but the advantages outweigh the extra cost in high-current applications.

Typical operating frequencies of switching regulators range from 10 to 50 kHz. However, there are some trade-offs. High frequencies reduce the ripple voltage at a price of decreased efficiency and increased radiated electrical noise. If the frequency is lowered, greater efficiency and less electrical noise will result, but larger coils and capacitors are needed. Also, a switching power supply operating at 10 kHz can become quite annoying to listen to after a while.

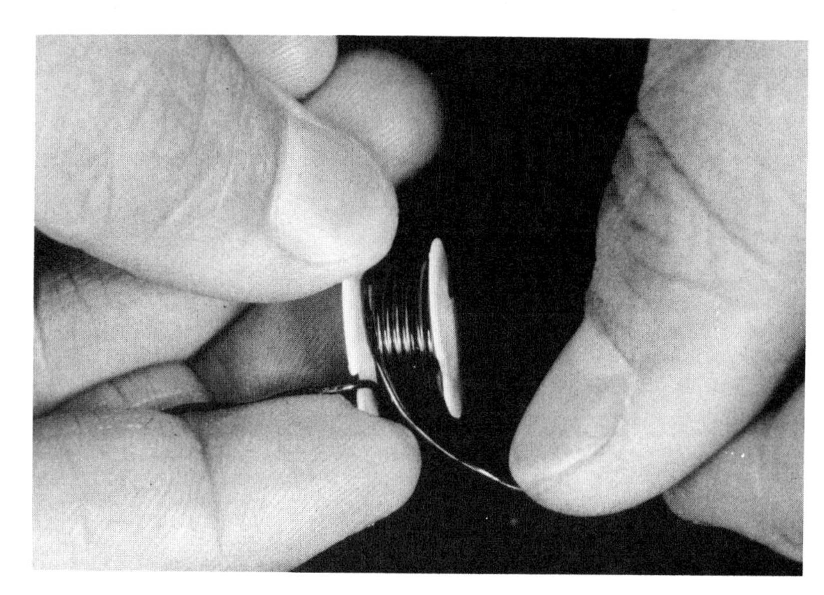

Photo 3: *Winding the inductor for the switching supply of figure 6a. The turns of wire are neatly wound around a plastic bobbin, evenly distributed.*

Photo 4: *Having been wound, the wire bobbin is placed between the two halves of the pot core, which are screwed together to form the finished coil.*

The most effective frequency range for optimizing efficiency and size with the components presently available is around 20 kHz. This is out of the range of human hearing yet low enough to be within the switching speeds of most inexpensive transistors and diodes. As switching speeds of newly developed high-current semiconductors increase and new ferrite components are introduced, practical operating frequencies will rise.

There are many kinds of switching regulator circuits that I could present. My previous article, "No Power for Your Interfaces? Build a 5 W DC to DC Converter" (reference 3), discussed the construction of a 5-watt converter which produced ±12 volts from a +5-volt input. This circuit was intended to facilitate powering linear interfaces and op amps (operational amplifiers) when the only source available was the 5-volt logic supply. If you have such a requirement, I refer you to that article.

This time, however, I'd like to present circuits to meet a different need. In the example of the conventional power supply with which I began this article, the input/output voltage differential was so great that more heat than useful energy was produced by the supply. In a small electronic package which requires several watts, dissipated heat can be difficult to remove if the enclosure has no vents. This points out a need for a more efficient 5-volt regulator. While we're at it, we might as well make one that can accommodate up to 30-volt inputs without significantly increased losses.

One power supply that seems to be in demand lately is one with a +25-volt output for programming EPROMs (erasable programmable read-only memories, as discussed in last month's Circuit Cellar; see reference 2). While a three-terminal regulator can be used for this voltage, I see this as a ripe opportunity to demonstrate the step-up variety of switching regulator. One circuit we'll look at will therefore be a +5-volt to +25-volt converter suitable for EPROM-programming use.

78S40 Switching Regulator

In recent years, monolithic (everything built on one semiconductor chip) linear voltage regulators have simplified power-supply design. Most systems have employed linear regulators because of their excellent reliability, low external parts count, and low cost. However, recent improvements in high-speed switching transistors and low-loss inductors have made switching supplies more attractive. The real breakthrough came with low-cost LSI (large-scale integration) monolithic switching regulators, which contain practically everything but the inductor on a single chip.

One of the many such integrated regulators available is the Fairchild 78S40, which is shown in the block diagram of figure 3. The 78S40 contains a current-controlled oscillator, current-limit sensor, voltage reference, high-gain comparator, high-current op amp, transistor switch, and power-switching diode. A single capacitor sets the frequency range (adjustable between 100 Hz and 100 kHz, but normally used at 20 to 30 kHz), and one external resistor provides current-limiting protection for the transistor and diode. Other than a few discrete resistors to set the output

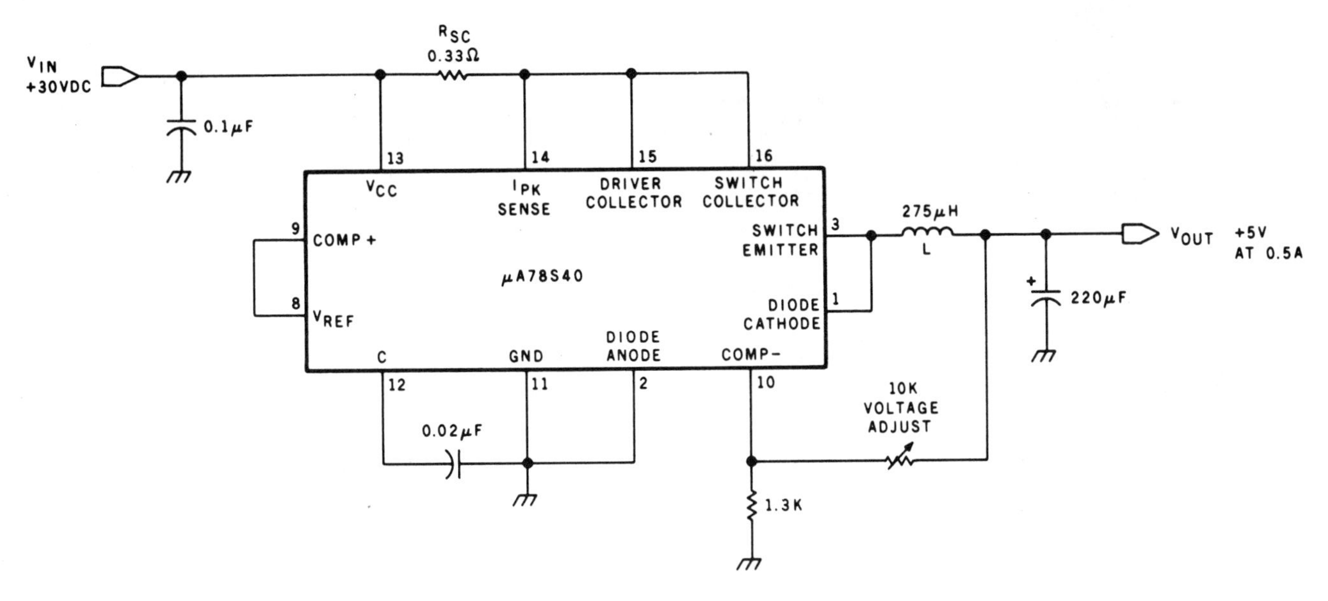

Figure 4: *Schematic diagram of a 5-volt, ½-amp step-down switching voltage regulator. For low currents, the 78S40's internal switching transistor and diode may be used.*

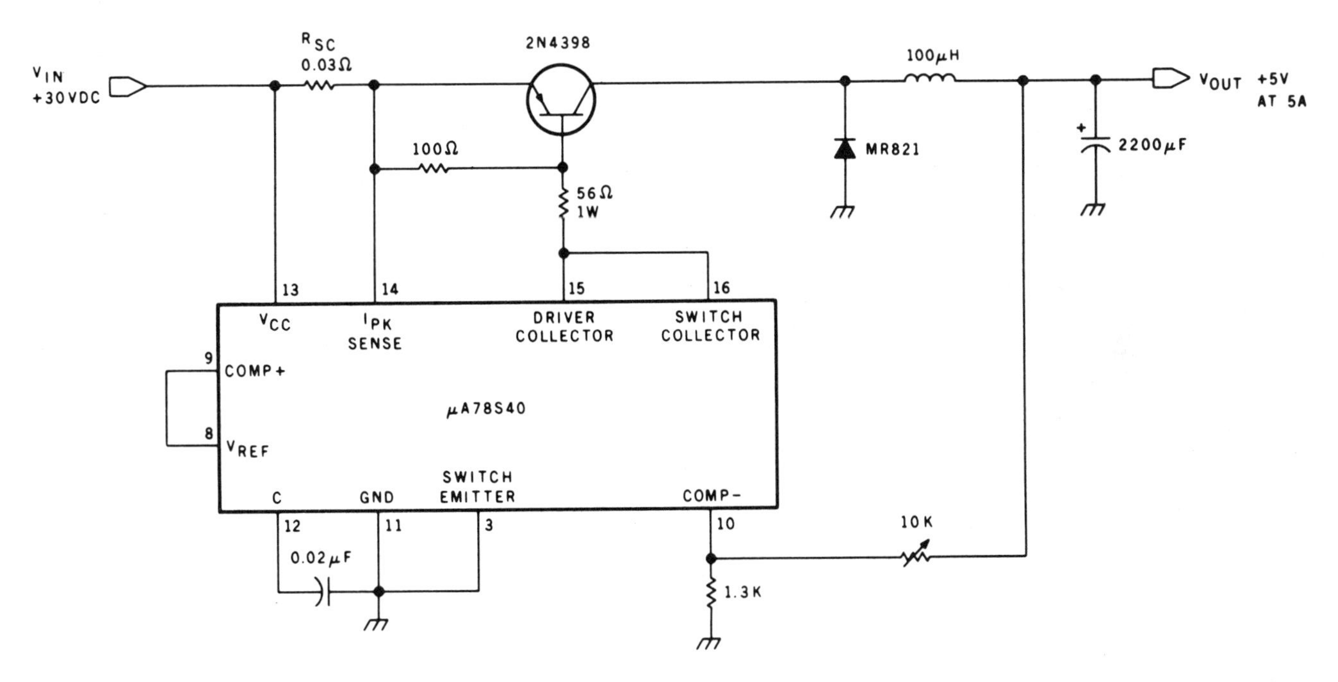

Figure 5: *Schematic diagram of a 5-volt, 5-amp step-down switching regulator. An external heat-sinked transistor and diode are needed for the higher current.*

voltage, only an inductor and capacitor are required to make a highly efficient switching power supply. The internal Darlington-configured transistor switch and diode are capable of handling up to 1.5 amps (peak) at 40 volts.

While virtually no calculation is required to design a circuit that uses a three-terminal linear regulator, you have to do some math to arrive at the correct values for the external components to use with switching regulators. A simple circuit to demonstrate the math is a standard step-down regulator, as shown in the schematic diagram of figure 4. This circuit is a 30-volt input, ½-amp, +5-volt regulator, with a computed efficiency of about 82%. The computation appears in the text box "Component-Value Calculation" on pages 140 and 141.

In applications where the peak current is greater than 1 amp or where you need voltages greater than 40 volts, an external diode and transistor should be used with the 78S40. If you want to experiment, figure 5 shows a regulator designed to provide +5 volts at 5 amps. The component values were selected using the same equations described in the text box. After I had actually built the unit, however, I was unable to get a current greater than 3 amps out of it. I attributed this to using thinner wire in my inductor winding than I should have.

Photo 5: *Close-up of the pot-core halves and the bobbin.*

Photo 6: *A prototype of the step-up switching regulator of figure 6a. This circuit can be adjusted to produce voltages between +8 and +28 volts.*

Another switching-power-supply circuit that might prove useful to you is shown in figure 6. Configured as a basic voltage-step-up regulator, this simple circuit converts +5 volts to +25 volts. With the voltage-adjustment pot shown, the circuit's output actually can be adjusted within a range of +8 to +28 volts. Set to +25 volts, it can be used as an EPROM-programmer supply. The design current rating is 200 milliamps, but this is dependent upon the output voltage. At 8 volts I measured 175 milliamps, but at 25 volts the maximum current was only 30 milliamps. Overall efficiency was about 70%. To achieve higher output currents at 25 volts, it is necessary to use an external switching transistor.

Winding the Inductor

The inductor coil in a switching regulator is designed for high power and large currents. Typically, the windings are around a ferrous core that is toroidal in shape, somewhat like a doughnut. The selection of a specific core size and core material for a desired inductance at a given steady current level is done through a set of iterative calculations using analytic curves showing permeability reduction versus DC magnetizing force (measured in oersteds).

The problem in selecting a core of suitable size and permeability is that all these parameters are interrelated. Rather than attempt to make everyone understand the curves and calculations, I prefer to take a little poetic license, by choosing a core which is optimal for the types of switching supplies I have just described. The remaining calculations are minor and less likely to discourage you from winding your own inductor.

The core I have chosen is the Ferroxcube 2213PA500-3C8, a pot-core set which differs from a pure toroid in that it consists of two cups which are secured together around a bobbin, around which, in turn, the windings are wound. Since the wire is wound on a bobbin rather than around the ferrite material itself, it is easy to change bobbins and experiment with various inductances produced by different numbers of turns and kinds of wire. Figure 7 outlines the characteristics of type-2213 pot core.

Two other factors are noteworthy in this selection. The type-2213 core is made in a variety of ferrite materials. I have chosen to use a type of material known as 3C8, which is a manganese-zinc ferrite substance with medium permeability and low losses. It is designed specifically for high-flux-density applications such as power supplies.

The 2213PA500-3C8 is a fixed-gap pot core. Flux saturation is avoided by introducing an air gap in series with the magnetic path. The effect of this 0.025-inch gap is to flatten the hysteresis loop, allowing greater cur-

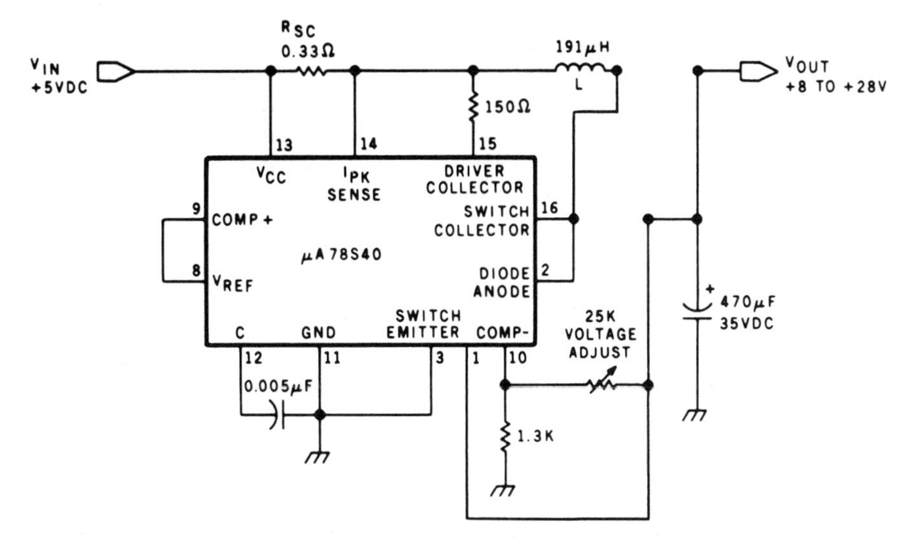

Figure 6a: *A step-up regulator that converts +5 volts to +25 volts, possibly for use in an EPROM programmer. The output voltage may be adjusted within the range of +8 to +28 volts. Current capacity depends on fine points of construction and on the voltage desired.*

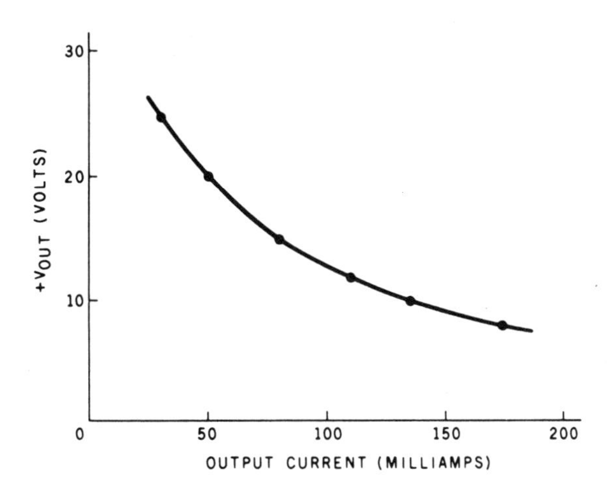

Figure 6b: *The measured curve of output voltage plotted against current for the step-up switching regulator of figure 6a.*

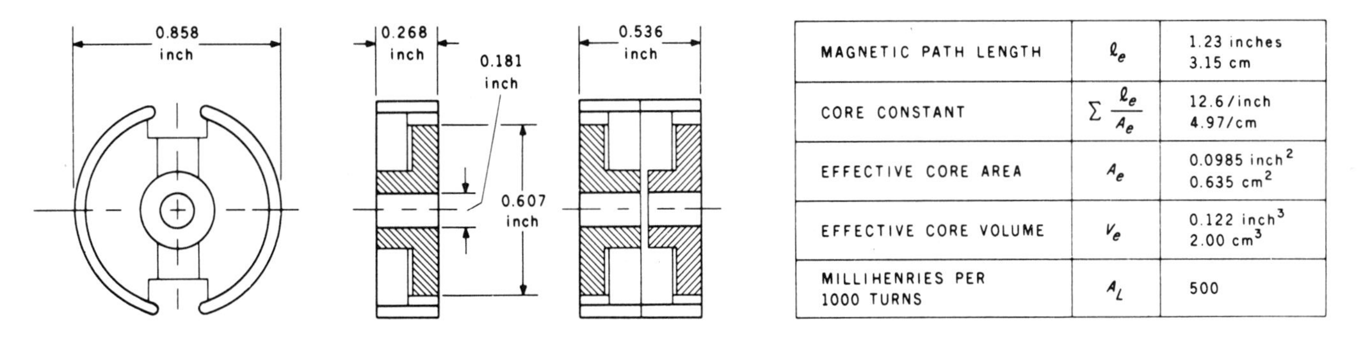

MAGNETIC PATH LENGTH	ℓ_e	1.23 inches 3.15 cm
CORE CONSTANT	$\Sigma \frac{\ell_e}{A_e}$	12.6/inch 4.97/cm
EFFECTIVE CORE AREA	A_e	0.0985 $inch^2$ 0.635 cm^2
EFFECTIVE CORE VOLUME	V_e	0.122 $inch^3$ 2.00 cm^3
MILLIHENRIES PER 1000 TURNS	A_L	500

Figure 7: *Physical and electromagnetic characteristics of the Ferroxcube 2213PA500-3C8 pot core. The coil turns are wound around a plastic bobbin that fits between the two core halves, easing experimentation with different numbers of turns and different gauges of wire.*

rent densities without saturation. An added plus is that it reduces the overall inductance per turn of wire and allows more turns of wire for a given inductance value. Pot cores work best when the windings completely fill the core.

The calculations for winding the specific inductors for the power supplies I have presented are relatively simple. For each core and material, the manufacturer specifies a proportionality factor A_L, which is the inductance in millihenries per 1000 turns. The inductance for any other number of turns N is then

$$L = A_L N^2 \times 10^{-9}$$

or

$$N = \sqrt{\frac{L \times 10^9}{A_L}}$$

where

N = number of turns
L = desired inductance in henries
A_L = millihenries/1000 turns

The 5-volt ½-amp regulator presented earlier required a 275-microhenry inductor. For the 2213PA500-3C8 core, $A_L = 500$ millihenries per 1000 turns. Plugging these values into the equation:

$$N = \sqrt{\frac{0.000275 \times 1000000000}{500}}$$

$$= \sqrt{550}$$

$$= 23 \text{ turns}$$

In addition to determining the required number of turns, we must consider the wire size and available space within the core. The objective is to fill the core, yet maintain a wire size that will carry the highest currents involved. There exist tables and charts of wire size, circular mils per ampere, and available-area specifications for exact determinations. As a rule of thumb, with designs like the ones presented here, the best tactic is to use the largest wire size that will fit for the required number of turns. If the pot core is not completely filled, the effect is not disastrous, but the true inductance could be as much as 10% less than calculated.

In Conclusion

I consider this subject a bit of a shot in the dark. I have found that as I build miniaturized electronic devices for my own use I can no longer tolerate the large volume, weight, and inefficiencies of linear-regulated power supplies. This article is a progress report on some of the circuits I have been experimenting with lately.

I don't anticipate any sudden shortage of ferrite cores as a result of the publication of this article. In my opinion, switching power supplies will come to have a very significant role, but mostly in commercially produced products. The greatest benefit of using this technology is the savings in energy and materials in large-scale production.

Component-Value Calculation

Conditions:

V_D (diode-saturation voltage) = 1.25 volts
V_{SAT} (transistor-saturation voltage) = 1.1 volts
V_{IN} = +30 volts
V_{OUT} = +5 volts
I_{OUT} (maximum) = 0.5 amps
$V_{RIPPLE} < 1\%$ = approximately 50 millivolts
V_{REF} = (internal reference voltage) = 1.3 volts

Calculate:

$$I_{PEAK} = 2I_{OUT} \text{ (maximum)} = 2 \times 0.5 = 1.0 \text{ amps}$$

The peak-current rating of the transistor is 1.5 amps.

Next, determine the value of the current-limiting resistor:

$$R_{SC} = 0.33/I_{PEAK} = 0.33/1 = 0.33 \text{ ohms}$$

Calculate the T_{ON}/T_{OFF} ratio:

$$T_{ON}/T_{OFF} = \frac{V_{OUT} + V_D}{V_{IN} - V_{SAT} - V_{OUT}}$$

$$= (5 + 1.25)/(30 - 1.1 - 5)$$

$$= 6.25/23.9 = \text{approximately } 0.26$$

Therefore:

$$T_{ON} = 0.26\ T_{OFF}$$

It is desirable to have the operating frequency of a switching regulator above 20 kHz (yielding a 50-microsecond period T), but neither T_{ON} nor T_{OFF} should be less than 10 microseconds for the 78S40 device. If we arbitrarily choose T_{OFF} to be 40 microseconds, then the values of the oscillator-timing capacitor (C_T) and inductor (L) are computed as follows:

References

1. Adamian, Andy. *Voltage Regulator Handbook*. Mountain View CA: Fairchild Camera and Instrument Corporation, 1978.
2. Ciarcia, Steve. "Build an Intelligent EPROM Programmer," October 1981 BYTE, page 36.
3. Ciarcia, Steve. "No Power for Your Interfaces? Build a 5 W DC to DC Converter," October 1978 BYTE, page 22. Reprinted in *Ciarcia's Circuit Cellar*. Peterborough NH: BYTE Books, 1979, page 1.
4. Hnatek, Eugene R. *Design of Solid-State Power Supplies*. New York: Van Nostrand Reinhold, 1980.
5. Pressman, Abraham I. *Switching and Linear Power Supply, Power Converter Design*. Rochelle Park NJ: Hayden Book Company, 1978.
6. Sevastopoulos, Nello, et al. *Voltage Regulator Handbook*. Santa Clara CA: National Semiconductor Corporation, 1980.
7. Spencer, John D, et al. *The Voltage Regulator Handbook*. Dallas TX: Texas Instruments, 1977.
8. *Switched-Mode Power-Supply Control Circuits*. Sunnyvale CA: Signetics, 1975.

$$C_T = 0.00045\ T_{OFF}$$
$$= 0.00045 \times 0.000040$$
$$= 0.018\ \mu F$$

Using a 0.02-microfarad standard capacitor value, T_{ON} and T_{OFF} really turn out to be:

$$T_{ON} = 11.6\ \mu s$$
$$T_{OFF} = 44\ \mu s$$
$$T = T_{ON} + T_{OFF} = 55.6\ \mu s$$
$$C_T = 0.02\ \mu F$$

For the inductor:

$$L = \frac{(V_{OUT} + V_D)\ (T_{OFF})}{I_{PEAK}}$$
$$= (5 + 1.25) \times 0.000044/1$$
$$= 275\ \mu H$$

The output-capacitor value C is calculated from the ripple requirements:

$$C = \frac{(I_{PEAK})\ (T)}{8\ (V_{RIPPLE})}$$
$$= 1 \times 0.0000556/8 \times 0.05$$
$$= 139\ \mu F \text{ (use standard 220 } \mu F)$$

Finally, compute the values of the two resistors required for the sampling network. Assuming that the comparator-input current is 1 milliamp (the comparator will work down to 100 microamps) then:

$$R1 + R2 = 5 \text{ k-ohms}$$
$$R2 = (R1 + R2)\ (V_{REF}/V_{OUT})$$
$$= 5 \text{ k-ohms} \times 1.3/5$$
$$= 1.3 \text{ k-ohms}$$

Select $R2$ = 1.3 k-ohms and use a 10-k-ohm potentiometer for $R1$.

14

Build a Computerized Weather Station

One of the few redeeming features of the weather here in New England is the abundance of wind. It may change directions five times a day, but there always seems to be a breeze.

For some time I have been thinking of installing a windmill at my house to provide supplemental electrical power. Maps and charts of my locale suggest that it might be feasible, but considering the complexities of the interactions of climate and terrain in Connecticut, I thought it might be worthwhile to gather more on-site weather data before pouring concrete.

The practical problem of collecting the data inspired this article. I started out by adapting a commercially available anemometer (wind-speed gauge) and wind vane for computer attachment. To simplify getting the data to the computer inside the house, I decided to convert the parallel output from the rooftop transmitter/sensor unit into serial format. Instead of stringing 200 feet of 12-lead cable from the rooftop unit to the computer, I could run a single two-conductor twisted-pair cable.

After this unpretentious start, I got a little carried away thinking how I could do away with even this one cable. But first let me describe the system as I initially built it, starting with the wind sensors.

Photo 1: *Wind-velocity measurements are taken by a cup anemometer and wind vane mounted high above any obstruction to air flow on a section of television-antenna mast.*

Weather Instrumentation

Devices capable of sensing and measuring wind speed and direction can be built from several different basic designs, but probably the most cost-effective wind-speed and direction sensors are the familiar cup anemometer and wind vane, shown in photo 1. The cup anemometer captures the moving air in cup-shaped air scoops that are attached via spokes to a shaft. The assembly spins at a rate proportional to the wind's velocity.

A wind vane looks and works like an arrow with a big tail. As the wind blows, the tail fin acts like a sail, causing the vane to align itself with the direction of the wind.

I briefly considered trying to design a homebrew cup anemometer and wind vane, but several factors argued against this.

In my application, survivability

and accuracy are important. To determine the economic feasibility of a windmill, measurements must be taken, for several months, from a location exposed to the full fury of the weather. An anemometer constructed from paper cups and a small permanent-magnet motor/generator would have been a kluge at best. It might have been capable of measuring wind speed for a little while, but it would not have survived exposure to the elements for very long. Also, I needed to have reliable accuracy to determine the potential power output of a windmill, which is a function of wind speed.

It is not easy to construct a reliable cup anemometer and wind vane. For weather instruments to work, they must survive the weather they are to monitor.

I prefer to concentrate on the applications of electronic technology rather than on techniques of fabrication or artistic excellence. Instead of attempting homebrew sensor designs, I decided to use the wind sensors from a commercially available weather-monitor kit, the Heathkit ID-1890 Digital Wind Computer, sold by the Heath Company, Benton Harbor, Michigan. This is a microprocessor-based unit that displays wind velocity and the date and time of peak gusts. The unassembled parts of the anemometer are shown in photo 2.

If you wish to duplicate my project, you can order the complete kit from Heath and use the appropriate parts. It is unlikely that the required parts will be available separately. (At the time of this writing, the ID-1890 Digital Wind Computer kit is on sale at $164.95, reduced from the regular price of $194.95.)

The required parts from the ID-1890 kit are listed in the text box. The ones unique to the kit are marked with an asterisk, while the rest are fairly common hardware or electronic parts.

The same wind vane and anemometer are used in the more complex ID-4001 Digital Weather Computer kit, which displays wind velocity, temperatures, barometric pressure, and the current date and time and stores weather data for future recall. The ID-4001 sells for $399.95. (In addition, the ID-4001 contains an output port designed to feed data into a Heath H-8 computer system for logging of weather conditions; it is likely that other computers could be connected through this interface as well.)

If you want to build an anemometer, you might try a different

Photo 2: *The anemometer and wind vane were constructed from parts used in the Heathkit ID-1890 Digital Wind Computer, shown here.*

Photo 3: *The partially assembled data encoder. The optical encoder disc is mounted on a shaft between the phototransistors and the LEDs. The opaque areas of the disc block the light path between appropriate phototransistor/LED pairs, producing a unique Gray-coded output value.*

measuring technique, such as the sonic anemometer described in BYTE several years ago by Neil Dvorak. His design used four ultrasonic transducers to measure wind speed, direction, and the temperature of the air. But due to the tight tolerances of the analog circuitry involved, I recommend the cup-anemometer approach.

Adapting the Wind Sensors

The output from the Heathkit cup anemometer and wind vane consists of encoded electrical impulses, which must be specially interpreted by the computer to derive information about wind conditions. Each of these wind-sensor units is not much more than a weatherproof mechanical housing for pairs of phototransistors and LEDs (light-emitting diodes) separated by an optical encoding disc.

As shown in figure 1, the anemometer and wind vane each have six basic components: the air-catching apparatus (the wind cup or vane), the top housing, two printed-circuit (PC) boards, the plastic optical encoder disc, and the bottom housing. The wind cup (or vane) and encoder disc are connected by a shaft supported by ball bearings. As the cup and shaft turn, the shaft rotates the encoder disc between the phototransistors, which are mounted on the top PC board, and the infrared LEDs, which are mounted on the bottom PC board.

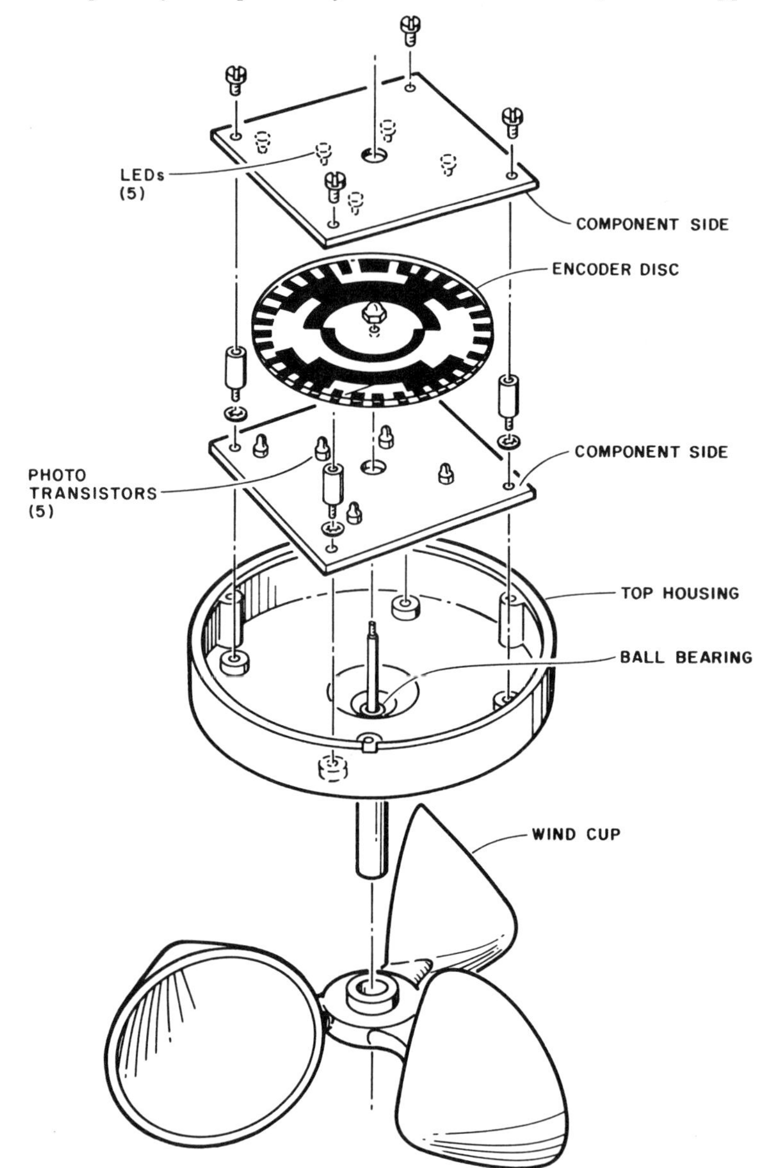

Figure 1: *Exploded mechanical diagram of the inverted Heathkit anemometer unit, showing the five LED and phototransistor positions on the two PC boards. The wind vane uses four LED/phototransistor sets, while the anemometer actually uses only one set.*

As the encoder disc turns, the opaque portions of its surface interrupt the light path between the LEDs and the phototransistors. A schematic diagram of the configuration is shown in figure 2.

There are five separate concentric bands on the encoder disc, as shown in figure 3. An identical disc is used in both the wind vane and the anemometer, but the two units use different portions. In the anemometer, the outside ring of the disc is positioned between a single LED/phototransistor pair. For each revolution of the cup shaft, 32 electrical pulses are generated as the 32 opaque disc areas pass the LED. The wind speed can be measured by simply determining the frequency of these pulses.

The wind vane uses four LED/phototransistor pairs to read the four inner tracks of the encoder disc. These four outputs form a 4-bit Gray-code value (interpreted in table 1), which defines the angular position to a resolution of 1 part in 16. Gray code is a modified binary code in which sequential numbers are represented by expressions that differ in only one bit position. This technique is preferable in slowly revolving encoders because "bit chatter" (oscillation between a 0 and 1 logic level at the point of transition) is less conspicuous than in simple binary or binary-coded-decimal (BCD) encoders. In such encoders, all four bits can change in certain positions (from 0111 to 1000, for example) with only a small change in angular position. Bit chatter can lead to ambiguous indications of direction.

A fairly simple circuit (shown in figure 4) provides a 20-mA (milliamp) current to the LEDs and conditions the output from the photo-transistors. The outputs of the 74LS04 inverter are TTL- (transistor-transistor logic) compatible and can be connected to any computer's parallel input port should you care to use the wind sensors as they are presently configured. Four LEDs connected to the vane output light up to aid calibration.

Calibrating the Wind Vane

Calibration of the vane for installation is simple and requires only a compass. Observe the state of the indicator LEDs with power applied to the vane. Rotate the housing and the vane until the indicators show all zeros. This setting of the vane should be oriented toward true north when the vane is installed. Be sure that the vane housing is secured so it won't rotate.

(In Connecticut there is a 14-degree difference between magnetic and true north, and the vane must be oriented 14 degrees from magnetic north to compensate. This sort of adjustment must be made in most of North America.)

Calibrating the Anemometer

Calibrating the anemometer is another story. The instructions that come with the kit make no mention of how many pulses are produced per second as a function of wind speed. The conversion of pulses to conventional units of speed (miles per hour [mph], kilometers per hour [kph], or knots) is handled by a microprocessor in the Digital Wind Computer, and this information is unnecessary for most users.

For me, however, it was essential. The only way to determine it was by empirically measuring the pulse rate in a known wind velocity. This can be accomplished by moving air across the anemometer, as in a wind tunnel, or moving the anemometer itself in still air. The indications should be the same.

As you can see in photo 6, I moved the anemometer in still air by hanging the anemometer out the side window of my car while driving down a side street near my house (I got some strange looks). As I drove, I measured the output frequency of the encoding mechanism.

Because it was inconvenient to use my frequency counter in the car while driving, I used a battery-operated audio-cassette tape recorder. Connecting it using the circuit of figure 5, which is a portable version of the conditioning circuit previously discussed, I simply recorded the tone produced as the cups spun. The frequency rose and fell as the relative wind velocity increased and decreased. After returning home, I played back the recording into the frequency counter.

I tried various speeds between 15 and 60 mph, and the results were fairly consistent. (I was unable to drive slower than 15 mph without creating a traffic jam.)

The results of my calibration runs are shown in figure 6. The output of this anemometer appears to be 11.6 pulses per second per mile per hour. A frequency of 600 Hz (hertz) corresponds to 50 mph. The curve is quite linear between 20 and 60 mph, but I suspect that readings below 10 mph might exhibit nonlinearities.

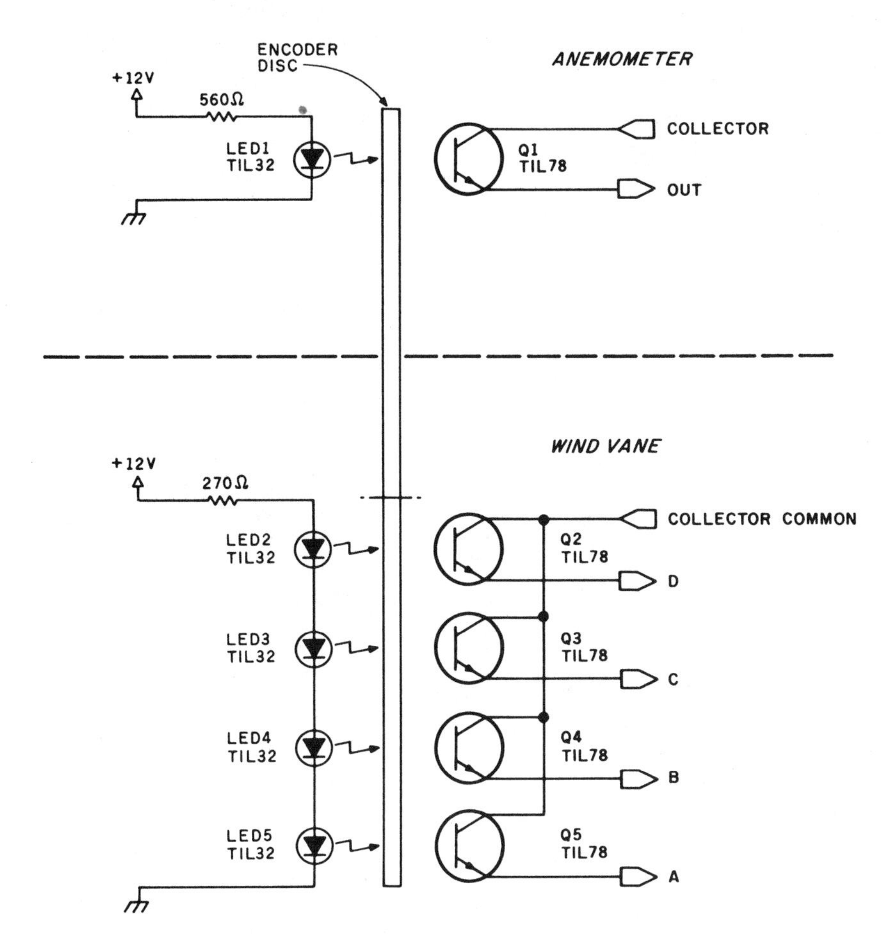

Figure 2: *Schematic diagram of the simple position-encoding circuitry inside the Heathkit wind-sensor units. The TIL32 LEDs and the TIL89 phototransistors operate in the infrared region.*

Decoding the reading of the anemometer with a computer can be accomplished most easily in software. The anemometer's pulse output can be measured by a machine-language subroutine that simulates a frequency

Figure 3: *The optical encoding disc uses a Gray code to eliminate ambiguity in angular position of the wind vane, while in the anemometer only the outermost ring is used as a sort of tachometer.*

counter; the algorithm for this will appear later in this article. The result is simply divided by 12 (close enough) to convert to miles per hour.

Adding a Digital Thermometer

With my scheme for measuring wind velocity well under way, I decided that I could easily upgrade the system to keep track of other weather conditions as well. While wind parameters were essential to my feasibility study, monitoring temperature provided an extra dimension to the data-gathering effort.

Most temperature indicators are analog in nature and require an A/D (analog-to-digital) converter to be read by a computer. This is not only an added complication, but it consumes more parallel-port resources to accommodate the A/D converter. A conversion resolution of 0.4 percent in parallel conversion requires 8 bits and generally occupies an entire 8-bit input port. Similarly, 0.002-percent converters use 16 bits.

Fortunately, parallel conversion is not a necessity in this application and others like it, which require modest accuracy but where input lines are at a premium. Here an analog-input-to-digital-frequency converter is more applicable. In my weather-monitoring system, I already had a digital frequency input from the anemometer. It was advantageous, therefore, to treat the temperature as a second frequency input and use the same software to measure it.

Figure 7 is the schematic diagram of a temperature-to-frequency converter suitable for this application. IC1 is an LM134 analog current source/temperature sensor with an operating range of −55 to +125°C (degrees Celsius). (You could substitute an LM334 to function within a temperature range of 0 to +70°C.) With a 230-ohm value set on the calibrating potentiometer (the R_{SET} value), the voltage from it will increase 10 millivolts per degree Celsius (mV/°C) from some nominal output. Through IC2, the rate is amplified to 100 mV/°C and the offset adjusted to a convenient value. IC3 is a type-2207 voltage-controlled oscillator that acts as a voltage-to-frequency converter. As configured, a 0- to 10-V input will result in a 0- to 10-kHz output. This output frequency is then measured by the computer.

To add excitement to the project, I decided to make my weather station talk.

Calibration is best established by immersing the temperature sensor (IC1) in ice water at 0°C and then in a liquid at a known elevated temperature. The calibration curve will be linear, but its slope is dependent on the particular components used to build the sensor. It's probably best to have a frequency of 2 kHz represent 20°C and 5 kHz represent 50°C. Conversion from Celsius to the Fahrenheit scale should be done by the host computer.

Serial Link to the Roof

Most wind sensors are located remotely from the recording devices. In the Heathkit units, a 150-foot 8-conductor cable is available for this connection. I don't like stringing any more wire than I have to, and I prefer to communicate digested rather than raw data.

The easiest way to condition the weather-sensor outputs and reduce the wiring is to attach a computer directly to the wind and temperature sensors. Any computer could be

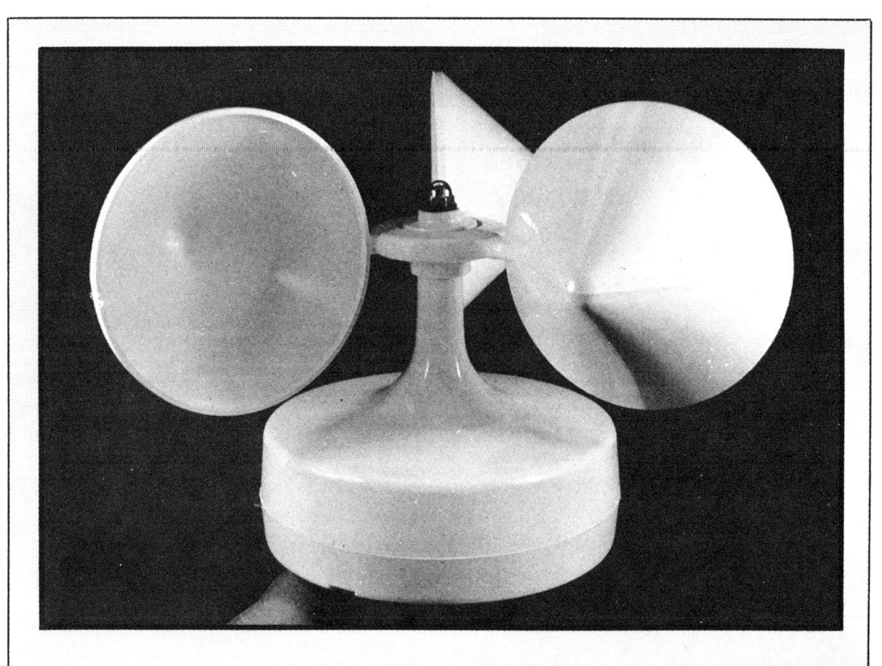

Photo 4: *Completed Heathkit anemometer assembly.*

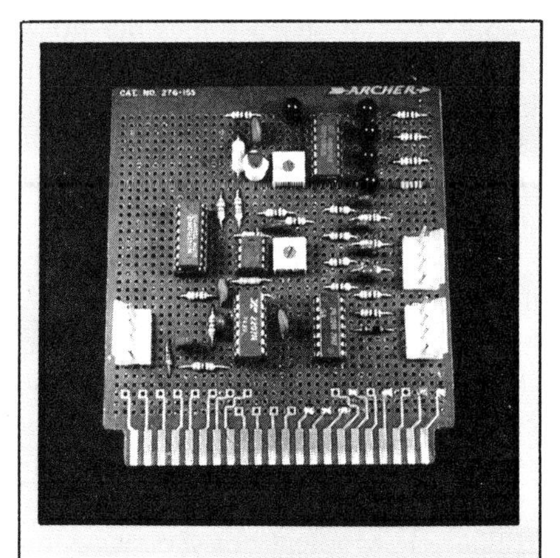

Photo 5: *Prototype of the wind-sensor signal-conditioning circuit board, which combines the input-conditioning and calibrating-display circuitry of figure 4 with the digital-thermometer circuitry of figure 7. The two 4-pin connectors on the right side connect to the wind vane, and the connector on the left goes to the anemometer.*

Number	Type	+5 V	GND
IC1	74LS04	14	7
IC2	7406	14	7

used, of course, but I decided that this was a natural application for the Z8-BASIC Microcomputer (which I described in the July and August 1981 issues of BYTE) used as a device controller and data concentrator, because it contains the necessary I/O (input/output) ports and can be programmed directly in BASIC.

I connected the Z8-BASIC Microcomputer/controller to the sensor units, ran my twisted-pair cable, and set up the computer/controller to use its RS-232C serial port to transmit the results to another computer inside the house for recording or for display on a video terminal.

A message sent down the serial link for recording need only consist of a header and the reduced data. A program running on the display computer could format the data as a compass diagram on the screen, or the Z8-BASIC Microcomputer could perform the formatting, given a more sophisticated program. In either case, the Z8-BASIC Microcomputer/controller board has the latent capability to reduce, record, and format the wind and temperature data as desired.

A Synthesized Weatherman

Having come so far in devising a versatile weather-monitoring system, how could I stop without giving it the ultimate in capability? Using serial communication for recording data was satisfactory, but dull. To add futuristic excitement to the project, I decided to make my weather station talk.

Exploiting as-yet-unused system resources, I connected a parallel-port Sweet Talker voice synthesizer (the subject of my September 1981 article) to port 2 on the computer/controller. I stored a simple phonetic vocabulary consisting of words like "wind," "velocity," and "temperature" in a table in the Z8-BASIC Microcomputer's memory and wrote a program to

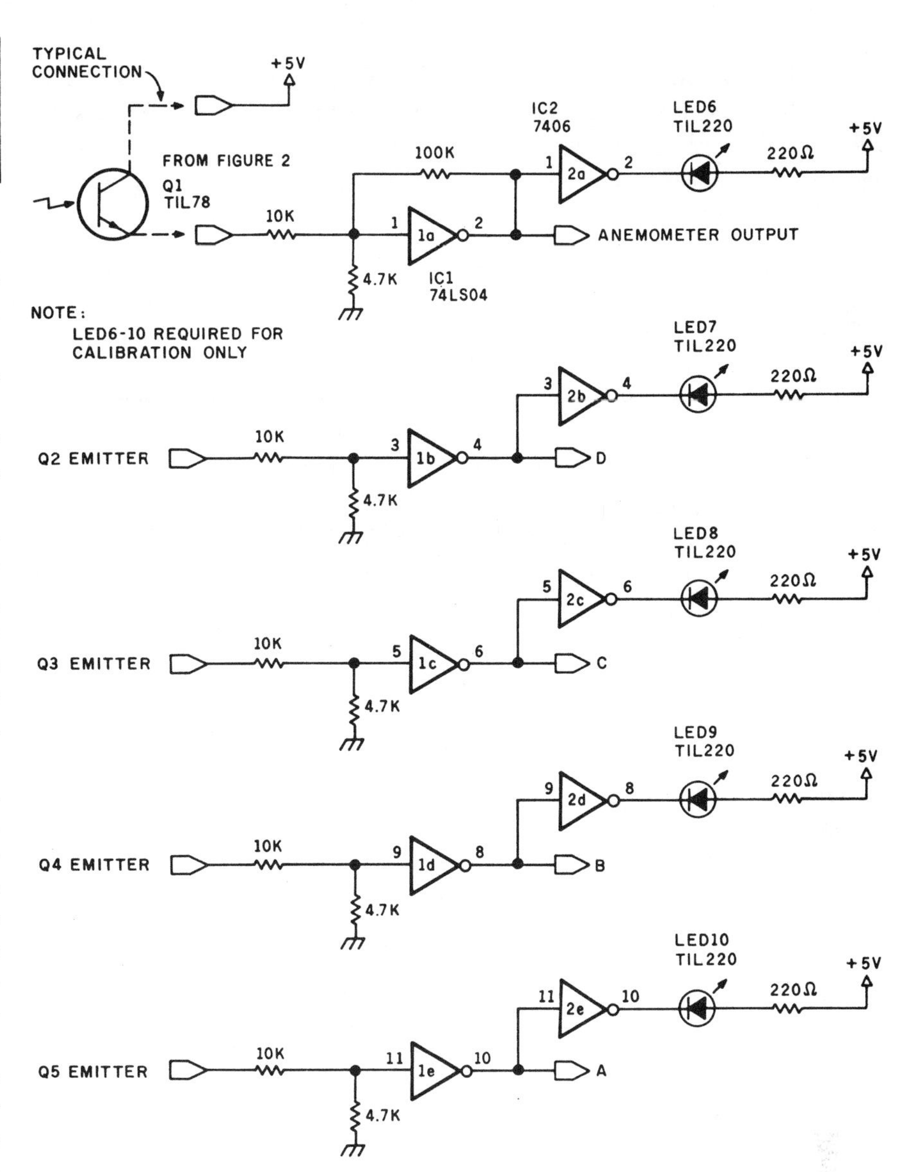

Figure 4: *Schematic diagram of the signal conditioner that accepts output from the phototransistors in the wind sensors and sends it to the controlling computer system. LED6 through LED10 are required only for calibration of the vane.*

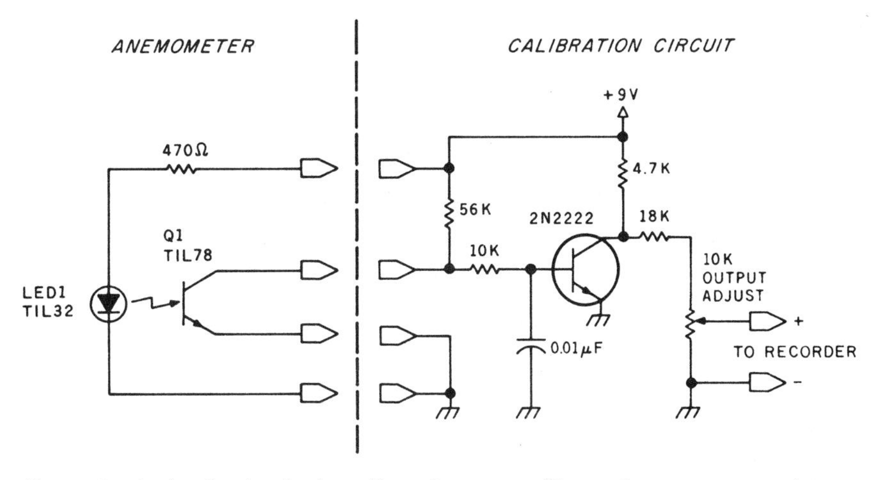

Figure 5: *A simple circuit that allowed me to calibrate the anemometer from my moving car by holding it out the window. The anemometer's output was fed through this circuit into a small, battery-operated cassette tape recorder, and the tape was later played back into a frequency counter.*

read the sensors and send appropriate word phonemes out the port to the Sweet Talker. (A list of appropriate words is contained in table 2.) Continuing along this line of thought to its logical conclusion, I connected the audio output of the Sweet Talker to the input of a low-power radio transmitter.

In the final configuration, the computer/controller board digests the weather-instrument data, the Sweet Talker converts it to English, and the transmitter transmits it to my radio. For up-to-the-minute weather data, I merely tune my radio to 98 MHz and listen to my own synthesized weatherman announcing, "Wind heading: north northwest at twenty miles per hour."

System Configuration

Figure 8 shows an outline of the connections in the completed system between the wind instrumentation, the temperature sensor, and the computer/controller board. The circuit boards are shown mounted on a con-

Compass Position	Gray Code D	C	B	A
N	0	0	0	0
N N W	0	0	0	1
N W	0	0	1	1
W N W	0	0	1	0
W	0	1	1	0
W S W	0	1	1	1
S W	0	1	0	1
S S W	0	1	0	0
S	1	1	0	0
S S E	1	1	0	1
S E	1	1	1	1
E S E	1	1	1	0
E	1	0	1	0
E N E	1	0	1	1
N E	1	0	0	1
N N E	1	0	0	0

Table 1: *Interpretation of the optical Gray code produced by the LED/phototransistor detectors inside the Heathkit wind-vane sensor unit.*

Word	Phonemes
anemometer	AE, N, AH1, M, AW1, AW2, M, I3, T, ER
average	AE1, EH3, V, R, I1, D, J
Celsius	S, EH1, L, S, I1, UH2, S
computer	K, UH1, M, P, Y1, IU, U1, T, ER
direction	D, I1, R, EH1, K, T, SH, UH3, N
east	E1, AY, S, T
Fahrenheit	F, EH1, R, I2, N, H, UH3, AH2, Y, ı
frequency	F, R, E1, K, W, EH3, N, DT, S, Y
hour	AH1, UH3, W, ER
kilometers	K, I1, I3, L, AW1, M, I1, T, ER, Z
maximum	M, AE1, EH3, K, PA0, S, EH3, M, UH2, M
miles	M, AH1, EH3, I3, UH3, L, Z
minimum	M, I2, N, I2, M, UH3, M
north	N, O2, O2, R, TH
peak	P, E1, AY, K
per	P, ER
south	S, AH1, UH3, U1, TH
temperature	T, EH1, EH3, M, P, ER, UH1, T, CH, ER
velocity	V, UH1, L, AW1, S, I1, T, E1, Y
west	W, EH1, EH3, S, T
wind	W, I1, I3, N, D, D

Table 2: *A list of words useful in describing weather conditions, with their Votrax phonemes. These phonemes can be transmitted to the Sweet Talker voice synthesizer by the controlling software running on the Z8-BASIC Microcomputer, in accordance with the prevailing weather.*

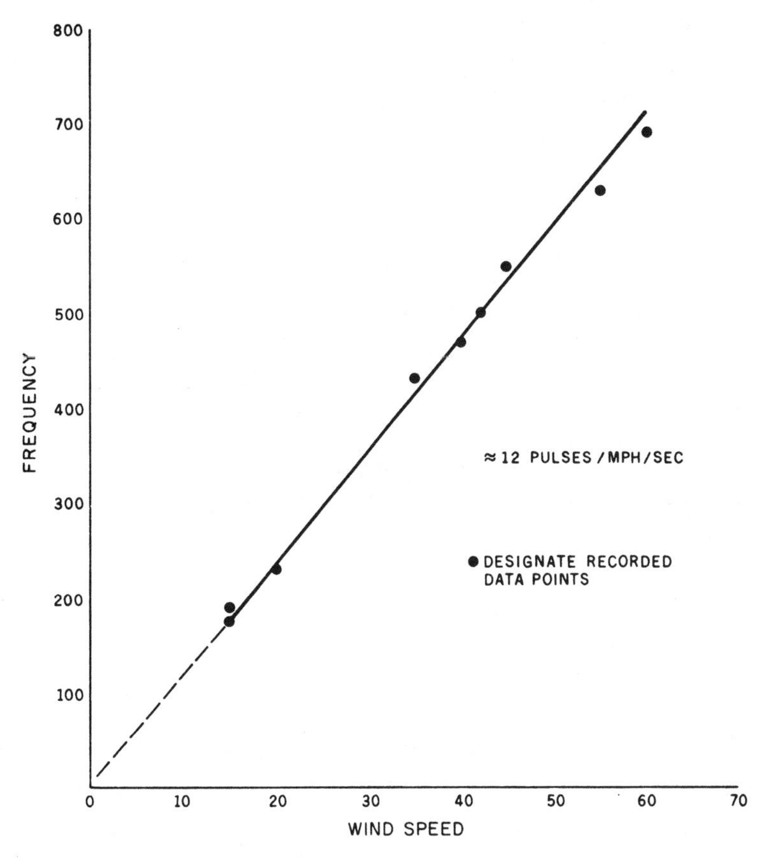

Figure 6: *Graph of anemometer-output voltage as a function of relative wind speed.*

Photo 6: *The anemometer was calibrated by moving it relative to still air; holding it out the window of a moving automobile worked quite well. Driving at a known speed, I used the circuit of figure 5 to record its pulses; the characteristic curve is shown in figure 6.*

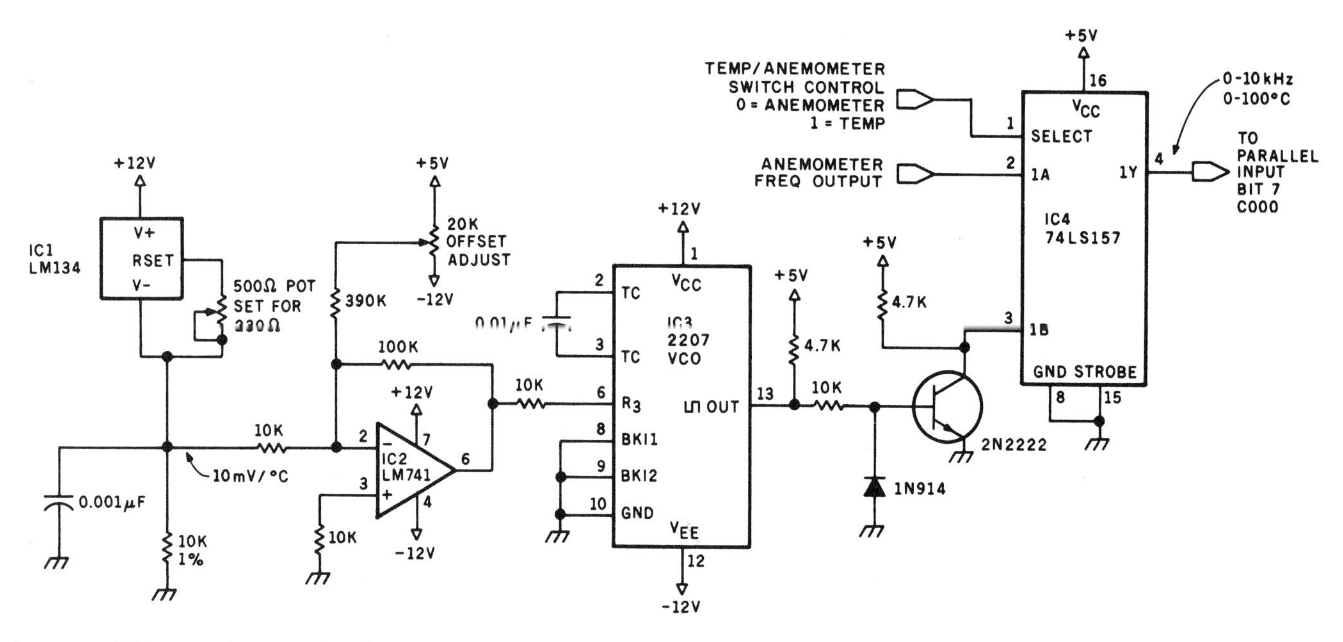

Figure 7: *Schematic diagram of a digital thermometer that varies its output frequency as a function of ambient temperature. The output can be read by the same frequency-counter software that interprets the wind-speed data from the anemometer.*

Component Sources

The following parts list is taken from the Heathkit ID-1890 Digital Wind Computer assembly manual. This list comprises the components necessary to build the wind-vane and cup-anemometer assemblies. Parts unique to the project are marked with an asterisk.

Part Number	Quantity	Description
250-235	8	6-32- by 1/4-inch stainless-steel screw
250-1168	6	#4 by 1-inch stainless-steel screw
254-25	8	#6 lockwasher
253-713	1	#6 rubber washer
252-80	1	6-32 cap nut
255-735	8*	short spacer
250-328	1	8-32 by 3/8-inch stainless-steel screw
250-43	2	8-32 by 1/4-inch setscrew
252-27	2	6-32 locking nut
253-1	2	#6 fiber flat washer
85-1982-1	4*	sensor printed-circuit board
412-635	5	TIL32 infrared light-emitting diode
417-919	5	TIL78 phototransistor
214-208-1	2*	top housing
214-209-1	2*	bottom housing
266-930	1*	wind vane
266-939	1*	wind cup
266-942	1*	wind vane cap
266-943	1*	counterweight
266-1032	2*	optical encoder disc
453-282	2*	1/8- by 3-inch shaft
253-712	4*	C-ring
455-643	4*	bearing
142-711	1	boom parts
142-712	1	boom
595-2399	1*	ID-1890 assembly manual
		miscellaneous hookup wire

necting mother-board in photo 8.

Figure 9 is a flowchart of a minimal application routine that reduces and transmits the resulting data down the serial communication line. Figure 10 is the flowchart of a frequency-counter subroutine written in Z8 machine language. This routine reads the inputs from the temperature sensor and anemometer and derives numeric values in hertz. The routine is stored in memory beginning at hexadecimal location 1500 (as presently assembled) and is invoked from the BASIC/Debug interpreter by the statement

A=USR(%1500)

The value returned in the variable A is the frequency. Listing 1 is the assembly-language listing.

If you wish to set up a radio weather station with a personal touch, as I did, you can use a low-power transmitter: either the AM (amplitude modulation) transmitter in figure 11a or the FM (frequency modulation) unit in figure 11b.

Ideas for Improvement

I have thought about enhancing the

Listing 1: *Assembly listing of the "Windy" routine in Z8 machine language. "Windy" is called by the BASIC statement A=USR(%1500). The frequency is read from bit 7 of the input port mapped into memory-address space at hexadecimal 1500, and the numeric value is returned to BASIC in the variable A. The routine "Windclk" is called in response to an interrupt that occurs every 0.01 seconds.*

```
         Op
Address  Code  D1   D2  Line   Label      Mnemonic         Comment

                               * Windy-   Count anemometer pulses coming in at hexadecimal
                               *          C000, bit 7 (pin K)
                               *
                               * Inputs-  None. Called as a "USR" routine from BASIC/Debug
                               *
                               * Output-  Count of number of pulses seen at location C000, bit 7
                               *          Result returned in registers R12 and R13
                               *
                               * Uses-    R12 - R13        Accumulate number of pulses
                               *          T1,T1 prescale   Set to provide 0.01-second interrupt clock
                               *          R32              Save old value of work-register pointer
                               *          R33              Counts the number of 0.01-second interrupts
                               *          R34 - 35         Indirect pointer to location C000
                               *          R36 - 38         Work registers. R37 becomes 'DONE' flag
                               *          LOC. 100F-1011   JP op code to vector the interrupt to my routine
                               *
                               * Calls-   None, but tests flag set by interrupt-
                               *          driven routine "Windclk"
                               *
                               * Notes-   All register notation is as follows:
                               *            RXX - Denotes full 8-bit register address
                               *            WX - Denotes work-register address
                               *            WPX - Denotes work-register-pair address
                               *            XX - Denotes hexadecimal data
                               *            ** All notation is in hexadecimal radix **
                               *            ** unless otherwise indicated **
                               *
                               *

1500     8F                    Windy      DI               Don't bother me 'til I'm set up
1501     E4    FD   32                    LD R32, RFD      Save current work-register pointer
1504     E6    FD   30                    LD RFD, 30       Point to my work registers
1507     E6    F3   03                    LD RF3, 3        Set up T, Prescale for mod-n, 64 count
150A     E6    F2   90                    LD RF2, 90       Set up T, to give 0.01-second interrupt
150D     E6    FB   20                    LD RFB, 20       Turn on IRQs I/R mask
1510     4C    C0                         LD W4, C0          Registers 34 and 35 point
1512     5C    00                         LD W5, 00            to the data-input address
1514     B0    12                         CLR R12          Clear registers 12 and 13. We
1516     B0    13                         CLR R13            will pass count in them.
1518     3C    00                         LD W3,00         Clear number of I/R's accumulator
151A     6C    10                         LD W6,10         Set up registers 36 and 37 to
151C     7C    0F                         LD W7,0F           store I/R vector for IRQ5
151E     8C    8D                         LD W8,8D         1st byte to store is JP op code
1520     92    86                         LDE WP6, W8      Move register 38 to address at registers 36 and
                                                           37
1522     7E                               INC W7           Step to next byte
1523     8C    15                         LD W8, 15        2nd byte is high byte of address
1525     92    86                         LDE WP6, W8      Store it.
1527     7E                               INC W7           Step to next byte
1528     8C    55                         LD W8, 55        3rd byte is low byte of address
152A     92    86                         LDE WP6, W8      Store this too
152C     46    F1   0C                    OR RF1, 0C       Initialization all done, start T1
152F     7C    00                         LD W7, 0         Clear register 37 to be used as flag
1531     9F                               EI               Turn on I/Rs to catch timer pops
                               *
                               *
                               *
                               *
```

Listing 1 continued:

Address	Op Code	D1	D2	Line	Label	Mnemonic	Comment
					*This is the main counting loop		
					*		
1532	76	37	80		Count	TM R37, 80	Test to see if we're done
1535	EB	17				JR NZ, Done	If bit on, we're through
1537	82	84				LDE W8, WP4	Load data at C000 into R38
1539	76	38	80			TM R38, 80	Is bit 7 at logic 1?
153C	6B	F4				JR Z, Count	If not, loop until it is
153E	76	37	80		Lowwait	TM R37, 80	Check to see if done just like before
1541	EB	0B				JR NZ, Done	If bit on, we're through
1543	82	84				LDE W8, WP4	Pick up data at C000 again
1545	76	38	80			TM R38, 80	Check bit 7 for transition to 0
1548	EB	F4				JR NZ, Lowwait	If not, wait for it
154A	A0	12				INCW R12	If yes, then high-to-low = 1 pulse
154C	8B	E4				JR Count	Do the whole mess over again
					*This is what we do when we're finished		
154E	56	F1	F3		Done	AND RF1, F3	Shut down T1 counter
1551	E4	32	FD			LD RFD, R32	Restore work-register pointer for BASIC/Debug
1554	AF					RET	Go back to BASIC pgm/monitor
					*		
					* This is the interrupt-driven routine that counts clock cycles		
1555	3E				Windclk	INC W3	Add 1 to number of cycles
1556	A6	33	64			CP R33, 64	have we done 100?
1559	1B	02				JR LT, More	No, do more
155B	60	37				COM R37	Turn all bits on in register 37
155D	BF				More	I RET	Issue Return-from-interrupt
					* That's all, folks!		
					*		
					*		

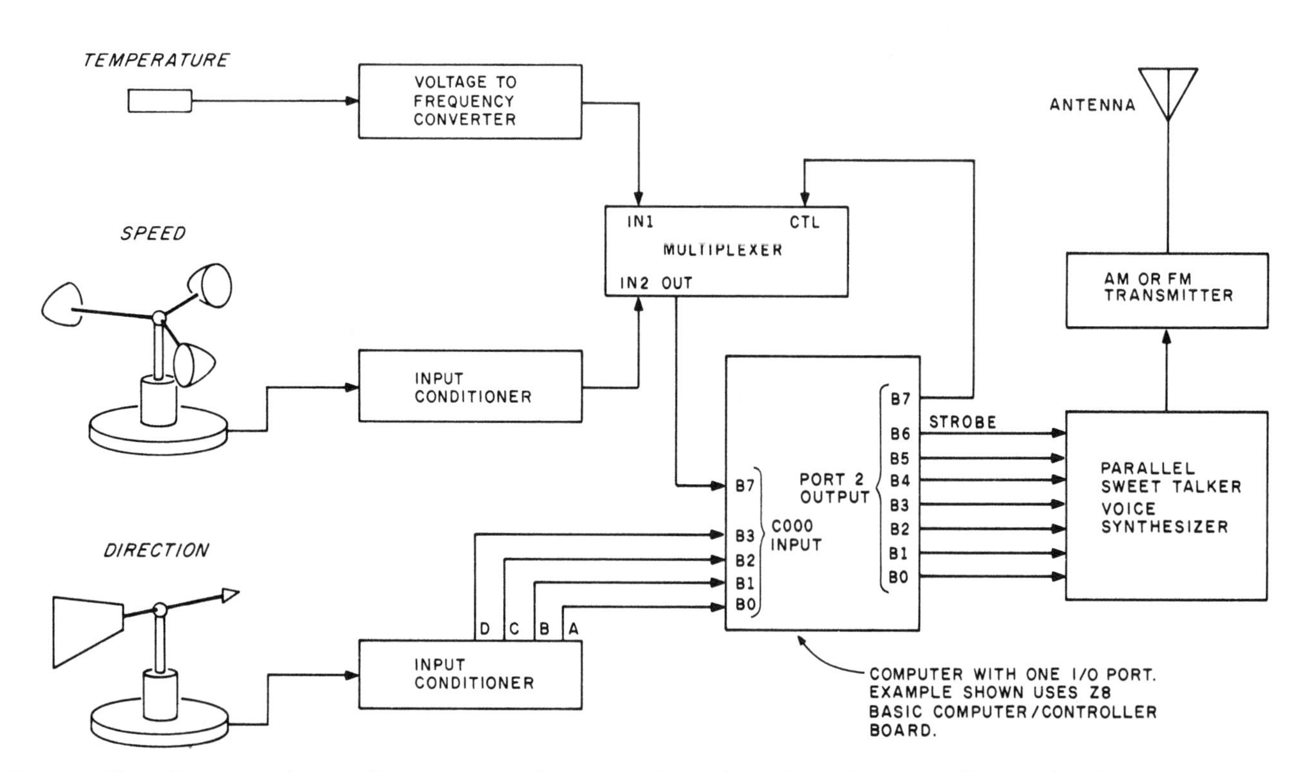

Figure 8: *Block diagram of the complete computerized, voice-synthesized weather radio station. The weather data may be directed to a host computer system for logging if radio transmission is not desired, or the output of the Z8-BASIC Microcomputer/controller could be sent directly to a printer or video terminal.*

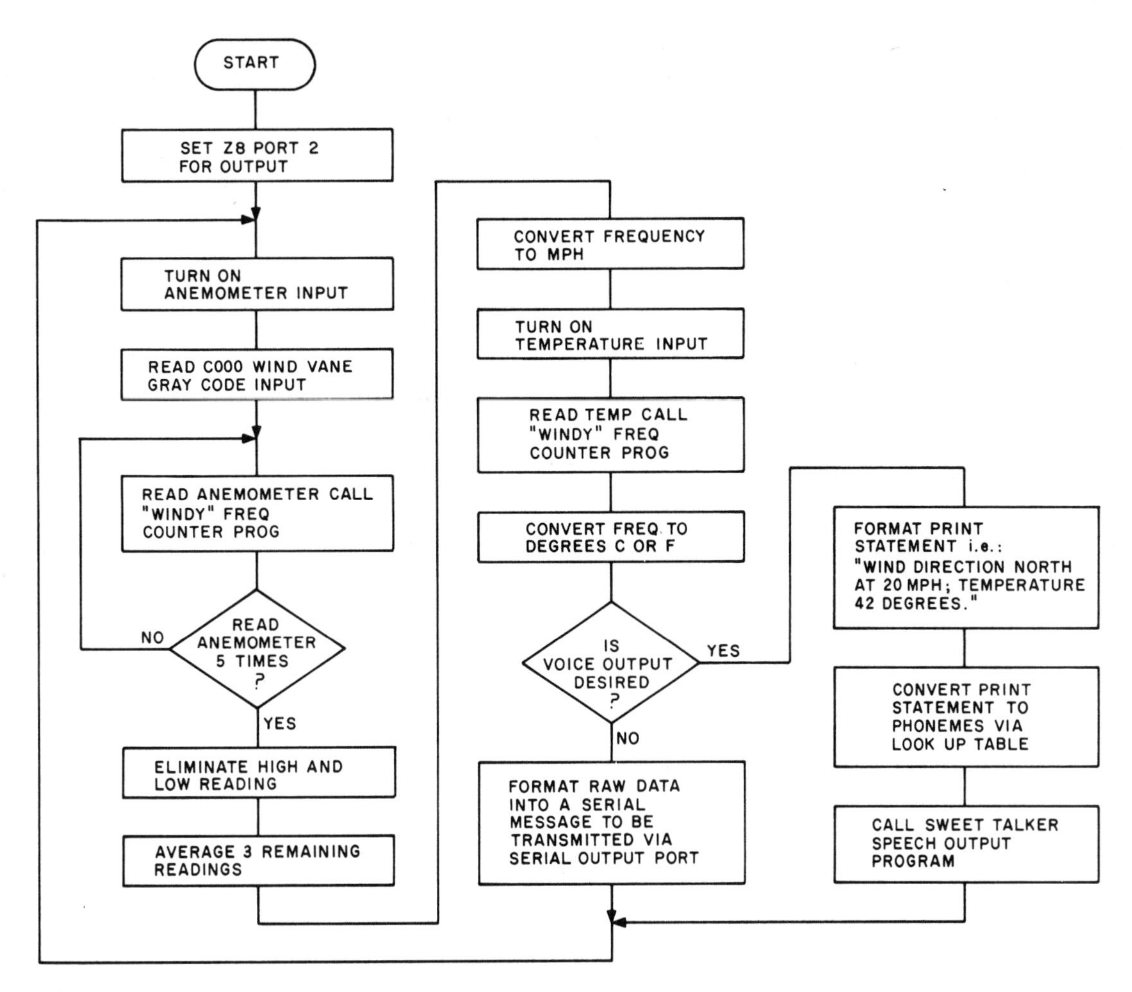

Figure 9: *Flowchart of the program that directs the Z8-BASIC Microcomputer to collect raw data from the wind sensors, digest it, and provide output either to the serial communication line or the Sweet Talker voice synthesizer.*

Photo 7: *The wind vane must be oriented in accordance with true north, which may vary from the magnetic north shown on the compass. Point the vane to the north and rotate the housing until the Gray-code value shown in the calibration display reads all zeros.*

system to measure barometric pressure in addition to the wind velocity and temperature. Conceivably, it could be accomplished with the hardware as presently configured plus one more sensor.

The method I thought might work was some sort of capacitance detector. The majority of modestly priced ($100) barometers are spring-and-bellows pressure detectors. The bellows contracts and expands with the changes in atmospheric pressure. Given the extremely short linear motion and low masses involved, a measuring technique that doesn't require mechanical sensing seems best.

One idea is to use the bellows as one side of a two-plate capacitor. As the pressure changes, the bellows contracts, changing the spacing of the capacitor plates and therefore the capacitance. This capacitor is in turn used to set the frequency of an oscillator. As the capacitance

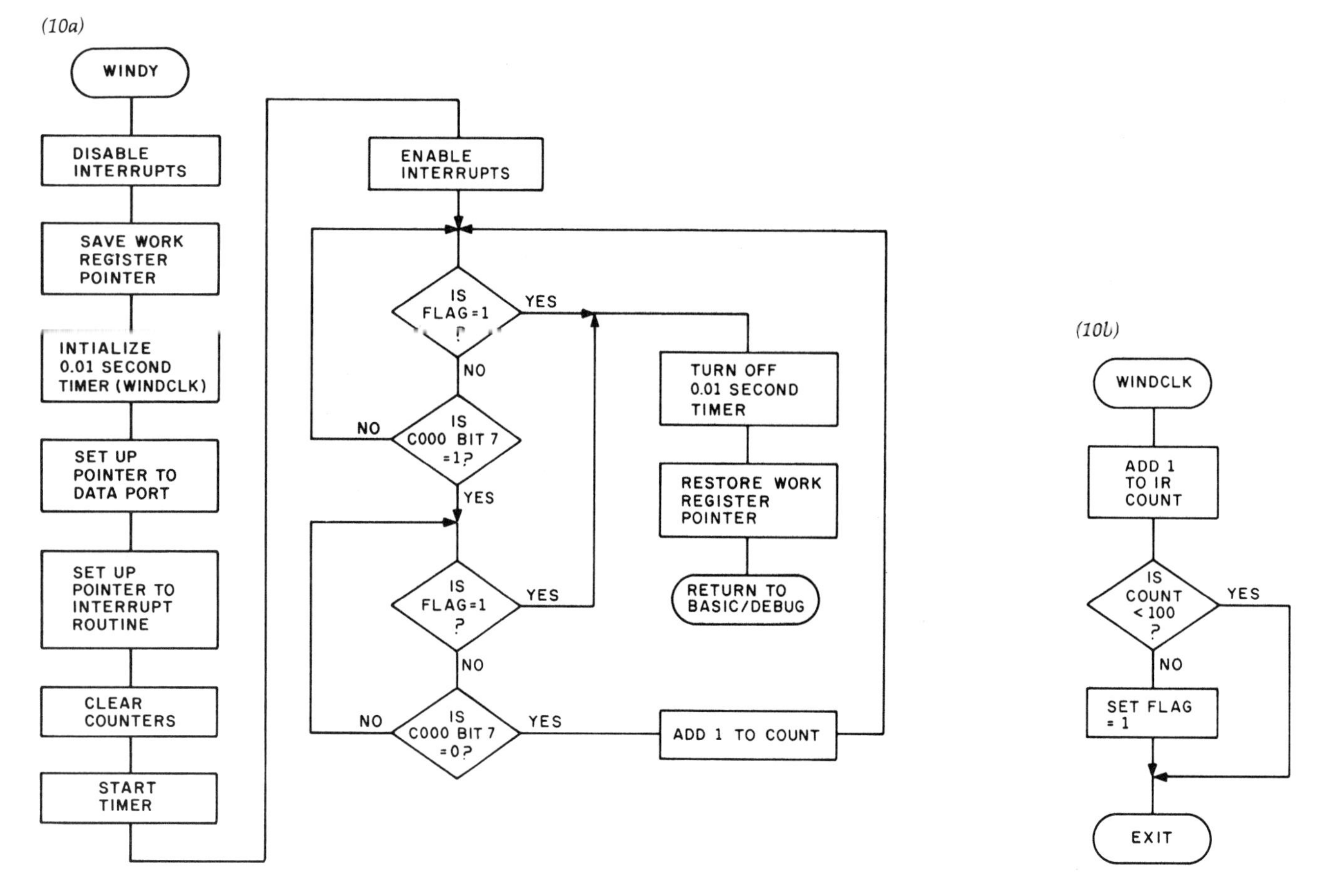

Figure 10: *Flowcharts of the machine-language routine "Windy" (figure 10a) and "Windclk" (figure 10b). The assembly-mnemonic listing is given as listing 1. "Windy" is called from the BASIC interpreter by the statement A = USR (%1500), while "Windclk" is called when the Z8 processor receives an interrupt from the real-time clock.*

changes, it varies the frequency. This output frequency can then be read by the computer/controller in the same way as the anemometer and thermometer.

Concluding Thoughts

I doubt that many of you will go to the extremes that I did to eliminate a few wires, but even directly attaching weather sensors to your computer is a satisfying project. In the process of reading about the specifics of my "synthesized weatherman," you may have seen an application for one of the subsystems. Or with this informa-

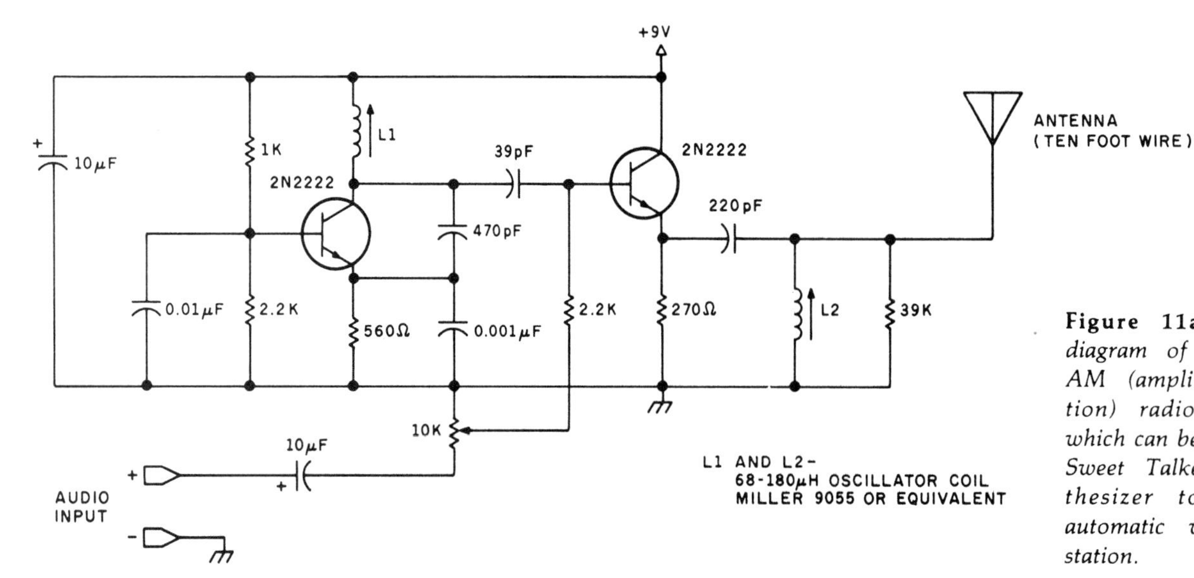

Figure 11a: *Schematic diagram of a low-power AM (amplitude modulation) radio transmitter, which can be used with the Sweet Talker voice synthesizer to create an automatic weather radio station.*

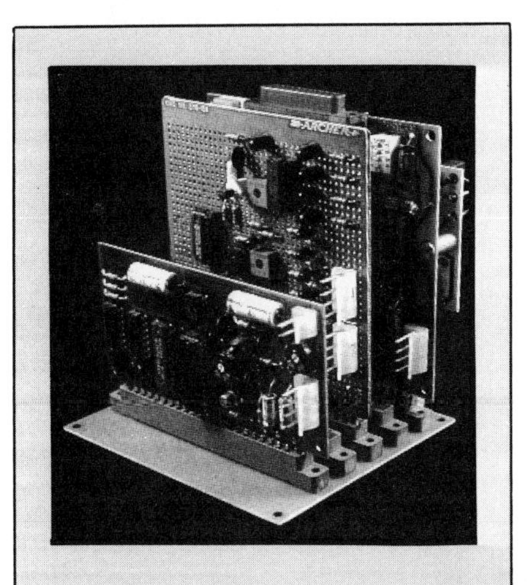

Photo 8: *The complete talking, broadcasting weather station is made up of the Z8-BASIC Microcomputer/controller board, in back, the input-conditioning and temperature board, in the center, and the Sweet Talker voice-synthesizer board, in front. The Z8-BASIC Microcomputer is based on the Zilog Z8 microcomputer-on-a-chip, and the Sweet Talker employs the Votrax SC-01.*

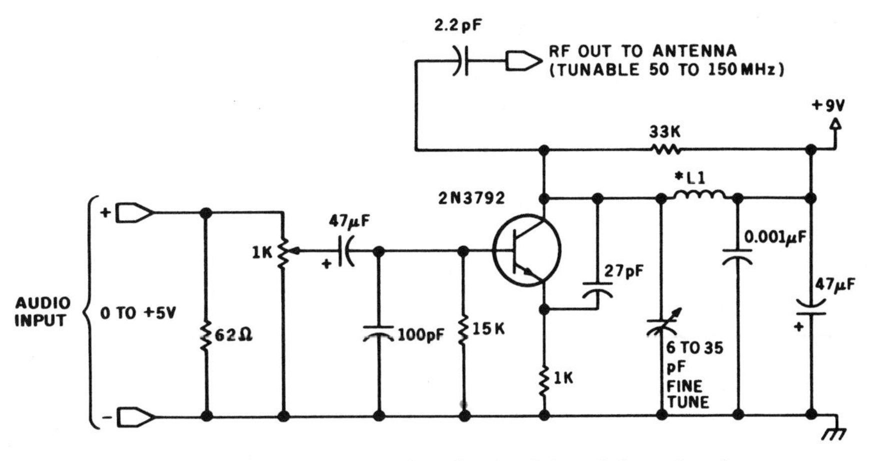

Figure 11b: *Diagram of a low-power FM (frequency modulation) radio transmitter, for use with the Sweet Talker voice synthesizer.*

tion you could easily configure your own custom weather station.

I think I'll listen to my voice-synthesized weatherman for a while before making modifications to the system. My only regret is that I won't be able to observe the expression on my neighbor's face the first time he tunes his radio across the dial. And I may never install a windmill after analyzing the accumulated data, but I will have the most personal weather reports in Connecticut.

See page 447 for a listing of parts and products available from Circuit Cellar Inc.

References

1. Ciarcia, Steve. "Build a Z8-Based Control Computer with BASIC, Part 1," BYTE, July 1981, page 38.
2. Ciarcia, Steve. "Build a Z8-Based Control Computer with BASIC, Part 2," BYTE, August 1981, page 50.
3. Ciarcia, Steve. "Build an Unlimited-Vocabulary Speech Synthesizer," BYTE, September 1981, page 38.
4. Cole, E. W. *Introduction to Meteorology*. New York: John Wiley and Sons, 1970.
5. Dvorak, Neil. "Sonic Anemometry for the Hobbyist," BYTE, July 1979, page 120.
6. Firth, Michael R. "Do It Yourself Weather Predictions," BYTE, December 1976, page 62.
7. Smith, Stephen P. "Graphic Input of Weather Data," BYTE, July 1979, page 16.
8. Viola, John T. and William E. McDermott. "A Recording Mercury Manometer," *Journal of Chemical Education*, October 1976, page 670.

Special thanks to Bill Curlew for his help in writing the software for the Z8 processor.

15

Use Voiceprints to Analyze Speech

Do you ever talk to your computer? I do. But it doesn't understand a word I say. That's just as well right now, because I talk to it mostly in moments of hardware-induced frustration.

Of course, the computer talks to me. If you've read my June and September 1981 Circuit Cellar articles, you know that my computers can talk using two different methods of voice synthesis. At present, a computer can synthesize speech much more easily than it can recognize speech.

Professional speech-recognition systems currently on the market can cost up to $100,000. Budget-priced systems for personal computers are available for about $500, but of course, they don't perform as well.

My mail has been full of requests from readers for a speech-recognition circuit. Most correspondents point out that such a project is a natural follow-up to my articles on voice synthesis. Unfortunately, designing a cost-effective voice-input speech-recognition system is a major project; it not only requires a complete understanding of the techniques involved but also necessitates skills in the design of filter networks and intricate data-comparison algorithms.

The basic concept of speech recognition is rather simple: have a computer digitize the analog voice waveform of each spoken word and compare it to a stored reference vocabulary. A basic block diagram is shown in figure 1.

First, the analog voice input is amplified, then it is digitized to form a *word template.* This template formatting can be done by various techniques that include bandpass filters, A/D (analog-to-digital) converters, zero-crossing detectors, or fast Fourier analyzers. The result, whatever the technique, is a digital representation of the word spoken into the microphone. In an inexpensive speech-recognition system, this word template might be 10 bytes long, whereas in a $100,000 system the template may have 10K bytes of data per word.

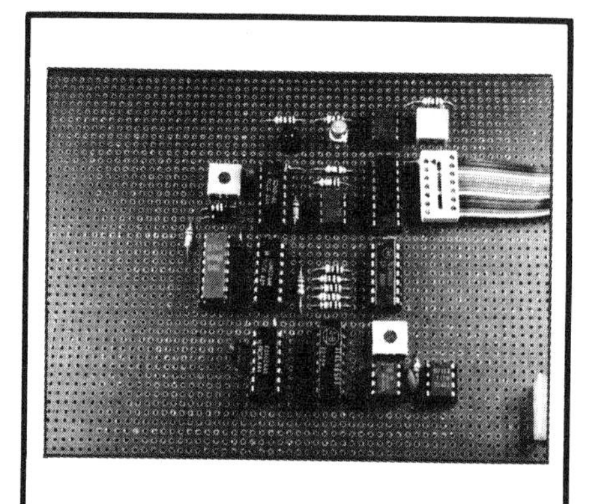

Photo 1: *The sample and scanning circuit of figure 5b, constructed on a breadboard.*

The input word template is then processed by a computer and compared to a series of templates stored in memory. The stored templates constitute the machine's vocabulary. A spoken word is deemed to be recognized when there is an exact or reasonably close match with one of the stored templates.

In practical speech-recognition systems, the size of the word template must be traded off against the amount of available memory or storage and the computing power of the processor. With a small template, the words are not very well defined, and there is a considerable possibility that the computer will confuse two different words. On the other hand, large templates, which more precisely define the words, take considerably more time for comparison as well as more storage space.

To achieve reasonably fast recognition with large templates, the computer must digest information at prodigious speed. In professional speech-recognition systems, a typical processor might perform 1 million 16-bit by 16-bit multiplications per second. Creating such a number cruncher is expensive.

To build a speech-recognition system on a low budget, using a microprocessor, we must make some compromise either in the time allotted for the computer to recognize a word or in the precision with which words are defined in the templates. There must be some amount of storage between 10K and 10 bytes that defines a word sufficiently well for our low-cost speech-recognition system to recognize it within a tolerable duration.

Preliminary Research

This article doesn't tell you how to build a speech-recognition system. We aren't ready for that yet. Instead, it describes a scheme to analyze the audible content of speech so that we can more accurately define a suitable template size.

A definition of just how much data is required can be determined only by carefully examining the spectral content of speech and analyzing the differences between the words we want to have the computer recognize. Just what is the audible difference between the numbers "six" and "eight"? Is there a unique set of data points that allows them to be easily differentiated?

In essence, the information we are looking for is a kind of fingerprint for speech, a *voiceprint.* (It may also be called a *spectrogram.*) By visually comparing the spectral voiceprints of words, we can perhaps come to understand details of definitive templates and the workings of comparison algorithms.

We may find that in a limited-vocabulary speech-recognition system the spectral differences between the words in the selected recognizable set may be so distinct that the template resolution can be reduced to perhaps less than 100 bytes. It is also possible that such an examination will demonstrate that a monumental effort must be exerted to distinguish between two words such as "seem" and "seen."

I hope to eventually write about a voice-response speech-recognition system. Such a project seems to lie within the scope of a Circuit Cellar article. For the present, however, I am still researching certain information about the significant differences between words, seeking to answer such questions as: Must data on amplitude as well as frequency be recorded? Must the input word be digitized in real time? Can the stored template data be compressed in some way? What frequencies are important and which can be ignored? Is there much variation between different utterances of the same word?

This month's hardware project, a spectral voiceprint display, should help answer some of these questions.

What Are Voiceprints?

When you speak, the sound that comes out of your mouth is composed of various frequencies blended together to create the tonal quality that is unique to your voice. If you attach a microphone to the input of an oscilloscope and speak into it, you can watch the frequency and amplitude changes. The bandwidth of meaningful sounds for most voices is about 4 kHz. (Not coincidentally, this is the passband of a voice-grade telephone line.)

Another method of looking at the various frequencies present in voices is to produce a graph of speech waveforms showing frequency as a function of time. An example of this is shown in figure 2. As the word "eight" is spoken, the majority of the energy is between 1 and 4 kHz for the

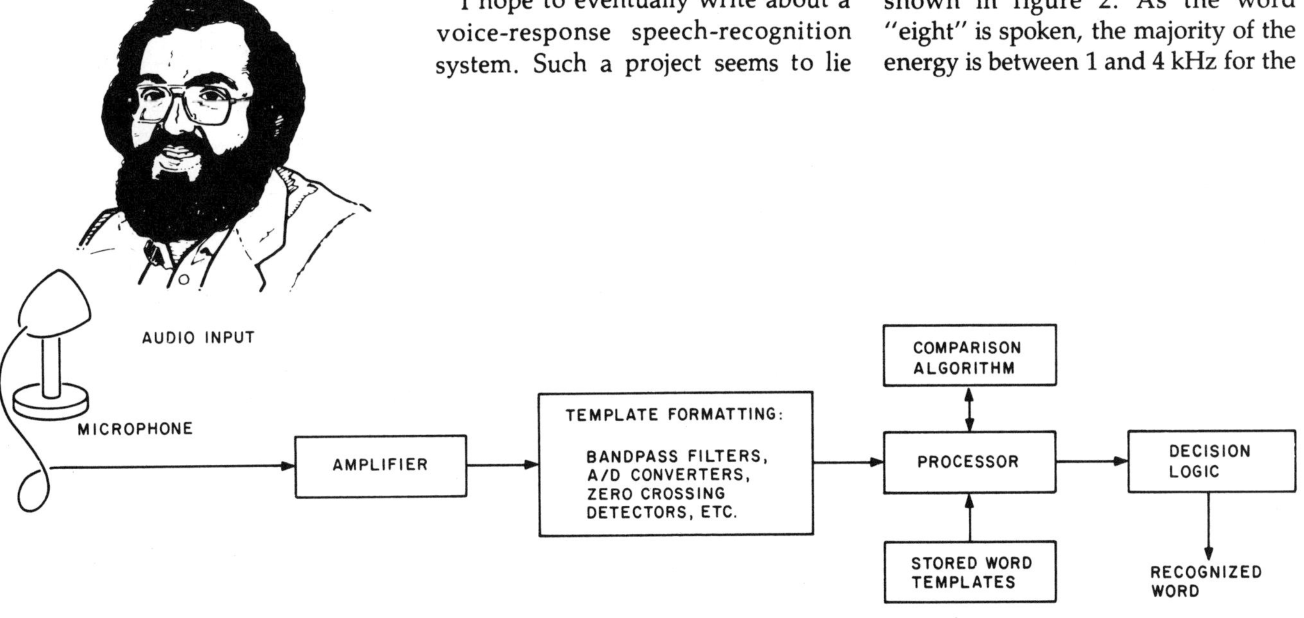

Figure 1: *Block diagram of a computer speech-recognition system using word templates.*

first 0.15 seconds, then a silent period is interrupted after another 0.15 seconds by a quick burst of energy at about 4 kHz. The first waveform group is the "eigh", and the final burst is the "t". A plot of the amplitude also provides significant information.

This sort of voiceprint or spectrogram shows a record of frequency and amplitude versus time.

Producing the graph shown in figure 2 requires an x,y plotter and a real-time spectrum analyzer. This equipment is costly and not generally available to the average experimenter, but with a little ingenuity we can obtain similar results with some simple bandpass filters and an oscilloscope.

Economy Voiceprint Display

The laboratory spectrum analyzer typically used to produce voiceprints often contains either a scanning filter or FFT (fast-Fourier-transform) processor. Such equipment has extremely high resolution (as well as cost) and allows the operator to resolve frequencies separated by only a few hertz (Hz). This is much more resolution than is required for our application, and a more cost-effective real-time spectrum analyzer can be substituted.

Figure 3 is a block diagram of the hardware I used to record voiceprints. It consists of an eight-octave bandpass filter connected to a microphone and some timing circuitry. The outputs of the circuit are connected to the x-axis, y-axis, and blanking (z-axis) inputs of an oscilloscope. The result is a three-dimensional view of the spoken word. The x axis represents time, the y axis represents frequency, and the z axis (brightness) represents amplitude.

The plot thus produced looks somewhat different from the spectrogram in figure 2, but it is equally representative of spectral content. The eight filter sections cover eight octaves from 31 Hz to 4 kHz. Concentrations of energy in the eight octaves appear as eight bands across the display.

For example, if there are any frequencies present around 1 kHz, the 1-kHz band on the display is illuminated, appearing as a stripe across the oscilloscope screen. The amplitude of these frequencies governs the intensity of the stripe. If this approximately 1-kHz signal is weak, the pattern will be dim; if it is strong, the pattern will be bright.

Figure 4 is an example of the kind of display produced by my interface circuit. This is approximately how the word "eight" appears when spoken. You'll note the grouping of energies corresponding to "eigh" and "t" as before. (There is also a shift in frequencies due to the fact that this display was produced by a different

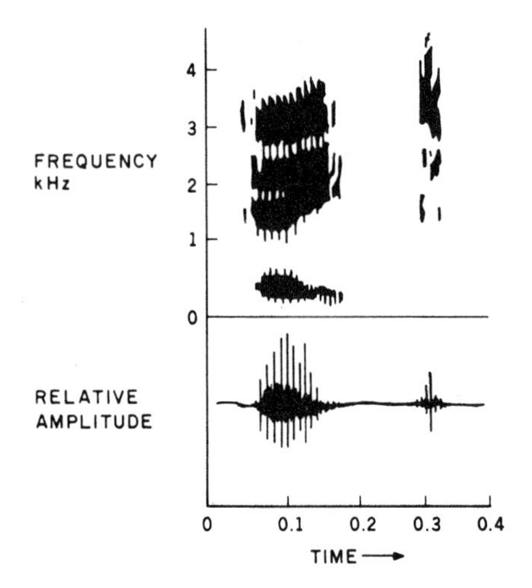

Figure 2: *A conventional voiceprint, or spectrogram, of a man saying the word "eight." Frequency is plotted as a function of time.*

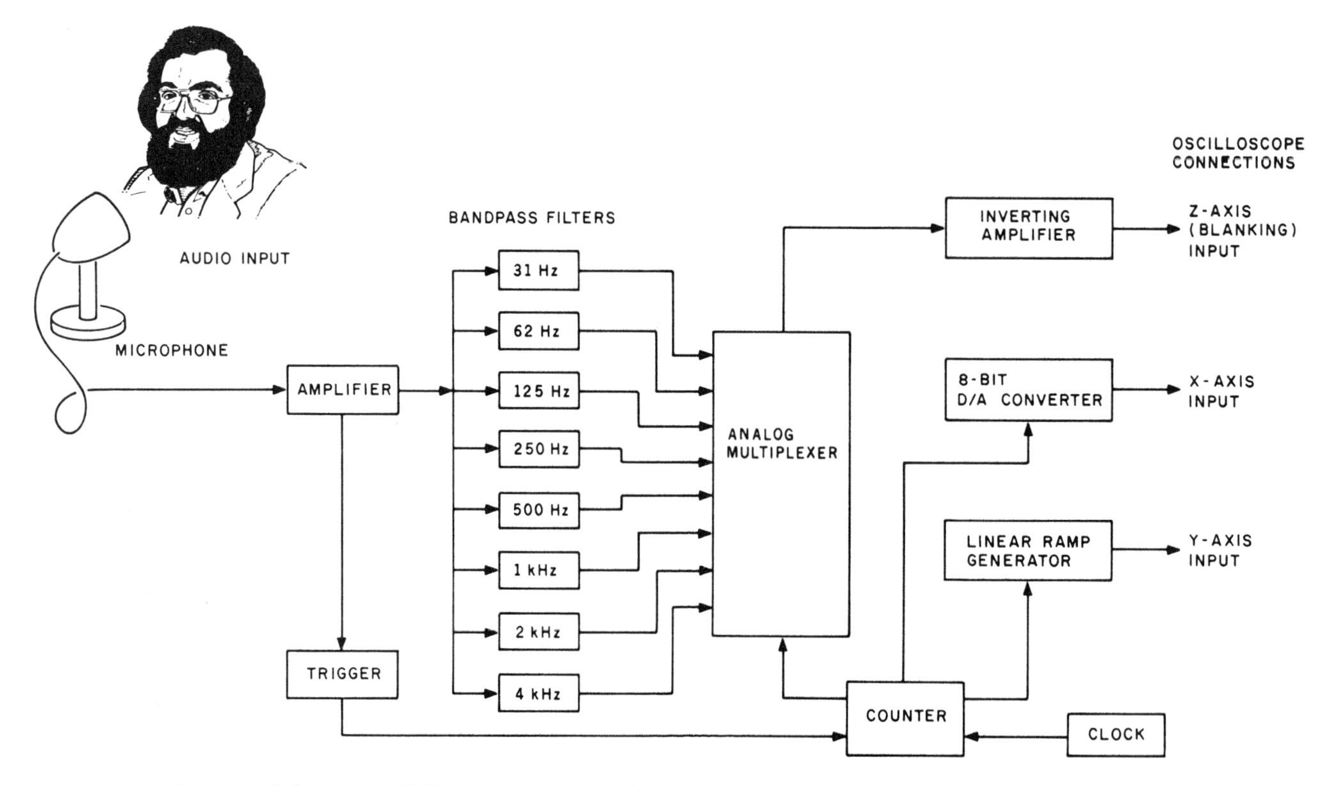

Figure 3: *Block diagram of the Circuit Cellar voiceprint-recording system.*

person speaking.) While unlike the ink-drawn spectrogram, it is equally detailed and unique.

On an 8- by 10-cm (centimeter) oscilloscope display, each frequency band occupies 1 cm on the vertical (y) axis. Time is recorded on the horizontal (x) axis where 1 cm corresponds to 0.05 seconds (all screen photos accompanying this article have these values). A complete word sample therefore represents sounds occurring during a one-half-second interval, consisting of 128 samples at each frequency. Changing the clock rate of the circuit can increase or decrease the scan time.

The scans appear as vertical lines on the screen. A full half-second sample consists of 128 vertical filter scans. Each vertical scan is divided evenly into eight sections corresponding to the eight filters. The bottom is 31 Hz, and the top is 4 kHz. The intensity of each segment of the scan line is determined by the output voltage of the particular filter: the more positive the output, the brighter the segment. If there is no output from a filter section during a segment interval, that portion of the segment will not be illuminated (it will be blanked). As configured, a half-second sample period scans the filters every 3.9 ms (milliseconds).

frequencies and produces a DC voltage output proportional to the average volume level. This output is fed to the voltage comparator IC9, which switches when the average input level is above a certain amplitude, thus triggering the sample period when pronunciation of the word begins.

Integrated circuits IC3 through IC6 are configured as eight separate bandpass amplifiers with center frequencies at 31 Hz, 62 Hz, 125 Hz, 250 Hz, 500 Hz, 1 kHz, 2 kHz, and 4 kHz. The filters are not particularly sharp, possessing a frequency rolloff of about 8 dB (decibels) per octave. The output stage of each filter contains an integrator that converts the pass frequencies into an average DC level.

The timing network appears more complicated than it is. For a half-second word sample the clock rate is set for 4096 Hz and is divided down through an 11-bit counter configured from IC14 and IC15. The reset lines of the counters are controlled by the trigger-level comparator IC9 and an RS (set-reset) flip-flop formed from IC7, sections c and d. When the circuit triggers, the reset line on the counters is raised to a logic 1, and they begin to count. After 2048 clock cycles, the flip-flop is reset and the scanning is stopped. A timing diagram is shown in figure 6 on page 164.

The 3 least significant bits of the counter control the address lines of an 8-channel analog multiplexer, IC16. The eight inputs of the multiplexer are the eight outputs from the filters, and the output of the multiplexer goes to the oscilloscope. When the multiplexer address is binary 000, the 31-Hz filter output is channeled through it to the scope blanking input, where it controls the oscilloscope-beam intensity. Similarly, binary 111 addresses the 4-kHz filter. (While this eight-cycle scan occurs every eight clock periods, it is displayed only at alternating scans.) The other 8 bits of the 11-bit counter set the 256 positions of the x axis (128 displayed and 128 blanked positions).

(The output level of the multiplexer should be set for the blanking range of your particular oscilloscope. This can be either a positive or negative voltage. My oscilloscope, a Tektronix model 2215, requires a negative blanking voltage, so I added IC10 as an inverting amplifier.)

Since an oscilloscope is an analog device, the digital counter outputs must be converted to analog voltages. Two different methods are employed in this circuit. The 8 most significant bits of the 11-bit counter drive an

How the Display Circuit Works

Figure 5 is a schematic diagram of the voiceprint-display system. It is basically divided into two sections: amplifier and filters (figure 5a) and the sample and scanning logic (figure 5b). A prototype of the sample and scanning logic is shown in photo 1. The ribbon cable leads off to the amplifier and filter board.

Integrated circuit IC1 is a two-stage microphone preamplifier (you could substitute a much simpler circuit; this just happens to be the one I used) feeding output into IC2b, which has a sensitivity adjustment potentiometer and an additional stage of amplification. IC2a is an average level indicator. While each filter responds only to its preset frequency passband, this portion of the circuit passes all

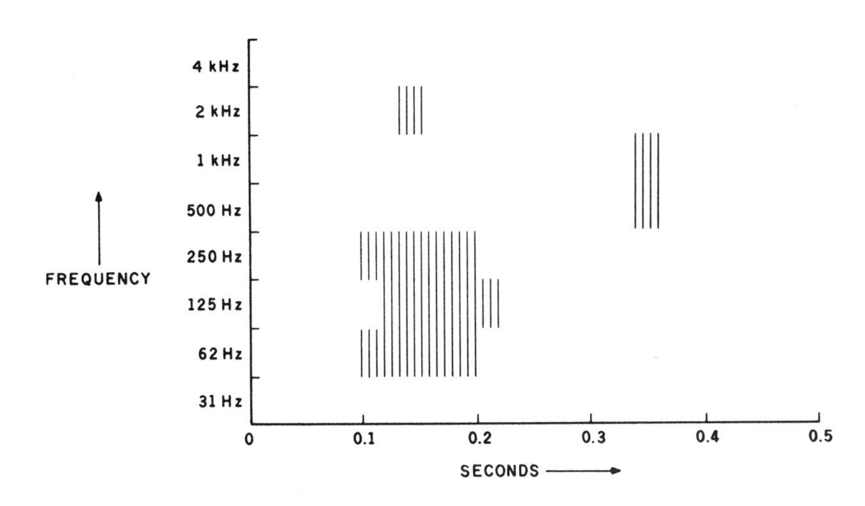

Figure 4: *Typical display produced by the Circuit Cellar voiceprint system, of the word "eight." Although dissimilar in appearance to the conventionally plotted spectrogram of figure 2, it contains the same kind of information, along with indications of amplitude through modulation of the intensity of the scanning beam.*

Photo 2: *Voiceprints, or spectrograms, of various words being pronounced by a Micromouth voice synthesizer, as recorded by the circuit of figures 5a and 5b attached to a Tektronix model 2215 oscilloscope. Eight frequency bands are defined in the vertical y axis, while the horizontal x axis gauges time elapsed during the sounding of the word. The amplitude of energy in the various frequency bands is indicated by the brightness of the oscilloscope trace.*

a. on

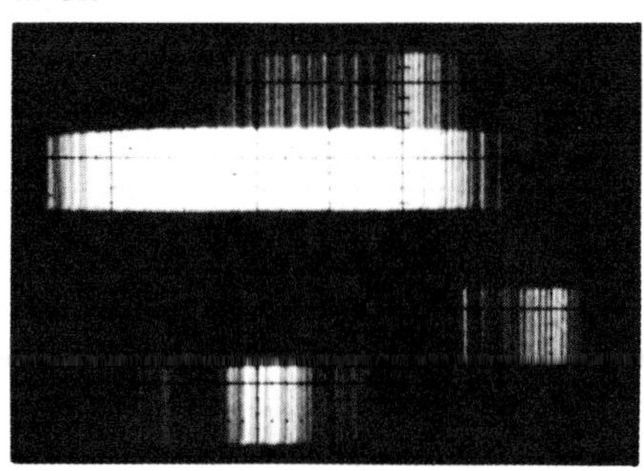

b. off

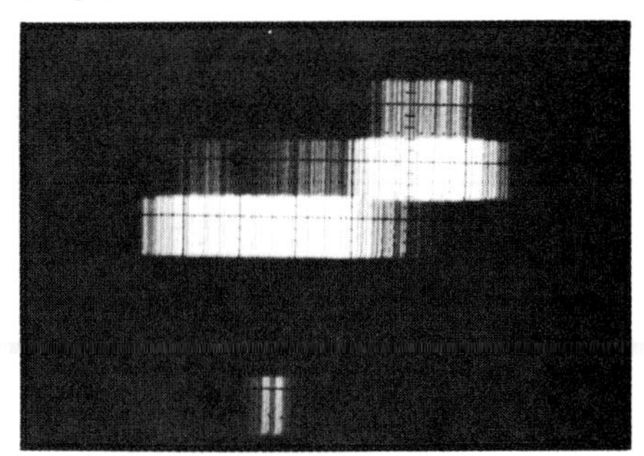

c. error

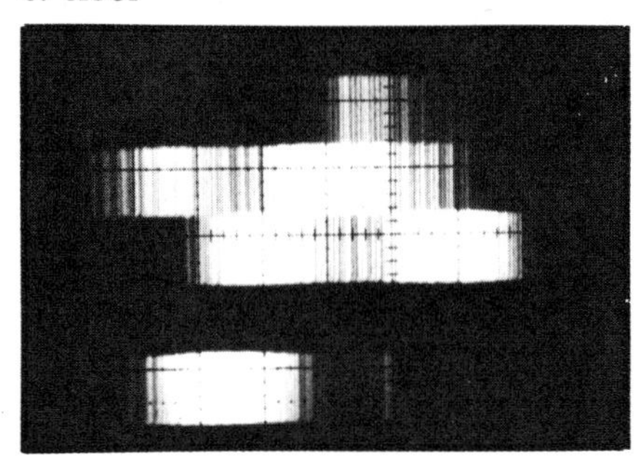

d. ready

e. stop

f. start

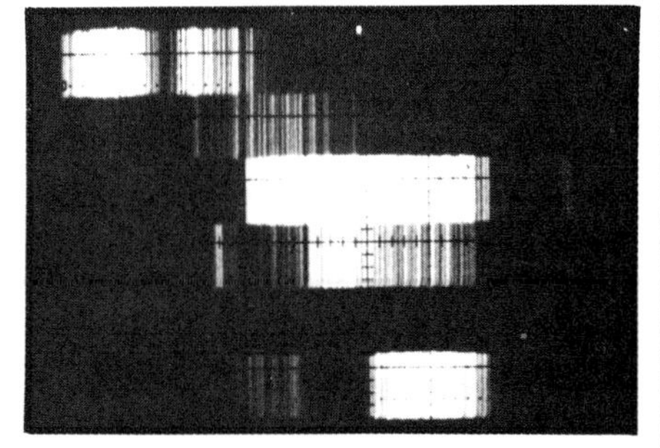

g. zero

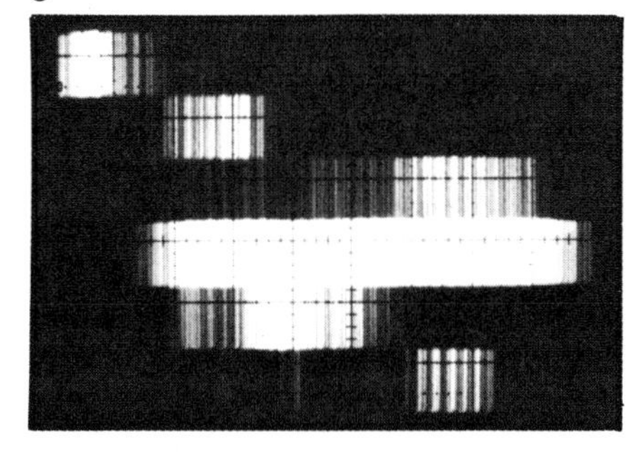

h. one

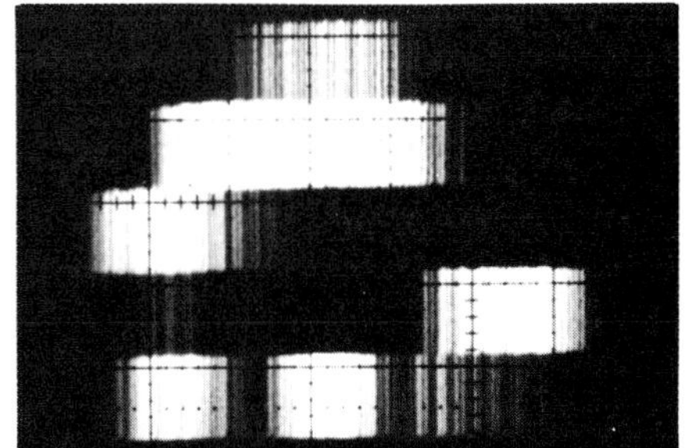

(Continued)

i. two

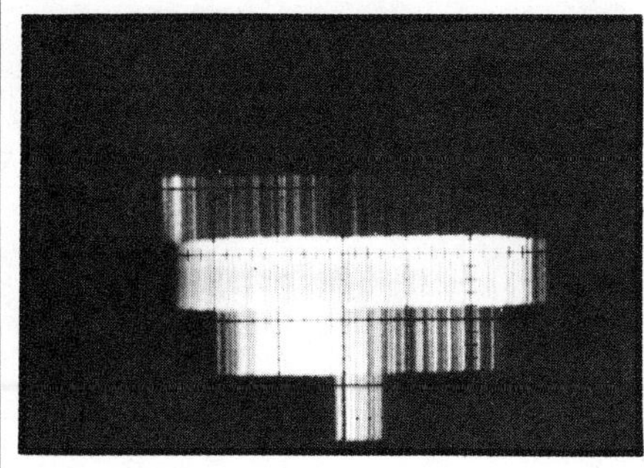

j. three

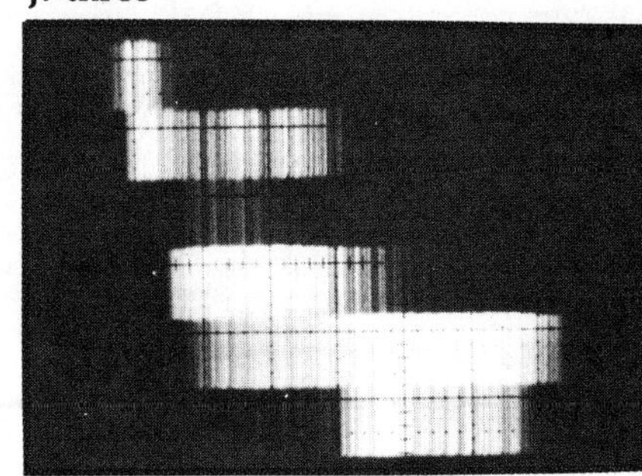

k. four

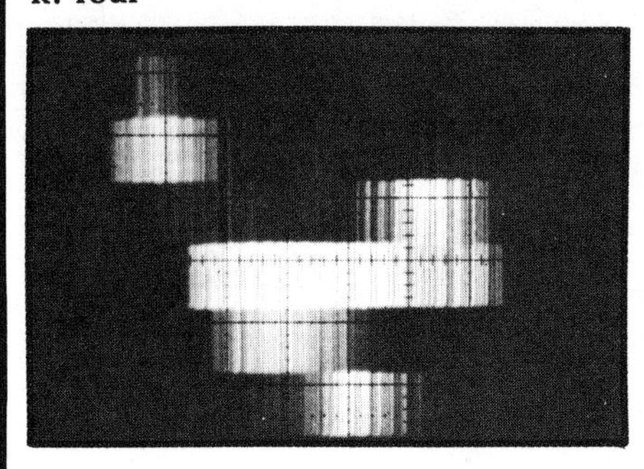

l. five

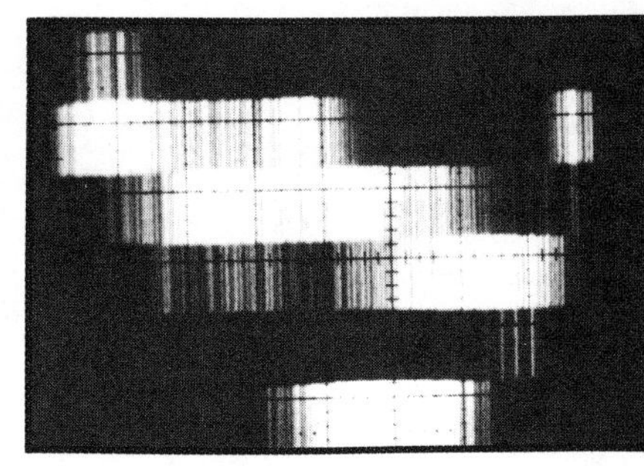

m. six

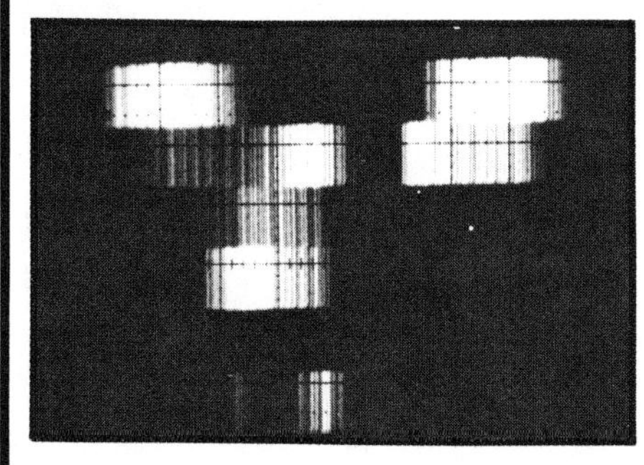

n. seven

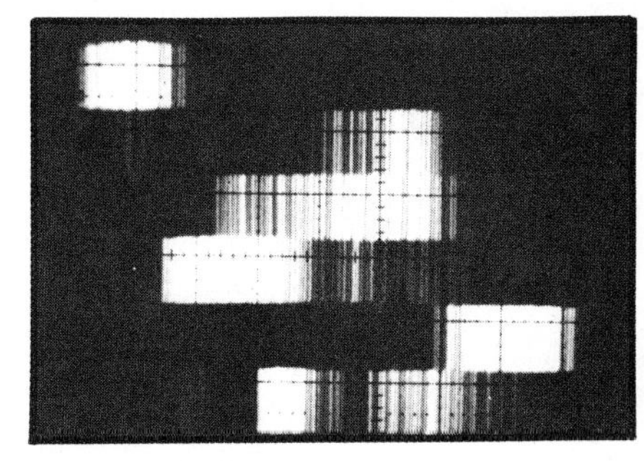

o. eight

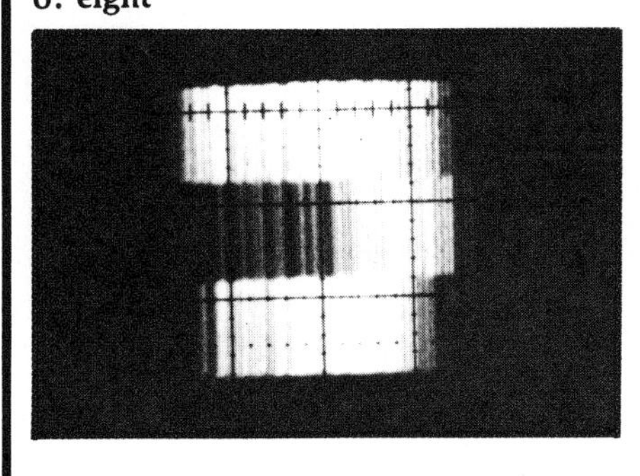

p. nine

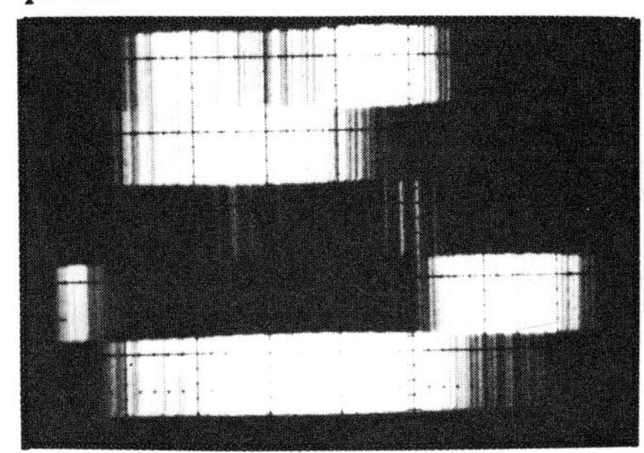

R-2R-ladder D/A (digital-to-analog) converter, which I discussed in my January Circuit Cellar article. With a count of 0, the output is 0 V (volts); with a count of 255, the output is +5 V. Therefore, each step is 19.5 mV (millivolts).

With a clock rate of 4096 Hz, the D/A converter increases its output voltage by an increment of 19.5 mV every 2 ms. With the scope set for x,y-vector display mode, the x-axis scope trace proceeds from the bottom left corner (0 V) to the bottom right corner (+5 V), taking half a second.

Initially I used a 3-bit D/A converter to increment the position of the y-axis beam. However, the 60-MHz bandwidth of the Tektronix scope was sufficient to cause each vertical scan to appear as eight dots rather than eight line segments. The scope was too fast. This was remedied by using a ramp-function generator configured from IC11 and IC17. IC11 is a positive-going integrator, and IC17 is a shorting switch connected across the integrating capacitor.

When the switch is closed, the output of IC11 is 0 V. This is the case during the odd-numbered scans, when the Q4 output of IC14 is high. On even scans the switch is open, and the capacitor is allowed to charge. As configured it charges linearly at a rate determined by the slope-adjustment potentiometer. This potentiometer should be set so that the output of IC11 (pin 6) goes from 0 V to 12 V during the 2-ms half period of the Q4 output. The clock rate affects this time period, so the slope will have to be readjusted if the clock frequency is changed.

Recording Voiceprints

After connecting the voiceprint-generating system to the scope, you can begin to experiment. Speak a word into the microphone. The beam will be triggered, and the trace will move from left to right across the screen. Slowly increase the input-sensitivity potentiometer until the background noise saturates the display. All filters will have some output, and the screen will be completely unblanked. Slowly back off

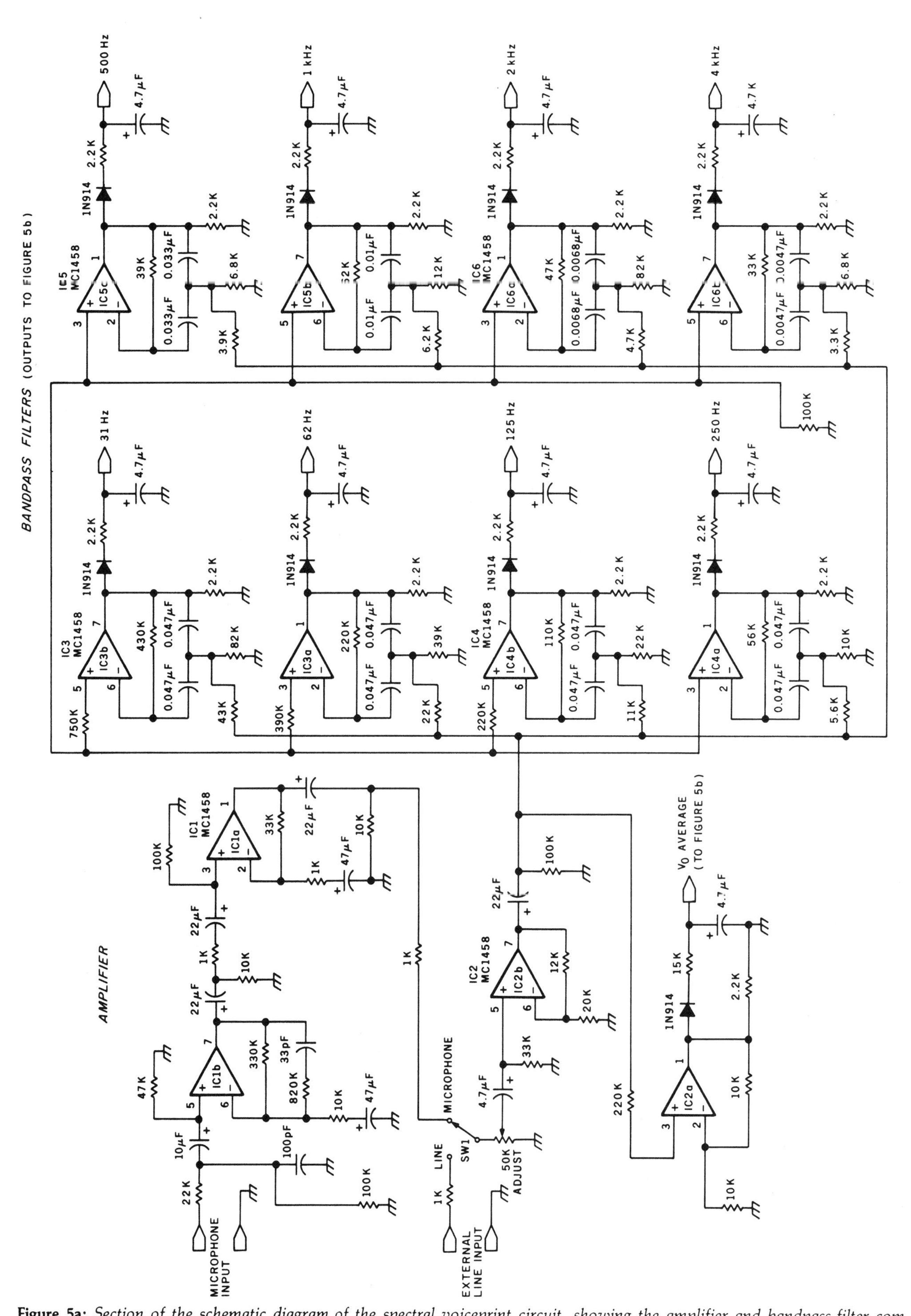

Figure 5a: *Section of the schematic diagram of the spectral voiceprint circuit, showing the amplifier and bandpass-filter components. Eight passbands are selected by the filter stages, with the output sent to the scanning and display section of figure 5b.*

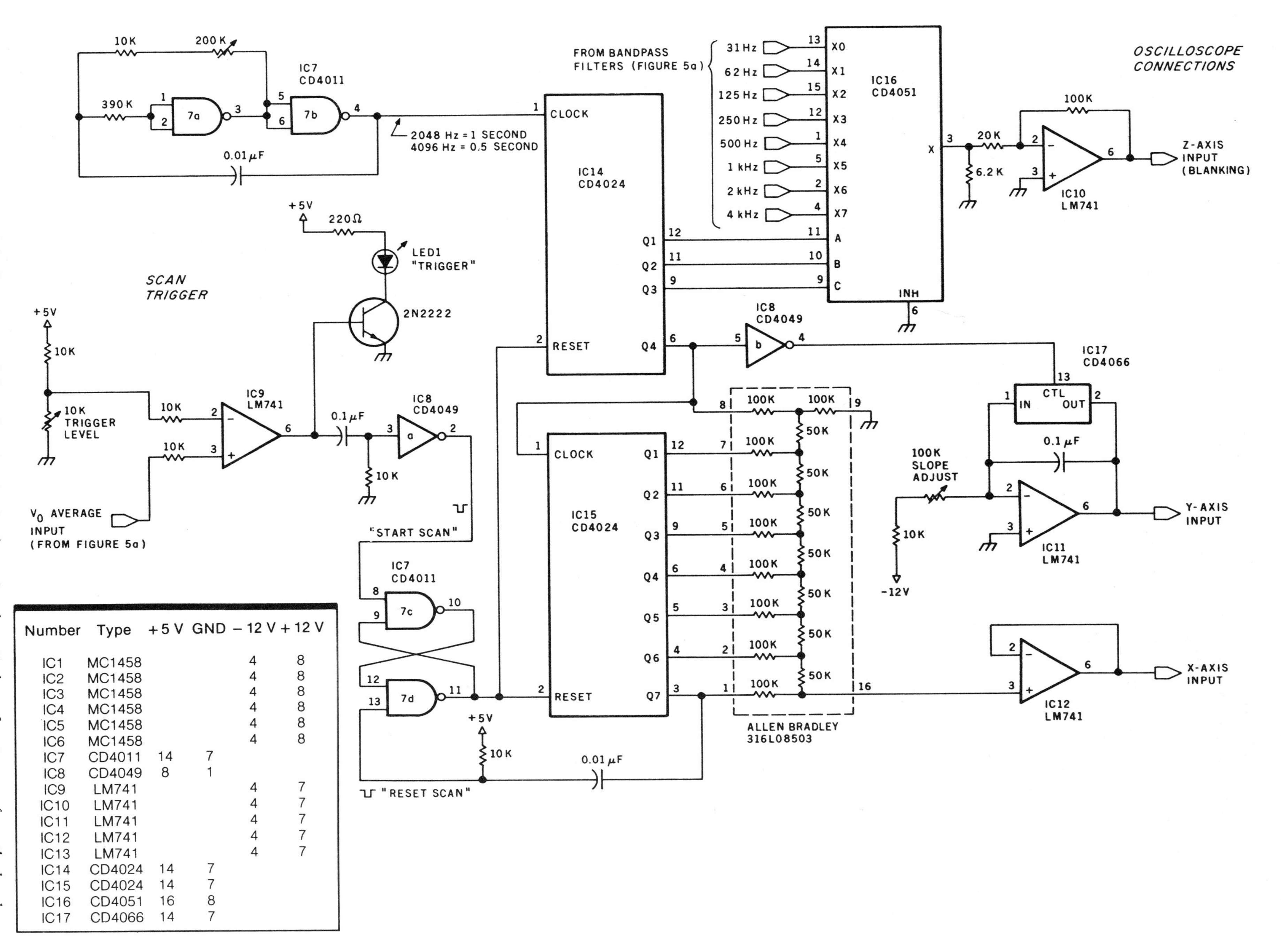

Number	Type	+5 V	GND	− 12 V	+ 12 V
IC1	MC1458			4	8
IC2	MC1458			4	8
IC3	MC1458			4	8
IC4	MC1458			4	8
IC5	MC1458			4	8
IC6	MC1458			4	8
IC7	CD4011	14	7		
IC8	CD4049	8	1		
IC9	LM741			4	7
IC10	LM741			4	7
IC11	LM741			4	7
IC12	LM741			4	7
IC13	LM741			4	7
IC14	CD4024	14	7		
IC15	CD4024	14	7		
IC16	CD4051	16	8		
IC17	CD4066	14	7		

Figure 5b: *The scanning and display section of the voiceprint circuit, shown in schematic form. Input comes from the bandpass filters, and output is sent to an oscilloscope. Discrete resistors may be used in place of the Allen-Bradley 316L08503 resistor package.*

the gain until the display appears to respond as you speak a word. Continue this adjustment until the display looks like the sample photos.

In bright ambient light the display will appear as a single vertical line moving across the screen. If you darken the room, the persistence of the phosphor screen will allow you to see the entire voiceprint.

To record the voiceprint for posterity you will need a camera. In a darkened room, simply set the camera on a tripod, open the shutter manually, allow one sample to scan on the screen, and then close the shutter. This is essentially the technique I used to produce the sample voiceprint photos that accompany this article. Unfortunately, since the Tektronix 2215 has no reticule illumination, no scale is reproduced in the photos. Keep in mind that there are eight vertical filter bands and that the x axis is half a second.

Examples of my own voiceprints wouldn't be especially helpful to you in trying to align your voiceprint system, so I have provided examples that can potentially be duplicated and compared. All the voiceprint photos here were produced using the output of a Micromouth voice synthesizer. The Micromouth, which I described in my June 1981 article, uses a National Semiconductor Digitalker speech-synthesis chip set. It has a limited vocabulary which is extremely intelligible and eminently reproducible. If you have a Micromouth, simply connect it up and compare your results to the various prints of words and numbers shown here.

Experimental Results

What can we learn from studying the results of our simple testing? First of all, the voiceprints of speech synthesizers and people are very different. While the words sound much the same to the ear, the frequency content is rather different. This difference should not bother a computer speech-recognition system so long as the word templates are set to recognize either synthesized or natural voices. But because of its repeatable speech, the synthesizer might provide a good way to initially test a speech-recognition system.

In general, there seem to be considerable spectral differences between the words in the minimum useful vocabulary I chose as examples. Because of the great differences, a speech-recognition system could use minimally precise template data to differentiate between these words.

Consider how a computer could store these voiceprints as word templates. An A/D converter could be used to read the filter values. Storing the output values from 128 scans of eight filters requires 1024 (1K) bytes for each word, assuming the use of an 8-bit A/D converter. The amount of memory required can be reduced by eliminating the dead air time at the beginning of words and between the sounds contained within a word.

Perhaps storing the output amplitude of the filters is unnecessary, and a simple threshold detector would be sufficient. A logic 1 could indicate that there is some spectral content in that frequency range while a logic 0 indicates none. The eight instantaneous filter outputs could then be stored in a single byte rather than eight. This translates into a memory requirement of 128 bytes per sample period. This presumes that information about the frequency content of speech with respect to time is more important than information about the amplitude of the energy in the different frequency bands. I think it will depend a lot upon the vocabulary chosen.

Finally, I saw little activity in either the 31-Hz or the 4-kHz band in speech both from my own voice and from the Micromouth. This may be a limitation of the hardware, but I think it would be safe to eliminate these passbands from any voice-response system. In my experience, the three frequency ranges that seem to always contain the most energy are about 60 Hz to 200 Hz, 200 Hz to 500 Hz, and 1 kHz to 2 kHz. I am at present unwilling to design a speech-recognition system with only three sampling passbands, but I'm still gathering data.

In Conclusion

I haven't yet decided how I will configure my speech-recognition system. I have only one major design criterion so far: because writing comprehensive software algorithms isn't among my greatest pleasures in life, I will attempt to do as much in hardware as I can.

Perhaps if I stall long enough a few inexpensive integrated circuits that can do it all will emerge from Silicon Valley. I have heard promising reports on a few such products. I know of the intense interest many of my readers have in the subject, and I intend to build a speech-recognition system as soon as I can make it cost-effective.

I hope that this article has at least helped you understand some of the first steps in speech recognition. If

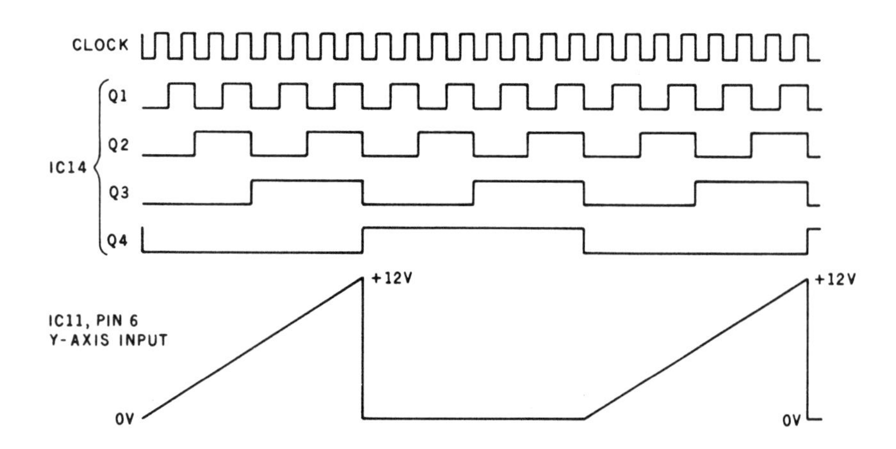

Figure 6: *Timing diagram of the voiceprint-recording system.*

you are talented in software, you may have been inspired with an idea that will make the process easy. But at any rate, I hope to have helped allay any suspicions that computerized voice response is a black art.

References

1. *Analog-to-Digital Conversion Handbook,* Norwood, MA: Analog Devices Inc., 1976.
2. Ciarcia, Steve. "Analog Interfacing in the Real World," BYTE, January 1982, page 72.
3. Ciarcia, Steve. "Build a Low-Cost Speech-Synthesizer Interface." BYTE, June 1981, page 46.
4. Ciarcia, Steve. "Build an Unlimited-Vocabulary Speech Synthesizer," BYTE, September, 1981, page 38.
5. Ciarcia, Steve. "Use ADPCM for Highly Intelligible Speech Synthesis," BYTE, June 1983, page 35.
6. *Digital-to-Analog Converter Handbook.* Billerica, MA: Hybrid Systems Corporation, 1970.
7. Doddington, George R. and Thomas B. Schalk. "Speech Recognition: Turning Theory to Practice," *IEEE Spectrum,* September 1981, page 26.
8. Levinson, Stephen E. and Mark Y. Liberman. "Speech Recognition by Computer," *Scientific American,* April 1981, page 64.
9. Wiggins, Richard H. and George R. Doddington. "Speech Recognition Spurred by Speech-Synthesis Success," *Electronic Design,* July 9, 1981, page 107.

16

Build an RS-232C Breakout Box

This diagnostic tool can help you make working serial connections.

Hundreds of exhibit-goers swirled around me as I elbowed wearily, yet warily, through the computer-convention crowd. Like Frodo Baggins approaching Mount Doom, I had to go to a place I feared to be: the BYTE booth. If I stayed anywhere near it, I was certain to be accosted by someone asking technical questions. (When your picture is plastered all over books sold at half a dozen booths, it's hard to remain incognito.)

Just as I started to take refuge behind the wall of 6000 boxed magazines in the BYTE booth, a voice behind me called out, "Steve, Steve! Help! I was going to send you an 'Ask BYTE' letter, but I need an answer right now on how to interface one of your projects to my computer."

I turned to respond as a man in his early thirties, carrying a briefcase and two shopping bags, hurriedly approached the counter. Panting a bit, he spoke with overtones of nervous anxiety combined with a note of triumph. I shrugged and responded, "Sure, I'll be glad to help if I can."

Most questions of this kind require minimal effort: at worst, digging out a few sheets of scratch paper and scribbling a schematic diagram or two. After all, most computers use similar parallel or serial interfaces.

"I have this computer. . . . " As the man started to speak, he cleared about four feet of counter space with one elbow, leaned over to one of his shopping bags, and started pulling out pieces of equipment and cables. "Here, hold this . . . um . . . somewhere in here are the monitor, recorder, and this other junk . . . and, wait, here it is! I want to connect it to this here Microvox of yours. Can you do it for me? I've made some homebrew modifications to my computer, but that shouldn't keep it from working."

I've been told that, to be perceived as a good teacher, you have to be able to leave everything as an exercise for the student. Here I was, in the middle of a convention with 50,000 people, and this guy had dragged in his whole computer system. Mind you, I've spent hours talking shop with computer hobbyists in public places, and I've designed as much circuitry on the back of napkins as you see in this column each month. This, however, was a new twist.

His computer? It was an off-the-shelf budget model that I'd used before but hadn't tried with the Microvox speech synthesizer. It had a serial output, so I presumed it would work easily. As we shall see, that assumption was a mistake.

On the surface, it seemed a simple task. The unit had a four-wire serial interface that was described as conforming to the RS-232C specification established by the Electronic Industries Association. The connections in the interface were defined for transmit, receive, status, and ground lines. The Microvox can be connected in that way or through various other handshaking protocols. Inside the Microvox are jumper connections to reverse the functioning of the transmit and receive lines so that the Microvox can be set up as either kind of RS-232C device—data terminal equipment (DTE) or data circuit-terminating equipment (DCE). Given that there were only four wires, it should have been duck soup, right?

My first surprise came in finding that my querist's computer used a DIN (Deutsche Industrie-Norm) connector rather than the DB-25 connector typically used for RS-232C links. But by referring to his computer's instruction manual, we were able to attach the Microvox, set up as a DCE device, and hard-wire the computer's status input so it would see a "ready" condition all the time. The computer was set up as a DTE device, receiving data on pin 2 and transmitting on pin 3.

To our disappointment, when we executed the program to make the synthesizer talk, his computer transmitted nothing that could be received by the Microvox. I grumbled, "OK, maybe there's at typo in the Microvox manual." We reversed pins 2 and 3, setting the Microvox as a DTE device, and tried again. No luck. Still bending down over the hobbyist's machine, I looked over the top of my wire-rimmed glasses and asked as politely as I could, "Are you sure the serial output on this thing works?"

The subsequent discussion revealed that one of the modifications he had installed was a 20-mA (milliampere) current-loop output. But a serial printer attached to it had worked just prior to his bringing the computer to the show.

Because he had serial data coming out, we decided that it must be the Microvox that wasn't communicating. But how to test it at the show? Hmmm. . . .

As I pondered that question, my gaze came to focus 20 feet across the aisle in another exhibit booth, on a computer system demonstrating a printer. As I looked closer, I saw an RS-232C cable connecting the computer and printer. Certain ideas came quickly into my mind. Gathering up the Microvox, my inquirer and I squeezed across the aisle and approached the booth's proprietor.

Fast footwork and a smooth tongue are necessities in such circumstances. You have to put off saying the phrases "have a problem," "test," and "your computer" as long as possible. So, as I slipped the Microvox onto his serial printer cable, I distracted the exhibitor's attention by saying something like, "Hi, we have a booth over there, and we saw that you were demonstrating this neat printer, and we were wondering if we could try something on your system. . . . Can I put this here? . . . Do you mind if I just plug this in your computer . . . to test just one little thing?"

I had the cable connected before the potential horror of what I was proposing had time to sink in. In fact, the exhibitor was a bit amused as the Microvox started speaking the text that was intended for the printer.

A half hour later, after explaining the theory of text-to-speech algorithms to the curious exhibitor, we went back to the BYTE booth and the original problem. Finding that the Microvox worked, and while not directly suggesting that my shopping-bag computer hobbyist's RS-232C output did not, I searched for a new strategy to make him happy.

Clearly, the problem was not simply one of making the right connections. I needed some way to take a better reading of what was happening. Even a mere bent paper clip and a light-emitting diode (LED) would have helped.

Again we went on the prowl for equipment. Three aisles away I found an exhibitor selling certain RS-232C diagnostic tools called breakout boxes who felt that he could trust me enough to lend me one of his less expensive models for a while. We took it back to the BYTE booth and set to work. We attached the breakout box to the serial output connector and watched for activity on the LED indicators. The transmit LED flickered as the data went to the Microvox, but I was still suspicious.

Back into the throng. Another aisle over I borrowed a DVM (digital voltmeter) and used it to check the voltage levels on the RS-232C output. Serial data appeared to be present, but it ranged from −0.7 to +6.8 V (volts) rather than from +12 to −12 V, as you typically find. The cheap ($150) breakout box was responding to the +6.8-V level, but it had no way of indicating negative voltage levels or open circuits.

Armed with this information, I checked some of the modifications that my comrade in debugging had made. Apparently the 20-mA current-loop circuit was at fault. It was optically isolated, as it should have been, but the optoisolators and other components were being powered directly from the RS-232C signals. The LEDs and protective shunting diodes were shorting out half the data! I disconnected the current loop, and two hours after the computer was first slipped out of its shopping bag, the Microvox spoke.

"Standard" RS-232C

Among the most exasperating experiences in any computer user's career is connecting two serial devices. I don't mean a terminal and a modem—making that connection is a piece of cake—but any other connection can be real trouble.

For instance, every time I buy a new piece of equipment, things seem to work this way: I spend five minutes reading the sales brochure, five minutes executing the financial transaction, and five hours trying to figure out how to make the new equipment communicate with my computer.

If you scan through product advertisements and equipment data sheets, frequently you see claims that products have "standard RS-232 interfaces." The people who perpetrate these lies have probably never connected two pieces of equipment together in their lives. What this phrase really means is that both ends of the communication line will probably be voltage-compatible and won't incinerate each other. Both will send and receive asynchronous or synchronous serial data following a specific timed sequence (bit rate) known by both ends. Beyond that, virtually nothing is reliably standard.

The only sure way to connect two serial devices is to examine the documentation for each and determine the proper connections of inputs and outputs, with lines crossing if need be. If the manuals are not available, then voltage-sensing and status-display devices must be used to examine individual lines. Depending on DTE or DCE convention, the same pin number may be either an input or output. And, generally speaking, in a fully implemented handshaking configuration, single-point signal monitoring is not enough. Frequently, multiple lines must be monitored to observe dynamic relationships.

One of the most useful pieces of test equipment is the RS-232C breakout box.

I don't intend to spend too much time discussing RS-232C handshaking or interfacing. That subject has been covered in numerous other articles in BYTE, most recently by Dr. Ian Witten (reference 7). In my experience, simply knowing the conventions

Pin	EIA	CCITT	Signal	Source
1	AA	101	Protective (Earth) Ground	
7	AB	102	Signal Ground	
2	BA	103	Transmitted Data (TD)	DTE
4	CA	105	Request to Send (RTS)	
20	CD	108.2	Data Terminal Ready (DTR)	
23	CH	111	Data Signal Rate Selector (DTE source)	
24	DA	113	Transmitter Signal Element Timing (DTE source)	
14	SBA	118	Secondary Transmitted Data	
19	SCA	120	Secondary Request to Send	
3	BB	104	Received Data (RD)	DCE
5	CB	106	Clear to Send (CTS)	
6	CC	107	Data Set Ready (DSR)	
22	CE	125	Ring Indicator (RI)	
8	CF	109	Received Line Signal Detector (or Carrier Detect—CD)	
21	CG	110	Signal Quality Detector	
23	CI	112	Data Signal Rate Selector (DCE source)	
15	DB	114	Transmission Signal Element Timing (DCE source)	
17	DD	115	Receiver Signal Element Timing (DCE source)	
16	SBB	119	Secondary Received Data	
13	SCB	121	Secondary Clear to Send	
12	SCF	122	Secondary Received Line Signal Detector	

Table 1: *RS-232C signals listed by function and source.*

hasn't helped one bit. I always have to study the manuals for each interface and often have to verify voltage levels with a meter. When all the handshaking signals are used, it can be a prolonged task.

Ultimately, successful serial interconnection depends more on your ability to measure, cross-connect, and jumper-connect signal lines until *some* combination works. And sometimes you have to deal with a case wherein the lines might be connected properly but, as in the example above, operating at the wrong voltage levels.

The point of this month's article is to describe some of the hardware involved in testing the RS-232C connection. We'll look at various circuits that make voltage measurements, including two kinds of breakout boxes. Finally, to allow dynamic testing of the data communication on the transmit and receive lines, we will see how to build a terminal simulator, which sends and receives ASCII (American National Standard Code for Information Interchange) characters at any standard data rate.

The RS-232C Breakout Box

One of the most useful pieces of test equipment is the RS-232C breakout box. This device is essentially a 25-line extension cable with apparatus attached that allows you to perform experiments on the serial link by inserting cross-connections, jumpers, and open circuits between the various signal lines. The most common use of the breakout box is to switch the Transmitted Data and Received Data lines (pins 2 and 3 in the commonly used connectors, abbreviated TD and RD). Another typical use is to make various combinations of connections between Request to Send (RTS, pin 4), Clear to Send (CTS, pin 5), Data Set Ready (DSR, pin 6), Carrier Detect (CD, pin 8—its official name is Received Line Signal Detector), and Data Terminal Ready (DTR, pin 20). The more expensive units can also monitor voltage levels or decode serial data. Table 1 identifies the RS-232C signals by source.

A variety of commercial breakout boxes, ranging in price from $100 to $1000, are on the market. The more expensive products include such additional features as bilevel or trilevel signal monitoring, absolute voltage sensing, and signal-injection capabilities. Some of these features may or may not be important in your application.

When you are choosing a breakout box, be sure you get your money's worth. In some of the $100 units, it appears that $99 of the cost is for the case. The mechanical features are similar at all price levels, and only the addition of electronic bells and whistles adds to the cost.

In any breakout box, you find two 25-pin type-D subminiature (DB-25) connectors, two sets of 25 connection

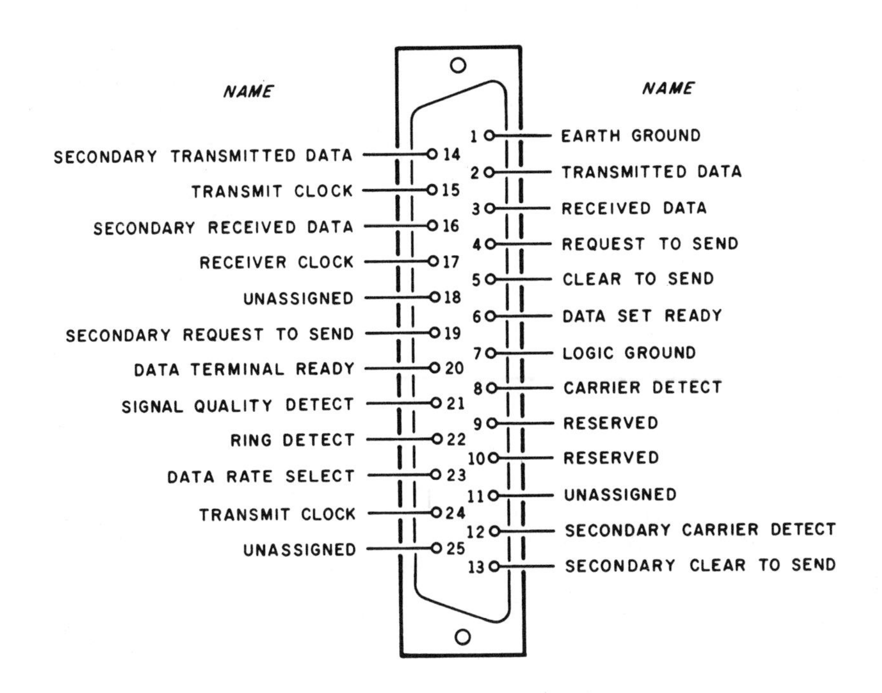

Figure 1: *Pinout specifications of the RS-232C interface. The popular DB-25 connector is most often used for RS-232C connections, but it is not required by the EIA standard.*

posts, and 24 or 25 switches. To begin testing, you connect the breakout box in series between the two pieces of equipment. The 25 lines on the first side go to 25 connection posts (often wire-wrap pins are used), and then 24 of the lines go into one side of three 8-pole DIP (dual-inline pin) switches. (Pin 1, the protective ground line, is directly routed to the output connector.) The 24 lines are directed through the DIP switches to another set of connection posts and then to the output connector. Figure 1 shows the pinout specification of the connectors.

To test the connection between two DTE-type RS-232C devices when no handshaking is involved, you merely insert the breakout box in the link; then, with all the switches closed except the two connected to pins 2 and 3, insert crossed jumpers between the connection posts to interchange the two signals. After you've made sure this works, you can permanently wire a cable to do the same job.

While most commercial breakout boxes connect and switch 24 of the 25 lines, most common RS-232C applications involving printers, low- and medium-speed modems, and video-display terminals generally use only connector pins 1 through 8, plus pin 20. For these applications, we can construct a bare-bones breakout box, such as the one shown in the schematic diagram of figure 2.

This simple construction will give you all the necessary features of a $150 box for about $15. Besides a signal-switching capability, this circuit incorporates passive voltage monitoring. Three components—a blocking diode, a constant-current diode, and an LED—indicate the presence of a potential of +4 V or greater on any signal line. The LED does not light for open circuits or negative voltages. Many low-priced RS-232C monitors use this type of circuit, and some cautions regarding its use are appropriate. Such an indicator draws its power from the signal itself (requiring about 4 mA), and in marginal situations it might kill the signal being measured.

For the RS-232C status signals, a line potential anywhere between +3 V and +15 V is logic 1, and a potential from −3 V to −15 V is a logic 0. If these signals are generated from low-power operational amplifiers (op amps) rather than RS-232C drivers, there may not be enough oomph to communicate and light LEDs simultaneously. (Op amps are frequently used as a cost-cutting measure in budget equipment.) Finally, this cheap monitor senses only positive voltages (unless the circuit is reversed in polarity and duplicated everywhere) and gives no indication of negative voltages or ground potential. For example, you couldn't tell the difference between a broken cable lead and a logic-0 (inactive) status signal. Remember that RS-232C is a bipolar signal. As much happens (or doesn't) at negative potentials as at positive.

A Functional, Low-Cost Breakout Box

The previous point may have seemed minor, but it leads us to a discussion of better voltage sensors and use of active, rather than passive, monitoring techniques.

Figure 3 is the schematic diagram of a much better indicator circuit for sensing RS-232C voltage levels. Using its own op amp to reduce signal loading, this bilevel monitor lights up in red for positive voltages and green for negative voltages. The feedback and gain-setting resistors are chosen so that the LEDs trigger at approximately +3 V and −3 V, respectively (internal current-limiting provides a relatively constant light level and protects the LEDs). Voltages between those values or open circuits light neither LED.

The indicator-circuit sections, one for every signal line, each use one section of an LM324 quad op amp. They present a virtually unnoticeable 100-kilohm load to the signal lines, can operate on any voltage with an absolute value between 4.5 V and 18 V, and can be powered easily from a battery of 6 AA cells for portability. I chose to use integrated two-color LEDs, but two regular LEDs (one red and one green) can be wired anode-to-cathode in parallel.

My idea of a reasonably useful breakout box is shown in figure 4. It combines the 24-switch-and-header portion of the commercial breakout boxes with the sensitive bilevel monitoring circuits just described.

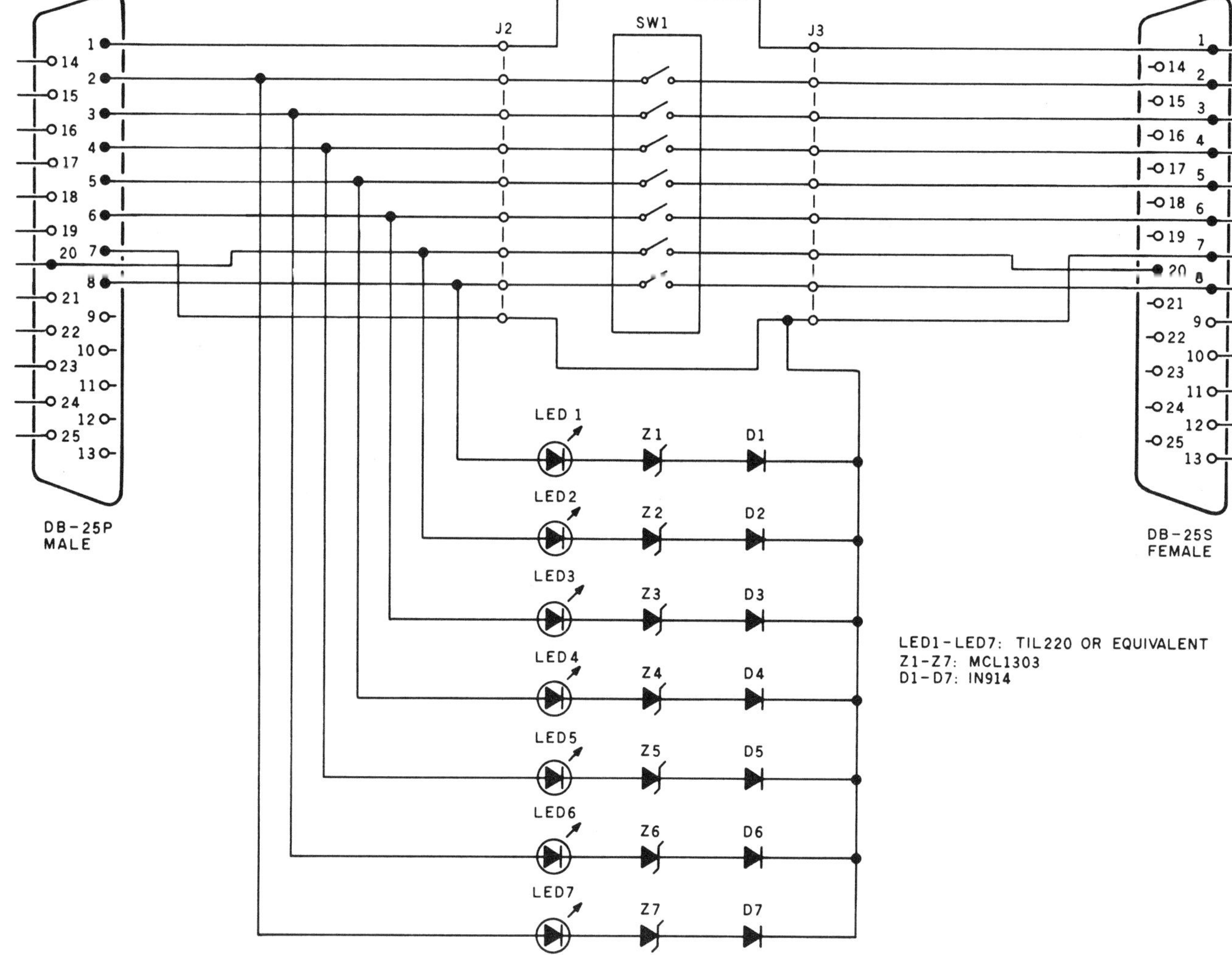

Figure 2: *A simple homebrew breakout box for diagnosing serial RS-232C communication links. The 9 lines connected here are the ones most often used.*

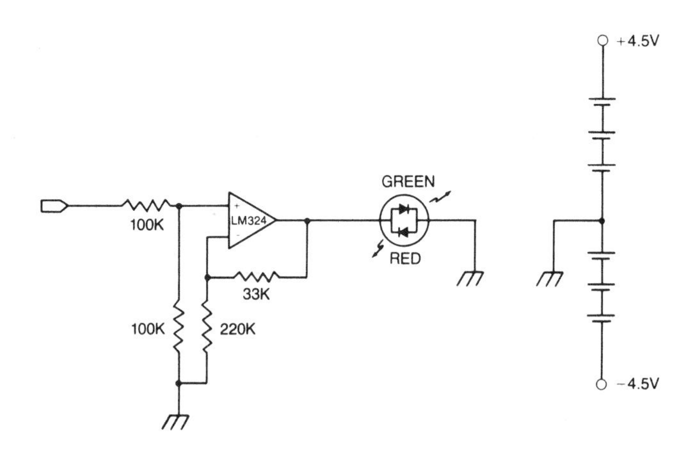

Figure 3: *A line-state status indicator that incorporates an op amp to keep line loading to a minimum. The dual-LED component will glow red for positive voltages and green for negative.*

Voltage indicators are hard-wired on the 11 most frequently used signal lines (2, 3, 4, 5, 6, 8, 15, 17, 20, 21, and 22), and the twelfth sensor can be connected to any line. A prototype of this circuit is shown in photo 1.

A Decade Voltage-Level Indicator

If you think back to my experience at the convention, you'll see that the problem was eventually detected as being voltage-related. It was undetectable with the cheap LED circuit, but it would have been caught by the circuit of figure 4. However, on some occasions even two levels are inadequate, and more precise measurements are necessary.

One extra indicator section was

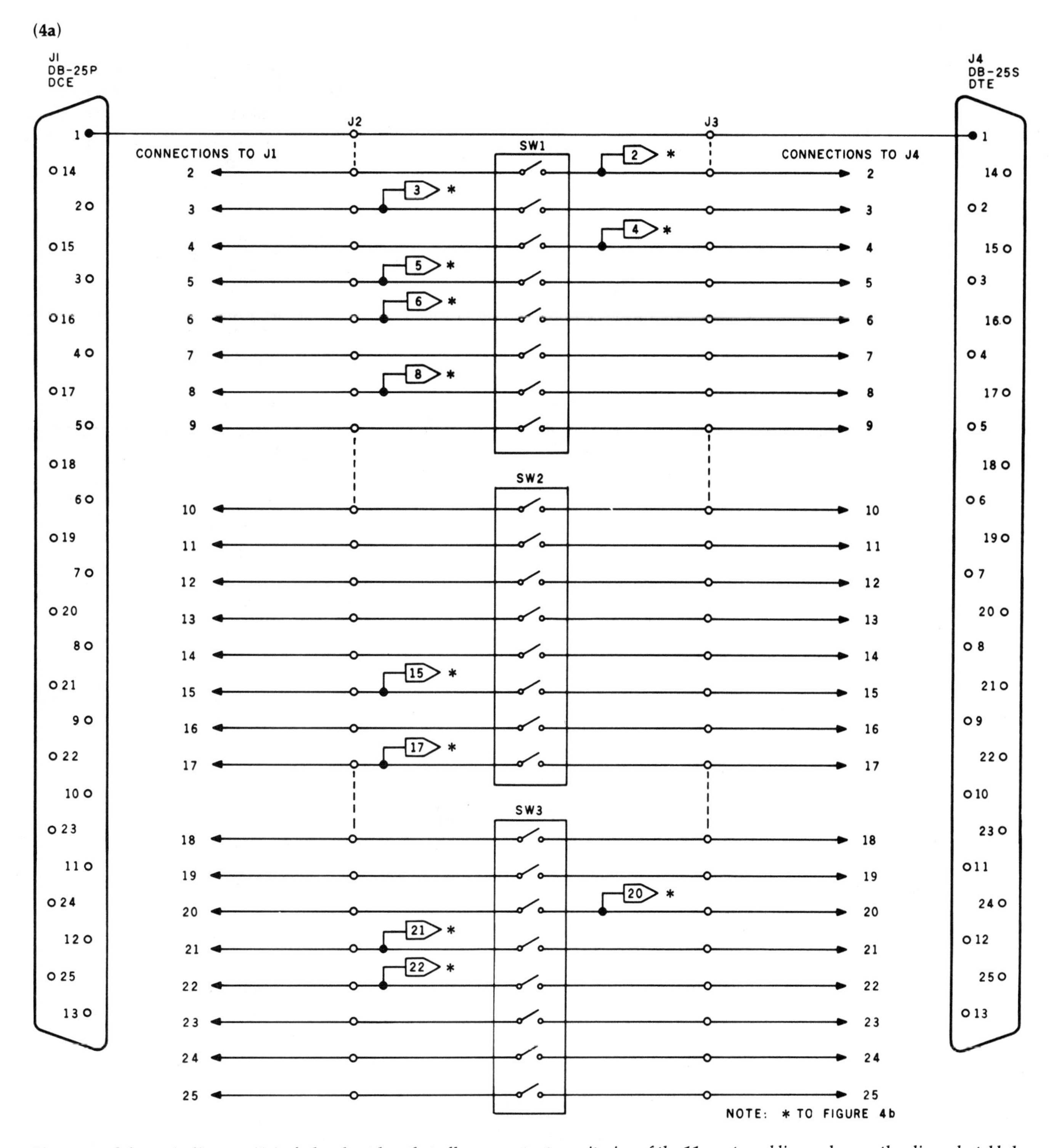

Figure 4: *Schematic diagram (4a) of a breakout box that allows constant monitoring of the 11 most used lines, plus another line selectable by a jumper connection. The indicator circuit of figure 4b is reproduced 12 times; three LM324 packages, each containing 4 op-amp sections are needed. The connections for each section are also shown in figure 4b.* (continued on the following page).

provided in the circuit of figure 4 so that it could be attached to any of the 25 input pins. If we expand this concept a bit further and replace this bipolar sensor with something having a little more resolution, we could take a better reading of voltage excursion. Granted, a DVM could be used, but such instruments are expensive and are much more elaborate than required. Because of the RS-232C voltage range of −15 V to +15 V (−3 V to +3 V being no-man's-land), the resolution need be only about 3 volts at best.

The circuit in figure 5 is a single-digit decade voltmeter. Designed for installation in the breakout box in place of the floating sensor, it has a higher resolution than the bilevel monitor, and it can be used in place of a DVM in less critical applications such as RS-232C level sensing. It uses an LM3914 LED dot/bar generator as a 10-step analog indicator. The 30-V (±15-V) input range is divided into 10 incremental

(4b)

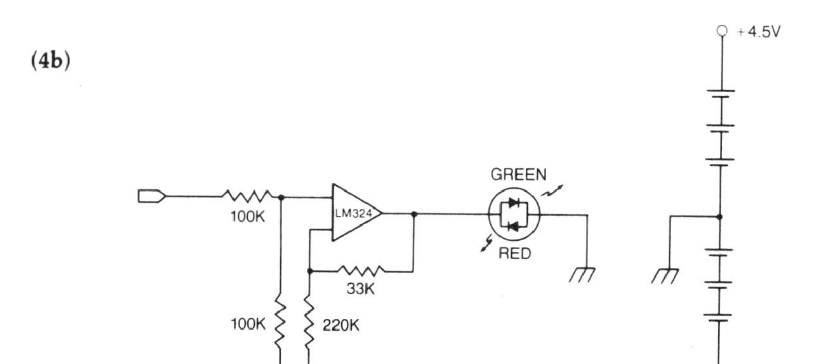

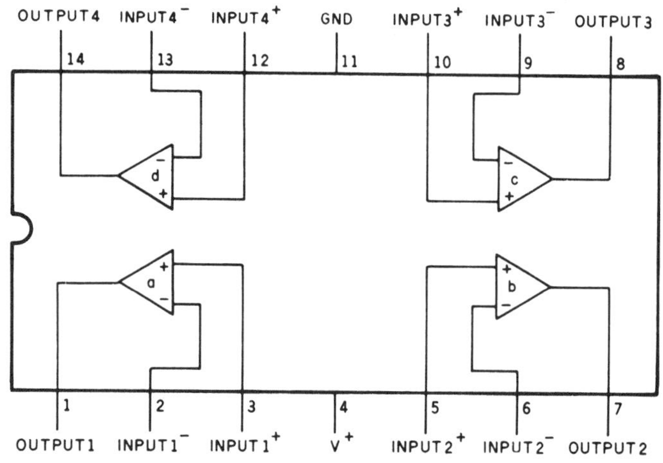

Figure 4 continued

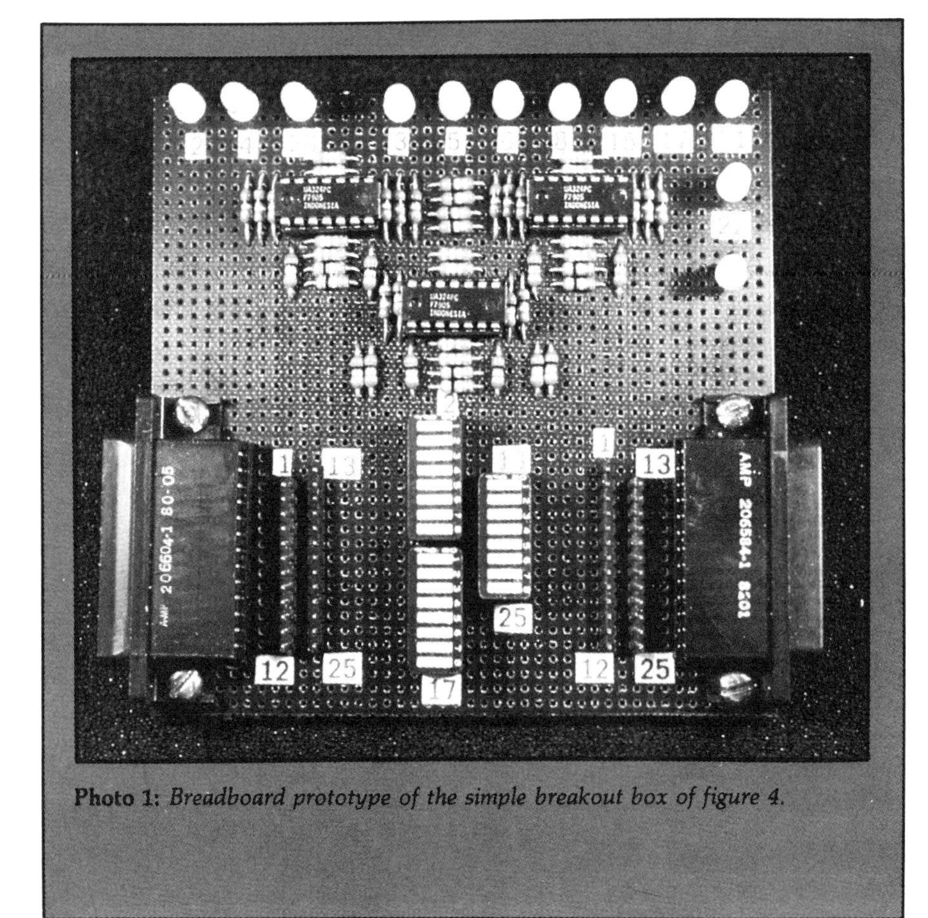

Photo 1: *Breadboard prototype of the simple breakout box of figure 4.*

steps of 3 V. Different-color LEDs provide a quick visual indication of relative magnitude and polarity of the line states. A prototype of the circuit is shown in photo 2.

The decade-voltmeter circuit uses 2 or 3 integrated circuits. IC1a (an LM324 op amp section) scales the input voltage range from ±15 V to ±2.5 V (adjust potentiometer R1 to set the gain for 1/6). Another section, IC1b, shifts the resulting ±2.5 V scaled signal to a 0-V to +5-V range at its output (adjust R2 for the proper shift magnitude). IC2, the LM3914 dot/bar generator, is configured to respond to a range of 0 V to +5 V.

When −15 V is applied to the circuit's input, LED1 lights; when +15 V is applied, LED10 lights. For an intermediate active voltage, such as +9 V, LED8 will light up. In the range of from −3 V to +3 V, which includes ground or open states, LED5 or LED6 will be lit. Using the same color convention as in figure 4, voltages greater than +3 V are shown on red LEDs, those less than −3 V are on green LEDs; in the inactive area between +3 V and −3 V, I used yellow. Any single input signal that swings between a red and green indicator is therefore at valid levels.

The decade-voltmeter circuit requires power at about +7.5 V (or greater) and −5 V (or less) for operation. Two 9-V battery cells (to provide ±9 V) work very nicely. Or using a single +7.5-V source and the ICL7660 voltage inverter (IC3) allows the unit to be battery-operated (for some reason, the 7660 didn't seem to like inverting +9 V when I tried it).

Build a Terminal Simulator

The final thing to check when testing an RS-232C serial link is the existence of properly formatted data bits coming over the wire at the correct data rate. Voltage-level indicators make only quantitative measurements. It is necessary to take a *qualitative* measurement if the integrity of the transmission is in question. Most often the quality of the link can be ascertained with the transmission and reception of a single character.

One way of testing a suspect communication line is to drag around a

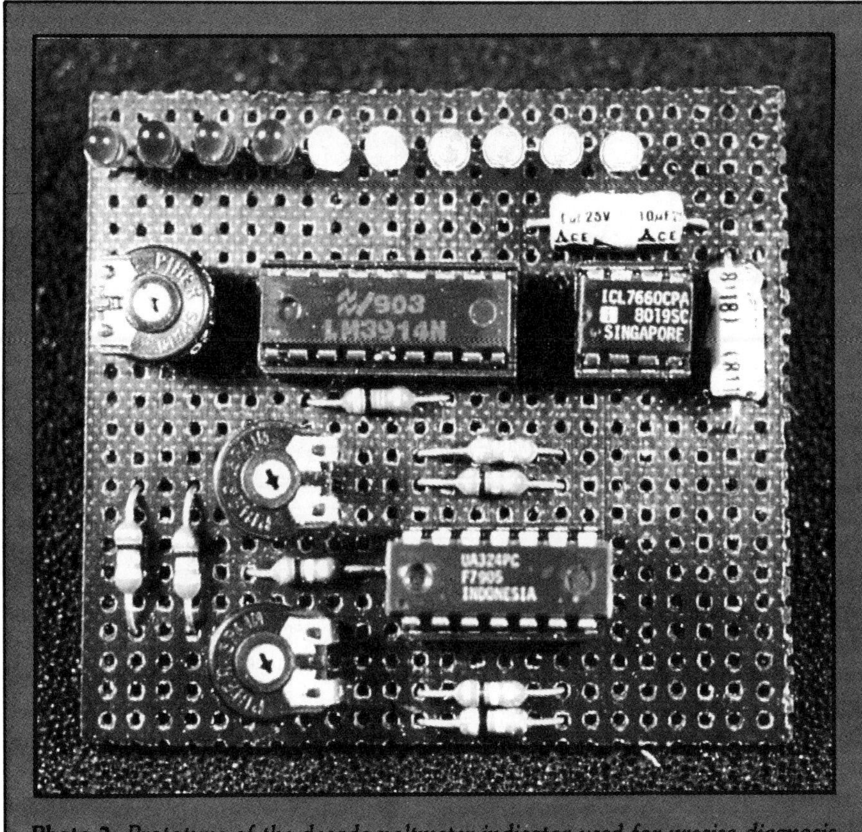

Photo 2: *Prototype of the decade-voltmeter indicator used for precise diagnosis of the state of an RS-232C line.*

big video-display terminal. Also the idea of adapting the single-chip terminal from the February article did occur to me. However, that was a relatively expensive item for a tester, and it couldn't cover all the data rates in common use, so I did some head-scratching and invented yet another circuit.

Figure 6 is the schematic diagram of my single-character terminal simulator. The prototype is shown in photo 3. Using 5 chips, it can send, receive, and display (in hexadecimal form) single ASCII characters at any of the 16 commonly used data rates between 50 and 9600 bps (bits per second). By switching pins 2 and 3, you can set it to appear on the line as either a DTE or DCE device. Have you got a suspect device on the line? Set your data rate and desired character code on the DIP switches and then press a button each time you want the character transmitted. Granted, a video-display terminal might be faster, but my simulator circuit is a lot lighter.

The terminal simulator uses a UART (universal asynchronous re-

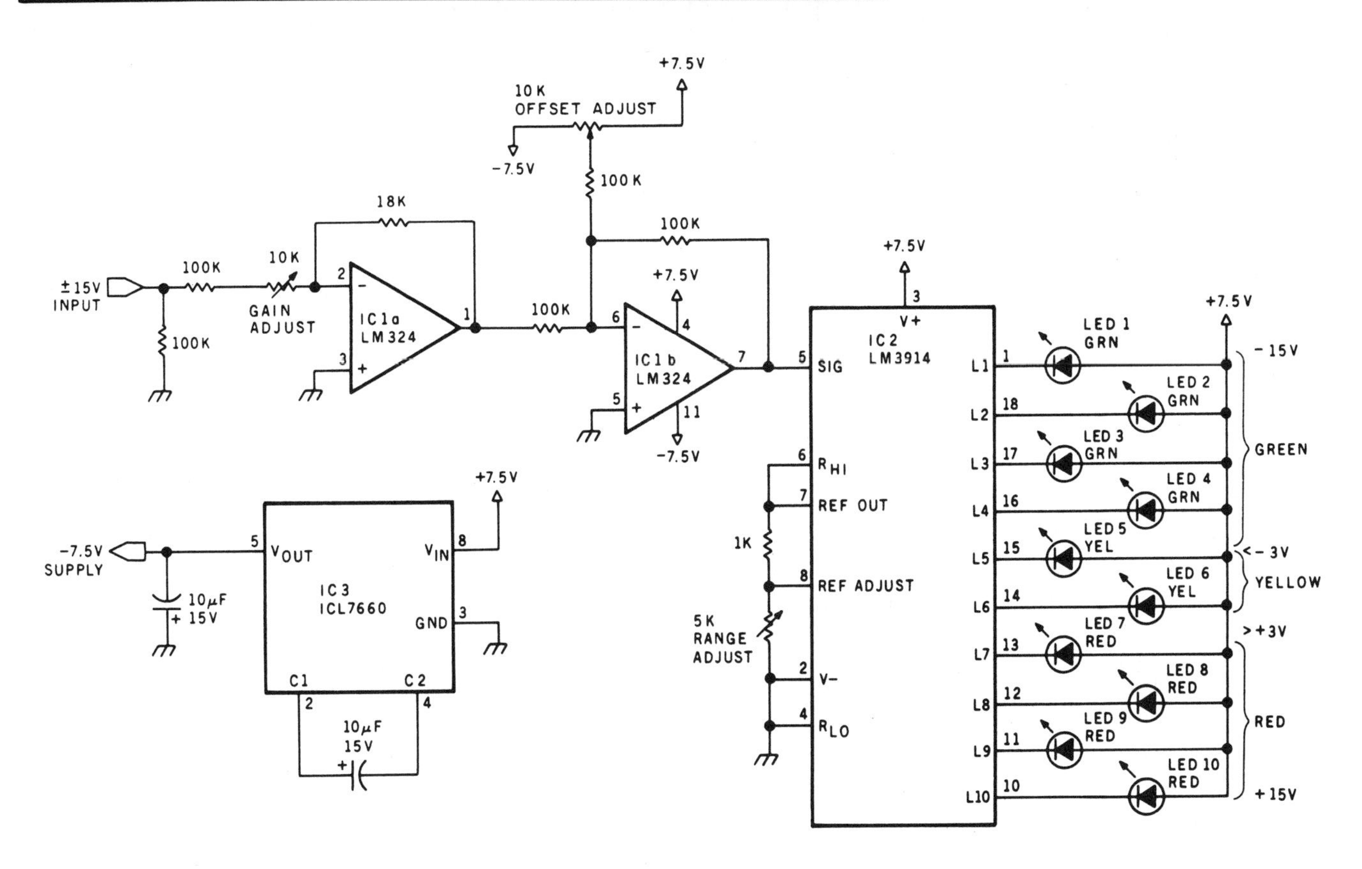

Figure 5: *A decade-voltmeter circuit. This more precise indicator may be substituted for the twelfth indicator in the breakout box of figure 4a. It lights up in one of three colors depending on the voltage present on its input.*

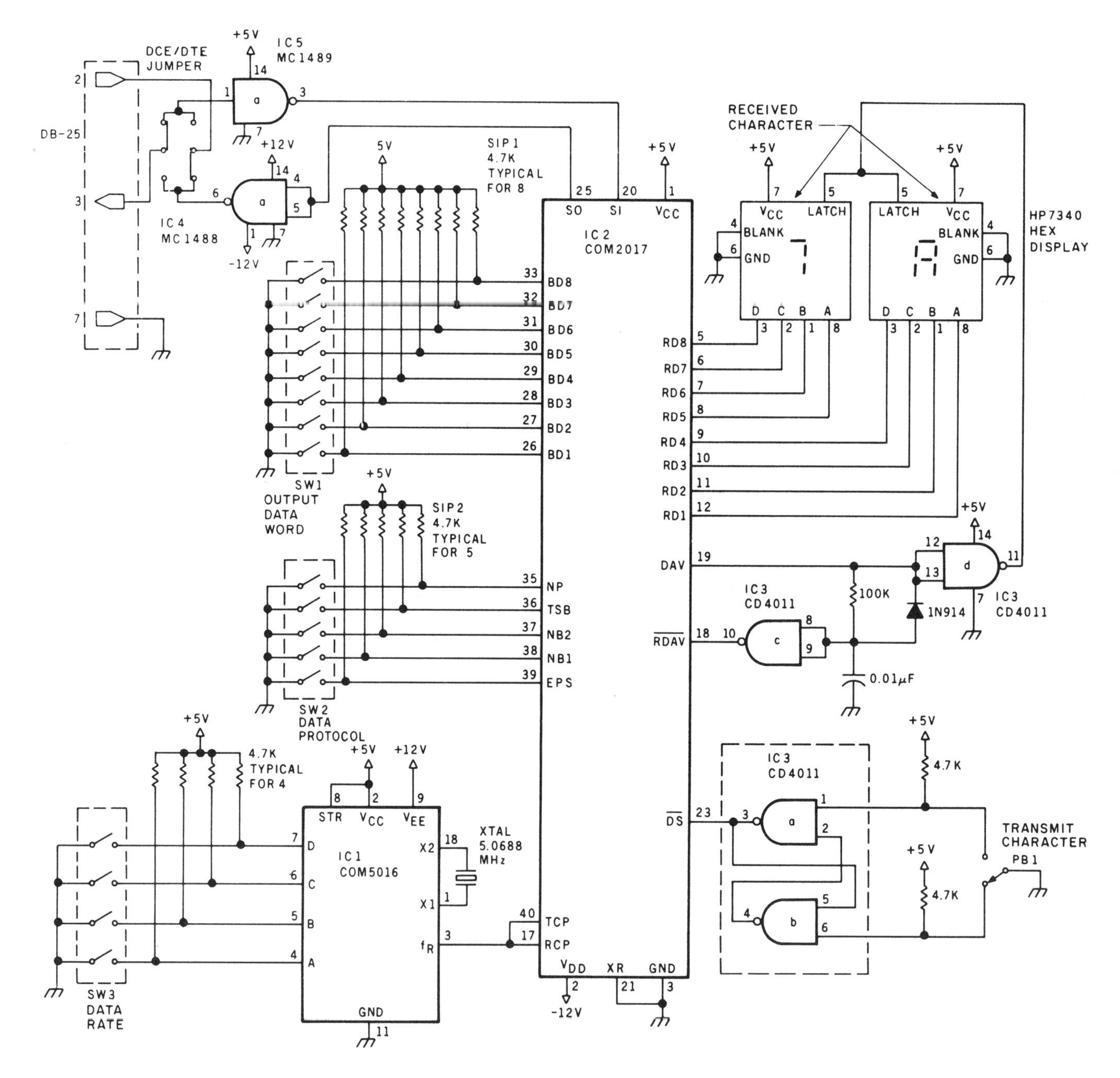

Figure 6: *A terminal-simulator circuit used to test the functioning of an RS-232C communication link. This simple-minded device transmits a single, switch-selected ASCII character on demand or receives a single character for display in hexadecimal form.*

ceiver/transmitter), IC2, to generate and accept data on the RS-232C line. (Because I have written elsewhere about the functioning of UARTs, I won't take the time now. See reference 2 for more information.) IC1 is a crystal-controlled data-rate generator. Depending on the settings of SW3, the clock rate of the UART will be set for one of the 16 data rates. Both the transmitter and receiver are set to the same rate.

The transmitter section is very simple. SW2 sets the status inputs to the UART for parity, word length, and stop bits. The character to be transmitted is set on SW1 (the most significant bit is BD8) in hexadecimal or binary format. It is transmitted each time pushbutton switch PB1 is pressed.

Data reception is just as easy. When a character has been received, the DAV (data available) output line of the UART goes high, and the received character can be read from the output lines RD1 through RD8. The DAV signal is delayed, inverted and used to reset itself through the $\overline{\text{RDAV}}$ input. This same logic signal is used to latch the 8 data bits onto a 2-digit hexadecimal LED display. Each time a character is received, the display will be updated, although you won't be able to read a continuous data stream.

Of course, more sophisticated commercial terminal simulators exist that incorporate some processing power,

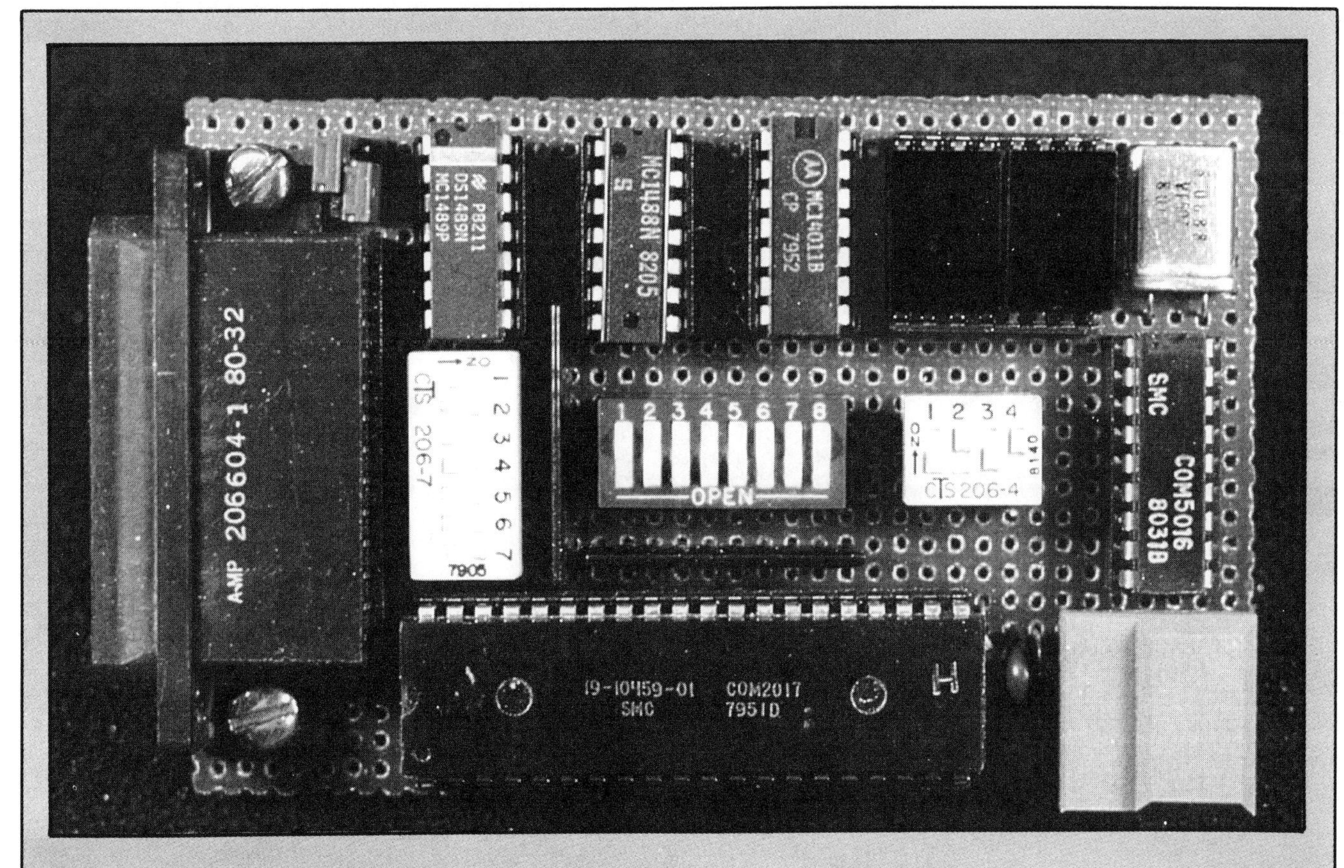

Photo 3: *Prototype of the terminal simulator of figure 6.*

but this one is simple enough to build yourself. Practically speaking, if I need more capability than this, I generally use a video terminal.

In Conclusion

This has been a rather long soliloquy on a rather oblique problem. I helped our shopping-bag computer hacker understand a little more about RS-232C, but a similar disaster might happen to him the next time he attempts to attach two serial devices.

As Dr. Witten has observed, most of the problems we find in using the RS-232C interface result from its being employed in ways never intended when it was invented.

For some of us, however, there's a light-emitting diode at the end of the tunnel. We can just whip out our little breakout boxes with all the pretty lights, decipher what's in or out, simulate all the handshaking signals, jumper everything to everything else, and make it communicate by brute force.

I'll probably still spend a few hours hooking up the Daisywriter 2000 I just got for my MPX-16. Nonetheless, maybe using some of the circuits I've presented will help cut down your frustration with RS-232C interconnections.

References

1. Ciarcia, Steve. "Build a Handheld LCD Terminal." February 1983 BYTE, page 54.
2. Ciarcia, Steve. *Build Your Own Z80 Computer.* New York: BYTE Books/McGraw-Hill Book Company, 1981. (See page 138).
3. *EIA Standard RS-232-C: Interface Between Data Terminal Equipment and Data Communication Equipment Employing Serial Binary Data Interchange.* Washington, DC: Electronic Industries Association Engineering Department, 1969.
4. Folts, Harold C. *McGraw-Hill's Compilation of Data Communications Standards*, 2nd edition. New York: McGraw-Hill Publications Company, 1982.
5. Leibson, Steve. "The Input/Output Primer, Part 4: The BCD and Serial Interfaces." May 1982 BYTE, page 202.
6. Liming, Gary. "Data Paths." February 1976 BYTE, page 32.
7. Witten, Ian H. "Welcome to the Standards Jungle." February 1983 BYTE, page 146.

17

Build an RS-232C Code-Activated Switch

This device will let you switch between several peripherals connected to one serial port.

Do you ever find yourself probing through a mass of tangled cables behind your computer? If your computer is like mine, you probably have only one serial port, to which you have to connect both a printer and modem. Of course, whenever you want to use a different peripheral, you have to unplug one and plug in the other.

The variety of peripheral devices necessary to gain full use of a personal computer can create a connection jam at the serial port. Many small computer systems have two I/O (input/output) ports intended to support a printer and modem. Usually the port intended for connection to the printer is a parallel port (although many printers require a serial port), while a serial port (perhaps called a "communications port") is provided for connection to a modem. For the typical user, this may be adequate.

Some of us, however, aren't typical users; we have more than one printer and one modem attached to our computer, or our printer uses a serial interface, not a parallel one. I have three serial printers and two modems, all of which I must connect to my workhorse computer system through a single RS-232C serial port. (It's not that my computer is a small configuration: it has eight parallel ports in addition to the one serial port. It's just that every new peripheral device I buy seems to be serially interfaced, and I can't fit any more serial ports inside the already crowded enclosure.

Recently, while juggling three cables and leaning over the computer, I began to wish for an easier way to switch between devices. I wondered if I could just put together a little box containing a multiple-pole rotary switch wired to a few DB-25 connectors as a workable compromise. But then I thought of a possible better solution as I remembered something I had seen in a catalog of data-communication products: a device called a code-activated switch.

Functional Analysis

The function we need here is the ability to multiplex—switch between as needed—several peripheral devices connected to a single I/O channel. A *communications multiplexer* performs this function. In essence, this device forms a bridge in the communication link between the master device (usually the host computer's I/O port) and one of several slave devices; transfer of data can proceed in either direction over this bridge. The physical linking of the input and output can be accomplished either mechanically or electronically.

Two Approaches to Switching

The simplest possible device for the purpose is a four-position mechanical switch box. Available commercially for about $150, a DB-25 switch box allows you to select one of four peripherals for output by turning a four-position multipole rotary switch. This manually activated switch is most frequently used where the peripherals, computer, and operator are in close proximity. Its major advantage is its relative low cost.

Unfortunately, mechanical switches are subject to deterioration from the elements and, of course, require a human operator to function. Harsh environments call for fully electronic switches. In situations where the communicating devices are at great distances from the computer, some

---

Z8 is a trademark of Zilog Inc.

Special thanks to Bill Curlew for his Z8-programming expertise.

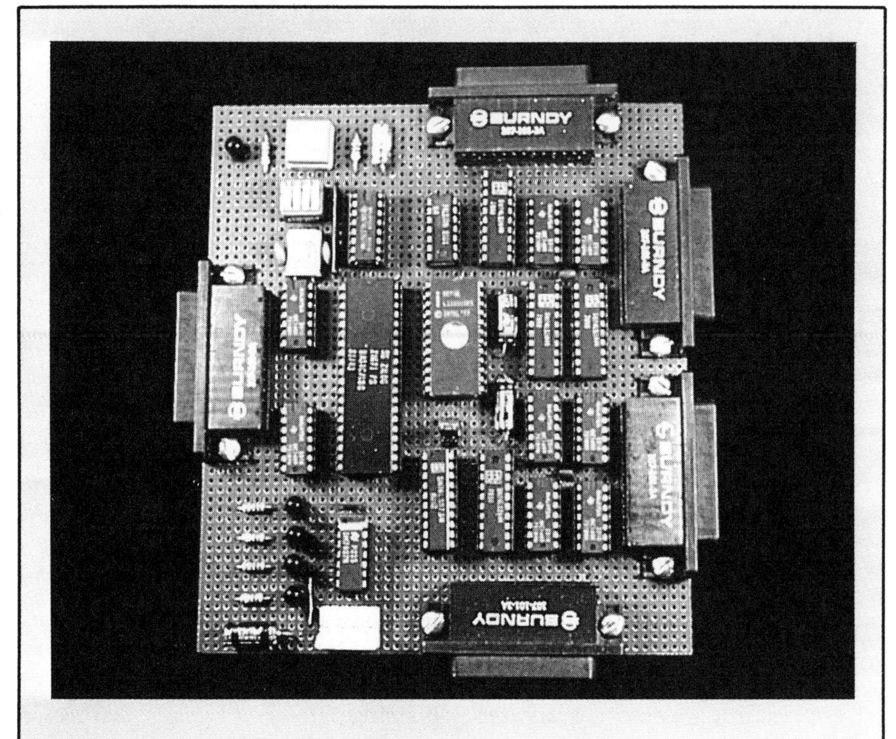

Photo 1: *Prototype of the Micromux four-channel code-activated switch, built around a Zilog Z8 single-chip microcomputer.*

form of remote-controlled switch must be used. An ordinary electrical relay can provide remote control, but the greatest flexibility and reliability are obtained from a fully electronic, software-controlled, code-activated peripheral-device switch. Let's look at some of the possibilities.

Electronic Switches

Figure 1a is a simple block diagram of a four-channel electronic RS-232C multiplexer switch. The master input on the left side is intended to be connected to a computer's I/O port and the four channels on the right side are intended to be connected to four peripheral devices. (In the following discussion, the assumption will be that data is being transmitted by the master device to one of the slave devices, although data can also move in the other direction.)

On the left side of the electronic multiplexer, the input serial data from the master device (the computer) is converted from RS-232C

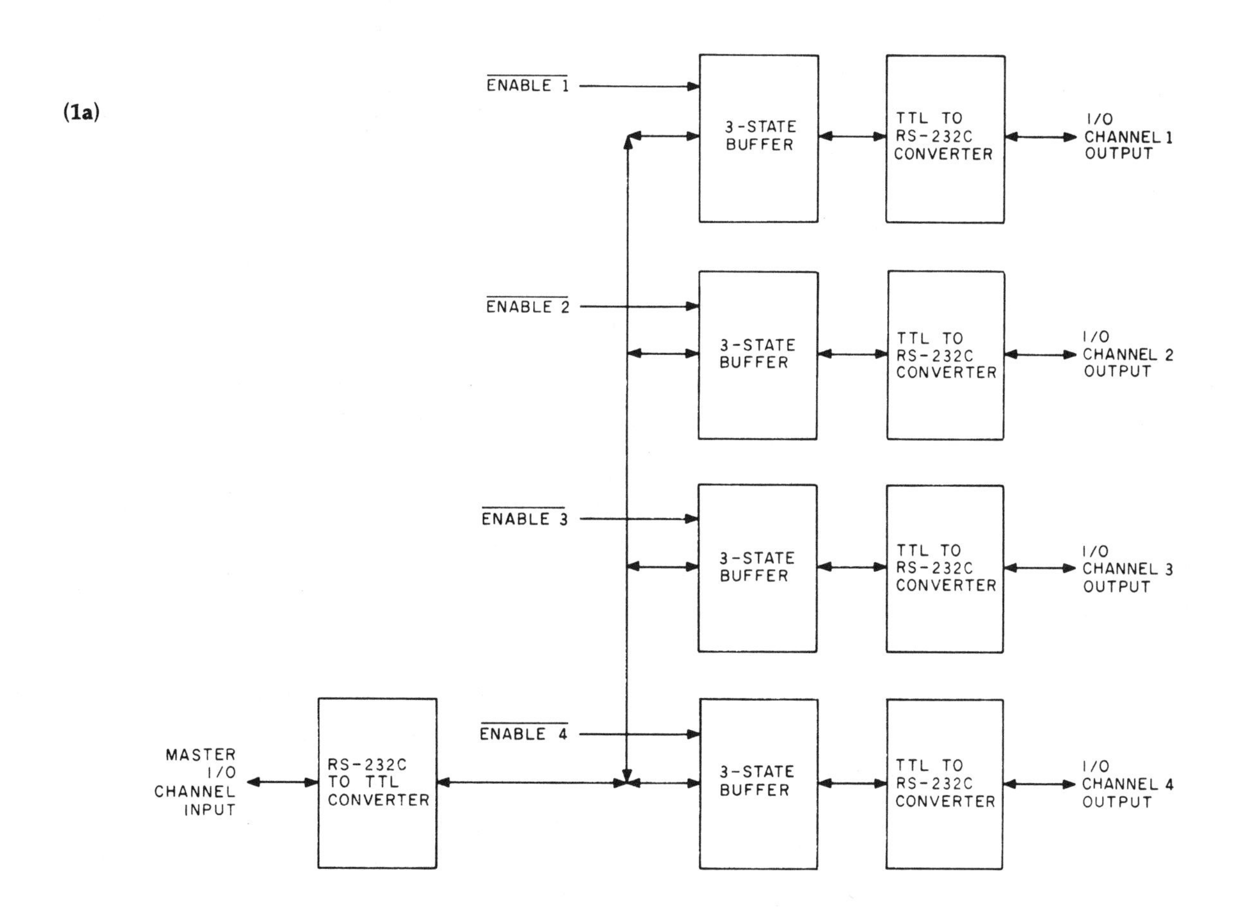

Figure 1a: *Block diagram of a four-channel peripheral switch for RS-232C serial communication, which may be controlled via wired remote activation or code activation.*

voltage levels (±15 V) to TTL (transistor-transistor logic) voltage levels (0 to +5 V) and directed to the inputs of four three-state signal buffers (also known as three-state switches) wired in parallel.

Each three-state signal buffer has an active-low enable input. An input voltage of 0 V on this enable input line allows a signal to pass through the buffer. When any one of the four buffers is enabled, the master device's input signal is allowed to pass through it. This signal is then reconverted to RS-232C levels and sent on to the peripheral. Changing the active-low enable signal from the first three-state signal buffer to another diverts the output to a different peripheral device. (In this particular configuration, only one of the four buffers should be activated at one time. Other setups could allow output to be sent to two, three, or all four of the output buffers.)

The method by which the enable lines of the three-state signal buffers are activated determines the complexity of the electronic switch. Two particularly important types of control methods are used by devices called remote-activated switches and code-activated switches.

Both remote- and code-activated switches are designed for hands-off operation. The difference between them is this: selection of an output channel is done in the former by decoding separately conveyed logic-signal inputs and in the latter by decoding signals conveyed as part of the data being transmitted through the multiplexer switch.

In the case of the *remote-activated switch*, wires for the remote-control signals must be provided in addition to the serial data connection. Furthermore, if the switch is located some distance from the computer, it may be necessary to add line drivers and receivers to these control lines. In the example control circuit shown in figure 1b, only three wires (plus ground) and a single type-74LS155 integrated circuit (a two- to four-line demultiplexer) are required to provide the enable signals to the four buffers. Two of the wires select one of the four control outputs; the third wire serves as a switch-enable line, selecting output from one buffer or none.

(1b)

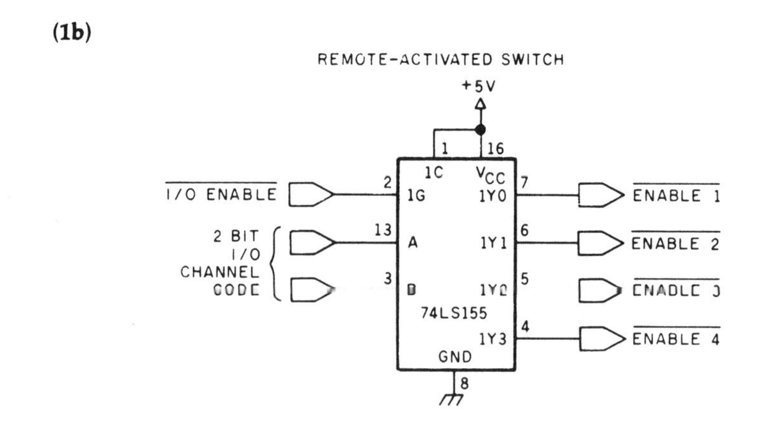

Figure 1b: *Functional diagram of a 74LS155 two- to four-line demultiplexer as it would be connected in a peripheral switch using wired remote activation.*

In principle, remote-activated switches are not much more complicated than their mechanical equivalents. They are generally cost-effective where high speed is essential or where there are tens of channels with various selection configurations. On the negative side, remote-activated switches require hardware and software control interfacing and are not easily adapted to different computers.

The Code-Activated Switch

The *code-activated switch* uses a microprocessor to analyze the characters in the data flowing through the switch. When a particular character or series of characters is received, the microprocessor turns output channels on or off. The only connection between the host computer and the multiplexer is the master input serial line.

Code-activated switches are available with various levels of complexity. The simplest ones merely switch channels upon recognizing a certain code sequence. More sophisticated units can accept incoming data at one data rate, collect it in a memory buffer, and send it to an output channel at another data rate. The most sophisticated units function more as message switchers and data concentrators than multiplexers, allowing party-line conversations and priority-interrupted communications.

Rather than confuse the issue by explaining all the various hardware categories, I've chosen as this month's project an "intelligent" (at least microprocessor-controlled) but rather simple code-activated switch that I call the Micromux. I hope you'll take the opportunity to build it and experiment with it. While useful as a printer or modem switch, this code-activated switch may perhaps find more demanding applications such as message channeling and data acquisition.

Build the Micromux

The intelligence of the Micromux, which is used to decode characters from the serial data, could have been provided by virtually any microprocessor. The only requirement is that the system contain a program-storage area and both a parallel and serial I/O port (the former to send enable signals to the three-state signal buffers, the latter to read and transmit serial data). While I could have chosen a general-purpose microprocessor such as the 6502 or Z80 and then used PIO (parallel input/output) and SIO (serial input/output) adapters, I chose to use the Z8671 variant of the Zilog Z8 single-chip microcomputer to reduce the complexity of the project. The Z8671 was the basis for my Z8-BASIC Microcomputer project (see reference 1) and the new Z8-BASIC System-Controller board available from The Micromint. The connections needed

(1c)

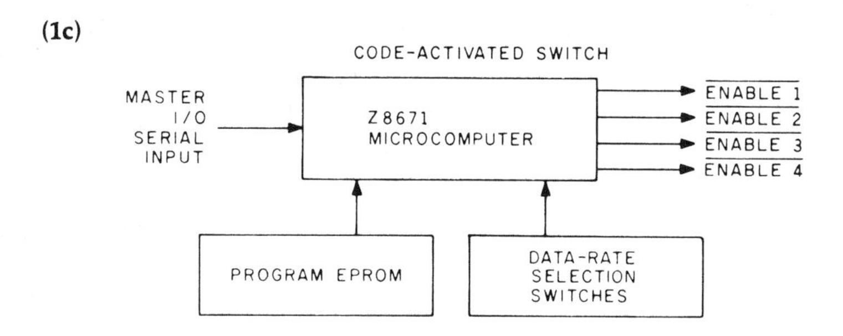

Figure 1c: *Functional diagram of a Z8671 microcomputer chip (variant of a Zilog Z8) as it would be used to control a four-channel code-activated peripheral switch.*

Number	Type	+5V	GND	−12V	+12V
IC1	Z8671	1	11		
IC2	2716/32	24	12		
IC3	74LS373	20	10		
IC4	74LS367	16	8		
IC5	74LS10	14	7		
IC6	74LS244	20	10		
IC7	MC1489	14	7		
IC8	MC1488		7	1	14
IC9	7407	14	7		

Table 1: *Power connections for the integrated circuits of figure 2.*

by the Z8671 to control the four-channel switch are shown in figure 1c.

Much of the hardware and many of the software subroutines required for the task are built into the Z8671 already. This Zilog product contains 256 bytes of RAM (random-access read/write memory), a serial port, two counter/timers, two parallel ports, and a 2K-byte tiny-BASIC (BASIC/Debug) interpreter within a single integrated-circuit package. Combined with a type-2716 EPROM (erasable programmable read-only memory) and a data-rate-selection switch, the five-chip Z8-BASIC Microcomputer system can be easily programmed to monitor RS-232C serial communications and switch channels on cue. The control program can be written in either BASIC or assembled machine language, as you will see.

Figure 2 is the schematic diagram of the Micromux. IC1 is the Z8671. Its serial input line (SI) is tied directly to the data input of the master input channel (RD). Four of the Z8671's port-2 output lines serve as the enable inputs to the three-state signal buffers. The program that controls the computer and analyzes the data transmissions has been written into the 2716 EPROM, IC2. IC3 is an 8-bit latch which holds the 8 low-order address bits (from the Z8's multiplexed outputs) during memory and I/O operations. IC5 is configured as a memory-mapped address decoder that enables IC4 when any address over hexadecimal C000 is accessed.

The data rate is selected by the switch settings on the input of IC4. When the system is powered up, these data-rate switches are read (as memory location hexadecimal FFFE) and used to set a counter/timer that divides the signal from the 7.3728-MHz crystal. The data rates that may be thus selected include 110, 150, 300, 1200, 2400, 4800, and 9600 bps (bits per second).

The integrated circuits IC6 through IC9 are type-74LS244 three-state signal buffers. Nothing flows through them unless their active-low enable lines, pins 1 and 19, are at logic 0. IC18 is a type-7407 buffer/driver, which lights one of four LEDs (light-emitting diodes) to show the enabled channel.

RS-232C drivers and receivers are provided, appearing in the schematic as IC10 through IC17. I chose to use only the six most frequently used RS-232C communication signals in this circuit. If you need additional signals, then you'll have to include more three-state signal buffers and driver/receivers. Conversely, if you need only Received Data and Transmitted Data (pins 2 and 3), then you'll need fewer chips in your code-activated switch. (Six signals—Received Data, Transmitted Data, Clear to Send, Ready to Send, Data Set Ready, and Data Terminal Ready—were as many as I was willing to wire by hand.)

Programming the Micromux

The Z8671 can be programmed in either tiny BASIC or assembly language. The primary difference is speed. With the Z8671 set to receive data at 9600 bps, only a machine-language program would execute fast enough to digest the data and make control decisions at that rate. But for slower data rates where there is more time for the processor to react, using BASIC makes the programming task easier.

For a multiplexer to switch quickly between four different output channels, four distinct character codes are required. These codes can be single ASCII (American National Standard Code for Information Interchange) characters such as the letters A, B, C, and D or sequences of characters such as "$%&1", "$%&2", "$%&3", and "$%&4". However, because the code-activated switch relies upon the data stream to contain its control codes, it's important that the channel-activation codes be different from any character sequences appearing in the data transmissions, or false channel selection may occur. Obviously, we would not choose the letter A as a practical selection code in most applications, so we choose some multicharacter sequence that would be unlikely to appear. For instance, you would probably never need to print the sequence "$%&1", and it could be used to designate a switch to a printer connected to output channel 1. Similarly, "$%&4" would enable channel 4.

Because the sequence-recognition time is dependent on the code length, an alternative is to use single *nonprinting* characters such as Control-A or Control-D. A single-character switch-control code is recognized faster than a three- or four-character code. When using a machine-language program running at 8 MHz the difference is hardly significant, but in BASIC the difference could be considerable.

Micromux Control in BASIC

At this point, let's look at how the

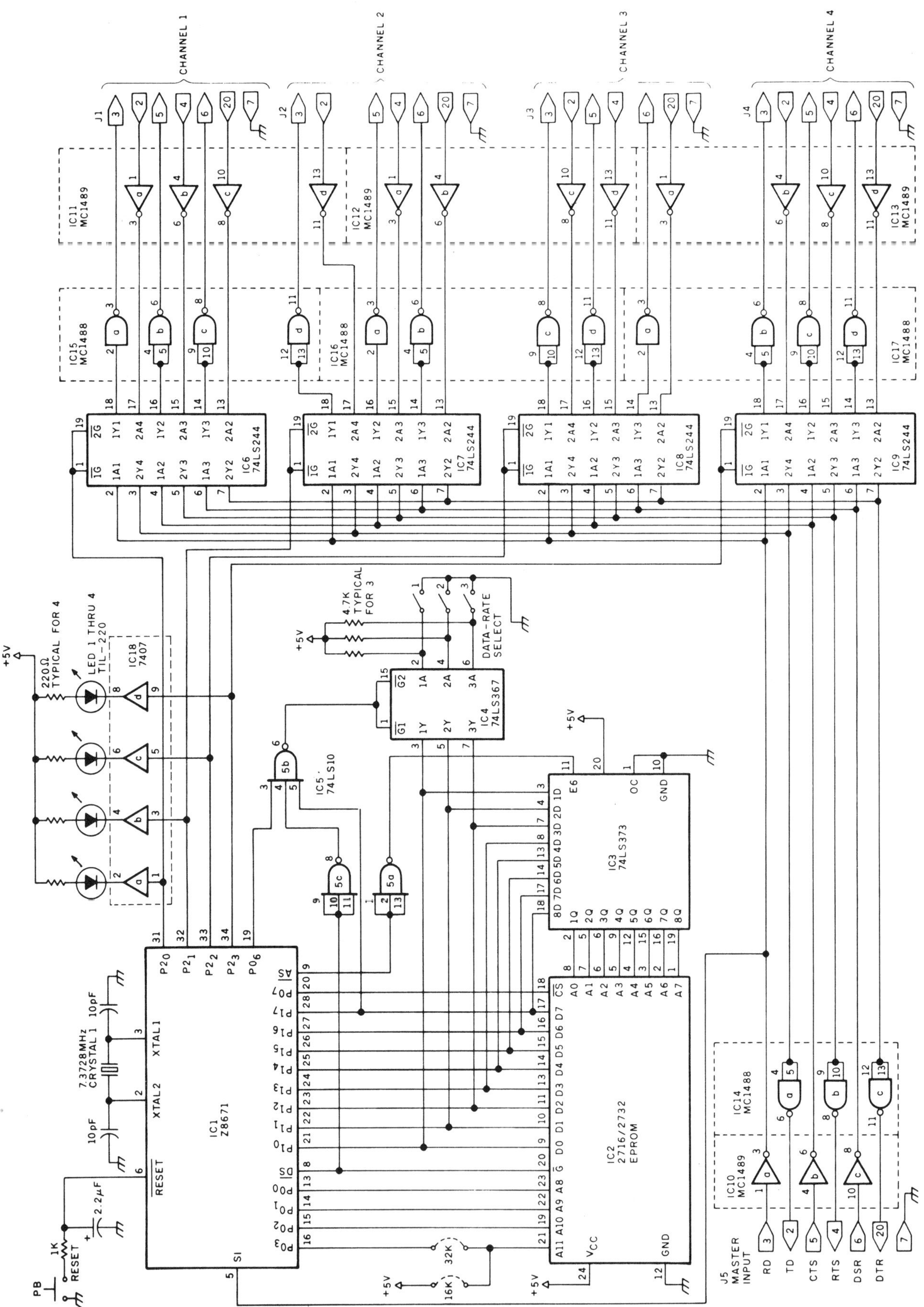

Figure 2: *Schematic diagram of the Micromux prototype four-channel code-activated switch. This design uses a Z8671 as a controller to monitor codes embedded in the transmitted data and activate the appropriate signal buffer for the channel. Only the six most used RS-232C signals are wired here; the addition of more buffer chips would allow switching of more RS-232C signals. Power connections are shown in table 1.*

Micromux might function with the Z8671 programmed in BASIC.

First, assuming that the switch is to be used only with printers and modems that use printable ASCII characters and commonly used control codes (Return, Backspace, Delete, etc.), I chose the four nonprinting characters Control-A, Control-B, Control-C, and Control-D to select the four output channels. For example, when the program sees a Control-C, it activates the enable line to the three-state signal buffer for channel 3.

Figure 3 is a flowchart of a control program written for the Z8-BASIC/Debug interpreter that obeys this convention. The program is seven lines long, as follows:

```
10 @246=0 : @2=255
20 X=@240 : IF X>4 THEN 20
30 IF X=1 THEN @2=254 : GOTO 20
40 IF X=2 THEN @2=253 : GOTO 20
50 IF X=3 THEN @2=251 : GOTO 20
60 IF X=4 THEN @2=247 : GOTO 20
70 GOTO 10
```

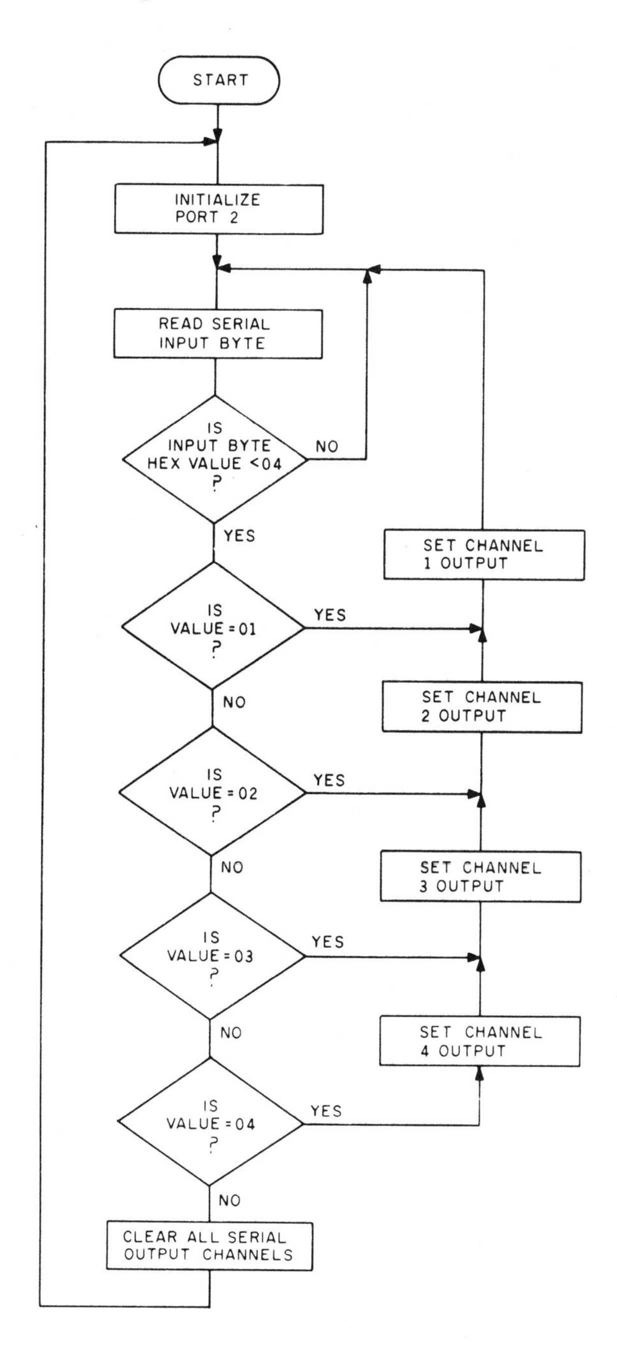

Figure 3: *Flowchart of a Z8-BASIC/Debug program to control the four-channel code-activated switch at low data rates.*

Line 10 configures port 2 for output and sets all enable lines to a logic 1 (not activated). Line 20 examines the serial input register of the Z8 and checks to see if its value is greater than 4. The only ASCII characters with values of 4 or less are the select codes mentioned above plus Control-@. Lines 30 through 60 set the output of the Z8's port 2 according to the control convention or, if a Control-@ is read, to disable all output channels. If the character coming in over the serial input line is not one of these control characters (that is, if its value is greater than 4), the character is sent on to the current output channel (if any), and the port is read again.

Because the read-and-analyze routine is a single program line which returns to itself, it operates fairly fast. The Z8 can be set for any of the 7 common data rates between 110 and 9600 bps. If the multiplexer channel is set before the transmission starts, or if the control code is the very last character in a particular message, then the data rate is irrelevant to the proper functioning of the program. It is only when the control code is embedded in a significant character stream that speed of execution is a consideration.

Regardless of the data rate, about 40 ms (milliseconds) are required to analyze a serial character using this program. As long as a delay of 40 ms is allotted after any of the 5 control codes is sent, the code-activated switch will respond properly. Another possibility is to put several of the same control codes in series so that the program will catch at least one of them.

Machine-Language Control

For the hasty folk who like to send and switch data at 9600 bps, we have to use a machine-language program. A flowchart of such a program is shown in figure 4. The increased processing speed using machine language allows us to use more characters in the switch-control code sequence and lessens the likelihood of confusion with data passing through the switch.

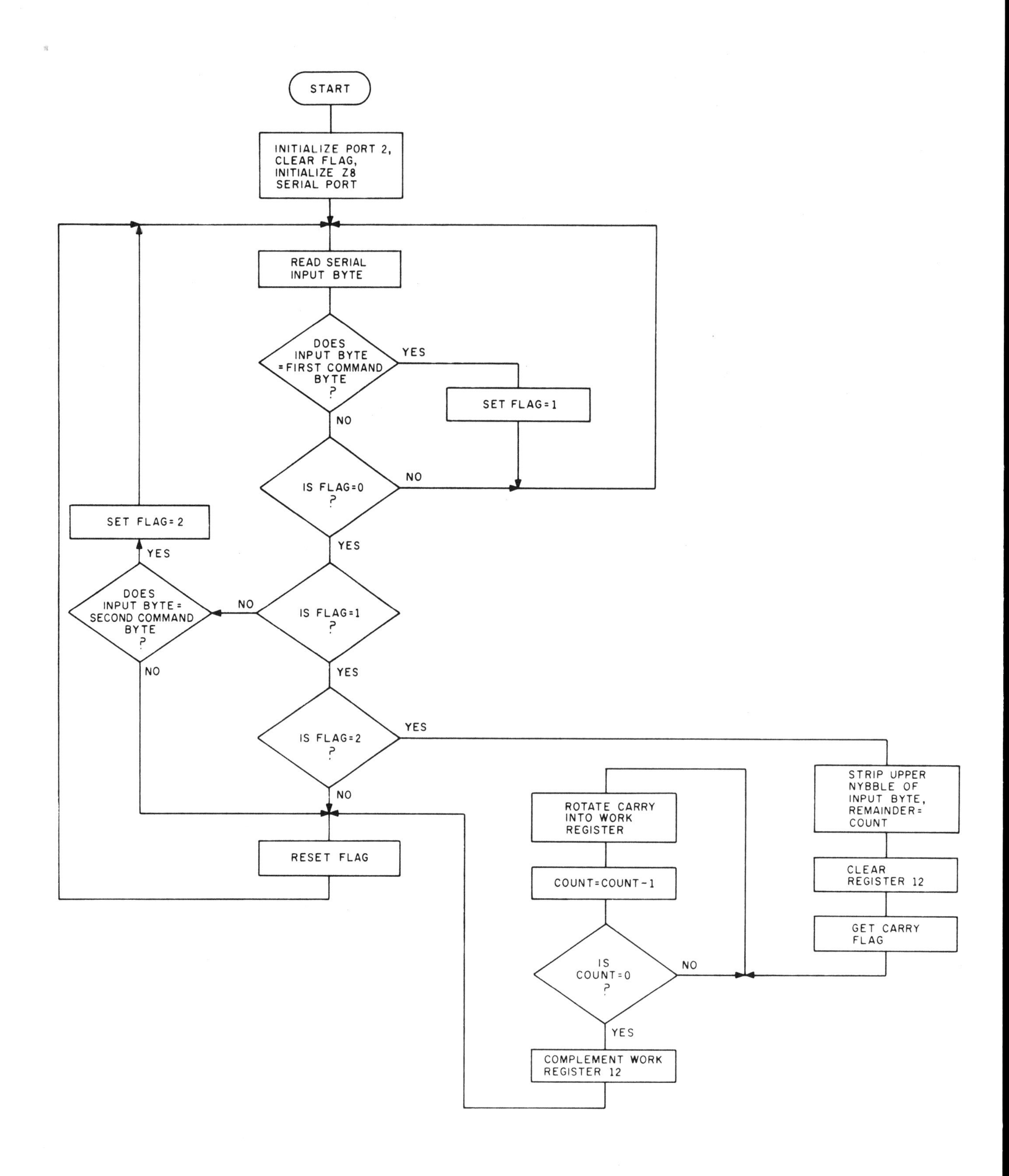

Figure 4: *Flowchart of a Z8 machine-language program that can control the code-activated switch at data rates up to 9600 bps.*

The final program, shown as assembly code in listing 1, uses many subroutines already available in the Z8671's ROM and requires only 88 bytes. It is designed to use a two-character control-sequence-recognition code, followed by a channel number. (I decided not to use unprintable control codes in this example, but any ASCII characters can be used.) The complete sequence is "@!x", where *x* is a value from 1 to 4. The code sequence "@!3", for example, would direct the output of the Micromux to channel 3, while the sequence "@!1" would send the output to channel 1. One handy feature of the read-and-analyze routine is that only the 4 low-order bits of the third character in the sequence are used. This allows us to use one of the Control-A, -B, -C, or -D output-selection characters (as used in the BASIC program) as the third character in the sequence in place of the digit.

Because the program of listing 1 runs entirely in machine language, channel changes may occur at any point in the transmission, at any of the available transmission speeds. If you plan on using this program with a printer attached to one of the output channels, you may want to change the "@!" part of the sequence to two nonprinting characters. Simply substitute two ASCII codes of your choice in the program.

In Conclusion

The primary purpose of this article is to present something the average computer hobbyist can successfully build and use. However, if you are an industrious programmer with a lot of experience, you might want to add some additional features, such as storing input in a memory buffer, trapping control-code sequences (which would filter the switching codes out of the data flowing to the output channels), and party-line communication (where data is sent to more than one output channel). Of course, all of these additions would have involved more complex hardware and software.

In my own case, the Micromux didn't end up attached to a printer. I'm using the prototype, consisting mostly of a modified Z8-BASIC Microcomputer, to communicate with four other Z8 boards that perform specific control and security assignments, none of which I had previously deemed important enough to merit tying up four separate serial ports.■

Listing 1: *Machine-language program for the Z8671 to control the code-activated switch at data rates up to 9600 bps.*

```
1000                   0010 ;
1000                   0020 ;
1000                   0030 ;  MULTIPLEXER SWITCHING PROGRAM FOR THE
1000                   0040 ;  ZILOG Z8 MICROPROCESSOR, P/N Z8671 WITH
1000                   0050 ;  ON-BOARD BASIC/DEBUG ROM
1000                   0060 ;
1000                   0070 ;  READ A TWO BYTE CODE AND THEN SET THE
1000                   0080 ;  PORT 2 CONTROL BITS AS REQUIRED BY
1000                   0090 ;  THE THIRD (DATA) BYTE.
1000                   0100 ;
1000                   0110 ;  USES ROM CALL TO READ BYTES ON THE SERIAL
1000                   0120 ;  PORT OF THE Z-8 COMPUTER BOARD.
1000                   0130 ;
1000                   0140 ;   TO CHANGE CONTROL BYTES, STORE FIRST COMMAND
1000                   0150 ;   AT 1041H, SECOND COMMAND AT 1055H. YOU MUST
1000                   0160 ;   USE DIFFERENT CODES FOR CMD1 AND CMD2, DUE TO THE
1000                   0170 ;   UNCONDITIONAL FLAG SET AFTER THE "FIRST" LABEL.
1000                   0180 ;
1000                   0190 ;
1000                   0200 ; FIRST SET THE FIRST 20H BYTES TO 0FFH. THESE ARE THE
1000                   0210 ; INTERRUPT VECTORS, AND ARE UNUSED BY THIS PROGRAM.
1000                   0220 ;
1000                   0230 ;
1000                   0240 FILL   EQU   1000H
1000 FF FF FF FF FF    0250        DB    #0FFH,#0FFH,#0FFH,#0FFH,#0FFH,#0FFH,#0FFH,#0FFH
     FF FF FF
1008 FF FF FF FF FF    0260        DB    #0FFH,#0FFH,#0FFH,#0FFH,#0FFH,#0FFH,#0FFH,#0FFH
     FF FF FF
1010 FF FF FF FF FF    0270        DB    #0FFH,#0FFH,#0FFH,#0FFH,#0FFH,#0FFH,#0FFH,#0FFH
     FF FF FF
1018 FF FF FF FF FF    0280        DB    #0FFH,#0FFH,#0FFH,#0FFH,#0FFH,#0FFH,#0FFH,#0FFH
     FF FF FF
1020                   0290 ;
1020                   0300 ;
1020                   0310 BASIC  ORG   1020H          ; PLACE A BASIC LINE AT 1020H
1020 00 0A 47 4F 20    0320        DB    #0,#10,'GO @%1030',#0,#0FFH,#0FFH,#0FFH
     40 25 31 30 33
     30 00 FF FF FF
102F                   0330 ;
102F                   0340 ;
102F                   0350 MULTI  ORG   1030H
1030 E6 F6 00          0360        LD    246,#0         ; SET UP PORT 2 AS OUTPUT
1033 E6 F7 C1          0370        LD    247,#0C1H      ; 7 BITS, ODD PARITY, 1 STOP, PORT 2 P
ULLUPS ON.
1036 E6 02 FF          0380        LD    2,#0FFH        ; SHUT DOWN ALL I/O AT START
1039 E6 30 00          0390        LD    30H,#0         ; SET FLAG TO 0
103C                   0400 ;
103C                   0410 ;
103C D6 00 54          0420 CMAND  CALL  #0054H         ; GET A BYTE FROM THE SERIAL PORT
103F A6 13 40          0430        CP    13H,#CMD1      ; IS THIS THE FIRST COMMAND CODE ?
1042 6B 05             0440        JR    Z,FIRST        ; YES, DO SET FLAG
1044 A6 30 00          0450        CP    30H,#0         ; IS FLAG 0 ?
1047 EB 05             0460        JR    NZ,PART2       ; NO, TEST FLAG SOME MORE
1049 E6 30 01          0470 FIRST  LD    30H,#1         ; SET FLAG TO 1
104C 8B EE             0480        JR    CMAND          ; AND GO BACK TO LINE
104E A6 30 01          0490 PART2  CP    30H,#1         ; IS FLAG A 1 ?
1051 EB 0A             0500        JR    NZ,PART3       ; NO, CHECK LAST POSSIBLE FLAG SET
1053 A6 13 21          0510        CP    13H,#CMD2      ; YES. IS THIS COMMAND BYTE 2 ?
1056 EB 1B             0520        JR    NZ,RESET       ; NO, FALSE ALARM. RESET FLAG BYTE
1058 E6 30 02          0530        LD    30H,#2         ; YES, UPDATE FLAG
105B 8B DF             0540        JR    CMAND          ; AND GO BACK TO LINE
105D A6 30 02          0550 PART3  CP    30H,#2         ; IS FLAG A 2 ?
1060 EB 11             0560        JR    NZ,RESET       ; NO. FALSE ALARM. RESET FLAG BYTE
1062 B0 12             0570        CLR   12H            ; CLEAR REG 12H
1064 DF                0580        SCF                  ; SET CARRY FLAG
1065 56 13 0F          0590        AND   13H,#0FH       ; MASK OFF H.O. BYTE OF DATA
1068 10 12             0600 LOOP   RLC   12H            ; ROTATE BIT ONCE.
106A 00 13             0610        DEC   13H            ; COUNT=COUNT-1
106C EB FA             0620        JR    NZ,LOOP
106E 60 12             0630        COM   12H            ; COMPLEMENT BECAUSE 0 = ENABLE TO CI
RCUIT
1070 E4 12 02          0640        LD    2,12H          ; WRITE BIT TO THE MUX CONTROL PORT
1073 E6 30 00          0650 RESET  LD    30H,#0         ; RESET COMMAND FLAG
1076 8B C4             0660        JR    CMAND          ; AND GO BACK TO LINE
1078                   0670 ;
1078                   0680 ;
1078                   0690 CMD1   EQU   '@'
1078                   0700 CMD2   EQU   '!'
1078                   0710 ;
1078                   0720 ZZZZ   END

SYMBOL TABLE

BASIC 1020   CMAND 103C   CMD1  0040   CMD2  0021   FILL  1000
FIRST 1049   LOOP  1068   MULTI 1030   PART2 104E   PART3 105D
RESET 1073   ZZZZ  1078
```

18

Build the RTC-4 Real-Time Controller

A 4-bit single-chip microcomputer from Texas Instruments comes preprogrammed for timed automatic control.

Bee-beep . . . bee-beep . . . bee-beep. I fumbled for the alarm button on my digital watch and briefly wondered what I had set it for.

"Oh yeah. Time to turn on the recorder."

The audio-cassette recorder in question was set up in one corner of the Circuit Cellar and connected to an FM radio. I had set it up to record a local news program, which is broadcast daily at 5:30 p.m. in my area. Because 5:30 is not usually a convenient time for me to listen to the radio, I often record the program so that I can listen to it later in the evening or the next day while I'm driving someplace.

So it was 5:28. Time to turn on the recorder. The trouble was I was outside working on my satellite dish antenna. As I dashed into the house and down the stairs, I thought, "There's got to be a better way to do this."

Once the recording was safely under way, I began to consider various automatic methods. I could use a mechanical timer, a simple digital alarm clock (with appropriate Circuit Cellar modifications), a BSR X-10 Home Control System timer, or my existing home-security and control system. None of these seemed completely satisfactory. Instead, I decided to look for a more universal solution.

The problem of timed activation pointed out the need for a generalized, cost-effective timer/controller that might, while solving my particular problem, have additional capabilities. As I was evaluating various circuits for real-time controllers, I came across some preprogrammed TMS1000-series 4-bit microcomputers-on-a-chip from Texas Instruments. One of these chips became the essential element in this month's project, the Circuit Cellar RTC-4 real-time controller.

The RTC-4 is a four-channel time-activated device or appliance controller that can be built for less than $100, complete with keypad, display, power supply, and relay outputs. It can be used in the home or laboratory for general time-dependent applications or, as in my case, to solve one particularly nagging problem.

Before jumping into how to build the RTC-4, let's take a look at the TMS1000 family tree and the particular branch of use in our project.

The TMS1000 Family

Texas Instruments (TI) makes a large family of single-chip MOS/LSI (metal-oxide semiconductor/large-scale integration) components called the TMS1000 series, which all contain 4-bit microprocessors. While the approximately 50 members of the family share a common subset of about 40 instructions, they differ in the varying amounts of read-only memory (ROM) and random-access read/write memory (RAM), and varying numbers of I/O- (input/output) control circuits. The family's members also differ in packaging and power requirements, but, in the 28-pin dual-inline package, the basic TMS1000 and TMS1100 are virtually the same with the exception of internal memory capacity. TI intends that the ROM in TMS1000-series components be mask-programmed for specific computing tasks; the devices are not general-purpose microprocessors, as are the familiar Z80, 6502, and 8088.

One major member of the product line is the TMS1100. This chip has 2048 bytes of internal ROM, 128 4-bit words (nybbles) of RAM, 4 input lines, and 19 output lines. (The basic TMS1000 contains less memory.) Figure 1 is a block diagram of its internal structure.

TMS1121C UTC

Somewhere on the 1100 branch of the TMS1000 family tree is an offshoot called the TMS1121C Universal

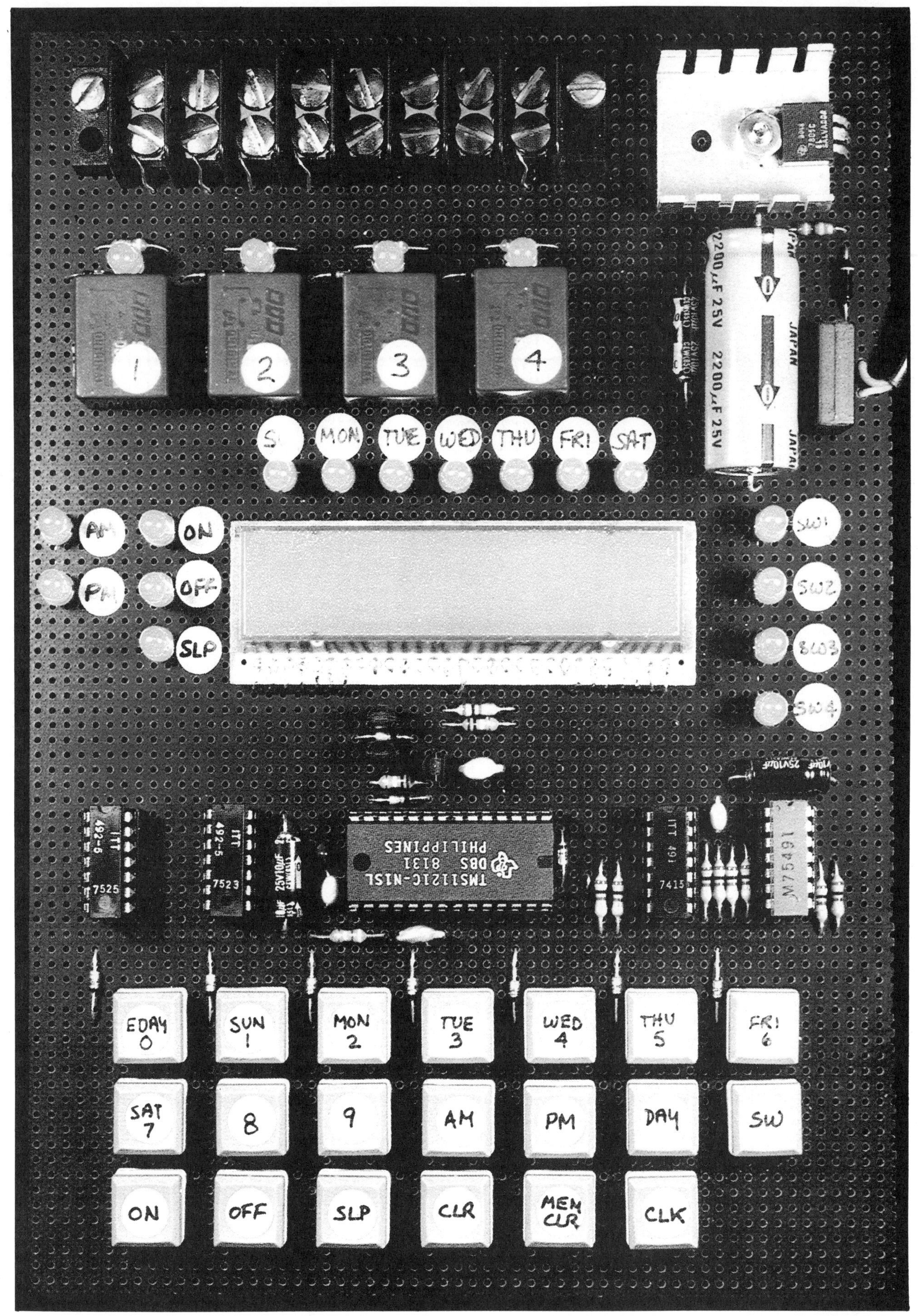

Photo 1: *Prototype of the Circuit Cellar RTC-4 real-time controller, a unit suitable for switching low-current external devices on and off according to immediate commands or stored time-lapse event settings.*

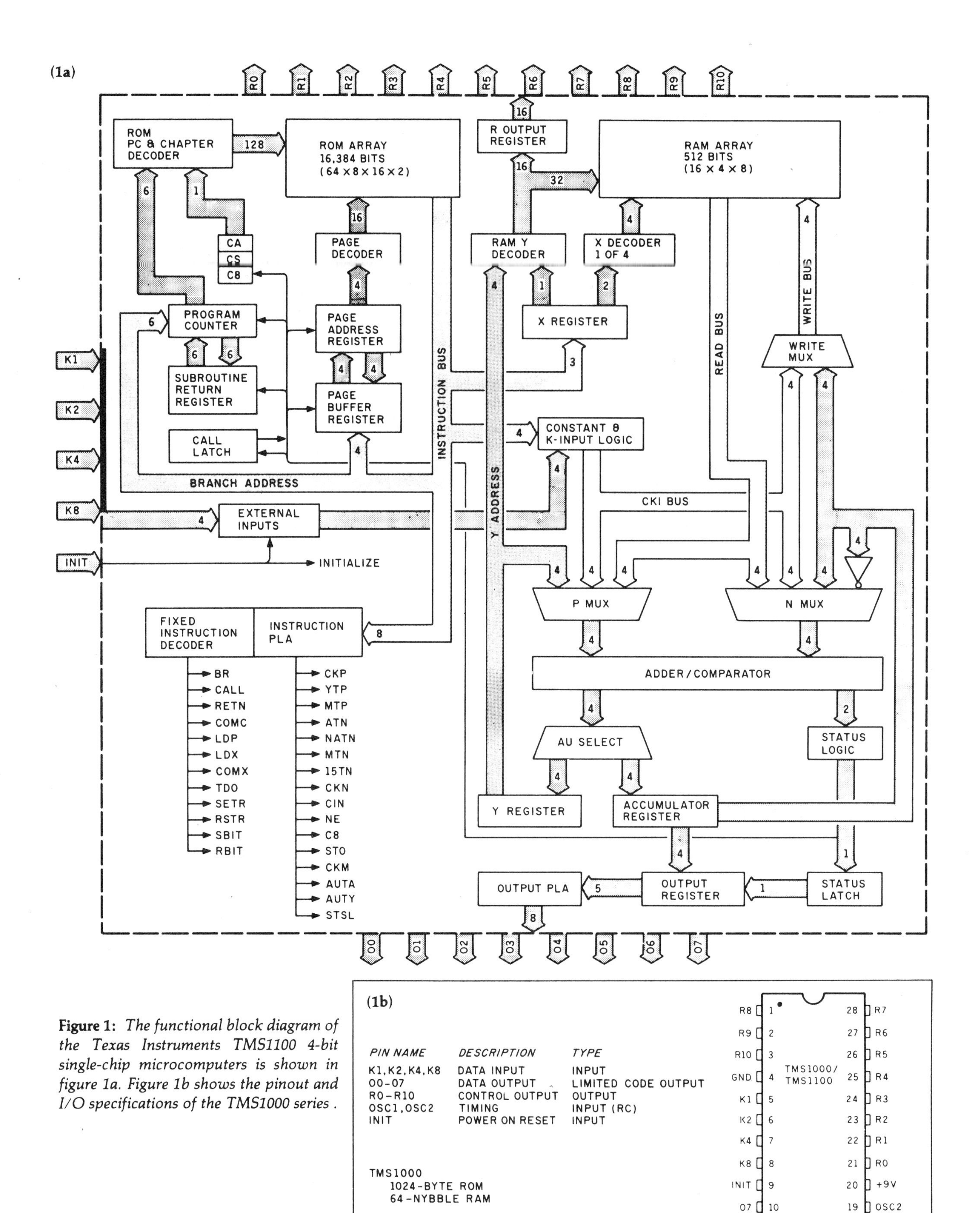

Figure 1: *The functional block diagram of the Texas Instruments TMS1100 4-bit single-chip microcomputers is shown in figure 1a. Figure 1b shows the pinout and I/O specifications of the TMS1000 series.*

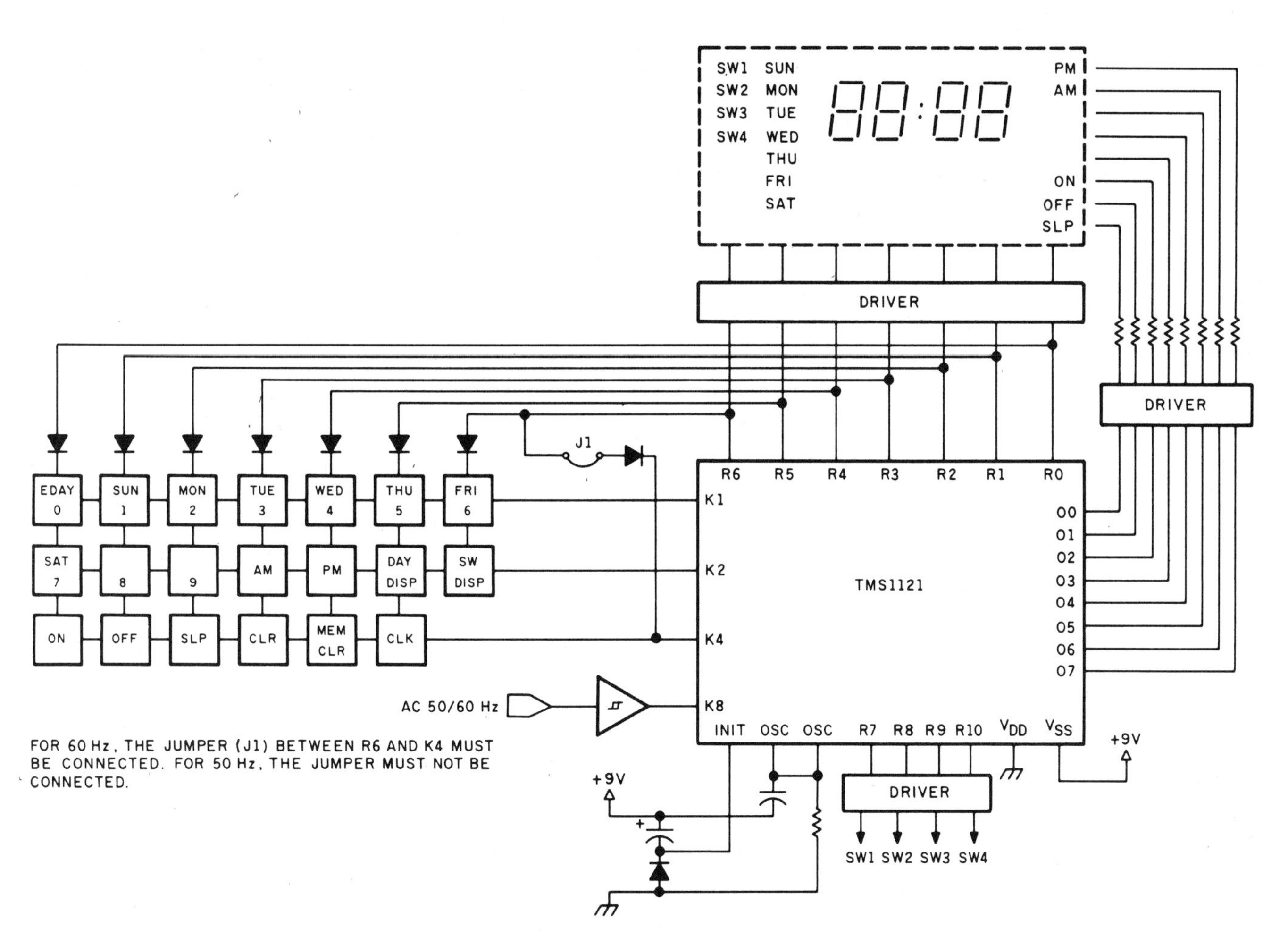

Figure 2: *Block diagram of the Circuit Cellar RTC-4 real-time controller, showing the TMS1121C 4-bit microcomputer and its associated user-interface components.*

Timer Controller. When I found out about the TMS1121C UTC, I realized that it was the perfect basis for the Circuit Cellar RTC-4 real-time controller. It is a variant of the 1100, designed and mask-programmed by Texas Instruments primarily to be a demonstration device for the TMS1000 product line. The 1121 is intended to provide a simple and inexpensive means for a prospective customer to become familiar with the workings of the 1000 series in an easily understood application that does not require costly custom programming. Fortunately for us, someone at TI did a little thinking and produced a rather neat (even considering its limitations) real-time programmable controller on a single chip.

RTC-4 Characteristics

The RTC-4 features a variety of control possibilities and operating aids: various status displays are available, many events can be programmed at once, and several options can be set up. Part of the RTC-4's task is merely to operate as a digital clock, displaying the time of day and the day of the week. The capabilities of the RTC-4 are essentially those of the TMS1121. A list of the unit's features is shown in table 1.

The RTC-4 contains additional circuitry necessary to interface the chip to the real world. In addition to the TMS1121, the unit includes an LED (light-emitting diode) display and associated drivers, a 20-key keypad, and four output relay switches.

The 20-key keypad enables you to enter instructions and data into the TMS1121. The instructions may be in the form of a stored program telling the RTC-4 what to do in the future, or they may be intended for immediate action. Using the pad, you can turn any of the output relays on or off directly without storing the commands in memory. Also, using the pad you can originate, change, inspect, or delete any timer programs.

The RTC-4 is capable of retaining up to 18 timer programs (on/off settings), which are entered through the keypad. Each of these programs can control one of the four output relay switches. The onboard output relays are small, low-current devices, so if

1. Four independent switch outputs with buffer
2. Display of time and day and status of its own switches
3. As many as 18 daily or weekly programmable setpoints
4. Memory display of programmed setpoints
5. Key entry for clock set and timer set
6. 50-Hz or 60-Hz operation

Table 1: *Capabilities of the RTC-4, which are essentially identical to those of the TMS1121C Universal Timer Controller.*

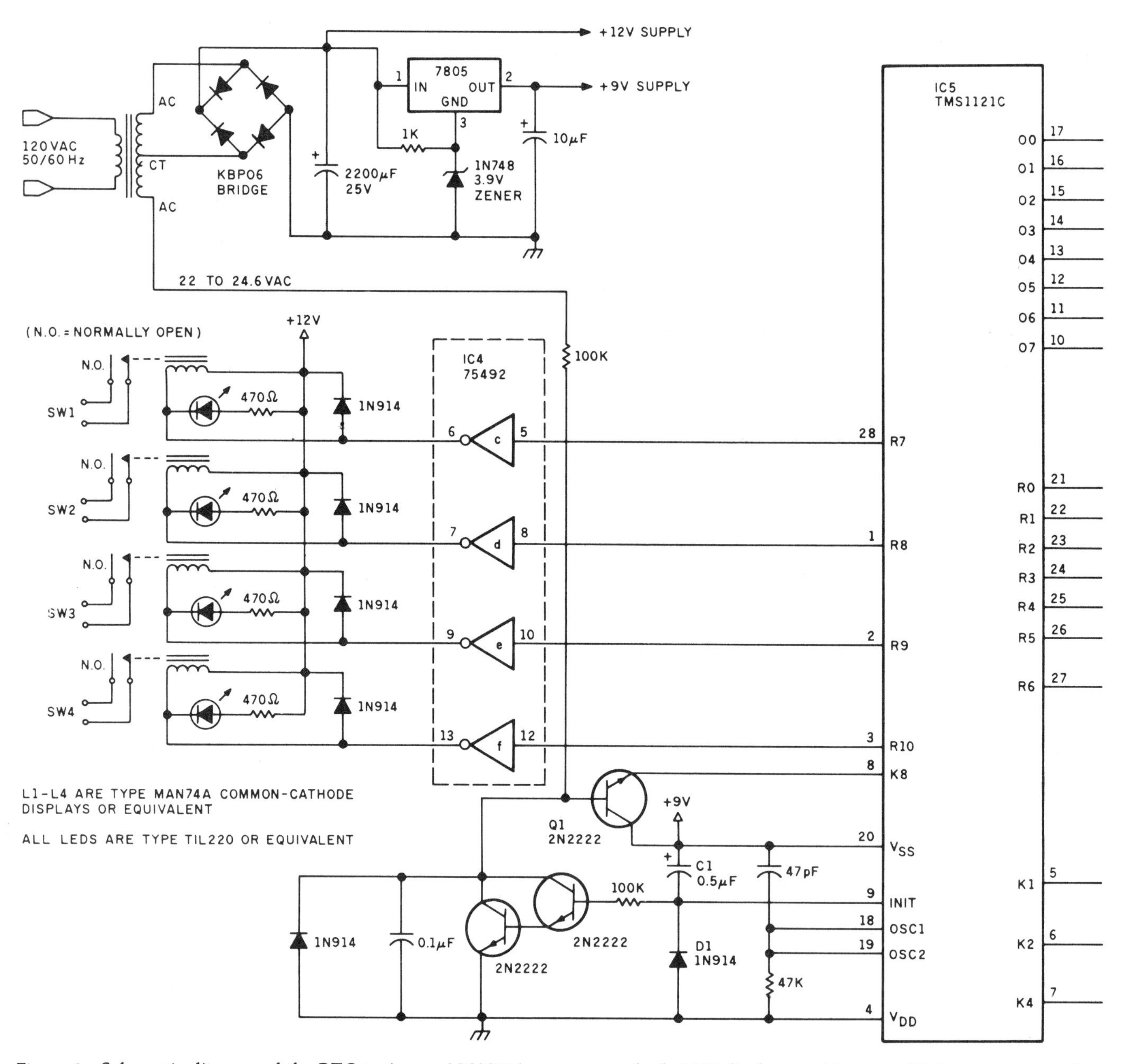

Figure 3: *Schematic diagram of the RTC-4. A type-MAN74A common-cathode LED display module is specified, a type somewhat different from the junk-box unit used in the prototype (see text box, page 192). Please note the position of the MSD (leftmost digit) and the LSD (rightmost digit) when wiring the unit.*

you want to control a toaster, you'll need to let the onboard relay control a larger outboard relay, which in turn operates the high-current device.

There are two kinds of timer programs: fixed programs and interval programs. Fixed-time programs toggle an output switch at a specific time of day, while interval programs toggle an output switch when a certain interval of time has elapsed since the previous toggling of the switch. Fixed-time programs are retained in memory and repeatedly executed. Interval programs are automatically deleted after execution.

Each program setting toggles only one switch, but a special function exists to combine on/off operations into one program sequence: the SLP (sleep) function is used to turn a switch on and then off 1 hour later. I'll tell you more about programming the RTC-4 later.

RTC-4 Hardware

Figure 2 is a simplified block diagram of the RTC-4, and figure 3 is the schematic diagram. Most of the circuitry of the RTC-4 is associated with the keypad and display and is multiplexed.

As with the other members of the TMS1100 group, the input lines on the TMS1121 are designated K1, K2, K4, and K8 (after their binary significance), and the two sets of output lines are labeled O0 through O7 and R0 through R10. The eight O output lines are configured during the mask-programming of the chip such that only certain combinations of their output values are attainable. Rather than a full range of 256 values, these eight lines can be set only to a

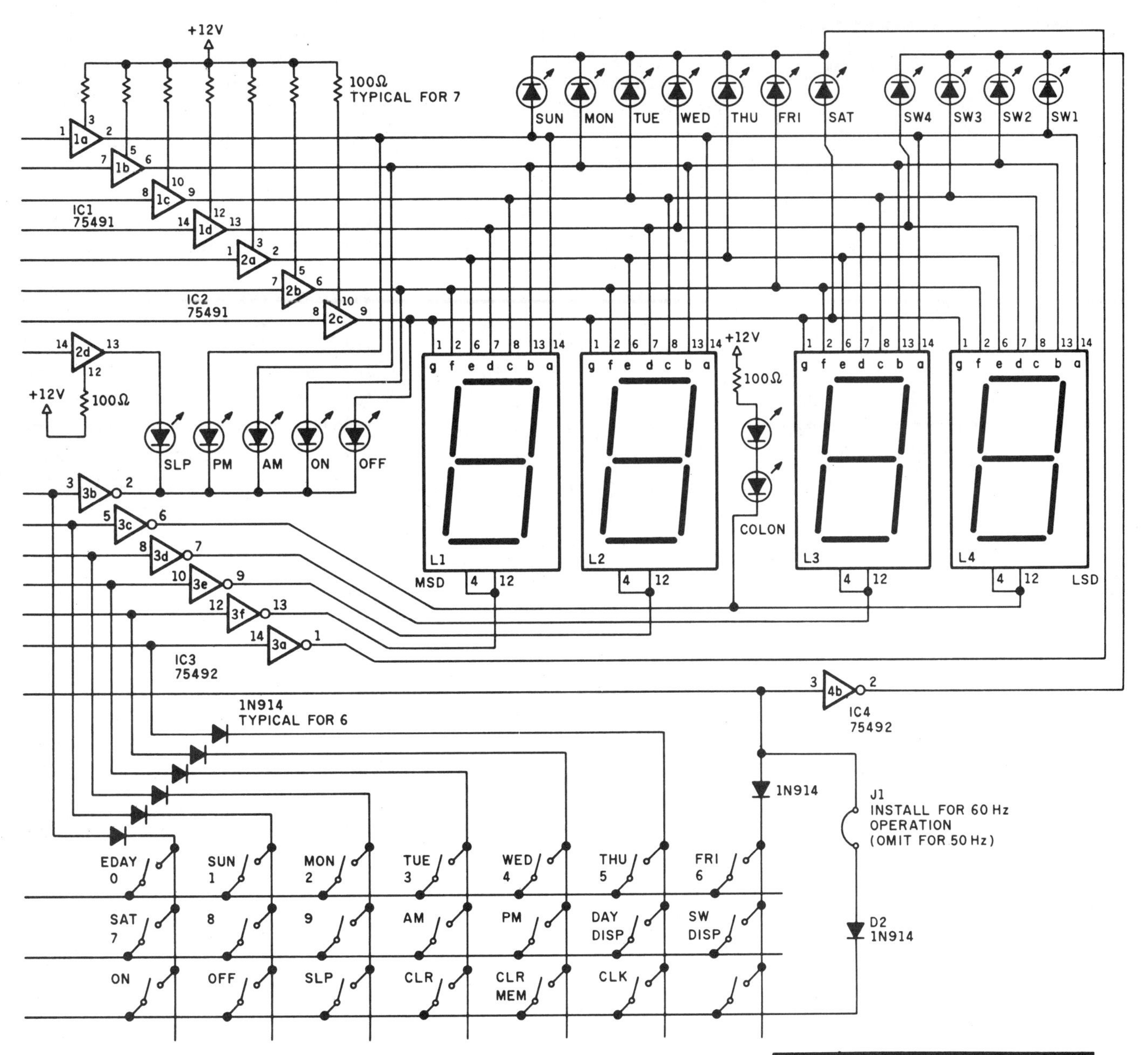

specific subset of combinations defined by the accumulator and status flags. Most frequently the O lines are used as display drivers or status indicators. In this application, very few combinations of the eight O lines are necessary. The 11 R-series output lines, on the other hand, are treated separately; each R line can be set or cleared individually.

The O0 through O7 lines are data-output lines used here to convey the mode, day, switch, and seven-segment codes (for the display) along a time-multiplexed bus. Each O line is buffered through a type-75491 MOS-to-LED-segment driver. Line O0 is segment A, O1 is segment B, and so on through O6, which is segment G. Line O7 is called the decimal-point segment even though no decimal points are used in this application; O7 is attached to the LED that indicates operation of the SLP (sleep) command.

Output lines R0 through R6 are the digit-select lines. Each of the four numeric digits in the LED display module is composed of a cluster of seven segments; the O lines control

Number	Type	+5V	GND
IC1	75491	11	4
IC2	75491	11	4
IC3	75492	11	4
IC4	75492	11	4
IC5	TMS1121C	20	4

the segments, while each of four of the R lines activates a digit cluster. Each of the digit-select lines is buffered through a type-75492 signal-inverting MOS-to-LED digit driver. Only one digit-select line is active and one display group illuminated at a time, but by sequencing (multiplexing) through all the digits rapidly, the system can make all the LEDs appear to be illuminated simultaneously.

The two MOS-to-LED drivers, the 75491 and 75492, are designed to

R0 — Mode	— AM, PM, ON, OFF, SLP
R1 — Tens of hours	— 0, 1, or 9 (error)
R2 — Hours	— 0 through 9
R3 — Tens of minutes	— 0 through 5 and 9 (error)
R4 — Minutes	— 0 through 9 and colon
R5 — Day of week	— SUN, MON, TUE, WED, THU, FRI, SAT
R6 — Switch	— SW1, SW2, SW3, SW4

Table 2: *Groups of LED status and numeric displays in the RTC-4, along with their corresponding select lines.*

drive common-cathode LED displays. The 4-digit numeric module is wired for common-cathode operation, and I set up the discrete LEDs in the three other display groups as common-cathode also. By impressing a logic high level (binary 1) on the O lines and logic low level (0) on one of the digit-select lines, the LEDs of each of the display groups can be lit. The seven display groups and their individual select lines are shown in table 2.

In addition to functioning as digit-select lines, the R0 through R6 outputs are used in combination with the K-group input lines in scanning the keypad matrix for user input. As each display group or digit cluster is selected in its turn by an R line, the same R line applies a logic 1 to a column of three keys (or two keys on the R6 line). If one of these three keys is pressed during the application of this scanning signal (thereby closing the circuit), the signal will flow into one of the three K input lines. The program in ROM determines which key has been pressed by reading the K lines in sequence and comparing the combined R and K addresses. This procedure includes use of a 10-ms (millisecond) software contact-debouncing subroutine.

The remaining four R output lines (R7 through R10) are the timer outputs. Buffered through some more sections of a 75492, these signals, in the case of the RTC-4, drive electromechanical relays; you could use solid-state relays also to control external equipment.

The rest of the RTC-4 consists of the power supply and timing sections.

The TMS1121 uses a resistor and a capacitor in a tuned circuit, rather than a crystal, to set its internal clock frequency. This is adequate because the real-time-clock function is synchronized to the 60- or 50-Hz AC power line, and the actual processor clock speed is not critical.

With the 47-picofarad/47k-ohm resistor/capacitor combination shown in the schematic, the clock frequency should be about 300 kHz. The 60- or 50-Hz timing signal is derived from one side of the power-supply transformer. This clock signal is made

The RTC-4 keyboard has 10 dual-function and 10 single-function keys; programming is relatively straightforward.

more square by the transistor Q1 and applied to the K8 input line. If the frequency source is 60 Hz, then you should install jumper J1 (it is omitted for 50-Hz operation). The INIT input pin, connected to diode D1 and capacitor C1, functions as a power-on reset line.

The power supply requires a center-tapped (CT) step-down transformer that has an output between 20 and 24 V (volts). I used a 300-mA (milliampere), 22-V CT unit (Micromint PITB-109), but a 450-mA, 24-V center-tapped transformer from Radio Shack (catalog number 273-1366) should also work. The RTC-4 cannot be battery powered because the power line's frequency is used for timing. The rest of the power-supply circuit consists of a standard type-7805 three-terminal voltage regulator configured for a +9-V output with a zener diode to raise its ground reference potential.

If you don't have any 4- or 5-V zener diodes for this kind of circuit, I recommend that you use an LM317 regulator instead. The power for the displays and relays need not be well filtered or regulated; it can be derived directly from the rectifier output when using a 22-V transformer. In the case of higher-voltage transformers, you might need to add a 7812 regulator to keep the LED drivers from dissipating too much power.

Keypad Programming

Programming the RTC-4 is relatively straightforward. The keyboard has 10 dual-function and 10 single-function keys, as shown in table 3.

When the RTC-4 is turned on, the display will automatically read Sunday at 12:00 p.m. if the unit is configured for 60-Hz operation (with jumper J1 installed). If the RTC-4 is set for 50 Hz, then you must press the CLK key to start it. Obviously, after you turn it on in this fashion, the first thing to do is set the correct time on the clock.

For instance, entering the following keypress commands:

SUN, DAY, 1, 1, 2, 4, PM, CLK

sets the clock to 11:24 p.m. Sunday evening.

The pattern for setting the clock is always the same. The day of the week is registered by typing, in this case, the SUN and DAY keys. The SUN key could be interpreted as meaning "1", but the entry of the DAY key immediately afterward leaves no ambiguity, and the software can determine your intent. (Texas Instruments uses the name WEEK for this key, instead of DAY, but since all entries involve a day setting, I felt that DAY was better than WEEK as the key legend.)

After the day of the week has been set, you tell the computer the hour and minutes, and whether these are antemeridian or postmeridian (a.m. or p.m.). Once the proper day and time have been selected, you press the CLK key to start the real-time clock from that setting.

Direct Switch Control

The RTC-4 controls external devices and appliances through four relay-switch outputs designated SW1,

SW2, SW3, and SW4. These switches can be individually turned on or off at specific times under program control, or they may be directly controlled from the keypad. To directly turn switch 2 on and then off, the commands would be:

2, SW, ON, . . . 2, SW, OFF

An alternative to separate on/off command sequences is the SLP (sleep) command. The sequence:

2, SW, SLP

will turn switch 2 on immediately and then automatically off 1 hour later. Any of the three basic functions, SLP, ON, or OFF, may be specified for any of the four switches. But the direct control sequences are not stored in RAM.

Fixed-Time Programs

Fixed-time programs change the state of the switch when the clock reaches a preset time. A typical sequence for entering a fixed-time program would be:

3, SW, MON, DAY, 9, 0, 0, AM, ON

This series would turn on switch 3 on Monday morning at 9:00 a.m. The first two keys, 3 and SW, indicate a switching function for the specific output channel 3. Next, you enter the day and time in the same manner as if you were setting the clock. Finally, you designate the action desired, ON.

As the key sequence is entered, the digital readout and LED indicators display the program settings. The day, time, and program function are automatically stored but will continue to be displayed until another sequence is initiated. To return the display to the current-time digital-clock mode, press CLK.

If the switch being controlled or day of the week differs in the next sequence from the preceding setting, the preceding key sequence must be repeated in its entirety with the new parameters. If, however, the switch number and day of the week are the same, a shortened sequence can be used:

1, 1, 4, 5, AM, OFF

If both of the above sequences are entered, the combined result would be to activate switch 3 on Monday at 9:00 a.m. and deactivate it at 11:45 a.m. the same day. In the case of my tape recorder, I want the action to take place every day, so I use the EDAY (every day) command to turn the recorder on every day at 5:30 p.m. and off at 7:00 p.m. as follows:

1, SW, EDAY, DAY, 5, 3, 0, PM, ON

7, 0, 0, PM, OFF

Exceptions to this can be added as separate program lines. Because the radio program I record runs for only 30 minutes on weekends, instead of

You are able to display stored timer settings by the day or by the switch channel. Status LEDs show pertinent information.

the usual 90 minutes during the week, I can shut the recorder off 60 minutes sooner on Saturday and Sunday:

1, SW, SAT, DAY, 6, 0, 0, PM, OFF

1, SW, SUN, DAY, 6, 0, 0, PM, OFF

Interval Programs

In an interval program, the function is performed after the specified time interval has passed. For example, in the sequence

4, SW, 2, 3, 0, ON

switch 4 would be turned on 2½ hours after the last key in the sequence is pressed. ON, OFF, and SLP commands can be used in interval programs. If SLP were substituted for ON in the above entry, switch 4 would have been turned on after 2½ hours and off again after 3½ hours from the time of programming. The maximum time length for any interval is 11 hours and 59 minutes. Interval programs can also be combined in the same manner as the fixed-time programs:

4, SW, 3, 0, ON

3, 1, OFF

These commands cause switch 4 to be activated after 30 minutes and deactivated after 31 minutes. The result is a 1-minute "on" time that starts 30 minutes after command entry.

Program Display and Errors

Stored timer settings can be displayed by the day or the switch channel. For example:

2, SW, SW, . . . SW, SW

would display all programs affecting switch 2. As each step is displayed, the status LEDs and numeric display show pertinent information such as the time, day, and what relays are set on or off. Entering

SAT, DAY, DAY, . . . DAY, DAY

Double-Function Keys	
EDAY/0	- Everyday or 0
SUN/1	- Sunday or 1
MON/2	- Monday or 2
TUE/3	- Tuesday or 3
WED/4	- Wednesday or 4
THU/5	- Thursday or 5
FRI/6	- Friday or 6
SAT/7	- Saturday or 7
SW/DISP	- Switch or Display switch-program memory
DAY/DISP	- Day or Display daily-program memory

Single-Function Keys	
8	- Numeric 8
9	- Numeric 9
AM	- AM time setting
PM	- PM time setting
ON	- Switch on
OFF	- Switch off
SLP	- Sleep — switch output on for 1 hour then off
CLR	- Clear entry or Error
MEM CLR	- Clear Program Memory
CLK	- Set or display clock

Table 3: *Functions of keys on the RTC-4's keypad; some have two functions, others only one.*

Prototype Construction Techniques

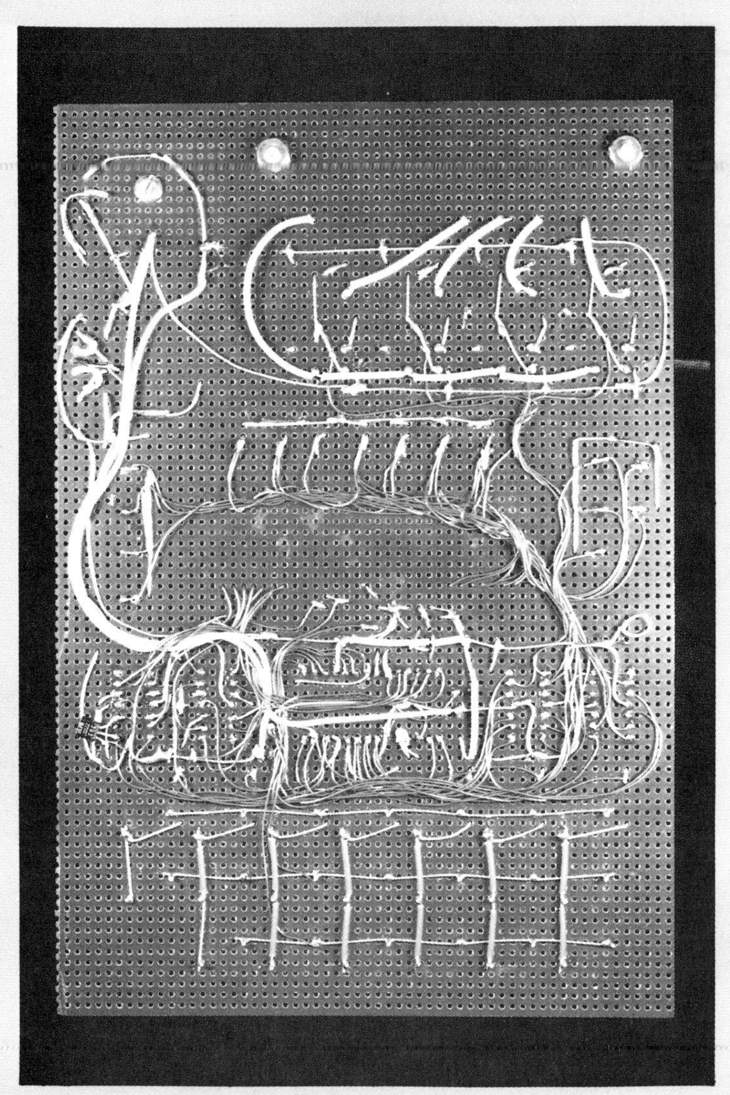

Photo 2: *Back side of the RTC-4 prototype.*

I receive at least half a dozen letters each month asking me how I put my project prototypes together and wire them. This month I'll reveal a few of the techniques I exercised in building the RTC-4.

To begin with, I don't recommend my techniques for everyone. I try to keep the circuit boards very small and lay the components out in an aesthetically pleasing arrangement. The results are not particularly easy to troubleshoot or modify.

I use standard integrated-circuit (IC) sockets (I prefer the Amp brand), and I hard-wire the power-supply and ground bus lines around the perforated project board using 22-gauge tinned bus wire. These soldered power connections fasten the IC sockets to the board.

Next, I insert the discrete components into the board and if possible directly route their leads to the appropriate pins on the IC sockets. When I absolutely must cross some leads, I use Teflon-insulated wire or sleeving. I finish by using 28- or 32-gauge wire-wrap wire to finish the connections, but I point-to-point solder them.

This may seem tedious, but I end up with a low-profile package, shown in photo 2, that looks very much like a commercially produced unit. It took me about 14 hours to assemble the RTC-4 prototype. Some of my other projects take much longer.

One detour I traveled during the RTC-4 project was the result of wanting large LED-display digits. I had some 0.3-inch type-MAN74A common-cathode single-digit display components, but I decided to adapt a 4-digit 0.7-inch display component I found in my junk box. Unfortunately, units of that type are not intended to be driven as multiplexed displays, and the digits are not wired individually. I had to disassemble and rewire it so that it functioned as 4 separate digits. Photos 3a through 3d show the process. It was so much trouble that I would just use the MAN74As if I were doing it over.

instead displays the programs for a particular day. You can also display those for every day (use EDAY).

Errors in command entry are indicated by the appearance of all nines ("99:99") in the display, which can be cleared by the entry-clear key, CLR, or the program-memory-clear key, MEM CLR. It is possible to selectively clear program segments. For example, entering

1, SW, MEM CLR

would erase all programs pertaining to switch 1. Similarly, entering

FRI, DAY, MEM CLR

erases any Friday programs.

Applications

Obviously, one of the more practical uses for the RTC-4 is to enhance your home's security by giving your house that lived-in look. If, for example, you have three lights and a stereo system plugged into the RTC-4, you could go away for weeks at a time and leave everything blinking on and off and sounding away. Because the RTC-4 can be programmed for sequences of one week, most normal home activities can be simulated. It is certainly more cost-effective and reliable than other timer-controllers I've seen.

A Previous Project's Woes

Some of you might have wondered why I didn't use the BSR X-10 Home Control System timer unit, which performs many of the same functions. Besides the obvious damper of losing a chance to experiment with some interesting components from TI, eliminating a good article topic, I've had less than consistent results with the X-10 devices over the years.

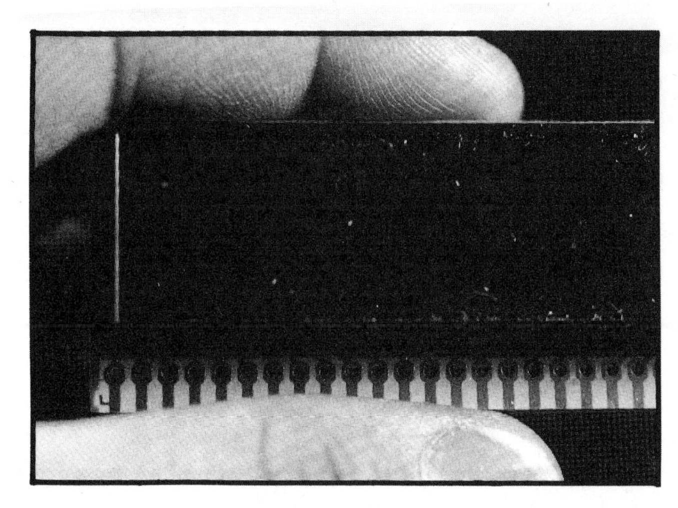

Photo 3a: *This typical National Semiconductor 4-digit LED display (which I believe to be one of the NSB7400 series) is most frequently used in digital clocks. It consists of three parts: a plastic circuit board with LEDs bonded to the circuit traces, a plastic reflective digit form, and a red lens cover.*

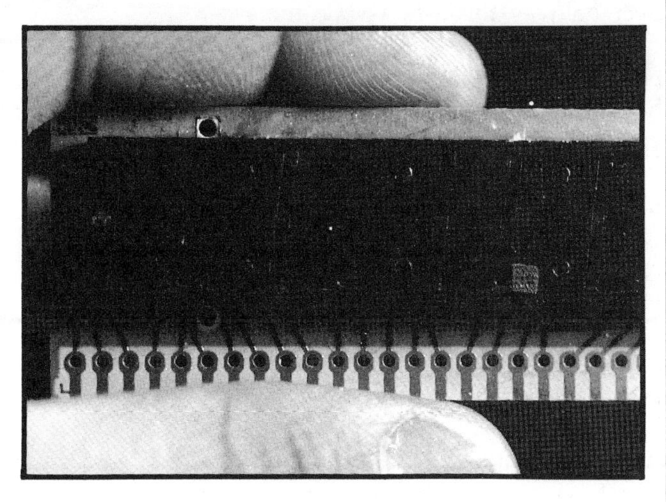

Photo 3b: *The LED display with the lens cover removed.*

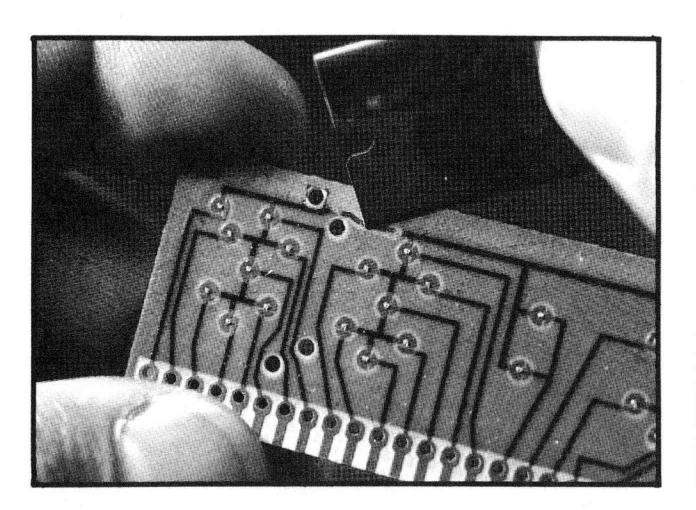

Photo 3c: *Because this display has one common cathode line for all 4 digits, I had to cut it at strategic points to isolate the digits.*

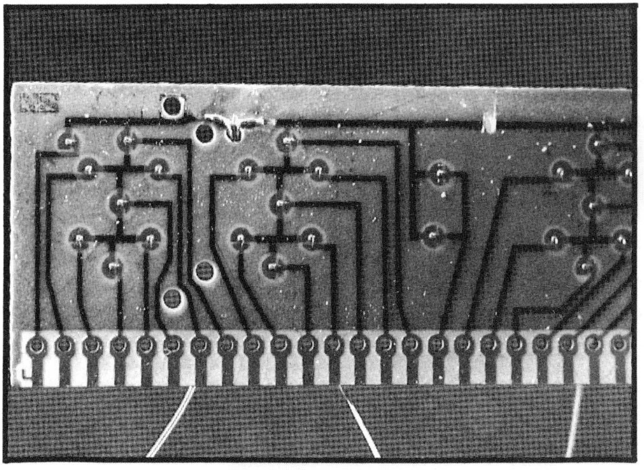

Photo 3d: *Finally, I drilled holes and soldered wires to the individual digit connections.*

At one time (shortly after my article on attaching an X-10 controller to a computer; see reference 1), most of my house was X-10-activated and remotely controlled. Frequently, however, I would find that the switches would reset or change state arbitrarily. I attributed this to power-line transients or electrical noise. Eventually I remedied this situation by retransmitting the intended status of each channel once every minute (obviously a tedious task suitable only for a computer).

Finally, I had a complete falling out with the BSR X-10 system shortly after a thunderstorm a few summers ago. While nothing else (computers, television sets, printers, etc.) in the house was affected, eleven X-10 remote receivers were blown out all at once. I expected to lose a couple receivers now and then, but I didn't expect to replace all of them. And I learned that when a light bulb burned out on a lamp connected to a lamp module, the module was often destroyed. In later production, BSR supposedly installed heavier SCRs (silicon-controlled rectifiers), but this has not completely eliminated the problem. Consequently, I've switched back to the old reliable copper wire and heavy-duty relays. But if I find any better methods I'll be sure to let you know.

A Trick Up My Sleeve

Some of you who are audiophiles might have been wondering how I have managed to record a 90-minute radio program unattended, due to a well-known property of ordinary Philips-type tape cassettes. Strictly speaking, a 90-minute program should just fit on a C-90 cassette. But, of course, a C-90 cassette can record only 45 minutes on one side; people are accustomed to turning the tape over halfway through. (While there do exist C-180 cassettes, they are expensive, and the thin tape doesn't handle the rigors of the automotive environment very well.)

But luckily, when I was still recording manually, I solved this problem. It was hard enough for me to turn the recorder on at all, let alone be there at 6:15 to turn the tape over, so after searching through most of the stereo shops in New England I eventually found a tape recorder that automatically reverses and records on both sides of the tape without turning the cassette over (a Pioneer CTF-750). Now, everything is automatic. But perhaps I should have taken this as an opportunity to build a robot.

In Conclusion

The RTC-4 owes its intelligence to the TMS1121C microcomputer chip. While I used it only in its off-the-shelf configuration, Texas Instruments would like you to know that there are other packaging and functional configurations that can be specifically mask-programmed for high-volume applications. For myself, I'll be satisfied with my somewhat-automatic tape recorder. Of course, to regularly record from radio broadcasts, I had a chat with the program director of the radio station, who gave me permission to record this particular news program. As an author, I'm very careful about copyright infringement.

Acknowledgments

Special thanks to Jeff Bachiochi for his help on this project.

Diagrams pertaining to the TMS1000-series devices are reprinted here through the courtesy of Texas Instruments Inc.

References

1. Ciarcia, Steve. "Computerize a Home." January 1980 BYTE, page 28. Reprinted in *Ciarcia's Circuit Cellar, Volume II*, page 137.
2. Staehlin, David C. "An 8080-Based Remote Appliance Controller." January 1982 BYTE, page 239.

19

Build a Power-Line Carrier-Current Modem

Communicate using electrical power wiring

"Jiggle the printer cable, Jeanette."

My assistant reached through the rat's nest of wires behind the computer and grabbed the one connected to the printer. As she moved it, I identified its other end from my cramped vantage point beneath the workbench and pulled it through a slot to attach it to my latest project. I was glad that what we were doing would keep us from having to run cables around the Circuit Cellar so often.

I have long had video terminals, printers, and other data-communicating equipment located at various places in the Circuit Cellar and around the upper stories of my home (see reference 3). Eventually the pain of rerouting cables whenever I moved a peripheral device got to me, so about a year ago I designed a communication system that would save having to string new wires every time. My system revolved around a *carrier-current modem,* which operates in much the same manner as the familiar telephone modem but sends its signals over electrical power wiring instead of over a telephone line.

After I pressed the carrier-current modems into service (with a little help), they served faithfully and I turned my attention to other projects, some of which have appeared in this column. But as of late more and more of my readers have written to me asking for help on how to send data through the AC power line. Apparently the widespread use of and media attention to the BSR X-10 Home Control System and similar products have given many people the idea of using the generally unexploited carrier-current modem for communication. Indeed, about five years ago I published a project on building a remote-control system that communicated through the AC power wiring of a building (see reference 4). It worked very much like the BSR X-10 as a carrier-current remote controller.

I hesitated to present the carrier-current modem as a Circuit Cellar project until now because I feel there is more to general-purpose carrier-current communication than meets the eye.

Simple on/off remote control is different. In most control applications, the communication is generally half-duplex or simplex; the transmission is limited to an intermittent tone or pulse burst that merely triggers a specific receiver into a binary control state. If the receiver is not activated properly by a single transmission because of interference, it's easy to send the control burst more than once. (Many computer control systems that use the BSR X-10 receivers send the same control code 10 times to make sure it is received.) But in general-purpose serial data communication, proper reception of every bit may be necessary, and errors in reception of the data may negate the usefulness of carrier-current operation.

To successfully use a carrier-current modem and the AC power wiring for data communication, we must either tolerate a dropped bit now and then or implement an intelligent protocol of error checking, redundant transmission, and handshaking. A really dependable power-line communication system has the physical link (AC-line transmission and reception) as only one of its components.

I was going to wait until I had perfected the control and error-checking protocols for use with the carrier-current modem, but the increasing interest indicated by my mail suggested that many experimenters might benefit from building a simple carrier-current modem; at least the physical part of the connection could be set up, even if the protocols and software are not ready.

This month's project, a modem for data communciation using the AC

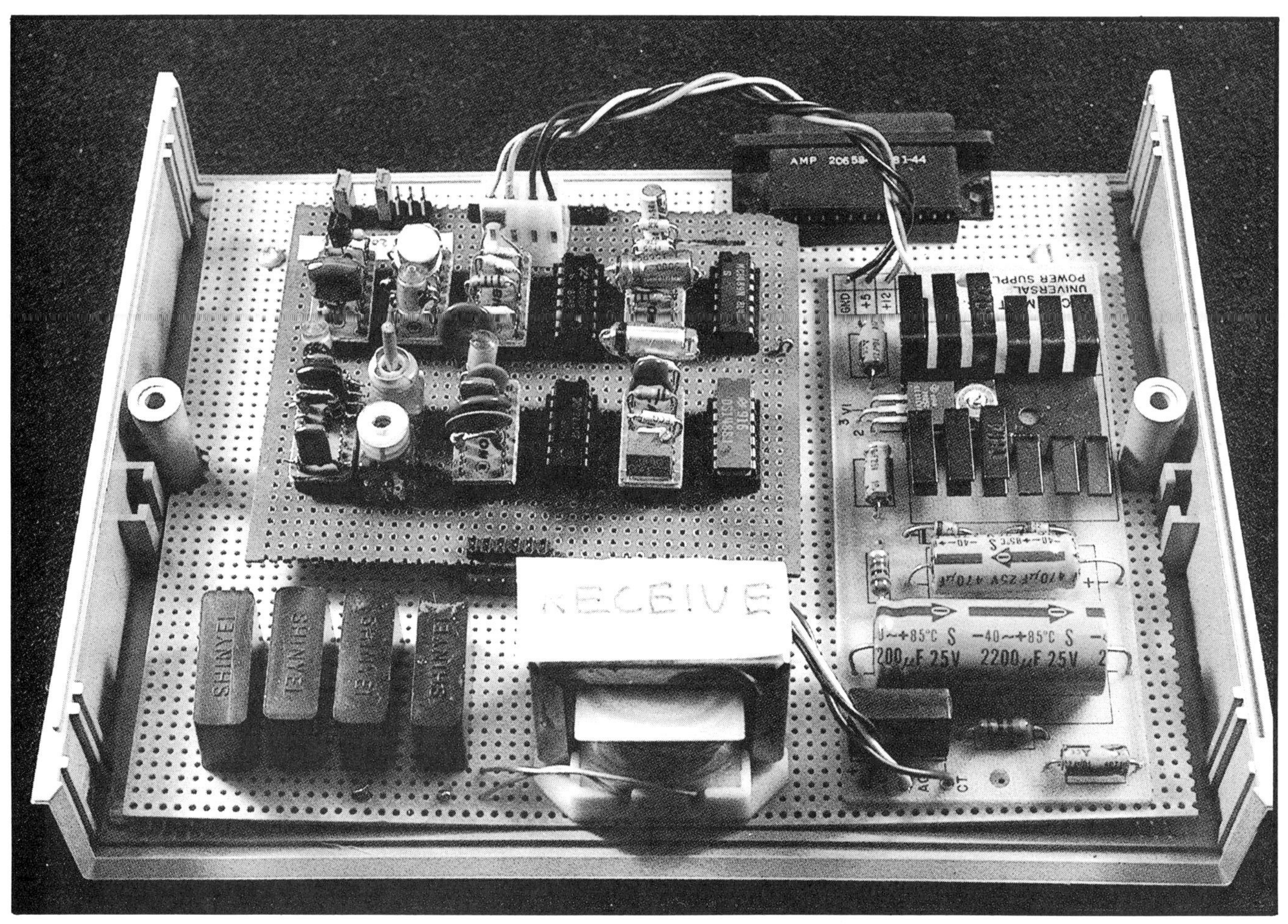

Photo 1: *Prototype of the Circuit Cellar CCM-1 carrier-current modem, which transmits serial data over the AC power line at 1200 bits per second. When in originate mode, the modem transmits mark signals at 90 kHz and space signals at 95 kHz; the answer mode transmits marks at 80 kHz and spaces at 85 kHz. The receive unit, shown here, differs from the transmit unit only in the frequency-selecting passive components.*

power line, is mostly an analog circuit. Successful operation of the modem, therefore, depends much more on tweaking and tuning the components than do digital computer-related projects. I am presenting this two-chip modem chiefly to discuss the principles involved, with some emphasis on selecting components for this application. Because the principles are susceptible to broad application, this knowledge should also be useful in understanding other modem designs as well.

All Modems Are Not Alike

The modem, named after a contraction of the words "modulator" and "demodulator," is a fairly common piece of computer equipment. You've probably seen modems built for sending data over telephone lines, and you may have read my March Circuit Cellar article about a low-cost modem (see reference 2). A modem allows two pieces of digital equipment to communicate with each other over long distances without having a direct hard-wired connection between them. With a telephone modem, the telephone lines form the communication path.

Modems of the usual type translate the voltage levels of the digital input signal (usually RS-232C levels) to tones at two frequencies, one of which signifies a logic 0, the other, a logic 1. The process of shifting the frequency of the output tone as the logic levels change is called *frequency-shift keying*, and the modems are called frequency-shift keyed, or FSK, modems.

To allow communication in two directions at once (full-duplex mode), rather than in only one (half-duplex), two pairs of frequencies are used, avoiding conflict when both ends of the connection talk at the same time. (By convention, one pair of tones is called the "originate" set, and the other is called the "answer" set. The two terms merely signify which set of frequencies each unit is using; no implication is intended regarding the content or origin of the data itself.) For compatibility, modems are built to adhere to certain standards of operation; the most common system in North America for low-speed modems was first used in the Bell System's Model 103 modem, so Bell-103-type modems abound.

Carrier-Current Systems

The AC power line is similar in some respects to the telphone line. One similarity is clear: we can send data through the power line by using an FSK modem.

Obviously, in addition to the data we want to transmit, the power lines

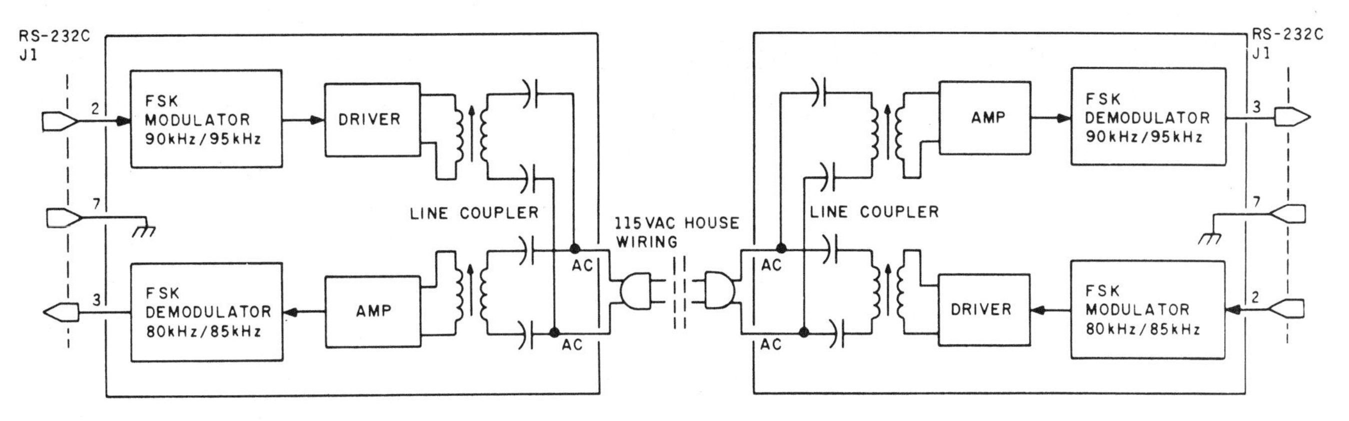

Figure 1: *Block diagram of a data-communication system employing carrier-current modems. The AC power wiring of the building is used to carry the frequency-shift-keyed transmission.*

must continue to carry power or we won't be able to operate the computer equipment. The carrier-current communication system superimposes a high-frequency signal on the 60-Hz power-carrying signal. On an oscilloscope, this is viewed as an additional small voltage carried on or riding atop the 115-V (volt) alternating current. At the receiving end, the modem filters out the 60-Hz signal and any other noise components on the power line, demodulating only the transmitted frequency. Unfortunately, the power line sometimes has an impedance less than 2 ohms, along with thousand-volt noise spikes that make it a hazardous environment and a less-than-optimal communication medium.

There is a price to be paid for the simplicity of this communication system. Unlike the complex digital carrier-current systems, which transmit around the zero-crossing interval of the power signal, the analog FSK carrier-current modems are more sensitive to line peculiarities and noise. However, the digital species is much more complex, and after all, my intention was to present a build-it-yourself project. Learning a little black art for the sake of simplicity can't hurt.

Carrier-Current Modem Circuit

Figure 1 is the block diagram of a carrier-current modem, which consists of three basic components: modulator/driver, amplifier/demodulator, and AC-line coupler. The simplest usable system consists of two modems, one attached to each of two pieces of data-communicating equipment. One of the two modems is arbitrarily designated as the originating modem and the other as the answering modem. As in the case of telephone communication, two sets of FSK frequencies are defined, although the power-line modems operate at much higher frequencies than the telephone-line type. The connections from the communicating equipment to the modulator and demodulator on each modem are through an RS-232C DB-25 connector. The driver and amplifier sections are in turn connected to the AC line through the coupler, the crucial component.

In a direct-connect telephone modem, the coupler is usually a 600-ohm isolation transformer, and the characteristics of the line are well defined. But in a carrier-current modem, the coupling transformer is very often a tuned circuit selected to resonate within the passband of the FSK tones to improve the signal-to-noise ratio in this particularly noisy environment. While tuned couplers are not aways used, most carrier-current driver circuits do employ them to increase the transmission range and receiver selectivity. For most experimenters, the driver and coupler are the hardest sections to construct because so much depends on selecting, balancing, and adjusting the components.

Taming the AC Line

Figure 2 shows two typical carrier-current driver-coupler circuits. Both consist of a transformer capacitively isolated from the AC line; 0.22-μF (microfarad) 600-V capacitors are recommended. Any 0.1-mH (millihenry) slug-tuned transformer will probably work, but I have had best success with standard low-Q (low-resonance) miniature IF (intermediate-frequency) transformers used in transistor radios. In practice, the circuit of figure 2b is less sensitive to component selection and more easily tuned.

If you have any doubts about whether a particular transformer will work, a few brute-force tests can help you tame the AC line and give you the confidence to build the rest of the modem circuit. *Just remember that working directly with the AC power line is dangerous if you aren't careful.*

Begin by building two couplers, one driver, and one receiver, using the component values and circuit layout in the schematic diagram of figure 3. (I'll get around to discussing it shortly.) Temporarily apply power to the driver section and attach it (carefully!) through its coupler to the AC line. The receiver should be powered and connected through its coupler across the AC line at some other nearby location.

Use an oscilloscope (connected to the AC line through an isolation transformer for safety) to monitor the signal present across the secondary coil of the receiver transformer (or from the collector of transistor Q2 to ground), while you inject a signal for transmission using a sine-wave func-

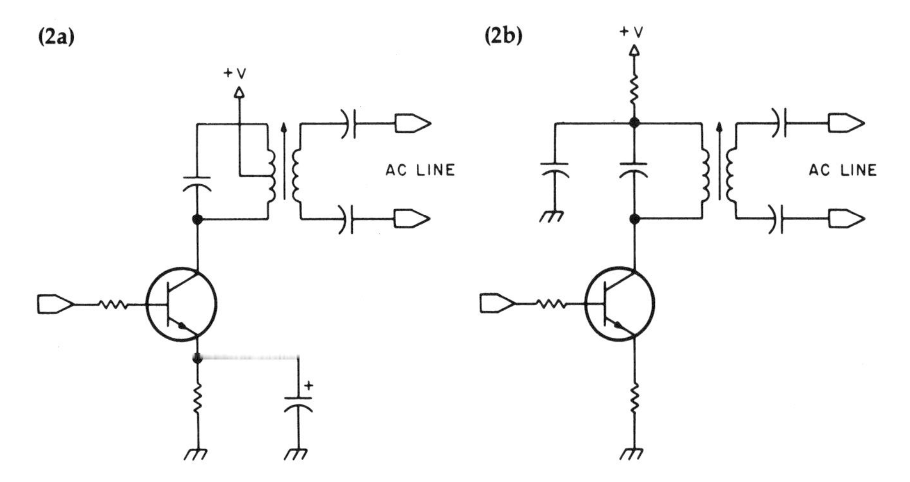

Figure 2: *Two possible schemes for coupling the modulator/driver portion of the carrier-current modem to the AC line. The circuit of figure 2b is the more stable.*

tion generator attached to the base of the driver transistor, Q1. Sweep the frequency between 50 kHz and 150 kHz until you detect the same frequency (at greater than 10 mV [millivolts]) at the receiver. Take care that you are receiving the fundamental frequency and not a harmonic. Don't be too surprised at the strange electrical noise you'll no doubt observe.

You can shift the detection band somewhat by adjusting the tuning slug in and out of the transformer windings or by changing the capacitor across the transformer secondary. The objective is to find the frequency band where the signal level at the receiver is highest. The band should be about 20 kHz wide; the frequency can go as high as 300 kHz (the upper frequency limit of the demodulator) if necessary.

In my case the best results were obtained between 80 kHz and 100 kHz, so I arbitrarily set two originate and answer frequency pairs within this band. One modem transmits on 90 and 95 kHz and receives on 80 and 85 kHz; conversely, the other modem transmits on 80 and 85 kHz and receives on 90 and 95 kHz. In a simple system, any originate and answer frequency pairs that work are acceptable because each frequency pair has its own tuned coupler. I recommend that the frequency separation between the mark and space tones be 5 kHz or less to facilitate easy demodulation. I only caution you not to set any frequency that is a multiple (or submultiple) of another one used in the system.

Remember that, in an analog FSK carrier-current communication system, success largely depends on your peaking the resonance of the coils and finding the proper transmission bands. I can't provide a parts list of components values that will be guaranteed to work because the behavior of parts in the list could be other than that predicted, due to performance and tolerance variations. The most important "component" in the coupler sections is your understanding the objective and knowing how to pursue it through testing and adjustment.

Fortunately, component selection in the FSK modulator and demodulator sections is much more straightforward and follows some basic formulas defined by the frequency and application. However, because it is possible that you might choose frequency pairs different from those in my design, I'll discuss the derivation of the component values rather than just the results.

Exar XR-2206 Modulator

First, let's consider the modulator. The XR-2206 is a function-generator integrated circuit, made by Exar Integrated Systems, which can produce sine, square, and triangular output waveforms at frequencies ranging from 0.01 Hz to 1 MHz. It is ideally suited for FSK applications because it can be set for two different time bases and digitally switched between them. A functional block diagram of the XR-2206 and typical FSK circuit is shown in figure 4.

The mark and space frequencies can be independently set by the choice of timing resistors R2 and R3 and the capacitor between pins 4 and 5. The FSK input signal is applied to pin 9. A high logic-input signal to pin 9 produces the frequency:

$$f_{high} = \frac{1}{R2 \times C}$$

and a low-level input signal produces the frequency:

$$f_{low} = \frac{1}{R3 \times C}$$

where R2 and R3 are in ohms and C is in farads.

R2 and R3 should be between 10 kilohms and 100 kilohms, and the capacitor should be polycarbonate, polystyrene, or Mylar for temperature stability. I chose to use a 0.001-μF capacitor, which produces the following resistor values for the frequency pairs I chose:

R2: 85 kHz, 11.76 kilohms
95 kHz, 10.53 kilohms

R3: 80 kHz, 12.50 kilohms
90 kHz, 11.11 kilohms

In the case of R2 and R3, you can use the nearest 1-percent-tolerance resistor or use a potentiometer in combination with the closest 5 percent fixed value.

You must also consider the settings of resistors R1 and R5, which adjust for minimum total harmonic distortion. In our case, where a few-tenths-percent distortion is irrelevant, pins 15 and 16 may be left open and R1 can be replaced by a fixed 200-ohm resistor. With R1 installed (same effect as closing switch S1), the output at pin 2 is a sine wave with an output impedance of 600 ohms and amplitude set by R4. The remaining components serve to stabilize operation and are the same for all frequencies.

XR-2211 Demodulator

The Exar XR-2211 is a phase-locked-loop (PLL) integrated circuit especially designed for data communication

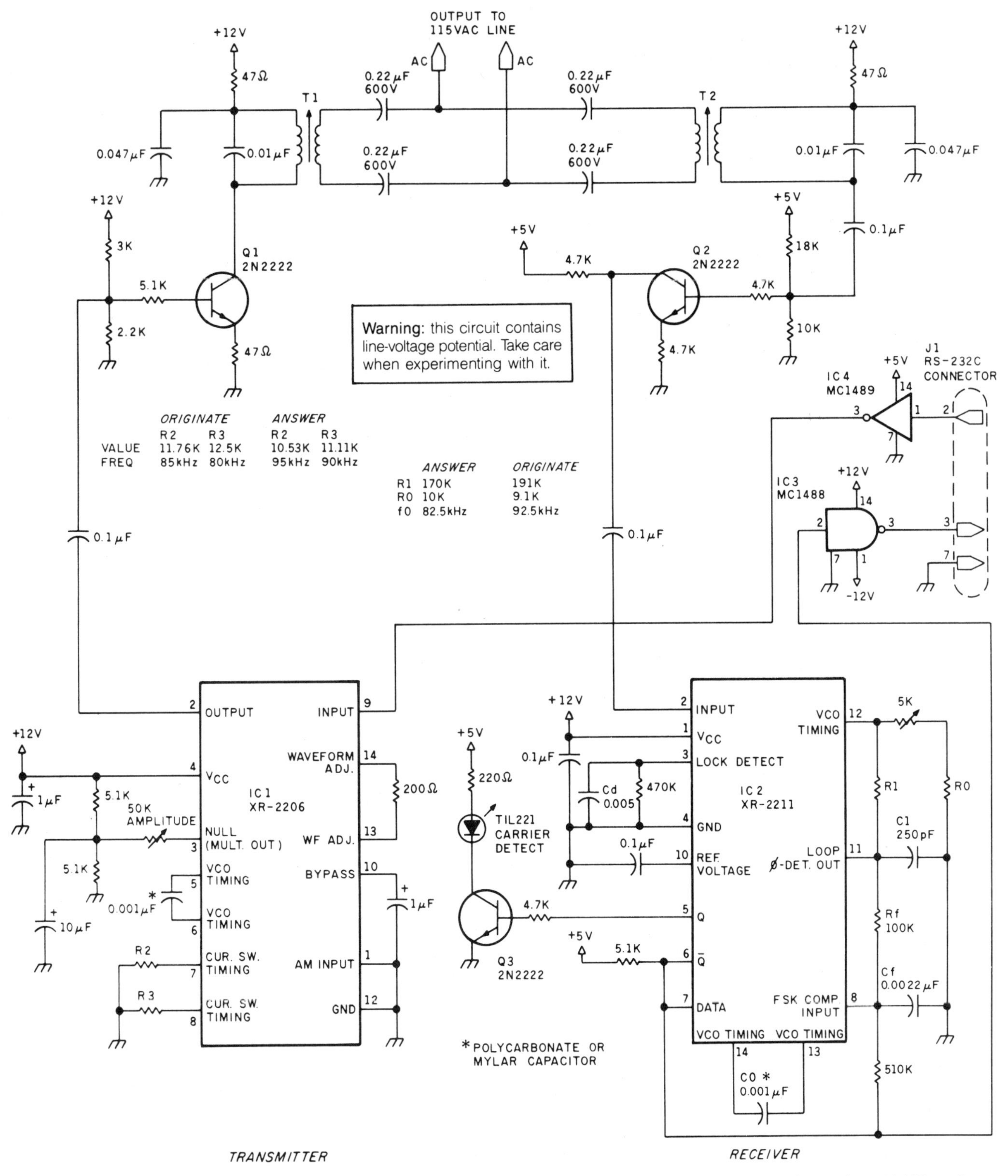

Figure 3: *Schematic diagram of a complete carrier-current modem. The originate-mode modem transmits mark signals at 90 kHz and space signals at 95 kHz; the answer modem transmits marks at 80 kHz and spaces at 85 kHz. This diagram shows the circuit for one end of the link; two such units are needed in the system with the proper component values differing between them. In one unit, the wiring of pins 2 and 3 of J1 should be reversed.*

and particularly suited for FSK applications. It operates over a frequency range of 0.01 Hz to 300 kHz and can accommodate analog input signals between 2 mV and 3 V. An XR-2211 functional block diagram and typical FSK demodulator circuit are shown in figure 5. Frequency-shift-keyed input signals fed into pin 2 of the XR-2211 must be capacitvely coupled through a 0.1-μF capacitor. The internal impedance is 20 kilohms, and the minimum recommended input signal is 10 mV.

The first order of business is to set

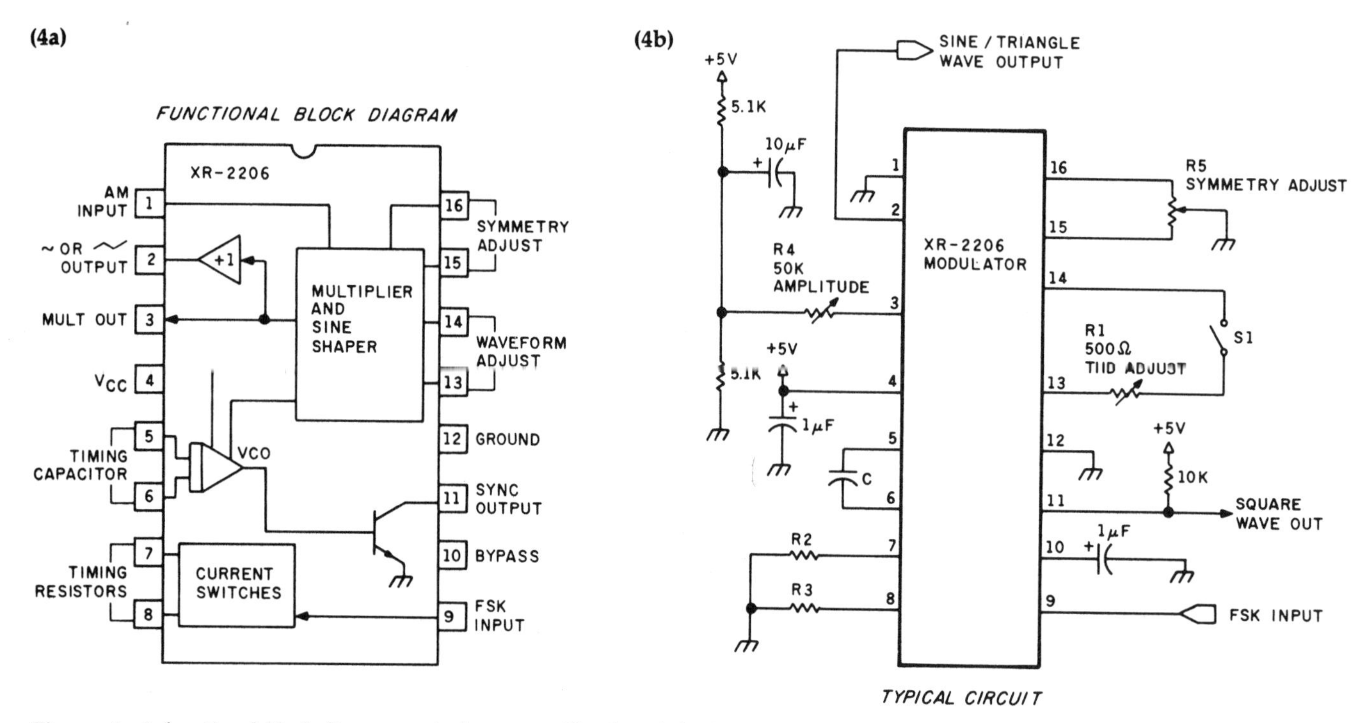

Figure 4: *A functional block diagram and pin-out specification of the XR-2206 (4a) and typical FSK circuit (4b).*

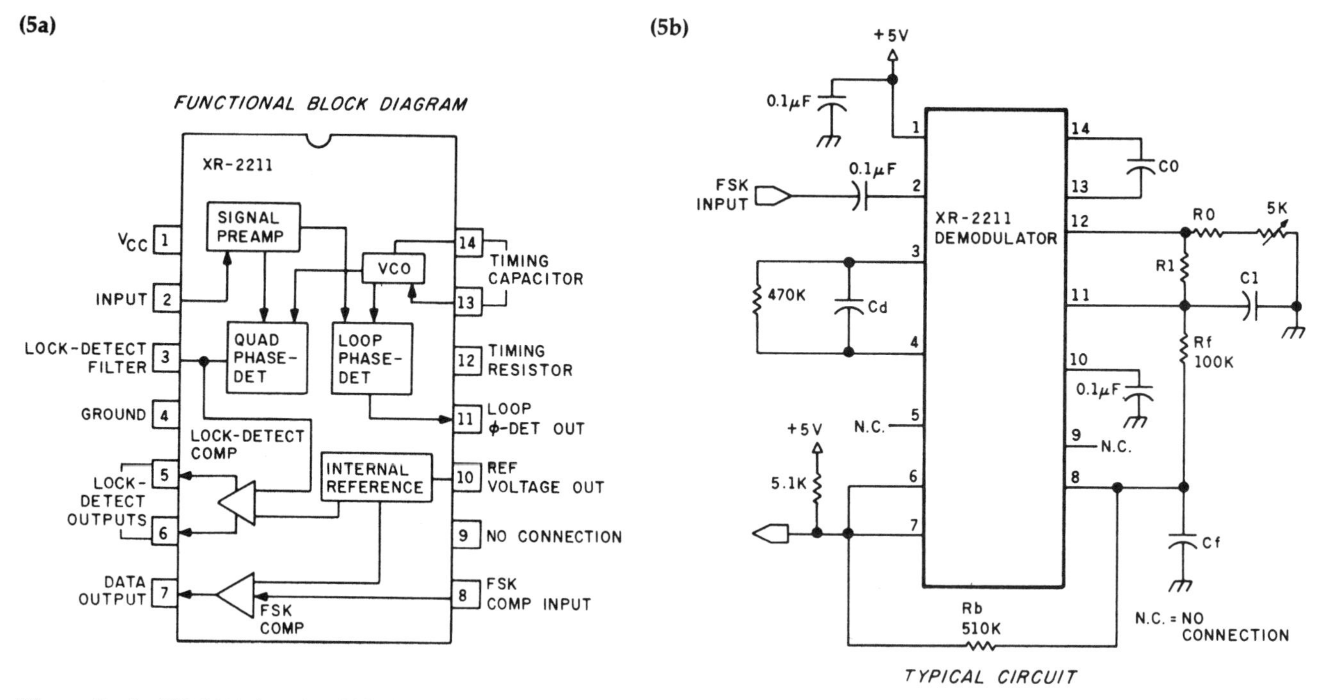

Figure 5: *An XR-2211 functional block diagram and pin-out specification (5a) and typical FSK demodulator circuit (5b).*

the center frequency of the demodulator passband at the center of the frequency band that we wish to detect. In my case, the passbands are defined by the tone pair at 80/85 kHz and the other pair at 90/95 kHz. The center frequencies for the two demodulators would then be 82.5 kHz and 92.5 kHz, respectively. The component values are computed as follows:

$$f_0 = \frac{1}{R0 \times C0}$$

where R0 is in ohms and C0 is in farads; f_0 is the center frequency.

Generally, R0 is in a range of 10 kilohms to 100 kilohms, but the choice is arbitrary. Often it is more convenient to choose a value for C0 and trim the value of R0 with an adjacent potentiometer. Using 0.001-μF value (Mylar, polycarbonate, or polystyrene) for C0, the computed R0 values are 12.12 kilohms (f_0 = 82.5 kHz) and 10.81 kilohms (f_0 = 92.5 kHz). With a 5-kilohm trim pot in series, more convenient resistors of 10 and 9.1 kilohms can be used instead.

R1 sets the system bandwidth and

C1 sets the loop-filter time constant and damping factor. The value of R1 is determined by the mark/space frequency difference:

$$R1 = \frac{R0 \times f_0}{(f_1 - f_2)}$$

The deviation is 5 kHz by design, and the values for R1 are 170 kilohms (f_0 = 82.5 kHz) and 191 kilohms (f_0 = 92.5 kHz).

While the equation for computing the loop-damping factor associated with C1 is complex, there is a convenient rule of thumb. The damping factor should be approximately ½, and a value of C1 = C0/4 will produce this. With C0 equal to 0.001 μF, C1 equals 250 pf (picofarads).

Resistor Rb provides positive feedback across the FSK comparator and facilitates rapid transition between output logic states. A value of 510 kilohms is used in most applications.

Cf and Rf form a single-pole post-detection filter for the FSK data output. Rf is most often set at 100 kilohms. Cf smooths the data output; its value is roughly calculated: Cf = (3/data rate in bits per second) where Cf is in microfarads. Because this modem is designed for operation at 1200 bps (bits per second), a value of 0.0022 μF or 0.0033 μF is acceptable.

The final area requiring calculation is the lock-detect section of the XR-2211, which is used here in a carrier-detect function. The open-collector lock-detect output, pin 6, is connected to the data output, pin 7. This will disable any output created by noise unless a carrier signal is present within the detection passband of the PLL. Presuming a parallel resistance of 470 kilohms, the minimum value of the lock-detect filter capacitor, Cd, is $16/(f_1 - f_2)/2$. In this case 0.005 μF is adequate.

Testing the Completed Unit

I built the complete Circuit Cellar carrier-current modem the way shown in figure 3, with component values for 80/85 kHz and 90/95 kHz tone pairs, but you may substitute other values as previously discussed. In addition to the three functional sections we have looked at, I have added a carrier-detect indicator and an RS-232C driver (IC3) and receiver (IC4).

To test the completed unit you need some source of serial data output. (I used a full-duplex video terminal.) The easiest test is a simple loop-back circuit. The terminal is connected to the originate modem and plugged into the power line. The answer modem is plugged in some distance away, with pins 2 and 3 jumpered together on J1, its RS-232C connector. As you type on the terminal, the data is transmitted to the answer modem where it is looped back through the jumper and retransmitted to the originate modem where it appears on the terminal's screen.

You should be able to place the modems anywhere within your home or office, or even an adjacent home or apartment. The ultimate range is limited by the power company's step-down transformer and the cross-coupling between the two 115-V legs of a multiphase 230-V distribution system. But you can arrange communication between the latter by attaching a fused capacitor between the two 115-V legs.

In Conclusion

Using this modem I was able to successfully communicate at 1200 bps for extended periods of time without loss of data. I've found FSK carrier-current communication to be fairly reliable; it's best at the lower data rates. Occasionally a few characters have been lost when my air-conditioner compressor or water pump turned on. These are occasions where an intelligent control system might be of significant help. I had intended that the intelligence necessary for error checking and redundant transmissions be part of this project, but as I explained, such a control system is much more involved than the modem itself. Given the excess computing power available in most personal computers, it would certainly be feasible in most cases for error-checking to be performed by applications software, perhaps using something like the well-known file-transfer protocol developed by Ward Christensen for use with his CP/M-based Modem-7 program.

Generally speaking, while I have detailed the hardware components of a complete system that works in the Circuit Cellar, it's important to recognize that AC line conditions differ significantly between locations. Complete frequency bands may be unusable due to interference produced by machinery, digital clocks, microcomputers, and fluorescent lights. For this reason, you should understand how the modem components and coupler are designed. Your ability to customize a basic modem design to the particular electrical environment of your home or office can make or break the project.

References

1. *Applications Data Book.* Sunnyvale, CA: Exar Integrated Systems Inc., 1981.
2. Ciarcia, Steve. "Build the ECM-103, an Originate/Answer Modem." March 1983 BYTE, page 26.
3. Ciarcia, Steve, "Come Upstairs and Be Respectable." May 1977 BYTE, page 50.
4. Ciarcia, Steve. "Tune In and Turn On: A Computerized Wireless AC Control System." Part 1, April 1978 BYTE, page 114. Part 2, May 1978 BYTE, page 97.
5. Edward, Harry J., Jr. *Residential Electrical Wiring: A Practical Guide to Electrical Wiring Practices in Residences.* Reston, VA: Reston Publishing Company, 1982.

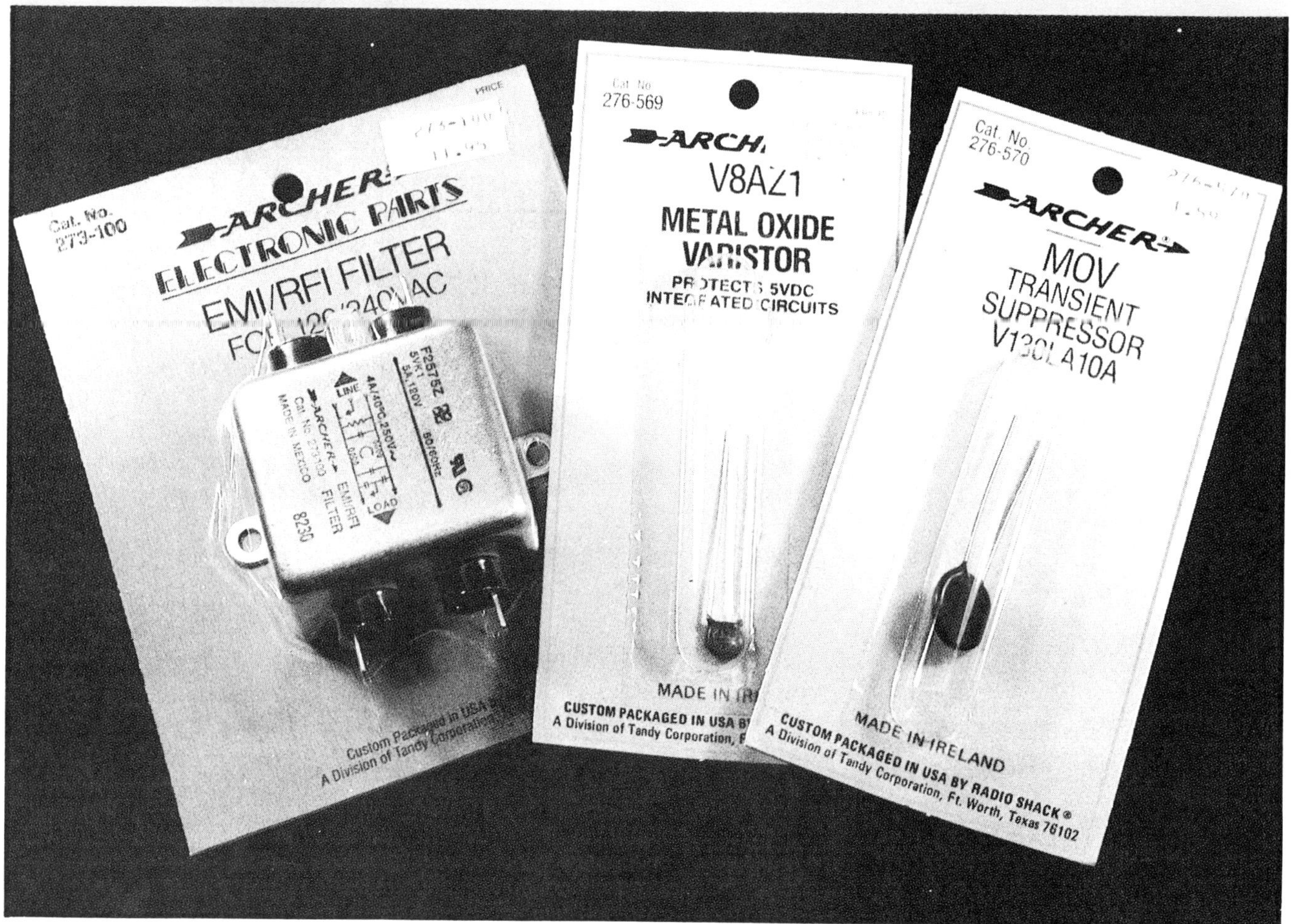

Photo 1: *Your computers and other electronic equipment are vulnerable to disturbances transmitted through the power line. Most of the components necessary for transient and noise suppression can be purchased from Radio Shack. Shown here are a commercial EMI/RFI line filter made by Corcom and two General Electric metal-oxide varistors.*

20

Keep Power-Line Pollution Out of Your Computer

A visitor once called the Circuit Cellar my mountaintop wilderness retreat. Since he lived in the center of Manhattan, the few oak and birch trees around my house seemed to him like a forest, and because he could view scenery further away than a block or two he must have felt like he was on Mount Whitney. Well, my area is one of the higher points in Connecticut, but that isn't very high. It's barely a prairie-dog mound to someone from Montana.

Life in a rural location has its special pleasures. I get to plow the snow from my own driveway, trim back the ever-encroaching foliage and rake the leaves, pile four cords of wood for the stove each winter (see reference 1), fight off the local animal population, and spend large sums of money repairing damage done to my electronic equipment by electrical disturbances.

This last item is the only one that really annoys me. Every year I can count on experiencing some equipment failure attributable to an external electrical impulse, usually coming in through the power line. For three years, just like clockwork, the first thunderstorm in June wiped out a DECwriter II terminal connected to one of my computers. After the first two times of spending a few hours replacing blown chips, I got smart and installed sockets. (Now I even know in advance which chips will be blown.) Last summer I kept the printer unconnected when I wasn't using it.

But the elements were not to be denied. During an August thunderstorm, lightning struck my house. I can't say for sure where the bolt actually hit (there were no burn marks or other visible clues), but I suspect the point of entry was the power line. I remember seeing an indistinct flash of light, hearing a tremendous crash, and then standing in darkness. My assistant Jeanette saw a bright blue glow behind one of the computers.

Such a tremendous power surge is not kind to semiconductor-based equipment. The casualties included one computer, one video camera, two video monitors, a microwave receiver, and probably several other assorted items I haven't found yet. The damage did not include the DECwriter (safely unplugged since May), but it was over $3000.

In December, thunderstorms are not an immediate threat, but as I write this in early September the memory is still fresh and I still have a month of potentially violent weather to contend with. I am forced to consider some defensive measures. Perhaps by relating my experiences I can save you from a similar fate.

Of course, lightning isn't the sole cause of electrical disturbances; you don't have to wait for a thunderstorm to be a victim. Many kinds of trouble can be ducted into your computer through the power line.

In the January 1981 Circuit Cellar article (reference 2), I wrote about electromagnetic interference (EMI) and radio-frequency interference (RFI). This month I'd like to pick up the saga by describing other forms of electrical pollution that occur on power lines. Afterward, I'll describe a few simple, inexpensive means of dealing with them.

The Power Line: A Hostile Environment

The lines leaving your local utility company's generating plant carry electrical power that in most respects is pure, smooth, and constant. However, as the power is routed through the distribution network, it comes under the increasing aberrant influence of external forces and the connection or shedding of electrical loads.

Your susceptibility to these aberrations depends on your location in the distribution system. If you are close to the power plant, you should have relatively few, with the low source-impedance of the generator and short distance of the transmission line limiting the influence of external forces. But rural customers at the end of the line usually experience the full effect. While the utilities try to distribute power evenly, the presence of a large-scale user of electrical power along the line between the generator and you can greatly affect the quality and quantity of the power you get.

If you own a personal computer, you should be concerned about the quality of the power you feed it. Power-line irregularities cause problems for computers and other digital equipment because certain kinds of extraneous electrical pulses can be interpreted as data or instructions, causing errors in operation. You face hazards every time you plug in a piece of electronic equipment, but there are certain precautionary measures that can protect your computer.

The degree of sensitivity depends somewhat on the type of equipment and the type of disturbances. As the operating speed of digital equipment increases, its tolerance to power-line pollution lessens. High-speed processors and memory components are susceptible to fast transients. (Dynamic memories, which must be periodically refreshed, are particularly susceptible.) Disk drives and displays, on the other hand, are more affected by lasting surges and sags in operating voltage.

Common Sources of Woe

Electrical power-line disturbances can come from either natural or man-made sources. Of the many ways the power line can be disturbed, the several varieties of voltage fluctuation most often cause problems with computer equipment. These fluctuations can be categorized by source and severity, as follows:

Blackouts. A blackout is a total power outage—the voltage goes to zero. Obviously if no alternate source of power is available as a backup, computer equipment will be severely affected, and data will be lost. Blackouts generally affect only a small number of utility customers (fewer than 5 percent) during a year and generally last less than 10 seconds.

Brownouts. A brownout is typically a corrective action taken by the utility when power demand exceeds generating capacity. The utility reduces the output voltage from a nominal 120 V (volts) by 5 to 15 percent. When the voltage is thus reduced, the resistive load presented to the generators by the distribution network consumes less power.

Generally speaking, most consumer and industrial equipment designed for use in North America functions properly when supplied with current within the range from 105 to 130 V. But when operating at either extreme, the equipment is more vulnerable to disruption from some other power-line anomaly. Fortunately, power companies rarely reduce the voltage by more than 7 percent.

Voltage transients. The phenomena of voltage transients include surges of voltage above the specified normal, voltage sags below, and instantaneous voltage spikes that leap far above the nominal levels.

Surges and sags are long-duration events occurring at some point in the distribution network when electrical equipment is routinely turned on or off nearby. The magnitude of the surge or sag depends upon the size of the load being removed from or placed on the network.

Sags are often produced by the turning on of electric motors, which have high starting currents. (You've probably noticed lights dimming

The important element in lightning protection is the lightning rod, a pointed shaft of copper to which a half-inch copper cable is fastened. The cable in run down the side of the building, where it is clamped to an 8½-foot copper-plated steel rod driven into the earth. The rod system pictured here costs $150.

How Lightning Strikes

A lightning flash is characterized by one or more strokes with typical peak currents of 20 kA (kiloamperes) or higher. In the immediate vicinity of the stroke's impact on the earth, hazardous voltage gradients exist. It is difficult to establish a definite grounding-conductance value necessary to protect equipment and personnel. The current in a lightning strike is so high that even 1 ohm of resistance can theoretically produce hazardous potentials.

When lightning strikes a building unprotected by a lightning rod, the stroke seeks out the lowest-impedance path to earth (most likely through the electric wiring or water pipes).

How It Starts

As the electric charge builds up in a cloud, the electric field in the vicinity of the charge center increases to the point where the air starts to ionize. A column of ionized air, called a pilot streamer, *begins to extend toward the earth at a velocity of about 100 miles per hour. After the pilot streamer has moved perhaps 100 feet to 150 feet, a more intense discharge called the* stepped leader *occurs. This discharge inserts additional negative charge into the region around the pilot streamer and allows the pilot streamer to advance for another 100 to 150 feet, after which the cycle repeats. As its name indicates, the stepped leader progresses toward the earth in a series of steps, with a time interval between steps on the order of 50 microseconds.*

In a cloud-to-ground flash, the pilot steamer does not move in a direct line toward the earth but instead follows the path through the atmosphere where the air ionizes most readily. Although the general direction is toward the earth, the specific angle of departure taken by each succeeding pilot streamer from the tip of the previous streamer is unpredictable. Therefore, each 100- to 150-foot segment of the stroke will likely approach the earth at a different angle. This changing angle of approach gives the overall flash its characteristic zigzag appearance.

As a highly ionized column, the stepped leader is at essentially the same potential as the charged area from which it originates. Thus, as the stepped leader approaches the earth, the voltage gradient between the earth and the tip of the leader increases. *The increasing voltage further encourages the air dielectric between the two regions to break down.*

Attracting Lightning

Objects extending above their surroundings are likely to be struck by lightning. Thin metallic structures, such as flag poles, lighting towers, antennas, and overhead wires, offer a very small cross-sectional area relative to the surrounding terrain, but ample evidence exists to show that such objects apparently attract lightning.

The ability of tall structures or objects to attract lightning serves to protect shorter objects and structures nearby. In effect, a tall object establishes a protected zone around it; within this zone, other structures and objects are protected against direct lightning strikes. As the height differen-

tial between the shorter surrounding objects and the tall one decreases, the protection provided to the shorter objects decreases. Likewise, as the horizontal distance between the tall and short structures increases, the protection afforded by the tall structure decreases.

Lightning Rods

A protective device that makes use of this phenomenon is the lightning rod, shown in photo. Generally just a sharp copper spike, the lightning rod is attached to the highest point on the structure to be protected. When lightning strikes, the current is shunted directly through a heavy copper wire from the rod to a grounding electrode buried in the earth.

Although the duration of a strike is typically less than 2 microseconds, the voltage generated is high enough to cause flashover strikes to conducting objects located as much as 14 inches away from the conducting path. For this reason, metallic objects in close proximity to down conductors should be electrically bonded to the conductors.

But circuits not in direct contact with the lightning discharge path can experience damage, even in the absence of overt coupling by flashover. Because the high current associated with a discharge builds up so fast, large inductively produced voltages are formed on nearby conductors. Experimental and analytical evidence shows that the surges thus induced can easily exceed the tolerance level of many components, particularly solid-state devices. Inductive surges can be induced by lightning current flowing in a down conductor or structural member, by a stroke to earth in the vicinity of buried cables, or by cloud-to-cloud discharges occurring parallel to long cable runs, either above ground or buried.

The Moral

The objective of all lightning-protection systems is to direct the high currents away from susceptible elements or limit the voltage gradients developed by the high current to safe levels. In a given area, certain structures or objects are more likely to be struck by lightning than others; however, no object, whether man-made or natural, should be assumed to be immune from lightning. The voltages that could be induced by such discharges present a definite threat to signal and control equipment, particularly equipment employing semiconductor components.

Power-Line Conditioner Sources

Cuesta Systems Inc.
3440 Roberto Court
San Luis Obispo, CA 93401
(805) 541-4160

Dymarc Industries Inc.
21 Governor's Court
Baltimore, MD 21207
(800) 638-9098
(301) 298-2629

Electronic Protection Devices
Division CNS Electronics Corp.
5-9 Central Ave.
Waltham, MA 02154
(800) 343-1813

Electronic Specialists Inc.
171 South Main St.
Natick, MA 07160
(800) 225-4876 (orders)
(617) 655-1532

Isoreg Corporation
410 Great Rd.
Littleton, MA 01460
(617) 486-9483

RKS Industries
4865 Scotts Valley Dr.
Scotts Valley, CA 95066
(800) 892-1342
(408) 438-5760

Sun Research Inc.
POB 210
Old Bay Rd.
New Durham, NH 03855
(603) 859-7110

Corcom Inc.
1600 Winchester Rd.
Libertyville, IL 60048
(312) 680-7400

Curtis Industries Inc.
8300 North Tower Ave.
Milwaukee, WI 53223
(414) 354-1500

Genisco Technology Corporation
18435 Susana Rd.
Rancho Dominguez, CA 90221
(213) 537-4750

Hopkins Engineering Company
12900 Foothill Blvd.
San Fernando, CA 91342
(213) 361-8691

The Potter Company
Division of Varian
POB 337
Wesson, MI 39191
(601) 643-2215

Siemens Corporation
8700 East Thomas Rd.
Scottsdale, AZ 85252
(602) 941-6366

Sprague Electric Company
87 Marshall St.
North Adams, MA 01247
(413) 664-4411

Stanford Applied Engineering
3520 De La Cruz Blvd.
Santa Clara, CA 95050-1997
(408) 988-0700

Power-Line Filter Sources

Cornell-Dubilier Electronics
Box B-967
New Bedford, MA 02741
(617) 996-8561

when an air conditioner comes on.) Surges are generally the result of network switching by the utility or of a sudden reduction in demand for power in the network; during the period necessary for the utility's electromechanical compensation system to function, an overvoltage transient condition can exist.

The most damaging power-line disturbance is the high-speed, high-energy *voltage spike*. People speaking loosely about "power-line transients" are probably talking about this type of event. Lasting usually less than 100 microseconds, spikes can be up to 6000 volts. Such high-energy transients are produced by the switching off of inductive loads by the opening of switch contacts, short circuits, or blown fuses; severe network load changes; or lightning. Inductive-load switching accounts for the majority of spikes.

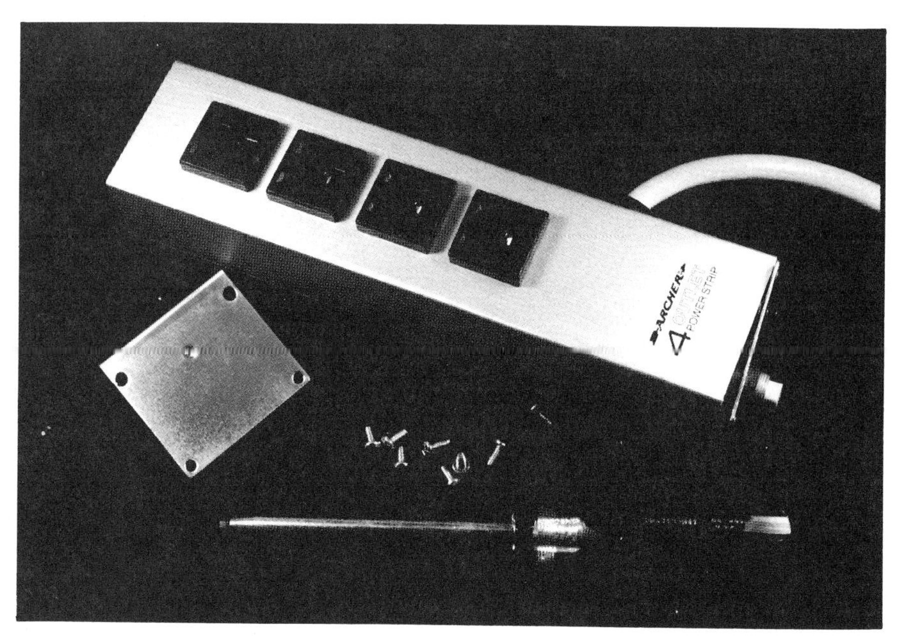

Photo 2: *You can save approximately $40 on the price of a transient-protected power strip by adding the protection yourself, as demonstrated on the Radio Shack Archer 61-2620 unit. First, unscrew the end plates.*

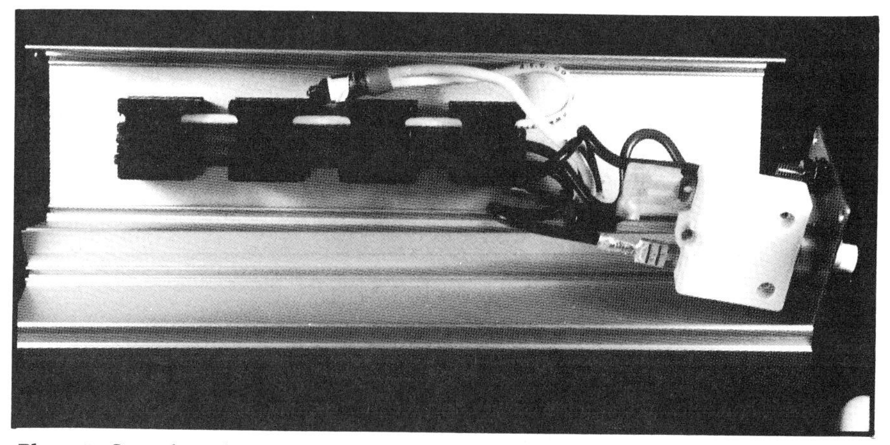

Photo 3: *Open the strip case, exposing the four receptacles and the white circuit-breaker block. The three wires conducting power run the length of the strip: black is the hot side, white is the neutral return, and the green wire is earth ground.*

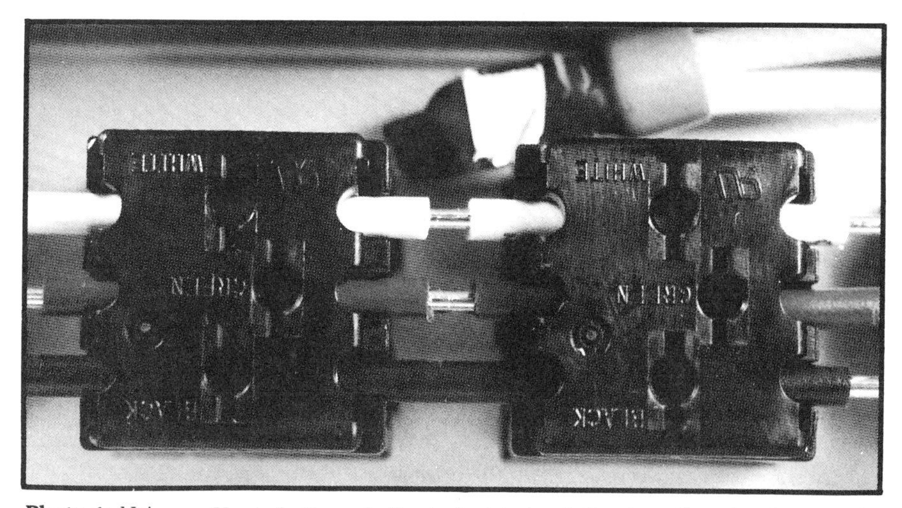

Photo 4: *Using an X-acto knife or similar tool, strip insulation from the wires between the receptacles (which I number 1 through 4, from left to right) according to the following system: between 1 and 2, strip the green and black; between 2 and 3, strip the green and white; between 3 and 4, strip the black and white.*

When the coil of an inductive load such as a transformer or motor is suddenly deenergized, the collapsing magnetic field must dissipate its energy, and it does this by placing a large voltage back into the circuit that energized it. Let's examine the process in detail.

As the circuit through the inductor is broken, current in the inductor continues to flow, charging the distributed capacitance in the windings. At some point, the charge voltage becomes sufficient to leap across the switch gap as a spark. This sudden shorting action discharges the winding's capacitive charge back into the circuit until the spark ceases. This process repeats in a cycle until there is too little energy left in the coil to create an arc across the contacts. The waveform of inductance-generated transients is oscillatory. For example, a contact opening while conducting 100 mA (milliamperes) in a 1-H (henry) inductance will produce a 3000-V spike, assuming about a 0.001-μF (microfarad) stray winding capacitance.

Whenever you plug in a vacuum cleaner, hair drier, or other appliance (even your computer), you could be creating some potentially serious transient disruptions for other equipment on the same power line. The equipment need not even be on the same wiring circuit. The capacitance of household wiring is often sufficient to couple a transient from one wire to another (differential mode) or from the wire to the ground (common mode).

Lightning is the most violent and most destructive source of transient energy. A direct lightning hit is catastrophic, but direct hits seldom occur. A more frequent danger is that a lightning strike on a power line miles away may result in a thousand-volt spike rushing throughout your home. Such hits happen frequently enough to cause much grief. (Because lightning is such a significant source of transients, I've explained it in detail in the text box "How Lightning Strikes." A secondary, and more widespread, effect of a lightning hit on a power line is a voltage sag over a large part of the network as the power company's safety circuits compensate for the spike.)

Electrical noise. Miscellaneous electrical noise is the final source of power-line disturbances. It is best understood as high-voltage high-frequency interference. Noise in the range from 10 kHz (kilohertz) to 50 MHz (megahertz) is the most common cause of computer failures. Because of its frequency, noise can be either broadcast through free space from its source or conducted directly through the power lines. Digital electronic equipment is a prime source of high-frequency noise.

Power-Line Protection

I'm not trying to make you afraid to plug your computer into the wall outlet. There are remedies for virtually all the problems I've mentioned, although some are more practical for some computer users than others.

If surges or sags are a constant problem for you, you can try having the power company change the tap on your local step-down transformer or installing a constant-voltage transformer on your premises. These measures, although expensive, are effective. If you are plagued by blackouts or have equipment that should never be shut down, I suggest that you consider obtaining an *uninterruptible power supply*, abbreviated UPS. Using a UPS gives you confidence in the quality of your power and effectively isolates your computer from damaging perturbations. However, a UPS is also quite costly.

In the case of electrical noise and EMI, there are filters and construction techniques that can be employed to reduce interference, but a better answer is to find the pollution at the source and eliminate it. My article in the January 1981 BYTE outlined most methods of filtration and preventive design. While I'll try not to belabor the point, a power-line filter is an important noise- and transient-suppression device.

The best answer to transients is to suppress their voltages to a harmless level, either with filters or a special category of components called *transient suppressors*.

Photo 5: *You can now solder a varistor between each of the stripped wire pairs, mounting it flat against the back face of the receptacle so that the case will fit together again.*

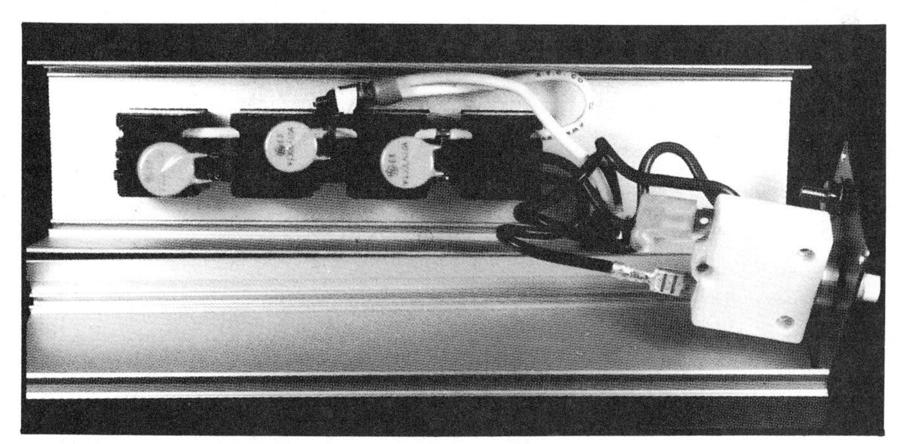

Photo 6: *The outlet strip with three MOVs installed provides both common-mode and differential-mode transient suppression. After you have finished soldering, carefully reassemble the power strip's enclosure and screw it back together.*

Power-Line Filters

A power-line interference filter is an electronic circuit used to control RFI and EMI conducted into and out of equipment. The filter is intended to provide unwanted interference signals with a high series impedance (into the vulnerable equipment) and low shunt impedance (to ground). It generally consists of a set of passive components that act as a mismatching network for high-frequency signals—a low-pass filter. The network attenuates RF energy above 10 kHz, while passing the 60-Hz power.

The simplest possible filter is a single capacitor wired in parallel or a single coil wired in series with the power line. More typically, several capacitors and/or coils are used together, connected into different configurations variously called L, π, and T filters.

Though containing only a few components, such passive bilateral networks have complex transfer characteristics that are extremely dependent upon the impedances of the source and load. Because you can't predict these impedances for all applications, it is not possible to unequivocally state that a specific filter configuration will work the same way in two different environments. But to allow

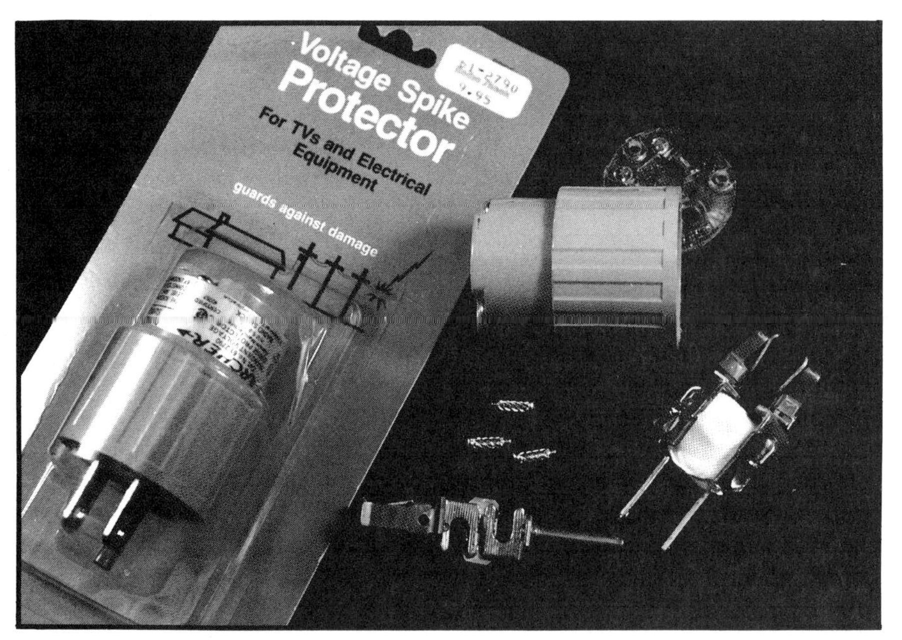

Photo 7: *For quicker and easier, though incomplete, protection, you can plug your computer into a simple voltage-spike protector such as the Radio Shack 61-2790. As you can see from the disassembled unit, the metal-oxide varistor (wrapped in fiberglass tape) is connected between only the hot and neutral lines (black and white). It has no varistor connection to the ground lead and therefore does not protect against common-mode transients.*

electrical specifications to be minimally compared, however, resistive source and load impedances of 50 ohms each are generally used.

Two similar power-line filters, even built with the same circuit topology and component values, may not perform identically; the mounting and wiring of the filter can be critical influences on its performance. A power-line filter is best installed at the point in your equipment where the power line comes inside the case rather than at the far end of a long cord. The filter's purpose is to attenuate high-frequency signals: this purpose is defeated if these parasitic signals can gain access to the equipment by capacitively coupling to the power cord at a point behind the filter.

It's not always possible to disassemble your computer to add a line filter, but the best location for a power-line filter is bolted to the chassis of the electronic equipment it protects, or at least in the immediate vicinity, such as at the power receptacle.

While you could construct a line filter using the formulas and designs from a magazine article, I heartily recommend that you buy a packaged unit instead. The selection is easier and much more controlled using commercial line filters (see the text box on page 205). So much depends upon component selection and layout that the only way to make sure power-line interference has been eliminated is to actually test the filter in your equipment. A circuit designed according to theory using a 50-ohm assumed impedance probably won't work as well as one empirically derived using the actual equipment and power line.

Transient Suppressors

Protection from the various kinds of line transients is obtained by suppressing or diverting them. The three types of circuits most often used for this are filters, crowbars, and voltage-clampers.

As I previously alluded, *filters* comprising inductances and capacitances are widely used for interference protection, including transients. Since most transient signals are high frequency, the suppression by a filter is often effective, provided it can withstand the associated high voltages.

Crowbar circuits use a switching action, such as turning on a thyristor or arcing across a spark gap, to divert transients. But crowbars that incorporate SCRs (silicon-controlled rectifiers) and triacs are much too slow to effectively suppress 100-μs (microsecond) transients. Most often they are incorporated in low-voltage DC power-supply output circuits where overvoltage conditions occur at more manageable speeds (milliseconds). Spark-gap devices, which include carbon blocks and gas tubes, are fast

Photo 8: *Some line filters are made to work in specific circumstances. This Radio Shack power-line-filter strip (stock number 26-1451) was devised to cure interference problems with the TRS-80 Model I computer; it contains two separate LC (inductance/capacitance) interference filters but no varistors. If you have this strip, I suggest you install some MOVs.*

and effective, but they trigger at relatively high voltages, making them unsuitable as the sole protection for semiconductor circuitry.

Voltage-clamping devices, on the other hand, have impedances that vary as a function of either the voltage across or the current through them. The circuit being protected is unaffected by the presence of the clamping device unless the incoming supply voltage exceeds the clamping level, as would be the case when a transient hits. The various kinds of high-speed voltage-clamping devices include selenium cells, zener diodes, silicon-carbide varistors, and metal-oxide varistors. Of these, the *metal-oxide varistors,* or MOVs, hold a significant price/performance advantage and are highly applicable in personal computing applications.

(1a)

TRANSIENT SUPPRESSOR

LINE IN

RECEPTACLES

FUSE 3A

MOV

MOV

MOV

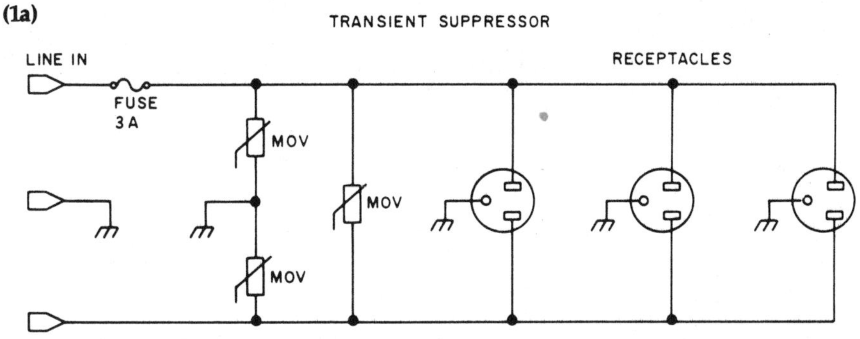

Figure 1a: *The Radio Shack four-outlet power strip can be easily modified to protect equipment from high-energy power-line transients. Three General Electric V130LA10A metal-oxide varistors (MOVs—Radio Shack number 276-570) are connected between the hot, neutral, and ground wires of the power line.*

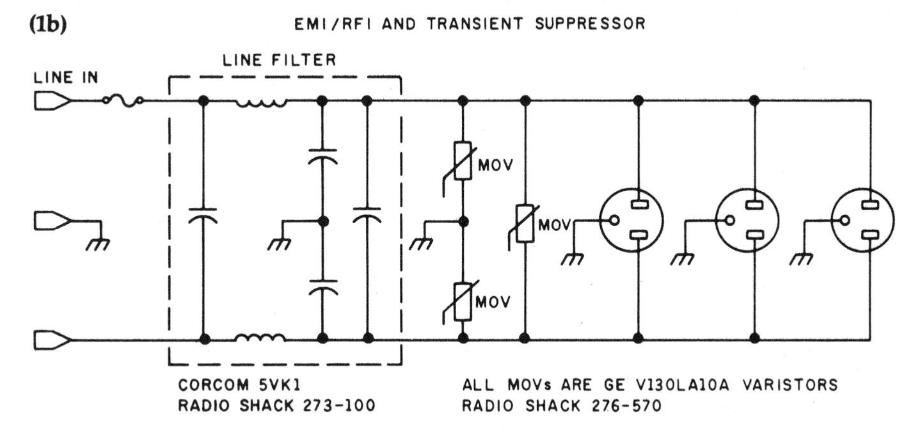

Figure 1b: *For added protection against low-energy electromagnetic and radio-frequency interference, the Corcom 5VK1 line filter (Radio Shack 273-100) can be installed in the circuit.*

MOVs to the Rescue

Metal-oxide varistors are voltage-dependent nonlinear devices that behave somewhat like a back-biased zener diode. When a voltage lower than its conduction threshold is applied across it, the MOV appears as a nonconducting open circuit. But if the applied voltage becomes greater than this set point (when a transient hits), the MOV begins to conduct, clamping the input voltage to a safe level. In effect, the MOV absorbs the transient and dissipates the energy as heat.

An MOV is made of zinc oxide combined with small amounts of bismuth, cobalt, and manganese. The individual zinc-oxide grains form many *p/n* (positive-doped/negative-doped) junctions that combine in a multitude of series and parallel arrangements. This diversity of microstructure causes its nonlinear semiconducting characteristics. An MOV is inherently more rugged than a single-junction semiconductor device (a zener diode, for example) because energy is uniformly absorbed throughout the bulk of the component.

The physical dimensions of the MOV determine its characteristics, its conduction-threshold voltage varying as a function of thickness, and its energy-dissipating capacity varying according to volume. MOVs are available in operating voltages from 6 to 2800 V, with peak current capacities of up to 50,000 A (amperes). MOVs respond to transients in only a few nanoseconds and are relatively inexpensive. The chief producer of MOVs is the General Electric Company.

Protect Your Computer

Large companies sometimes solve power-line problems by producing their own power. In the home or small office, it's more practical to protect your computer and peripherals through comprehensive application of filtering and transient suppression.

Most of the commercially available filtered power strips contain MOVs as their primary suppression device. Even those costing $50 or $75 rarely contain more than $5 worth of transient protection. By purchasing the suppression components separately and installing them yourself, you can save a lot of money.

The majority of the projects I've presented in Circuit Cellar articles can be built for $50 to $2000, but the project this month wins hands down for economy. For the most part, line filters and MOVs are available off the shelf, and adequate transient suppression for your computer might cost as little as $1.59!

You can take two approaches in installing suppression. If you are interested in protecting only a few items of equipment, MOVs can be wired across the AC line where it enters the enclosures. You can find the General Electric V130LA10A MOV component at Radio Shack for $1.59 (stock number 276-570). This device is ideally suited to 120-VAC applications. It has an energy rating of 38 joules (watt-seconds) and will clamp to 340 V at 50 A within 35 ns (nanoseconds). Its peak-current rating is 4500 A. (For heavier duties, you'll need to use V130LA20A or V130PA20A MOVs.)

(As a rule, if you are going to be

working inside the equipment you should also install line filters. You can buy Corcom type-5VK1 5-A RFI power-line filters at Radio Shack for $11.95 (stock number 273-100). These units, like the one shown in photo 1, are adequate for most consumer applications and fit in very nicely with existing equipment.)

The easier alternative is to modify a regular power strip to include transient suppression. Radio Shack's 4-outlet strip (number 61-2620, costing $15.95) is perfect for this application. Merely open it up and install three MOVs, as demonstrated in the series of photos 2 through 6, connected as shown in figure 1. One MOV is installed directly between the black (hot) and white (neutral) leads, the second MOV is connected from the black lead to the green (ground) wire, and the third from the white to the green. While you might squeak through by installing one MOV across the line, complete common-mode and differential-mode suppression requires three MOVs. (Photo 7 shows a commercial adaptation of the simplified scheme.) The price for all the parts of the protected power strip is $20.72. If you were to buy a larger power strip or build your own distribution box, you could also add a power-line filter. And if you have a filter strip already on your computer, you might want to check its degree of transient protection (see photo 8).

An Ounce of Prevention . . .

This project may not seem very exciting. I didn't find the idea very exciting, either, until the flash and subsequent smoke coming out of my favorite article-writing computer provided all the excitement I'll need for months. Most of the $3000 worth of damage I had was for equipment plugged into a single circuit, some of it on the same power strip. I had always known the protective value of MOVs, but I thought it wouldn't happen to me. A few dollars' worth of parts could have saved a lot of aggravation.

Voltage spikes and power-line disturbances aren't always the result of storm activity. Transient-caused equipment failures can happen anytime. The events I've described just served as a catalyst for presenting the subject. And even if lightning never hits you, you should know that many of the new computers I have been evaluating this year have shown an increased sensitivity to external interferance, including power-line glitches. You wouldn't want to find your new computer rebooting suddenly at a critical point or discover the memory to be scrambled after you plug in a printer on the same outlet. Transient suppression constitutes an ounce of prevention. You can spend thousands for the cure.

References

1. Ciarcia, Steve. "A Computer-Controlled Wood Stove." February 1980 BYTE, page 32.
2. Ciarcia, Steve. "Electromagnetic Interference." January 1981 BYTE, page 48.
3. Roberts, Steven K. *Industrial Design with Microcomputers.* Englewood Cliffs, NJ: Prentice-Hall, 1982.

21

A Musical Telephone Bell

Personalizing the sound of your telephone

About a month ago, I was visiting an IC (integrated circuit) manufacturer that produces a line of communication and voice-synthesis chips, among others. I was there to get some firsthand information on some new chips that will be the heart of a low-cost Circuit Cellar voice-recognition project coming later this year.

The visit started like most business meetings I'm used to: it included a tour of the facility, discussions with the technical staff, and lunch. After returning from lunch, a group of us were standing in the corridor adjacent to a large divided office area deciding who should be part of the next meeting when a phone rang in the middle of the room. Everyone immediately stopped talking and looked toward their section of the office as the phone rang again.

They glanced at each other again as one said, "Is that my phone or yours?"

The consensus was, "It must be yours, it doesn't sound like mine."

I watched as all four started walking toward their offices. About halfway there, each one stopped dead and turned around. With a somewhat exasperated sigh, one of them said, "It was George's phone."

"It gets so frustrating. All the bells sound alike, and with the acoustics in here it's impossible to tell whose phone it is unless you're within 10 feet of it."

The others shared the same expression of annoyance as the discussion shifted from electronics-related topics to an area of more immediate concern: Why can't telephone manufacturers make different-sounding bells for phones? It would seem that in a free-enterprise system, adept at producing pet rocks and Cabbage Patch dolls, a custom telephone bell would be trivial.

While I didn't make any pledge to solve this problem, I recognized legitimate concerns and decided to intervene electronically. Therefore, projects on talking robots, automatic houses, and rainmaking machines will have to be put off for another month as I try to build a better mousetrap.

I hope that this month's project will solve the auditory confusion in an office or at least add a little spice to an otherwise boring telephone. Rather than just ring, the Circuit Cellar Whimsi-Bell plays the first few bars of 25 preselected tunes. Instead of hearing an annoying metallic clamor, you can be greeted by the theme to *Star Wars*, or perhaps you would prefer the "William Tell Overture."

Cleverly disguised in this whimsical project is a discussion of the telephone system. My intention is to help you understand characteristics and specifications that govern the telephone products you purchase or the telephone interfaces you might build. It also sets the stage for future Circuit Cellar projects dealing with telephone lines.

Reading between the lines, however, you'll soon realize that the central theme is not musical phones but rather ring detection and auto-answering the phone by computer. It may seem like a trivial consideration, but the environment is hostile and connection restrictions abound. First, a little about the phone system.

The Status Quo Telephone

Essentially, the characteristics of the telephone have been unchanged for 90 years. It originally used a carbon microphone and electromagnetic earphone with a capacitively coupled electromagnetic ringer triggered from a hand-turned magneto. Today's Western Electric phones incorporate many of the same materials, and new electronic phones merely simulate their archaic predecessors. For example, the characteristic impedance of the early Edison phones was 600 ohms, and today's electronic units must still abide by this specification.

The design and use of telephone equipment are dominated by line resistance. When you wish to answer or initiate a telephone call, the only requirement is to place a load across the phone lines (between tip and ring). The handset, or the data-access arrangement (DAA) in either your modem or your auto-dialer, will cause a DC current flow of approximately 25 to 30 milliamperes (mA). A current-sensing relay at the telephone company then signals the system that you are "off hook." If it is an outgoing call, you will receive a dial tone; if it is an incoming call, the ringing will stop and you will be connected to the incoming party.

The on-hook voltage between tip and ring of the telephone line is approximately 48 volts (V) DC. It is generally supplied by a battery from the telephone office and can range from 42.75 to 52.5 V. Tip and ring have nothing to do with the telephone ring itself; they refer to the plugs that the operators used to connect callers many years before automatic dial exchanges. The original system had large arrays of connection jacks with operators who would physically insert patch-cord jumpers between initiating and receiving calls. The two conductor patch cords made their electrical connection to the tip-and-ring portion of the plug. The tip connection was usually ground. (Anything designed for connection to the telephone line should not be polarity sensitive. Polarity is sometimes reversed, and 200-V test voltages are sometimes placed on the line.)

Once the line is captured, an off-hook situation exists and a dial tone will be heard in approximately a second. The dial tone actually consists of two tones: 350 Hz and 440 Hz. If you are contemplating building a computer-activated automatic dialer, a tone decoder should be incorporated to signal the computer that dialing can commence. It will also signal you when the call has terminated if it was not initiated at your end. (If you want to get elaborate, the awful sound that the phone company blares at you to attract your attention when you forget to hang up a phone consists of four tones: 1400 Hz, 2060 Hz, 2450 Hz, and 2600 Hz. Pulsed at a rate of 5 Hz, it is called the receiver off-hook tone.)

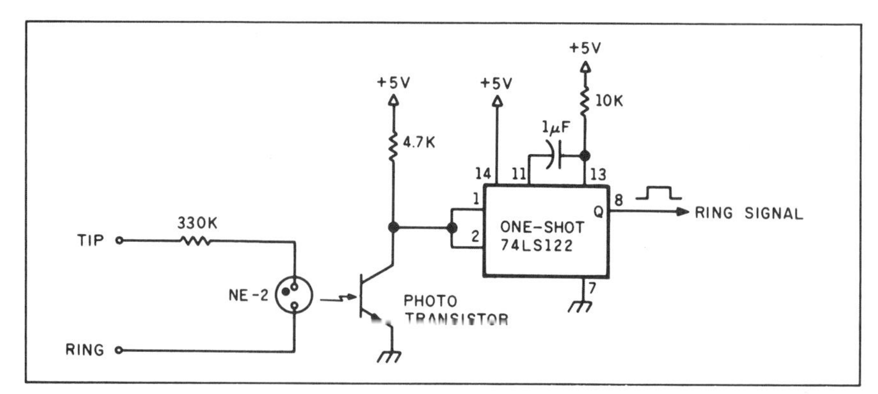

Figure 1: *A simple ring-detector circuit.*

Dialing

Most residential telephones in this country still use mechanical pulse dialers. When you turn and release the telephone dial, the current in the telephone line through the off-hook relay is interrupted by the number of times indicated on the dial position. These dial pulses are issued at approximately 10 pulses per second (pps)—some exchanges will accept up to 21 pps—with about a 60/40 percent on-/off-hook duty cycle. Nine breaks in succession are interpreted as a "9" digit, and three breaks define a "3" digit.

The separation of digits is determined by the *interdigit time*. Any succession of pulses occurring more than 750 milliseconds (ms) after the last pulse are considered part of the next digit. For example, if you send five pulses at 10 pps, wait 200 ms and send three more pulses, the telephone company will interpret this as the single digit "8." If you had waited 750 ms between transmissions, they would have been interpreted as the two digits "5" and "3."

The latest innovation in the telephone system is Touch-Tone, the registered trademark for AT&T's dual-tone multifrequency (DTMF) communication. The pulses go only as far as the local exchange office. From there, and throughout the rest of the telephone system, DTMF tones are used to direct calls.

The advantages of DTMF are increased dialing speed and, more important, the ability to transmit data. DTMF tones are sent with a minimum duration of 50 ms and an interdigit time of 45 ms. A seven-digit number on a pulse-dial system would take 11 seconds versus less than 1 second for DTMF.

I won't dwell on DTMF because it is covered in detail in my "Build a Touch Tone Decoder for Remote Control" on page 42 of the December 1981 BYTE.

Photo 1: *Whimsi-Bell attaches to the tip and ring wires in parallel with your phone.*

The Busy and Ring Signals

Again, if you intend to build an auto-dialer at some point, you should incorporate some means to recognize a busy signal. The busy signal consists of two tones, 440 Hz and 620 Hz, that are on for 0.5 second, then off for 0.5 second. Either the tones themselves can be recognized, or the unique 50/50 0.5-second duty cycle can be monitored.

When your telephone rings, it is the result of a high AC voltage being ap-

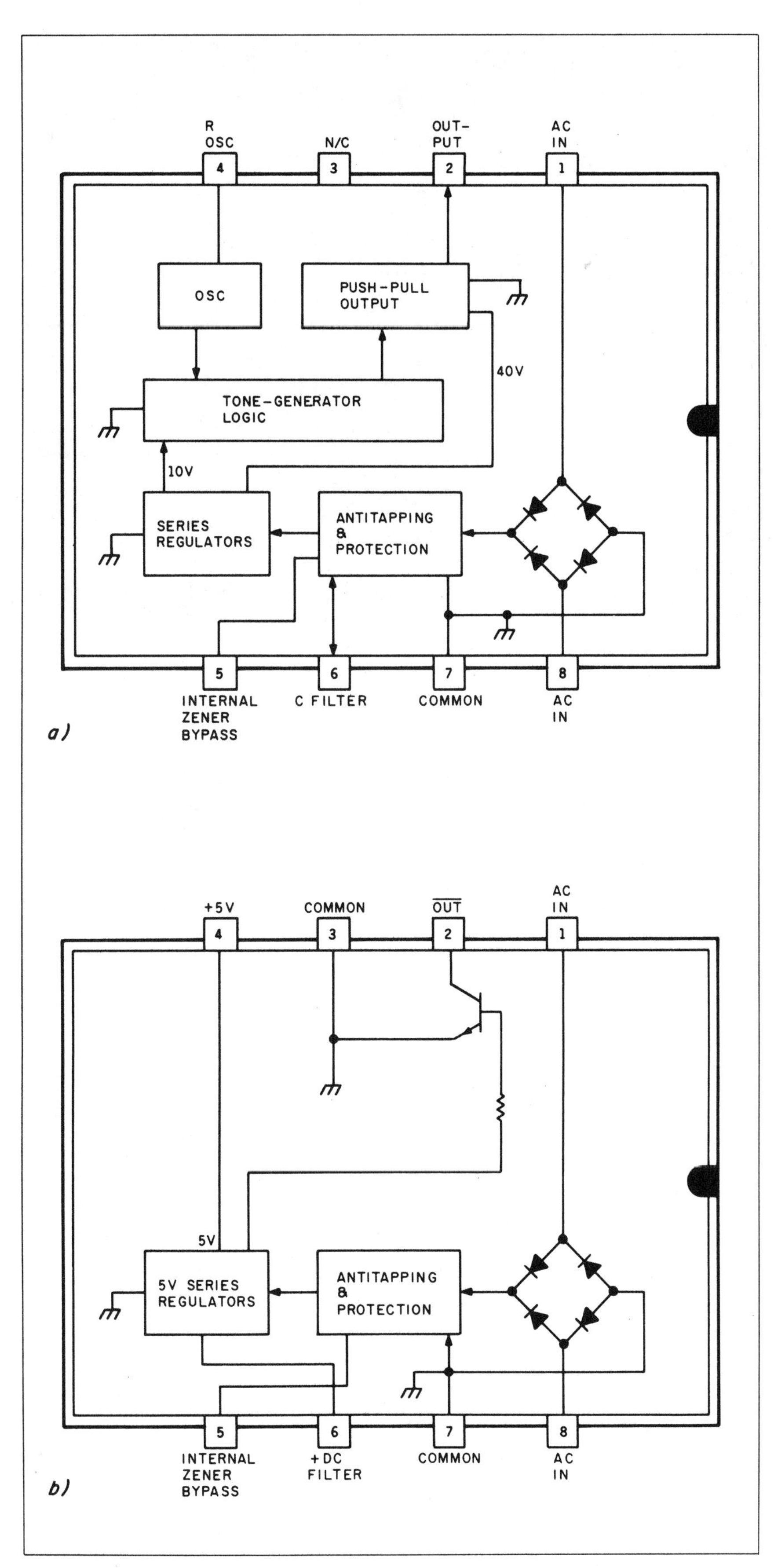

Figure 2: *Texas Instruments* CMOS *ring-detector chips.* (2*a*) *Block diagram of the* TCM1501A, TCM1505A, TCM1506A, *and* TCM1512A *versions.* (2*b*) *Block diagram of the* TCM1520A *version.*

If you are building a computer-activated automatic dialer, use a tone decoder to signal the computer.

plied to the telephone lines. The voltage is capacitively coupled to the electromagnetic bell. When you are initiating a call and you hear a ringing in the earpiece, you are not hearing the ringing voltage. Instead, you are hearing a pair of tones, 440 Hz and 480 Hz, used by the phone company for signaling. The on/off duty cycle depends on the exchange being dialed. When the ringing stops, it usually means the call has been completed or that you are irrevocably lost among the trunk lines.

Ring Detection

The solution to the problem described earlier is a ring detector. Rather than try to change all the bells in an office, it is a relatively simple matter to monitor the telephone line in parallel with the existing phone (with its bell turned low or off) and generate a different sound corresponding to the incoming ring signal. This new sound can be a buzzer, a slightly different bell, or an entirely new electronic signaling device. The actual alerting mechanism is secondary. It is all triggered by a circuit called a ring detector.

The incoming ring is the highest non-test voltage encountered on the telephone line. The normal on-hook condition is a high-impedance state with approximately 50 V DC between tip and ring. When the phone rings (usually 2 seconds on and 4 seconds off), it is because an additional 86 V AC of ringing voltage has been superimposed on the line. This 20 Hz ± 3 Hz signal is passed through a capacitor to the telephone bell, causing it to ring. While 86 V AC is nominal, the ringing voltage can vary from 65 to 130 V AC, and the DC component can appear as much as 70 V negative.

Because the ringing voltage is so different from other telephone signals, a ring detector is simple in theory to construct. The simplest ring-detector circuit (shown in figure 1) consists of a neon lamp and a phototransistor. Neon lamps such as the NE-2 have a turn-on threshold of about 65 V and therefore would respond only to the higher ringing voltages. When the neon lamp lights, it in turn causes the transistor to conduct and triggers the one-shot.

Variations on this circuit employing LED (light-emitting diode) optoisolators and level-detection circuits are available, but a price is paid for simplicity. The telephone line is not an ideal environment and contains many aberrations that can lead to false triggering by crude ring detectors. Just the action of going on hook or off hook (also called tapping) generates local line transients that are sufficient to cause a neon bulb or LED to fire. We can add more components or compensate for these peculiarities in our communications software, but, fortunately, alternatives exist.

Specially Designed Ring-Detector Chips

Texas Instruments produces a line of CMOS (complementary metal-oxide semiconductor) ring-detector chips that offer all the necessary features. (See figures 2a and 2b.)

The normal installation of the ring-detector chip uses a capacitor and a 2.2K-ohm resistor connected between the detector and the line. The network formed by the DC-blocking capacitor, current-limiting resistor, and full-wave bridge rectifier supplies power to the IC from the phone lines. The rectified AC signal is filtered by a 10-microfarad (μF) capacitor attached between pins 6 and 7. The value of this capacitor determines the minimum input voltage and turn-on time of the ring detector. It is also used to suppress response to dial tapping (tapping is a false triggering of the bell due to transient-producing pulses on the phone line, usually from rotary-dial phones).

In use, the ring-detector IC stays in standby mode until the incoming

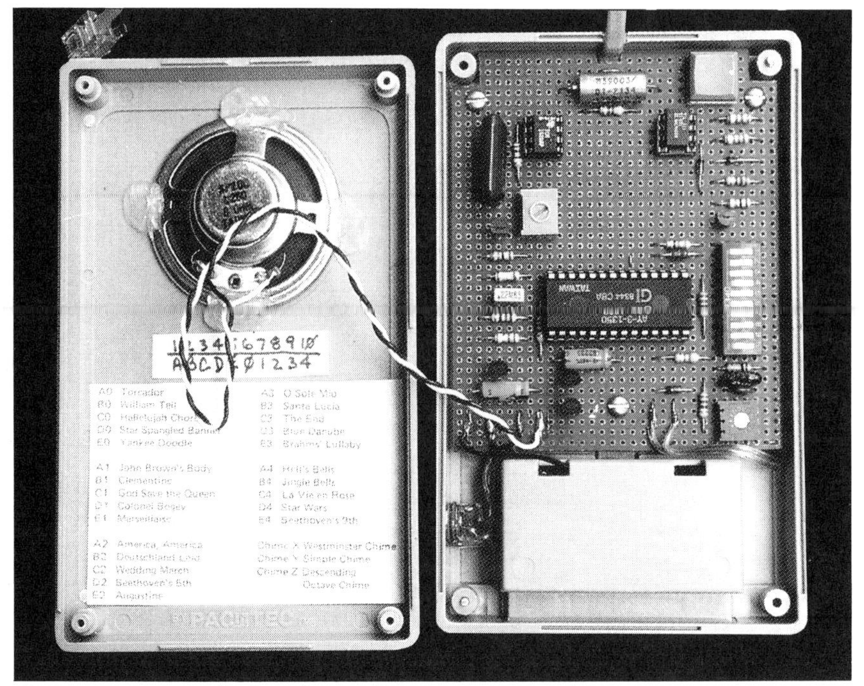

Photo 2: *A view inside the Whimsi-Bell prototype. The doorbell synthesizer chip can play 25 different tunes.*

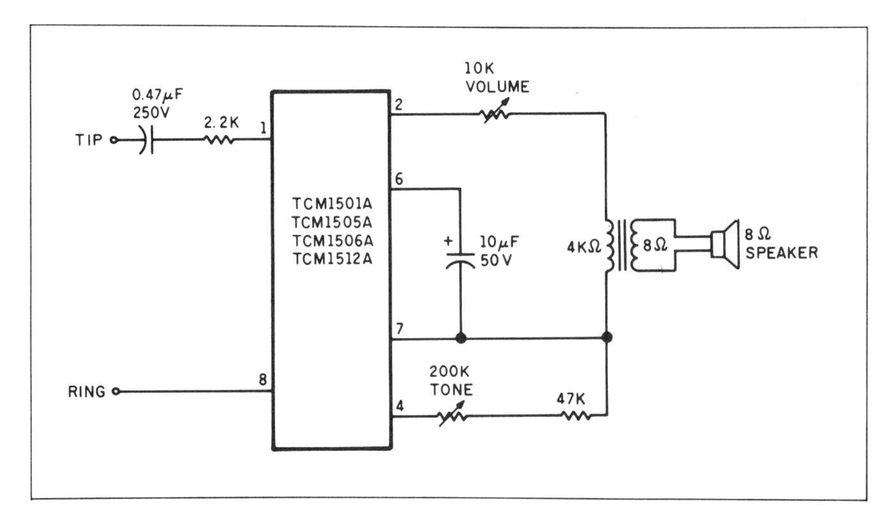

Figure 3: *A typical ring detector using the TI ring-detector chip.*

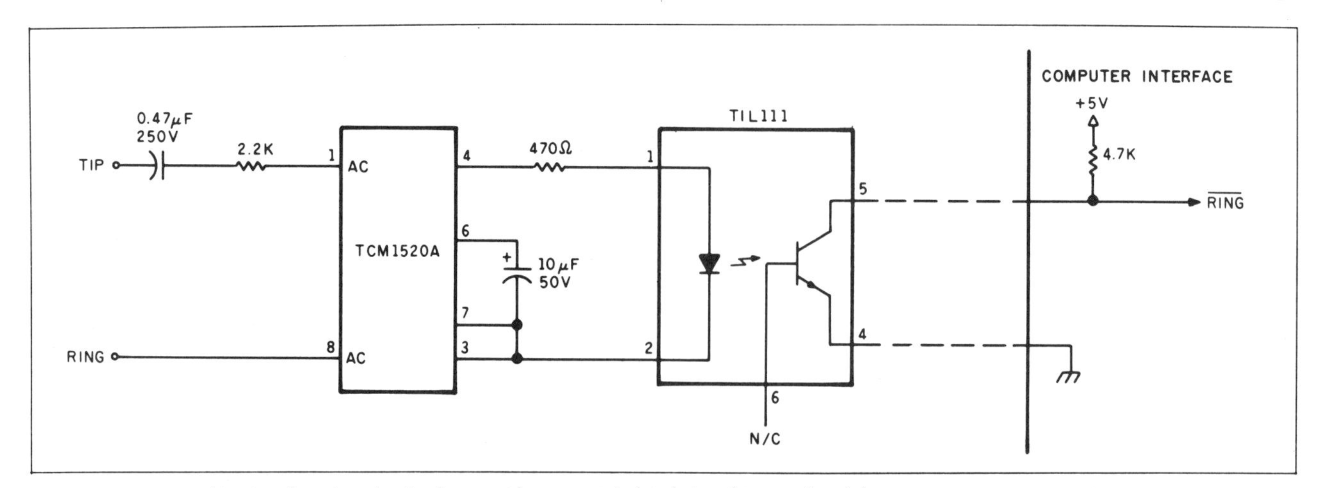

Figure 4: *A two-chip ring-detection circuit that provides an optoisolated ring-detector signal for your computer.*

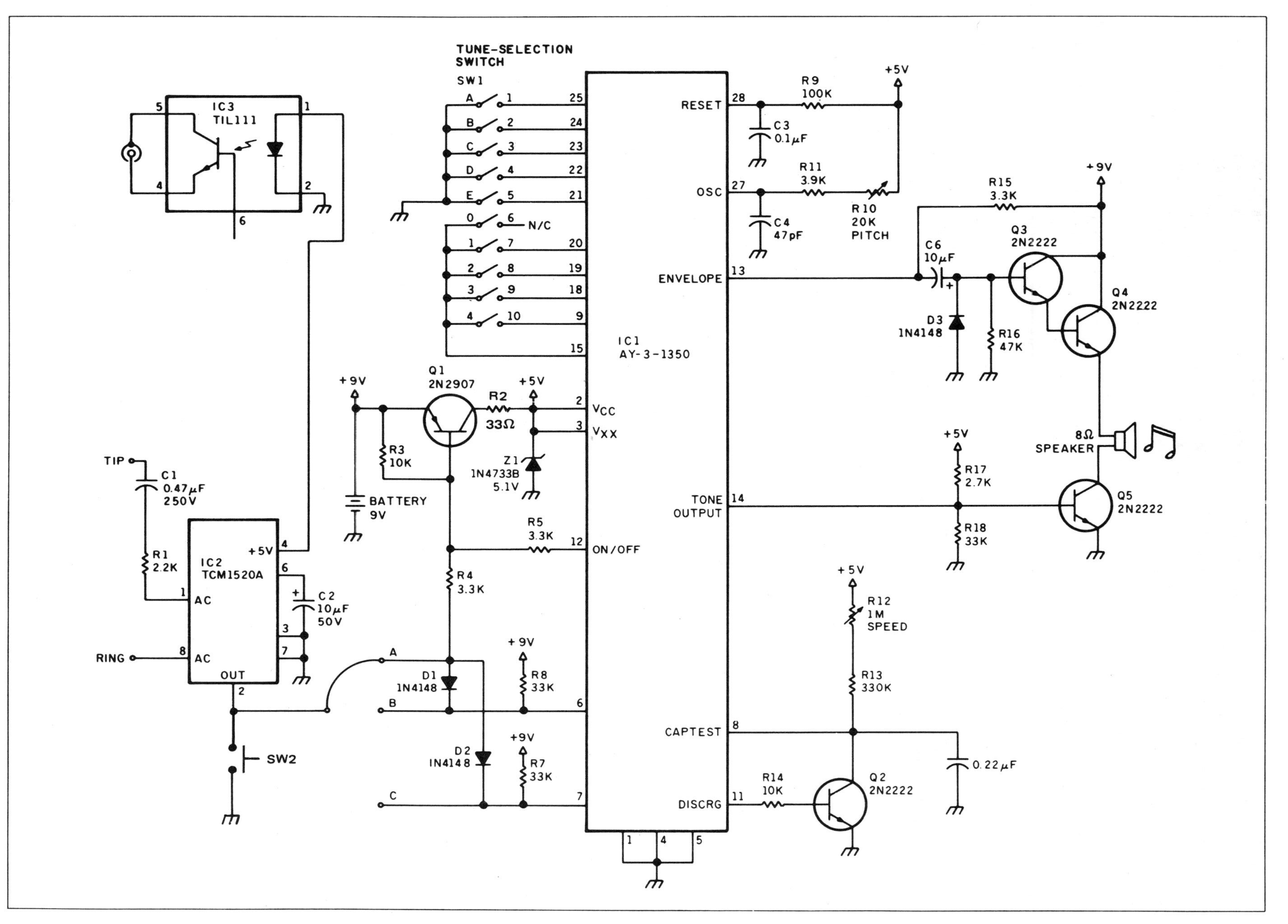

Figure 5: *The schematic of the* Whimsi-Bell.

voltage between pins 1 and 8 exceeds 8.9 V (remember, the off-hook condition is a DC voltage on the line, and the series DC-blocking capacitor prevents the ring detector from seeing this voltage). While in standby mode, the impedance is approximately 1 megohm.

When the voltage exceeds 8.9 V, the IC begins to conduct. This energy is not transferred to the load, however, until the input reaches 17 V. Should the input voltage continue to rise beyond a predetermined limit (a transient instead of a true bell signal), an internal high-current SCR (silicon-controlled rectifier) is triggered. The excess energy is dissipated in the 2.2K-ohm series resistor.

Two versions of the ring detectors are supplied by Texas Instruments. Models TCM1501A, TCM1505A, TCM1506A, and TCM1512A are detector tone drivers intended for use as electronic alternatives to the standard electromagnetic telephone bell. These chips incorporate an oscillator and a power audio-generator section to drive a piezoelectric transducer or speaker. Figure 3 demonstrates a typical circuit.

The TCM1520A is a ring detector only. It has a +5-V output signifying the ring signal rather than an audio output. The TCM1520A is best utilized in auto-answer modems. Figure 4 demonstrates a two-chip circuit that provides an optoisolated ring-detection signal that can be connected to a computer. When a ring signal is applied, the 10-μF capacitor charges until it passes the 17-V threshold. At that point, pin 4 outputs +5 V, which in turn drives the LED side of an optocoupler. The illuminated LED causes the transistor portion of the optocoupler to saturate, providing a low true signal to the computer. The computer, on recognizing a valid ring, gives

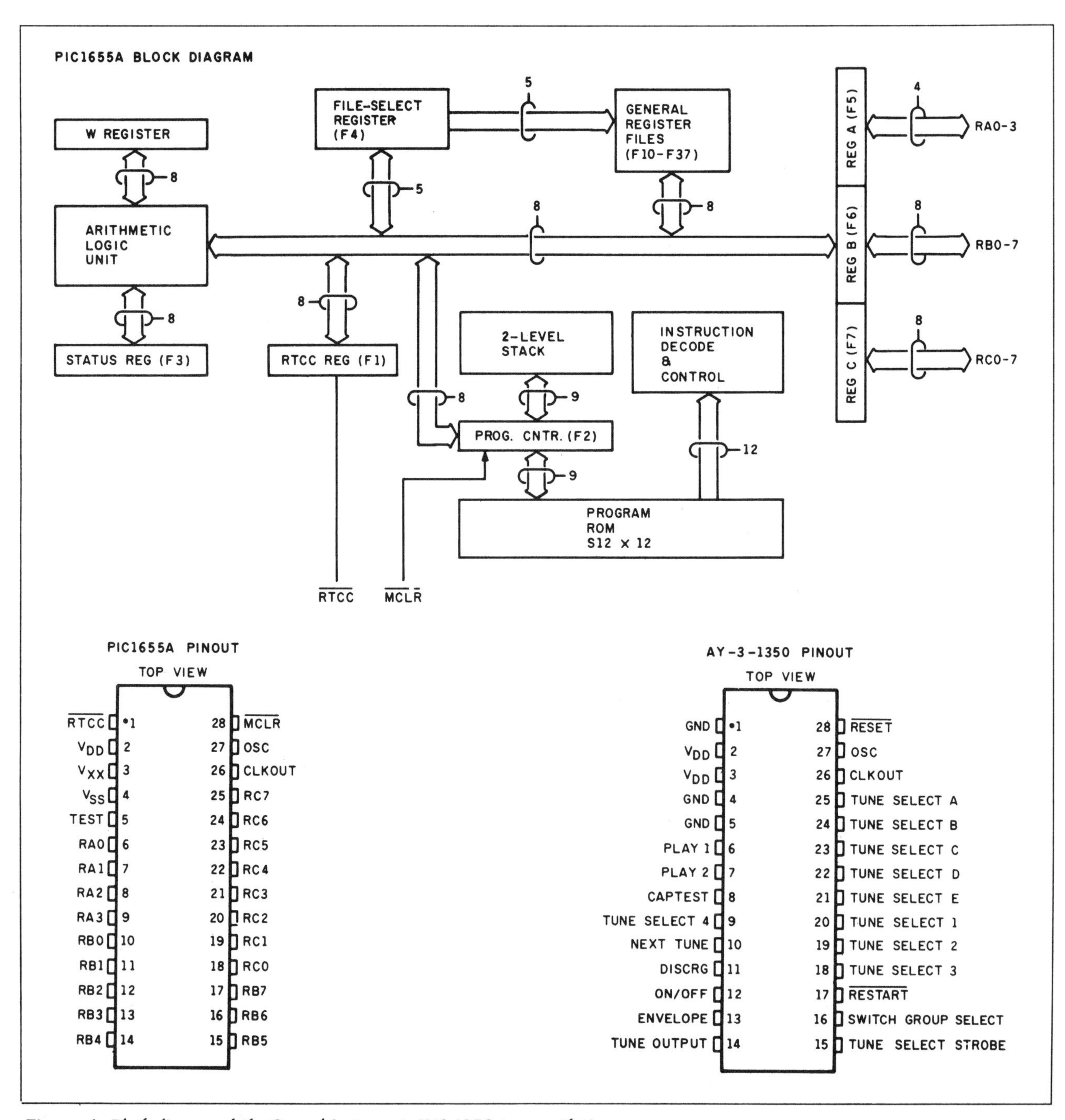

Figure 6: *Block diagram of the General Instrument AY-3-1350 tone synthesizer.*

an off-hook signal to the DAA or the modem to answer the phone.

Back to the Initial Problem

The office situation described earlier can be easily solved for one or two individuals with the one-chip circuit shown in figure 3. Here, a TCM1512A is used to drive a speaker. The circuit is connected in parallel to an existing phone that has had the bell disconnected or turned down. R1 sets the frequency of the "bell"; R2 sets its volume. When the phone rings, a different kind of bell is heard.

Such a circuit might be useful in some applications. Beyond a few installed in a large room, however, the same confusion would arise. Instead, a more varied signaling device must be employed.

A variety of alternatives comes to mind. Once you've detected the ring, virtually any triggerable event can occur. Initially, I thought it might be cute to attach a voice synthesizer or other equally nonstandard signaling device. However, I'd rather reserve voice synthesizers for more serious purposes. Instead, in the name of cost-effectiveness and mass production, I opted for Whimsi-Bell.

Whimsi-Bell

The Circuit Cellar Whimsi-Bell, shown in photos 1 and 2 and schematically diagramed in figure 5, combines a ring detector and a sophisticated microprocessor-controlled doorbell synthesizer. The phone is connected to the Whimsi-Bell through a Y-junction phone-line connector. When the phone rings, it plays the first few bars of one of 25 tunes or three chimes. The switch-selectable repertoire is listed in table 1. The circuit also includes an optoisolated output so you can connect the device to a computer as well.

The heart of Whimsi-Bell is the General Instrument AY-3-1350 tone synthesizer, which is actually a PIC 1655A 8-bit microcomputer that has been specifically mask-programmed as an electronic doorbell chime. Block-diagramed in figure 6, the PIC 1655A includes thirty-two 8-bit RAM (random-access read/write memory) registers, 512 by 12 bits of program ROM (read-only memory), an on-board oscillator and real-time clock, and 20 I/O (input/output) lines on a single chip. The chip runs on +5 V.

The Whimsi-Bell is powered by a 9-V battery and consumes virtually no power until it is triggered by a ring-detect signal from IC2. This signal turns power on to IC1 through Q1. Q1 is sustained for the musical period regardless of the length of the ring by the on/off signal, IC1 pin 12. The chip shuts itself off when it has concluded playing.

The ring-detect signal is attached to one of three points in the circuit (A, B, or C), depending upon the combination of tones you wish. All 25 tunes are accessible if the connection is to point A. The chimes are selected by opening all the selection switches and connecting the ring-detect input to one of the three input terminals. If A, it will play the Westminster chime; if B, it will play the simple chime; if C, it will play the descending-octave chime.

With the ring-detector connection at point A, where all 25 tunes are available, a particular tune is selected on the 10-position DIP (dual-inline package) switch SW1. SW1 positions 1 through 5 are assigned to selection codes A through E; SW1 positions 6 through 10 are assigned to selection codes 0 through 4. To select the *Star Wars* theme, D4, positions 4 and 10 would be closed (all others open). Every time the phone rings, that tune will be played.

The 25 tunes average nine notes each. The tunes are stored internally as a series of notes with a stop code following the last note in each tune. Each 8-bit note comprises a 5-bit pitch value and 3-bit duration. Two-and-a-half octaves of notes can be accommodated, with the durations ranging from a sixteenth note to a whole note. (I mention this only because the AY-3-1350 can be configured for external ROM should you care to play the entire "Star-Spangled Banner" when the phone rings. Contact General Instrument for an application note describing this procedure.)

There are two potentiometer adjustments. R10 varies the processor clock speed about 1 MHz to set the pitch. The other adjustment potentiometer, R12, varies the charging time on C7 to set the speed at which the notes are played.

The output of IC1 is a combination of signals. The actual tune output comes from pin 14, while pin 13 serves as an envelope generator to control the volume. A three-transistor amplifier directly drives an 8-ohm speaker.

In Conclusion

This month's project is a little less taxing than the 99-chip Trump Card from the last two months. Nonetheless, ring detection and the telephone system are important subjects for discussion.

While I don't think everyone is going to want a Whimsi-Bell attached to their phone, all the people who have heard my prototype have been amused enough to want one, if only as a unique conversation piece. More important, I'm encouraged by all the experimenters who have built one of the Circuit Cellar modem projects and see this as a means to add auto-answering.

Table 1: *Tunes and switch settings for the Whimsi-Bell.*

Switch	Tune	Switch	Tune
A0	Toreador	A3	O Sole Mio
B0	William Tell	B3	Santa Lucia
C0	Hallelujah Chorus	C3	The End
D0	Star-Spangled Banner	D3	Blue Danube
E0	Yankee Doodle	E3	Brahm's Lullaby
A1	John Brown's Body	A4	Hell's Bells
B1	Clementine	B4	Jingle Bells
C1	God Save the Queen	C4	La Vie en Rose
D1	Colonel Bogey	D4	Star Wars
E1	Marseillaise	E4	Beethoven's Ninth
A2	America, America		
B2	Deutschland Leid		
C2	Wedding March		
D2	Beethoven's Fifth		
E2	Augustine		

22

AN ULTRASONIC RANGING SYSTEM

Build the SonarTape

Those of you who have followed the Circuit Cellar projects over the years will recognize that I have discussed sonar and ultrasonic ranging before. November 1980's "Home In on the Range! An Ultrasonic Ranging System" used the Polaroid Ultrasonic Ranging System Designer's Kit. This $150 kit, based on a modified, custom-manufactured sonar-ranging circuit board from Texas Instruments for Polaroid's SX-70 camera, greatly simplified the circuitry normally associated with ultrasonics. The November 1980 project described circuit additions that facilitated connection of the ranging kit to a computer's I/O (input/output) port.

To my knowledge, these kits are still available from Polaroid if you are interested in producing the stepper-motor-driven scanning system described in the original article. However, if you are looking for a cost-effective distance sensor for your computer, you'll be happy to know that LSI (large-scale integration) technology did not stand still in the interim. Texas Instruments introduced a new sonar-ranging module that is both cost-effective and simple to integrate into computer-based systems.

I'll first describe this new module in detail and then demonstrate how easily you can attach it to a computer or use it independently with an LCD (liquid-crystal display) to create an electronic tape measure.

THE SONAR-RANGING MODULE

The latest Texas Instruments sonar-ranging module, the SN28827, is actually an updated and higher-functioning version of the original SX-70 module, both of which are shown in photo 1. The newer unit has similar performance characteristics but requires far less support circuitry and interfacing hardware. It is designed to drive a 50-kHz 300-volt (V) electrostatic transducer, which Polaroid is still manufacturing. The module and the transducer are the only components necessary to measure distances from 1.33 to 35 feet with an accuracy of ±2 percent.

In operation, a pulse is transmitted toward a target and the resulting echo detected. The elapsed time between the transmission and echo detection is a function of the distance to the target. Basically, this distance in feet is simply the elapsed time in seconds (actually milliseconds) multiplied by the speed of sound in feet per second

Specifications

Minimum Transmitting Sensitivity at 50 kHz 300 V AC peak to peak, 150 V DC bias (dB re 20 μPa at 1 meter)	110 dB
Minimum Receiving Sensitivity at 50 kHz 150 V DC bias (dB re 1 V/Pa)	−42 dB
Suggested DC Bias Voltage	150 V
Suggested AC Driving Voltage (peak)	150 V
Maximum Combined Voltage	400 V
Capacitance at 1 kHz (typical) 150 V DC bias	400–500 pF
Operating Conditions	
Temperature	32°–140° F
Relative Humidity	5%–95%
Standard Finish	
Foil	Gold
Housing	Flat Black

Figure 1: *Specifications for the Polaroid electrostatic transducer.*

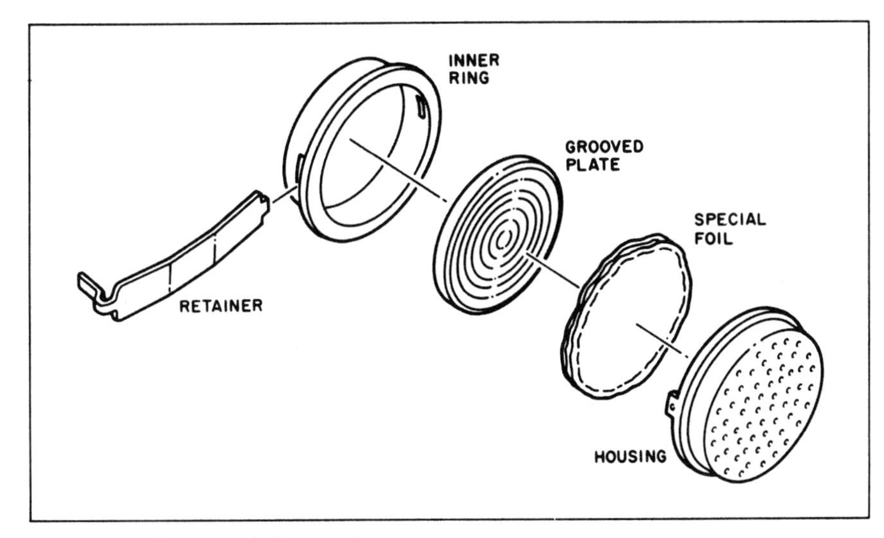

Figure 2: *The parts of the transducer.*

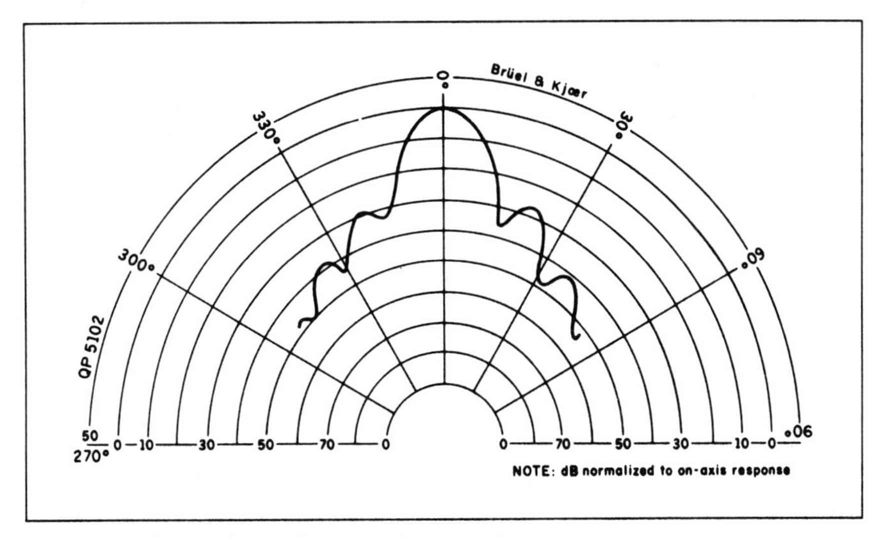

Figure 3: *The transducer's beam pattern at 50 kHz*

(approximately 1100 ft/sec). More specifically, the rate is 1.78 milliseconds (ms) per round-trip foot. It takes 3.55 ms for a pulse to leave the transducer, strike a target 2 feet away, and return to the transducer.

The transducer most frequently used with this module is the instrument-grade Polaroid electrostatic transducer (see figure 1 for its specifications), which acts as a speaker in the transmit mode and a microphone in the receive mode. The transducer (shown in photo 2 and disassembled into its component parts in figure 2) is 1.5 inches in diameter and consists of a 3-millimeter gold-plated foil stretched over a concentrically grooved aluminum disc.

The foil, electrically insulated yet bonded closely to the metallic backplate, forms a capacitor. The foil is the moving element in the transducer that converts electrical energy into sound and the returning echo into electrical energy.

The Polaroid sensor is larger and, in my opinion, considerably more expensive than other 50-kHz transducers. However, it is designed this way for a specific purpose. (I've tried 50-kHz transducers from two other manufacturers with no success.) The diameter of the transducer determines its directional sensitivity. The Polaroid unit is extremely directional, as indicated in the graph of acoustical-signal strength shown in figure 3.

Operating the Module

The sonar-ranging module (pictured in photo 3 and diagramed in figure 4) is a two-chip, 2- by 2-inch module. It is a 12-component, custom-manufactured module built around the Texas Instruments TL851 and TL852 sonar-ranging controller/receiver chip set. In small quantities, the assembled module is far less expensive than the individual components, and it eliminates the ever-present aggravation of finding the correct coils and chokes. Electrical connection is made through an eight-conductor flat ribbon cable or direct solder attachment to the connector base, which contains three output lines, three input lines, power, and ground.

The two basic modes of operation for the module are single echo and multiple echo. The single-echo mode

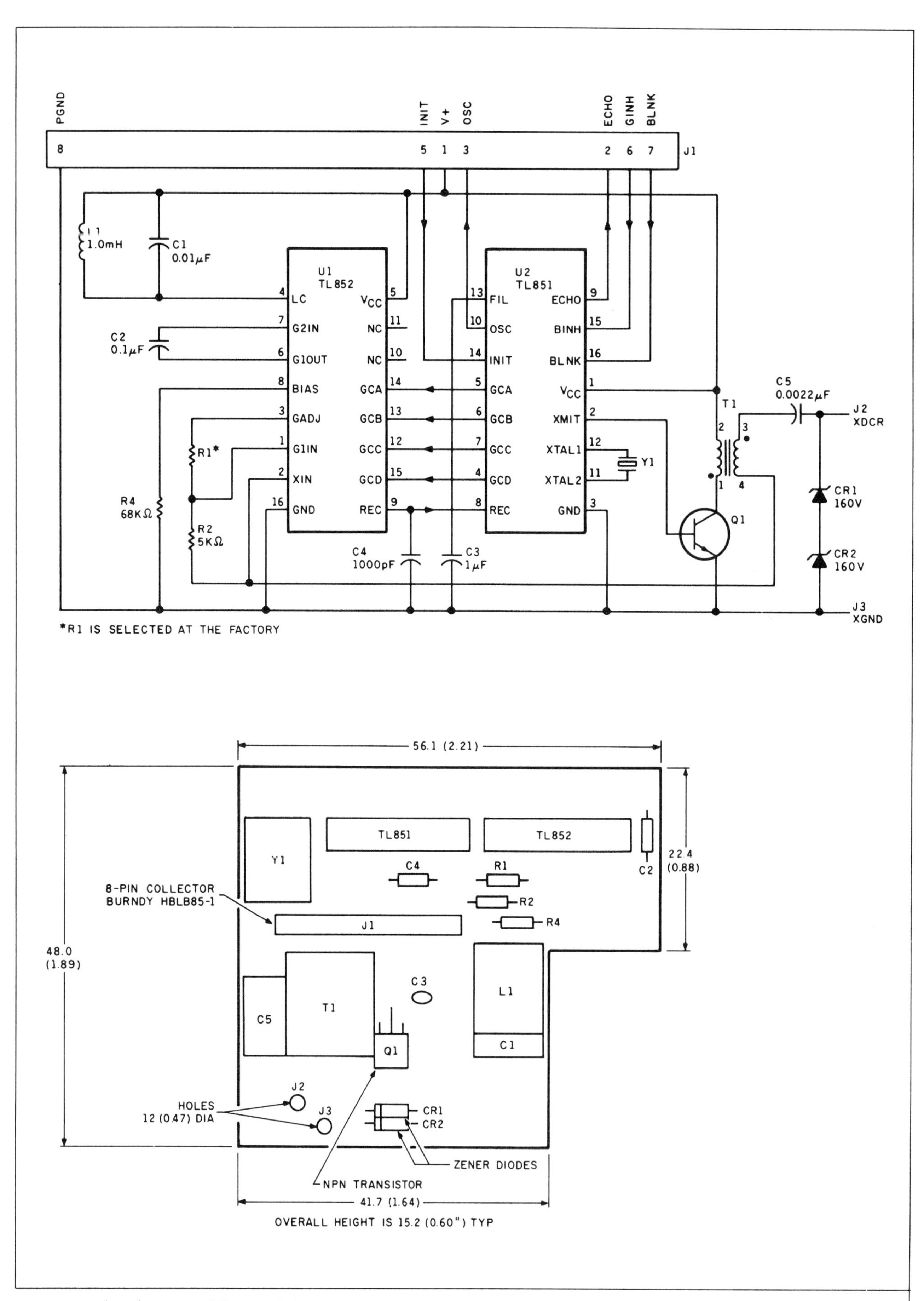

Figure 4: *The schematic and layout of the Texas Instruments sonar-ranging module.*

Photo 1: *The original Texas Instruments ultrasonic ranging module (left) provided in the experimenter's kit from Polaroid compared to the new TI ranging module (right) that is the basis of this article.*

Photo 2: A *close-up of the Polaroid ultrasonic transducer.*

Photo 3: A *close-up of the TI ranging module.*

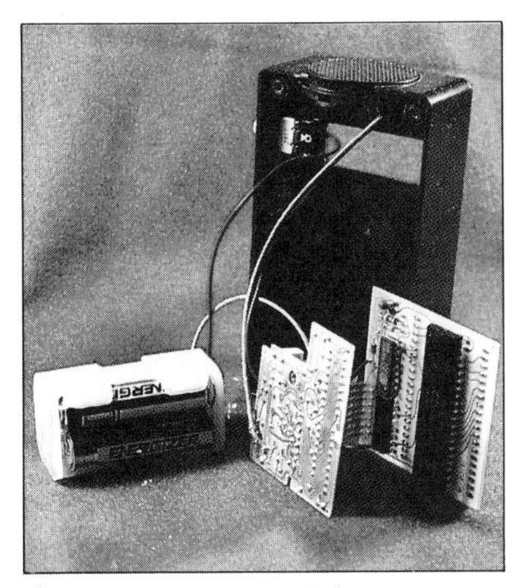

Photo 4: A *rear view of the disassembled tape measure.*

Photo 5: A *front view of the disassembled tape measure.*

implies that only one target exists and that a single ranging value is desired. In the multiple-echo mode, the echo monitoring time is extended to "hear" the echo from objects farther away than the closest target. Differentiating among these echoes can be a data-reduction nightmare, so I will limit my discussion and the use of the ranging module to the single-echo mode.

Figure 5 is the basic timing diagram of the ranging module. Within 5 ms of applying power (at 4.5 V to 6.8 V), the internal circuitry has reset and stabilized and the module is ready. The distance-measuring sequence is activated by raising the INIT input line to a logic 1 state. This enables the 420-kHz on-board ceramic resonator, and 16 cycles of a 300-V 49.4-kHz (420 kHz divided by 8.5) signal are generated. The 16 cycles are passed through a driver/transformer combination that boosts the signal's magnitude to 300 V at the transducer. At the end of the 16 cycles, a DC bias of 150 V remains on the transducer for optimum operation, and the oscillator output steps up to 93 kHz (420 kHz divided by 4.5), where it remains as long as INIT is high.

In order to prevent any ringing in the transducer from being detected as a return signal, the receive input of the ranging control IC (integrated circuit) is inhibited by an internal blanking signal for 2.38 ms. This blanking interval corresponds to 1.33 feet, which is the minimum distance that the ranging module can sense without external control intervention. To detect objects closer than 1.33 feet, the blanking inhibit line, BINH, can be taken high prior to 2.38 ms to enable the receiver.

In the single-echo mode, all that must be done is to wait for the echo from the target. After INIT, the transmitted pulse travels at a rate of 0.9 ms per foot. When the ranging module hears the echo, the ECHO output line goes high. The difference in time between INIT and ECHO both going high is a measure of the distance to the target. This elapsed time can be measured through a timing loop in a computer or gated with the oscillator output to increment a counter and drive a display. I will explore both of these techniques.

One final note on the ranging-module electronics. If you have experimented with ultrasonics, you've probably found, as I have, that fixed-distance reception is far easier to accomplish than variable-distance reception. Sound intensity decreases geometrically proportional to increases in distance. If you have designed a transmitter-and-receiver system that works well at 6 feet, it may not function at 12 feet without substantially increasing the receiver sensitivity (gain) to account for the reduced echo amplitude. Leaving the sensitivity at a high level and reducing the distance again invites other interference and false-echo-detection problems. To adequately compensate for changing distances, the sensitivity setting must also be adjustable with distance.

The TI ranging module dynamically tailors the receive amplifier's sensitivity. Lower amplification is needed for close objects, higher amplification for distant echoes. Twelve gain steps within the range of 0 to 35 feet are automatically incremented as the time between INIT and ECHO lengthens. If the distance is 6 feet, the receiver will be at its second gain setting. At 20 feet it will be at its sixth level. The twelfth level is sufficient for unaided reception of echoes from 35 feet. The addition of a direction cone to the transducer, which will further improve sensitivity, will facilitate measuring distances beyond 35 feet.

Computer Connection

Figure 6 is the schematic for connecting the ranging module to the parallel port of a computer. The entire single-echo-mode interface requires only 1 input bit and 1 output bit. The output bit connects to the INIT line; the input bit connects to the ECHO line. To measure distance, the computer merely raises the INIT line and measures the time until ECHO goes high. The repetition rate depends on the distance. The cycle is repeated when INIT is lowered and raised again. This can occur at any point after ECHO. For short distances, a 100-Hz repetition rate is possible. To allow a full 35-foot range, however, the repetition rate should be limited to 10 Hz.

Unlike the Polaroid Designer's Kit, which is specified to run on 6 V, the TI ranging module can function between 4.5 and 6.8 V. If you use a 5-V supply, the ranging module I/O is TTL

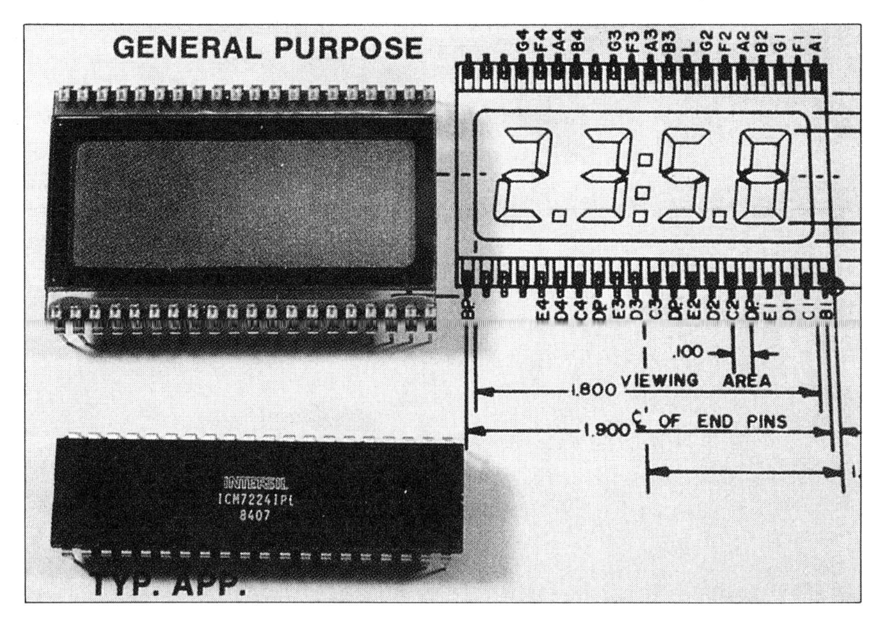

Photo 6: *With the addition of the 4-digit LCD, such as the UXD unit shown, and a 4½-digit 7224 LCD counter/decoder/driver chip from Intersil, a hand-held tape measure can be easily constructed.*

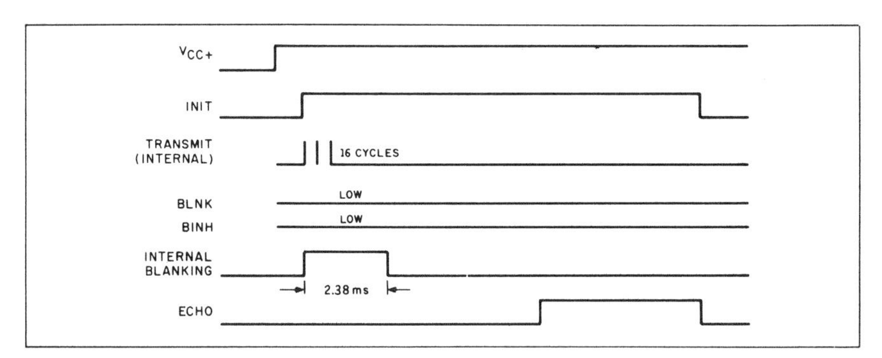

Figure 5: *A timing diagram of the sonar-ranging module.*

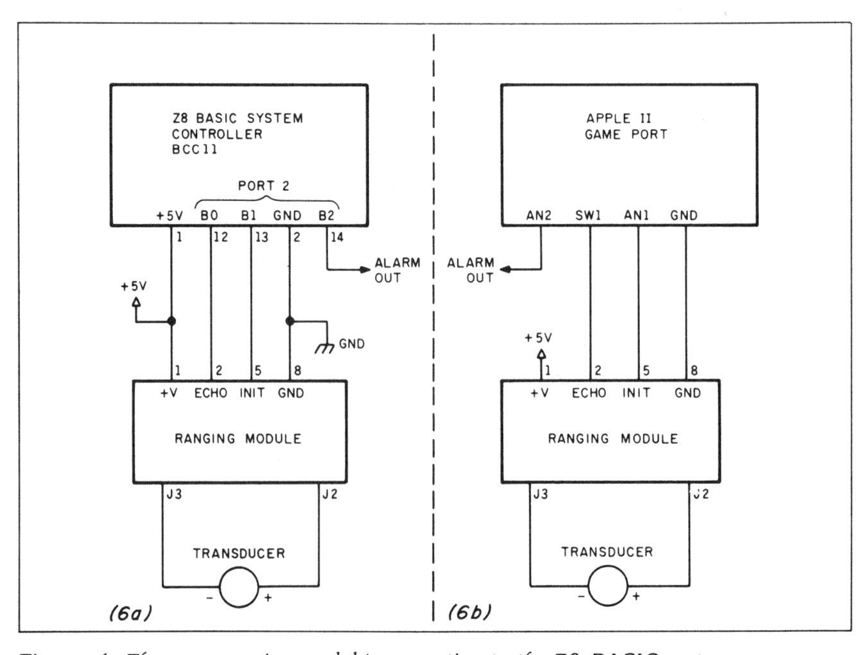

Figure 6: *The sonar-ranging module's connection to the Z8 BASIC system controller (6a) and to the Apple II system controller (6b).*

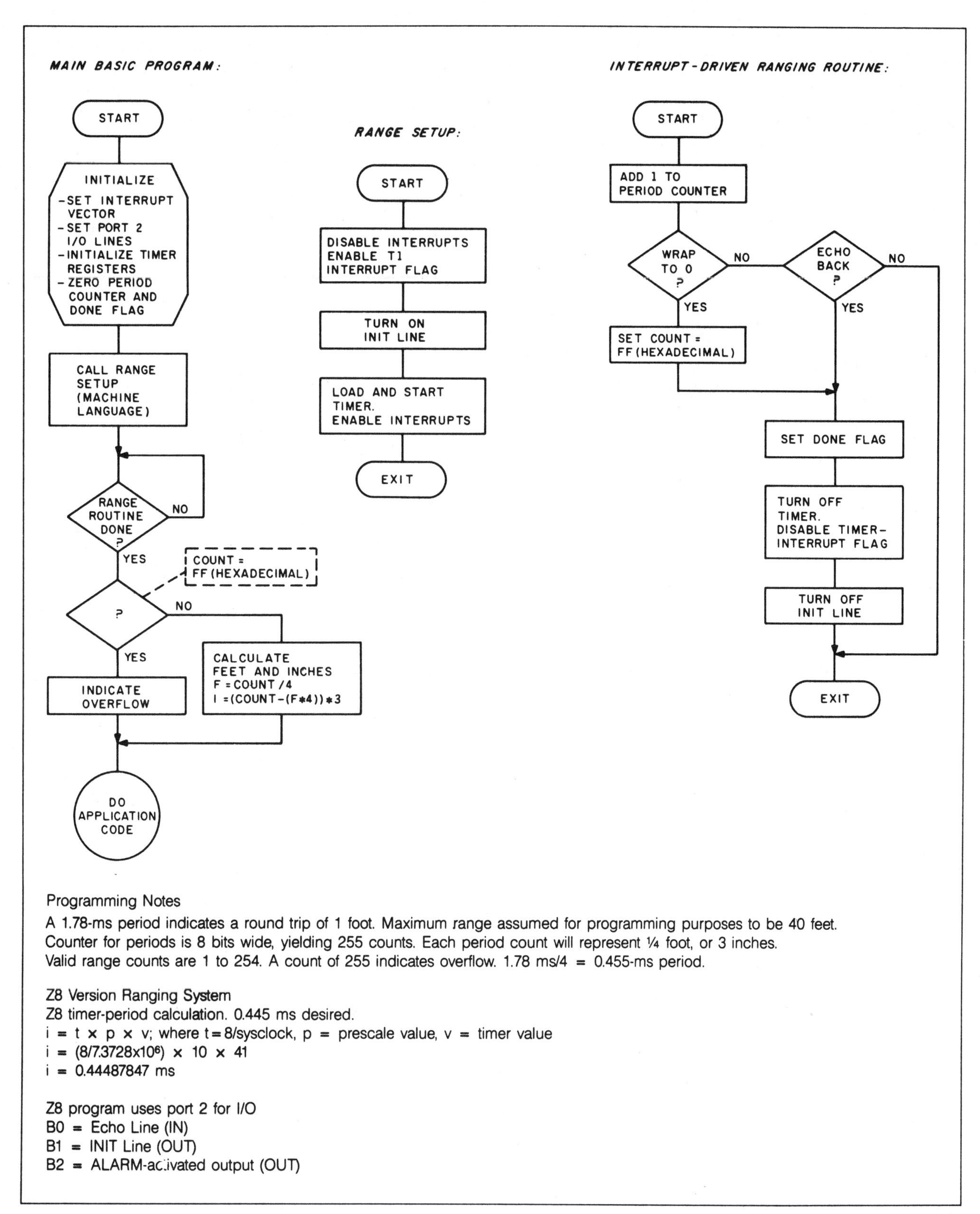

Figure 7: A *flowchart of the range-finder program for the Z8 controller.*

(transistor-transistor logic) compatible and can be connected directly to most computers. The module normally requires about 100 milliamperes (mA) except during the 16-cycle transmission period, when it can reach 2 amperes (A). Any power source intended for use with the module should have an intermittent power rating this high. For portable operation, you can use either standard AA or larger alkaline batteries or Polaroid's Polapulse high-current batteries.

For the purposes of this article, I chose to demonstrate connection of the ranging module to an Apple II and the Z8 system-controller board, which is an improved descendant of the original Z8-based computer controller presented in the June and July 1980 Circuit Cellar articles. The description of connecting the ranging module to the Z8 system controller through a parallel I/O port, measuring the elapsed time in a callable machine-language routine, and manipulating the results within the tiny BASIC interpreter are transferable in principle to any computer. (I have since had a Z8 FORTH chip produced that is interchangeable with the Z8671 tiny BASIC chip. The FORTH chip could also have been used.) By using the Z8 board, however, I can produce a dedicated measuring system that can sense and average a number of readings, make decisions, activate specific control outputs, and communicate readings to larger host systems serially.

Figure 7 is a block diagram of the necessary steps to initialize the module and measure distance with the Z8. Figure 8 is a block diagram of the Apple II version of the same code. Figure 9 is the block diagram of a typical proximity-detector application with alarm outputs. Listings 1, 2, and 3 show the code to perform these tasks.

Electronic Tape Measure

While describing the electrical characteristics and computer connection of the ranging module might ordinarily suffice as a Circuit Cellar project, I was intrigued by the simplicity of using the ranging module and decided to build the electronic tape measure shown in photos 4 and 5. The circuit shown in figure 10 required only three additional CMOS (complementary metal-oxide semiconductor) chips and can be constructed as a hand-held device.

The Circuit Cellar Sonar Tape Measure, hereafter called SonarTape, consists of the TI ranging module, a CD4049 inverter, a CD4029 counter, and an Intersil ICM7224 4½-digit LCD counter/decoder/driver chip (shown in photo 6). When pointed at an object or a wall and activated, the SonarTape transmits once and holds its reading (so that you can turn the unit and see the display if it wasn't already in view). The LCD indicates the distance in feet and tenths of feet.

The SonarTape is activated by turning on the power (6 V provided by

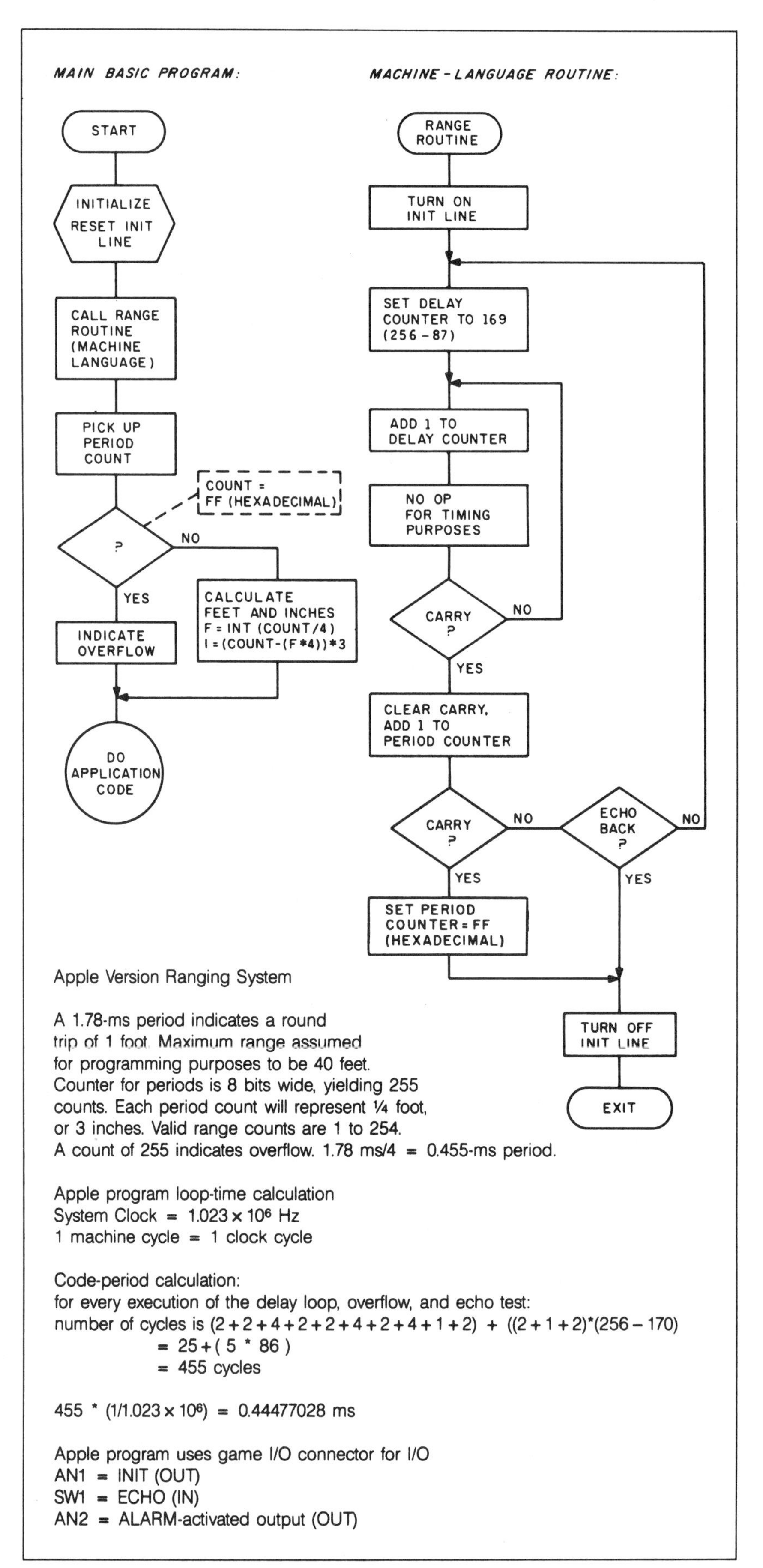

Figure 8: *A flowchart of the range-finder program for the* Apple II.

four AA alkaline cells) to the circuit through a momentary push button. A resistor/capacitor timing circuit connected to V_{cc} and attached to pin 3 of IC1 resets IC3, presets IC2, and keeps the INIT input line to the ranging module low until the 5-ms power-on stabilization time has passed. When the capacitor charges up to a logic 1 level, the INIT line goes to a logic 1, allowing the ranging module to transmit.

Once the INIT line goes high, the clock output, CLOCK, from the ranging module is enabled, and a 93.333-kHz clock (after the first 16 cycles at 49.4 kHz) is presented to the clock input of IC2. Configured as a divide-by-16 counter, the carry output, pin 7, will be at 5.83333 kHz. This frequency is connected to the count input of the counter/decoder/driver chip, IC3, which continues to count input pulses (started when INIT went high) until the ECHO line, connected through an in-

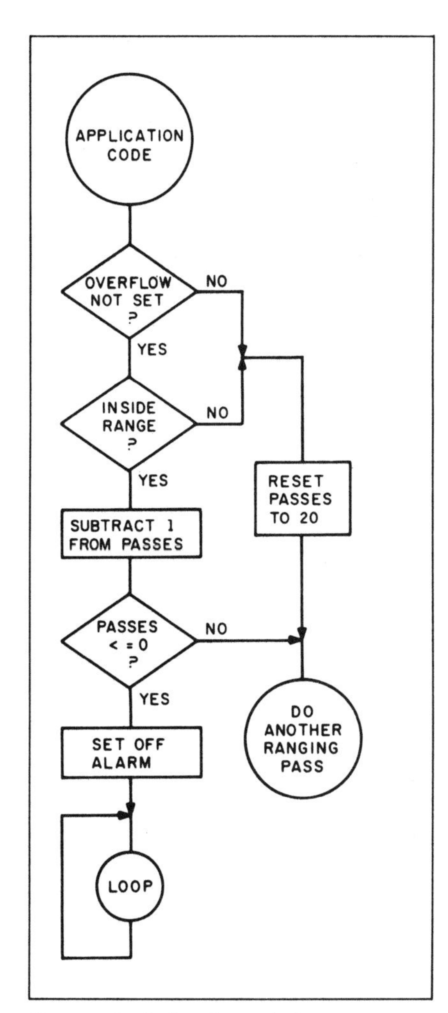

Figure 9: *A flowchart of the alarm-monitor program.*

Listing 1: *The Z8 programs to initialize the ranging module, measure the distance, and print the results. The BASIC program (1a) initializes the module and calls the machine-language routine, represented here in assembly language (1b). The machine-language program oversees the distance measurement. Finally, the BASIC program prints the results (1c). Remember to delete the REM statements.*

(1a)

```
1 REM Z8 BASIC PROGRAM FOR RANGE FINDER
10 @%100F=%8D:@%1010=%15:@%1011=%0C   :REM SET UP INTERRUPT VECTOR
20 @246=%F9:@2=0                      :REM SET UP PORT 2 I/O. RESET INIT LINE
30 @243=%2B:@242=41                   :REM SET UP TIMER REGISTERS
40 @%30=0:@%31=0                      :REM ZERO COUNTER AND DONE FLAG
50 GO @%1500                          :REM CALL RANGE-SETUP ROUTINE
60 REM LOOP UNTIL DONE FLAG SET
70 IF @%31=0THEN60                    :REM TEST DONE FLAG AND LOOP
80 C=@%30                             :REM PICK UP COUNT VALUE
90 F=0:I=0:O=0                        :REM CLEAR FEET, INCHES, OVERFLOW
100 IF C=%FF THEN O=1:GOTO 130        :REM SET OVERFLOW IF COUNT=%FF
110 F=C/4                             :REM CALCULATE FEET
120 I=(C-(F*4))*3                     :REM AND INCHES
130 REM APPLICATION CODE STARTS HERE
140 IF O=1 THEN PRINT "OVERFLOW":GOTO 40
150 PRINT F; " FEET ";I;" INCHES"
160 PRINT C; " .445-MS PERIODS COUNTED"
170 GOTO 40
:
:
:
:NOTE: DO NOT ENTER THE :REM STATEMENTS, OR THE PROGRAM WILL NOT RUN.
```

(1b)

```
0010 ;
0020 ; Z8 VERSION RANGE-FINDER ROUTINES
0030 ;
0040 ; DRIVE INIT LINE AND SET UP CLOCKS
0050 ;
0060        ORG    1500H
0070 INIT   EQU    $
0080        DI                   ; SHUT DOWN INTERRUPTS
0090        OR     IMR,#20H      ; ENABLE IRQ 5 (TIMER 1)
0100        LD     02H,#02       ; TURN ON INIT LINE
0110        OR     TMR,#0CH      ; LOAD AND START TIMER 1
0120        EI                   ; ENABLE INTERRUPTS
0130        RET                  ; RETURN TO BASIC
0140 ;
0150 ;
0160 ; THIS IS THE INTERRUPT-DRIVEN RANGING ROUTINE
0170 ;
0180 ;
0190 RANGE EQU     $
0200        INC    30H           ; ADD 1 TO PERIOD COUNTER
0210        JR     NZ,ECHO       ; IF NO WRAP, CHECK ECHO
0220        LD     30H,#0FFH     ; SET COUNT TO %FF
0230        JR     DONE          ; AND SHUT DOWN RANGE ROUTINE
```

(continued)

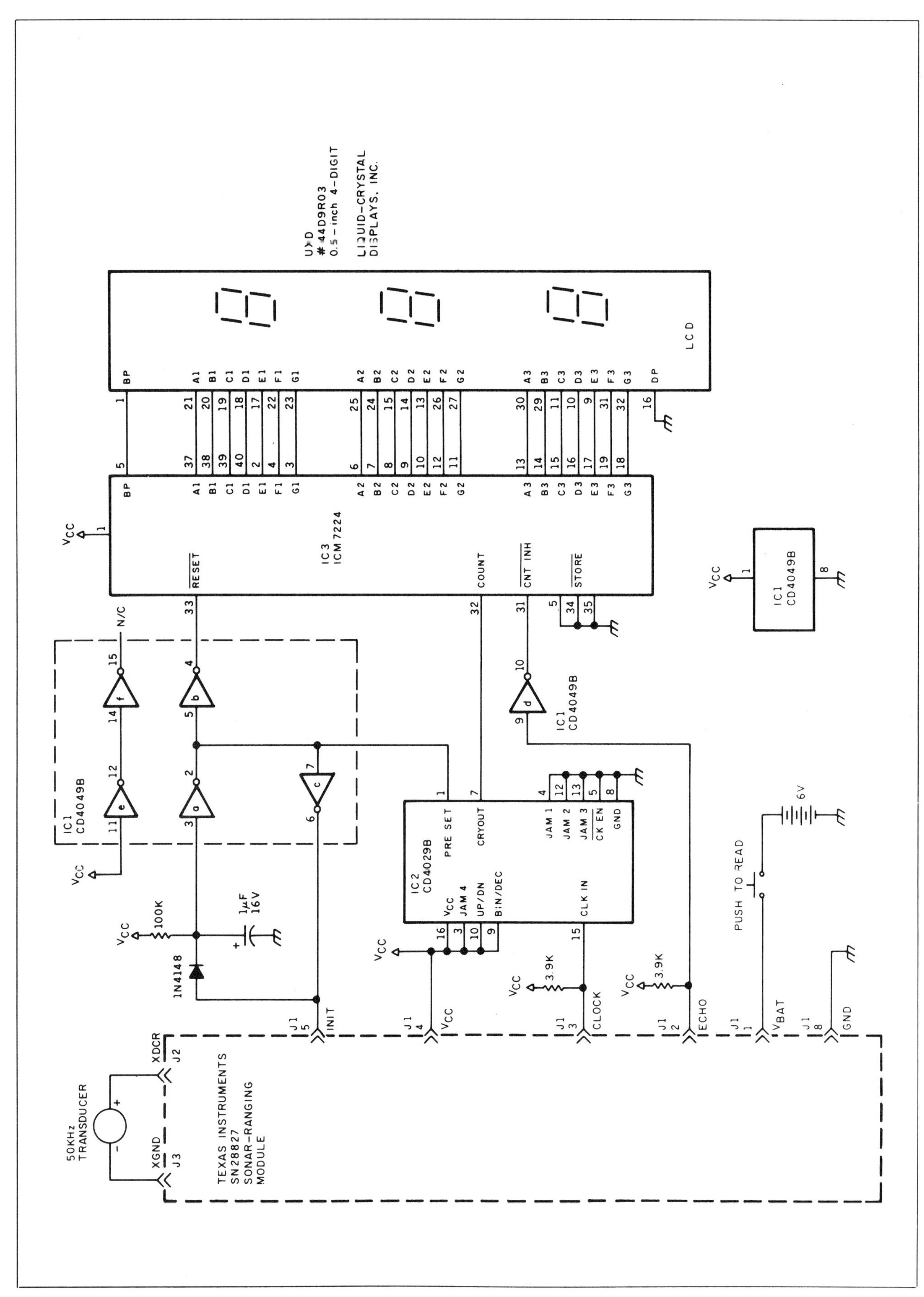

Figure 10: *The schematic of the SonarTape.*

A sonar sensor would let a sightless person "hear" a picture of the environment.

verter to the count-inhibit line on IC3, goes high, indicating reception of the echo. Clock pulses will therefore have been counted during the elapsed time period between INIT and ECHO.

If the elapsed time were 6 ms, about 34 clock cycles would have been counted by IC3. With the decimal point permanently set between the least and next most significant digit, the indication on the LCD would have been 3.4 feet. Similarly, 31 ms of elapsed time would allow about 175 cycles to occur, which would indicate 17.5 feet.

In Conclusion

I did think about commercial applications for SonarTape. Just because I'd like one, however, doesn't mean it has any potential. The high cost of the additional circuitry, packaging, and LCD make it either a high-priced novelty or cost-effective only in mass production. I'll stick with my prototype and wait for the next development iteration so that I can cover this subject in another four years.

One additional application that I started prototyping but discontinued as not completely germane to this article was a blind sonar sensor. Such a device (for the blind or any sightless application, such as firemen entering a smoke-filled room) would conceivably allow a sightless person to "hear" a picture of the environment by sweeping a hand-held ranging unit in front as he or she walked. Rather than a visual display, the distance measurement would be a tone whose frequency was a function of distance. My initial design consisted of continuously triggering the INIT line with a 5-Hz oscillator and applying the INIT and ECHO signals to an AND gate to produce a single pulse output whose pulse width was the elapsed time between INIT and ECHO. Next, using an integrator circuit, or even a simple RC (resistor-capacitor) combination, the pulse width is converted to a DC voltage level. The longer the pulse width, the higher the DC level.

```
0240 ECHO   EQU  $
0250        TM   02H,#01      ; IS ECHO BIT ON?
0260        JR   Z,EXIT       ; NO, LET ANOTHER PASS OCCUR
0270 DONE   EQU  $
0280        LD   31H,#0FFH    ; TURN ON DONE FLAG
0290        AND  TMR,#0F3H    ; TURN OFF T1 TIMER
0300        AND  IMR,#1FH     ; DISABLE IRQ 5 INTERRUPT
0310        LD   02H,#0       ; SHUT OFF INIT LINE
0320 EXIT   EQU  $
0330        IRET              ; RETURN FROM INTERRUPT
0340 ;
0350 ; END OF MACHINE-LANGUAGE ROUTINES
0360 ;
0370 ZZZZ   EQU  $
```

(1c)

```
 RUN
32 FEET 3 INCHES
129 0.445-MS PERIODS

32 FEET 3 INCHES
129 0.445-MS PERIODS
```

Listing 2: *The Apple II versions of the programs in listing 1. The output is the same as for the Z8.*

(2a)

```
1 REM APPLE II VERSION RANGING SYSTEM
10 POKE -19293,0: REM SHUT OFF INIT LINE
30 PRINT CHR$ (4);"BLOAD RANGE.0": REM LOAD OBJECT PROGRAM
40 POKE 12328,0: REM CLEAR COUNT
50 CALL 12288: REM CALL RANGING SUBROUTINE
60 C = PEEK (12328): REM GET COUNTER VALUE
70 F = 0:I = 0:O = 0: REM CLEAR FEET, INCHES, OVERFLOW
80 IF C = 255 THEN O = 1: GOTO 120: REM SET OVERFLOW IF
   COUNT=$FF
90 F = INT (C / 4): REM CALCULATE FEET
100 I = (C - (F * 4)) * 3: REM AND INCHES
120 REM APPLICATION CODE STARTS HERE
130 IF O = 1 THEN PRINT "OVERFLOW": GOTO 40
140 PRINT F;" FEET ";I; " INCHES"
150 PRINT C;" .455-MS PERIODS": PRINT
160 GOTO 40: REM AND MAKE ANOTHER PASS
```

(2b)

```
0010 *----------------------------------------------------------------
0020 * APPLE II  VERSION RANGING SYSTEM
0030 *----------------------------------------------------------------
0040 *
0050        .OR  $3000
0060 RANGE  .EQ  *
0070        LDA  $C05B      * TURN ON INIT LINE
0080 DELAY  .EQ  *
0090        LDA  #184       * SET DELAY COUNT TO 184
0100 LOOP   .EQ  *
0110        ADC  #1         * ADD 1 TO DELAY COUNTER
0120        NOP             * NO OPERATION FOR TIMING
0130        BCC  LOOP       * LOOP UNTIL CARRY
0140        CLC             * CLEAR CARRY FLAG
0150        LDA  COUNT      * PICK UP OLD COUNT
0160        ADC  #1         * ADD 1 TO IT
```

(continued)

```
0170          STA  COUNT    * AND SAVE IT
0180          BCC  ECHO     * IF NO WRAP, CHECK ECHO
0190 * WRAP OF PERIOD HAS OCCURRED. SET OVERFLOW.
0200          LDA  #$FF     * GET A $FF
0210          STA  COUNT    * WRITE TO COUNTER
0220          JMP  EXIT     * AND LEAVE THIS ROUTINE
0230 ECHO .EQ  *
0240          LDA  $C061    * PICK UP ECHO BYTE
0250          AND  #$80     * TEST ECHO BIT
0260          BEQ  DELAY    * IF NO ECHO, DO ANOTHER LOOP
0270 EXIT  .EQ  *
0280          LDA  $C05A    * SHUT OFF INIT LINE
0290          RTS           * RETURN TO BASIC
0300 *----------------------------------------------------------------
0310 COUNT     .BS $1       * COUNTER DATA BYTE
0320 *----------------------------------------------------------------
0330 *----------------------------------------------------------------
0340 * END OF RANGE SUBROUTINE
0350 *----------------------------------------------------------------
0360 ZZZZ  .EQ  *
```

Listing 3: *To modify the Z8 program in listing (1a) for the alarm application, add the* BASIC *statements in (3a), except for the* REM *statements. To modify the* Apple II *program in listing (2a) for the alarm application, add the* BASIC *statements in (3b).*

(3a)

```
5 R=120:P=20                           :REM SET RANGE AND
                                        NUMBER OF PASSES
140 IFO=1THENP=20:GOTO40               :REM RESET PASSES IF
                                        OVERFLOW
150 IF ((F*12)+I)>R THEN P=20:GOTO 40  :REM RESET PASSES IF
                                        OUTSIDE RANGE
160 P=P-1                              :REM PASSES=PASSES-1
170 IF P>0 THEN 40                     :REM IF PASSES NOT DONE,
                                        TRY AGAIN
180 REM OBJECT WITHIN RANGE FOR NUMBER OF PASSES. ALARM
190 @2=4                               :REM SET ALARM BIT
200 REM LOOP
210 GOTO 200                           :REM LOOP FOREVER
:
:
:
:AS BEFORE, DO NOT ENTER THE :REM STATEMENTS
```

(3b)

```
2 REM TO MODIFY THE APPLE PROGRAM FOR THE APPLICATION, ENTER
  THE FOLLOWING.
3 R = 120:P = 20: REM SET RANGE TO 10 FEET, PASSES TO 20
20 POKE -16291,0: REM SHUT OFF ALARM
80 IF O = 1 THEN P = 20: GOTO 40: REM RESET PASSES IF OVERFLOW
90 IF ((F * 12) + 1) > R THEN P = 20: GOTO 40:
   REM RESET PASSES IF OUTSIDE RANGE
100 P = P - 1: REM PASSES = PASSES -1
110 IF P > 0 THEN GOTO 40: REM IF NOT THROUGH ALL PASSES, DO
    ANOTHER
120 POKE 16292,1: REM TURN ON ALARM
130 REM LOOP HERE
140 GOTO 130
```

When ECHO goes high, the integrator output would be sampled and held (until the next ranging sample) with a sample-and-hold circuit. The output of the sample-and-hold circuit is connected to a voltage-controlled oscillator such as the XR4151 or XR2206. Short distances produce low tones; long distances result in high tones. A little bit of experience recognizing tone patterns should allow any of us to walk without seeing.

I chose not to pursue the design at this time. If any of you build a working unit, however, I'd like to know about it. Among the hundreds of letters I receive each month are some from readers who might benefit by the information.

Finally, I'm not ending this article by discussing all the possible applications for the TI ranging module. The number is so great that they are impossible to list. Now that you know the unit exists and how to attach it to a computer, perhaps you'll demonstrate some novel uses. For me, it's on to the next project.

See page 447 for a listing of parts and products available from Circuit Cellar Inc.

23

UNDERSTANDING LINEAR POWER SUPPLIES

Proper design brings simplicity and reliability

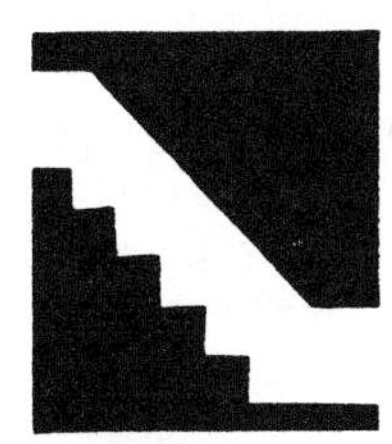

Sometimes it is the more trivial aspects of electronics that create the greatest problems. With all the concern about 16-bit versus 32-bit processors and multitasking operating systems, who would think that a simple linear power supply could cause the demise of a company? Even with a board of directors full of venture capitalists, it's often too late when they look beyond their spreadsheet projections and ask whether the product they are financing actually works.

While looking over someone's shoulder is not my favorite consulting activity, I recently was involved in such a situation. One of the founders of a venture-capital-funded company was getting very nervous because his engineering department was seriously overdue on two products. Since his responsibility was sales, and about $800,000 in pending orders was riding on cost-effective delivery of these products, it was no wonder that he was concerned. One of the products was way over budget, and the other seemed to have a "heat problem." My job was to determine if there was a problem and help rectify it if possible.

The first product was a speech synthesizer that attached to a parallel printer port. Its problem was "engineering buzzword injection phenomenon." Inexperienced engineers try to impress management by designing microprocessors into products that don't need them.

The synthesizer chip required parallel data and a strobe. It signified that it needed more data with a single ready line. Instead of merely attaching the chip directly to the printer port and pretending it was a printer (attaching printer busy to the ready line), the engineer had added a mask-programmed microprocessor, external character-buffer memory ("In case the programmer wanted it," he said.), external program memory (in case the mask-programmed chips didn't arrive in time and they had to use EPROMs [erasable programmable read-only memories]), and a parallel port for the synthesizer chip. Direct connection to the printer port (without the microprocessor) afforded a 75 percent cost reduction.

The second product, which I'll call E, was a stand-alone speech-and-music synthesizer board that communicated serially with the host computer. To make a long story short, I was called in to look at E after 5000 sets of components and printed-circuit boards had been purchased but nothing had been shipped. It was now four months overdue.

I agreed on the basic design method. Because of the data in serial format and multiple peripheral chips, this device did require a microprocessor. Unfortunately, it suffered from another common ailment among inexperienced designers: "three-terminal-regulator narcosis." This occurs when you read the manufacturer's spec sheets on a three-terminal regulator and use the information without understanding it or the other elements in the power supply.

My first experience with the E product almost burned my nose. I leaned over an operating prototype to make a closer inspection and sensed intense heat rising from the power-supply section. According to the designer, everything was within the manufacturer's specifications. The board needed +12 and +5 volts (V) at 0.5 ampere (A), which was regulated down from an 18-V (plus 10 percent ripple) rectifier output. The 90° Celsius (C) case temperature was "warm," but the engineer hotly contested that everything was okay. When further queried about added heat once the unit was enclosed, he assured me that it still wouldn't exceed the manufacturer's specified limit of 150°C (apparently he didn't know the difference between junction and case temperature).

I come from the school of design that says, "If you can't touch it, you've got big problems." Eventually, my greatest fears were realized. E boards were installed in ABS (acrylonitrile-butadiene-styrene) thermoformed plastic cases and allowed to burn in (aptly named) overnight. When inspected in the morning, the tops of many cases had melted and deformed. In addition, many of the regulators had failed and were now inoperable. At this point, the bad design could not be hidden from the venture capitalists. Even marketing concurred, "We can't ship an incendiary device to every kid with a computer!"

Linear Power Supplies

Virtually all electronic equipment operates on a DC power supply. This DC voltage can come from a battery or can be converted from an energy source such as the AC power line. The two commonly used conversion methods are switching and linear.

The advent of easy-to-use three-terminal regulators has given designers a false sense of security. Because of the wide operating limits and built-in protection of many of these monolithic regulators, brute force and a rule-of-thumb design technique can still result in functional power supplies. It takes resourceful and knowledgeable designers who understand the interrelationships of power-supply components to produce efficient and cost-effective products. In the case of E, larger heatsinks or lower input voltages would have offered some after-the-fact relief. Some better understanding of linear power supplies and a bit of initial computation would have resulted in a shippable design in the first place, however.

My Circuit Cellar projects range from the esoteric to the instructive. I do, however, presume that the builders of these projects have a certain level of basic understanding and that many hours of construction won't go up in smoke because of a poorly designed power supply. This recent experience has made me hesitate to be quite so presumptuous, and I will now add linear power supplies to my periodic tutorial subject list (with speech synthesis, home control, etc.).

This month I'll go back to basics and analyze the construction of linear power supplies. I'll describe transformer selection, input-filter design, regulator selection and connection, heatsinking, and layout. I will particularly emphasize the filter, heatsinking, and layout. Most articles seem to

CONFIGURATION	CIRCUIT	V_{IN}	$V_{O(PEAK)}$	AVERAGE V_O DC	PEAK INVERSE VOLTAGE PER DIODE	FUNDAMENTAL OUTPUT RIPPLE FREQUENCY 60Hz	OUTPUT WAVEFORM
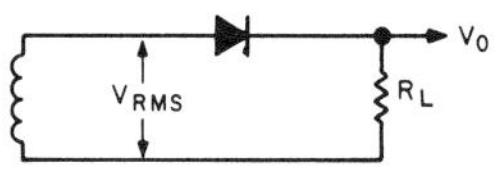 V_{RMS}, R_L, V_O	SINGLE-PHASE HALF WAVE	V_{RMS}	$1.41V_{RMS}$	$\frac{1}{\pi}V_{O(PEAK)}=0.45V_{RMS}$	$1.41V_{RMS}$ ($2.82V_{RMS}$ WITH FILTER)	$1F_L$	0VDC
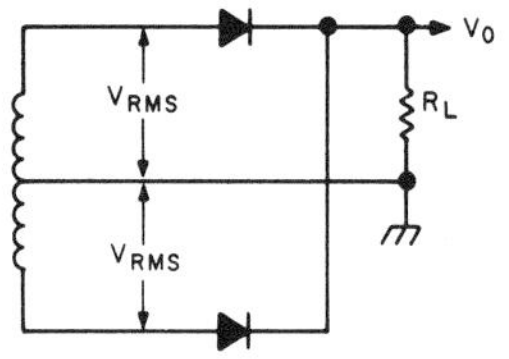 V_{RMS}, V_{RMS}, R_L, V_O	SINGLE-PHASE CENTER-TAP FULL WAVE	V_{RMS}	1.41_{RMS}	$\frac{2}{\pi}V_{O(PEAK)}=0.90V_{RMS}$	2.82_{RMS}	$2F_L$	0VDC
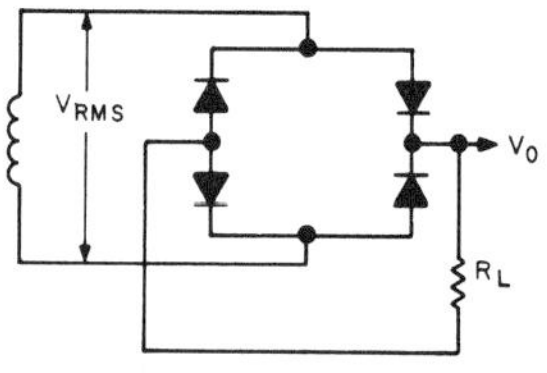 V_{RMS}, R_L, V_O	SINGLE-PHASE BRIDGE FULL WAVE	V_{RMS}	$1.41V_{RMS}$	$\frac{2}{\pi}V_{O(PEAK)}=0.90V_{RMS}$	$1.41V_{RMS}$	$2F_L$	0VDC

Figure 1: *Three single-phase transformer-rectifier configurations.*

overlook these items while they discuss various regulator configurations. If, after building a supply from such a slanted article, the end product is in thermal shutdown most of the time because of naive filter design, you are better off reading comic books. I believe that the construction of power supplies isn't difficult, but perhaps no one has ever described how to do it. Hopefully, the process will become easy after reading this article.

STARTING WITH THE BASICS

Generally speaking, a basic single-phase linear power supply consists of little more than a transformer, rectifier, and filter. Where it is necessary to accurately maintain the output potential, a voltage-regulator circuit is added. More precisely, the four components function as follows:

1. A *transformer* isolates the supply from the power line and reduces the input voltage (120 or 220 V AC) into usable low-voltage AC.
2. A *rectifier* converts the AC to a DC waveform and satisfies the charging-current demands of the filter capacitor.
3. A *filter* capacitor maintains a sufficient voltage level between charging cycles to satisfy the minimum voltage requirements of either the load directly or a voltage regulator attached to the load.
4. A *regulator* maintains a specific output voltage over various combinations of input voltage and load.

The three basic forms of single-phase transformer-filter circuits are half-wave, full-wave bridge, and full-wave center-tap. The terms "half-wave" and "full-wave" refer to the AC-input waveform. In a half-wave circuit, only half of the 360-degree input voltage is applied to the load. In a full-wave circuit, the full 360 degrees is usable. Figure 1 shows their configurations and relationships.

The first consideration to be made in the transformer choice is the type of circuit configuration: full-wave center-tap or full-wave bridge. The half-wave rectifier is generally used only for low-current or high-frequency applications since it requires twice the filter capacitance to maintain the same ripple as a full-wave rectifier.

Both the center-tap and bridge configurations have their own merits. The center-tap circuit dissipates less power, requires less space, and is potentially more economical than the bridge because it uses only two (as opposed to four) rectifier diodes. Using only two diodes, it has a lower impedance than a bridge circuit. However, for the same DC output voltage, the diodes must have twice the PIV (peak inverse voltage) rating. And since diodes are inexpensive, there is less real economy in using center-tap transformers. Their selection often results more from finding available transformers with the proper secondary voltages for a particular application.

A 120-V AC RMS (root mean square) sine wave is applied to the primary winding of the transformer. A similar lower-voltage waveform is produced at the secondary windings. This AC output voltage is then applied to a full-wave bridge rectifier of the form described in figure 1.

Since we are dealing with actual components and not theoretical examples, it is important to note that different output voltages will be produced from bridge and center-tap circuits, even though they may start with the same secondary potential. If you observe a full-wave rectifier output on an oscilloscope, you will note a period of nonconduction at every zero crossing. Real diodes have an intrinsic voltage drop across them and dissipate power. For most low-current applications, this threshold voltage is about 0.6 V. At 5 A or more it is closer to 1.1 V.

Depending upon the configuration, one or two diode drops may be between the transformer and the filter capacitor. (Figure 2 shows the current flow through a bridge rectifier.) The voltage regulator requires a certain minimum DC input level to maintain a constant output voltage. Should the applied voltage drop below this point, output stability can be severely de-

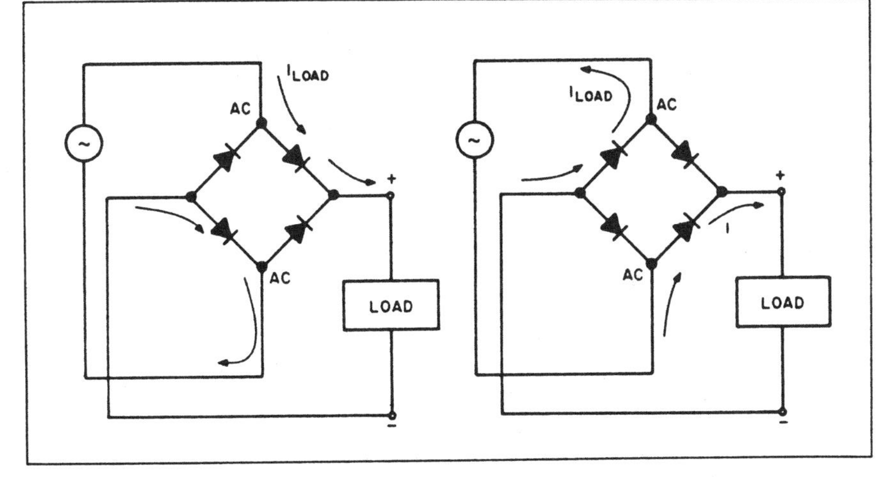

Figure 2: *The current flow through a full-wave bridge. Two diodes are conducting current at any one time. They present a voltage drop of 1 V each.*

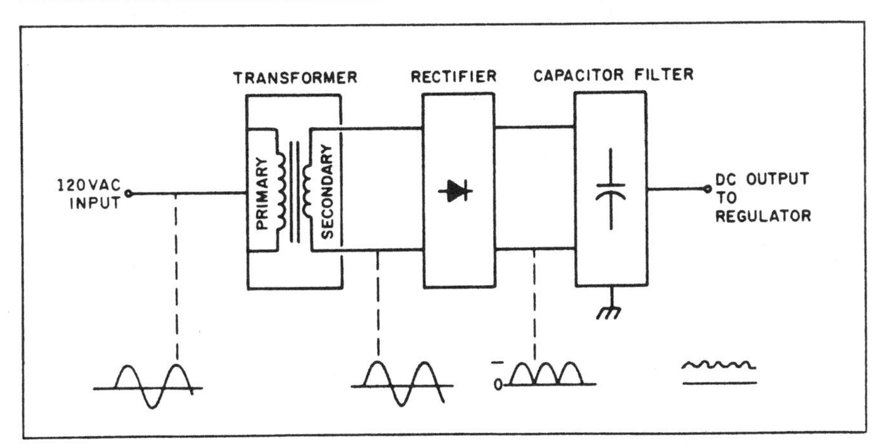

Figure 3: *The block diagram of a filtered power supply.*

graded. Where efficient low-dissipation designs are involved, these diode drops can be significant. In full-wave bridge designs, two diodes are in series at all times. The 2.2-V loss through the bridge is an important consideration that should be reflected in the calculations.

To smooth the rectifier output and help maintain a minimum input level to the voltage regulator, a filter capacitor is used (see figure 3). When the diodes are conducting, the capacitor stores enough energy to maintain the minimum voltage until the next charging cycle. With a 60-Hz transformer and a full-wave rectifier, the charging cycle has a frequency of 120 Hz. The capacitor must charge up in 8.3 milliseconds (ms) and maintain a sufficient level until the next charging cycle, 8.3 ms later.

The peak-to-peak magnitude of this periodic charge/discharge cycle is called the ripple voltage (V_{ripple}). The highest level of voltage including the ripple is called peak voltage (V_{peak}). Shown in figure 4, the ripple voltage should never be more than 10 percent of the steady-state voltage (V_c); V_c should never be less than the minimum input required by the regulator. Selecting the values of these components requires some calculation.

Learning by Doing

The best way to understand the interrelationships of the components in a linear power supply is to design one. For purposes of this discussion, let us design a whopping 12- to 13.2-V 15-A supply using LM338K regulators.

This choice is not arbitrary. Most articles on linear power supplies present low-current circuits that are relatively idiot-proof. At such low currents, simple rule-of-thumb practice is acceptable. It is only at high power levels that design knowledge, layout principles, and proper component selection are mandatory. Somewhere in between the 5-watt (W) bench supply you build from an article for hobbyists and the 198-W supply we will discuss is a gray area where experimenters who rely on luck and rule of thumb will be out of one or both.

I am presenting such a big supply for another reason. For quite some time I've been receiving letters regarding uninterruptible power supplies. While I can't guarantee that I can design one that is more cost-effective than a commercial unit, I intend to investigate the options. To aid in this task, I need a high-current DC supply for the initial experiments. Rather than dragging in car batteries (a "12-V" car battery actually produces 13.2 V) and chargers, I thought I'd build a high-current supply that demonstrates something for this project and can be put to good use later.

198 Watts!

You might think that specifying the transformer's secondary voltage would be the first consideration when building a power supply. Yes and no. While it is important, the choice of components in a power supply is interrelated. Too great an emphasis on one component over another can greatly influence cost and performance. The approximate secondary voltage can be determined by certain logical rules, but the exact requirements are deduced only by a thorough analysis that begins at the final power-supply output voltage and proceeds backward. In practice, the advantages gained in laborious transformer calculations are of benefit only to those designers capable of specifying custom-wound transformers. The majority of designers will have to rely upon readily acquired transformers with standard output voltages. For greatest efficiency, the standard voltage should be as close to the calculated value as possible.

Given an understanding of the basic filter components at this stage, we can proceed to the case at hand: a 13.2-V 15-A supply. The regulator, which I'll discuss later, uses the LM338K chip. These units are variable-voltage output devices, in contrast to fixed-voltage output devices such as the LM340T-5 (5 V) or 7812 (12 V). For them to properly operate over a temperature from 0 to 70°C, the input voltage to the LM338K must be 3 V greater than the output set point. (Fixed-voltage output regulators by contrast can suffice with a 2- to 2.5-V difference.) To be on the safe side, I always plan a minimum of 3.5 V. If this I/O (input/output) differential is less than 3 V, regulation becomes unstable. For a 13.2-V output, therefore, V_c has to be at least 16.7 V. For a 5-V output, V_c should be 8.5 V minimum. (Too much input voltage creates a different problem. Any $V_{out}-V_{in}$ difference greater than 3 V simply generates heat and should also be avoided.)

Whatever the magnitude of V_{peak} and V_{ripple}, V_c must not drop below 16.7 V or the regulator may not work. If this supply, which operates at 115-V AC input, is to still function at 105 V AC, we must make sure that V_c is 16.7 V at 105 V AC. The 8.5 percent voltage rise to 115 V AC, however, will make V_c 18.2 V (and 20.5 V at 130 V AC). Going much above these values, while still satisfying the input criteria, will increase power dissipation substantially.

Thus far, we have calculated or assumed the following at 25°C and 115-V AC input:

V_c = regulator input voltage at 115 V AC = 18.2 V

V_{ripple} = 10 percent of V_c maximum = 1.8 V

$V_{peak} = V_c + V_{ripple} = 20$ V

I_{out} = output current = 15 A

V_{rect} = voltage drop across diode bridge = 2.2 V (two diodes)

Choosing the Transformer

We have determined the voltage drops across the various components and the minimum regulator input voltage. These values can be used to calculate the required RMS secondary output voltage as follows:

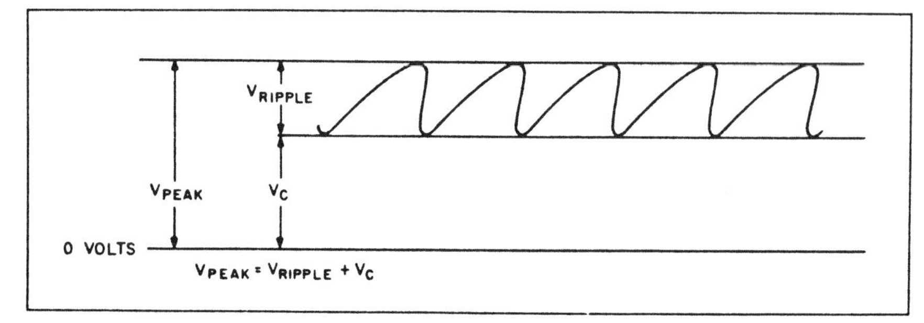

Figure 4: *The components of peak voltage,* V_{peak}, *are the steady-state voltage,* V_C, *and the ripple voltage,* V_{ripple}.

$$V_{sec,rms} = \frac{V_c + V_{ripple} + V_{rect}}{\sqrt{2}}$$

$$= \frac{18.2 + 1.8 + 2.2}{1.414}$$

$$= 15.70 \text{ V AC}$$

In practice, 15-A 15.70-V transformers aren't available off the shelf, but many 15- and 18-V units are available. While 16 V is the proper transformer secondary-winding selection, 15- and 18-V 10-A transformers will work fine if you can live with 10 A rather than 15 A. (One of these is part number F-62U from Triad-Utrad, 1124 East Franklin St., POB 1147, Huntington, IN 46750.) The 15-V unit will not give complete operation to 105 V AC, however, and the 18-V transformer will increase power dissipation in the regulator by 12.5 percent. (I don't recommend using 25.2-V center-tap transformers.)

I don't like presenting an optimized design and then apologizing for not taking my own advice. The calculations might seem a bit rigged when you find that I just happened to use a 16-V 15-A transformer to complete this project. I have to confess that I had it custom wound to meet the application. While I don't expect the average experimenter to resort to such expensive tactics, when I build a piece of test equipment, I want reliable and consistent operation. The final transformer secondary winding is 16 V AC at 15 A, and the secondary resistance is 0.04 ohm. (A 2-A 18-V center-tap transformer such as the Radio Shack 273-1515A by contrast is about 0.6 ohm.) With the additional wiring and connections between the transformer and the bridge, the source resistance is about 0.1 ohm.

Using a 16-V transformer, the true, as opposed to calculated, voltage levels are

V_{sec} = 16 V AC
V_{rect} = 2.2 V
$V_{peak} = (V_{sec} \times 1.414) - V_{rect} = 20.4$ V
V_c = 18.6 V
V_{ripple} = 1.8 V
R_s = transformer secondary resistance and resistance of connecting wires = 0.1 ohm

Sizing the Filter Capacitor

When the supply is turned on, a 120-Hz rectifier output is applied to the capacitor. The capacitor is large enough so that it can supply the full load current with only a negligible drop. If the capacitor is very near peak when the next charging cycle occurs, as would be the case with light loads and large capacitors, the diodes conduct for a very short time. The exact time during which the capacitor supplies current is fixed by the permissible peak-to-peak ripple voltage. This time:

$$T_{conduction} = (\theta/180)\ (8.33) \text{ ms}$$

where $\theta = [90 + \arcsin(V_c/V_{peak})]$ degrees.

For ripple voltages equivalent to 10 percent of V_{peak}, the filter-capacitor conduction time is 7.14 ms rather than 8.33 ms. For simplicity, however, it is often assumed that the capacitor must carry current for the full half cycle and 8.33 ms is used in the calculations. The capacitor value is chosen as follows:

$$C = (T_c/V_{ripple})\ I$$

where C = capacitance in farads (F) = ?, I = continuous output current = 15 A, T_c = charging time of capacitor = 8.33 ms, and V_{ripple} = allowable ripple voltage = 1.8 V. Plugging in the values:

$$C = \frac{(15)\ (0.00833)}{(1.8)}$$

$$= 0.069417 \text{ F}$$

or

$$= 69{,}000 \text{ microfarads } (\mu\text{F})$$

In the nearest commercial value, C = 75,000 μF at 25 V. Generally available commercial electrolytic capacitors have a tolerance of +50 percent and −10 percent. I chose to use a General Electric 86F543 75,000-μF 25-V (V_{peak} is 20 V) unit, but any capacitor of similar size will work. The ripple-current rating on capacitors of this size is also adequate.

Choosing the Bridge Rectifier

The four considerations when choosing a bridge rectifier are surge current, continuous current, PIV rating, and power dissipation. These parameters are generally ignored in rule-of-thumb designs because the 3- and 10-A diode bridges (which are generally available, coincidentally) have ratings

When I build a piece of test equipment, I want reliable and consistent operation.

that protect against bad designs. With 15-A power supplies, however, we should take nothing for granted. These specifications are not inconsequential and must be considered.

When a power supply is first turned on, the filter is totally discharged and for an instant appears as a dead short to the diode bridge. In this condition, the only thing that limits the current flowing through the bridge is the resistance in the secondary windings and the connecting wires. This sudden inrush is called surge current and is computed as follows:

$$I_{surge} = \frac{V_{peak}}{R_s}$$

$$= 20.4/0.1$$

$$= 204 \text{ A}$$

The time constant of the capacitor is

$$\tau = (Rs)\ (C)$$

$$= (0.1)\ (0.075000)$$

$$= 7.5 \text{ ms}$$

Generally speaking, power surges will not damage the bridge if the surge is less than its surge-current rating and if the time constant, τ, is less than 8.33 ms, which it is. A readily available bridge rectifier that fits the bill is the Motorola MDA990-2, which is rated at 30 A continuous and 300 A surge. Its PIV rating is 100 V, which is significantly in excess of our 22.6-V secondary peak.

One final consideration on the bridge is power dissipation. Since diodes exhibit voltage drops when current flows through them, they dissipate power just as the regulators do. The rule of thumb says that if I_{out} is 15 A and V_{rect} is 2.2 V, power dissipation (PD) is 33 W.

A possibility exists, however, that

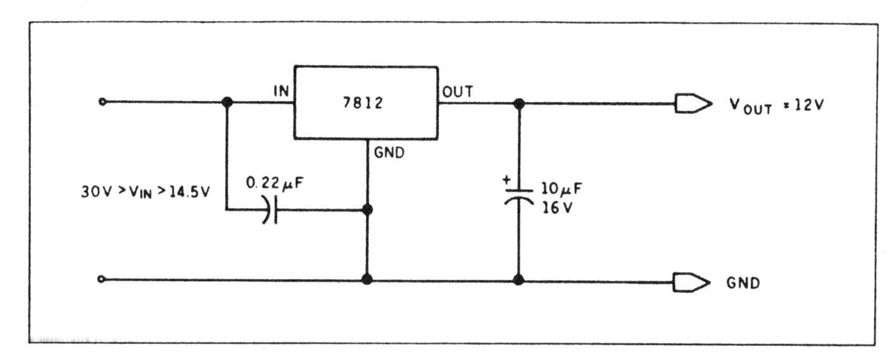

Figure 5: A *typical +12-V fixed-voltage output regulator.*

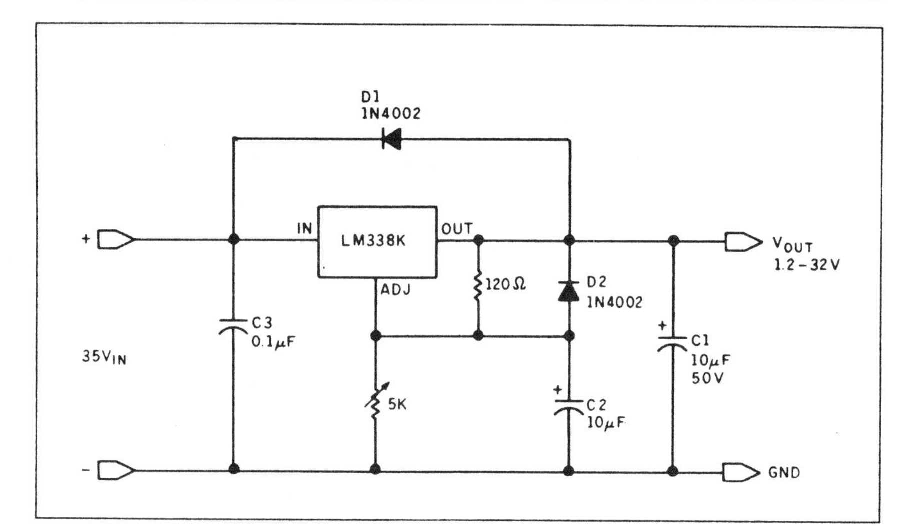

Figure 6: A *typical variable-voltage three-terminal regulator.*

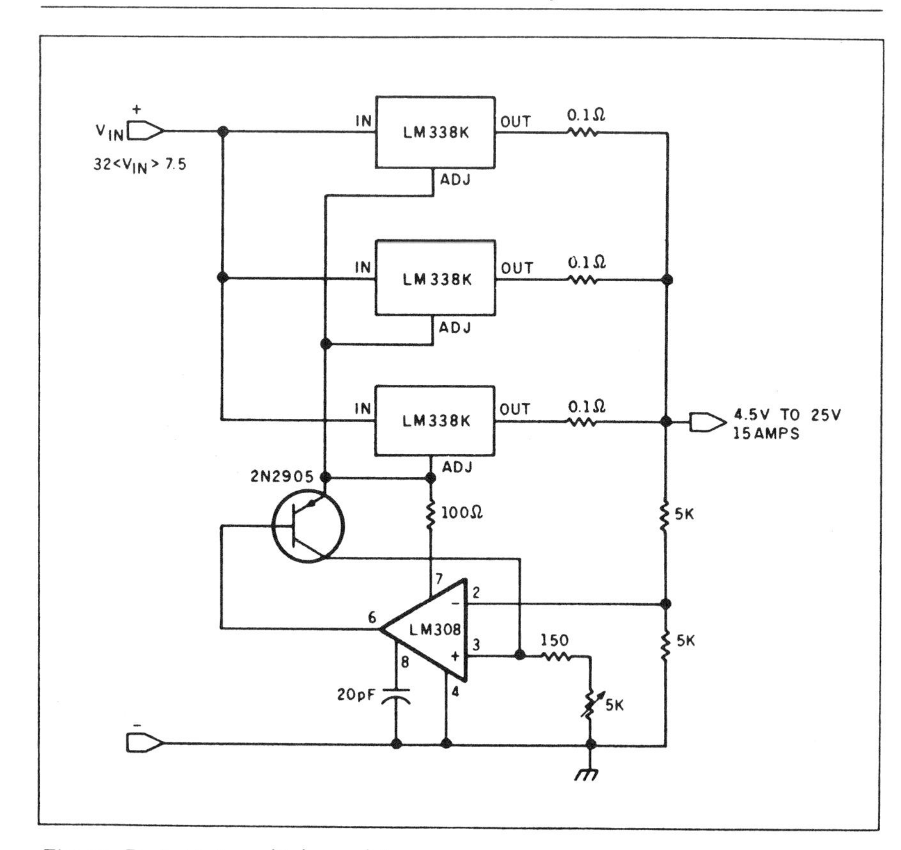

Figure 7: A 15-A *multiple regulator.*

the true (as opposed to the rule-of-thumb) PD could be significantly different and may or may not need extraordinary measures for heat removal. At the very least, you might like to know how to calculate real dissipation.

First, the only time the diode in the bridge conducts for a full cycle is immediately after turn-on. After that, it conducts only during that period when the input voltage is greater than V_c. In our supply, V_c is 18.6 V and V_{peak} is 20.4 V.

You'll remember that I previously said that the capacitor supplied power for 7.14 ms of every 8.33-ms charging cycle. The diode, therefore, conducts for the remaining 1.19 ms. (The time depends on the amplitude of the ripple). The average current flowing during this 1.19 msec (this current will not be constant) multiplied by 1.19 msec gives the charge passed by the diodes each half cycle. This charge must equal the output current times 8.33 msec. In other words, all the charge that passes through the load in a steady stream flows through the diodes in large pulses. Thus the peak current is

$$I_p = (8.33/1.19)15\ A = 105\ A$$

This high peak current would not change the average diode power dissipation if the voltage drop across the diode were unchanged; there would just be higher dissipation for a shorter time. The correction to the rule-of-thumb comes in because there will be a higher voltage drop across the diodes at this higher current. The relationship between two currents and the associated voltage drops for a power diode is roughly given by

$$V1 - V2 = 50\ mV|\ln(I1/I2)|$$

where ln indicates the natural logarithm.

In our case I1 = 105 A and I2 = 15 A, which leads to V1 − V2 = 0.1 volts. Thus the diode voltage drop we should use in the dissipation calculation is 1.2 volts per diode rather than 1.1 volts. So we find a power dissipation of 15 A multiplied by 2.4 V equals 36 W, or 9 W per diode.

Nine watts isn't too much. Simply mounting the bridge to the metal power supply enclosure should provide enough cooling. To know for sure, however, look up the bridge's rated junction temperature, keep the PD value in mind, and calculate the cooling requirements when we get to heatsinking.

Voltage Regulators

Once the filter section is configured, our next consideration is the voltage regulator. All linear regulators perform the same task: convert a given DC input voltage into a specific, stable, DC output voltage and maintain it over wide variations of input voltage and output load.

Entire books have been written on regulation circuits, and I think the subject material is adequate. The best sources are, in fact, the data manuals from the regulator-chip manufacturers. These manuals specify the I/O voltages and other specifications important to the power-supply designer. (Rather than go into the history and successive milestones in regulator evolution, I'm going to presume that you know a lot about this and want me to quickly get back to building a real supply.)

Three-terminal regulators come as fixed- or variable-voltage output devices (fixed-voltage output regulators can be configured to provide variable-voltage outputs). A typical +12-V fixed-voltage output regulator is shown in figure 5. The regulator has three terminals: in, out, and ground reference. In a 7812 regulator, $V_{out} - V_{in}$ should be 2.5 V; therefore, V_c should be 14.5 V for proper operation. The maximum input, disregarding power dissipation as a limiting factor, is 30 V.

If you want a 13.2-V supply, you would substitute a variable-voltage three-terminal regulator such as the LM338K shown in figure 6. Here, the three terminals become in, out, and voltage adjust. A potentiometer in the adjust line sets a reference level to the chip that determines its output voltage. This circuit also contains diodes to protect the regulator.

While manufacturers would like you to think otherwise, three-terminal regulators are not indestructible and can fail. One source of failure is the discharge of external capacitors through the regulator. For example, if the regulator output is shorted, C2 will discharge through the voltage-adjust pin. A diode, D2, diverts the current around the regulator protecting it. If the input is shorted, C1 can discharge through the output of the regulator, possibly destroying it. Diode D1 shunts the current around the regulator, protecting it. While such protection is merely insurance on hefty devices such as the LM338K, it is a necessity on lower-current regulators.

The LM338K can be adjusted for

Three-terminal regulators have been known to fail.

outputs from 1.2 to 32 V at 5 A, and devices can be paralleled to provide increased output current. Figure 7 outlines a circuit composed of three LM338Ks configured as a 15-A regulator that will satisfy the regulation requirements of the supply we are building. With a V_c input of 18.6 V at 115-V AC input, the supply is adjustable from about 4.5 to 15.6 V. Figure 8 is the final schematic of the unit.

Layout Is Important

Three-terminal regulators employ wideband transistors to optimize response. Unfortunately, stray capacitance and line inductance caused by poor layout can introduce oscillations and unstable operation into these circuits. Keeping lead lengths short, as shown in photo 1, and adding external bypass capacitors will limit the problems caused by the regulator. Builder-introduced problems are another matter entirely.

Figure 9 illustrates a typical three-terminal-regulator supply layout, in-

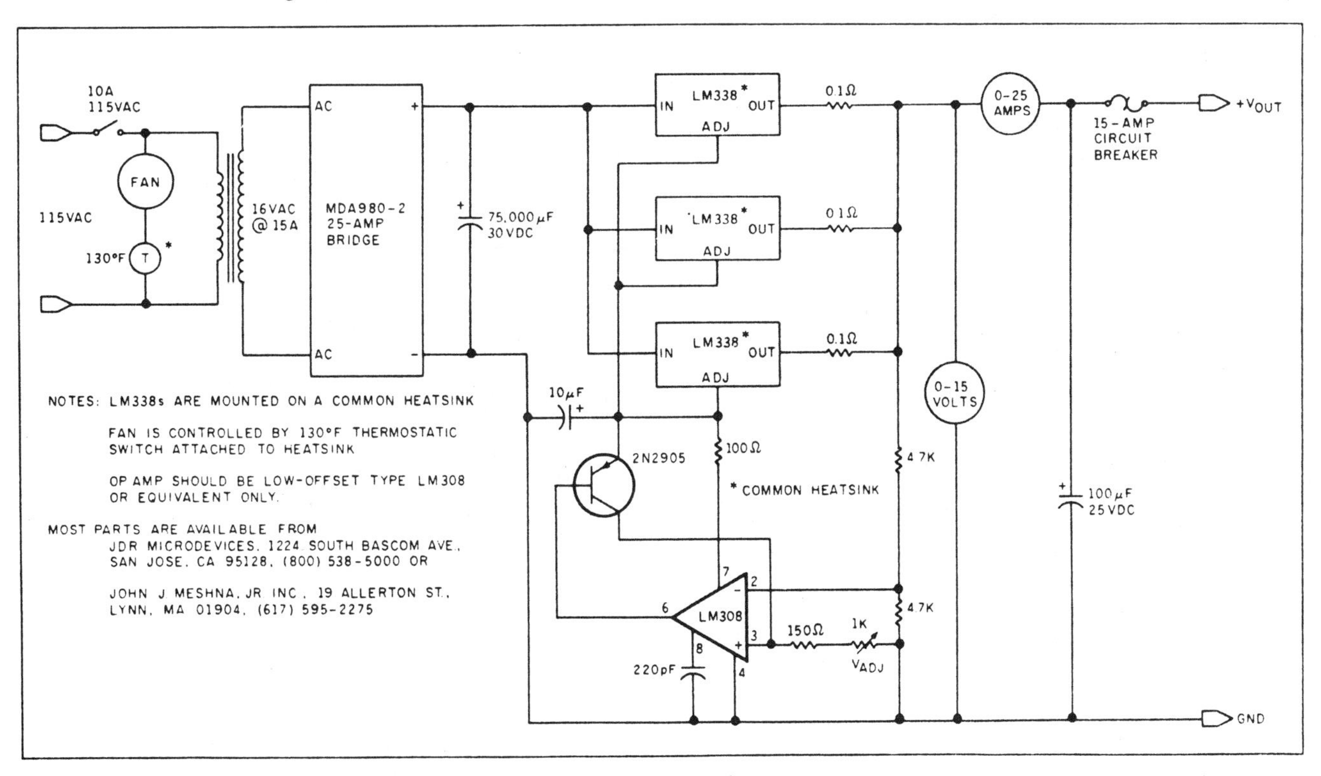

Figure 8: A 12- to 13.2-V 15-A *power supply that uses readily available components.*

cluding the areas that can cause problems. All wires and connections within a power supply have resistance. In the case of high-current supplies such as ours, small resistances can introduce major errors. For example, a 0.1-ohm resistance at 15 A drops 1.5 V. Heavy wire should be used, and it is important to separate the charging-current and the output-current paths.

As demonstrated in figure 9, improper placement of the input capacitor can induce unwanted ripple on the output voltage. This occurs when the charging current to the filter capacitor influences the common ground or voltage-adjustment line of the regulator. As previously mentioned, the peak currents in the filter circuit are in excess of 100 A. The voltage drop across R2′ will cause the output to fluctuate as if the voltage trim were being adjusted.

The output-current loop is also susceptible to layout. In a three-terminal fixed-voltage output regulator, the output voltage is referenced between the output pin and the common line of the chip. Because the load current flows through R2′, R3′, and R4′ as well as R_{LOAD} there may not be the correct voltage across the load due to accumulated voltage drops in the wiring. Also, while points B and C are both ground, they are at different voltages depending upon the resistance of and the current flowing through R3′. Similarly, resistance R4′ in the output lines continually reduces the output voltage as the current increases. This serves to negate the purpose of the regulator.

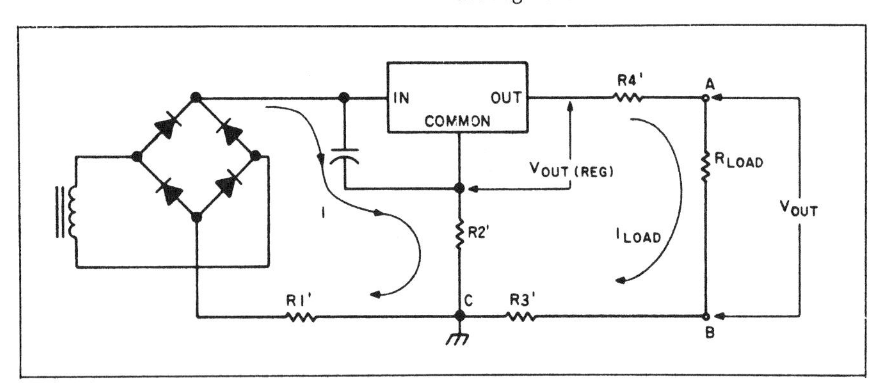

Figure 9: *Sources of layout-induced errors in a typical three-terminal-regulator power supply.*

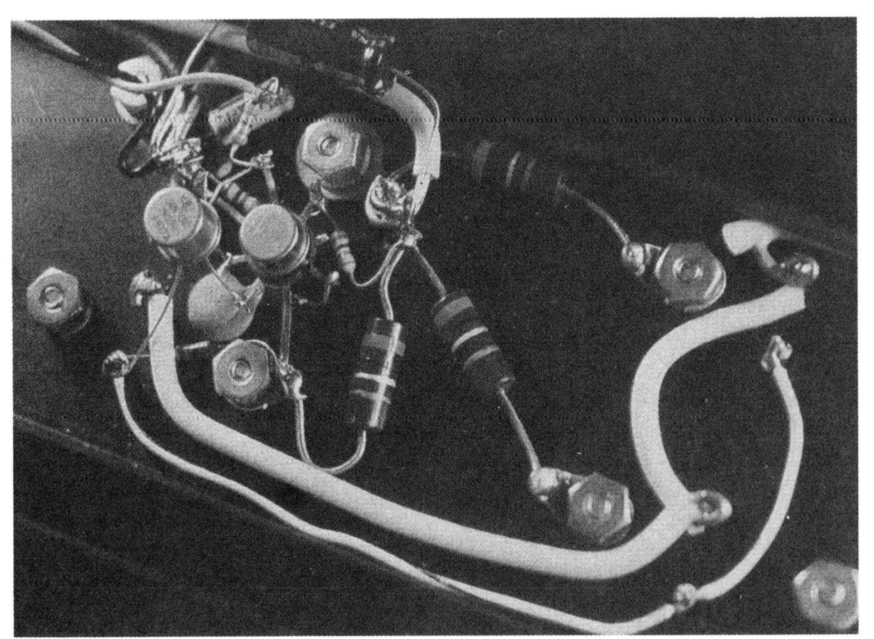

Photo 1: *Short leads in the regulator sections limit noise pickup and add to overall stability. In my prototype, I connected the op amp and other discrete components directly to the regulators on the bottom side of the heatsink.*

Figure 10 is a diagram of the proper layout. In the layout, all high-current paths should use heavy wire to minimize resistance, and the input-filter and output-load circuits are effectively separated. Most important, the wires from the transformer go directly to the bridge and then to the filter capacitor. Power to the rest of the circuit should come directly from the terminals of the capacitor and not from any point between the bridge and the capacitor. The result is two sets of wires (input from the bridge and output to the regulator) connected to the capacitor—but it is absolutely necessary. Mixing current paths is the most common problem in experimenter-built supplies.

The last layout consideration is the concept of a single-point ground. One point in the power supply must be designated as the ground, and the ground connections of the other circuit sections are connected to it. In practical terms, this is often just a metal strip or busbar called a ground bus. There should be virtually no measurable voltage between any two points on this bus. Don't be afraid to use thick wire!

HEATSINKING

The final consideration is heatsinking. Generally speaking, linear power supplies, while easy to build, are grossly inefficient. A 45 percent-efficient design is good (the usual range is 40 to 55 percent). Before you start thinking of this as both a 5- and 15-V 15-A supply, remember the old saying, "what goes in, comes out." With a $16\text{-}V_{rms}$ 15-A input, we are putting in 16 × 1.414 × 15 = 339 W and taking out 5 V × 15 A = 75 W. The other 264 W is dissipated in heat. Power is simply $V_{out} - V_{in}$ times the current. If you are going to want a 5-V supply, you should not start with a V_c of 18.6 V but rather something like 9 V. (The best transformer/filter for a 5-V supply is an 18-V center-tap configuration.) In the 13.2-V supply we are building, the maximum power dissipation is

$$PD_{max} = ((V_c + V_{ripple}/2) - V_{out}) \times (I_{out})$$
$$= ((18.6 + 1.8/2) - 13.2) \times 15$$
$$= 94.5 \text{ W}$$

For linear supplies of this magnitude, 95 W is relatively cool. Nonetheless, it must be dissipated properly through a device called a heatsink, as shown in photo 2.

Basically, the entire process of calculating factors such as dissipation, temperature rise, and junction temperatures is to determine a quantitative value of absorbable power for a given set of physical conditions. For a predetermined rise in heatsink temperature, you will be able to calculate the maximum power dissipation of the circuit to maintain that limit or, vice versa, to calculate the junction and heatsink temperatures given the input power.

Heatsink ratings and heat transfer through component mountings are stated in terms of thermal resistance: °C/W. For a particular application, it is necessary to determine the thermal resistance that a cooler must have to maintain a junction temperature that sustains adequate semiconductor performance. The basic relationship is

$$PD = \frac{\Delta T}{\Sigma R_\theta}$$

where PD = power dissipated in the semiconductor, ΔT = difference in temperature between ambient and the heatsink, and ΣR_θ = the sum of the thermal resistances of the heat flow path across which ΔT exists. In elaboration:

$$PD = \frac{T_j - T_a}{R_{jc} + R_{cs} + R_{sa}}$$

where T_j = the maximum junction temperature as stated by the semiconductor manufacturer (°C), T_a = ambient temperature (°C), R_{jc} = thermal resistance from junction to case of semiconductor (°C/W), R_{cs} = thermal resistance through the interface between the semiconductor case and the heatsink (°C/W), and R_{sa} = thermal resistance from heatsink to ambient air (°C/W).

The best way to understand this is to look at an example. First, we have a 7805T in a TO-220 case that is dissipating 5 W, and we select the proper heatsink. Given:

PD = 5 W
R_{jc} = 4°C/W (from the manufacturer)
T_j = 125°C maximum for TO-220 package
T_a = 50°C ambient
R_{cs} = 1°C/W insulator with heatsink grease

We use the equation for PD to solve for R_{sa}.

$$R_{sa} = \frac{125 - 50}{5} - (4 + 1)$$

$$R_{sa} = 10°C/W$$

Thermalloy part number 6299B has a 50° rise in temperature for a 12-W input. Therefore:

$$R_{sa} = 50/12 = 4.16°C/W$$

The 6299B is more than adequate for the task and will, in fact, heat up only 4.16 × 5 = 20.8°C over ambient in this example.

Getting back to the supply we are building, the minimum power dissipation is at maximum output voltage and vice versa. If the supply will be used in the range of 12 to 14 V, the heatsinking must accommodate the worst-case conditions. When the output is set for 12 V, the power dissipated in the regulator section will be 112.5 W (at 15 A). Each of the three regulators will be dissipating 112.5/3 or 37.5 W. The minimum R_{sa} is as follows. Given:

PD = 37.5 W
R_{jc} = 1.0°C/W
T_j = 125°C
T_a = 40°C
R_{cs} = 0.28°C/W (anodized washer and heatsink grease)

The R_{sa} minimum is thus

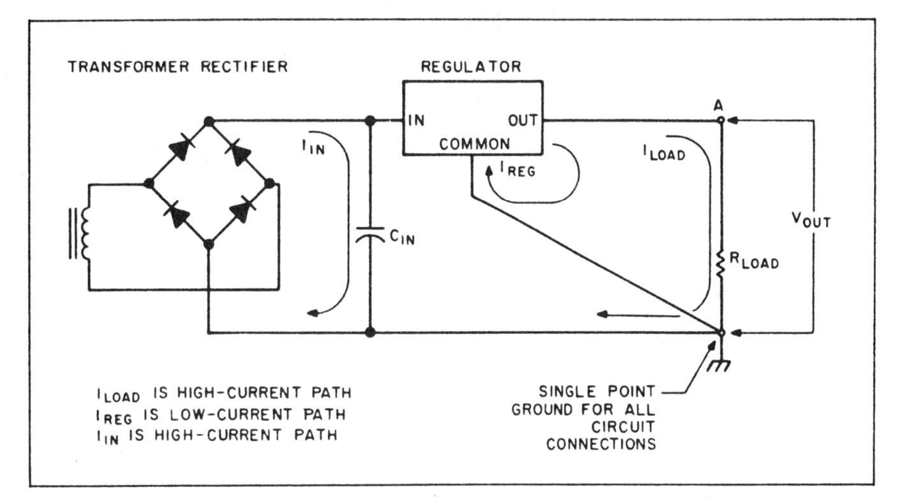

Figure 10: *The proper power-supply layout.*

Photo 2: *Any high-current linear power supply needs large heatsinks to carry away the heat dissipated in the regulators. When more than one regulator is to be mounted on the same heatsink, special insulated mounting kits must be used.*

Photo 3: *The completed power supply has little extra internal space. You'll notice that the filter capacitor has two sets of connecting wires to conform to the layout guidelines I've described. While the heatsink is sized properly for ambient air installations, I have added a fan to compensate for the insulating effect of the enclosure. To limit noise, the fan is controlled by a 130°F thermostat attached to the heatsink. The thermostat turns on the fan only when necessary.*

$$R_{sa} = \frac{125 - 40}{37.5} - (1 + 0.28)$$

$$= 0.99°C/W$$

A 3-inch piece of Thermalloy part number 6560 has a 0.70°C/W R_{sa} value. It would operate satisfactorily for an individual LM338K. A 7-inch piece of the same material would have an effective R_{sa} of 0.42°C/W.

To accommodate the full 125 W, forced-air cooling is recommended. From the data I have at hand, it appears that R_{sa} is reduced approximately by half with 600-cubic-feet-per-minute forced convection. With a fan on the heatsink, 112.5 W should be adequately dissipated while maintaining a low ambient temperature within the supply case. Photo 3 shows the complete supply with fan. I added a 130°F thermostatic switch on the heatsink to turn on the fan only when needed.

In Conclusion

The product failure described in the beginning was not a result of misunderstanding three-terminal-regulator specifications but instead ignorance of the supporting circuitry. I could have discussed a lot more, but much of it relates to experience, and it might sound as if I were a proponent of rule-of-thumb design. Instead, I would hope that you no longer take linear power supplies for granted. Even in today's VLSI (very-large-scale integration) world we continue to depend on tried-and-true, even if somewhat ancient, designs. Linear power supplies have a definite place in our world of electronics.

I don't expect venture capitalists to get excited about power-supply design, but the next time the words "meltdown" and "incendiary" are mentioned, I know a few who will be listening more closely.

This article is dedicated to Kram Nurtam and the E product. May that great heatsink in the sky cool any thoughts he might have of designing another linear power supply.

24

BUILD THE BASIC-52 COMPUTER/ CONTROLLER

A single-board problem solver with great potential

One of the most popular Circuit Cellar projects was the Z8 BASIC computer/ controller presented in July and August of 1981. Since then, thousands of Z8 controller boards have found their way into end-user and OEM applications.

I specifically designed the original Z8 controller because I hate programming. Generally speaking, if the program has fewer than 100 lines I'll grin and bear it. Any longer than that, however, and I lose interest and call in a programmer. To ease the pain, I generally use high-level languages like BASIC. Most people understand BASIC, and it excuses me from wasting time on tedious bit manipulations merely to demonstrate a hardware peripheral device. (My favorite programming language is solder.)

I don't try to justify using BASIC, I just get results. While others are arguing the merits of Pascal and C, I've plugged in my single-board computer/controller and am plinking away in BASIC to solve the problem. I've learned enough about other programming languages so that I know when to nod appreciatively at a programmer's description of a random-number seed generator written in some obscure programming dialect.

This "plug and program" approach has been adequately satisfied by the Z8, but I find that I purposely avoid applications involving floating-point calculations or trigonometric functions that would otherwise force me to resort to assembly-language programming (ugh!). In an effort to forestall my inevitable defection from BASIC, I am continually on the lookout for cost-effective performance boosters that I can package as single-board problem solvers (that execute in BASIC, naturally). And I just found another one!

What I have found is the Circuit Cellar BASIC-52 computer/controller (BCC-52) board. It uses the new Intel 8052AH-BASIC microcontroller chip that contains a ROM (read-only memory)-resident 8K-byte BASIC interpreter. The BCC-52 board includes the 8052AH, 48K bytes of RAM/EPROM (random-access read/write memory/erasable programmable ROM), a 2764/128 EPROM programmer, three parallel ports, a serial terminal port with automatic data-transmission-rate selection, a serial printer port, and is bus-compatible with the BCC-11 Z8 system/controller and all the BCC-series expansion boards I've already designed. Figure 1 is a block diagram of the hardware.

BASIC-52 is particularly suited for process control, providing IF. . . THEN, FOR. . .NEXT, DO. . . WHILE/UNTIL, ONTIME, and CALL statements among its broad repertoire of instructions (figure 2 lists the software features). Calculations are handled in integer or floating-point math and are fully supported with trigonometric and logical operators. Because of its low system overhead it is extremely fast and efficient.

I'll get into the system configuration and the design details momentarily, but I first have to mention an interesting aspect of BASIC-52. While I considered using EEPROMS (electrically erasable programmable ROMs) and other nonvolatile storage techniques, the sophisticated EPROM programming capabilities of BASIC-52 justified eliminating them simply on the basis of cost and board real estate. Unlike most one-shot EPROM programmers that fill the entire contents of an EPROM regardless of the application program's size, BASIC-52 treats the EPROM as write-once mass storage.

When a BASIC application program is saved to EPROM, it is tagged with an identifying ROM number and stored only in the amount of EPROM required to fit the program (plus header and EOF [end of file]). Additional application programs can be stored to the same EPROM and recalled for execution by requesting a particular ROM number. A 27128 EPROM provides 16K bytes of mass-storage space. When it is full (a nondestructive EPROM FULL error will tell you), simply erase the present EPROM or insert another. Finally, since this pseudo–mass storage exists in directly addressable memory space rather than cassettes or disks, it runs at full processor speed and stored application programs are instantly accessible.

BASIC-52 bridges the gap between expensive, intelligent control capabilities and hard-to-justify, price-sensitive control applications. BASIC-52's full floating-point BASIC is fast and efficient enough for the most complicated tasks, while its cost-effective design lets it be considered for many new areas of implementation.

I'm bullish on the BCC-52 board, and you can expect to see it in future Circuit Cellar projects. With so much power and convenience, I can accomplish quite a bit in a few lines of code—especially since that's all I may ever write.

The BCC-52 Board

The BCC-52 is a single-board controller/development system. Shown as a prototype in photo 1 and as a schematic in figure 3, this 17-chip circuit fits in a compact 4½ by 6½ inches (the same size as the Term-Mite smart terminal [see photo 2], if you want a two-board complete system—see my columns in the January and February 1984 issues of BYTE). It contains RAM/EPROM, an EPROM programmer, three parallel ports, and two serial ports.

The BCC-52 board has five main sections: processor, address decoding and memory, parallel I/O (input/output), serial I/O, and EPROM programmer.

The BCC-52 board is based on the 8052AH-BASIC chip, a preprogrammed version of Intel's 8052AH microcontroller (see figure 4). The 8052AH is the newest of Intel's 8-bit microcontroller-chip series, also known as the MCS-51 family.

The 8052AH contains 8K bytes of on-chip ROM, 256 bytes of RAM, three 16-bit counter/timers, six interrupts, and 32 I/O lines. In the 8052AH-BASIC chip, the ROM is a masked BASIC interpreter, and the I/O lines are redefined to address, data, and control lines. Figure 5a illustrates the 8052AH-BASIC chip pinout.

The 8052AH-BASIC chip has a 16-bit address and an 8-bit data bus (the 8 least significant address bits [AD0–AD7] and the data bus [D0–D7] are multiplexed together, similar to

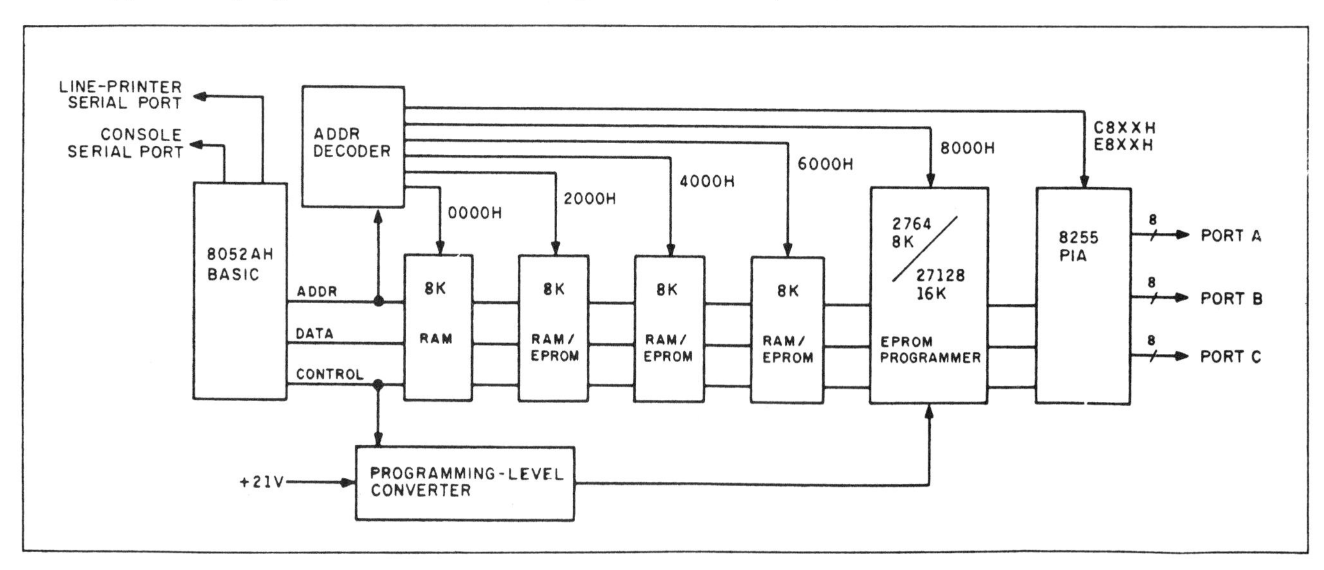

Figure 1: *Block diagram of the Circuit Cellar* BASIC-52 *computer/controller board.*

Command	Function
RUN	Execute a program
CONT	Continue after a stop or Control-C
LIST	List program to the console device
LIST#	List program to serial printer
NEW	Erase the program stored in RAM
NULL	Set null count after carriage return/line feed
RAM	Evoke RAM mode, current program in read/write memory
ROM	Evoke ROM mode, current program in ROM/EPROM
XFER	Transfer a program from ROM/EPROM to RAM
PROG	Save the current program in EPROM
PROG1	Save data-transmission-rate information in EPROM
PROG2	Save data-transmission-rate information in EPROM and execute program after reset
FPROG	Save the current program in EPROM using the intelligent algorithm
FPROG1	Save data-transmission-rate information in EPROM using the intelligent algorithm
FPROG2	Save data-transmission-rate information in EPROM and execute program after reset, use intelligent algorithm

Statement	Function
BAUD	Set data-transmission rate for line-printer port
CALL	Call assembly-language program
CLEAR	Clear variables, interrupts, and strings
CLEARS	Clear stacks
CLEAR1	Clear interrupts
CLOCK1	Enable real-time clock
CLOCK0	Disable real-time clock
DATA	Data to be read by READ statement
READ	Read data in DATA statement
RESTORE	Restore read pointer
DIM	Allocate memory for arrayed variables
DO	Set up loop for WHILE or UNTIL
UNTIL	Test DO loop condition (loop if false)
WHILE	Test DO loop condition (loop if true)
END	Terminate program execution
FOR-TO-{STEP}	Set up FOR... NEXT loop
NEXT	Test FOR... NEXT loop condition
GOSUB	Execute subroutine
RETURN	Return from subroutine
GOTO	GOTO program line number
ON GOTO	Conditional GOTO
ON GOSUB	Conditional GOSUB
IF-THEN-{ELSE}	Conditional test
INPUT	Input a string or variable
LET	Assign a variable or string a value (LET is optional)
ONERR	ONERR or GOTO line number
ONTIME	Generate an interrupt when time is equal to or greater than ONTIME argument; line number is after comma
ONEX1	GOSUB to line number following ONEX1 when INT1 pin is pulled low
PRINT	Print variables, strings, or literals, P. is shorthand for print
PRINT#	Print to software serial port
PH0.	Print hexadecimal mode with zero suppression
PH1.	Print hexadecimal mode with no zero suppression
PH0.#	PH0.# to line printer
PH1.#	PH1.# to line printer
PUSH	Push expressions on argument stack
POP	Pop argument stack to variables
PWM	Pulse-width modulation
REM	Remark
RETI	Return from interrupt
STOP	Break program execution
STRING	Allocate memory for strings
UI1	Evoke user console input routine
UI0	Evoke BASIC console input routine
UO1	Evoke user console output routine
UO0	Evoke BASIC console output routine

Operator	Function
CBY( )	Read program memory
DBY( )	Read/assign internal data memory
XBY( )	Read/assign external data memory
GET	Read console
IE	Read/assign IE register
IP	Read/assign IP register
PORT1	Read/assign I/O port 1 (P1)
PCON	Read/assign PCON register
RCAP2	Read/assign RCAP2 (RCAP2H:RCAP2L)
T2CON	Read/assign T2CON register
TCON	Read/assign TCON register
TMOD	Read/assign TMOD register
TIME	Read/assign real-time clock
TIMER0	Read/assign TIMER0 (TH0:TL0)
TIMER1	Read/assign TIMER1 (TH1:TL1)
TIMER2	Read/assign TIMER2 (TH2:TL2)
+	Addition
/	Division
**	Exponentiation
*	Multiplication
−	Subtraction
.AND.	Logical AND
.OR.	Logical OR
.XOR.	Logical exclusive OR

Stored Constant	
PI	PI − 3.1415926

Operators—Single Operand	
ABS( )	Absolute value
NOT( )	One's complement
INT( )	Integer
SGN( )	Sign
SQR( )	Square root
RND	Random number
LOG( )	Natural log
EXP( )	"e" (2.7182818) to the X
SIN( )	Returns the sine of argument
COS( )	Returns the cosine of argument
TAN( )	Returns the tangent of argument
ATN( )	Returns the arctangent of argument

Figure 2: *Detailed description of the Intel 8052AH BASIC-52 programming language.*

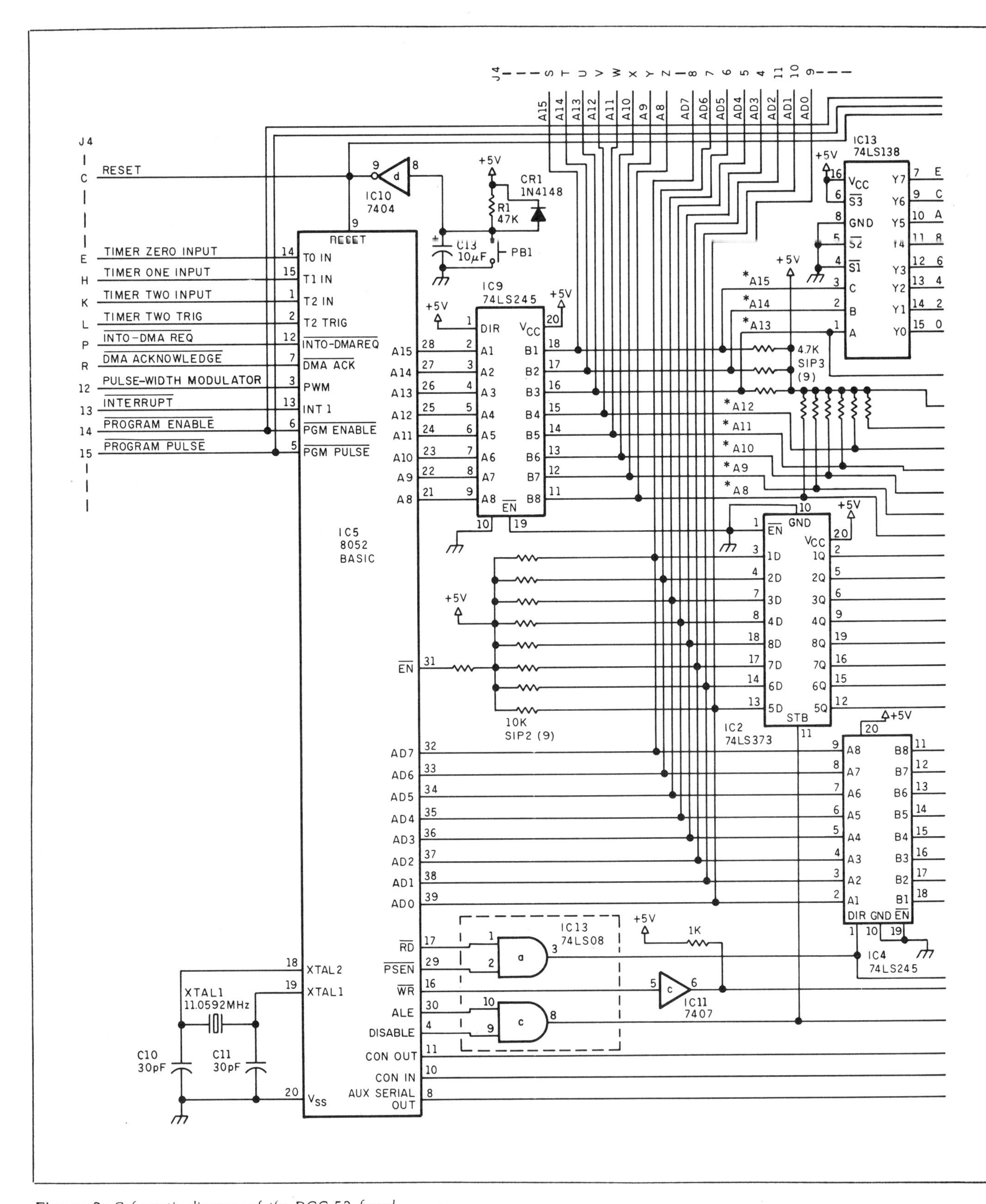

Figure 3: *Schematic diagram of the BCC-52 board.*

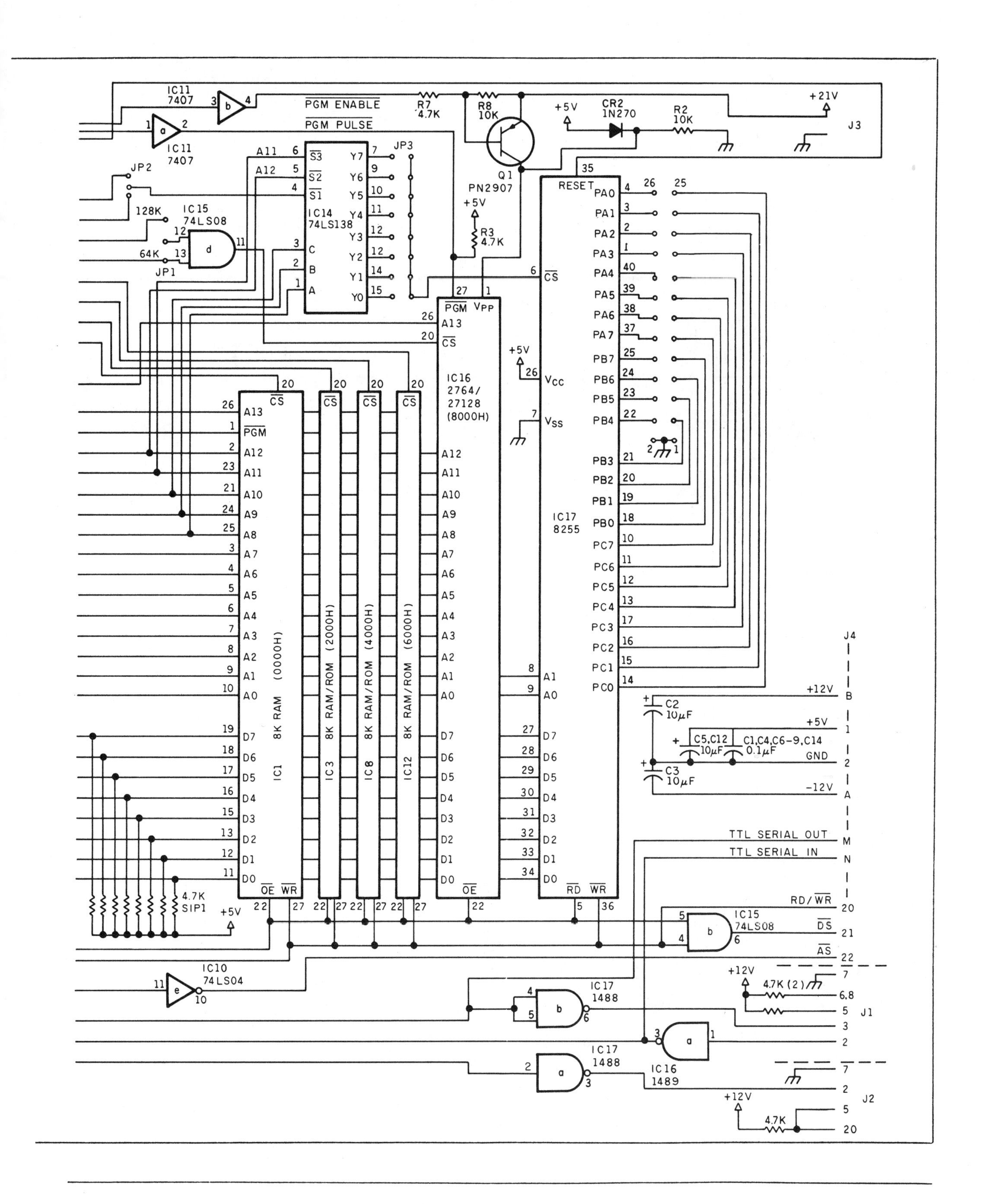

IC11 7407
IC15 74LS08
IC14 74LS138
PGM ENABLE
PGM PULSE
R7 4.7K
R8 10K
Q1 PN2907
+5V
CR2 1N270
R2 10K
+21V
J3
JP2
JP3
JP1
128K
64K
R3 4.7K
IC16 2764/27128 (8000H)
IC17 8255
RESET
8K RAM (0000H)
IC1
8K RAM/ROM (2000H)
IC3
8K RAM/ROM (4000H)
IC8
8K RAM/ROM (6000H)
IC12
4.7K SIP1
J4
+12V
C2 10µF
C5,C12 10µF
C1,C4,C6-9,C14 0.1µF
GND
C3 10µF
-12V
TTL SERIAL OUT
TTL SERIAL IN
RD/WR
DS
AS
IC10 74LS04
IC17 1488
IC16 1489
4.7K (2)
J1
J2
4.7K

the 8085 and Z8). When the chip is powered up, it sizes consecutive external memory from 0000 to the end of memory (or memory failure) by alternately writing 55 hexadecimal and 00 to each location. A minimum of 1K bytes of RAM is required for BASIC-52 to function, and any RAM must be located starting at 0000. [*Editor's note: For the remainder of the article, all addresses and data values will be hexadecimal unless otherwise specified.*]

Three control lines, $\overline{RD}$ (pin 17), $\overline{WR}$ (pin 16), and $\overline{PSEN}$ (pin 29), partition the address space as 64K bytes each of program and data memory. However, user-called assembly-language routines and EPROM programming are unsupported in data memory. For that reason, the BCC-52 board as I've designed it is addressed completely as program memory (RAM/EPROM mode), both for RAM and I/O. The addressing logic is as follows:

Photo 1: *The* Circuit Cellar BASIC-52 *computer/controller prototype.*

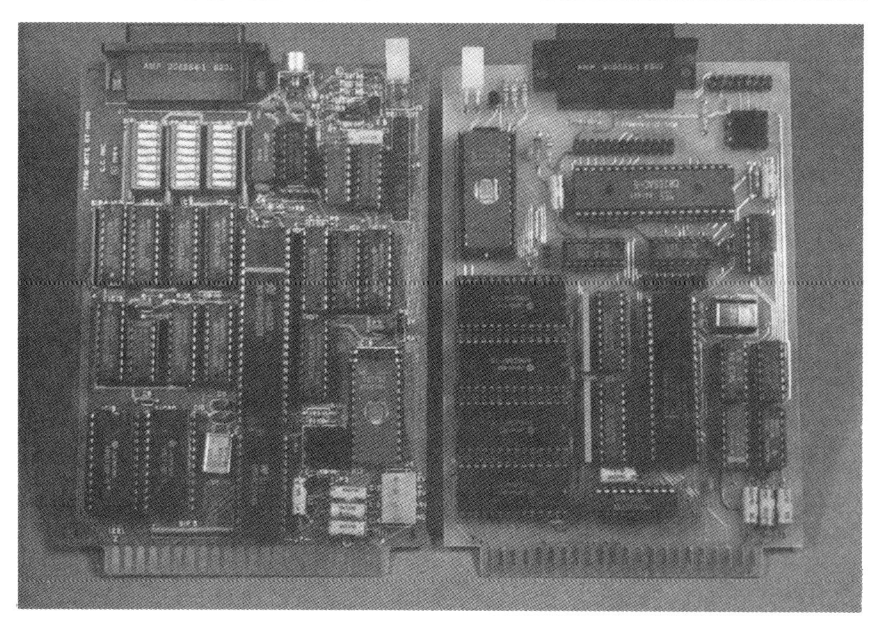

Photo 2: *On the right is the* BCC-52 *prototype; on the left is the* Circuit Cellar BCC *Term-Mite smart-terminal board (see the January* 1984 *Circuit Cellar). With the addition of a video monitor and keyboard, the two boards constitute a complete computer system suitable for software development or installed use.*

1. The $\overline{RD}$ and $\overline{WR}$ pins on the 8052AH chip enable RAM from 0000 to 7FFF. Addresses are used to decode the chip select ($\overline{CS}$) for the RAM devices, and $\overline{RD}$ and $\overline{WR}$ are used to enable the $\overline{OE}$ and $\overline{WE}$ (or $\overline{WR}$) pins, respectively.
2. $\overline{PSEN}$ is used to enable EPROM from 2000 to 7FFF. Addresses are used to decode the $\overline{CS}$ for the EPROM devices, and $\overline{PSEN}$ is used to enable the $\overline{OE}$ pin.
3. Between 8000 and 0FFFF, both $\overline{RD}$ and $\overline{PSEN}$ are used to enable either EPROM or RAM. $\overline{RD}$ and $\overline{PSEN}$ are applied as inputs to AND gate IC15, a 74LS08. The $\overline{WR}$ pin on the chip is used to write to RAM in this same address space.

BASIC-52 reserves the first 512 bytes of external data memory to implement two software stacks: the control stack and the arithmetic or argument stack. Understanding how the stacks work is necessary only if you want to link BASIC-52 and 8052 assembly-language routines. The details of how to do this are covered in the assembly-language linkage section of the *MCS BASIC-52 User's Manual.*

The control stack occupies locations 60 (96 decimal) through 0FE (254 decimal) in external RAM. This memory is used to store all information associated with loop control (i.e., DO. . . WHILE, DO. . .UNTIL, and FOR. . . NEXT) and BASIC subroutines (GOSUB). The stack is initialized to 0FE and "grows down."

The argument stack occupies locations 12D (301 decimal) through 1FE (510 decimal) in external RAM. This stack stores all the constants that BASIC is currently using. Operations like add, subtract, multiply, and divide always operate on the first two numbers on the argument stack and return the result to the argument

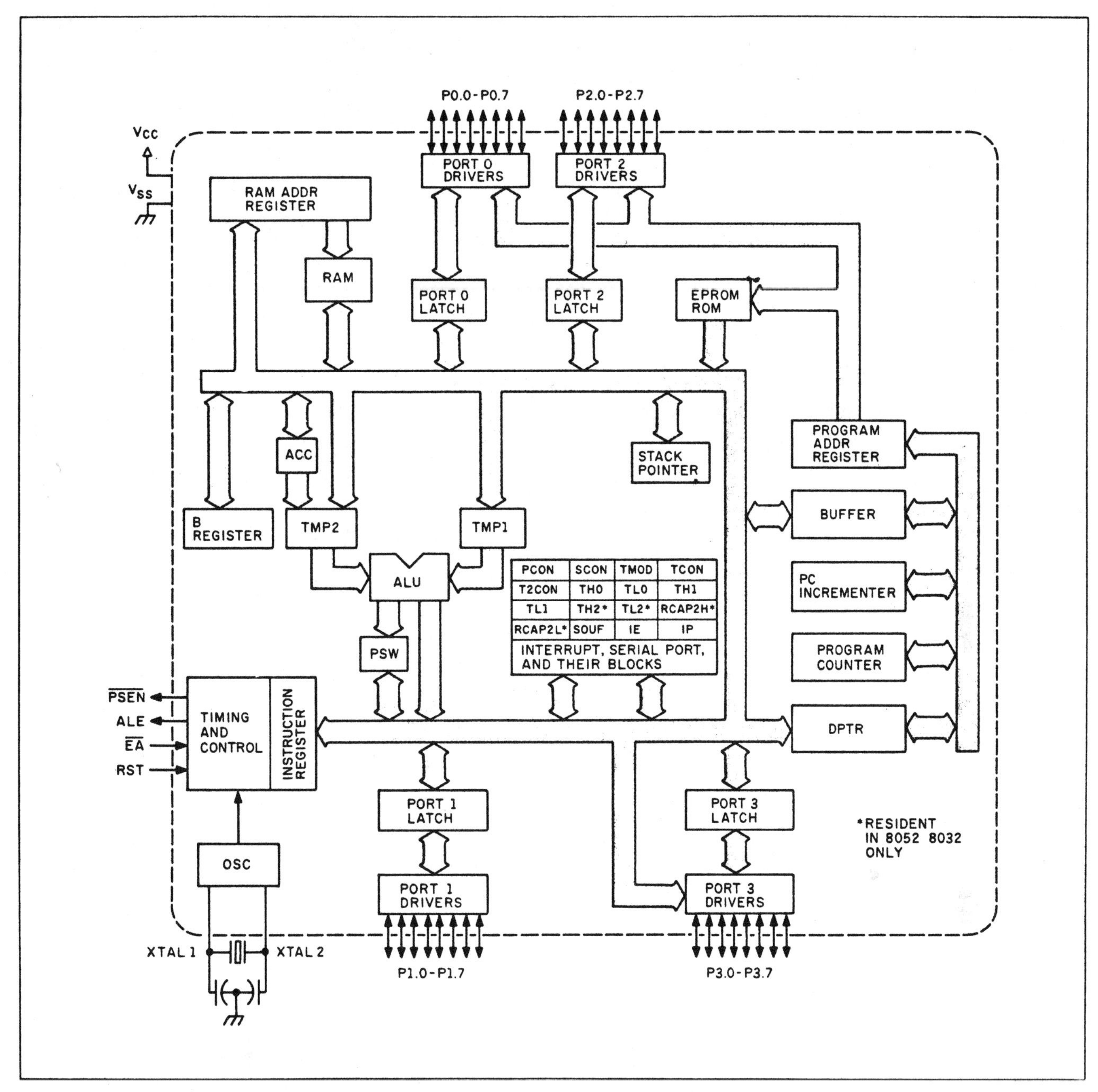

Figure 4: *Block diagram of the Intel 8052AH-BASIC chip.*

stack. The argument stack is initialized to 1FE and "grows down" as more values are placed on it. Each floating-point number placed on the argument stack requires 6 bytes of storage.

The stack pointer on the 8052AH (special-function register, SP) is initialized to 4D (77 decimal). The 8052AH's stack pointer "grows up" as additional values are placed on the stack.

Address Decoding

The three most significant address lines (A13–A15) are connected to a 74LS138 decoder chip, IC13, which separates the addressable range into eight 8K-byte memory segments, each with its own chip select (Y0–Y7). The four least significant chip selects are connected to 28-pin, 64K-bit (8K by 8) memory devices, either 2764 EPROMs or 6264 static RAMs. IC1, addressed at 0000, must be RAM in order for BASIC-52 to function. IC locations 3 (2000–3FFF), 11 (4000–5FFF), and 12 (6000–7FFF) can use either RAM or EPROM. IC16 (8000–

9FFF or BFFF) is an EPROM programming socket intended for 2764 or 27128 EPROMs (see figures 5b and 5c).

Altogether, you have 48K bytes of memory on the BCC-52 board if you use four 6264 RAMs (as ICs 1, 3, 8, and 12) (see figure 5d) and a 27128 EPROM in IC16. The memory and I/O can be further expanded through the expansion bus using BCC-series Z8-system expansion cards.

A second 74LS138 decoder, IC14, partitions either C800–CFFF or E800–EFFF as eight 256-byte I/O blocks. Rather than simply using the available C000 or E000 strobes from IC13 alone, which would occupy a 2000 address space for a single PIA (peripheral interface adapter) chip, IC14 allows many peripheral devices to share the remaining address space by using only a 256-byte address range. This addressing convention is consistent with other expansion boards I've designed, and it is easy to configure a 64-channel A/D (analog-to-digital) or 128-channel power I/O system using this board with a number of peripheral cards.

Parallel I/O

The BCC-52 board contains an 8255A-5 PIA (IC17) that provides three 8-bit I/O software-configurable parallel ports. The three I/O ports, labeled A, B, and C, and a write-only mode-configuration port occupy four consecutive addresses in one of the

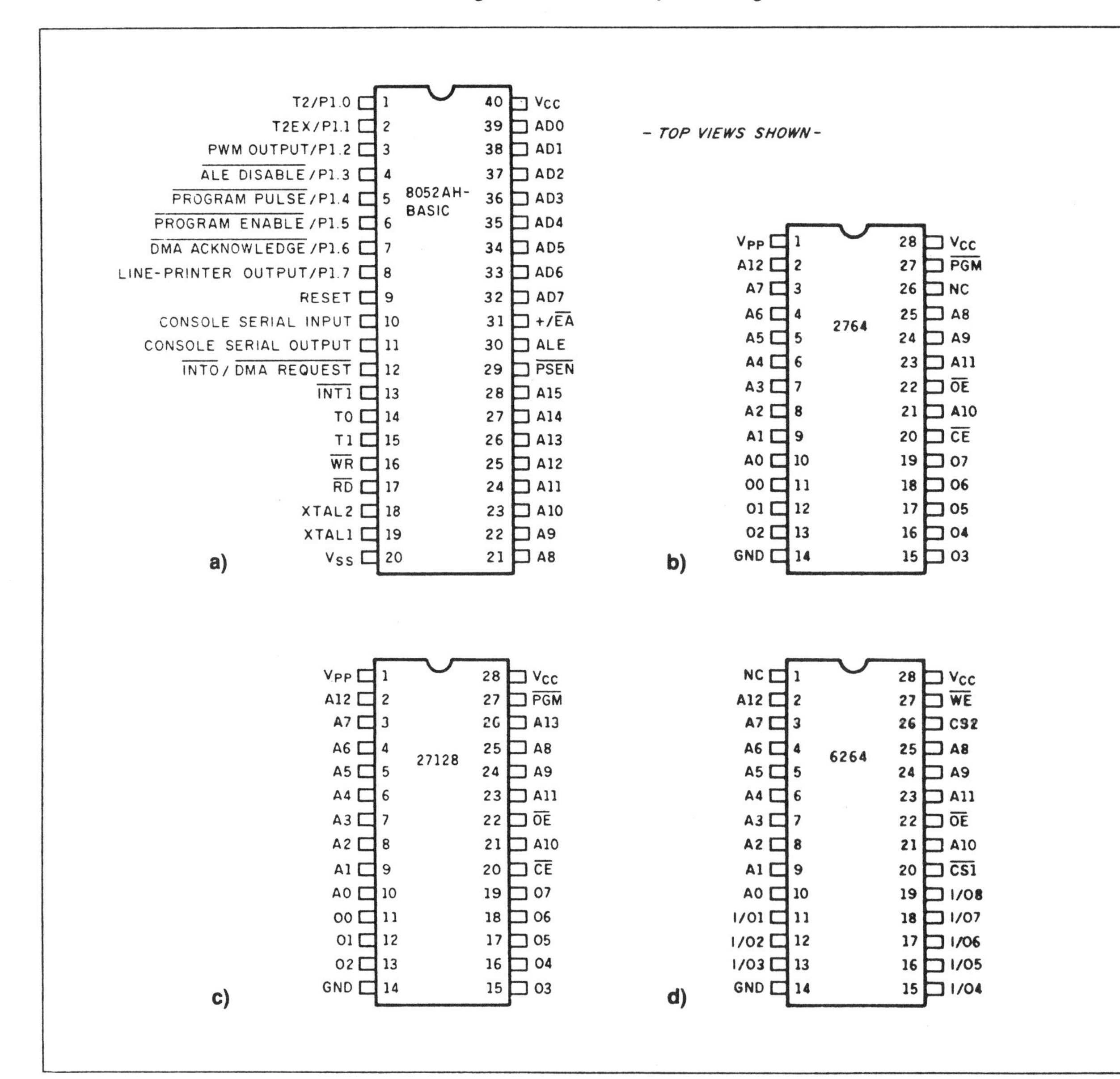

Figure 5: *Pinouts for (a) the* 8052AH-BASIC *chip, (b) the* 2764 8K-*byte* EPROM, *(c) the* 27128 16K-*byte* EPROM, *and (d) the* 6264 8K-*byte* RAM.

eight jumper-selectable I/O blocks. With C000 selected and pin 17 of IC14 (Y0) jumpered to pin 6 of IC17 (at JP3), the range would be C800–C803. Using the XBY() operator in BASIC, data can be written to and read from this PIA. (You are probably more familiar with PEEK and POKE. PEEK (C802H) is accomplished with XBY (C802H), and POKE C902H,A is XBY(C802H) = A.) I won't belabor the discussion on the 8255. I have used it many times in Circuit Cellar projects and refer you to the manufacturer's data sheets.

The three parallel ports and ground are connected to a 26-pin flat ribbon-cable connector. The outputs are TTL (transistor-transistor logic)-compatible.

SERIAL I/O

Two serial ports are found on the BCC-52 board. One is for the console I/O terminal (IC5 pins 10 and 11); the other is an auxiliary serial output (IC5 pin 8) frequently referred to as the line-printer port. When using an 11.0592-megahertz (MHz) crystal, the console port does automatic data-transmission-rate determination on power-up (a preset data-transmission rate can alternatively be stored in EPROM as well). I've used it at 19,200 bits per second (bps) with no degradation in operation.

The BAUD[expr] statement is used to set the data-transmission rate for the line-printer port. In order for this statement to properly calculate the data-transmission rate, the crystal (special-function operator—XTAL) must be correctly assigned (e.g., XTAL = 9000000). BASIC-52 assumes a crystal value of 11.0592 MHz if no XTAL value is assigned.

The main purpose of the software line-printer port is to let you make a hard copy of program listings and/or data. The command LIST# and the statement PRINT# direct outputs to the software line-printer port. If the BAUD[expr] statement is not executed before a LIST# or PRINT# command/statement is entered, the output to the software line-printer port will be at about 1 bps, and it will take a long time to output something. It is necessary to assign a data-transmission rate to the software line-printer port before using LIST# or PRINT#. The maximum data-transmission rate that can be assigned by the BAUD[expr] statement depends on the crystal, but 4800 bps is a reasonable maximum rate.

MC1488 and 1489 level shifters (ICs 6 and 7) convert the TTL levels from the console and line-printer ports to RS-232C. (The TTL serial lines are also connected to the bus to allow use of the Term-Mite smart-terminal board without RS-232C voltages.) The BCC-52 board requires only about 200 milliamperes (mA) at +5 volts (V) to function. The voltage required for external RS-232C communication is ±12 V; that required for EPROM programming is +21 V.

EPROM PROGRAMMER

One of the more unique and powerful features of the BCC-52 board is its

■Mode Selection (goes with figure 5b)

Mode / Pins	$\overline{CE}$ (20)	$\overline{OE}$ (22)	$\overline{PGM}$ (27)	V_{PP} (1)	V_{CC} (28)	Outputs (11–13, 15–19)
Read	V_{IL}	V_{IL}	V_{IH}	V_{CC}	V_{CC}	Dout
Standby	V_{IH}	×	×	V_{CC}	V_{CC}	High Z
Program	V_{IL}	×	V_{IL}	V_{PP}	V_{CC}	Din
Program Verify	V_{IL}	V_{IL}	V_{IH}	V_{PP}	V_{CC}	Dout
Program Inhibit	V_{IH}	×	×	V_{PP}	V_{CC}	High Z

■Mode Selection (goes with figure 5c)

Mode / Pins	$\overline{CE}$ (20)	$\overline{OE}$ (22)	$\overline{PGM}$ (27)	V_{PP} (1)	V_{CC} (28)	Outputs (11–13, 15–19)
Read	V_{IL}	V_{IL}	V_{IH}	V_{CC}	V_{CC}	Dout
Standby	V_{IH}	×	×	V_{CC}	V_{CC}	High Z
Program	V_{IL}	×	V_{IL}	V_{PP}	V_{CC}	Din
Program Verify	V_{IL}	V_{IL}	V_{IH}	V_{PP}	V_{CC}	Dout
Program Inhibit	V_{IH}	×	×	V_{PP}	V_{CC}	High Z

■Mode Selection (goes with figure 5d)

$\overline{WE}$	$\overline{CS_1}$	CS_2	$\overline{OE}$	Mode	I/O Pin
×	H	×	×	Not Selected (Power Down)	High Z
×	×	L	×		High Z
H	L	H	H	Output Disabled	High Z
H	L	H	L	Read	Dout
L	L	H	H	Write	Din
L	L	H	L		Din

×: don't care

ability to execute and save programs in an EPROM. The 8052AH chip actually generates all the timing signals needed to program 2764/128 EPROMs. Saving programs in EPROMs is a much more attractive and reliable alternative to cassette tape, especially in control and/or noisy environments.

The entire EPROM programming circuitry consists of two 7407 open-collector drivers and a single transistor circuit that switches between +5 V and 21 V (CR2, connected to the collector of the transistor should be a germanium diode like a 1N270).

Port 1, bit 4 (IC5 pin 5) is used to provide a 1- or 50-millisecond (ms) programming pulse. The length of the pulse is determined by whether we are programming Intel fast-program EPROMs or generic 2764s and 27128s. BASIC-52 calculates the length of the pulse from the assigned crystal value. The accuracy of this pulse is within 10 processor clock cycles. This pin is normally in a logical high (1) state. It is asserted low (0) to program the EPROMs.

Port 1, bit 5 (IC5 pin 6) is used to enable the EPROM programming voltage. This pin is normally in a logical high (1) state. Prior to the EPROM programming operation, this pin is brought to a logical low (0) state, and it is used to turn on the high voltage (21 V) required to program the EPROMs on or off.

BASIC-52 saves several programs on a single EPROM. In fact, it can save as many programs as the size of the EPROM permits. The programs are stored sequentially in the EPROM, and any program can be retrieved and executed. This sequential storage of programs is referred to as the EPROM file. The following commands permit you to generate and manipulate the EPROM file.

RAM and ROM[integer] tell the BASIC-52 interpreter whether to select the current program (the one that will be displayed during a LIST# command and executed when RUN is typed) out of RAM or EPROM. The RAM address is assumed to be 200 (512 decimal), and the EPROM address begins at 8010 (32,784 decimal).

When RAM is entered, BASIC-52 selects the current program from RAM. This is usually considered the normal mode of operation and is the mode that most users employ to interact with the command interpreter.

When ROM[integer] is entered, BASIC-52 selects the current program out of EPROM. If no integer is typed after the ROM command (i.e., ROM(cr)), BASIC-52 defaults to ROM 1. Since the programs are stored sequentially in EPROM, the integer following the ROM command selects which program you want to run or list. If you attempt to select a program that does not exist (i.e., you type in ROM 8 and only six programs are stored in the EPROM), the message Error: Prom Mode will be displayed. The error is nondestructive, and you can retype the correct command.

BASIC-52 does not transfer the program from EPROM to RAM when the ROM mode is selected, and you cannot edit a program in ROM. Attempting to do so will result in an error message.

Since the ROM command does not transfer a program to RAM, it is possible to have different programs in ROM and RAM simultaneously. You can flip back and forth between the two modes at any time. Another benefit of not transferring a program to RAM is that all the RAM can be used for variable storage if the program is stored in EPROM. The system-control values, MTOP and FREE, always refer to RAM.

The XFER (transfer) command transfers the currently selected program in EPROM to RAM and then selects the RAM mode. After the XFER command is executed, you can edit the program in the same manner any RAM program can be edited.

The PROG command programs the resident EPROM with the current program (this is the only time that the +21-V programming voltage needs to be applied). The current program can reside in either RAM or EPROM. After PROG is typed, BASIC-52 displays the number in the EPROM file the program will occupy.

Normally, after power is applied to

the BASIC-52 device, you must type a space character to initialize the 8052AH's console port. As a convenience, BASIC-52 contains a PROG1 command. This command programs the resident EPROM with the data-transmission-rate information. The next time the MCS BASIC-52 device is powered up, i.e., reset, the chip will read this information and initialize the serial port with the stored data-transmission rate. The sign-on message will be sent to the console immediately after the BASIC-52 device completes its reset sequence. The space character no longer needs to be typed.

The PROG2 command does everything the PROG1 command does, but instead of signing on and entering the command mode, the BCC-52 board immediately begins executing the first program stored in the resident EPROM.

By using the PROG2 command, it is possible to run a program from a reset condition and never connect the BCC-52 board to a console. In essence, saving PROG2 information is equivalent to typing ROM 1 and RUN in sequence. This is ideal for control applications, where it is not always possible to have a terminal present. In addition, this feature lets you write a special initialization sequence in BASIC or assembly language and generate a custom sign-on message for specific applications.

Powering Up the Board

The best way to check out the BCC-52 board is to run it with the minimum hardware first. With only ICs 1, 2, 4–7, 9–11, 13, and 15 installed, we have an 8K-byte RAM-only system. After applying power, BASIC-52 clears the internal 8052AH memory; initializes the internal registers and pointers; and tests, clears, and sizes the external memory.

BASIC-52 then assigns the top of external RAM to the system-control value (MTOP) and uses this number as the random-number seed. BASIC-52 assigns the default-crystal value, 11.0592 MHz, to the system-control value (XTAL) and uses this default value to calculate all time-dependent functions, like the EPROM programming timer and the interrupt-driven real-time clock. Finally, BASIC-52 checks external memory location 8000 to see if the data-transmission-rate information is stored. If the data-transmission rate is stored, BASIC-52 initializes the data-transmission-rate generator (the 8052AH's special-function register, T2CON) with this information and signs on. If not, BASIC-52 interrogates the serial-port input and waits for a space character to be typed (automatic data-transmission-rate detection).

If you have entered nothing on the console device, BASIC-52 will appear inoperative to the uninitiated. Simply type a space, and the console device should display the following:

```
*MCS-52(tm) BASIC Vx.x*
READY
>
```

To see if the processor is operating correctly, we type the following:

```
>PRINT XTAL, TMOD, TCON,
 T2CON
```

BASIC-52 should respond with the control and special-function values:

```
11059200 16 244 52
>
```

A Word About the BASIC

As I mentioned earlier, BASIC-52 is oriented toward process control and is significantly more powerful than a tiny BASIC. Since most of you are familiar with BASIC, I will not describe individual instructions like DO. . . WHILE and FOR. . .NEXT. Instead, I'd like to point out the pertinent features that demonstrate the exceptional small-package performance of the BCC-52 board.

MCS BASIC-52 contains a minimum-level line editor. Once a line is entered, you cannot change the line without retyping it. However, it is possible to delete characters while a line is in the process of being entered. This is done by inserting a rubout or delete character (7F). The rubout character will cause the last character entered to be erased from the text input buffer. Additionally, pressing Control-D will cause the entire line to be erased.

Variables and Expressions

The range of numbers that can be represented in BASIC-52 (in decimal) is +1E–127 to +0.99999999E+127.

It has eight digits of significance. Numbers are internally rounded to fit this precision. Numbers can be entered and displayed in four formats: integer, decimal, hexadecimal, and exponential, for example, 129, 34.98, OA6EH, 1.23456E+3.

Integers are numbers that range from –32,768 to +32,767 decimal. All integers can be entered in either decimal or hexadecimal format. A hexadecimal number is indicated by placing the letter "H" after the number. When an operator like AND requires an integer, BASIC-52 will truncate the fraction portion of the number so that it will fit the integer format. All line numbers are integers.

A variable can be either a letter (e.g. A, X, I), a letter followed by a number (e.g., Q1, T7, L3), a letter followed by a one-dimensioned expression (e.g., J(4), G(A+6), I(10*SIN(X))), or a letter followed by a number followed by a one-dimensioned expression (e.g., A1(8), P7(DBY(9)), W8(A+B)). Variables with a one-dimensioned expression are called dimensioned or arrayed variables. Variables that involve only a letter or a letter and a number are called scalar variables.

BASIC-52 allocates variables in a static manner. Each time a variable is used, BASIC-52 allocates 8 bytes specifically for that variable. This memory cannot be deallocated on a variable-by-variable basis. If you execute a statement like Q=3, later on you cannot tell BASIC-52 that the variable Q no longer exists and free up the 8 bytes of memory that belong to Q. You can clear the memory allocated to variables with a CLEAR statement.

Relative to a dimensioned variable, it takes BASIC-52 much less time to find a scalar variable. That's because a scalar variable has no expression to evaluate. If you want to make a program run as fast as possible, use dimensioned variables only when you have to. Use scalar variables for intermediate variables, then assign the final result to a dimensioned variable.

An expression is a logical mathematical term that involves operators (both unary and dyadic), constants, and variables. Expressions can be simple or quite complex, e.g., 12*EXP(A)/100, H(1)+55, or (SIN(A)*SIN(A)+COS(A)*COS(A))/2. A stand-alone variable [var] or constant [const] is also considered an expression.

REAL-TIME OPERATION

After RUN is typed, all variables are set equal to zero, all BASIC-evoked interrupts are cleared, and program execution begins with the first line number of the selected program. The RUN command and the GOTO statement are the only ways you can execute a program in the command mode. Program execution can be terminated at any time by typing a Control-C on the console device.

Unlike some BASIC interpreters that allow a line number to follow the RUN command (e.g., RUN 100), BASIC-52 does not permit such a variation on the RUN command. Execution always begins with the first line number. To obtain the same functionality as the RUN[ln num], use GOTO[ln num] in the direct mode.

The CLOCK1 statement enables the software real-time clock in BASIC-52. The special-function operator time is incremented once every 5 ms after the CLOCK1 statement has been executed. The CLOCK1 statement uses timer/counter 0 in the 13-bit mode to generate an interrupt once every 5 ms. Because of this, the special-function operator time has a resolution of 5 ms.

BASIC-52 automatically calculates the proper reload value for timer/counter 0 after the crystal value has been assigned (i.e., XTAL = value. If no crystal value is assigned, MCS BASIC-52 assumes a value of 11.0592 MHz). The special-function operator time counts from 0 to 65,535.995 seconds. After reaching a count of 65,535.995 seconds, time overflows back to a count of 0.

The interrupts associated with the CLOCK1 statement cause BASIC programs to run at about 99.6 percent of normal speed. That means that the interrupt handling for the real-time-clock feature consumes only about 0.4 percent of the total processor time. This is small interrupt overhead. The CLOCK0 statement disables or turns off the real-time-clock feature.

The TIME statement is used to retrieve and/or assign a value to the real-time clock after the CLOCK1 statement enables it. TIME = 5 presets the real-time clock to 5 seconds, while ONTIME 30,100 causes the program to jump to line 100 when the real-time clock reaches 30 seconds.

Finally, PWM might be useful to literally add bells and whistles to your next control application. PWM stands for pulse-width modulation. It generates a user-defined pulse sequence on IC5 pin 3.

The statement appears as PWM 50,50,100. The first expression following PWM is the number of clock cycles the pulse will remain high. A clock cycle is equal to 1.085 microseconds (11.0592-MHz crystal). The second expression is the number of clock cycles the pulse will remain low; the third expression is the total number of cycles you want to output. All expressions in the PWM statement must be valid integers, and the minimum value for the first two expressions is decimal 20.

These are only a few of the 103 commands, statements, and operators in BASIC-52. The *User's Manual* describes them in detail.

IN CONCLUSION

This was a hard article for me to write, but not for any of the reasons you might think. So much is built into this compact board that I am impatient to use it, and it was hard to sit down and write. Unfortunately, documentation is the drudge work side of engineering.

It won't take long to put the BCC-52 board into some serious applications. It might be a single-board computer, but its configuration does not stop with a single board. The BCC-52 is BCC-series Z8-bus-compatible and can be expanded using many of the

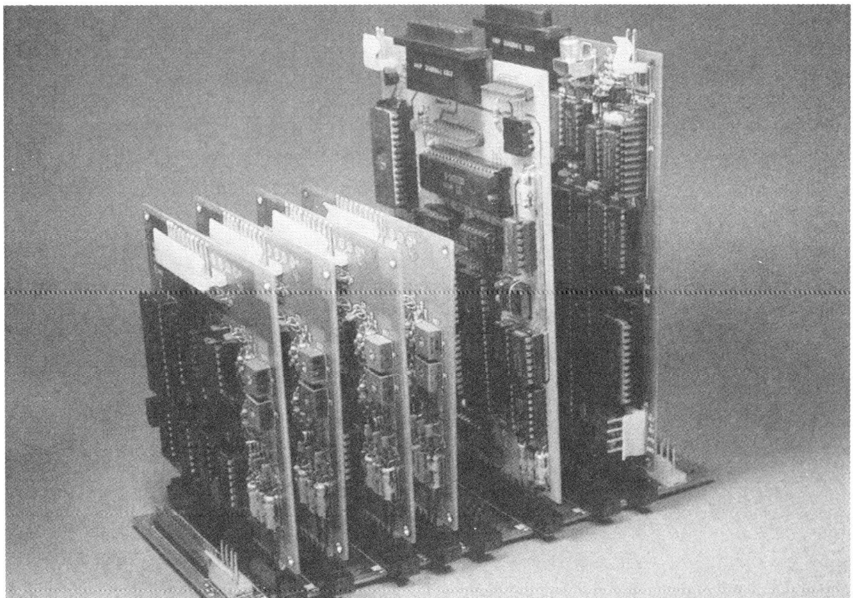

Photo 3: *The* BCC-52 *and* Term-Mite *boards can be combined with other* BCC-series *peripheral devices to create control and data-acquisition systems. Here, they are combined with four* BCC-13 8-*channel,* 8-*bit* A/D *converter boards to make a* 32-*channel data-acquisition system.*

projects and boards I've already designed. For example, monitoring temperatures, controlling motors and heaters, and reporting events are adequately handled by existing power I/O, serial and parallel expansion, and A/D converter boards (see photo 3).

This BASIC-52 project has just started. Because of its power, I am inspired to further develop applications and peripheral support devices. While a specific time has not been chosen, I'll be back in a few months with the next chapter on the BCC-52.

Diagrams and data pertinent to the 8052AH-BASIC *chip are reprinted courtesy of the* Intel *Corporation.*

See page 447 for a listing of parts and products available from Circuit Cellar Inc.

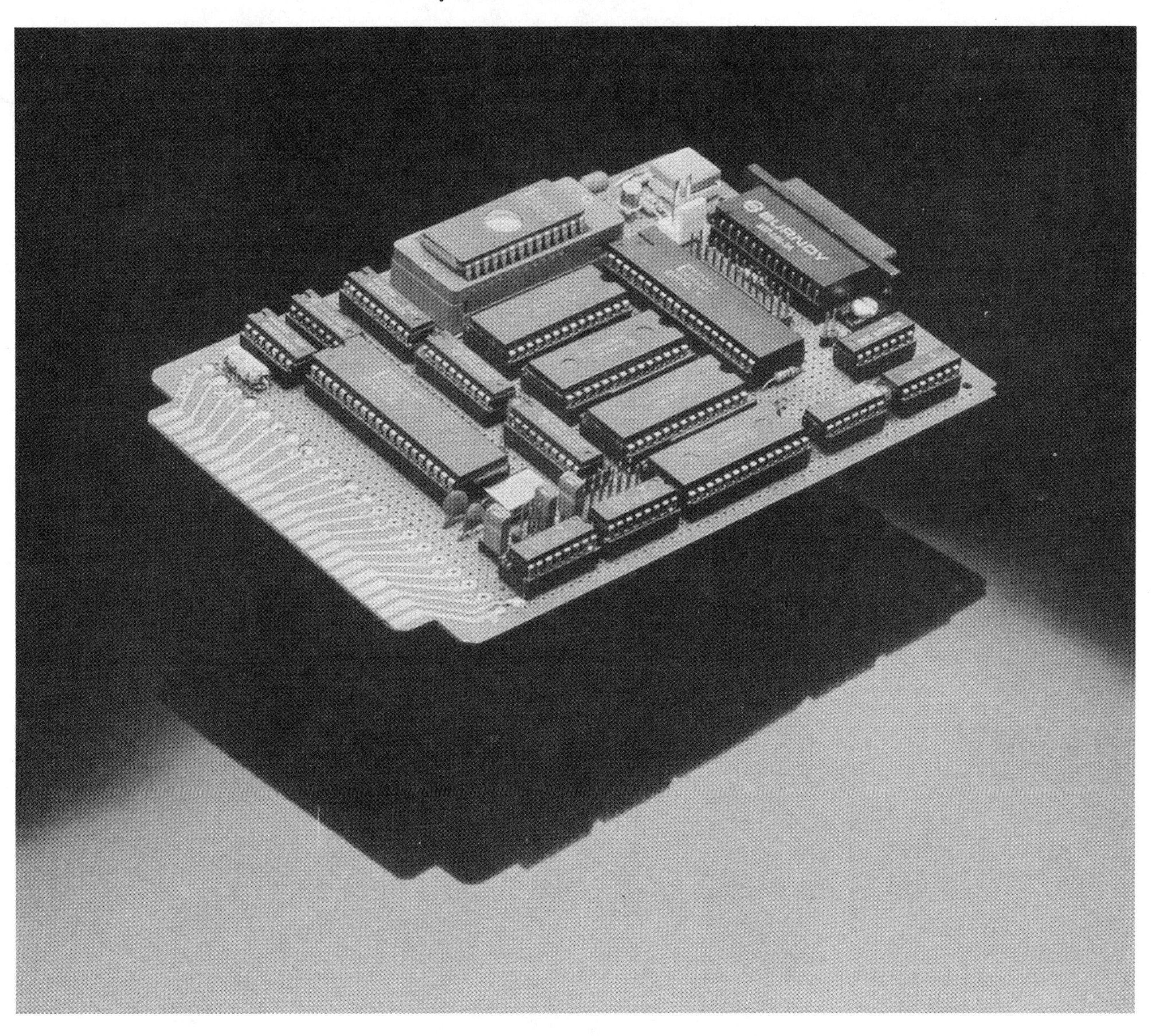

PHOTOGRAPHED BY PAUL AVIS.

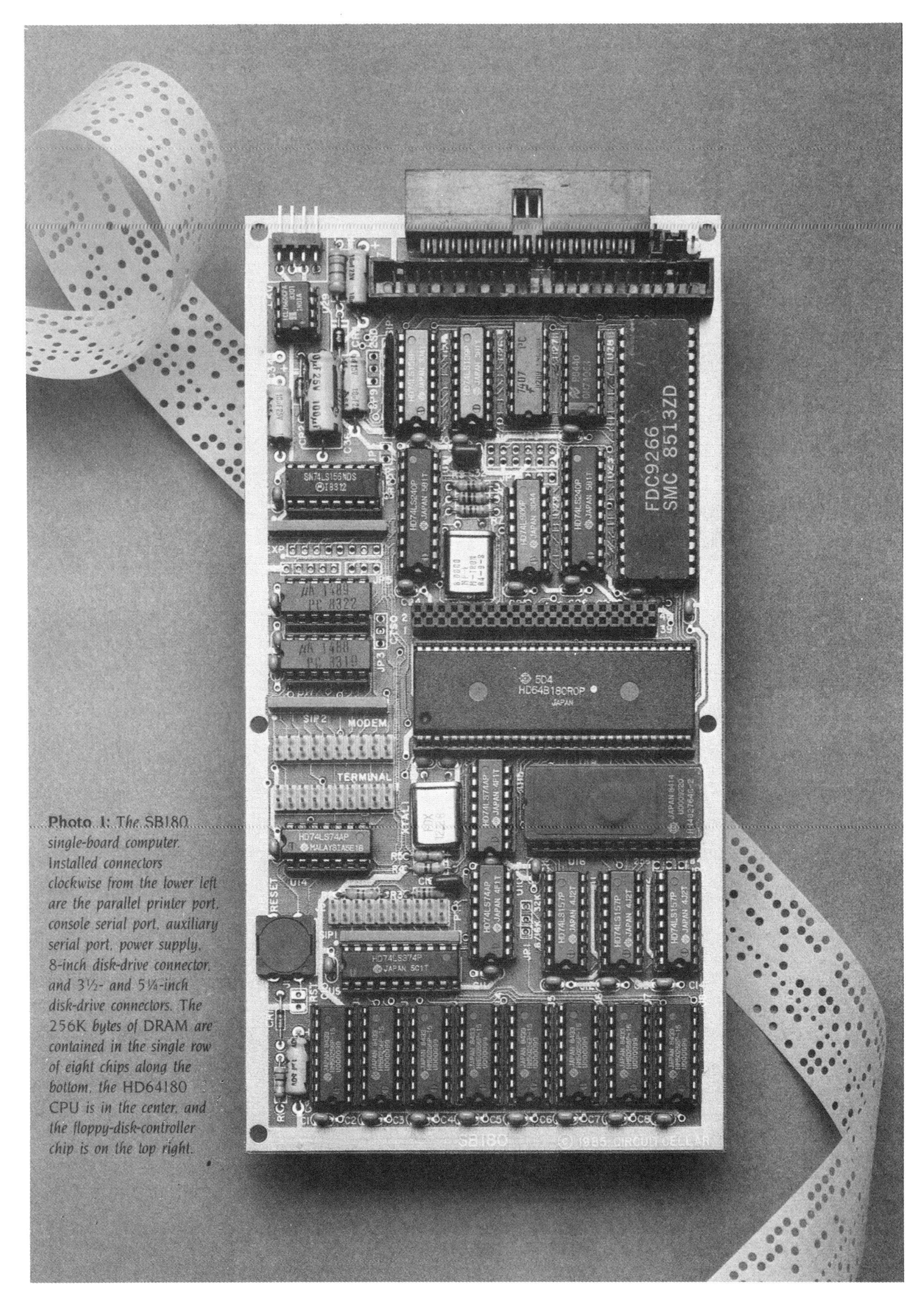

Photo 1: *The SB180 single-board computer. Installed connectors clockwise from the lower left are the parallel printer port, console serial port, auxiliary serial port, power supply, 8-inch disk-drive connector, and 3½- and 5¼-inch disk-drive connectors. The 256K bytes of DRAM are contained in the single row of eight chips along the bottom, the HD64180 CPU is in the center, and the floppy-disk-controller chip is on the top right.*

PHOTOGRAPHED BY PAUL AVIS

25

BUILD THE SB180 SINGLE-BOARD COMPUTER

PART 1: THE HARDWARE

This computer reasserts 8-bit computing in a 16-bit world

Newer, faster, better. These words are screamed at you in ads and reviews of virtually every new computer that comes to market. Unfortunately, many of the proponents of this rhetoric are going on hearsay evidence. While advertising hype has its place in our culture, a more thorough investigation may lead you to alternative conclusions.

Generally speaking, quotes of increased performance are basically comparisons of CPU (central processing unit) instruction times rarely involving the operating system. The 68000 is indeed a more capable processor than the 6502, but that doesn't necessarily mean that commercial application programs always run faster because the CPU has more capability. People owning 128K-byte Macintoshes have discovered this.

The bus size of the processor is only one factor in the performance of a computer system. Operating-system design and programming styles contribute much more to the overall throughput of a computer. It is not enough to simply compare 8 to 16 bits or 16 to 32 bits. For example, the Sieve of Eratosthenes prime-number benchmark runs faster in BASIC on the 8-bit 8052-based controller board presented in Chapter 6 than it does on a 16-bit IBM PC.

The ultimate performance of a computer is the sum of its subsystem interaction times and not just the CPU execution speed. Using a simple database-management program involves interaction among the user input device, operating system, disk drives, firmware, system memory, and user output devices. Slow communication or a bottleneck between any two of these subsystems can degrade the performance of the entire computer.

In my opinion, the processor/operating system connection contributes most to user satisfaction. In the days before the IBM PC, the de facto computer standard was the 8080/Z80 and CP/M 2.2. Unfortunately, software developers considered it difficult to use, especially in turnkey applications. The frustration of having to warm-boot the system merely to change disks and the lack of a PATH command created many ready-and-willing PC-DOS customers. If only there were an 8-bit operating system with the capabilities and friendliness of MS-DOS.

The Z80 is still considered a high-performance CPU. In reality, the 8-bit processor is a cost-effective and efficient choice for personal computers and industrial controllers. (Remember that peripheral support chips such as PIAs [peripheral interface adapters] and floppy-disk controllers are 8-bit devices, as are many memory chips.) One problem is that the Z80 has an address limitation of 64K bytes, which discourages the use of 50K-byte BASIC interpreters and 100K-byte integrated spreadsheet programs common to IBM PC users. While additional memory has been added through hardware bank switching, it has never been properly integrated into the CP/M operating system, and its function is kludgy. If only someone would design a Z80 chip that directly addresses more than 64K bytes.

THE CIRCUIT CELLAR SB180

Hitachi has recently developed a CMOS (complementary metal-oxide semiconductor) Z80-code-compatible processor, designated the HD64180, that directly addresses 512K bytes. Echelon Systems has developed an operating system for the processor that is an amalgam of CP/M, MS-DOS, and UNIX. This operating system is called the Z-System. Using the HD64180 and the Z-System, I have produced a computer that reasserts 8-bit computing in a 16-bit world and outperforms a standard IBM PC by 20 to 100 percent.

In this two-part chapter, I take a look at the Hitachi HD64180 and the evolution of CP/M as embodied in the Z-System. The hardware project is a single-board computer (4 by 7½ inches) called the SB180, which is the 29-chip equivalent of a large S-100 system (see photos 1 and 2). Refer to figure 1 for a block diagram of the SB180 board.

The Circuit Cellar SB180 has the following capabilities:

- 256K-byte on-board RAM (random-access read/write memory), which can be partitioned as 64K-byte system memory and 192K-byte RAM disk or as paged system memory
- 32K-byte EPROM (erasable programmable read-only memory)
- full ROM-based monitor with system-disk boot
- two RS-232C serial ports, one with automatic data-transmission-rate detection
- a Centronics parallel printer port
- single/double-density programmable floppy-disk controller capable of handling a mix of up to four 3½-, 5¼-, or 8-inch drives; different-size drives can run concurrently
- requires only +5 volts (V) and +12 V for operation (+12 V only for RS-232C)
- I/O (input/output) expansion bus
- fits on a 4- by 7½-inch board that mounts directly to a 5¼- or 3½-inch floppy-disk drive

Disk-based software includes the Z-System DOS (disk operating system), a debugger, and HD64180 assembler.

While this is in fact a hardware project, true functionality would be an exercise left to you readers were it not for the operating system and BIOS (basic input/output system) specifically written and adapted to the SB180. The combination of the HD64180 and Z-System is what gives this tiny computer so much power. Much of the project description therefore has to be the software that uniquely sets performance levels far above traditional 8-bit computer designs. I don't want to diminish the significance of the hardware, but I am a realist.

In Part 1, I'd like to describe the HD64180 chip and give an overview of the changing evolution of CP/M as it pertains to this project. After that, I'll describe the design of the SB180. Part 2 will be dedicated to the operating-system software and BIOS.

Photo 2: *The* SB180 *shown beside a* 3½*-inch disk drive and an* 8*-inch disk drive.*

CP/M AND BEYOND

Anyone who has been involved with microcomputers for more than two years acknowledges the tremendous impact of CP/M upon the history of personal computing. While hobbyists were still debating whether the "standard" tape format should be Kansas City CUTS (cassette user tape system) or Tarbell, Gary Kildall made CP/M available at a reasonable price. It quickly became the de facto standard DOS for 8080- and Z80-based microcomputers. With a "standard" operating system and a "standard" 8-inch disk format, entrepreneurial programmers saw the opportunity to write serious business applications that needed disk capability to be viable and that could be sold to thousands of CP/M machine owners.

The first major commercial release of CP/M was version 1.4. In an effort to fix some bugs and improve its file-handling capabilities and limits, it was upgraded first to version 2.0 and quickly to version 2.2. Version 2.2 has been a stable product that is familiar to most readers of BYTE. Version 3.0, or CP/M Plus, has been available for about two years, but it hasn't matched the popularity of version 2.2. While CP/M Plus does offer some advanced features, it is not significantly better than version 2.2.

CP/M has many quirks and shortcomings. Having to warm-boot the system when changing disks is a major inconvenience, and named directories would be more convenient than trying to remember disk designators and user-area numbers. Almost no file or system security is provided. CP/M 2.2 does not support redirection of I/O devices, as MS-DOS and UNIX do. Even the MS-DOS batch facility beats CP/M's more primitive Submit and XSUB two ways from Sunday; CP/M lacks conditional testing at the command level.

Many system integrators decry CP/M's lack of a good menu utility to integrate stand-alone, executable programs. Command-line editing is primitive, and multiple commands on one line would be appreciated; creating a single-name "alias" for multiple commands would be even better. Since CP/M has no search path (à la PC-DOS and UNIX), you must either keep multiple copies of often-used programs in many different user areas (wasting disk space) or patch CP/M with a software kludge to make user area 0 accessible from all user areas (even that doesn't help if you need to access a file from user area 3 while you are in user area 7). Many CP/M users supplement their system utilities with more useful public-domain programs like XDIR, DU, CRC, Help, Unerase, and Diff. It's too bad that CP/M wasn't upgraded more regularly so that it could have evolved into a more mature product.

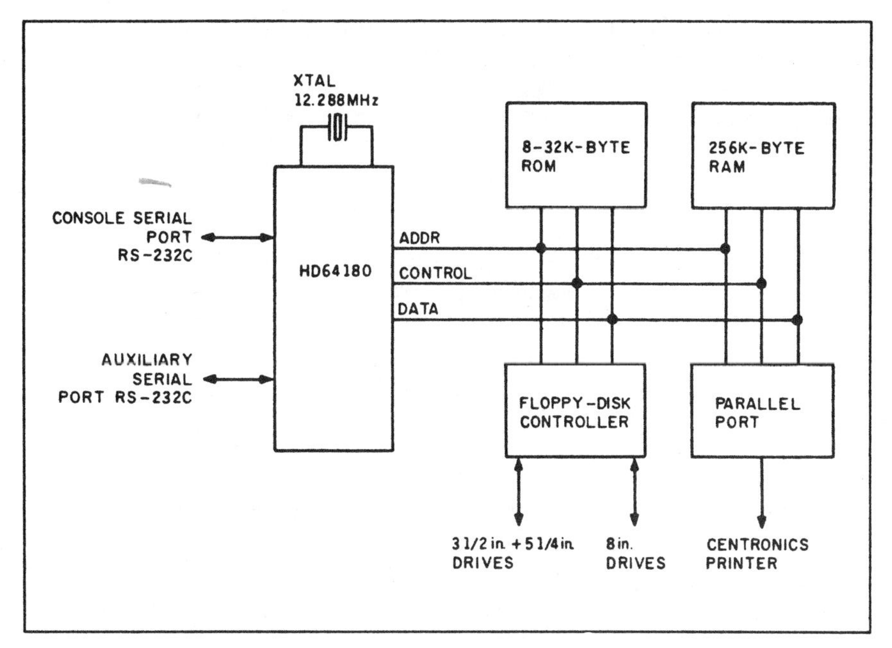

Figure 1: *Block diagram of the* SB180 *single-board computer.*

These observations are certainly not unique. When millions of people use a computer and its operating system in an attempt to become more productive in some way, obstacles to efficiency and productivity are bound to show up. Computer users who have had experience with many different operating systems will miss certain features to which they had become accustomed. And creative individuals of all types always can think of a "better way" to perform some function already provided.

One of the best-known attempts to improve CP/M was ZCPR (Z80 command processor replacement), developed under the direction of Rick Conn. ZCPR was written to replace the console command processor (CCP) supplied with CP/M. This public-domain software featured scaled-down features of UNIX appropriate for a single-user CP/M system. However, since it was designed to fit into the same 2K-byte space as the CCP, it was limited in what it could offer. It did make user area 0 "public" so that executable programs there could be invoked from other user areas. It also changed the prompt so that it indicated the user area as well as the drive currently logged in. As a result of user feedback, ZCPR2 evolved as an extended version of ZCPR. However, it required much more technical sophistication to install it into a CP/M system, and it started patching into and replacing parts of CP/M itself.

Within the past year, Echelon Inc. released ZCPR3, a much-improved version of ZCPR2, which provides solutions to all the problems with CP/M 2.2 and adds more features. Echelon has also developed a complete replacement for CP/M 2.2 in the form of the Z-System. It is composed of two major sections: ZCPR3 and ZRDOS. ZRDOS completely replaces CP/M's BDOS (basic disk operating system).

Written entirely in Z80 assembly language, the Z-System offers the benefits derived from the expanded Z80 instruction set (CP/M is 8080 code only) and from fixing bugs in CP/M 2.2 itself. It is downward-compatible with all CP/M software and takes advantage not only of the Z80 instruction set but of the Hitachi HD64180 CPU as well. Echelon provides both the operating system and a macro-relocating assembly language with linker and librarian, translators, and debuggers. The operating-system utilities, based on many good public-domain utilities, have all been rewritten to have a consistent user interface and make optimum use of ZCPR3 facilities.

The Z-System will be discussed more completely in next month's article. Suffice it to say, it has significantly greater utility and functionality than MS-DOS or plain CP/M. Table 1 gives a comparison of the three operating systems.

THE HITACHI HD64180

The Hitachi HD64180 is based on a microcoded execution unit and advanced CMOS manufacturing technology. It provides the benefits of high performance, reduced system cost, and low-power operation while maintaining complete compatibility with the large base of standard CP/M software. The HD64180 can be combined with CMOS VLSI (very-large-scale integration) memories and peripheral devices to form the basis for process-control applications requiring high performance, battery-power operation, and standard software compatibility.

Performance is derived from a high clock speed (6 MHz now, 9 MHz in the near future), instruction speedup, and an integrated memory-management unit (MMU) with 512K bytes of memory address space. The instruction set is a superset of the Z80 instruction set; 12 new instructions include hardware multiply, DMA (direct memory access), I/O, TEST, and a SLEEP instruction for low-power operation.

The HD64180 requires operation at specific frequencies in order to generate standard data-transmission rates. The standard operating frequency for the SB180 is 6.144 MHz (12.288-MHz crystal). Other frequencies that maintain standard data-transmission rates are 3.072 MHz, 4.608 MHz, and (later) 9.216 MHz.

System costs are reduced because many key system functions have been included on this 64-pin chip. Table 2 compares the HD64180 with other 8-bit processors.

The block diagram in figure 2 shows that the HD64180 CPU is composed of five functional blocks:

- Central processing unit: The CPU is microcoded to implement an upward-compatible superset of the Z80 instruction set. Besides the 12 new instructions, many instructions execute in fewer clock cycles than on a standard Z80.
- Clock generator: This generates the system clock from an external crystal or external clock input. The clock is programmably prescaled to generate timing for the on-chip I/O and system-support devices.
- Bus-state controller: This controller performs all status/control bus activity, including external bus-cycle wait-state timing, $\overline{\text{RESET}}$, DRAM (dynamic RAM) refresh, and master DMA bus exchange. It generates "dual-bus" control signals for compatibility with

Table 1: *A comparison of the Z-System, CP/M 2.2, and MS-DOS.*

Feature	Z-System	CP/M 2.2	MS-DOS
Software-compatible with CP/M 2.2	yes	yes	no
No warm boot required when changing disks	yes	no	yes
Multiple commands per line	yes	no	no
Named directories	yes	no	yes
Password protection for directories	yes	no	no
Dynamically variable user privilege levels for commands	yes	no	no
Searching of alternate directories for invoked programs and files	yes	no	partial
Terminal-independent video capabilities	yes	no	no
Input/output redirection	yes	no	yes
Conditional testing and execution at the operating-system level (IF/ELSE/ENDIF)	yes	no	no
Shells and menu generators with shell variables	yes	no	no
Tree-structured on-line help and documentation subsystem	yes	no	no
512-megabyte file sizes, 8-gigabyte disks	yes	no	no
Complete error trapping with recovery, customizable messages, and prompts	yes	no	no
Screen-oriented file manipulation and automatic archiving and backup	yes	no	no
Full-screen command-line editing with previous-command recall and execution	yes	no	no

Table 2: *A comparison of some 8-bit processors.*

	HD64180	8080/85/Z80	NSC800	Z800	80188
Process	CMOS	NMOS	CMOS	NMOS	NMOS
Power	100 mW	1 W	100 mW	2 W	2 W
Maximum clock	10 MHz	8 MHz	4 MHz	10 MHz	8 MHz
Address space	512K bytes	64K bytes	64K bytes	512K bytes	1 megabyte
UARTs	2-ch.	no	no	1-ch.	no
DMAC	2-ch.	no	no	4-ch.	2-ch.
Timers	2-ch.	no	no	4-ch.	2-ch.
Clocked SI/O	yes	no	no	no	no
CS/wait logic	yes	no	no	yes	yes
DRAM refresh	yes	yes (Z80)	no	yes	no

Note: The availability of the Zilog Z800 is unknown, and specifications on it are subject to change.

both 68xx and 80xx family devices.

• Interrupt controller: The interrupt controller monitors and gives priorities to the four external and eight internal interrupt sources. A variety of interrupt-response modes are programmable.

• Memory-management unit: The MMU maps the CPU's 64K-byte logical-memory address space into a 512K-byte physical-memory address space. The MMU organization preserves software object-code compatibility while providing extended memory access and uses an efficient common area/bank area scheme. I/O accesses (64K bytes of I/O address space) bypass the MMU.

The integrated I/O resources comprise the remaining four functional blocks:

• Direct-memory-access controller (DMAC): The two-channel DMAC provides high-speed memory-to-memory, memory-to-I/O, and memory-to-mem-

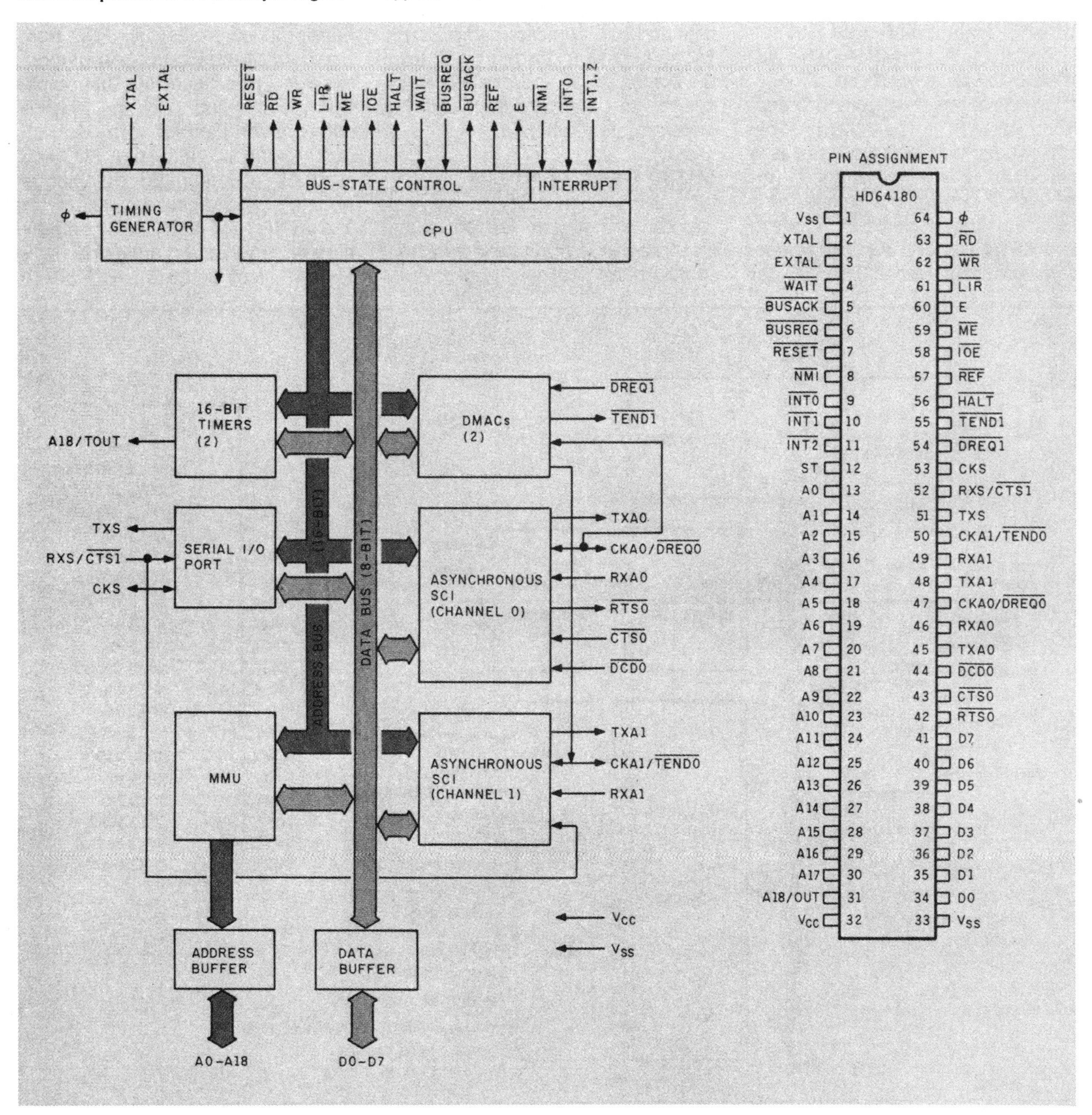

Figure 2: *Block diagram and pin-out of the Hitachi* HD64180.

ory-mapped-I/O transfer. The DMAC features edge or level sense-request input, address increment/decrement/no change, and (for memory-to-memory transfer) programmable-burst or cycle-steal transfer. In addition, the DMAC can directly access the full 512K bytes of physical-memory address space (the MMU is bypassed during DMA) and transfers (up to 64K bytes in length) can cross 64K-byte boundaries. At 6 MHz, DMA is 1 megabyte per second.

- Asynchronous serial communication interface (ASCI): The ASCI provides two full-duplex UARTs (universal asynchronous receiver/transmitters) and includes a programmable data-transmission-rate generator, modem-control signals, and a multiprocessor communication format. The ASCI can use the DMAC for high-speed serial data transfer, reducing CPU overhead.
- Clocked serial I/O port (CSI/O): The CSI/O provides a half-duplex clocked serial transmitter and receiver. This can be used for high-speed connection to another microcomputer.
- Programmable reload timer (PRT): The PRT contains two separate channels, each consisting of 16-bit timer data and 16-bit timer-reload registers. The time base is divided by 20 from the system clock, and one PRT channel has an optional output allowing waveform generation.

SB180 Design Criteria

With all this functionality on one chip, you can see why so few additional chips are needed to implement a truly sophisticated 8-bit single-board computer in such a small space (less than 30 square inches). In terms of the original Altair microcomputer of 10 years ago, the functionally equivalent machine would have taken about 35 S-100 boards for a total of 1750 square inches (and that's using 8K-byte memory boards!).

In order to reduce chip count further, I decided to use an enhanced floppy-disk-controller chip (FDC) from Standard Microsystems Corporation, the 9266 (see figure 3). This 40-pin DIP (dual in-line package) chip is software-compatible with the industry-standard NEC 765A FDC and is actually an integrated combination of SMC's 9229 digital data separator and an NEC 765A FDC. It is compatible with single- and double-sided 3½-, 5¼- (40- and 80-track), and 8-inch

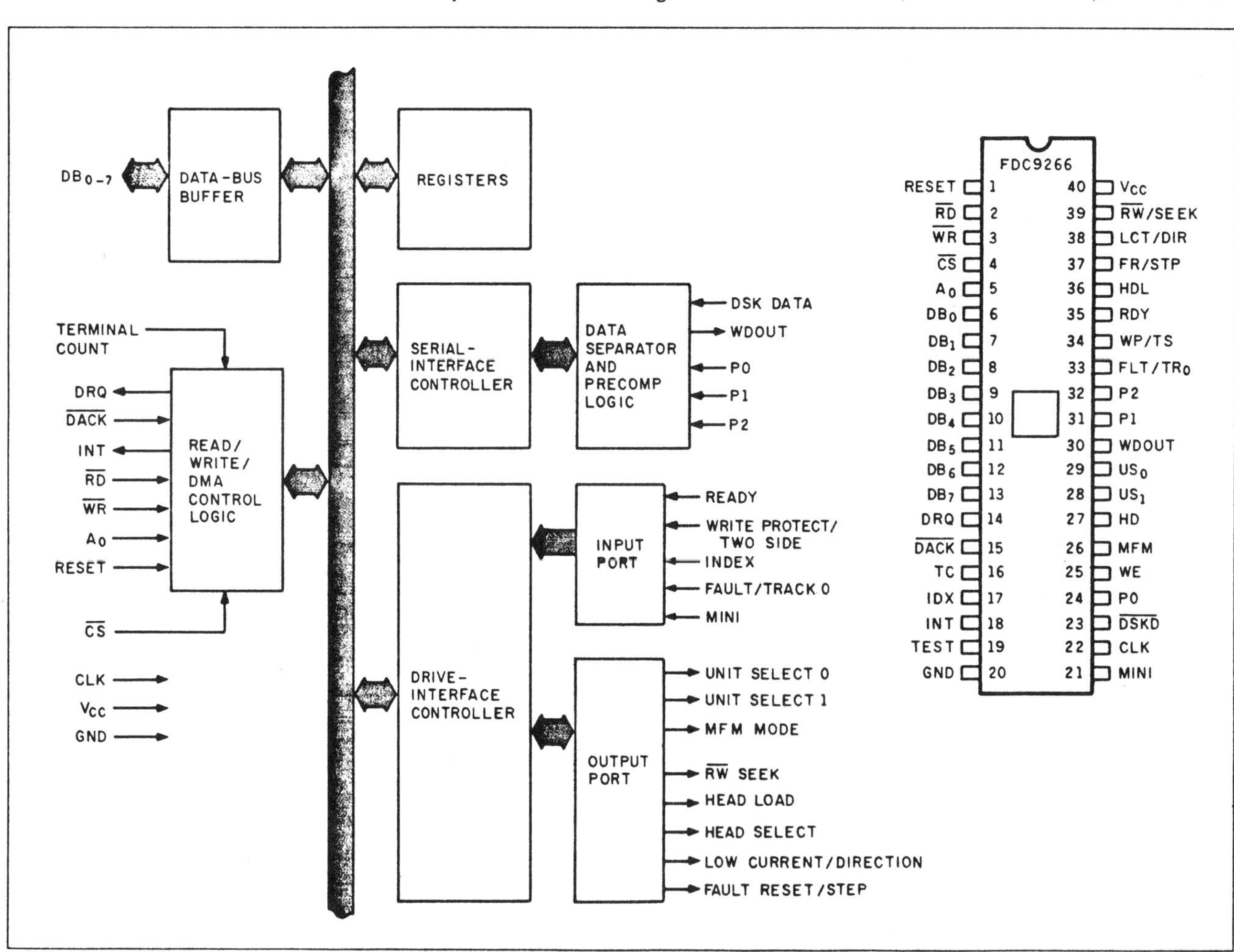

Figure 3: *Block diagram and pin-out of* SMC's 9266.

drives; the data separator handles both single- and double-density data. This means that it can be programmed to read and write most soft-sectored CP/M disk formats.

With the HD64180's two-channel ASCI built in, two serial ports were included in the design automatically. Provision was made for a Centronics parallel printer port as well. Since 256K-bit DRAM chips are now plentiful and inexpensive, eight were used for memory (64K-bit DRAMs can also be used). Because only 64K bytes is used for the logical-memory address space, you can optionally designate the other 192K bytes as a RAM disk in the operating system, or you may prefer to use if for other purposes (such as implementing banked memory for CP/M Plus).

The SB180 Hardware

Figure 4 is the schematic diagram of the SB180 computer. Its design is primarily characterized by the high-performance, high-density MOS (metal-oxide semiconductor) devices, including 256K-bit DRAMs.

In addition to the CPU elements previously discussed, the SB180 system design implements the following functional blocks: RS-232C interface, memory interface, Centronics printer interface, floppy-disk interface, XBUS expansion bus, and power supply.

RS-232C Interface

The HD64180 ASCI two-channel UART is connected to 1488/1489 RS-232C line drivers/receivers to provide two separate ports. ASCI channel 1 is used for the console; ASCI channel 0 is used for auxiliary RS-232C devices like printers, plotters, and modems. This distinction is made because modems require the extra handshake signals that are available with ASCI channel 0, while terminals do not. All primary RS-232C parameters (data-transmission rate, handshaking, data format, interrupts) are software-programmable.

Memory Interface

The SB180 incorporates a 28-pin boot-ROM socket that can be jumpered to hold 8K by 8-bit, 16K by 8-bit, and 32K by 8-bit memory devices. The boot ROM (contains disk boot and ROM monitor) occupies the bottom 256K bytes of the HD64180 physical-memory address space since it is selected whenever A18/TOUT is low. (Note: The TOUT timer output function is not used.) Thus, the boot-ROM contents (whatever the size) are simply repeated in the lower 256K bytes. The boot-ROM output (OE) is enabled by the HD64180 ME (memory enable) signal. (As configured, the maximum RAM on the SB180 is 256K bytes. To support larger memories, additional address decoding would be required to designated RAM and ROM areas in the current 256K-byte boot-ROM space.)

ROMs of 200 nanoseconds (and, marginally, 250 ns) can operate with one wait state.

At RESET, the HD64180 begins execution at physical address 00000, the start of the boot ROM. [*Editor's note: All addresses are in hexadecimal.*]

256K-bit Dynamic RAM

Standard 256K-bit 150-ns DRAMs, requiring 256 refresh cycles (8-bit refresh address) every 4 milliseconds (ms) are used. These DRAMs occupy the top 256K bytes of the HD64180 512K-byte physical-memory address space.

The interface is quite straightforward. Complete DRAM refresh control is provided by the HD64180 in conjunction with control logic IC14 and IC22 and address multiplexers IC11, IC12, and IC13.

The HD64180 WR output directly generates DRAM WE. The HD64180 ME output directly generates RAS. During normal read/write cycles (A18 high, REF high), CAS goes low at the next rising edge of ϕ following the rising edge of E (enable). This provides plenty of setup time for the address multiplexers since the rising edge of E switches the address multiplexers from row to column addresses.

RAS-only refresh is used. The HD64180 generates the refresh addresses. During refresh cycles (REF low), ME generates RAS, while CAS is suppressed at IC22.

The HD64180 can be programmed to generate refresh cycles every 10, 20, 40, or 80 ϕ cycles as well as selecting a refresh every two or three clock cycles. Since the DRAM requires a refresh cycle every 15.625 microseconds (μs) (4 ms/256), the HD64180 is programmed for 80-cycle refresh request since 80 $\times$ (1/6.144 MHz) = 13.02 μs. Two-cycle refresh is also programmed. Thus, refresh overhead is only 2.5 percent (2 cycles every 80 cycles).

Centronics Printer Interface

The Centronics printer interface is composed of the 8-bit latch IC9 and flip/flop IC10. The Centronics port is decoded at I/O address 0C0 by IC26. To write to the printer, the following sequence is used:

Write data to port 0C1

This sets up the data to the printer and asserts STB low. The following sequence:

Write data to port 0C0

deasserts the printer STB signal high.

When the printer has processed the data, it will return the ACK signal, which generates an external interrupt (INT 1) to the HD64180. The interrupt handler clears the interrupt by performing a dummy output to port 0C0. This clears the INT 1 interrupt request.

The printer interface is not buffered, so compatibility with all printer/cable setups cannot be guaranteed. In practice, however, problems should be rare since the software scheme provides adequate data-setup and -hold times. Also, note that this printer interface is interrupt-driven, which allows high-performance operation. In a more primitive polling design, excessive overhead limits acceptable performance in applications like background print spooling.

Floppy-Disk Interface

SMC's 9266 FDC manages almost all details of the drive interface, including data separation and (with external logic IC22 and IC23) programmable

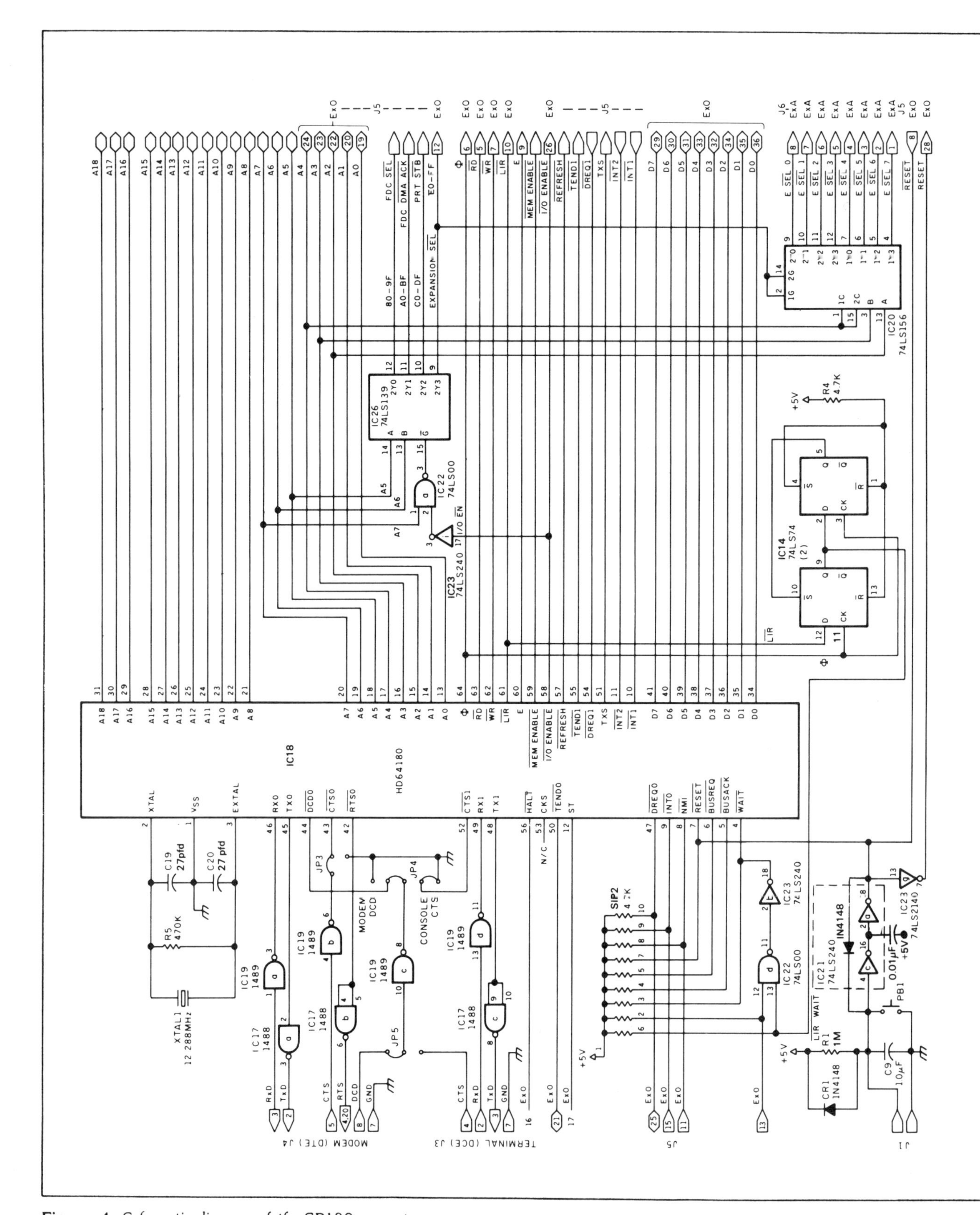

Figure 4: *Schematic diagram of the SB180 computer.*

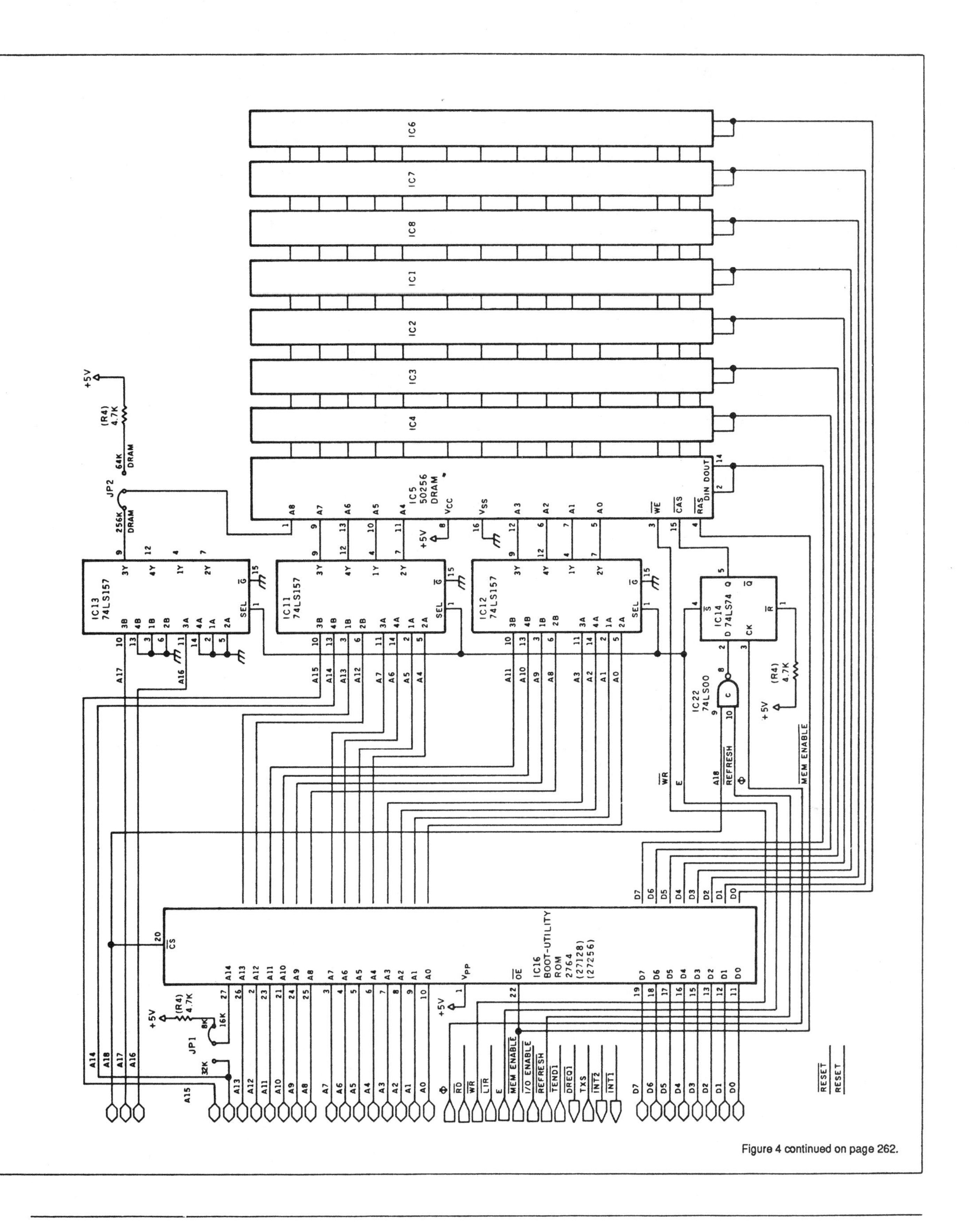

Figure 4 continued on page 262.

Figure 4 continued

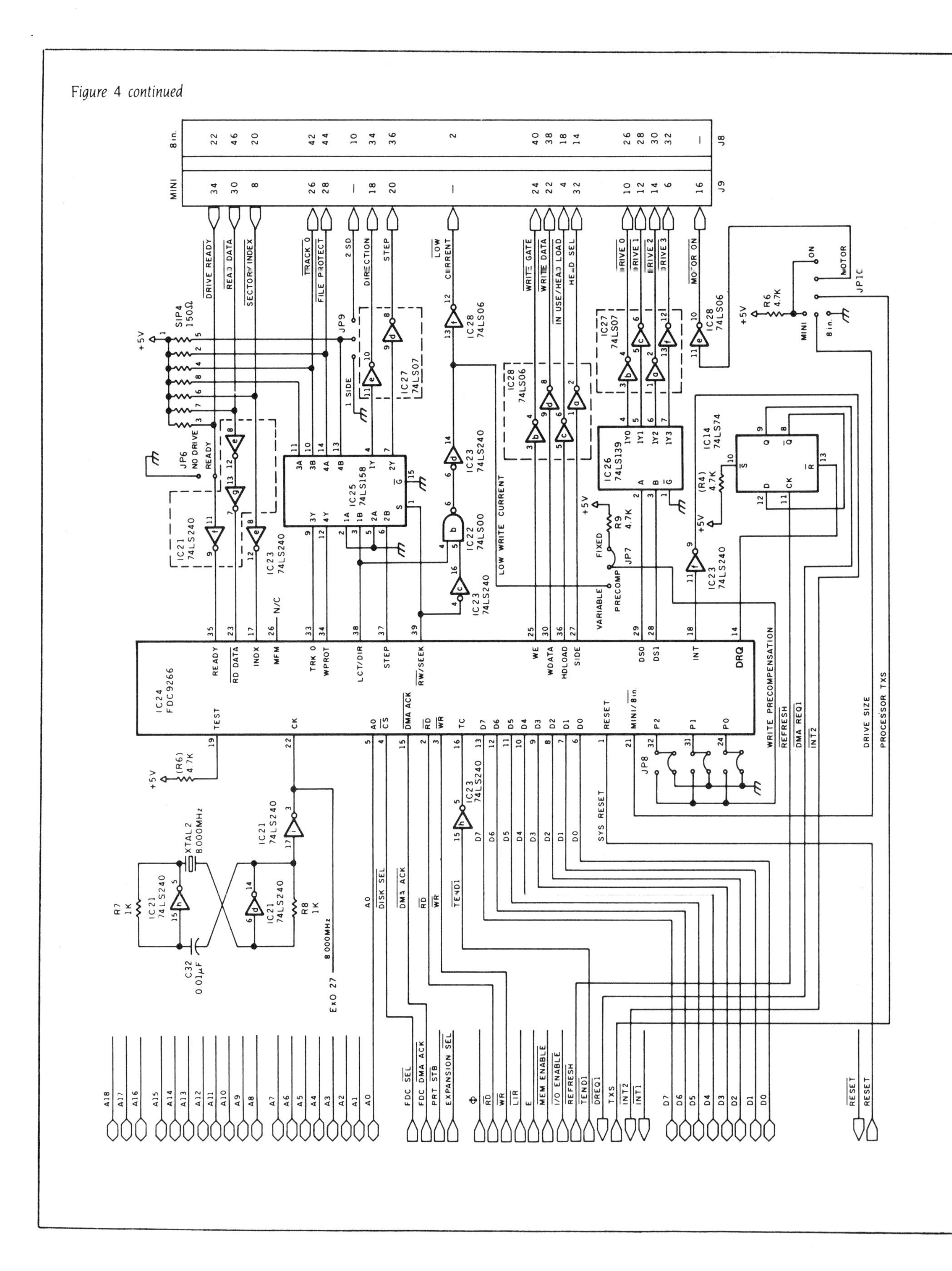

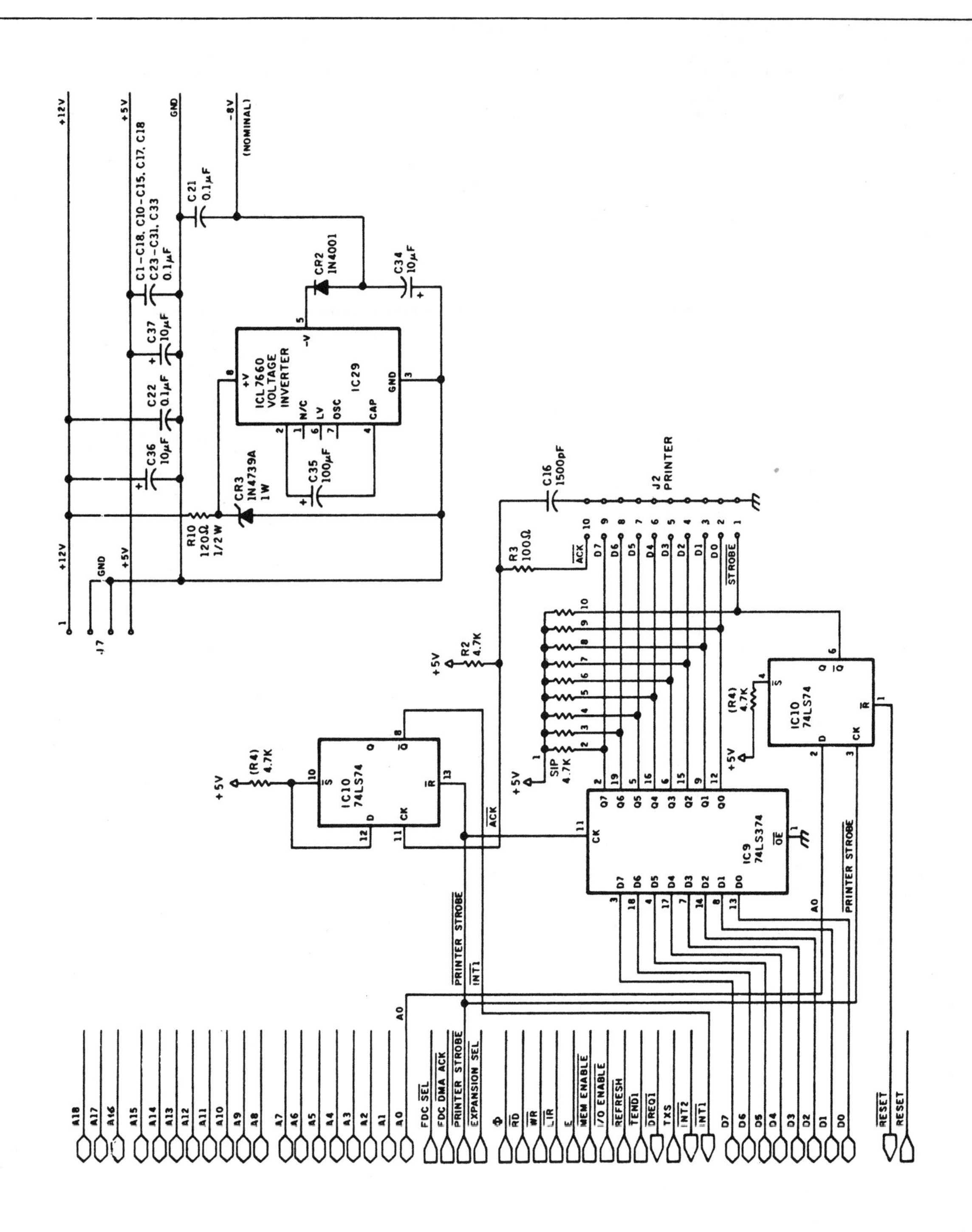

write precompensation. The 9266, as mentioned earlier, combines the NEC 765A/Intel 8272 FDC with SMC's 9229 digital data separator. Thus, from the host CPU side, the 9266 looks just like these devices, including hardware and software compatibility.

The 9266 clock is generated by an 8-MHz oscillator composed of a crystal and IC21. Jumpers are provided to select write precompensation and allow 8-inch floppy-disk drives to be interfaced.

On the CPU side, the key requirements are interfacing the 9266 with both programmed I/O ($\overline{\text{CS}}$), for things like initialization and status check, and with DMA (DRQ, $\overline{\text{DACK}}$), for data transfer.

Programmed I/O is straightforward, with $\overline{\text{CS}}$ generated for I/O address 80, and $\overline{\text{RD}}$ and $\overline{\text{WR}}$ directly generated by the HD64180. This is the same scheme used to interface with other 80xx family peripheral devices.

DMA is a little more involved. First, DMAC channel 1 is used for the FDC since dedicated handshaking lines ($\overline{\text{DREQ1}}$, $\overline{\text{TEND1}}$) are provided on the HD64180. Since DMAC channel 0 control lines are multiplexed (with ASCI clocks), DMAC channel 0 is used for memory-memory DMA. This means that the ASCI clock functions are available, although they are not currently used in this design.

For disk DMA, the 9266 asserts DRQ, which in turn causes HD64180 $\overline{\text{DREQ1}}$ assertion. The HD64180 performs DMA reads/writes to I/O address 0A0, which causes the 9266 $\overline{\text{DACK}}$ to be asserted, completing the transfer cycle. After the DMAC's programmed number of reads/writes has completed, the HD64180 $\overline{\text{TEND1}}$ output is asserted and, after inversion, causes the 9266 TC (terminal count) input to be asserted, completing the DMA operation. This is typically followed by the 9266 generating an HD64180 $\overline{\text{INT 2}}$ external interrupt. This interrupt-service routine can read the 9266's status to determine if errors occurred, etc.

However, one "gotcha," fixed by flip/flop IC14, conditions the 9266 DRQ output. It turns out that if 9266 DRQ directly generates HD64180 $\overline{\text{DREQ1}}$, the HD64180 may respond too quickly. This is because HD64180 $\overline{\text{DREQ}}$ input logic was designed to minimize latency, and $\overline{\text{DREQ}}$ can thus be recognized at a machine-cycle breakpoint. Unfortunately, the 9266 requires that at least 800 ns elapse from the time it asserts DRQ before the DMA transfer ($\overline{\text{DACK}}$) actually occurs. In other words, when the 9266 "asks" for service, it really doesn't want it . . . yet! To prevent accessing the 9266 too quickly after DRQ, DRQ from the 9266 is delayed at IC14 before issuing the $\overline{\text{DREQ1}}$ to the HD64180. DRQ is delayed by one $\overline{\text{REF}}$ cycle time.

Minifloppy double-density data transfers occur at a 250-kilohertz (kHz) data rate. Thus, each byte must be read within 32 μs. The disk-driver software reprograms the refresh-

Table 3: *Power connections for figure 4.*

		+5 V	GND	+12 V	−8 V
IC1	50256 DRAM 150 ns	8	16	—	—
IC2	50256 DRAM 150 ns	8	16	—	—
IC3	50256 DRAM 150 ns	8	16	—	—
IC4	50256 DRAM 150 ns	8	16	—	—
IC5	50256 DRAM 150 ns	8	16	—	—
IC6	50256 DRAM 150 ns	8	16	—	—
IC7	50256 DRAM 150 ns	8	16	—	—
IC8	50256 DRAM 150 ns	8	16	—	—
IC9	74LS374	16	8	—	—
IC10	74LS74	14	7	—	—
IC11	74LS157	16	8	—	—
IC12	74LS157	16	8	—	—
IC13	74LS157	16	8	—	—
IC14	74LS74	14	7	—	—
IC15	74LS74	14	7	—	—
IC16	2764 monitor ROM 200 ns	28	14	—	—
IC17	1488	—	7	14	1
IC18	HD64180	32	1,33	—	—
IC19	1489	14	7	—	—
IC20	(74LS156) optional	16	8	—	—
IC21	74LS240	20	10	—	—
IC22	74LS00	14	7	—	—
IC23	74LS240	20	10	—	—
IC24	FDC 9266	40	20	—	—
IC25	74LS158	16	8	—	—
IC26	74LS139	16	8	—	—
IC27	7407	14	7	—	—
IC28	7406	14	7	—	—
IC29	7660 regulator	—	3	—	—

Jumpers

JP1	(1x3)	ROM size
JP2	(1x3)	RAM size
JP3	(1x3)	CTS0 disable
JP4	(1x5)	RS-232C gate output
JP5	(1x3)	RS-232C gate input
JP6	(1x2)	Drive without ready
JP7	(1x3)	Write precomp mode
JP8	(2x6)	Write precomp value
JP9	(1x3)	Single-/double-sided drive
JP10	2 – 1x3	TXS function/drive size

request rate from every 80 ϕ cycles to every 40 ϕ cycles prior to disk DMA and reassigns it back to 80 ϕ cycles after the disk DMA is completed. The 9266 DRQ is delayed from between 40 ϕ cycles to 79 ϕ cycles. This is about 6 to 14 μs. Therefore, the 800-ns delay and 32-μs data-transfer constraint are both met. Note that 8-inch floppy double-density data transfer is twice as fast (500 kHz) and requires a refresh-rate increase to every 20 ϕ cycles.

Expansion Bus

The spare $\overline{CS}$ from address decoder IC26 (I/O addresses 0E0 to 0FF) and all major buses (address, data, control) are routed to the expansion bus. This allows an I/O expansion-board capability. The full complement of HD64180 control signals ($\overline{IOE}$, E, $\overline{RD}$, $\overline{WR}$, etc.) allows easy interface to all standard peripheral LSI (large-scale integration) devices, including the 80xx, 68xx, and 65xx families. Example expansion boards could include a hard-disk controller, 1200-bits-per-second (bps) modem, or LAN (local-area network) interface.

Power Supply

The SB180 requires +5-V and +12-V power. A negative voltage is generated on board, which is used only by the RS-232C driver. The negative voltage is obtained by using a Zener diode to obtain +9 V from +12 V, which is then inverted using an Intersil 7660 converter. The +12-V power is used only for the RS-232C driver. Thus, the SB180 uses only significant power from the +5-V supply. Typically, this may be from 0.3 to 0.6 ampere (A) (depending on the proportion of the TTL [transistor-transistor logic] and memory devices that are CMOS)—about the same as a 5¼-inch floppy disk.

The SB180 ROM Monitor

While my initial discussion espoused the merits of the Z-System, every good single-board computer needs a ROM monitor that should be more useful than just booting the operating system from a floppy disk. The SB180 8K-byte ROM monitor includes commands for everything you need—from A to Z. It is supplied on an 8K-byte EPROM. (The SB180 also supports 16K- and 32K-byte ROMs, so additional commands or application programs can be supported.)

The SB180 ROM monitor provides commands to assist the design and debugging of SB180-related hardware and software. Also, it serves as a stand-alone training vehicle for the HD64180 CPU.

The monitor supports the following I/O devices:

CON: Console RS-232C serial port
AUX: Auxiliary RS-232C serial port
CEN: Centronics parallel printer port
DSK: Floppy-disk storage devices

The monitor supports one to four 5¼-inch, 48- or 96-tpi (tracks per inch), 40- or 80-track, double-sided double-density disk drives. During initial system checkout, such a drive should be connected to verify operation of the disk interface.

At $\overline{RESET}$, the monitor first checks for fatal RAM failure. If the RAM is bad, the monitor waits for a carriage return to be typed on the console in order to determine the console data-transmission rate and then prints 8 binary digits—each corresponding to one DRAM chip. A chip associated with a 1 is faulty. The RAM-check se-

A ASCII table: Prints an ASCII table
B Bank select: Selects a 64K-byte memory bank
C Copy disk: Systems with 256K bytes of RAM can perform single-drive copies
D Display memory: Displays memory in hexadecimal and ASCII
E Emulate terminal: Console keyboard is echoed to the auxiliary RS-232C output, and the RS-232C input is echoed on the console display
F Fill memory: Any portion of memory is filled with a data byte
G Go to program: Starts program execution at specified address and optionally includes a breakpoint
H Hexmath: Prints the 20-bit sum and difference and 32-bit product of the two arguments
I Input port: Prints the 8-bit data input from specified port
K Klean disk: Formats a specified drive
M Move memory: Moves a block of memory
N New command: Enables new commands from extended ROM space
O Output port: Byte is output to specified port address
P Printer select: Toggles printer selection between the Centronics parallel port and the auxiliary RS-232C port
Q Query memory: Searches memory for pattern of 1 to 4 bytes
R Read disk: Reads specified sectors from drive into memory
S Set memory: Displays memory contents and allows new data to be entered
T Test system: Tests various system devices
U Upload hexadecimal file: Uploads Intel hexadecimal file from auxiliary or console serial port
V Verify memory: Compares two blocks of memory
W Write disk: Writes specified sectors to disk from memory
X eXamine CPU registers: Displays main and alternate CPU registers and prompts for modification of main registers
Y Yank I/O registers: Displays the HD64180 on-chip I/O registers
Z Z-System boot: Boots the Z-System from disk

Figure 5: *A list of the monitor commands.*

quence is repeated each time a carriage return is typed.

If the RAM is okay, the stack is set up, and the monitor checks to see if a disk is loaded in drive #0. If so, the DOS boot-load sequence (same as the Z command) is started.

The monitor commands are shown in figure 5.

THE LUNCHBOX COMPUTER

The only things you need to turn the SB180 into a full-fledged stand-alone microcomputer are a +5-V and +12-V power supply (or a 12-V battery and 5-V regulator), at least one 5¼-inch floppy-disk drive (initially), and a serial terminal.

You can mount such a small computer in many ways (see photos 3 and 4). If you use a half-height drive, the SB180 will fit on top of the drive inside a single full-height disk-drive chassis including power supply! Since the console serial port can automatically detect the terminal's data-transmission rate, you can carry your SB180 with you and connect it to any terminal (or computer emulating a terminal) running at 300, 1200, 9600, or 19,200 bps (other rates are optionally selected).

It is possible to fit the SB180, power supply, and two 3½-inch floppies into a lunchbox. If you want to get fancy, you could fit the SB180, power supply, one or two half-height floppies, the Micromint Term-Mite terminal board, and a keyboard into a small attaché case. Use any handy video monitor, and . . . voilà! You have just out-Osborned Adam Osborne!

I can imagine many unusual ways to package the SB180. When good LCDs (liquid-crystal displays) come down in price a bit more, I think the SB180 can form the basis of a functional notebook computer. The SB180 makes minimal use of the +12-V supply (and only for RS-232C operation) and can use less than 1 A at +5 V, so battery power is a real possibility. With just one floppy disk, the SB180 can become a super data logger. While my Z8 BASIC or FORTH single-board computer might ordinarily be used for such an application, not everyone likes to program in FORTH or BASIC. With a CP/M-compatible computer, a developer can now choose Pascal, C, FORTRAN, or even PL/I for applications. Since the data is already in CP/M format, it makes data analysis convenient.

The Z-System is provided with the complete SB180 board and software package. While it comes with a utility to read several other common 5¼-inch formats (like Kaypro and Osborne), its native format is identical to the well-optimized 386K-byte

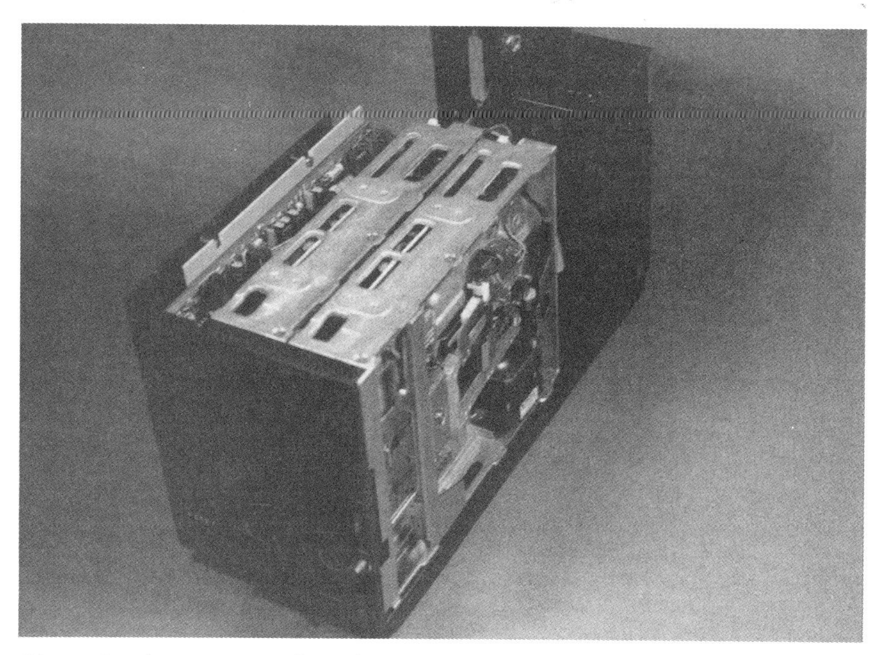

Photo 3: *The* SB180 *can be packaged in many off-the-shelf enclosures advertised in* BYTE. *It fits comfortably to the left of a pair of half-height 5¼-inch disk drives in a case available from* Disk Plus. *This enclosure also includes the power supply.*

Photo 4: *The enclosure from photo 3 is shown with the cover on. The only additional hardware necessary to use the system is a serial terminal.*

double-sided double-density format of the Little Board from Ampro Inc. Provisions have been made to support 8-inch drives as well. Of course, it is possible to implement CP/M 2.2, CP/M Plus, MP/M II, TurboDOS, or Oasis if you prefer.

Although simple benchmarks are not to be taken as the last word in describing a computer's performance, I did run the BYTE Sieve of Eratosthenes prime-number program on the SB180 and the MPX-16 (8088-based with a 4.77-MHz clock, same as the IBM PC) using appropriate versions of Microsoft BASIC. For one iteration, the MPX using GW-BASIC took 203 seconds, while the SB180 using MS-BASIC took 147 seconds. An empty FOR . . . NEXT loop with 20,000 iterations took 27 seconds on the MPX and only 18 seconds on the SB180. Don't forget that these results are based on the 6-MHz implementation of the HD64180; with a 9-MHz clock, the results will be spectacular—not quite the speed of an IBM PC AT, but close!

Special thanks to Tom Cantrell, Merrill Lathers, and Bob Stek for their contributions to this project.

See page 447 for a listing of parts and products available from Circuit Cellar Inc.

Photo 1: A *standard lunchbox can be converted into a powerful computer with the* SB180.

PHOTOGRAPHED BY PAUL AVIS

25

BUILD THE SB180 SINGLE-BOARD COMPUTER

PART 2: THE SOFTWARE

This computer reasserts 8-bit computing in a 16-bit world

The SB180 computer system represents the state of the art in 8-bit systems (for detailed specifications of the SB180, see pages 252 to 267). It also elevates the power-per-square-inch ratio to a new high (see photos 1 through 6). However, much of the hardware's potential would be wasted if the software were not as advanced. In Part 2 of Chapter 7, I'll continue my discussion of the SB180 with emphasis on the DOS (disk operating system).

I began with some general ideas about what the software should do. I wanted a DOS, but it had to accommodate the new 3½-inch disk drives as well as older 5¼- and 8-inch units. A primary requirement was that it needed to be compatible with the most widespread "8-bit" DOS, CP/M 2.2. However, it needed to be free of the many restrictions and quirks of CP/M 2.2 and should represent a step forward in the logical development of operating systems.

The operating system and its utilities should not be separately developed and then stitched together, like a crazy quilt, but they instead should be developed concurrently so that they use a consistent command structure. Finally, since it is a Circuit Cellar project, the system must facilitate a high degree of user customization. It must be flexible enough to operate at 100 percent of the system's potential in one application, yet it must allow a terminal to be connected and a user to interact with it even if no disk drives are connected in another application.

ADVANCED FEATURES OF THE HD64180

From a programming point of view, the HD64180 microprocessor resembles its predecessor, the Z80, but also executes 10 additional instructions. The mnemonic names of these instructions are SLP, MLT, INO, OUTO, OTIM, OTIMR, OTDM, OTDMR, TSTIO, and TST.

The SLP instruction puts the microprocessor into a "sleep" mode that uses little power; it would not be used in a DOS situation but is available for use in a user's own programs.

The MLT instruction is an impressive feature of the HD64180. It multiplies two 8-bit quantities and results in a 16-bit product. Again, this instruction is not usually used in operating-system software.

The remaining instructions perform functions like block output with increment, decrement, and repeat; input or output of any register to an immediate I/O (input/output) port address; and nondestructive AND logic operations on the various registers, I/O ports, and immediate data.

This last group of eight instructions would be convenient for use in a DOS based on an HD64180 microprocessor. Most of these instructions could be used in accessing the on-chip peripheral hardware like the asynchronous RS-232C ports or the memory-management unit (MMU).

Photo 2: A *section of an aluminum cake pan is cut and bent to form a support for a small switching power supply.*

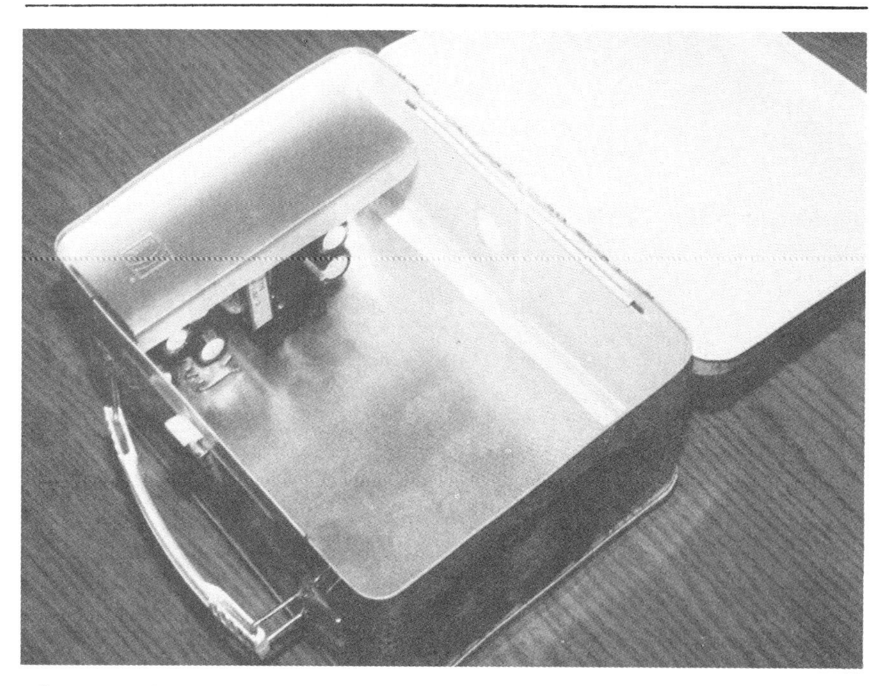

Photo 3: *The power supply is installed in the corner of the lunchbox.*

MATING HARDWARE AND SOFTWARE

I eventually found an operating system that met these challenging requirements in the Z-System from Echelon Inc. of Los Altos, California. The Z-System is compatible with programs that run under CP/M 2.2 and contains a multitude of improvements. The Z-System and its utilities were developed together in a common environment and share a common command structure. The system utilities can be used together in different ways to create new, powerful commands.

The system is adaptable, with several of the utilities being menu-driven. It even has menu processors that allow you to personalize the entire operating system to whatever level of sophistication you require. Best of all, extensive amounts of the source code to the system will be made available to those users who have the knowledge and desire to customize the system at the fundamental levels.

As I explained in pp. 252-267, better and faster microprocessors mean absolutely nothing if they are bound by inefficient operating systems. The SB180 with the Z-System is an unbeatable combination that seriously challenges the advertising-hyped credibility of 16-bit computers.

INTRODUCTION TO THE SOFTWARE

The most visible component of the Z-System is the command processor, called ZCPR3. It is the most visible component because ZCPR3 acts as the user interface to the rest of the operating system: interpreting commands, loading programs that are to be executed, and more.

ZCPR3 is more than just a command processor; it is also more than 70 utility programs, all of which make use of the special features of the environment provided by ZCPR3. These utilities can be used together in many ways to create new system commands to accomplish more powerful tasks. It is similar to UNIX in the way that individual programs can be combined to make a single new program.

DOS ARCHITECTURE

First, I'll define some terms. An *application program* is one that is not an intrinsic part of the operating system. Examples of this are WordStar or the ZCPR3 utilities. A *user area* is a way of partitioning the storage capacity of a disk and was originated in CP/M 2.2. Up to 32 user areas are on each disk drive; only user areas 0 through 15 are usually accessible. *Transient program area* (TPA) is the segment of memory, beginning at address 100 hexadecimal, where application programs are loaded by the command processor. A *file type* refers to the file's intended use and is indicated by the last 3 characters of a filename. In the Z-System, a filename has a total of 11 characters. (As an example, the filename FILENAME.TYP has the file type TYP.)

Figure 1 outlines the memory map of the SB180 software environment.

ZCPR3

To understand ZCPR3, you need an understanding of what a command processor is. A minimum definition is that it acts as the interpreter between you and the rest of the operating system. A command processor is the component of the operating system that prompts you for a command and then attempts to execute it.

ZCPR3 does everything mentioned above and much more. One of its most powerful features is its ability to act as an interpreter for application programs that want to generate operating-system-level commands, not just for the user. This means that programs like the ZCPR3 utilities can generate new commands that are then fed to the operating system and processed as though you had typed the command at the console.

Another aspect of ZCPR3 that needs to be understood is the concept of system segments. Six system segments are in a fully implemented ZCPR3 system. A system segment is a file that is loaded into a predetermined area of memory.

The ZCPR3 command processor and utilities can call upon a system segment to perform a function or provide information. Memory-resident segments can be overlaid with a new segment at any time, providing the ZCPR3 command processor and utilities with extended functions.

ENVIRONMENT DESCRIPTOR

The first, and most important, segment is the Environment Descriptor

Photo 4: *Another section of cake pan is bent into a "U" shape and drilled to support two 3½-inch disk drives.*

Photo 5: *The disk drives are installed in the cake-pan housing.*

(ENV), which occupies addresses 0FE00 through 0FEFF hexadecimal in the memory map (see figure 1). [*Editor's note: The addresses in this article are in hexadecimal.*] Because of the many possible choices in exactly how ZCPR3 can be implemented, ENV "describes" how that particular ZCPR3 implementation is configured. The ZCPR3 command processor and utilities use the information supplied by the ENV to determine the CPU (central processing unit) clock rate, number of disk drives installed in the system, where the other system segments can be found in memory, and more. When the Environment Descriptor segment is stored in a disk file, the file type is ENV.

NAMED DIRECTORY

The next segment is the Named Directory (NDR). In the memory map, NDR occupies addresses 0FC00 through 0FCFF. This segment also supplies information to the ZCPR3 command processor and utilities, as ENV does.

The NDR segment assigns symbolic names to disk drives and user areas of the system. This means that a name like BASIC may become associated with a particular user area on a disk drive. The ZCPR3 command processor and utilities will then refer to that disk drive and user area whenever a command is executed that contains the directory name BASIC. This gives you the ability to assign names to specific sections of your disk drives, which you can easily reassign by loading a new NDR segment. The file type of a file containing a Named Directory segment is NDR.

Photo 6: *The side of the lunchbox is cut out so that the drives protrude, and the drive assembly is bolted to the power-supply bracket. The SB180 is installed underneath, and cables are run for power and disk drives.*

Listing 1: *An example of the flow commands in a possible* STARTUP.COM *alias. Only if the operator responds with a "Y" in step 4 will the* MDSK /I *command in step 5 be executed (which will initialize the* RAM *disk). This is handy for rebooting the system after a program crash and preserving the contents of the* RAM *disk.*

Command	Comments
1 LDR SYS.RCP,SYS.FCP,SYS.NDR,SYS.Z3T;	| Load system segments
2 WHEEL SYSTEM;	| Set wheel byte
3 ECHO SHOULD THE RAM DISK BE INITIALIZED?;	| Ask question
4 IF INPUT;	| Get yes or no
5 MDSK /I;	| If yes, initialize RAM disk
6 FI	| Terminate IF (ENDIF)

RESIDENT COMMAND PACKAGE

The Resident Command Package (RCP) segment is a collection of subroutines that extend the intrinsic commands of the operating system. An intrinsic command is a routine that resides in memory and that can be executed without disturbing the TPA. As an example in the CP/M 2.2 environment, the DIR command is an intrinsic command, but the STAT command (which loads the STAT.COM program into the TPA and therefore disturbs the TPA) is not.

In the software supplied with the SB180, the intrinsic commands that reside in the ZCPR3 command processor are GO, SAVE, GET, and JUMP. The intrinsic commands added by the RCP are CP, ERA, TYPE, LIST, P (PEEK), POKE, PROT, and REN. As you see, the RCP adds many commands. Additional commands are available within an RCP, but the RCP segment must fit within a 2K-byte area of memory. The above commands make it just a few bytes short of this limit.

If you include another command, you would have to disable one of the existing commands in order for the RCP to fit into its assigned area. In the memory map, the RCP occupies addresses 0F200 through 0F9FF. The file type of a file containing a Resident Command Package segment is RCP.

FLOW CONTROL PACKAGE

The Flow Control Package (FCP) resides between addresses 0FA00 and 0FBFF. It is unique to ZCPR3; no comparable feature is found in any other microcomputer operating system. The FCP adds conditional testing to operating-system-level commands.

An example of conditional testing is the IF. . .THEN . . . ELSE statement in high-level languages like BASIC. The FCP gives the ZCPR3 command processor this testing capability. It is not usually used while you are entering commands at the console, but you can take full advantage of it in batch-processing operations. The file type of a file containing a Flow Control Package segment is FCP.

To understand how the flow commands are useful, you must first know about the ZCPR3 flow state, which is either true or false. While the flow state is true, the ZCPR3 command processor will execute all commands. If the flow state is false, the ZCPR3 command processor will ignore all commands except ELSE and FI.

The IF command is capable of setting the flow state to either true or false. The IF command can evaluate a number of tests: the existence of a file on disk, whether or not a file is empty, the state of the wheel byte (explained in detail later), and more.

A good example of how to use the IF command is shown in listing 1. ZCPR3 allows you to nest the IF/ELSE/FI (ENDIF) flow commands up to eight levels deep.

INPUT/OUTPUT PACKAGE

The Input/Output Package (IOP) segment acts as a traffic cop in routing input and output to and from peripheral devices. The print spooler supplied with the SB180 is an example of an IOP. Other IOPs let you set up programmable function keys or capture characters in a disk file that are normally sent to the console or list device.

The IOP occupies addresses 0EC00 through 0F1FF in the memory map, and the file type of a file containing an Input/Output Package segment is IOP.

TERMINAL CAPABILITIES

The Terminal Capabilities (TCAP) segment is actually contained within the

```
 -------------------------------------  3FFFF Hex
 | RAM Disk (192K bytes)              |
 -------------------------------------  0FFFF Hex
 | ZCPR3 External Stack               |
 -------------------------------------  0FFD0 Hex
 | ZCPR3 Multiple Command             |
 | Line Buffer                        |
 -------------------------------------  0FF00 Hex
 | ZCPR3                    TCAP      |
 | Environment            ------------  0FE80 Hex
 | Descriptor Segment                 |
 -------------------------------------  0FE00 Hex
 | ZCPR3 Wheel Byte                   |
 -------------------------------------  0FDFF Hex
 | ZCPR3 External Path                |
 -------------------------------------  0FDF4 Hex
 | ZCPR3 External File                |
 | Control Block                      |
 -------------------------------------  0FDD0 Hex
 | ZCPR3 Message Buffers              |
 -------------------------------------  0FD80 Hex
 | ZCPR3 Shell Stack                  |
 -------------------------------------  0FD00 Hex
 | ZCPR3 Named Directory              |
 | Segment                            |
 -------------------------------------  0FC00 Hex
 | ZCPR3 Flow Control                 |
 | Package Segment                    |
 -------------------------------------  0FA00 Hex
 | ZCPR3 Resident Command             |
 | Package Segment                    |
 -------------------------------------  0F200 Hex
 | ZCPR3 Input/Output                 |
 | Package Segment                    |
 -------------------------------------  0EC00 Hex
 | SB180 BIOS (Basic Input/           |
 | Output System)                     |
 -------------------------------------  0DA00 Hex
 | ZRDOS Disk Operating               |
 | System                             |
 -------------------------------------  0CC00 Hex
 | ZCPR3 Command Processor            |
 -------------------------------------  0C400 Hex
 | TPA (Transient Program             |
 | Area)                              |
 -------------------------------------  00100 Hex
 | Page 0 Buffers and                 |
 | Reserved Locations                 |
 -------------------------------------  00000 Hex
```

Figure 1: *The memory map of the* SB180's *software system, as initially configured. You can modify the system to meet your own needs.*

Environment Descriptor segment, although it can be loaded independently. It resides at addresses 0FE80 through 0FEFF in the memory map. Information stored here describes characteristics of your terminal, specifically the strings that invoke the terminal's clear screen, cursor addressing, highlight on/off, and other functions.

Also stored in this segment are the codes generated by any arrow keys on the terminal. The ZCPR3 command processor and utilities use this information to enhance interaction with you, by offering flashy displays and using the arrow keys for various functions. The important thing to understand about the TCAP segment is that it is easily changed if you attach a different terminal. Because the ZCPR3 utilities refer to the segment for their information, they do not need to be changed as well. This feature can be described as terminal independence. The file type of a file containing a Terminal Capabilities segment is Z3T.

Other ZCPR3 Concepts

The path, originally incorporated in ZCPR2, is a ZCPR3 concept that provides a tremendous amount of flexibility. The path lets ZCPR3 search other directories (disk drives and user areas) if the program or file to be invoked is not in the active directory.

An example: The path is set up for ZCPR3 to search (in the following order) the current drive and user area, drive A/user area 0, and drive A/user area 15. When you issue a command that is not an intrinsic command, the ZCPR3 command processor begins searching for the file of that name in the current drive and user area. If the file is found, it is loaded and executed. However, if the file is not found, the path instructs the ZCPR3 command processor to continue the search at drive A/user area 0. Again, if the file is found, it is loaded and executed. If the file is not found, the ZCPR3 command processor searches drive A/ user area 15. Once again, if the file is found, it is loaded and executed. The SB180 software allows up to five levels of search.

The flexibility derives from the fact that the path is, like the system segments, changeable at any time. What this means to you is that your frequently invoked programs can be stored in a specific drive and user area, usually A15, and can be invoked from any currently active drive and user area without needing to specify the disk drive, as long as the path points to the appropriate directory.

The path is also used by many of the ZCPR3 utilities. For example, the Help utility will search along the path when looking for HLP files.

The Wheel

The last important component of ZCPR3 that I should describe is the wheel byte, which resides in the distribution software for the SB180 at address 0FDFF. If the RAM (random-access read/write memory) at that address contains a zero value, the wheel byte is considered reset (off); if the address contains a nonzero value, the wheel byte is considered set (on).

The wheel byte functions as a security system. All the intrinsic ZCPR3 commands can be set up so that they check the status of the wheel byte before they execute. This is ideal for situations where a security function is necessary, like a public computerized bulletin-board application. If "dangerous" commands like ERA (erase files) are set up to check the wheel byte, and a user who does not have wheel privileges attempts to use ERA, all that will happen is that the message "No Wheel" will be displayed.

Several ZCPR3 utilities will function only if the wheel byte is set on; otherwise, they abort immediately. Also, both intrinsic commands and a utility program let you manipulate the status of the wheel byte; both require a password to operate.

The ZCPR3 Utilities

All ZCPR3 utilities (there are more than 70) are included in the full SB180 software package (a subset is included with the boot disk only). See the "Z-System Utilities" text box on page 275 About 20 percent of the utilities correspond to intrinsic commands. Consequently, if you elect to omit, for example, ERA as an intrinsic command, you can use the ERASE.COM utility to perform the same function.

ZCPR3 utilities all share many common features, the most significant of which is that they reference the Environment Descriptor segment to determine information about the system configuration, e.g., to determine the location of the Terminal Capabilities segment or the Named Directory segment. However, since the Environment Descriptor segment is not necessarily located at the same addresses in every ZCPR3 configuration, the ZCPR3 utilities must be installed for that particular configuration. This is an easy task because, as you might surmise, a ZCPR3 utility will do it for you! (Note that the software supplied with the SB180 does not require this installation; it is preinstalled for the memory map of the SB180 default configuration.)

Finally, if you get lost, you can always find help. A help screen for any of the ZCPR3 utilities can be called by invoking it with a command-line parameter of //, so that LDR // as a command calls up a help screen for the LDR utility. Other command-line options are usually preceded by a single slash or a space character.

Shells and Aliases

To understand the concept of a shell, think of your computer system as an onion. It is made up of various layers of software and hardware, with the microprocessor at the very center. The outermost, and visible, layer would be an application program like WordStar. When WordStar is executing you are presented with its displays, and the computer will process your input in accordance with the commands of WordStar. When you exit WordStar, the outermost layer of the onion is removed and you see the next inner layer, which is the ZCPR3 command processor. If you looked deeper, you would see ZRDOS as the next layer, then the BIOS (basic input/output system).

While each layer has its own appearance and commands, it has to rely on deeper layers to execute these commands. A shell is an additional layer that fits between the application-program layer and the ZCPR3 command processor. A shell can present its own displays and process your input in accordance with its own commands, and it relies on the ZCPR3 command processor (the next deeper layer) to actually execute the commands. A common use of a ZCPR3 shell is that of a translator between you and the ZCPR3 command processor that converts single keystrokes you type to command sequences that are then executed by the ZCPR3 command processor.

Finally, a ZCPR3 shell can be nested. Multiple layers of shells can be simultaneously active, with only the outermost layer visible to you. This gives one shell the capability of invoking another shell. After the second shell is finished executing, control is returned to the shell that invoked it. A shell knows how and when to return to itself.

An alias is a .COM (executable binary program code) file created by the Alias program that contains one or more commands that are to be placed in the multiple command line buffer and then executed. Command strings can be built into an Alias program and then invoked by a single command. This is an impressive and powerful feature of ZCPR3.

As a simple example, you create numerous files with file types of BAK, HEX, and SYM, which are no longer required. Rather than repetitively entering the commands ERA *.BAK, ERA *.HEX, and ERA *.SYM, an alias named CLEANUP.COM is created. CLEANUP.COM contains the commands ERA *.BAK, ERA *.HEX, and ERA *.SYM. When CLEANUP.COM executes, it places its command string into the ZCPR3 command line buffer, and then the ZCPR3 command processor executes those commands, erasing the files specified.

Aliases can be considerably more powerful than this example because they are the technique used under

Z-System Utilities

The following is a list of some of the more interesting Z-System utilities. Keep in mind that there are more than 70 utilities in all, so this is just a small sample to give you an idea of the capabilities available.

AC: This stands for archive/copy. This utility copies a file from one directory to another, with the option of copying only files that have been modified since last archived.

CLEANDIR: Clean directory removes all deallocated references to files on the disk and sorts the remaining active filenames in either ascending or descending order. Used often, and you're nearly guaranteed a successful **UNERASE** (see below).

CONFIG: This menu-based utility is used to configure BIOS parameters like I/O port speeds, set up the printer as serial or parallel, alter the number of CPU wait states, and more. (This utility was written specifically for the SB180.)

DPROG: This is a device-programming utility that is capable of sending predefined byte sequences to peripheral devices like printers and terminals.

FINDF: Find file searches for a file or files in all disks and user areas of the system and reports their location(s).

HELP: Invokes the help subsystem. Entering HELP ZCPR3 will invoke the ZCPR3 on-line documentation. Other help information can be created with a text editor and displayed using this utility. (This utility uses the TCAP system segment to enhance its displays.)

HELPCHK: This utility checks files to be used with the **HELP** program for proper structure and syntax.

MENU: This utility invokes the menu subsystem under ZCPR3. Menu files can be created with a text editor according to the instructions in the help file. This is a Z-System shell. (This utility also uses the TCAP system segment.)

PAGE: This utility sends the contents of a file or files to the console for viewing. The data is "paged," filling only one screen at a time and then waiting for the operator to strike a key. (Also uses TCAP.)

PWD: The print working directories utility shows currently available named directories.

SHOW: This is a menu-oriented display of the status of the ZCPR3 environment, which includes all system segments, the path, and more. (Also uses TCAP.)

TCMAKE: This menu-oriented utility allows creation of TCAP system-segment files in case your terminal is not already handled by the SB180 system. (An associated utility, TCCHECK, checks TCAP system-segment files for errors in structure or syntax.)

UNERASE: Does just what its name implies; it allows the recovery of accidentally deleted files if run immediately after the deletion. Usually successful if **CLEANDIR** is run frequently.

VFILER: This is an extremely useful utility to manipulate files in various ways, such as sending contents to printer, displaying on console, copying, unsqueezing, etc. Command entry occurs merely by pointing to the filename and selecting a command. **VFILER** can be personalized with up to 10 additional user-determined commands.

Z3INS: Use this utility to install all ZCPR3 utilities if any changes are made in the location or structure of the ZCPR3 system. The supplied utilities are already installed for the SB180 environment but will need to be reinstalled if the system is reconfigured in any way.

ZAS: A relocating macro assembler.

ZCPR3 to create new commands, using whatever other programs that may exist on disk in different ways to add easily invoked powerful functions. Aliases support nesting of other aliases within their command string and also support parameter substitution so that programs invoked by an alias can be fed parameters specified when the alias is invoked, in a fashion similar to Digital Research's SUBMIT.COM utility. This parameter substitution allows the command sequence contained in the alias to operate on different filenames or with different options.

ZRDOS AND BIOS

ZRDOS is the core of the SB180 operating system. It occupies space in the memory map from 0CC00 to 0D9FF. ZRDOS, like the CP/M 2.2 BDOS (basic disk operating system), creates the standard virtual machine that application programs are written for. This lets software vendors write one version of a software package, which will execute on more than one type of hardware configuration. The virtual-machine environment is provided via standardized system functions, such as sending a character to the console or checking the status of the list device. The CP/M 2.2 BDOS contains 39 functions; ZRDOS provides these same functions, thereby maintaining compatibility, and adds four more.

Two aspects of ZRDOS are visible to you in comparison with CP/M 2.2. The major significant difference of ZRDOS is that when a new disk is placed in a disk drive, it is not necessary to type Control-C to log in the new disk, as in CP/M 2.2. The other difference is improved error messages. Instead of the cryptic Bdos Err on A:, you see Read Error on A:, which is much more meaningful. ZRDOS also includes file archive handling compatible with CP/M Plus and MP/M, which can be used to make automatic backup copies of a file that has been changed. ZRDOS also recognizes what are known as wheel-protected files and does not allow modification to those files unless the wheel byte is set.

The BIOS for the SB180 handles several important functions not found in most computers and uses the on-chip hardware of the HD64180 to the fullest. It was written specifically for the SB180, with emphasis on rapid and efficient code. Disk operations are extremely quick in comparison with other machines.

The SB180 BIOS resides between addresses 0DA00 and 0EBFF in the memory map. Like the BIOS in any microcomputer, its function is to act as the interface between the software and hardware. Another way of saying this is that, because different hardware configurations may be used, a certain part of the operating-system software exists that is specially customized for that hardware configuration. Because of the customization of the BIOS, the same DOS can function on many different computer types, despite the fact that there may be significant differences between the machines.

Table 1: *A list of programs tested and known to be compatible with the SB180's operating system.*

MicroPro:	WordStar 3.0, WordMaster, MailMerge, StarIndex, SuperSort
Microsoft:	Multiplan, Macro-80, BASIC-80, BASIC compiler
Digital Research:	MAC, SID, ZSID, CB-80, Pascal/MT+
Sorcim:	SuperCalc2
Ashton-Tate:	dBASE II
Borland International:	Turbo Pascal 2.0
Manx:	Aztec C 1.05g
CompuView:	Vedit
Others	T/Maker III, The Word Plus, Punctuation and Style, MIX, Modem7, MDM7, MEX, BDS C, C/80

The BIOS is responsible for interfacing all peripheral devices like floppy disks and video terminals that are used on a computer to the standard virtual-machine environment created by the operating-system software.

For example, an application program requests ZRDOS to send a character to the console. ZRDOS is the same no matter what machine it is operating on and has no way of knowing which I/O port address the console might be found at, or anything else about the console. Yet it does know that the request has to do with I/O, and ZRDOS passes this request to the BIOS to send the character to the console. Finally, the BIOS is the component that actually transmits the character to your terminal, because it has been configured to know that the "console" is actually a terminal connected to I/O port number two. This example shows how the peripheral devices attached to the machine (console attached to I/O port number two) are interfaced to the software of the machine (request to output a character to the console).

The SB180 BIOS incorporates this type of standard software/hardware interfacing and several special features. The most important feature is the integrated RAM disk. The upper 192K bytes of the memory can be set up by the BIOS to be used as an extremely fast file-storage device, which, to application programs and the operating system, looks like disk drive M. This integrated RAM disk is a powerful tool, one that gives the SB180 an incredible performance advantage. One of the two direct-memory-access controllers (DMACs) of the HD64180 is dedicated to the RAM disk, providing the best performance possible.

Table 1 lists programs that have been tested to run in the software environment of the SB180. The programs listed are not the only programs that will run; they represent all that was available to be tested.

It is important to understand that, although the Z-System is compatible with programs designed to run under CP/M 2.2, it is a case of upward compatibility. This means that these programs usually cannot use most of the advanced features of the Z-System but simply perform as they would in the environment they were intended for. Programs written for CP/M 2.2 can benefit from some aspects of the Z-System, but other aspects of the Z-System cannot be utilized.

STARTING THE SYSTEM

Two modes of operation are available when starting the system. The first mode is the SB180 monitor. It will be invoked if the computer is powered on without floppy-disk drives connected or if disk drives are connected but no disk is in drive A. The monitor has its own command set to perform such functions as examining and changing the contents of memory, transmitting and receiving byte values to and from the I/O ports, and more.

The monitor software resides in the on-board system EPROM (erasable programmable read-only memory) and is used for debugging the system hardware and as a bare-bones operating system for SB180 users who have not added floppy-disk drives. If the system is used in this way, simply strike the Return key so that the monitor can determine the console's data-transmission rate. (A listing of the monitor commands is provided on page 265.)

The second mode of operation involves attaching one to four floppy-disk drives (3½-, 5¼-, or 8-inch) to the appropriate drive connector on the SB180 and simply placing a disk that has the operating system on it into drive A and powering on, or resetting, the system.

This is referred to as cold booting and loads the operating system into the SB180's memory. Several messages are displayed in the process of cold booting, most of which are originated by the ZCPR3 utility LDR.COM. When the system has completed the cold-boot process, you are presented with the system prompt, which is A0:BASE> if you are using the distribution software. The system is now ready to accept your commands.

The cold-boot process has many stages. An important one is the execution of the STARTUP.COM program. The SB180 operating system searches for the STARTUP.COM program as the last step of the cold-boot process, and if it is found, it is executed. STARTUP.COM is created by the Alias program, and its major role in the cold-boot process is to load the ZCPR3 system segments by placing the command LDR SYS.ENV,SYS.RCP, SYS.FCP,SYS.NDR into the multiple command line buffer.

This is not the only role of the program. You can easily customize STARTUP.COM so that whenever the computer goes through the cold-boot process, an additional series of commands are executed automatically. Because each disk can have its own personalized STARTUP.COM, it is possible to create turnkey systems for specific applications. Disks could be set up specifically for word processing by using STARTUP.COM to automatically load and execute WordStar, or for a turnkey database operation. An unattended remote-access computerized bulletin board could have a STARTUP.COM set up so that if power failed and then was restored, all the needed commands to start the system again would be executed. The STARTUP.COM concept gives you a great deal of flexibility and convenience.

RAM-DISK INITIALIZATION

The RAM disk is an exciting feature of the SB180. It will improve system performance many orders of magnitude when used. Like all RAM disks, it has some characteristics that should not be overlooked. The RAM disk, unlike a floppy disk, does not retain its contents when power is removed, so you must be sure to make floppy-disk copies of files used in the RAM disk. It is quite possible to use aliases to make the process convenient, so that when you edit a file that resides in the RAM disk, it is automatically copied onto a floppy disk at the conclusion of the edit session.

The SB180 BIOS is written to *not* initialize the RAM disk when a cold boot is performed. This is so that when the Reset button of the computer is activated, the RAM-disk contents are retained. You may need to use the Reset button if a buggy program goes into an endless loop, and it is nice to reset the computer without losing the contents of the RAM disk!

A utility program called MDSK (the command is MDSK /I) is used to initialize (format) the RAM disk. It is important that MDSK /I not be used in the STARTUP.COM alias, because this will destroy the contents of the RAM disk whenever the Reset button is used. An alternative is to use the flow commands of ZCPR3 to query the user, so that the STARTUP.COM alias may perform the command or not, depending on the user's response.

CHANGING THE DRIVE/USER AREA

The Z-System accepts the generic commands found in both MS-DOS and CP/M for changing the currently active disk drive. For example, to select the B drive as active, simply

Table 2: *The different floppy-disk formats that can be processed with the* SB180 BIOS *and their associated storage capacities. You do not need a conversion program to format, read, or write these formats (tpi=tracks per inch; dsdd=double-sided double-density; ssdd=single-sided double-density).*

Format	tpi	Sides	Capacity per disk
SB180 native	96	2	782K bytes (dsdd)
SB180 native	48	2	386K bytes (dsdd)
Hitachi QC-10	48	2	286K bytes (dsdd)
Kaypro 2	48	1	191K bytes (ssdd)
Ampro	48	1	188K bytes (ssdd)
Osborne 1	48	1	183K bytes (ssdd)

type B:. To select the integral RAM disk, type M:. When a drive is selected as active, it becomes the default drive. In other words, it is where programs and files will be searched for first, unless specified otherwise by the user or application program.

These generic commands are actually a subset of the commands available under the Z-System to select the active disk and user area. Two major types of commands are available under the Z-System for this purpose: the DU: form and the DIR: form. Both forms are recognizable by the trailing colon character.

The SB180 BIOS can automatically recognize different floppy-disk formats, so exchange of disks with dissimilar formats becomes easy. See table 2 for a list of supported formats. To change to a disk of different format, you may have to type Control-C.

The DU: form is made up of two components. D is a disk drive, and the acceptable range for it is A through P. U is a user area, and its acceptable range is 0 through 15. Thus, DU: forms of A0:, B6:, and M15: are allowed, but forms like Z0: or G33: are invalid. So, to move to the A2: drive/user area, simply type A2:. Another way of using the DU: form is to realize that the D and U are optional. This means that forms such as 3: and A: are valid.

The DIR: form is derived from the Named Directory system segment, which associates a symbolic name with a specific drive and user area. A directory name is up to eight characters in length. For instance, if the symbolic name ROOT is defined in the Named Directory segment as being associated with A15, and if you type ROOT:<cr> at the system prompt, A15 will be selected as active.

To have the list of defined directory names printed for you, use the PWD (print working directories) utility. The default configuration of the SB180 system software allows 14 directory names to be defined, but this number can be changed (see the "User Customization" section).

A characteristic of the DIR: form is that a password can be associated with a directory name. The password can be up to eight characters in length and is also defined in the Named Directory segment. If you type a DIR: form that is passworded, a prompt of PW? is presented, and if the proper password is not entered, you are not allowed to enter that directory. (Note that the DU: form, if enabled, allows password-free access to that directory. To make the DIR: passwords come into full force, reconfigure the system to not accept the DU: form.)

SECURE SYSTEMS

The SB180 system software has been configured with a minimal amount of security. This was done because most users will probably not be in a situation where public access to their SB180 will be allowed.

However, the system does possess extensive options for security. If you intend to use the SB180 as a public-access computerized bulletin board, you can reconfigure the software into a more secure system. For example, you could deny use of the ERA command to "ordinary" users.

I previously discussed how the DU: form of changing the active disk drive and user area bypasses any passworded Named Directory entries. In a secure system, the DU: form may be disabled, so that the only way to refer to any other area would be to use the DIR: form. Then, because the DIR: form requires a password if one has been defined, critical areas of the system can be password-protected.

These changes are implemented by editing the source code supplied with the system to set the new options and then merging the changes into the operating-system software.

USER CUSTOMIZATION

You can customize a number of other areas in the SB180 system software. The first is the BIOS, using the CONFIG utility.

The BIOS source code is included in the full software package to make easy installation of significant changes in the hardware-configuration information. The assembler, ZAS, lets you regenerate the operating system into new and different forms. An example is if you want to rewrite your RCP so that some of its commands will respond to the wheel byte. Another example is expansion of the Named Directory system segment to handle more than 14 directory names.

Although some of the above examples require programming knowledge, a great degree of customization can be done without specific programming knowledge through the use of Alias programs and the shell utilities.

The STARTUP.COM alias is a likely candidate for personalization. It is normally executed only when the system does a cold boot. When using the SB180 for word processing, I use a STARTUP.COM command sequence that automatically copies the files I will be using to the RAM disk. Once the files I am using are copied into the RAM disk, execution of commands is nearly instantaneous. Having STARTUP.COM automatically place the appropriate files in the RAM disk to gain the benefit of the speed of access can be used in many other areas besides word processing.

SUPPORT

A full set of manuals are included with the SB180 operating-system software. This includes manuals for ZAS, ZDM, EDIT, ZRDOS, and some SB180 utilities. Documentation for the ZCPR3 utilities and ZCPR3 itself are included in the HLP files.

EXPERIMENTERS

As always, I try to support the computer experimenter by rewarding diligence. If you build the SB180 from scratch, send me a picture and I'll send you a copy of the BIOS and the ROM monitor on disk (SB180 double-sided double-density format) at no charge, provided it is for your personal use.

If you build, buy, or otherwise assemble an SB180 system, I'd like to know about it. I will be designing expansion boards for the SB180 (the first one is a 300/1200-bps modem) and can notify you of them in advance of publication. In addition, having your name will greatly simplify the organization of any users groups that might arise.

The SB180 is a fully supported Circuit Cellar project. I have arranged for the hardware to be available in kit or assembled form, and I contracted with Echelon Inc. to write the BIOS and integrate it into the operating system. Echelon has telephone technical assistance available and also has affiliations with more than 40 public remote-access bulletin-board systems, called Z-Nodes. Z-Nodes are located throughout the nation, and there may be one in your area. You can find ZCPR3 utility-program updates and informative newsletters about the Z-System on a Z-Node for the price of a phone call.

In Conclusion

This has been a big project, even though it's only a small board. Sports cars have a lbs/HP rating. It's too bad there isn't an MPS/sq. in. rating that could be used to truly compare the capabilities of the SB180 to other computers.

Like the Z8 and 8052, the HD64180 has joined the Circuit Cellar preferred processor list, and you can expect it to be well supported with a variety of expansion peripheral devices. A hard-disk controller and modem are currently in the works. As soon as I figure out how to design them and gather together the software people who know how to glue it together, I'll let you know. Until then, I'll just keep plinking along with simple projects like home controllers, intelligent terminals, and voice-recognition systems. My PC board designer needs a vacation.

Special thanks to Tom Cantrell, Frank Gaude', Merrill Lathers, Dave McCord, Joe Wright, and the people at Custom Photo and Design Inc. and Tech Circuits Inc. for their contributions to this project.

See page 447 for a listing of parts and products available from Circuit Cellar Inc.

26

BUILD AN ANALOG-TO-DIGITAL CONVERTER

A 16-*channel* 12-*bit* *high-speed* A/D *converter*

It is evident that many applications for computer controls, including energy management, security, and environmental monitoring, require measurement inputs and control outputs in quantities not easily expressed in the 0- and +5-volt TTL (transistor-transistor logic) levels present in your computer.

An energy-management system, for example, may need to monitor a temperature range of 0 to 100° C with a resolution of 0.1 degree. The thermocouple sensing this temperature range might generate only 1 or 2 millivolts per degree. A proportional-drive pump motor in the same system might require a 2.40-V set-point control input to produce the proper flow rate throughout the system.

Continuous analog systems like these are in the real world, outside the binary logic-0 and logic-1 domain of digital computers. For the computer to interact with the real world, we need some scheme for translating analog measurements to and from quantized binary equivalents.

This is not the first time I have touched upon analog-to-digital and digital-to-analog conversion. I try to cover this topic every three or four years so that new readers can be brought up to speed on the basics. For the old-timers, however, I spice up the project with the latest whiz-bang conversion interface that can be cost-effectively produced.

The previous projects have all used 8-bit converters. However, the overwhelming response to the BASIC-52 computer/controller (BCC-52) presented on pages 105–117 has created a demand for something more challenging. Thousands of BCC-52 industrial and end users are applying computer control to applications that ultimately require greater accuracy of measurement.

Presently, an 8-channel 8-bit A/D converter (10,000 samples per second, 0 to 10 V or −5 to +5 V, P/N BCC-13) is available for the BCC-52, but many measurements require more resolution. Therefore, it's time to dust off the old theoretical explanations and present an up-to-date, high-speed, high-resolution A/D interface for the BCC-52.

First, because one is an integral component of the other, I'll outline the basics of D/A conversion and then go on to A/D conversion. After a few circuit examples, I'll get into the heavy stuff. Ultimately, this month's project is the design of a 16-channel (8-channel differential input) 12-bit plus sign bit, −5 to +5 V, 10,000 samples/sec, BCC-52/Z8 bus-compatible A/D converter board. In a few months, after I have presented a few more essential peripheral devices, I will demonstrate the configuration and application of a full-fledged data-acquisition and control system based on the BCC-52.

D/A Conversion

The D/A converter can be thought of as a digitally controlled programmable potentiometer that produces an analog output voltage. This output voltage (V_o) is the product of a digital signal D, a multiplier constant K (usually 1), and an analog reference voltage V_{ref}, related by the following equation:

$$V_o = KDV_{ref}$$

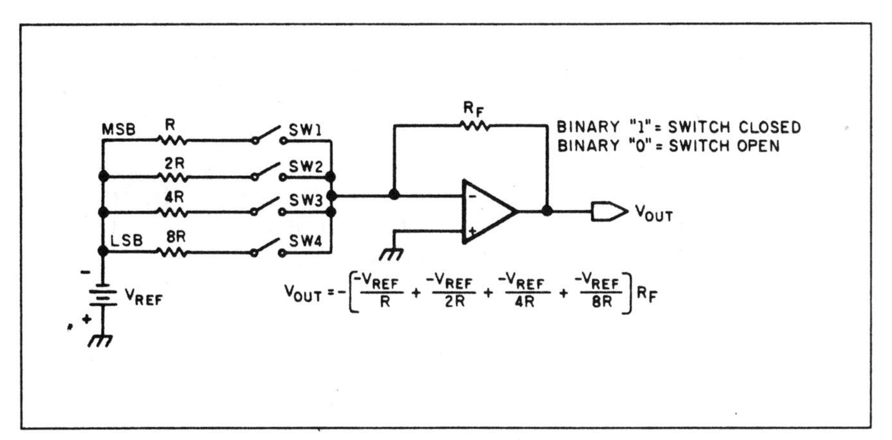

Figure 1: *A 4-bit weighted-resistor D/A converter. A 4-bit word is used to control four single-pole single-throw solid-state switches. Each switch is in series with a resistor. The resistor values are related as powers of 2. The other sides of the switches are connected together at the summing point of an op amp. Currents with magnitudes inversely proportional to the resistors are generated when the switches are closed. They are summed by the op amp and converted to a corresponding voltage.*

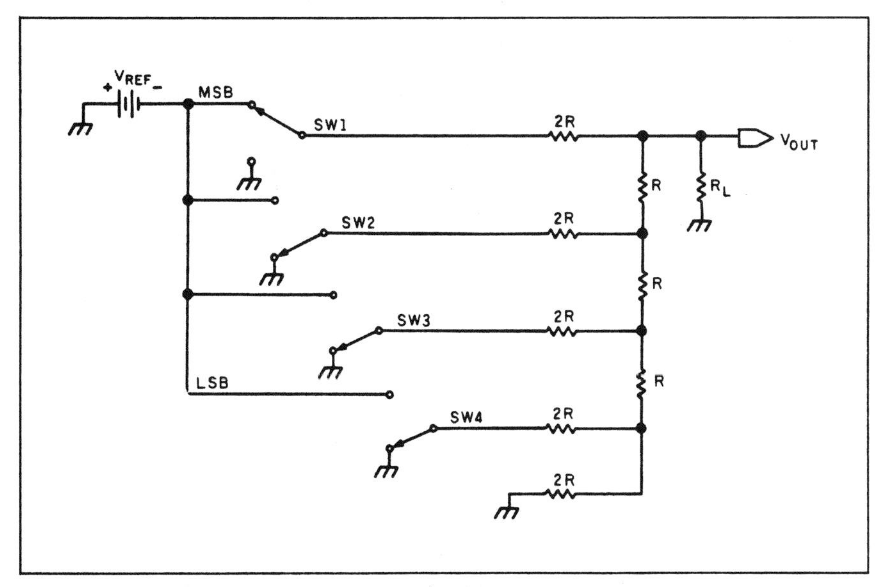

Figure 2: *A 4-bit R-2R-type resistor-ladder D/A converter. The topology of this network is such that the current flowing into any branch of a three-branch node will divide itself equally through the two remaining branches. Because of this, the current will divide itself in half as it passes through each node on its way to the end of the ladder.*

The binary value transmitted to the D/A converter by the computer is a binary fraction representing what portion of the full output voltage is emitted. The fraction is multiplied by a reference voltage, which can be either fixed or variable. D/A converters with variable reference voltages are often referred to as multiplying D/A converters, although all D/A converters can be said to multiply.

In finite binary fractions, the most significant bit (MSB) has a value of 1/2 (that is, 2^{-1}), the next most significant bit is 1/4 or 2^{-2}, and the least significant bit (LSB) is $(1/2)^n$ or 2^{-n}, where n is the number of bits in the binary fraction. If all the bits in the fraction are added, the sum approaches 1; the more bits in the fraction, the closer the sum is to 1. The difference between 1 and the approach to 1 is the quantitation error of the digital system. I'll discuss this later.

Different implementations of D/A and A/D converters use different formats for representing the binary digital quantities. One basic difference is how systems represent negative binary numbers and negative voltages; some can, and some can't. Analog interface systems that can manipulate positive and negative numbers and voltages are called bipolar converters; systems that can handle only positive voltages and quantities are called unipolar.

Unipolar converters chiefly use straight binary and binary-coded-decimal (BCD) representations of digital quantities. Bipolar converters use a variety of representations, including offset binary, one's- and two's-complement formats, and Gray code. For brevity, I will limit this discussion to converters using straight-binary and offset-binary representations. Later, I will get into two's-complement representations since the converter chip used in this project represents negative numbers in two's-complement form.

Offset binary differs from straight binary only slightly. In offset binary, a number consisting of all zeros represents the most negative possible quantity. The most obvious consequence of this is that the MSB acts as a sign bit, 0 for negative values and 1 for positive. For instance, in offset notation, the bit string 01000000 represents −64, while the bit string 11000000 stands for +64.

Frequently, offset notation is referred to as a resolution value plus sign bit, i.e., 12-bit plus sign converter. The sign bit, while performing as a thirteenth bit in bipolar operation, should not be confused with a 13-bit converter. The sign bit can be used to indicate only quantities above 0 V (in this case, sign bit=0) or below 0 V (sign bit=1) and not shifted in scale. Between −5 V and +5 V on a 12-bit plus sign converter, there will be 8192 divisions (13 bits). However, if the converter were to measure inputs only in the range of 0 to 5 V, only 4096 divisions (12 bits) can be represented. In this project, the A/D is set for −5 to +5 V and is therefore indistinguishable from a 13-bit converter between these limits and would be 1 bit better than a straight 12-bit converter used to measure the same range.

The translation of digital values to proportional analog values is performed by either of two basic D/A-conversion circuits: the weighted-resistor circuit or the R-2R circuit. The weighted-resistor converter is by far the simpler and more straightforward. This parallel decoder requires only one resistor per input bit.

In the weighted-resistor D/A converter, solid-state switches are driven directly from the signals that represent the digital number *D*. Individual currents with voltage magnitudes related by powers of 2 (magnitudes of 1/2, 1/4, 1/8, . . . , 2^{-n}) are generated and summed by connecting a network of resistors with values of R, 2R, 4R, . . . , 2^nR between the reference voltage $-V_{ref}$ and the summing point of an operational amplifier (op amp) by means of the set of electronic switches. After being summed, the various currents are converted to a voltage by the op amp, as shown in figure 1.

While this may appear to be a simple answer to an otherwise complex problem, this method has some significant drawbacks. The accuracy of this type of converter is a function of the combined accuracies of the resistors, switches (all switches have some resistance), and the op amp. In D/A-conversion systems of greater than 10 bits resolution, the values of the resistors become extremely large, and the resultant current flow is reduced to such a low value as to be lost in circuit noise.

For example, in an 8-bit D/A converter with R (the value of the resistor for the MSB) set to 10 kilohms, the value of the resistor for the LSB turns out to be 1.28 megohms. With a reference voltage of 10.00 V, only 7.8 microamperes would flow into the op amp. This current is significantly below the response threshold of most low-cost op amps and would not be detected. Lowering the value of R to 100 ohms creates the opposite problem. At a reference voltage of 10.00 V, the input current to the op amp would be 100 milliamperes, more than most op amps can handle.

A reasonable alternative to the weighted-resistor D/A converter is the R-2R D/A converter, often referred to as a resistor-ladder converter. This type is more widely used, even though it uses more components than the weighted-resistor type. A simple R-2R design is shown in figure 2, including the reference voltage, a set of binary switches, and an output amplifier. The basis of this converter is a ladder network constructed with resistors of two values: R and 2R.

In each bit position of the network, one resistor (2R) is in series with the bit switch, and the other (R) is in the summing line, so that the combination forms a pi network. This suggests that the impedances of the three branches of any node are equal, and that a current *i*, flowing into a node

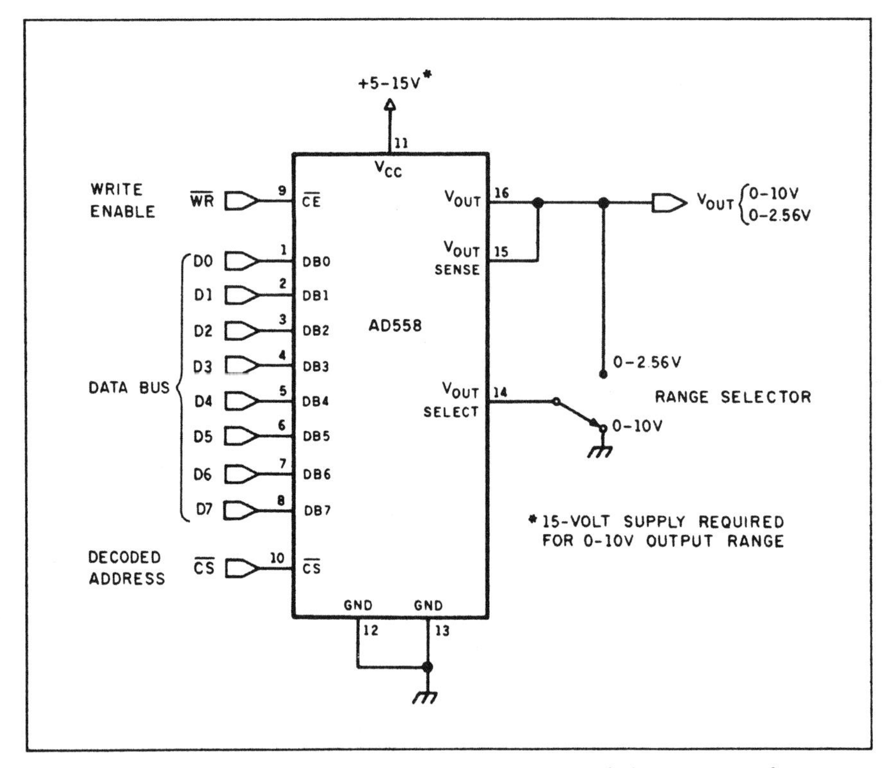

Figure 3: *A block diagram outlining a typical connection of the* AD558 *8-bit multiplying* D/A *converter.*

through one branch, flows out as $i/2$ through the other two branches. In other words, the current produced in the network by closing a bit switch is cut by half as it passes through each node on the way to the end of the ladder. Simply stated, the position of a switch with respect to the point where the current is measured determines the binary significance of the particular switch closure.

The R-2R D/A converter is easy to manufacture because only two resistor values are needed. The component stock can be reduced to one resistor value if two are used in series for each bit. Keeping matched resistor values that have the same temperature coefficients contributes to a stable design. Certain trade-offs are required between ladder resistance values and current flow to balance accuracy and noise.

One form of the R-2R ladder circuit is found in the multiplying D/A converter. This type of converter, which utilizes external-variable analog reference voltages, produces outputs that are directly proportional to the value of the digital input multiplied by this reference. Functionally, this type of converter is available as current- or voltage-output types. The current-output devices are faster and less complex because they do not include additional output-amplifier stages. Therefore, they cost less than voltage types.

An economical 8-bit multiplying D/A converter is the Analog Devices AD558. Shown in figure 3, it contains an 8-bit latch, R-2R ladder network, reference-voltage source, and output amplifier. The AD558 can run on a +5- to +15-V power supply and can be jumper-selected for 0- to 2.56-V or 0- to +10-V ranges. Using a separate op amp, you can configure an offset converter or modify the output of the range.

The AD558 can be used as a transparent D/A converter by holding the chip-enable and chip-select lines constantly low. However, it was primarily designed to be bus-operated and appear as a write-only location in memory or I/O (input/output) address space. Typical connections consist of a decoded address strobe, a write-enable signal, and the 8-bit data bus.

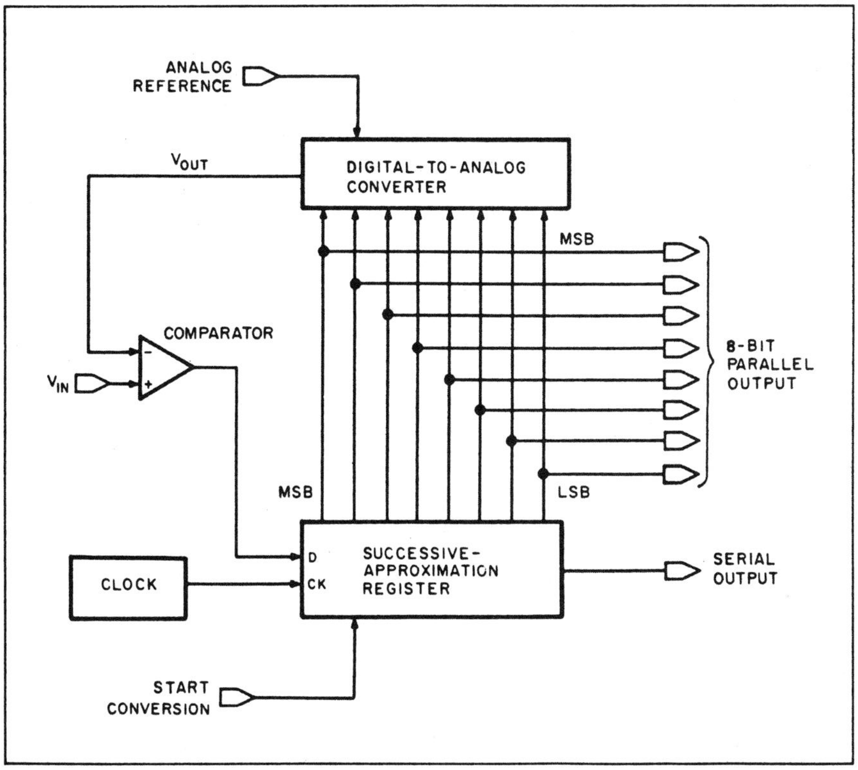

Figure 4: A *block diagram of a typical 8-bit successive-approximation* A/D *converter.*

A/D Converters

Virtually all high-resolution A/D converters incorporate a D/A converter as an integral component. That is why, even though our ultimate aim is A/D, I always discuss D/A converters first. Hopefully I have made you aware of the binary-conversion process, and you can appreciate the concepts of resolution and accuracy.

An A/D converter changes an analog voltage into a digital representation compatible with the computer's input needs. Akin to the 8-bit D/A converter, an A/D converter is subject to the same conversion rules. If you are trying to read a 10-V signal with an 8-bit converter, resolution is 1/256 of 10 V (approximately 40 mV), and accuracy will be ± 1/2 the LSB.

For greater resolution, more conversion bits are necessary. The number of bits does not set the input-voltage range of a converter; it only determines with what precision the output value is represented. An 8-bit converter (either A/D or D/A) can be set up just as easily to cover a range of 0 to +1 V as it can be to cover 0 to +1000 V. Often, the same circuitry is used with only a final amplification stage or resistor-divider network changed.

Note, however, that an 8-bit converter with a range of 1000 V has a resolution of only 4 V (1000/256), and it would be useless to measure 0- to 10-V signals. You can solve this problem in a number of ways. The easiest solution is to use a converter with more bits. A 16-bit converter, which has 65,536 steps instead of 256, would cover the same 1000-V range in 15-mV increments.

As a practical matter, though, a reasonable price/performance ratio is often more important than wide-range capability. A/D conversion is considerably more expensive than D/A conversion, and price is directly related to resolution and accuracy. If you intend to read 0- to 5-V input signals and you have to be accurate within only 35 mV, it hardly makes sense to use a 1000-V range 16-bit converter (probably costing $5000) when an 8-bit 0- to 5-V range unit ($150) would more than suffice.

The rule in choosing an A/D converter boils down to "be realistic." Assess the quality of the signal source (noise, rate of change of input, ground referenced or differential, etc.) when you choose your converter. Installing a converter with 1-microvolt (μV) resolution to measure an input signal buried in 200 mV of noise is pointless.

An A/D converter that scans thermistor probes and controls the ambient temperature in a large supermarket cannot encode video information from an optical scanner. A/D converters, much more than D/A converters, are specifically tailored to an application. Speed, accuracy, and resolution are variables in any converter design, but the blending of these choices can greatly affect the cost in A/D conversion.

Most confusing is the variety of A/D-converter designs. They range from very slow, inexpensive techniques to ultrafast, expensive ones. You get what you pay for. The two fastest techniques are flash conversion and successive approximation.

The flash converter is just that. It consists of a separate analog-input comparator for each incremental voltage it is to measure. An 8-bit flash A/D converter has 256 comparators with gating logic that outputs the binary code corresponding to the comparator triggered by the input voltage. Flash converters are very fast (1 million–100 million samples/sec), but they are also very expensive.

A somewhat slower (1000–1 million samples/sec) and more cost-effective alternative is the successive-approximation converter. Shown in figure 4, this type—like the binary-ramp-type A/D converter—uses a D/A converter in the feedback loop to compare a calculated D/A voltage to the unknown input voltage. In this implementation, the binary counters are replaced with a special successive-approximation register (SAR).

Initially, the outputs of the SAR and the mutually connected D/A converter are at a zero level. After a start-conversion pulse is received, the SAR enables its bits one at a time starting with the MSB. As each bit is enabled, the comparator gives an output signi-

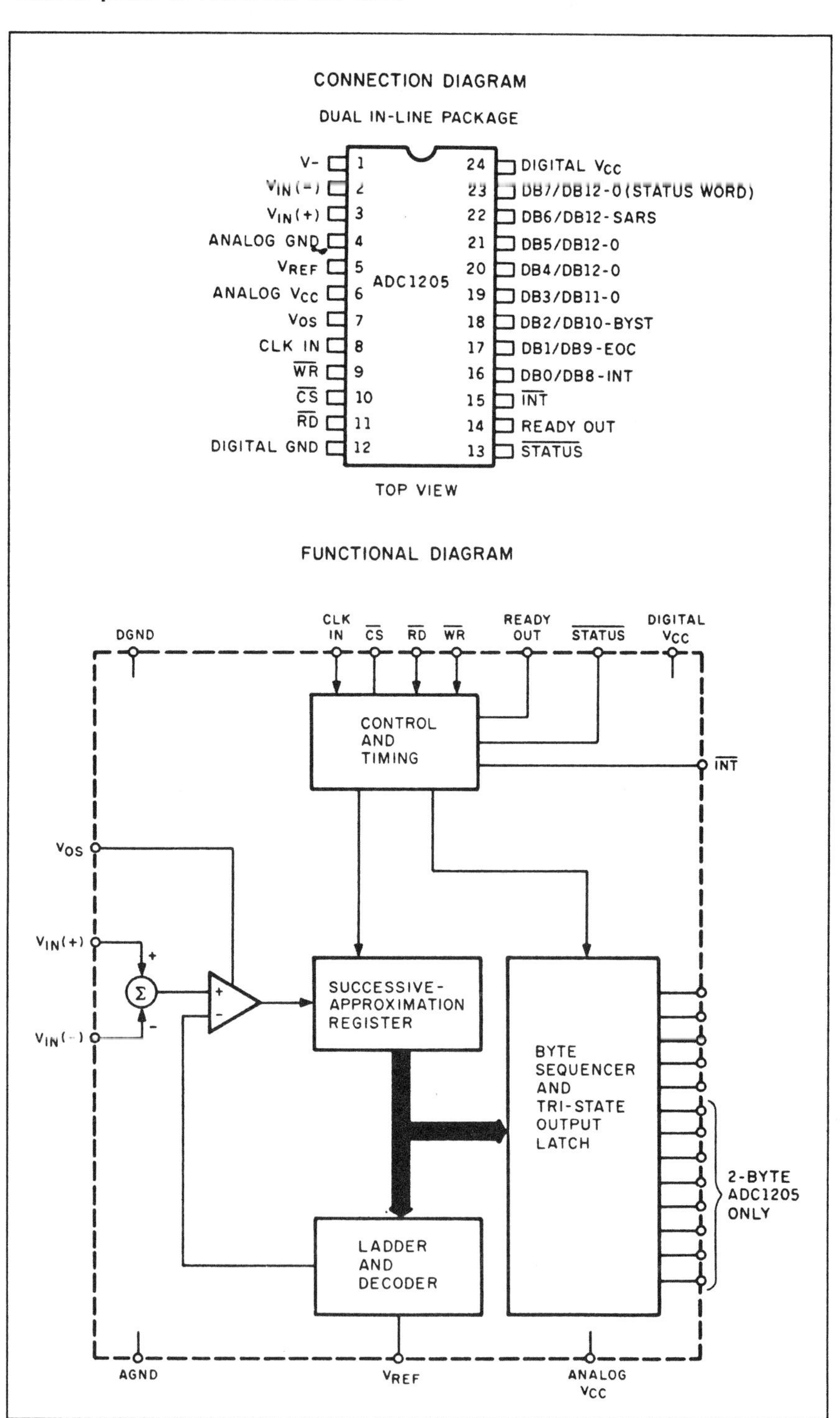

Figure 5: *Pin-out and block diagram of National Semiconductor's ADC1205CCJ 12-bit plus sign A/D converter chip.*

fying that the input signal is greater or less in amplitude than the output of the D/A converter. If the D/A output is greater than the input signal, a 0 is set as the value of the corresponding output bit. If the D/A output is less than the input signal, the circuit sets the corresponding bit to a 1. The register successively moves to the next bit (retaining the settings on the previously tested bits) and performs the same test. After all the bits have been tested, the conversion cycle is complete. An 8-bit successive-approximation A/D converter takes only eight clock cycles to complete a conversion.

This one-to-one relationship between conversion resolution and SAR clock counts is generally true only for discrete-component SAR-based A/Ds. In higher-resolution integrated-circuit A/D converters, the clock cycle/conversion bit times are less distinct due to extensive housekeeping circuitry. Like many microprocessors with high clock-crystal frequencies, the actual system clock is much slower.

The ADC1205

Figure 5 is the pin-out and block diagram of the National Semiconductor ADC1205CCJ 12-bit plus sign A/D converter chip. It operates on a single +5-V logic supply and 5.000-V reference input to provide a 12-bit conversion on 0- to 5-V inputs. With a 1.08-megahertz clock frequency, the ADC1205 will do 10,000 conversions per second (108 microseconds per conversion).

If an additional −5- to −15-V supply is connected to V− (pin 1), the ADC1205 will convert −5- to +5-V inputs using a thirteenth output bit. This MSB is the sign bit. It is a logic 0 for positive values and logic 1 for negative values.

Figure 6 shows the output characteristics of the converter. For 0- to 5-V inputs (sign bit=0), the codes range from binary 0000000000000 to 0111111111111, respectively. In a 5-V range, each bit represents 0.0012 V, or 1.2 mV resolution! If the output of the converter were binary 0000010111100 (hexadecimal 000BC), this would be (188)*(0.0012)=0.2256 V. Similarly, binary 0110101111000 (hexadecimal 00D78) is +4.1376 V.

Negative inputs are represented in two's-complement binary. For 0- to −5-V inputs (sign bit=1), the codes range from binary 1111111111111 to 1000000000000, respectively. The output code for negative values is represented as the magnitude of the difference from the unknown input to −5 V and not its distance from zero. An output code of 1000010111100 (100BC) is −((5.00)−(188)*(0.0012))= −4.7744 V. Similarly, 1110101111000 (10D78) is −((5.00)−(3448)*(0.0012))= −0.8624 V.

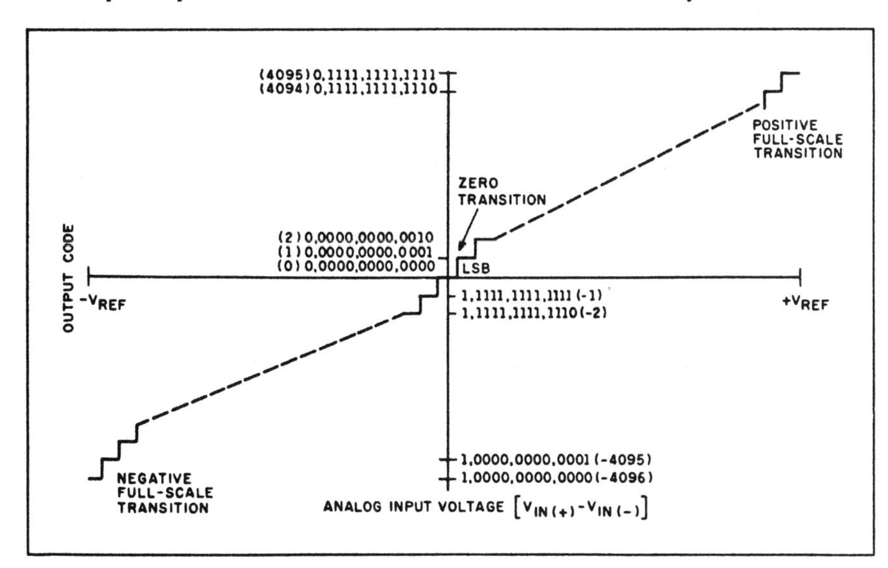

Figure 6: ADC1205 *output characteristics.*

Under computer control, the conversion is relatively easy. At each reading, determine the absolute value of the 12-bit number by multiplying it by 0.0012 V. If the sign bit is a 0, add a plus sign to your calculation, and you have a positive output of that magnitude. If, on the other hand, the sign bit is a 1, subtract that value from 5.0 V and append a minus sign. You can see that watching the sign bit is important, and this is not as simple as offset binary.

One further consideration before presenting the entire schematic is the concept of single-ended and differential inputs. There is a significant difference between them. Most low-cost multichannel A/D converters have single-ended inputs.

All converters have a V_{in+} and a V_{in-} input. In a single-ended A/D converter, the V_{in-} line is connected to ground. Therefore, all measurements are referenced to a common ground. Even if an 8-channel multiplexer switches inputs to the V_{in+} line, all readings are referenced to a single ground, and voltages from two different systems cannot be monitored simultaneously unless their grounds are connected. This is often not the case, and conditions called ground loops result. Many of you no doubt remember "smoking" an early-generation oscilloscope by accidentally viewing the hot side of the AC line while referenced through the line cord to the other side (even today I still use an isolation transformer on my scopes).

Another consideration is trying to measure voltages that are not necessarily relative to ground. Perhaps resistor R_{tc} in figure 7 is a thermistor, and we wish to read the voltage drop across it to determine temperature. A single-ended A/D converter could not be connected directly across R_{tc} if both the circuit and the A/D converter have the same ground without shorting out one of the resistors. To read the thermistor, you would have to separately read the voltages at points B and C and subtract them. Furthermore, unless you manually move the probes, the only way to do it is to increase the number of channels on the A/D converter. Hence, the proliferation of multichannel single-ended A/D converters.

Unfortunately, measurements referenced to ground often contain noise and power fluctuations from other components in the circuit. It is far better in some applications to simply measure the voltage between two points in a circuit irrespective of ground. Such a measurement is termed "differential." For lack of a better example, think of this as the two probes on a digital voltmeter (DVM). If the meter is battery-operated, it is completely isolated from ground, and the two probes measure absolute potential between them. Only when the V_{in-} probe is physically connected to the circuit ground are the readings then single-ended and ground-referenced.

The ADC1205, while being powered from ground-referenced power supplies, has analog input lines that are isolated from ground. These two lines are like the two probes on the DVM. In a multichannel single-ended A/D converter, only the V_{in+} line is multiplexed. The V_{in-} line is attached to ground. In a differential-input multichannel A/D converter, both the V_{in+} and V_{in-} lines are multiplexed, and neither is tied to ground. To read across R_{tc}, the V_{in+} line is attached to point B, and the V_{in-} line is connected to point C (in industry parlance, V_{in+} is V_{in} High and V_{in-} is V_{in} Low).

The ADC1205 is a 12-bit converter designed to attach directly to an 8-bit microcomputer bus. The system communicates with the chip as memory-mapped I/O through the $\overline{CS}$ (chip-select bar) and $\overline{RD}$ (read bar) $\overline{WR}$ (write bar) signals. An additional $\overline{STATUS}$ (status bar) line is used as a signal to start conversion or check conversion progress.

The 12 bits and sign are read as 2 successive bytes. Data is right-justified with the most significant byte presented first (the 4 MSBs of the first byte all have the value of the sign bit). A second read to the chip automatically presents the least significant byte. The three possible interactions are given in table 1.

Communicating with this chip may look complicated, but it is much less so than you might think, especially if you are operating the converter in BASIC. I will demonstrate it shortly.

THE BCC-30 16-CHANNEL A/D CONVERTER BOARD

When you invent things, you get to name them. I called the BASIC-52 board the BCC-52. Since this A/D converter board is BCC-bus-compatible, I've decided to call it the BCC-30 (other more appropriate numbers are unfortunately taken). See photo 1. The schematic of the BCC-30 is shown in figure 8.

The configuration of the BCC-30 is as a bus-compatible peripheral device to the BCC-52 and the BCC-11 Z8-based computer/controller redesigned from the original presentation in July 1981. See photo 2. Both units and a number of expansion boards I've designed over the years share a common 44-pin bus sometimes called

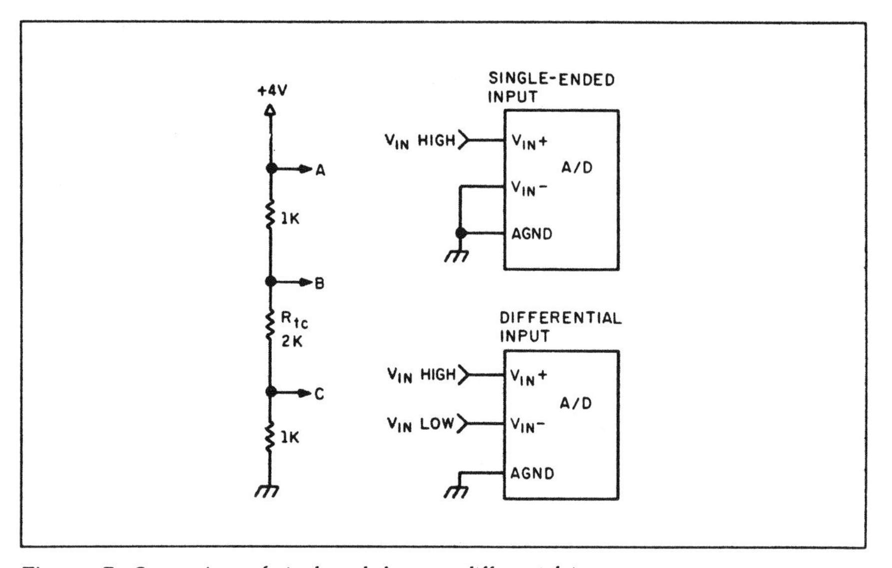

Figure 7: *Comparison of single-ended versus differential input connections.*

Table 1: *The three possible interactions with the AC1205 A/D converter chip.*

$\overline{CS}$	$\overline{RD}$	$\overline{WR}$	$\overline{STATUS}$	Function
0	1	1	0	Reset data-byte counter and start conversion.
0	0	1	1	Read data. First byte is sign and 4 MSBs; second byte is 8 LSBs.
0	0	1	0	Read status word.
			Status-word format:	Bit 0 — High indicates conversion complete and data ready. Bit 1 — High indicates conversion complete. Bit 2 — High indicates next byte is 8 LSBs. Low indicates next byte is sign and 4 MSBs. Bit 6 — High indicates conversion still in progress.

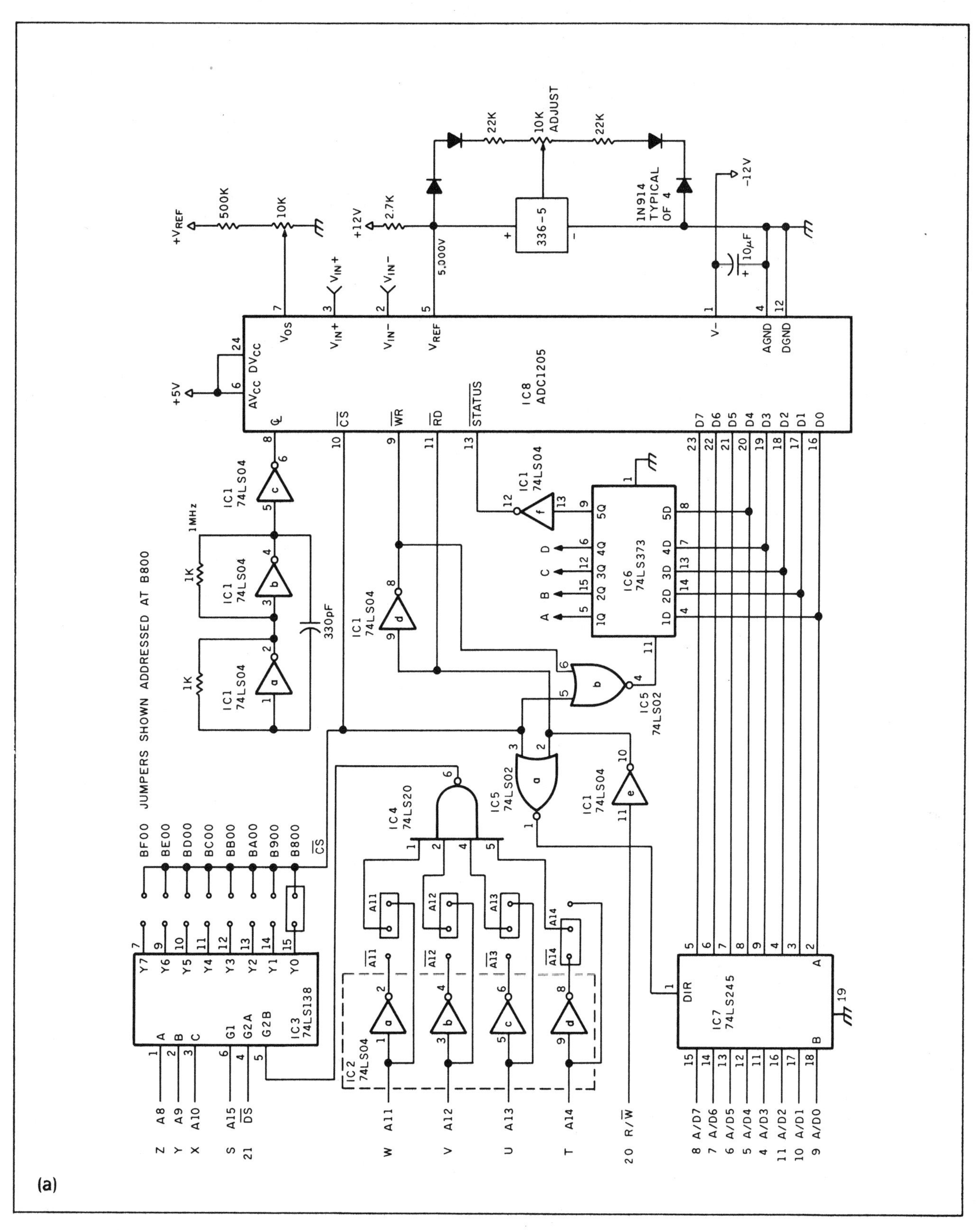

Figure 8: *Schematic diagram of the* BCC-30 16*-channel* A/D *converter board* ***(continued on the following page).***

(continued on the following page)

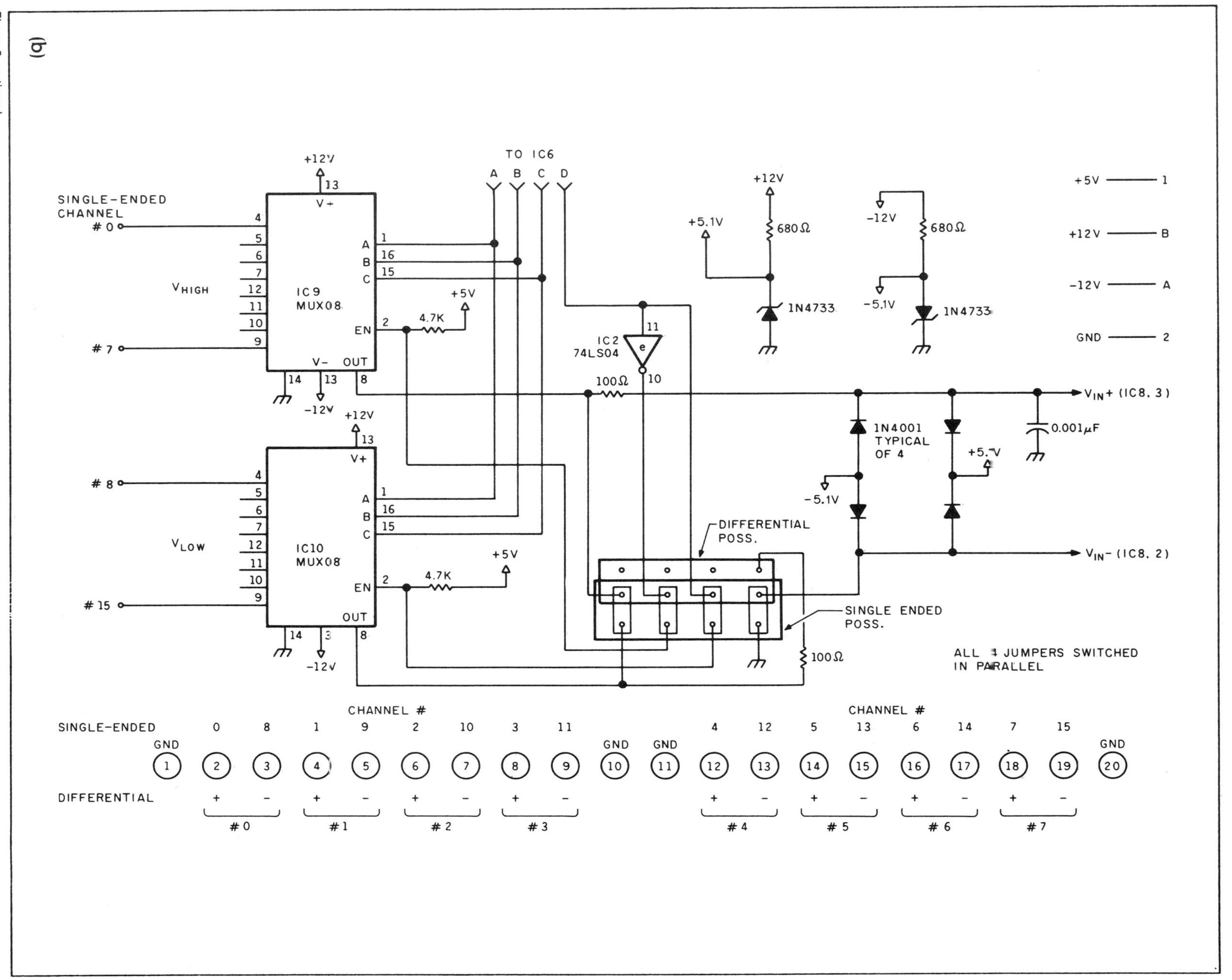

Figure 8 continued

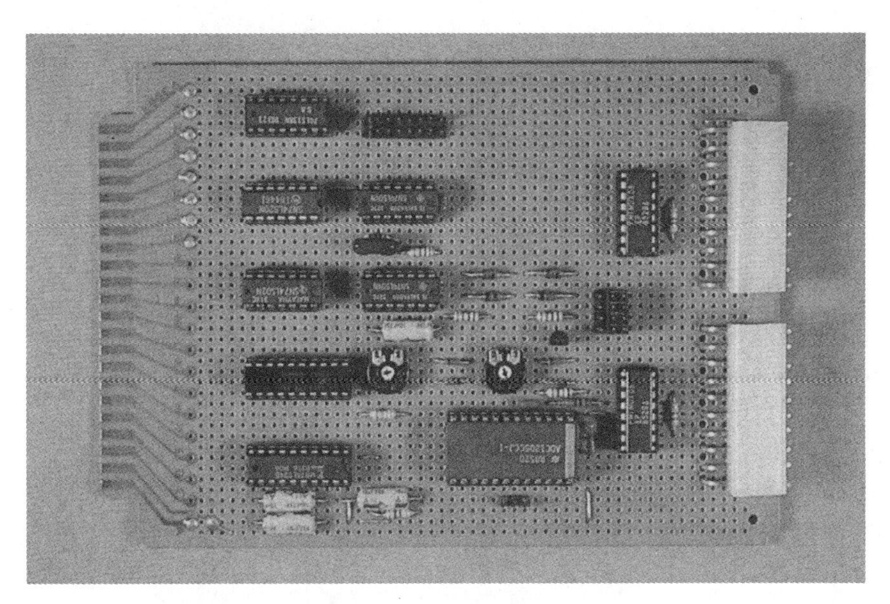

Photo 1: *Prototype of the Circuit Cellar 16-channel 12-bit plus sign A/D converter board.*

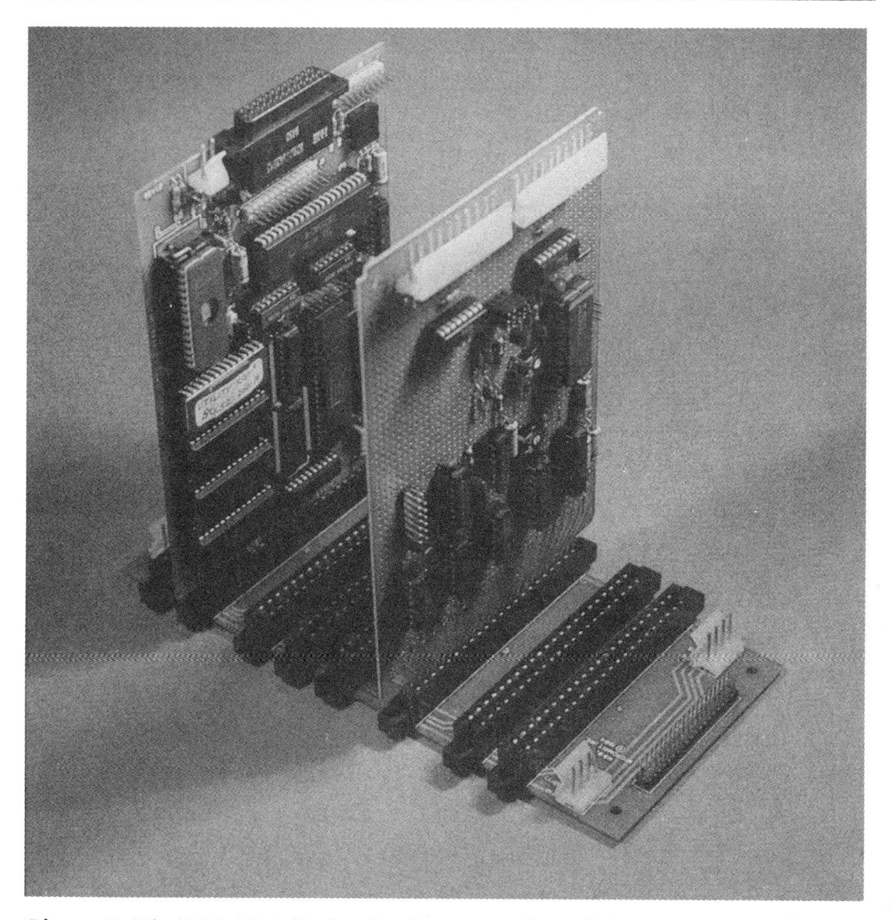

Photo 2: *The BCC-30 A/D board is bus-compatible with the BCC-52 controller board. The BCC-30 A/D prototype and BCC-52 are shown plugged into a backplane for use together.*

the "Z8 Bus" but more properly called the MMZ8 bus. Nothing is unique about the signals on this bus except perhaps their pin designations. It is a multiplexed address/data/control bus primarily oriented to 8-bit computers (16-bit address and 8-bit data).

The BCC-30 A/D board looks to the computer as a single address at any one of 128 predefined (jumper-selectable) locations. It can be configured either as 16 independent single-ended-input channels or 8 differential-input channels. Single-ended or differential operation is determined by the placement of jumpers JP1–4 and is therefore not under program control. The data byte sent by the computer to the board address defines which channel the input multiplexer is set for.

ICs 2, 3, 4, and 5 decode A8–A14 address bits to produce $\overline{CS}$ for the ADC1205 chip and latch data directed through buffer IC7 into the multiplexer address latch (IC6). The jumper positions selected in the schematic locate this address at B800 hexadecimal (47104 decimal). The 4 LSBs of this register control the input multiplexer while the fifth bit (b4) sets the $\overline{STATUS}$ level control line to the ADC1205.

Running the A/D board in BASIC is straightforward and consists of four sequential operations: set multiplexer address and reset A/D, start conversion, read most significant byte, and read least significant byte. While the status of the A/D is available as an output, a conversion takes only 100 microseconds and therefore could never be seen in BASIC (reading the status will be necessary if you are taking 10,000 samples/sec in an assembly-language program, however). It is simple enough to start the conversion and then go back immediately and read it since it will always be completed.

Executing an XBY(47104) = 18 in BASIC will load hexadecimal 12 into the address latch (the XBY( ) command in BASIC-52 is like PEEK and POKE in other BASICs). This corresponds to a multiplexer address of 2 and a status bit set to a logic 1 (reset). Resetting the status bit starts the conversion with an XBY(47104) = 2. The 2 bytes are then read as A1 = XBY(47104):A2 = XBY(47104). The most significant byte/least significant byte counter automatically increments on the successive reads. Summarizing, to read channel #2 (board address B800 hexadecimal), we execute code as outlined in figure 9.

A1 and A2 can then be combined to produce the desired output. I refer you to listing 1 for that procedure.

As mentioned earlier, four jumpers (JP1–4) decide whether the function of the A/D is 16-channel single-ended or 8-channel differential. All four jumpers are moved together, and all must occupy either the single-ended or differential jumper positions together. Each MUX08 (IC9 and 10) multiplexer is an 8-channel JFET-type analog switch. While CMOS (complementary metal-oxide semiconductor) switches might function in the circuit (and be about a tenth the cost), their I/O-transfer characteristics are not adequate for a 12-bit converter. The variations in resistance with input signal level would surface as measurement errors and instability. JFET multiplexers are specifically designed for this application and have very flat response curves.

Four bits from the multiplexer address latch (IC6) are directed through the jumpers to the multiplexer control lines. In the single-ended position, V_{in-} of the ADC1205 is physically grounded, and the two MUX08s sequentially address 16 input signals through it to the V_{in+}. When they are in the differential position, however, address line D is disabled, V_{in-} is removed from ground, and both V_{in+} and V_{in-} are switched through the input multiplexers. A differential input on channel #2, for example, would have V_{in} high on IC9 pin 6 and V_{in} low on IC10 pin 6 (setting channel #10 when using differential mode will enable channel #2 instead).

The remaining areas worth commenting about are the reference voltage and input protection. For a 12-bit A/D to be worth anything, it must have a precise, stable reference voltage for its internal D/A. In the BCC-30, the 5-V reference is supplied from an LM336-5 voltage reference chip. Additional diodes and a trim pot allow it to be precisely set at 5.000 V with virtually no temperature drift. Only a positive reference is required, even though the converter measures negative voltages as well.

The only "gotcha" in using the ADC1205 is input protection. While it measures +/− 5-V inputs, levels above or below +/− 5.3 V may damage the device. One method of protecting the inputs is through clamping diodes and current-limiting resistors. Using these techniques, I have connected V_{in+} and V_{in-} to a voltage source that will shunt damaging inputs away before they exceed 5.3 V. Unfortunately, if these diodes are connected to +/− 5 V, they will not begin conducting until +5.6 V and −5.6 V, respectively (germanium diodes with similar speed and power capabilities are much more expensive). I have chosen the least painful alternative by providing +/− 4.7-V Zener-generated sources to the clamping diodes that will start conducting at 5.3 V.

Presently, only a 100-ohm series

BASIC Command	Function
XBY(47104) = 18	Set multiplexer channel #2 and set status line high to reset A/D converter.
XBY(47104) = 2	Retain multiplexer channel setting and set status line low to start conversion.
A1 = XBY(47104)	Read first (most significant) byte.
A2 = XBY(47104)	Read second (least significant) byte.

Figure 9: *Series of* BASIC-52 *statements used to read channel #2 of the* BCC-30.

Listing 1: A *sample* BASIC-52 *program to read and display channels 0–7 on the* BCC-30.

```
10    CLEAR
20    REM  READ AND DISPLAY A/D CHANNEL 0-7
30    REM  SINGLE-ENDED OR DIFFERENTIAL
40    REM  -5- TO +5-VOLT INPUT
50    REM
60    REM
70   N=47104 : REM BOARD ADDRESS
80    REM  STATUS BIT IS B5 - LOGIC 1 IS RESET
90    FOR A=0 TO 7 : REM  DO ALL CHANNELS 0-7
100   GOSUB 160 : REM  READ A CHANNEL
110   NEXT A : REM  NEXT CHANNEL
120   PRINT CHR(18),CHR(27),"Y" : REM  TERMITE — HOME AND CLEAR SCREEN
130   REM  DISPLAY ARRAY HOLDING CHANNEL 0-7 READINGS
140   PRINT USING (#.###),A(0),A(1),A(2),A(3),A(4),A(5),A(6),A(7),"VOLTS"
150   GOTO 20 : REM  DO IT ALL AGAIN
160  XBY (N)=A+16 : REM  RESET A/D AND SET MULTIPLEXER CHANNEL
170  XBY(N)=A : REM  CLEAR STATUS BIT TO READ DATA
180  D1=XBY(N) : D2=XBY(N) : REM  READ 12 BITS AS TWO SUCCESSIVE WORDS
190  R=0.0012207 : REM  VOLTS PER COUNT
200   IF D1>=240 THEN GOTO 230
210  A(A)=R*((D1*256)+D2) : REM  SAVE POSITIVE READING IN ARRAY
220   RETURN
230  D1=255-D1 : D2=255-D2 : REM  ADJUST D1 & D2 FOR
     TWO'S COMPLEMENT
240  A(A)=-1*R*((D1*256)+D2) : REM  SAVE NEGATIVE READING IN ARRAY
250   RETURN
```

resistor is used in each input line to dissipate any input overvoltage. The ADC1205 converter has about a 50-kilohm input impedance so this extra resistance is unnoticeable. The series resistance can be increased further for more protection, but the temperature drift of this resistor adds errors to the system. The quantity of error depends upon the signal source impedance.

Frankly speaking, I would much rather have added a clamped-output op-amp stage, but it would have been very expensive. Remember, we are talking about 1-mV signals and 20 parts per million maximum permissible temperature drift. It hardly makes sense to add an op amp with a 30-mV offset and 200 ppm temperature drift combined with piles of who-knows-what discrete components moving in all different directions. Low-drift, low-offset, high-speed op-amp circuits are expensive.

It would be easy for me to simply provide an untried schematic of a typical protection circuit, but, as a practical matter, a properly designed and tested circuit with no offset or drift would have been a bigger project than the whole A/D board. I suggest that you simply try to limit your input range to +5 to −5 V. Half-watt 100-ohm series resistors will protect the inputs up to +/− 12 V.

While faster diodes might eventually be required in the clamping circuits shown, they are reasonably priced and adequate protection for normal use. No one wants to pay what it would take to guard against all possible circumstances. Only an idiot would try to measure the voltage across the tips of an arc welder with this board.

CONCLUSION

The price/performance of A/D converters is a balance of speed and resolution. There are $200 4-bit 100 million samples/sec A/D chips and $9.95 12-bit 2 samples/sec units (I won't bother to tell you how much 12 bits at 100 million samples/sec would cost). In environmental systems that have slowly varying conditions, speed is not as important as accuracy. Room temperature, for example, doesn't change so fast that you need to sample it 500 times a second. The accelerometers on a shake table, however, may need to be sampled 20,000 times a second for accurate G-force event records.

The BCC-30 has more than enough performance for most data-acquisition situations and will be finding a home in industrial control applications along with the BCC-52.

See page 447 for a listing of parts and products available from Circuit Cellar Inc.

27

REAL-TIME CLOCKS: A VIEW TOWARD THE FUTURE

One of these clocks also provides nonvolatile RAM

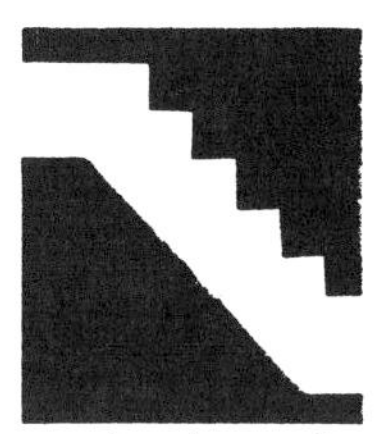

Ever have one of those occasions when everything is progressing smoothly and then you hit a big snag? Here I was, standing on a 6-foot ladder with screwdrivers in my back pockets, electric drill in my right hand, hammer in my belt, mouth full of screws, a pencil over my ear, and the audio/video multiplexer that I described in the previous chapter balanced in my left hand. A precarious climb if ever there was one, I assure you.

I had solved the ever-present do-it-yourself problem of forgetting some needed tool by dragging everything along with me up the ladder. I cursed as I banged my head on an overhead heating duct and swore that the next time I set aside an area to mount all this home-control junk, I'd know everything I was going to put on it in advance and leave plenty of room. Instead of a convenient shoulder-high location where life would be easy, I was halfway up under the floor joists sniffing concrete dust and soot. (Installing this stuff is getting about as pleasurable as writing software for me these days. Back when all I had was 48-instruction processors and lots of empty wall space, everything was copacetic.)

The only saving grace to this temporary agony was that it would soon be relieved by the enhanced automatic living afforded through the intelligent audio/video multiplexer (AVMUX). Using the BCC-52 computer and a smart terminal board from a couple of previous Circuit Cellar projects, the AVMUX would take inputs from a variety of audio and video sources and channel them to any of a number of specific outputs. My intention was to enhance my present level of automatic living to include programmed light, sound, and music.

I wiped my brow after turning the last mounting screw for the AVMUX and instinctively dodged another floor joist as I looked down at the card cage containing the computer that I still had to mount. I had included all the necessary cabling and mounting hardware between it, the multiplexer, and the Home Run Control System (HCS) already in operation. My control program was already written and saved in EPROM (erasable programmable read-only memory) so that it would automatically start when I powered up the system. Depending upon the day, time, direct signals from the HCS, and a yet-to-be-designed remote-control interface, specific or dynamically variable I/O (input/output) configurations would be mapped into the AVMUX. By retaining the different I/O maps in RAM (random-access read/write memory), the controller could switch among different command situations easily and rapidly.

Time? Memory maps? Uh-oh. It's terrible getting revelations on a ladder. Instead of the normal reaction—stand up, slam my hand on the desk, and yell "That's it!"—I had to be cognizant of the limited headroom and the dust I would raise by yelling.

In actuality, I had discovered nothing—nothing except a realization that what I had remembered was what I had forgotten to incorporate in my controller design. While the BCC-52 had a real-time clock/calendar within BASIC, it lacked the capability of retaining the I/O maps in RAM or the time if a power failure occurred.

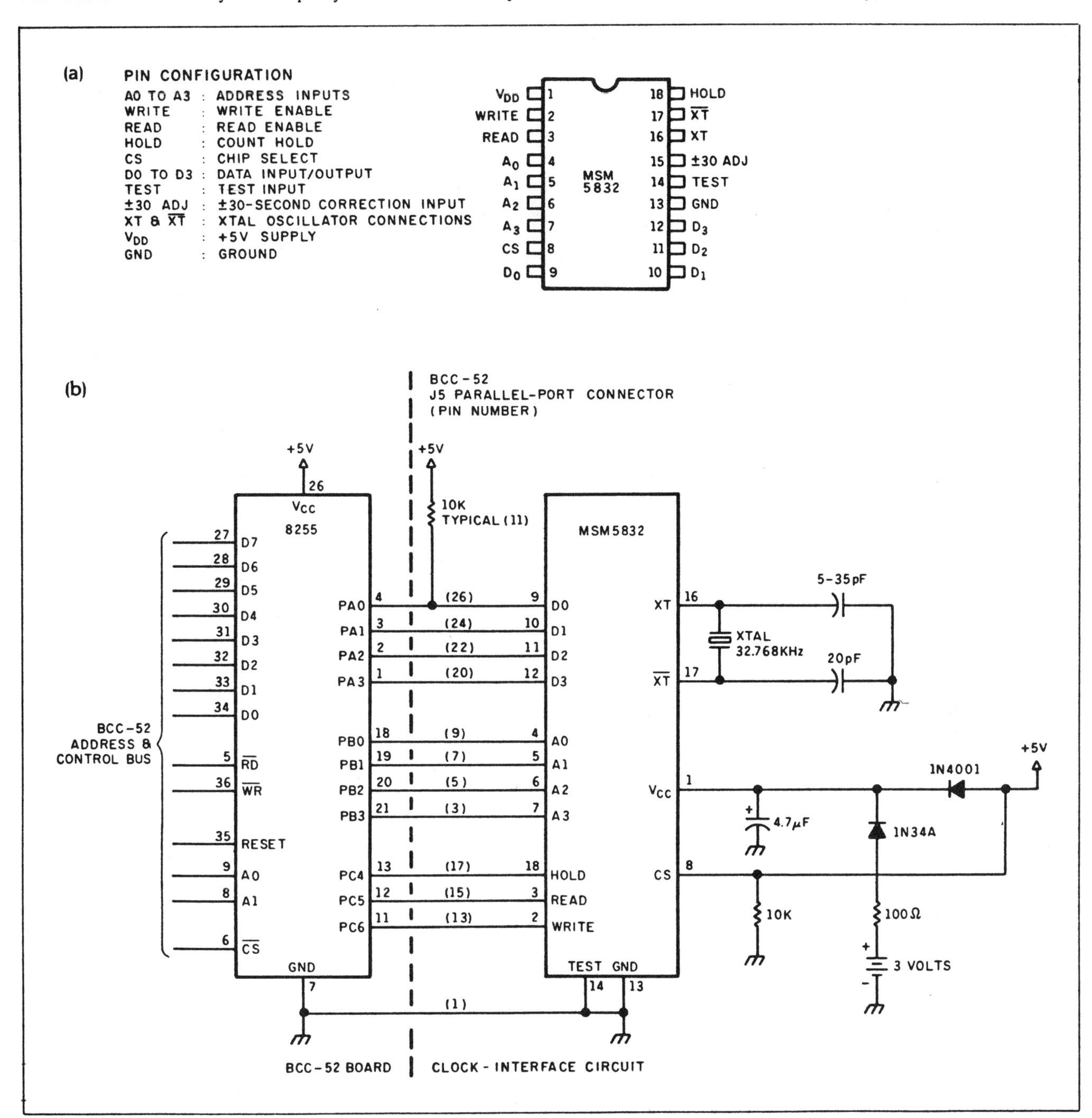

Figure 1: *(a) Pin-out of the Oki MSM5832 clock chip. (b) Schematic diagram of a real-time clock circuit using the chip from (a) and an 8255 PIA (already on the BCC-52).*

The EPROM-resident control program was self-initiating and would not be destroyed. But it would start up with a time and date of zero and RAM cleared.

My first reaction was to consider a UPS (uninterruptible power supply) or backup battery power for the whole computer. Either option was expensive and would consume a lot of circuit-board real estate that I didn't have. Even if successful, what good is powering the whole computer system unless all the peripherals that it controls are operational as well? In reality, all I needed to power was less than 1K byte of RAM and whatever component or system functions as the real-time clock/calendar. Of course, powering the whole computer would indeed provide nonvolatile memory and an uninterruptible clock/calendar.

Another trek up this ladder to mount a half-dozen rechargeable batteries didn't appeal to me. I needed a self-contained, battery-backed real-time clock that operated independently of the BCC-52 yet could be interrogated periodically regarding the time. Along the way, I'd look into making some portion of RAM nonvolatile.

Within seconds, I was down the ladder and shuffling through my junk box: piles of unopened news releases and an eyebrow-high pile of trade publications. The dimensions of this pile are usually affected only by an earthquake or a desperate search for some new technical idea around article time. Fortunately, my search was rewarded, and I came upon two solutions. One of them uses conventional technology; the other one is rather innovative.

I'll describe two real-time clock/calendars here and let you be the judge of which one you want to implement in your application. While both circuits are applicable to any computer, I was specifically looking to attach a clock and nonvolatile RAM to the BCC-52. Also, since I have covered the basics of real-time clocks and the specific attributes of the BCC-52 computer in previous Circuit Cellar articles, I refer you to them for additional information: "Everyone Can Know the Real Time" (*Byte*, May 1982, page 34) and "Build the BASIC-52 Computer/Controller" (Chapter 24 of this volume).

The first device uses a CMOS (complementary metal-oxide semiconductor) clock/calendar chip that attaches to the computer through parallel ports. Easy to exercise in software but consumptive of I/O, this conventional approach appeals to many of us who simply want a quick resolution of a problem and are not concerned about the I/O costs or having to manufacture thousands of them. The

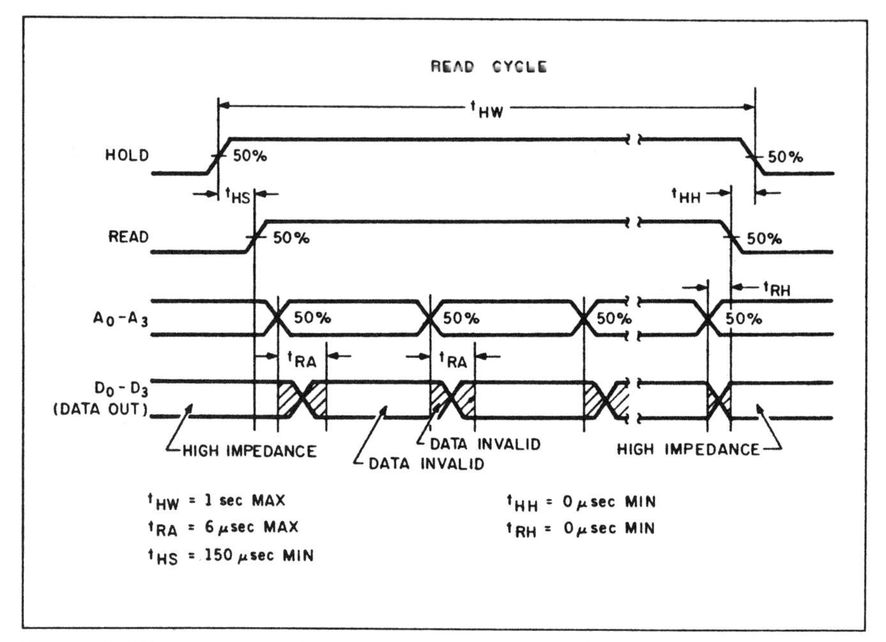

Figure 2: *The read-cycle timing diagram for the* MSM5832.

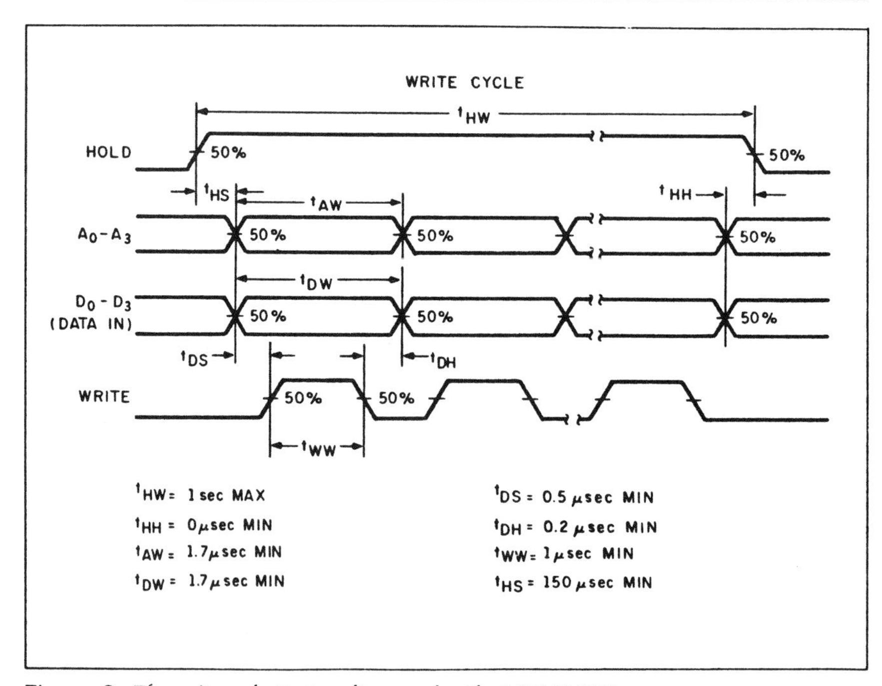

Figure 3: *The write-cycle timing diagram for the* MSM5832.

second approach uses a new clock "socket" from Dallas Semiconductor that requires no independent interfacing and merely plugs in with a static-RAM chip that might already be resident in the computer system. While software-intensive compared to the conventional clock chips, the concept is unique and will surely prevail long into the future. First, a little tradition.

A Single-Chip Clock

Without chopping traces and gluing a clock chip directly onto the BCC-52 bus lines, the only logical way I could add a hardware real-time clock was to build another interface card (with its associated bus drivers and address-decoding logic) or attach it to the parallel I/O connector (three parallel I/O ports are generated by an on-board 8255 PIA [peripheral interface adapter] and attached to a 26-pin connector designated as J5). For simplicity and ease of programming, I chose the latter route.

Figure 1 is the schematic of a single-chip battery-backed real-time calendar clock. As far as clock circuits go it is relatively low-tech, but it does work and was easy to build. It uses an Oki MSM5832 clock chip that functions as follows:

- Read cycle (see figure 2): The HOLD line goes high first, followed by the 150-microsecond hold setup time (t_{HS}). After this, READ is set high, and the specific MSM5832 registers are sequentially (or randomly) addressed. There is a 6-μs read-access time delay (t_{RA}) between setting the address and reading the data. This procedure is repeated to read all 16 registers. The HOLD line is then returned to a logic 0.
- Write cycle (see figure 3): The write cycle starts by raising the HOLD line, followed by a 150-μs t_{HS}. Next, the MSM5832 register address and the data to be loaded into it are presented on the address and data lines. While these lines are set, the write-input line is strobed high for a minimum of 6 μs. Concluding writing to the 13 counter/registers and three flags, the HOLD line is set again to a logic 0.

I chose the 5832 because it is probably the simplest clock on the market to use in a parallel-port connection. While the timing constraints just described make the 5832's use as a bus-connected clock chip a more difficult connection, they are of little consequence when using it with a PIA, especially one operated in BASIC. Many clock chips are available to match the speeds of today's processors. By connecting them through a PIA, however, this increased speed is filtered by the PIA and is of no extra value. Specific bits on the PIA function to simulate the computer bus, but the range of control is greater, and bus timing constraints are eliminated.

The 5832 functions by counting pulses from a 32-kilohertz crystal. These counters can be individually preset (setting the time) or read (reading the time) under program control. The chip is CMOS, and while it operates on 5 volts, it retains its status and continues keeping time with a power supply of only 2.2 V. Two hearing-aid batteries (1.5-V alkaline) are soldered together to produce 3 V. (While lithium batteries are preferred, all the lithium batteries I had were larger than I cared to use on this clock.) When the computer is powered, 5 V is applied to V_{CC} and the clock chip-select line, enabling all I/O and programming functions. With the absence of power, the chip continues to function on 2.8 V supplied through a germanium diode, but I/O and programming functions are inhibited because the chip-select line will be at logic 0.

Five volts is not normally available on the J5 parallel I/O connector, so I had to add a jumper to the back of the board. Rather than try to reroute the circuit connections of J5, I wired +5 V to pin 10 of the J2 serial-printer header and attached a jumper wire with a 2-pin header to connect this power to the prototype board. This pin is not normally used on a serial printer and will not obstruct its use. If you want both the real-time clock and a printer to function together, however, you'll have to make an alternate connection point.

The entire clock is constructed on a 3-square-inch piece of prototyping board that piggybacks on the BCC-52 and plugs directly into the J5 parallel-port connector. Four bits of port A (on the 8255) function as the clock chip's bidirectional data bus; 4 bits of port B serve as the address bus. Three ad-

Function	MSM5832 Hex Address	Hex Entry Range
X1 Seconds	00	0–9
X10 Seconds	01	0–5
X1 Minutes	02	0–9
X10 Minutes	03	0–5
X1 Hours	04	0–4
X10 Hours	05	0–11
Bit 0 and Bit 1 — X10 Hours		
Bit 3 — AM(0)/PM)1)		
Bit 4 — 12(0)/24(1) Hours Format		
Day of Week	06	1–7
X1 Day of Month	07	0–9
X10 Day of Month	08	0–7
Bit 0 and Bit 1 — X10 Days		
Bit 2 — 28(0)/29(1) Day Leap Year		
Bit 3 — N/C		
X1 Month of Year	09	0–9
X10 Month of Year	0A	0–1
X1 Years	0B	0–9
X10 Years	0C	0–9

Figure 4: A *map of the* MSM5832*'s registers.*

ditional bits of port C provide control signals that gate data into and out of the clock chip. The addresses and functions of these clock registers are shown in figure 4.

As listing 1 demonstrates, the 5832 is easy to set and read in BASIC. To set the time, the three ports are all configured as outputs by loading 80 hexadecimal into the 8255's control register (P4). Since I was writing the program (and you know how I feel about programming), I didn't get too fancy with PRINT statements in the clock-setting routines. Considering that the clock should have to be set only once forever, I think you can do a little hand calculation on the bit configurations and enter each of the presets with a simple prompt of which register is being loaded. If it's August 8th, for example, I expect that you could translate that into an 8 for register 9 and a 0 for register 10. As each entry is made, the port C (P3) control lines are activated in this sequence—hold on, write on, write off, hold off—thus functioning as a simple write strobe.

Reading the device is much simpler. Port A is set as input; ports B and C are set as outputs by loading 20 hexadecimal into the 8255's control register. The 13 clock registers are read by sequentially setting an address on port B and storing the value presented to port A. Because we are doing this all in BASIC, timing is noncritical. Even the hold command, normally enabled whenever the 5832 is read or set, can be ignored if counter-update ripple-through is not critical. After reading all the registers, the values are multiplied by the appropriate constants and added. My software is bare bones, and 24-hour timing was chosen for simplicity. (I realize I could have done it with 11 bits on only two ports—and probably never have finished the software.)

Listing 1: A BASIC *program to set and read the* MSM5832 *real-time clock chip.*

```
100    DIM N(200) :  DIM M(200)
110    REM
120    REM     REV 1.5 11/8/85
130    REM  5832 REAL TIME CLOCK FOR BCC-52 I/O PORT
140    REM
150   P1=51200 : P2=51201 : P3=51202 : P4=51203
155    REM SET 8255 PORT A AS INPUT AND B&C AS OUTPUT
160   XBY(P4)=90H
170    REM   PORT B IS ADDRESS AND PORT C IS CONTROL BUS
180    PRINT "ENTER 0 TO SET TIME OR 1 TO READ TIME",
       : INPUT A
190    ON A GOSUB 350,220
200    GOTO 180
210    GOTO 145
220    REM   READ 13 5832 REGISTERS
230   XBY(P3)=20H :  REM      SET READ MODE
240    FOR A=0 TO 12
250   XBY(P2)=A : N(A)=XBY(P1)
260    NEXT A
270    REM  DISPLAY CONTENTS
280    PRINT "DATE ",
290    PRINT N(10)*10+N(9),"/",N(8)*10+N(7),"/",N(12)
       *10+N(11)
300    PRINT "TIME ",
310    IF N(5)>=8 THEN N(5)=N(5)-8
320    PRINT (N(5)*10)+N(4),"  :  ",(N(3)*10)+N(2),"  :  ",
       (N(1)*10)+N(0)
330    PRINT
340    RETURN
350    REM SET TIME
360   XBY(P4)=80H :  REM   SET PORTS A,B,&C AS OUTPUT
370    REM MSB OF REG 5 12(0)/24(1) HRS & MSB-1 AM(0)/PM(1)
380    FOR A=0 TO 12
390    PRINT "REGISTER",A, :  INPUT X
400   XBY(P2)=A : XBY(P1)=X
405    REM WRITE STROBE
410   XBY(P3)=10H : XBY(P3)=50H : XBY(P3)=10H : XBY(P3)=00H
420    NEXT A
430   XBY(P4)=90H :  REM    RESTORE READ PORT SETTINGS
440    PRINT
450    RETURN
```

A Real-Time Clock with No Interface Required

As I wrote the above subhead, I tried to imagine how you would interpret it. Would the skeptical among you say, "Who's he trying to kid?" Or would you take it on faith that I've had the answer up my sleeve all along?

While the previous circuit works very well, and many of you no doubt will build it, it is I/O-intensive. The value of a single-board computer like the BCC-52 is that it is often desirable to implement it as a single-board control solution. As such, the parallel ports might be needed for the application and not be available for the clock interface I've described.

Not using the PIA means that the clock chip must be connected to the bus and addressed as memory or additional I/O. Short of using one of the already-decoded 8K-byte memory-block address strobes and building the clock to plug it in in place of a memory chip, it would appear that the only alternative is to attach the clock circuit through the external expansion-bus connector. That, however, would be the traditional answer to the problem.

Using one of the RAM-chip locations sounds like the most logical approach, until you realize that you are sacrificing 25 percent of the available on-board RAM (the BCC-52 holds 32K bytes of RAM in four 8K-byte chips on board) for less than 10 bytes of clock data. It would be much better if both

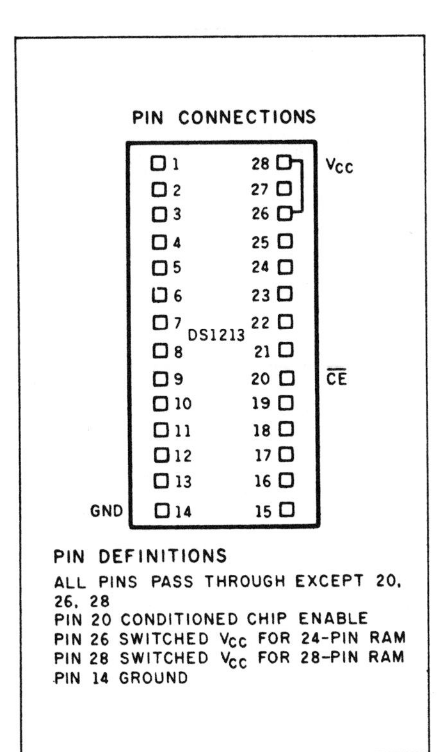

Figure 5: *A pin-out diagram of the SmartSocket.*

the clock and the RAM could share the memory-block address and any locations not those of the clock could be RAM. Better yet, why not keep all the RAM and attach the clock as a phantom interface that is there only when you need it?

Just as I was cleaning off the soldering iron to build the ultimate real-time clock, I discovered that such a device was just introduced by Dallas Semiconductor (4350 Beltwood Parkway South, Dallas, TX 75244, (214) 450-0470). Called SmartWatch, it does everything I had hoped in a single package and has some startling side benefits, like 8K bytes of nonvolatile memory. This latter revelation necessitates starting at the beginning and discussing a few additional Dallas Semiconductor products that are incorporated in the SmartWatch. The most notable of these is the DS1213 SmartSocket.

Nonvolatile RAM

The DS1213 is a 28-pin 0.6-inch-wide DIP (dual in-line package) socket with a built-in CMOS controller circuit and lithium battery (see figure 5). It accepts either a 28-pin 8K by 8 or 24-pin 2K by 8 lower-justified JEDEC (Joint Electronic Device Engineering Council) byte-wide CMOS static RAM. When the socket is mated with a CMOS RAM, it makes the RAM contents nonvolatile by automatically switching the RAM to battery operation and write-protecting it upon any occurrence of power interruption.

The SmartSocket performs five circuit functions necessary for implementing battery backup on a CMOS memory. First, a switch is provided to direct power from the battery or V_{CC} supply, depending on which is greater. This switch has a voltage drop of less than 0.2 V. The second function is power-fail detection at input voltages less than 4.75 V. The DS1213 constantly monitors the V_{CC} supply. When V_{CC} falls below 4.75 V, a precision comparator detects the condition and inhibits enabling of the RAM chip.

The third function accomplishes write protection by holding the chip-

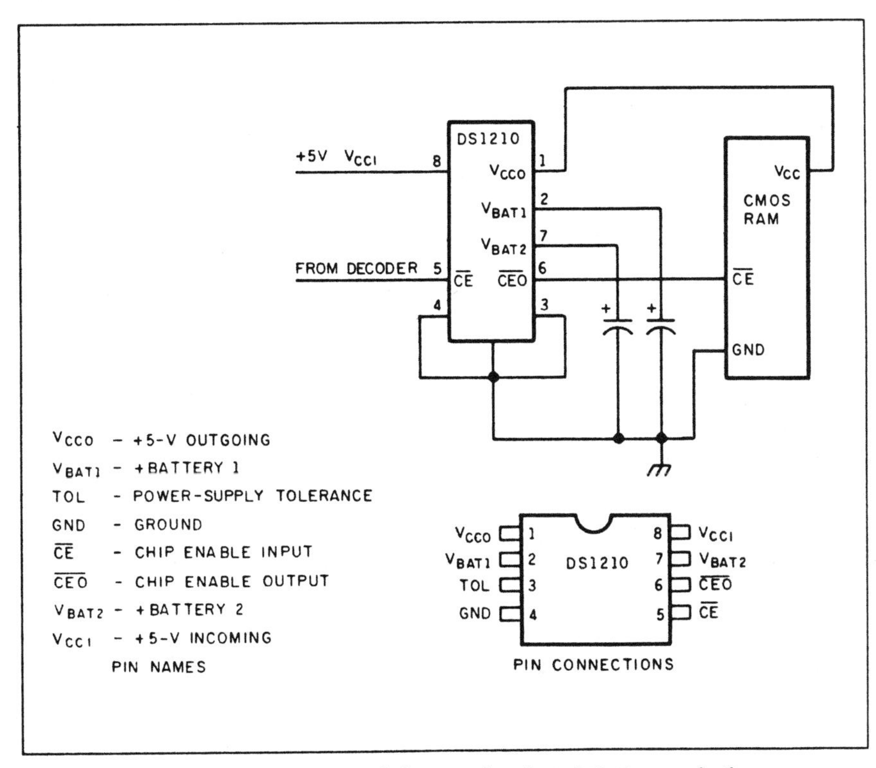

Figure 6: *The pin-out and functional diagram for the DS1210 nonvolatile-RAM controller.*

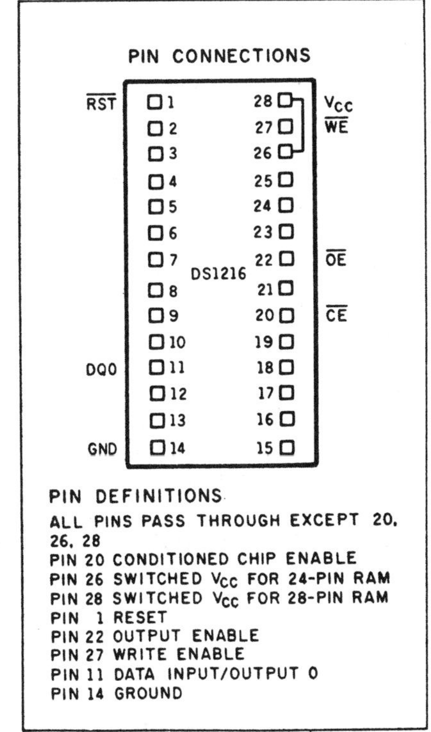

Figure 7: *A pin-out for the DS1216 SmartWatch.*

enable signal to the memory to within 0.2 V of V_{cc} or battery supply. If the chip-enable signal is active at the time power-fail detection occurs, write protection is delayed until after the memory cycle is complete to avoid corruption of data. During nominal power-supply conditions, the memory chip-enable signal will be passed through to the socket receptacle with a maximum propagation delay of 20 nanoseconds.

The SmartSocket's fourth function is to check battery status and warn of potential data loss. Each time that V_{cc} power is restored to the SmartSocket, the battery voltage is checked with a precision comparator. If the battery supply is less than 2.0 V, the second memory cycle is inhibited. Battery status can, therefore, be determined by performing a read cycle after power-up to any location in the memory, recording the contents of that memory location. A subsequent write cycle can then be executed to the same memory location, altering the data. If the next read cycle fails to verify the written data, the contents of the memory are questionable because the battery may not have retained it.

The fifth function is battery redundancy. In many applications, data integrity is paramount. The DS1213 SmartSocket has two internal batteries. During battery-backup time, the battery with the highest voltage is selected for use. If one battery fails, the other automatically takes over. The switch between batteries is transparent to the user. A battery status warning occurs only if both batteries are less than 2.0 V. Each of the two lithium cells contains 35 milliampere/hour capacity, making the total 70 mA/hr.

If you are contemplating a new design and want the benefits of nonvolatile static memory, the essential ingredients of the SmartSocket are available in chip form. Designated as DS1210, DS1224, and DS1212, they coordinate and perform the above described backup and write-protect functions for banks of 1, 4, or 16 in-

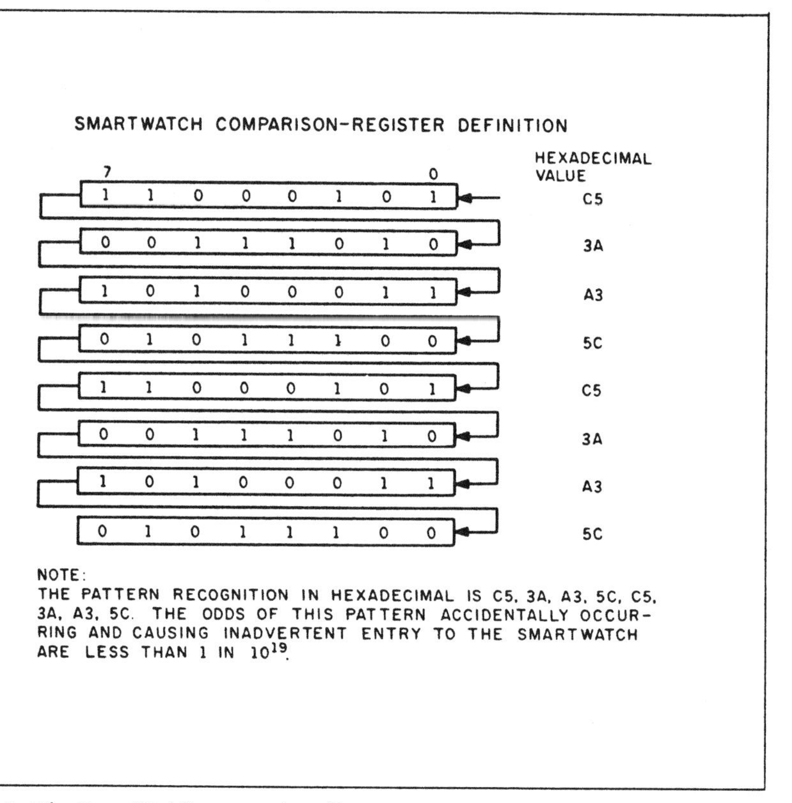

Figure 8: *The SmartWatch's comparison bit pattern.*

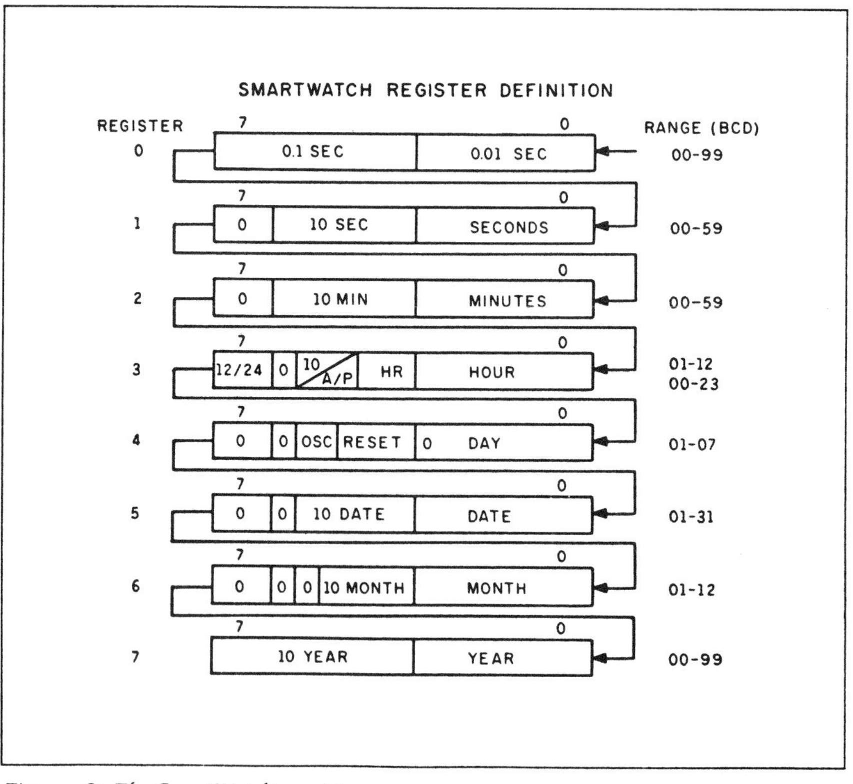

Figure 9: *The SmartWatch's registers.*

dividual CMOS RAM chips. (Contact Dallas Semiconductor directly for data sheets.) The DS1210 is shown in figure 6.

THE DS1216 SMARTWATCH

A new socket-style device called the DS1216 SmartWatch retains the non-volatile-RAM capability of the Smart-Socket and adds a calendar time function (see figure 7 for the pin-out). The SmartWatch includes its own crystal time base and maintains time information, including hundredths of seconds, seconds, minutes, hours, day of week, day of month, month, and year. The date at the end of the month is automatically adjusted for months with fewer than 31 days, including correction for leap year. Hours of the day can be tracked in both 12- and 24-hour formats.

Communication with the Smart-Watch is established by pattern recognition on a serial bit stream of 64 bits that must be matched by executing 64 consecutive write cycles containing the proper data on DQ0. All accesses that occur prior to recognition of the 64-bit pattern are directed to memory. After recognition is established, the next 64 read or write cycles either extract or update data in the SmartWatch; memory access is inhibited.

Data transfer to and from the time-keeping function is accomplished with a serial bit stream under control of chip enable ($\overline{CE}$), output enable ($\overline{OE}$), and write enable ($\overline{WE}$). Initially, a read cycle to any memory location using the $\overline{CE}$ and $\overline{OE}$ control of the Smart-Watch starts the pattern-recognition sequence by moving a pointer to the first bit of the 64-bit comparison register. The next 64 consecutive write cycles are executed using the $\overline{CE}$ and $\overline{WE}$ control of the SmartWatch. These 64 write cycles are used only to gain access to the SmartWatch. Therefore, any address to the memory in the socket is acceptable. However, the write cycles generated to gain access to the SmartWatch are also writing data to a location in the mated RAM. The preferred way to manage this requirement is to set aside one

Listing 2: A BASIC *program that sets and reads the SmartWatch.*

```
10      REM     APPLICATION PROGRAM USING ONLY BASIC
                TO DEMONSTRATE
20      REM     SMARTWATCH REAL TIME CLOCK ON BCC52 COMPUTER
                CONTROLLER BOARD
30      CLEAR
40      STRING 200,15
50     $(1)="SUNDAY"
60     $(2)="MONDAY"
70     $(3)="TUESDAY"
80     $(4)="WEDNESDAY"
90     $(5)="THURSDAY"
100    $(6)="FRIDAY"
110    $(7)="SATURDAY"
120     REM
130     REM ********* MAIN MENU ********
140     REM
150     PRINT "0=READ DATE/TIME    1=ENTER NEW DATE/TIME ?"
160    G=GET
170     GOSUB 1350 :  REM    GET NUMBER 0-9
180     PRINT CHR(18),CHR(27),"Y"  : REM CLR & HOME TERMITE
190     IF G=0 THEN  GOSUB 790 : REM READ & DISPLAY
        DATE/TIME INFO
200     IF G=1 THEN  GOSUB 250 : REM GATHER & SAVE
        NEW DATE/TIME INFO
210     GOTO 150
220     REM
230     REM    ********* GATHER $ SAVE NEW DATE/TIME INFO
240     REM
250    J=XBY(4000H) : REM SAVE BYTE LOCATED IN 4000H
       TO REPLACE WHEN DONE
260     GOSUB 1420 : REM SEND PATTERN RECOGNITION CODES
270     PRINT "ENTER DATE     MMDDYY"
280    G=GET
290     FOR Z=6 TO 8 :  REM USE G(6) FOR MM, G(7) FOR DD,
        G(8) FOR YY
300     GOSUB 1350 :  REM   GET NUMBER 0-9
310     PRINT G, :  REM   ECHO NUMBER 0-9
320    H=G*16 :  REM   STORE NUMBER IN UPPER NIBBLE
330     GOSUB 1350
340     PRINT G,
350    G(Z)=H+G :  REM   COMBINE NUMBERS 1 IN UPPER NIBBLE,
        1 IN LOWER NIBBLE
360     NEXT Z
370     PRINT
380    G=G(6) :  REM
390    G(6)=G(7) :  REM    SWAP 6 & 7, NOW 6,7,8 IN DD/MM/YY
400    G(7)=G : REM
410    G(1)=0 : REM SET TENTHS & HUNDREDTHS OF A SECOND = 0
420     PRINT "DAY OF THE WEEK SUN=0 MON=1 TUE=2 WED=3
        THU=4 FRI=5 SAT=6 ?"
430    G=GET
440     GOSUB 1350
450     PRINT G
460     PRINT
470    G(5)=G.OR.10H :  REM   OR BIT4 TO IGNORE
       RESET FROM PIN 1
480     PRINT "ENTER TIME    HHMMSS"
490    G=GET
500     FOR Z=4 TO 2 STEP -1 :  REM   USE G(4) FOR HH,
        G(3) FOR MM, G(2) FOR SS
510     GOSUB 1350
520     PRINT G,
530    H=G*16
540     GOSUB 1350
550     PRINT G,
```

Listing 2 continued

```
560   G(Z)=H+G
570    NEXT Z
580    PRINT
590    PRINT "IS THE TIME IN   0=24 HOUR FORMAT
       1=12 HOUR FORMAT  ?"
600   G=GET
610    GOSUB 1350
620    IF G<>1 THEN 680 :  REM   IF NOT 1 THEN JUMP
630   G(4)=(G(4).OR.80H) :  REM   OR BIT7 TO
      INDICATE 12 HOUR FORMAT
640    PRINT "IS IT   0=AM   1=PM  ?"
650   G=GET
660    GOSUB 1350
670    IF G=1 THEN G(4)=(G(4).OR.20H) :  REM      OR
      BIT5 TO INDICATE PM
680    REM   HOLD FOR TIME SYNCHRONIZATION
690    PRINT "HIT '0' TO GO SET THE NEW DATE/TIME"
700    GOSUB 1350
710    IF G<>0 THEN 700
720   GOSUB 1530 : REM STORE DATE/TIME INFO TO SMARTWATCH
730   XBY(4000H)=J : REM REPLACE BYTE TO 4000H
740   G=0
750    RETURN
760    REM
770    REM   ********* READ & DISPLAY DATE/TIME
780    REM
790   J=XBY(4000H)
800    GOSUB 1420 :  REM   SEND PATTERN RECOGNITION CODES
810    GOSUB 1230 :  REM   READ SMARTWATCH REGISTERS
820    PRINT "TODAY IS ",$((G(5).AND.7H)+1) :
       REM   STRIP OFF DAY OF WEEK
830   $(8)="  /  /  " :  REM   INITIALIZE DATE STRING
840   Z=7 :  REM   USE G(7) MM REGISTER
850   X=1 :  REM   PLUG CHARACTERS INTO STRING STARTING
      AT POSITION 1
860    GOSUB 1630 :  REM   GET 2 CHARACTERS FROM G(Z)
       AND PLUG INTO STRING $(8)
870   Z=6
880   X=4
890    GOSUB 1630
900   Z=8
910   X=7
920    GOSUB 1630
930   $(9)=$(8) :  REM   SAVE IT IN $(9) FOR ANY FUTURE USE
940    PRINT $(9)
950   $(8)="        .        " : REM INITIALIZE TIME STRING
960   G(9)=G(4)
970    IF (G(4).AND.80H)=0 THEN 1020 :
       REM   IF BIT7=0 THEN 24 HR FORMAT, JUMP
980    IF (G(4).AND.20H)=0 THEN ASC($(8),13)=41H :
       REM      IF BIT5 = 0, PLUG A
990    IF (G(4).AND.20H)=20H THEN ASC($(8),13)=50H :
       REM      IF BIT5 SET, PLUG P
1000  ASC($(8),14)=4DH :  REM    PLUG M
1010  G(9)=(G(4).AND.1FH) :
      REM   STRIP OFF FORMAT FROM HOUR REGISTER
1020  Z=9
1030  X=1
1040   GOSUB 1630
1050  ASC($(8),3)=3AH : REM PLUG IN THE CHARACTER FOR COLON
1060  Z=3
1070  X=4
1080   GOSUB 1630
```

(continued)

address location in RAM as a SmartWatch scratchpad.

When the first write cycle is executed, it is compared to bit 1 of the 64-bit comparison register. If a match is found, the pointer increments to the next location of the comparison register and awaits the next write cycle. If a match is not found, the pointer does not advance, and all subsequent write cycles are ignored. If a read cycle occurs at any time during pattern recognition, the present sequence is aborted, and the comparison-register pointer is reset.

Pattern recognition continues for a total of 64 write cycles until all the bits in the comparison register have been matched (this bit pattern is shown in figure 8). With a correct match for 64 bits, the SmartWatch is enabled, and data transfer to or from the timekeeping registers can proceed. The next 64 cycles will cause the SmartWatch to either receive or transmit data on DQ0, depending on the level of the $\overline{OE}$ pin or the $\overline{WE}$ pin. Cycles to other locations outside the memory block can be interleaved with $\overline{CE}$ cycles without interrupting the pattern-recognition sequence or data-transfer sequence to the SmartWatch.

The SmartWatch information is contained in eight registers of 8 bits each, which are sequentially accessed a bit at a time after the 64-bit pattern-recognition sequence has been completed. When updating the SmartWatch registers, each must be handled in groups of 8 bits. These read/write registers are defined in figure 9.

Data contained in the SmartWatch registers is in BCD (binary-coded decimal) format. Reading and writing the registers are always accomplished by stepping through all eight registers, starting with bit 0 of register 0 and ending with bit 7 of register 7. A few of the significant bits are the following:

- AM-PM/12/24 mode: Bit 7 of the hours register is defined as the 12- or 24-hour mode select bit. When high, the 12-hour mode is selected. In the

Listing 2 continued

```
1090  ASC($(8),6)=3AH
1100  Z=2
1110  X=7
1120   GOSUB 1630
1130  Z=1
1140  X=10
1150   GOSUB 1630
1160   PRINT $(8)
1170  XBY(4000H)=J
1180  G=0
1190   RETURN
1200   REM
1210   REM     ******** READ SMARTWATCH REGISTERS
1220   REM
1230   FOR Z=1 TO 8
1240  G(Z)=0
1250   FOR X=1 TO 8
1260  G=(XBY(4000H).AND.1) :  REM   G = BIT0
1270   IF G=0 THEN 1290 :
       REM   BIT = 0, DON'T ADD ANYTHING TO REGISTER BYTE
1280  G(Z)=G(Z)+(2**(X-1)) :
      REM   BUILD REGISTER BYTE FROM BITS RECEIVED
1290   NEXT X
1300   NEXT Z
1310   RETURN
1320   REM
1330   REM   ******** GET NUMBER 0-9
1340   REM
1350  G=GET
1360   IF G<48.OR.G>57 THEN 1350
1370  G=G-48 :  REM   ASC TO 0-9
1380   RETURN
1390   REM
1400   REM  ******** INITIALIZE PATTERN RECOGNITION CODES
1410   REM
1420  G(1)=0C5H
1430  G(2)=3AH
1440  G(3)=0A3H
1450  G(4)=5CH
1460  G(5)=0C5H
1470  G(6)=3AH
1480  G(7)=0A3H
1490  G(8)=5CH
1500   REM
1510   REM    ******** SEND REGISTERS TO SMARTWATCH
1520   REM
1530   FOR Z=1 TO 8
1540   FOR X=1 TO 8
1550   IF (G(Z).AND.(2**(X-1)))<>0 THEN G=1 ELSE G=0 :
       REM   STRIP OFF BIT
1560  XBY(4000H)=G :  REM   SEND BIT TO SMARTWATCH
1570   NEXT X
1580   NEXT Z
1590   RETURN
1600   REM
1610   REM  ******* GET 2 CHARACTERS FROM G(Z) REGISTER
1620   REM  ******* PLUG $(8) @ X
1630  G=INT(G(Z)/16)
1640  ASC($(8),X)=G+48
1650  ASC($(8),X+1)=G(Z)-(G*16)+48
1660   RETURN
1670   REM
1680   REM   ************ END
```

12-hour mode, bit 5 is the AM/PM bit with logic high being PM. In the 24-hour mode, bit 5 is the second 10-hour bit (20–23 hours).

- Oscillator and reset bits: Bits 4 and 5 of the day register are used to control the reset and oscillator function. Bit 4 controls the reset pin (pin 1). When the reset bit is set to logic 1, the reset input pin is ignored. When the reset bit is set to logic 0, a low input on the reset pin will cause the SmartWatch to abort data transfer without changing data in the watch registers. Bit 5 controls the oscillator. This bit is shipped from Dallas Semiconductor set to logic 1, which turns the oscillator off. When set to logic 0, the oscillator turns on, and the watch becomes operational.
- Zero bits: Registers 1, 2, 3, 4, 5, and 6 contain one or more bits that always read logic 0. When writing these locations, either a logic 1 or 0 is acceptable.

Reading and Setting SmartWatch in BASIC

While it is ultimately smarter to exercise the SmartWatch through an assembly-language routine, the universality of BASIC suggests that it would be a better tool for demonstrating the intricacies of communicating with SmartWatch. Listing 2 is a BASIC program that sets and reads a SmartWatch installed at 4000 hexadecimal on a BCC-52 computer/controller board. This program is more involved than the 5832 clock-chip program described earlier, primarily because it has more reporting features and communicates with the operator through menus. Embedded among all the REM and PRINT statements are the essential time-setting and read routines that you can translate from BASIC-52 to any other BASIC.

An Assembly-Language Firmware Utility

The assembly-language interface to the SmartWatch is also done on the bit level. The device requires that a particular pattern of 64 consecutive bits be written to data bit 0 in order to access the time and date registers of the SmartWatch. Then, 64 more consecutive reads or writes are required to examine or set the watch.

The firmware EPROM consists of a set of routines that take care of manipulating the SmartWatch and conversion of the SmartWatch data from BCD to binary format (I call this firmware SmarTime). This allows the use of BASIC to directly read the data in memory with no conversion code necessary in the BASIC-52 program. Year, month, date, day of week, hours, minutes, seconds, and hundredths of seconds are made available in the BCC-52 data memory for reading by the user program. SmarTime uses 24-hour military time (00:00–23:59) in order to eliminate the need for an AM/PM indicator. If AM/PM time is required, you can convert it to the more standard format in BASIC.

The SmarTime system is contained within an EPROM at location 6000 hexadecimal (for this demonstration) and occupies 300 (hexadecimal) bytes of memory (the SmarTime routines can be reassembled to run in any available 300 (hexadecimal)-byte EPROM space on the board). The rest can be used by your application if desired, but be careful to put your data into the EPROM from the top down.

SmarTime stores its register save and load areas and date/time information in the area directly above BASIC-52's current MTOP pointer. Because of this, the MTOP address must be adjusted down by 30 bytes prior to calling the initialization routine at location 6000 hexadecimal. A memory map appears in figure 10.

Three basic functions are found in SmarTime (figures 11–14 outline the logic flow of these programs). The first is a routine that sets up the memory environment for SmarTime to use. This routine, executed with a CALL 6000H, creates a load table of information in the memory above MTOP. The table contains the base address of the SmartWatch device, as well as a pointer to where the time information is stored.

A second routine, invoked with a CALL 6003H, uses the binary data stored in the time and date fields to set the SmartWatch. The required control bits for establishing 24-hour time and for turning on the SmartWatch internal oscillator are added to the data prior to its being written to the SmartWatch device. The routine also does the binary-to-BCD conversion.

The third function of the software is a routine for reading out the SmartWatch date and time information, converting it from BCD to binary format, and storing it in the memory area above MTOP. It is executed with a CALL 6006H.

Using the SmartWatch with SmarTime is easy. With the DS1216 installed at address 4000 hexadecimal and SmarTime at 6000 hexadecimal, simply enter and run the program

SmarTime uses 24-hour military time to eliminate the need for an AM/PM indicator.

shown in listing 3.

Finally we are back to simple BASIC programs with the help of a little firmware tucked away in an EPROM. As this program runs, you should see the seconds location of the clock/calendar printed out once per second. If it is correctly incrementing, the EPROM is installed properly at location 6000 hexadecimal and the Smart-

Program Memory	6000H–6300H		SmarTime EPROM-resident software
External Data Memory	Offset Above MTOP Value		Function
	DEC	HEX.	
	24	18	YEARS (00–99)
	23	17	MONTHS (01–12)
	22	16	DATE (01–31)
	21	15	DAY (01–07)
	20	14	HOURS (00–23)
	19	13	MINS. (00–59)
	18	12	SECS. (00–59)
	17	11	HUNDREDTHS of SECS (0.00–0.99)
	16	10	BANK 3 REG 7 LOAD (RESERVED)
	15	0F	BANK 3 REG 6 LOAD (RESERVED)
	14	0E	BANK 3 REG 5 LOAD (RESERVED)
	13	0D	BANK 3 REG 4 LOAD (RESERVED)
	12	0C	BANK 3 REG 3 LOAD (TIME AREA LOW)
	11	0B	BANK 3 REG 2 LOAD (TIME AREA HIGH)
	10	0A	BANK 3 REG 1 LOAD (RESERVED)
	09	09	BANK 3 REG 0 LOAD (WATCH BASE)
	08	08	BANK 3 REG 7 SAVE AREA
	07	07	BANK 3 REG 6 SAVE AREA
	06	06	BANK 3 REG 5 SAVE AREA
	05	05	BANK 3 REG 4 SAVE AREA
	04	04	BANK 3 REG 3 SAVE AREA
	03	03	BANK 3 REG 2 SAVE AREA
	02	02	BANK 3 REG 1 SAVE AREA
MTOP += >	01	01	BANK 3 REG 0 SAVE AREA

Figure 10: *The SmarTime firmware memory map.*

Watch at location 4000 hexadecimal. You can now write your own BASIC-52 programs to use SmarTime.

IN CONCLUSION

Either real-time clock I've presented is applicable and valuable in control applications. Which you use depends primarily upon the application. It took only a few hours to build and test the 5832 circuit, and it proved an immediate success for a one-shot problem. In the long run, however, the SmarTime system incorporating the SmartWatch and nonvolatile RAM is a more useful BCC-52 peripheral that can be easily duplicated, especially now that the software is written.

Speaking of software, the BASIC listings and a file of the SmarTime executable code (to run at 6000 hexadecimal) discussed in this article are available for downloading from the Circuit Cellar BBS at (203) 871-1988.

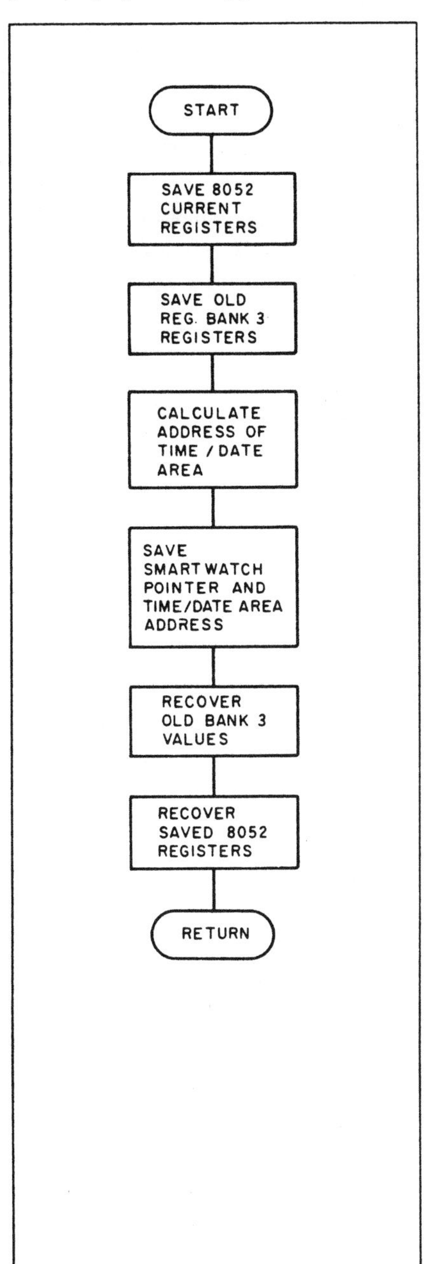

Figure 11: *The SmarTime firmware flowchart—the initialization routine.*

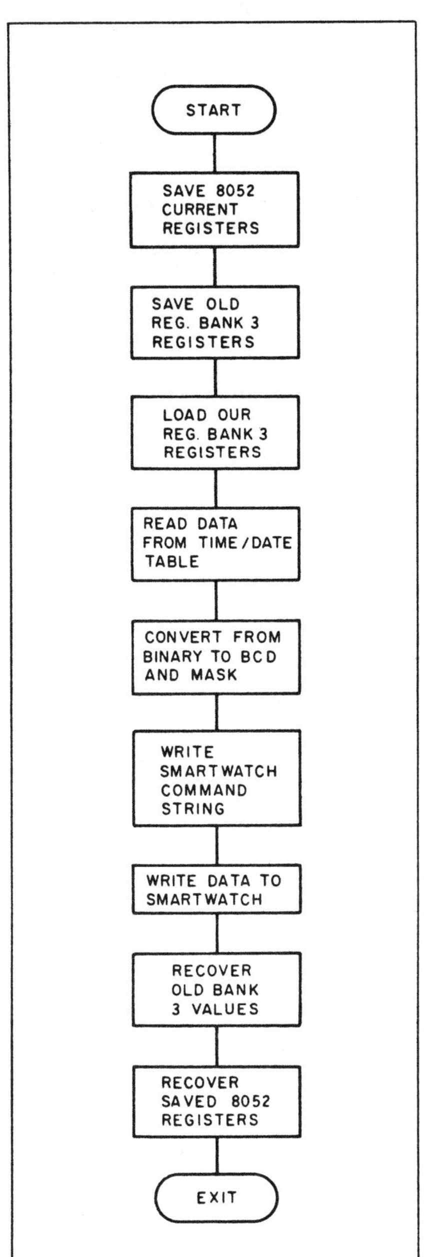

Figure 12: *The SmarTime firmware flowchart—a routine to set the SmartWatch.*

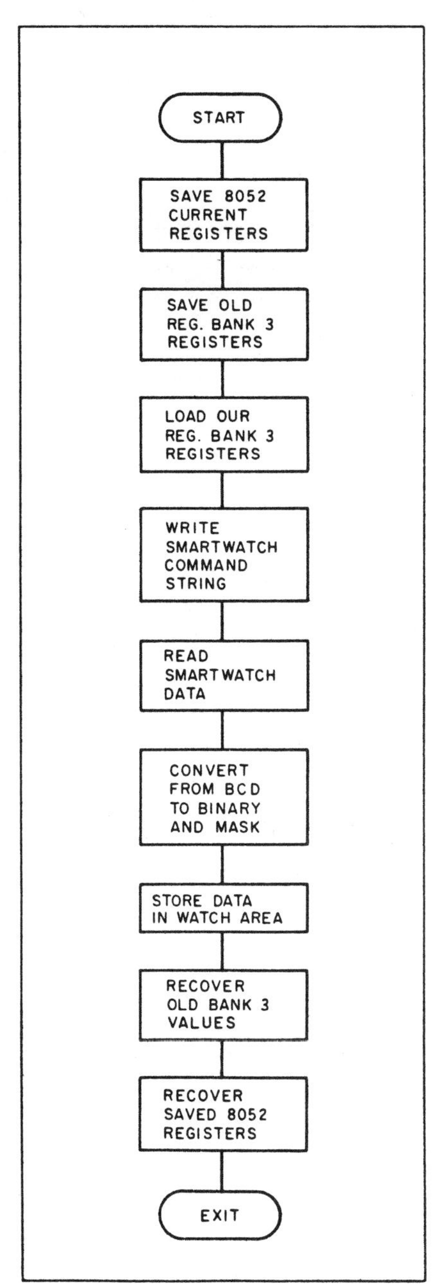

Figure 13: *The SmarTime firmware flowchart—a routine to read the SmartWatch.*

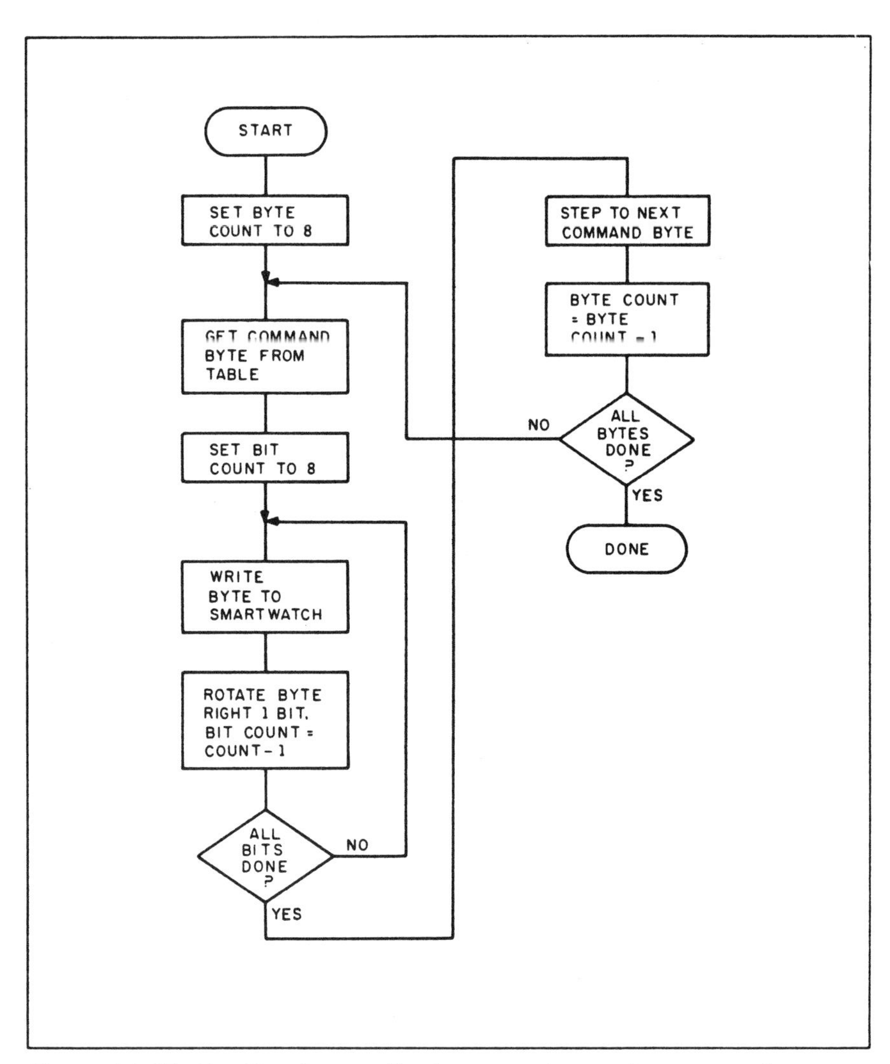

Figure 14: *The SmarTime firmware flowchart—a routine to write a command string or data to the SmartWatch.*

Listing 3: *This BASIC program uses the SmarTime firmware to update a seconds counter on screen. Note that comments in parentheses next to the code should not be entered as part of the program.*

```
10 MTOP = MTOP - 30   (RESET MTOP POINTER)
20 DBY(18H)=040H  (ASSUME SMARTWATCH AT 4000H)
30 CALL 6000H    (INITIALIZE THE SYSTEM)
40 REM NOW SET SMARTWATCH TIME
50 FOR X=MTOP+24 TO MTOP+17 STEP -1
60 READ C
70 XBY(X)=C
80 NEXT X
85 REM ZZ/01/85  14:25:00.00
90 DATA 85,11,01,05,14,25,00,00
100 CALL 6003H  (WRITE THE VALUES)
110 CALL 6006H  (READ THE VALUES)
120 PRINT MTOP+18 (SECONDS COUNTER)
130 GOTO 110
```

Special thanks to Jeff Bachiochi and Bill Curlew for their software expertise.

See page 447 for a listing of parts and products available from Circuit Cellar Inc.

28

COMPUTER ON GUARD!

A tale of overzealous computer security

Editor's note: Steve Ciarcia has been working with and writing about home-control systems for many years. This article shows what could happen when a sophisticated system runs amok.

"Merrill, you gotta help me!"

The feeling of panic was coming over me as I beat on Merrill's back door. I needed help, and Merrill was the only person I could trust—the only person that would understand that I wasn't crazy. As I knocked on the door, I glanced over either side of my shoulders to make sure no one else was around.

"Merrill, you gotta help me!"

I stood next to the door in a shadow that the moonlight failed to illuminate. It was a cool spring evening. While the stars shown brilliantly in their quiet elegance, I couldn't help but fear that this would be the last quiet moment of the evening if I failed.

"Steve? What are you trying to do? Can't you just ring the bell and wait 30 seconds like everyone else?"

Merrill wasn't really mad, just startled at my wild-eyed look and disheveled appearance. I often visited him but usually announced myself by some means other than beating down his back door. He waited a few seconds. He realized that this wasn't a social call and changed his tone to one of concern.

"What's wrong? You look terrible."

"I locked myself out!"

For any other person in the world, that would not be a catastrophic occurrence. In fact, the words sounded a bit absurd as I said them. I only hoped that Merrill valued our friendship enough to listen to me.

"You locked yourself out? Didn't you once give me a key to hold just in case this ever happened?" Merrill was becoming increasingly curious as to why I should be so distraught. I should have known he had a key.

I nervously glanced at my watch and answered, "That was when one needed a key to get in my house." Such a statement obviously would lead to all kinds of conjecture, but I didn't have time to explain.

"What do you mean, no key? How do you get into your house? Whistle?" Merrill seemed a bit disturbed that I was playing guessing games.

"I don't use a key anymore. I use a digital code. I really don't have time to explain.

Please, just put on some dark clothes and help me."

His help-thy-neighbor attitude took five giant steps back when I mentioned the necessity for dark clothes. Glancing at my watch once again to see how much time we had left, I determined that a portion of it had to be allocated for explanation. I stepped into the doorway and moved past Merrill.

"You see, Merrill, I've locked myself out of the house, and I have a soufflé in the oven."

Merrill looked at me like I was some kind of nut. He walked over to the kitchen sink and opened the cabinet doors beneath it, revealing a toolbox. "Look, we'll zip over and pull the hinges on one of the doors. It's a cinch."

Before he could pass me any tools, I interrupted him. "Merrill, it's not that easy. You don't understand. Let me explain."

The expression "Please do" was painted all over his face and needed no verbalization. As he sat down in the overstuffed chair, he extended and crossed his legs on the footstool and stroked his gray beard nervously. The little bit of fuzz on the top of his balding head seemed to bristle like a cat. To further the impression that he was ready for a real fish story, he took out a briar pipe and nonchalantly started to clean it. Between the sounds of tapping the pipe on the ashtray and blowing through the stem to clear it, he extended his hand toward me and said, "Do begin, please."

"Merrill, you gotta help me!"

The delay was excruciating. It was critical to act soon. The soufflé was irrelevant. It was the chain of events that could be accidentally touched off that I was worried about. My only hope was to talk fast.

"Merrill, my house doesn't use a house key anymore because it has a computerized environmental and security control system!"

He puffed on the pipe and interjected, "Fair enough. But what's that got to do with the soufflé?"

"This isn't just any home-control and security system. I designed it! An advanced sensor system tied directly into my computer makes it about the most sophisticated home burglar alarm in the world. I got thinking one night that I needed a burglar alarm. Since practically all the lights and appliances were already connected to the control system, I just extended its capability a little. But I got a little carried away on the engineering, and I'm not sure I know how to get in without setting it off."

Merrill was amused. Every time he and I had spoken lately, it had something to do with computers. He no longer thought I was completely crazy, just a little. There was still that one burning question, "What has that got to do with the soufflé?"

"There's a soufflé in the oven, and let's see . . . it should be done in 30 minutes. But the oven timer only buzzes. It doesn't shut the oven off. I know you're only an engineer and not Betty Crocker, but even you can guess that it wouldn't be more than another 20 to 30 minutes before it starts to burn."

I spoke rapidly. We were eating up precious seconds. "When the smoke from the burning soufflé hits the smoke detectors on my alarm system, all hell is going to break loose on this street."

"Wow! What does it do, call the police?"

Most people are familiar with the standard smoke and burglar alarms that automatically dial the fire department. While the end result was the same, my method was quite different. The sophistication of my home-control computer was unmatched by anything that commercial companies had to offer. That, in combination with the mind of your average, everyday mad scientist, can produce startling results.

"Well," I started rather sheepishly. It isn't often one has to explain the limits of his paranoia. "It isn't every day you have a fire in your house. When you do, you want action fast so you can reduce the damage and get people out in time. This system is predicated on everyone acting fast. When a fire or smoke is detected, it first sets off the alarm horns mounted outside next to the garage. I've never tried them, but they're war surplus air-raid sirens.

"Mathematically, the sound level coming out ought to be high enough to break about half the windows on the street. Mrs. Picker, who lives directly across from my house, will probably have her whole house moved back about two feet when they go off.

"Secondly, there are four xenon strobe aircraft-landing lights mounted on the corners of the house that will start flashing with about 2 million candlepower each. That was just in case the fire trucks had trouble finding the house.

"Then come the automatic telephone calls out on the three telephone lines. Remember, Merrill, my computer has a voice synthesizer, so I don't need a tape recorder. It definitely doesn't sound like a recording, so it should

ILLUSTRATED BY ELLIOTT BANFIELD

prompt immediate action. The first call is to the fire department. It also is simultaneously transmitted on CB channel 9. Then a whole bunch more. The end result is more cars and trucks than we can fit on this street."

The pipe in Merrill's mouth drooped lower and lower as I conveyed the consequences of my alarm going off. It was hanging down to his chin when he muttered, "Why don't you add me to the list of calls in case I miss the initial shock wave."

"Don't worry, Merrill! You're the ninth call!" Merrill definitely had a concerned expression. As I expanded upon the next step, it turned to terror.

"Merrill, you gotta help me break into my house and shut the alarm off before the soufflé burns."

The pipe fell out of his mouth, and the ashes formed a line down the front of his shirt. He barely noticed them as he exclaimed, "Are you crazy? Break into your own house?"

"Look, Merrill, I designed that system to prove I could do it. Now that I can count the seconds before I know it's going to go off, I recognize it as pure overkill. I'll replace it later with something more sane, like six Doberman pinschers and a minefield. But right now we have to stop it! Will you help me?"

"But what's that got to do with the soufflé?"

Merrill brushed the ashes off his lap and jumped up. "Do I need dark clothes?"

"Yes. I'll explain later. And wear a dark sweatshirt with a hood or something to cover your head."

The evening newspaper fell to the floor as it was sucked off the table by the vacuum created as Merrill ran to change. I could detect a cold sweat forming as I checked my watch repeatedly. It was only 10 minutes since we had first started talking, but now it was only 20 minutes before the soufflé would be done.

I could picture in my mind the progression of events that would follow. First, the soufflé would blacken and crack. Then, as it shriveled, some of the exterior sections would have dried enough to be combustible. The first whiffs of smoke would go unnoticed, but eventually a billowing cloud would spew forth from the oven. When it reached the smoke detectors, the computer would go into action. Our only hope was to get inside in time to stop the computer. If we failed, we had better make sure we were not standing next to one of those sirens when it blew. Further thoughts were interrupted as Merrill burst into the room fully dressed for action.

"I'm ready. Let's go."

Merrill looked like a cat burglar. The solid-black sweatshirt had a hood that completely covered his balding head, and, while his gray beard still showed, it aided the camouflage. His pants were equally dark and skintight. All reflective surfaces like belt buckles and key chains were carefully omitted. Black track shoes completed the outfit. I only hoped we didn't have to do too much running with all the rope, and as we jogged up the street toward my house, Merrill turned and asked, "You sure you know how to get in?"

The details of the computer alarm design flashed through my mind. I knew every wire, every sensor. Yes, I knew what the components of the system were. But the computer had far greater speed than I at analyzing the data received from them. A pressure switch activated in the wrong sequence, a heat sensor detecting human presence, any number of things could activate the alarm. I had let my inventive genius run "open loop." The tiny credit card that now lay on the coffee table in the living room had been my only control over the potential Frankenstein I had created. True, it would foil a burglar or call the fire department, but the ends to which I had gone in devising the system were aimed more at instant incineration of any perpetrator than protection of property.

To fully answer Merrill's question was impossible. I didn't know whether I could beat myself at my own game. "I don't know, Merrill. I hope so."

We stopped in front of my house. Almost magically a floodlight switched on to illuminate the area before us. Music could be heard from inside. A light in one room switched off, and another turned on. I didn't wait for Merrill to ask since I knew he was curious.

"Most of the AC outlets in the house are remote-controlled. The computer can control almost any light or appliance in the house, except the stove. The computer knows that something or someone is out here from microwave motion sensors planted in front of the house. No one is in the house, but it is simulating habitation by playing music and making it appear as though people are moving from room to room. Just for good measure, it turned the floodlight on to tell you that it knows you're here too."

Merrill started toward the front walk. I grabbed his arm to stop him.

"Forget it. The only way into the house is through some window that doesn't have any sensors attached. They're in the back of the house. Possibly one of the bathroom windows would be the best to try."

"Hey, Steve, before I lay my life on the line to save your soufflé, do you mind telling what happens if we set off the burglar alarm while trying to break in. You already told me about the fire alarm."

My reputation had preceded me. The fire alarm was only part of the system. The burglar alarm was equally devastating.

"Well, there's a bunch of stuff I'll explain as we go along. It's too complicated to explain in detail. But the end result is that the computer determines the location of the perpetrator and then tries to lock him in the area where he has been detected and calls the police."

"If that's all, you can explain the accidental phone call to the police. They often get false alarms from automatic dialers."

"*Mathematically* . . ."

"Wait, you didn't let me finish. Then, it sets off all the sirens and lights, just for good measure. And, oh yeah, there's a very loud noise source inside the house that's triggered, which is supposed to temporarily disable the perpetrator. Then it does all the same telephone calls, explaining there is a break-in instead of a fire."

Merrill looked at me in amazement. The adventurer in him wanted to go full speed ahead and tackle the Mount Everest of electronic obstacle courses, while the quiet engineering instinct suggested that he go home and check his medical insurance first. He shook his head as he said facetiously, "Why didn't you just use tear gas?"

"Oh, I considered it. It's just too hard to get the smell out of the Oriental rugs."

This unexpected response was too much for Merrill. As we stood there in the moonlight, I could see the sweat forming above his brow.

To this point, he had been aiding an eccentric neighbor. Though it had taken a long time and not through any direct explanation, Merrill was ready to admit that this computer alarm had to be stopped. There was no animosity that I had created it, just a realization of the full consequences of its being.

He, too, looked at his watch and sensed the seconds ticking away. No longer was he along for the ride. Now he was a committed participant. "Let's go."

I knelt down next to a sandy area at the corner of the lot. Merrill looked over my shoulder. Grabbing a short stick to draw in the soft soil, I started to lay out the attack plan. "Here's the house, the property line, and key obstacles. There's only one way to approach the house from the rear and not be detected. We have to go over the side-yard fence, along through the brush to the pine trees behind the house, and then across the back lot. Have you done any pole-vaulting recently?"

"Pole-vaulting? Are you kidding? I just about have enough energy to go from the couch to the refrigerator for another beer. What are you talking about?"

His eyes opened wide and projected a common expletive. The general translation was, "Hey man, I agreed to break in a house with you, but I ain't pole-vaulting over no fence."

That was the easiest way, but I had to agree with Merrill. The years in the cellar being a mad inventor rather than a tennis pro had taken their toll. I wasn't about to pole-vault over any fence either.

"We've got to find a way over the fence without actually climbing on it. There are vibration sensors in the vertical supports that are meant to detect anyone climbing over it. Tripping it won't set the whole alarm off, but it will start a timer where the computer treats perimeter events more seriously. If during that period the computer senses too many motion and vibration inputs, it will treat it as a threat and react accordingly." I didn't elaborate on the latter.

We stood next to the fence. It was constructed of heavy wire mesh attached to metal supports. Trying to vault over such a fence and missing would be like putting your body into a cheese grater. It was only about 5 feet high though, so there had to be an easy way over it.

Merrill looked at the situation. I could see his engineering mind going to work. Pictures of levers, fulcrums, balances, and pulleys were flashing through his mind. Walking over to the tree adjacent to the fence, he started coiling a length of rope in one hand. With one mighty swing, he threw the coil of rope over a 20-foot-high tree limb hanging directly over and parallel to the fence. Now the rope hung down and touched the top of the fence.

"Come here, Steve," he said.

I was still a little puzzled, even as he looped the rope around under my arms and tied a knot at my chest. Only when he pulled on the other end and hoisted me off the ground did I realize how he intended to get us over the fence.

"Gee, Steve, why don't you lose a little weight for the next break-in?"

I felt like a side of beef hanging on a rope 6 feet off the ground. When he started swinging me from side to side, I thought I was going to get seasick. The amplitude of the swing got longer and longer until the arc carried me over the fence to the other side. The realization of what the next part of the sequence would be came a fraction too late for me to protest. As the arc carried me over the fence, Merrill let go of his end of the rope. Logically, I should have expected that this was the only way, but the experience of being swung on the end of that rope hadn't any semblance of logical reasoning on my part. My far-too-late protest started something like a "whoop" and concluded with the tonal equivalent of Tarzan merrily swinging through the jungle and suddenly missing the last vine.

"Gee, Steve, why don't you lose a little weight for the next break-in?"

The fall was only 6 feet, but it felt like a hundred stories. I thought that if that was a sample of things to come, maybe I should take my chances with the alarm. It didn't help matters when I landed sitting down. The ground was quite moist, and my clothing sucked up the water like a sponge. When I put my hand down to reorient my position, I felt the cold spring mud ooze between my fingers. The totality of my situation and the immediate sensations at hand were summed up with the single word, "Yech!"

As I turned to check on Merrill, I caught a glimpse of him sailing through the air. Rather than be hoisted, he had secured one end of the rope and tied large knots in the other to aid climbing. Once at the 6-foot level, he swung out over the fence as I had and let go. Even though he came down feet first, the momentum was too great for the terrain. It took only a fraction of a second for two skid marks to form behind his heels, and Merrill came crashing down in the same sitting position next to me. His first word was, "Yech!"

I glanced at my watch and realized there were only 10 minutes left on the oven timer. I said, "Come on, Merrill, we can't sit here like two idiots. There's not much time left. We have to head for the brush on the right and then crawl toward the pine trees."

"Crawl? Why do we have to crawl?"

"I'll explain when we get there. Right now, pull your hood up over your head like this. Whatever you do, don't look at the house as you run past the brush into the pines, or the computer will see you."

"What is this, a science-fiction movie or something? What do you mean see us?" Merrill's nervousness was evident by the shrillness of his voice. He should have believed me when I said it was the most sophisticated alarm installed in a home.

"Just that. See that small box on the corner of the porch roof?" I pointed to a small black rectangular enclosure suspended below the corner of the roof line. About every 10 seconds a small red light flashed, giving it the appearance of being activated.

"There's a digital television camera in that box that scans this section of the yard between a height of 3 and 7 feet. When that light flashes, it starts a scan and looks for changes in light patterns from one scan to the next. With our dark clothes, by running just ahead of the scan we should go unnoticed."

The 30 seconds it took while we watched the blinking light until we could anticipate the next scan seemed like an eternity. When the precise moment came, I yelled, "Head for the pines. Go!"

Running with both hands in our pockets to shield our skin from detection made trying to run at full speed rather awkward. It was more like a high-speed waddle than the statuesque gait of a long-distance runner. We had 5 seconds to make it to the pines before the camera would start to retrace its path and compare the new image to that of the preceding scan. It was barely 120 feet, but it took all our effort to achieve it in time.

As I was about to dive under the first pine for concealment, I remembered something vitally important. I crouched under instead. "Merrill, watch out where you walk. There is where my dogs . . . Oh, I see you just found out. Sorry about that, Merrill."

Merrill was apparently just mentally chalking it up on his list of reasons to strangle me when the escapade was over—which it wasn't. Standing out there in no-man's-land was not accomplishing the task. Pointing to his watch, he said, "We have 5 minutes. What's this about crawling?"

"Don't worry about it, just crawl. Remember, we have to stay below 3 feet high. Don't stand up or we're dead. Ready? Go!"

Merrill still didn't understand why he was on all fours, crawling toward my house at ten o'clock at night. Life used to be so much simpler.

We were neck-and-neck about halfway across the yard when the computer spotted us. Two bright floodlights came on illuminating the area where we lay. Merrill, exercising reflex actions learned from years in the Marines, instinctively dove into a prone position, as though he anticipated an imminent artillery barrage. At the same time the lights came on, the tumultuous roar of many vicious, snarling dogs filled the yard.

Frozen in his position, Merrill yelled, "What have you got, a pack of hungry timber wolves in the basement? What do you need an alarm for?"

"Don't talk. Just bark!"

"Bark?" Merrill looked at me and shook his head.

"Bark," I said. "Like this. Arf! Arf! Arf!"

"Bark," I said. "Like this. Arf! Arf! Arf!"

Soon we were both barking and woofing up a storm. My two Scotties would be proud of us. We kept it up for about 25 seconds, until the lights and the ferocious dogs stopped as miraculously as they had started.

Speaking very softly and not waiting for questions, I said, "Hey, you can stop barking. There's a laser perimeter intrusion detector in this corner of the yard. It sensed our presence below the 3-foot level. It turned on the floodlights and the recorded sounds of barking dogs to see what it was or try to scare it off.

"Now, here's what the computer is great for. After all that was triggered, the computer turned on a microphone to listen out here at the same time. When it heard us barking the same as any real dog would do upon hearing the recording, it shut off the alarm sequence. You see, Merrill, the computer thinks we are just a dog that wandered through the yard and not an intruder. A real burglar, smart enough to see the different sensors and trying to crawl as we have been doing, wouldn't know enough to bark back at the computer. Neat, huh? Now we can finish crawling to the house. It won't bother us again."

Merrill rolled his eyes and put his muddy palm to his sweat-laden forehead. As straight-faced as one could be, all things considered, he said, "Steve, you're crazy."

Not wishing to argue, since time was running out, I merely responded, "Genius is never appreciated until it's too late."

"Steve, tell me why I'm going through all this. What do you have in your house that is so valuable that you installed a system designed to counter an invasion?"

"Well, if I really think about it, I guess the computer control system and all the alarm sensors are probably worth the most."

Merrill didn't know how to respond to that information. We computer hackers design things sometimes just for the challenge. Unfortunately, this particular challenge was getting out of hand, and time was very short.

The remaining distance across the lawn was far less wasteful than the first. We encountered no land mines, bear traps, or quicksand. We finally found ourselves resting against the house just below the bathroom window. Reaching the next objective was not as bad as the preceding events. The window lock was easily pried open with a pinch bar. I warned Merrill not to make any noise once he was inside the house. Then I hoisted him up to the window. Grabbing the top of the window frame for support, he lifted himself off my shoulders and knelt on the window ledge. Next, trying to be as graceful as he could in such an awkward position, Merrill swung his body around so that he now sat on the sill, with the trunk of his body hanging outside the window and his legs projecting inside. Once in that position, it was easy to swing into the bathroom and land squarely on the floor.

In a gymnasium, Merrill would have executed it perfectly. A small bathroom was quite another story. One foot came down squarely on the carpeted floor, as it should have. The other foot came down squarely into the open toilet, as it shouldn't have. Remembering that I had warned of excessive noise, he cussed very quietly as he extracted his foot from the toilet.

As he leaned out the window to help pull me up, he said, "Hey, Steve, I hear some kind of buzzer in the house."

I quickly glanced at my watch and responded, "That's the oven timer. It's running a little faster than I thought. Now the soufflé is overcooking. Help me up. We haven't got much time."

Merrill leaned out the window and grabbed the shoulders of my sweatshirt as I jumped up to the window ledge. My entrance was far less graceful than his. I had no alternative but to go through the window head first. I'd swear that Merrill directed my flight toward the toilet on purpose, but I have no proof. At the last instant, I was able to extend an arm and apply a force opposite to that of my trajectory. The result was a dull, rolling thud on the bathroom floor.

Our totally disheveled appearances lent no levity to the situation. But we were inside the house, and the stove was just 20 feet away. If we could get to the soufflé in time to stop it from burning, we would have all the time in the world to shut off the rest of the alarm. "Merrill, don't say anything louder than a whisper. There are mikes planted around the house, and the computer is listening for loud noises." I extended a forefinger against my lips to dramatize what I had said.

"Steve, I just saw something outside. Outside near the fence in the backyard!" Merrill was looking out the window, and after a few moments he excitedly pointed over my shoulder toward a dining room window, visible from the bathroom even though it was on the other side of the house. "There it is again!"

I jerked around in time to detect motion from an unknown object. "What do you think it is, Merrill?"

Before he had time to answer, a human form stood for a second in front of the window. Extending from an arm was a long, slender object. For a moment, Merrill and I just stood with our mouths open watching the proceedings. The figure turned suddenly. The slender object exhibited a metallic gleam in the moonlight. Then the figure was gone, as quickly as it had appeared.

We looked at each other. Our eyes were wide open as we whispered in unison, "I think that was a gun!"

"I think we have a real prowler, Steve. What are we going to do? He has a gun, too!"

"Don't ask me! Remember, I'm stuck here, too."

"Suppose. Now just suppose he was able to get by all the alarms and got into the house. And just further suppose he fills his pillowcase and is about to leave when he decides to go to the bathroom. Voilà, us two looking down the barrel of that gun!"

"Shhh, Merrill. Don't be an alarmist. Nobody can get through my alarm system."

Simultaneously, as Merrill spoke them, I thought of the exact same words, "But we did!"

I'd swear that Merrill directed my flight toward the toilet on purpose, but I have no proof.

The situation presented a problem. Should I leave the oven on and purposely trigger the alarm to bring help and catch the prowler? Or should we still try to finish what we had started and then hope the perpetrator wasn't smart enough to make it through my alarm?

I looked at my watch. The soufflé had to have been over-cooking about 10 minutes. The stove timer was buzzing relentlessly in the background. I sniffed the air. What had previously smelled freshly baked now had the scent of being overdone. It would still take a few minutes before smoke would be produced that the computer could smell. We were in a real dilemma. We were caught between our protector and the prowler.

"Steve, look again!" Merrill pointed toward the dining room window. "It's a woman!"

The figure was in full moonlight in front of the window. The features were easily discernible, and I recognized the person immediately. The metallic glint previously thought to be a gun was the stainless-steel tip of a walking cane. I grabbed Merrill's arm tightly and said, "That's no woman. That's Mrs. Picker from across the street."

"Is that bad?" Merrill had little experience with Mrs. Picker. He could not fully comprehend the grave position we were now in.

"That's worse than any prowler with 10 guns. She probably saw us and thinks we are the prowlers."

"Boy, that must really take guts to confront two prowlers single-handedly." Merrill still didn't understand what I was trying to tell him.

"That feisty old lady might be 80, but I wouldn't put it past her to climb over the fence after us if she discovered the rope. What I'm really worried about is that while she's looking for us, she'll probably set the alarm off. When the law arrives, guess who is wearing the cat-burglar costumes and covered with mud?"

Merrill looked down at his clothes and back up at me. His eyes pleaded with me to act fast. We were in the worst possible combination of circumstances to be caught in. The only solution was to try to turn off the system before Mrs. Picker triggered it.

"Let's go," I said. "We still have to turn the oven off."

Merrill agreed. We had no other choice. Extra time to shut off the system was gone. First, we had to get to the stove. Motioning to Merrill that he should follow in my exact footsteps and mimic my every motion, like a childhood game, we started the ordeal.

"Merrill, see those two holes on either side of the door molding? Those are photosensors. The computer can tell if we pass through the door and in what direction we are going. Fortunately, they are only 18 inches off the floor."

At the doorway of the bathroom, I lifted my right leg very high and extended it out over the other side. Shifting my weight to the now firmly planted foot outside the bathroom, I retracted the other leg by reversing the process. Merrill followed suit. We stood in the back hallway outside the bathroom.

"Every doorway we go through, we will have to follow the same procedure. Got it?" Merrill nodded affirmatively as I continued to whisper, "Now step over this area and these other two. There are pressure switches under the rug that will go off if you step on them. Try not to make too much noise jumping. Remember the mikes!"

Mrs. Picker

Ordinarily, all these sensors and switches caused the computer to turn on lights and direct the stereo system to the appropriate rooms as I walked through the house, all in the name of convenience. Now, however, the feeling we had was like being in combat. We were in the middle of a minefield directing those behind to follow in our footprints. While the sensation of stepping on a mine could not be exactly equaled by my computer, the heart attack following the first sound of that air-raid siren could be just as lethal. We silently high-stepped and hopscotched our way through the house until we reached the stove.

As I extended an arm to turn the oven off, I could see the blackened soufflé through the oven door's window. It was very disconcerting to see a creation of one's hand and mind shriveled and destroyed. But the realization that we were still at the mercy of another such creation prompted a fast exit. We had not gotten to the stove too soon. Inside it was filled with smoke. While not so dense as to obscure total view, I dared not open the oven door. The smell was of burned baked goods, but it was not dense enough for the computer to get excited about—yet.

Our final objective was the cellar, where the computer was headquartered. It was quicker to go there than try to find and insert the digital card in the reset mechanism in the front hallway. The motion sensors in that area of the house were not as easily overcome as the simpler variety that we had thus far defeated. The cellar door was but 5 feet and one pressure switch from the stove. We made it to this objective as easily as we had the others. There was no sensor on the door. I opened it slowly so that the squeaking of hinges would not reach an appreciable volume level.

When we opened the door, my two Scotties looked up at us. "No time to play now, guys," I said.

I went bounding down the stairs with Merrill in close pursuit. "When the alarm is activated, the dogs are kept in the cellar. So there are just a few sensors down here. We're home free now!"

Merrill and I stood in front of the computer control system. This computer did not have the usual panel full of flashing lights. That was old hat. The new stuff all had cathode-ray-tube displays. The monitor attached to the control system displayed a matrix of control parameters and on/off state. Peripheral sensors, not directly used to determine specific alarm conditions and still experimental, scanned the grounds like radar and displayed their activity around an outline of the house on another monitor. A dot flashed on the screen next to the outline. It slowly moved around the periphery of the house.

"That's Mrs. Picker," I pointed out to Merrill. "The computer knows she's out there. It has turned on the lights, but it will ignore her unless she goes over the fence into the backyard. See, she's moving in that direction now. I'll need about 3 minutes to enter the disarm commands."

Merrill looked around the cellar at all the equipment. Spying a refrigerator, he started to walk toward it. "Hey, Steve, why can't you just pull the plug on the computer?"

"That wouldn't do anything. In case of a power failure, the computer has battery backup and all kinds of redundancy."

I started to type in the first abort code. Merrill, who finally felt relaxed again, stood at the refrigerator and said, "Boy, all this work has really made me thirsty. Do you have any beer in here?"

He opened the refrigerator door. The fact that the refrigerator contained refreshment became immediately irrelevant. Suddenly a small speaker next to the computer started to emit a loud, repetitive sound: beep . . . beep . . . beep . . . beep.

"Merrill! You triggered the alarm! It's going to go off in 10 seconds!"

My mind raced with the thoughts of things that were about to happen. Everyone but the National Guard would be here in 10 minutes. Large jetliners approaching the nearby airport would be distracted by the brilliant flashing lights and start to circle the house instead. They would find Merrill and me in a state of partial rigor mortis from the loud horns that would now go off inside the house. The computer had sensed an intruder. Finally, and most important, there was Mrs. Picker. If she was standing next to one of those sirens when it started, it would be curtains!

"Give me a minute or two to fix it, and I'll show you that the sirens really do go off."

Merrill's eyes bulged with terror. Internally, he screamed, how could this be happening? Vocally, he yelled, "I thought you said that there were very few sensors down here because of the dogs! Why did it go off?"

Simple, yet true. We were done for, but he still had to know. "Dogs don't open refrigerator doors. That's why."

The 10 seconds had almost elapsed. My final words were, "Hit the deck! Cover your ears!" That was exactly what we did. It was a hard-tiled cement floor, but we dove under one of my workbenches and covered our heads with our arms. Almost immediately the beeping stopped. Then, silence. . .and more silence. . .and more silence. After about 15 seconds I peeked out. At 30 seconds we got up and walked over to the computer.

"I don't understand," I said. "It should have gone off. At the end of the beeping, it should have started the sirens and lights and everything. I don't understand."

I walked over to the console and started to list the program on the display. "There must be a program bug or a loose wire in the back here someplace. Otherwise, it would have gone off." I busily typed on the keyboard as I spoke. "Gee, Merrill, that's a lousy demonstration of my talents. I'm a better programmer than that.

"Merrill, wait a few minutes, and let me see if I can fix it. Don't think this was all a waste of your time. I want you to know that this thing really works. Give me a minute or two to fix it, and I'll show you that the sirens really do go off."

Merrill didn't wait. He gave me a fierce glance and took off up the staircase. I yelled, "Where are you going? Don't you believe this will work?"

Merrill yelled down the stairs. "I'll be right back. I'm just going to borrow Mrs. Picker's cane!"

This story was adapted from the chapter called "Computer ON Guard" in "*Take My Computer. . . Please!*" by Steve Ciarcia and is reprinted by permission of Scelbi Publishing.

Autographed copies of my hardbound book "*Take My Computer. . . Please!*" can be obtained for $10 including shipping (add $6 outside continental U.S., $4 for Canada). Send check or money order directly to me.

29

BUILD A GRAY-SCALE VIDEO DIGITIZER

PART 1: DISPLAY/RECEIVER

An imaging system with remarkable features for the price

Video technology has always interested me. One look at all the monitors, TVs, and displays around my house suggests that it goes deeper than interest. Freudian views aside, this is not the first time I have covered video technology in a Circuit Cellar project. In previous articles, I have described high- and low-resolution video display systems and even a low-cost digital camera. However, the one project I've always wanted to do has eluded me. Until now, I have had to hold off on the presentation of a cost-effective general-purpose high-performance gray-scale video-digitizing "frame grabber."

I'll explain all this later, but the key terms for the moment are "gray scale" and "frame grabber." Such terms usually indicate commercial units costing thousands of dollars.

While some video digitizers are designed as peripherals for specific computers, virtually all digitizers endent and involve significant trade-offs in performance to maintain low cost. Generally, their digitizing speed is significantly less than the rate necessary to capture a video image as it is transmitted in real time (1/30 to 1/60 second). Instead, they must repeatedly sample many sequential video frames. Digitizers like these—sequential field scanning digitizers—can deal only with stationary objects in front of a camera and can take as long as 30 seconds to scan and record an image. Such digitizers are useless if you are working with moving objects.

Another factor to consider is how a digitizer represents the intensity of each pixel. Most low-cost digitizers meet the minimum video display capabilities of their host computer and digitize each element only as black or white. Some allow a limited gray scale. Higher-performance digitizers offer 64 or more levels of gray scale as well as a high digitizing speed.

Fortunately, both static memory and video integrated circuit technology have progressed to the point where I can finally offer a project that attempts to meet the level of perfection I have outlined. In parts 1 and 2 of Chapter 1, I will describe a complete digital video system. You can use it independently as a video camera digitizer and display (to implement a video telephone, for example), or you can connect it to any personal computer for tasks like image processing, character recognition, and desktop publishing graphics.

ImageWise consists of two separate boards: a digitizer/transmitter and a display/receiver. Each board can be used independently, or they can be connected to form a complete digitizer/transmitter/receiver system (see photo 1).

In contrast to other digitizers, ImageWise is a true frame grabber that takes only 1/60 second to capture an image. It accepts the video signals from devices like a standard TV camera (either monochrome or color), VCR, laserdisc player, and camcorder, and it then stores the picture as 244 lines of 256 pixels with 64 levels of gray scale—256 by 244 by 6 bits (see photo 2). The ImageWise digitizer/transmitter board converts the stored video image to RS-232 serial data that can be transmitted to any computer or to the ImageWise display/receiver board. Transmission rates are selectable from 300 bits per second up to 57.6k bps.

The ImageWise display/receiver board has a serial RS-232 input and a composite video output. It receives serial data directly from the digitizer/transmitter board or transmitted from a file downloaded from your computer and converts this data back into a picture on a composite video input black-and-white monitor (adding a pair of modems lets you send the images over telephone lines). The displayed image is an interlaced 256 by 244 by 6-bit gray-scale picture. The following specifications for the display/receiver board sum it up better:

- Resolution: The three selectable resolutions are 256 by 244, 128 by 122, and 64 by 61. All resolutions support 64 levels of gray scale (each picture element is represented by 6 bits). Note that, regardless of resolution, the system displays all pictures as interlaced full-screen images. Lower-resolution images are composed of larger pixel blocks.
- Video output: 75-ohm, 1.5-volt peak-to-peak composite video.
- Serial input: RS-232, 8 bits, 1 stop bit, no parity. Transmission rate is selectable from 300 bps to 28.8k bps.
- Hardware: 8031 microprocessor, Telmos 1852 video D/A converter, 64K bytes of static video RAM.

The specifications of the digitizer/transmitter board are

- Resolution: Same as above.
- Video input: 1-V peak-to-peak, black and white or color, 75-ohm termination.
- Serial output: RS-232, 1 start bit, no parity. Transmission rate is selectable from 300 bps to 57.6k bps.
- Hardware: 8031 microprocessor, RCA CA3306 6-bit flash A/D converter, 64K bytes of static RAM.

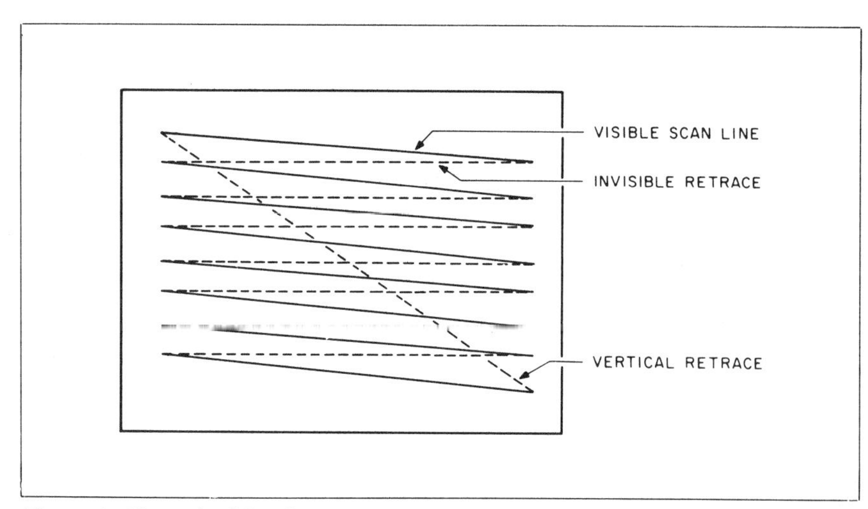

Figure 1: *The path of the electron beam during a video scan.*

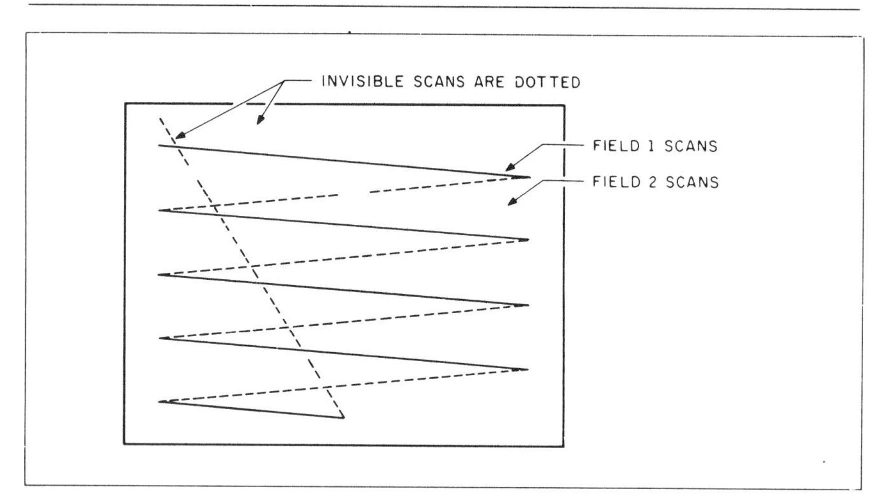

Figure 2: *The path of the electron beam for an interlaced video scan.*

Photo 1: *The ImageWise video digitizer system consists of two boards that can be used together or independently. The display/receiver board, shown on the right, reassembles and displays images that are flash-digitized and serially transmitted by the digitizer/transmitter board on the left.*

The ImageWise video system is designed to be intelligent as well as functional. You can control many of its features—such as digitizer resolution and picture update—remotely from the receiver or another computer. In addition, the system incorporates various compression techniques, including run-length encoding, to considerably reduce image-transmission time (essential with slow modems).

ImageWise is no small project. Consequently, I will present it in two parts, beginning with the display/receiver section. (If you built the digitizer/transmitter section first, you would have no way of displaying a gray-scale picture other than converting it to a dot-dithered black-and-white-only image on your computer—and what would that prove?) You can check out the receiver using its internal test patterns and by downloading picture files from a computer or my bulletin-board system. It's much easier to verify that the transmitter is sending the correct data after you have a working receiver to show any problems.

Before I delve into the hardware and software, however, I think it's a good idea to review what goes into a "standard" TV signal. With that in mind, it will be easier to see how the ImageWise transmitter digitizes the video and the receiver reconstructs it. As you'll discover, there is a lot more to video than just another pretty picture.

Today's Class: TV Basics 101

Although there may be a few folks tucked away in odd corners of the country who don't have a TV set, I think it's safe to say that everyone who reads BYTE has at least seen a TV picture at one time or another. While most TV is color TV these days, I'll describe only monochrome (black-and-white) TV signals because that's what ImageWise uses. The circuitry required to digitize and reconstruct color TV signals is considerably more complex than seemed reasonable for this project. Fortunately, the color video standards include monochrome as a subset, so we can use color cameras and monitors as well.

Figure 1 shows a simplified diagram of the process used to build an image on a TV monitor. An electron beam is scanned horizontally across the screen, starting from the upper left corner. It is moved downward after each line, and the result is a set of lines scanned left to right and top to bottom filling the screen. After scanning the last line, the beam is returned to the upper left corner to begin scanning the next screen.

The faceplate of the screen is covered with a phosphor that glows when struck by the electron beam and continues to glow even after the beam passes on. Because the entire screen is scanned rapidly enough to get the beam back to each spot before the phosphor glow fades out, the entire screen seems to be illuminated at once.

One key difference between a TV display and most computer CRT displays is that the electron beam in a TV set can take on a wide variety of intensities, ranging from completely off (black) through shades of gray to completely on (white). A computer display may allow only black, one shade of gray, and white. We'll see what this difference means a little later on. Most composite video input amber or green monitors also have some gray-scale (or should I say green-scale?) capability.

As you might expect, the actual details are a bit more involved. A TV screen is scanned twice for each image, with the two sets of scan lines interlaced on the screen (computer displays are generally noninterlaced). This allows the whole screen to be scanned in half the time a noninterlaced scan would take, without reducing the number of lines in the image. Figure 2 shows how interlaced scanning paints lines on the screen. Each vertical scan is called a *field*, and two matched fields make up a *frame*.

One field is completed in 1/60 second. There are 262.5 lines in each field (see figure 2 to locate the half lines), so each line must be scanned in 63.5 microseconds (1/60 second divided by 262.5). Figure 3 illustrates the signal voltage sent to the display for a single scan line, along with the allowable times for each part of the waveform. (Note: By strict definition, ImageWise is a field rather than frame grabber, since it digitizes only the first 244-line field of an interlaced frame. However, since the term frame grabber has come to mean a digitizer with the speed to digitize within the time period of a video frame, I shall continue to use the term frame grabber.)

The horizontal sync pulse occurs every 63.5 μs and tells the monitor to end the current line, return to the left edge, and begin another line. Surrounding each sync pulse is a blanking voltage that ensures that the track of the retrace will not be displayed on the screen. After allowing time for retrace and blanking, about 52 μs are left for the actual video picture on the horizontal scan line.

A vertical sync signal tells the monitor when to end the current field, return to the top of the screen, and begin the next field. Because the two fields in each frame are offset by exactly half a line, the timings at the end of each field are slightly different. Figure 4 diagrams the analog voltage required to display a complete field. To keep the size of the diagram down, only the first and last lines of active video are shown. The horizontal scale is distorted so that you can see the details.

You should note that the horizontal

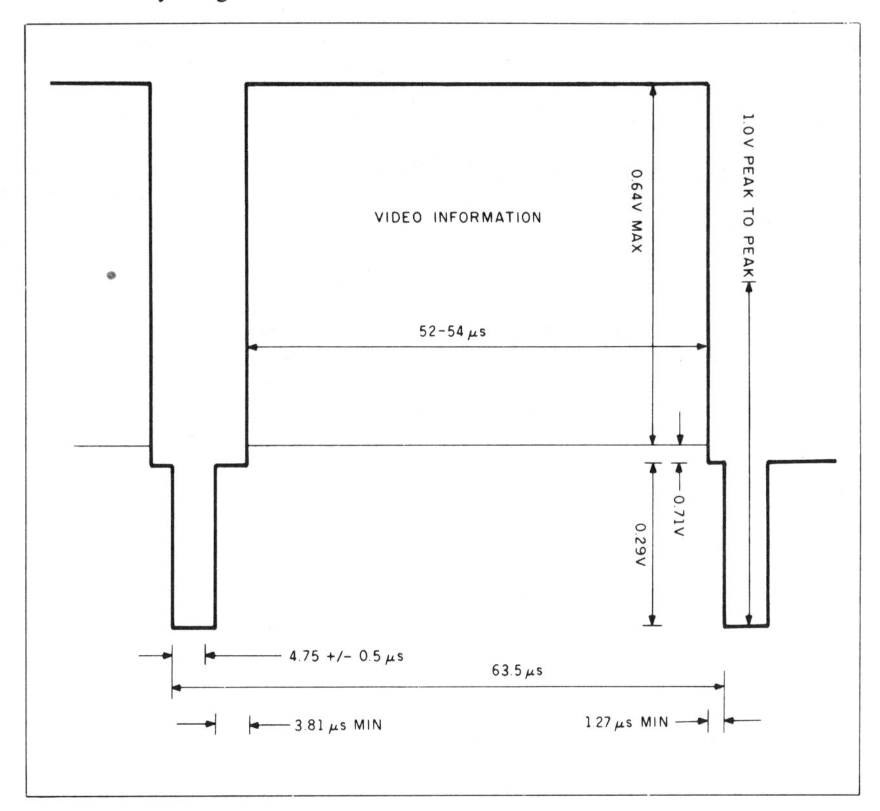

Figure 3: *A profile of the signal voltage sent to the display during a single scan line. Note the bracketing 63.5-microsecond horizontal sync pulses.*

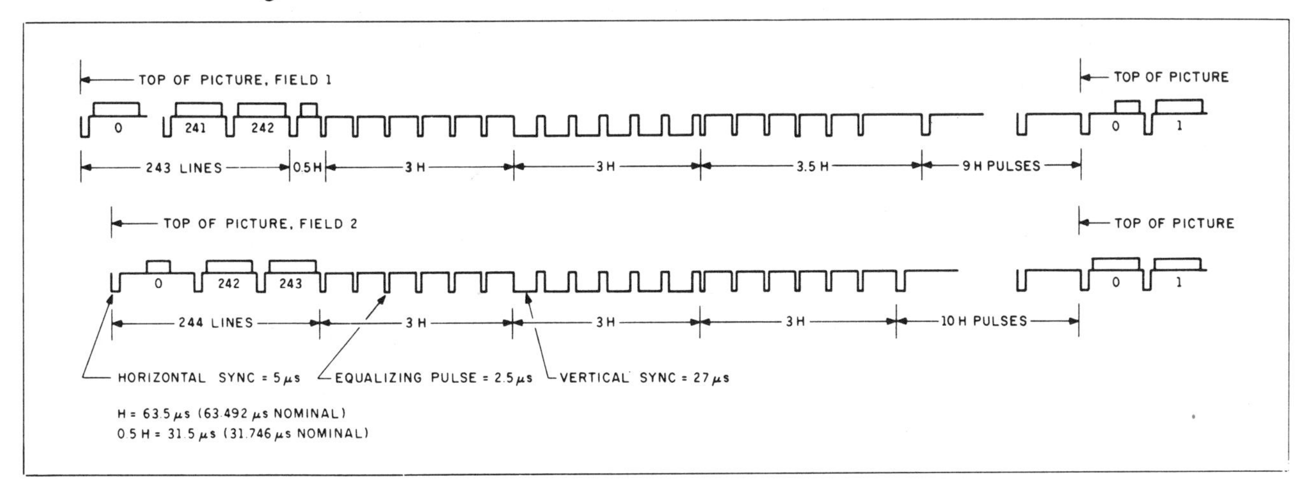

Figure 4: *Signal voltage timing diagram for a complete field.*

sync pulses do not stop during the vertical sync and retrace. In fact, to make sure that the monitor switches smoothly across the half line, sync pulses occur every 31.75 μs during most of the vertical sync period.

A blanking voltage surrounds the vertical sync pulse itself to ensure that the vertical retrace is not visible on the screen. Each field contains 244.5 visible lines, which you can verify by examining the timing diagram in figure 4.

Although video monitors can tolerate small variations in the number of displayed lines or the exact line timings, any errors will be immediately visible as jitter or distorted images. The sync voltages and timings must be exact to ensure a stable picture. The worst offense is to have timings that vary "just a little" from field to field; this will cause the picture to jitter annoyingly.

Variations on a Theme

Now that you're acquainted with standard video, I can explain some of the basic design criteria for ImageWise. As with any project, there are trade-offs between "the ultimate system" and "the one that got built." I will try to explain why I made the decisions I did.

The most basic question was one of resolution: How many picture elements (pixels) should appear on each line? A single pixel corresponds to the smallest unit of video information handled by the system. Computer monitors typically have 300 to 1000 pixels per line. With about 50 μs available in each line to display those pixels, a 1000-pixel line requires a new pixel every 50 nanoseconds. Since typical dynamic RAMs have a cycle time of 300 ns, allowing a few nanoseconds for the other circuitry would require about eight banks of RAM to ensure that a pixel was ready every 50 ns. Using 50-ns static RAMs, while feasible, would be very expensive.

I decided to look at it the other way: How many pixels will fit on a line with affordable hardware? We find that 32K-byte static RAMs are increasingly affordable, and they are considerably faster than DRAMs. Since even the "slow" ones have a cycle time of about 130 ns, one pixel every 200 ns is reasonable. That allows 256 pixels in 51.2 μs. Be-

Photo 2: *The ImageWise frame-grabber (or "freeze-frame digitizer") captures a video signal in 1/60 second; fast enough to digitize live TV broadcast signals as easily as those from a stationary camera.*

Photo 3: *The high-quality gray-scale images of the ImageWise digitizer can be used in security, pattern recognition, a video telephone, and image database applications. (Phone me sometime; I may be looking back at you through the camera.)*

Photo 4: *Teletransfer of pictures for purposes of identification or verification is a legitimate application for ImageWise. Simply hold the part in front of the camera and transmit the picture to everyone.*

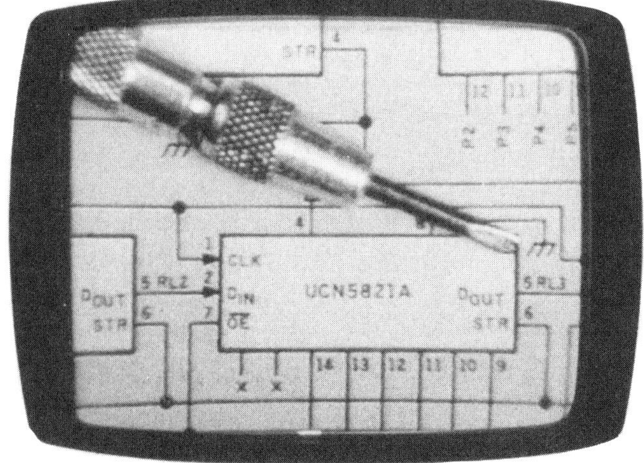

Photo 5: *The 6-bit gray scale of ImageWise adds significantly more to the perceptible resolution of an image, whether it be a black-and-white schematic or a silver screwdriver. (The schematic and the screwdriver could not be represented accurately without gray scale.)*

cause 256 is a "magic" number, I knew I was on the right track.

As I mentioned earlier, each field has about 244 visible lines. Therefore, a 64K-byte buffer could hold one field with some room left over. Two fields could be contained in 128K bytes. With two fields, however, the vertical resolution (488) is twice the horizontal resolution (256). This seemed excessive. Fortunately, because both fields often contain redundant information, I decided to keep the amount of RAM within reasonable bounds and digitize only a single field. But how would a 256- by 244-pixel picture look compared to the original?

All my experience with 320 by 200 computer displays suggested that I might not like the results and be forced to go back to expensive plan A. However, seeing is believing, so I figured I'd build it and decide then. (Often it is easier and faster to build a prototype and take a look at the results than to argue about what might be.)

Photo 3 shows the quality of the image I got with a 256 by 244 display. (So much for my prior experience with computer displays!) There's a good reason why I was wrong, and if you're as surprised as I was, here's the explanation.

You see jagged diagonal lines or "jaggies" on low-resolution computer displays because each pixel can have only a few levels of brightness. The jaggies can be reduced only by increasing the number of pixels on each line. Depending upon the subject material, resolutions of 640 pixels per line and 350 to 400 lines per screen are required to see noticeable improvement.

But there is another way to reduce the jaggies: If each pixel can take on many levels of brightness, the sharpness of the edges can be reduced. ImageWise uses 6 bits to represent each pixel, allowing 64 shades of gray. Real-world images don't have crisp, computer-generated edges, so each pixel tends to shade into the adjoining ones. The effect is a rather smooth picture that has more "effective" resolution than you'd expect (see photos 4, 5, and 6). This is why pictures shown on color displays that incorporate palette D/A converters often look better. Look closely at a line boundary and see if there is some gradual shading.

As an example, you're probably aware that a standard TV does not make a good computer monitor. Trying to fit more than 40 characters on a line results in an unreadable display. However, newspaper headlines displayed on a TV are easily readable even though the characters are very small, simply because each pixel can take on many brightness levels. Watch your TV carefully and see.

Finally, why does ImageWise use 6 bits per pixel and not more if gray scale is such a good idea? Again, it is a cost tradeoff. We have to digitize and determine the gray-scale value of 256 data points in 50 μs, or one every 200 ns. This requires a fast A/D converter called a flash A/D converter. The price of one is directly related to the number of bits it resolves. Eight-bit models are considerably more expensive than 4- or 6-bit chips. The device I ultimately chose was the RCA 3306 6-bit flash A/D converter, which can operate at 12 million to 16 million samples per second (our sample rate is 5 megahertz). I'll talk more about this next month.

Display/Receiver Hardware

The receiver has two main functions: It accepts data from the RS-232 serial port and displays the resulting picture on a monitor. Figure 5 shows the receiver hardware.

As in many recent Circuit Cellar projects, the receiver uses an Intel 8031 single-chip microprocessor to control the rest of the hardware. A 2764 EPROM stores the 8031's program. An Intel 8254 Programmable Interval Timer (PIT) produces the sync pulses. The video field data is held in a pair of 32K-byte static RAMs and is converted to an analog voltage by a specialized video D/A converter. The MC145406 converts RS-232 voltages into TTL levels for the 8031's serial port.

It was tempting to use the 8031 to produce the sync pulses directly, but a little study showed that there was no way to get the precise timings required for a stable picture. The 8254 is connected to produce repetitive pulses, so the 8031 need only program the appropriate values into the 8254's registers when a change is required. The 8254 uses a 500-ns (2-MHz) clock divided from the 10-MHz crystal oscillator. Figure 6 shows the 8254 pulses for a normal video line.

Often it is easier and faster to build a prototype and take a look at the results than to argue about what might be.

The Telmos 1852 is a specialized video D/A converter that accepts up to 8 video data bits, a blanking input, and a sync input. The analog output conforms to the standard video specifications. Using this D/A converter eliminates a lot of hardware that would otherwise be required to combine the video, blanking, and sync signals to produce the right output voltage with enough power to drive the monitor. The 8031 ensures that the 2 unused bits (the low-order ones) are always 0.

The 16-bit address required by the 64K-byte field buffer is divided into two parts: a high byte supplied by the 8031 and a low byte that can come from either the 8031 or an 8-bit counter. Normally, the counter steps through the 256 pixels on each line, and the 8031 counts out the lines in the high byte. Both bytes are supplied by the 8031 when it reads or writes buffer data.

A pair of LS244s isolate the field buffer's data bus from the 8031's data bus,

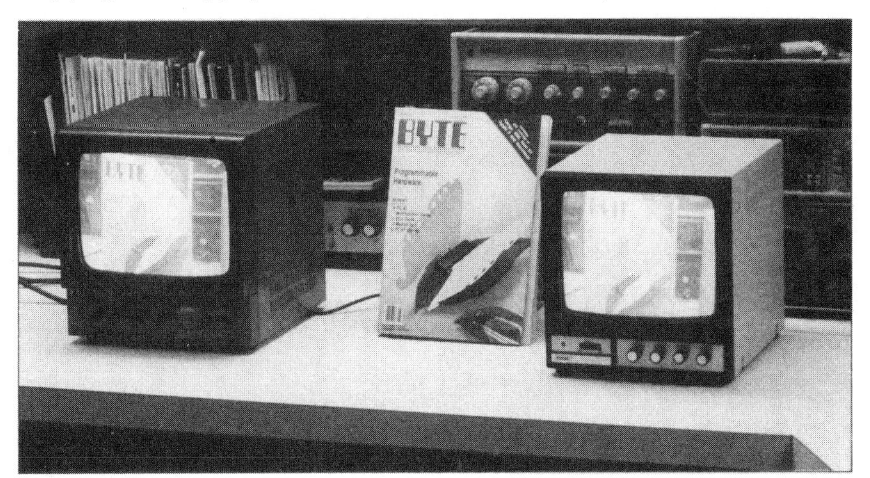

Photo 6: *An in/out comparison. The camera is pointed at the magazine in the middle. The monitor on the left displays the image seen by the camera and the digitizer/transmitter. The monitor on the right shows the digitized image received by the display/receiver board.*

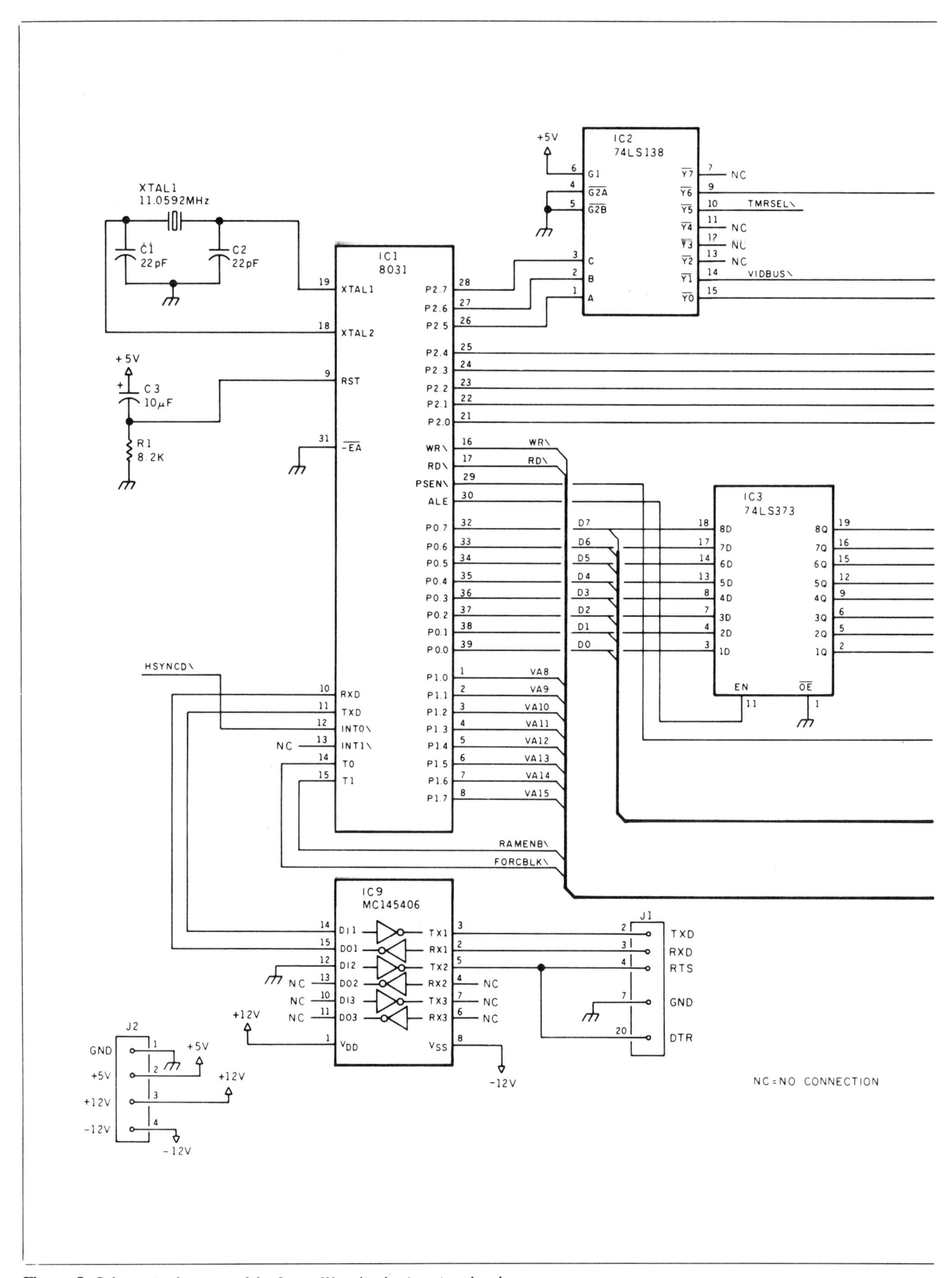

Figure 5: *Schematic diagram of the ImageWise display/receiver hardware.*

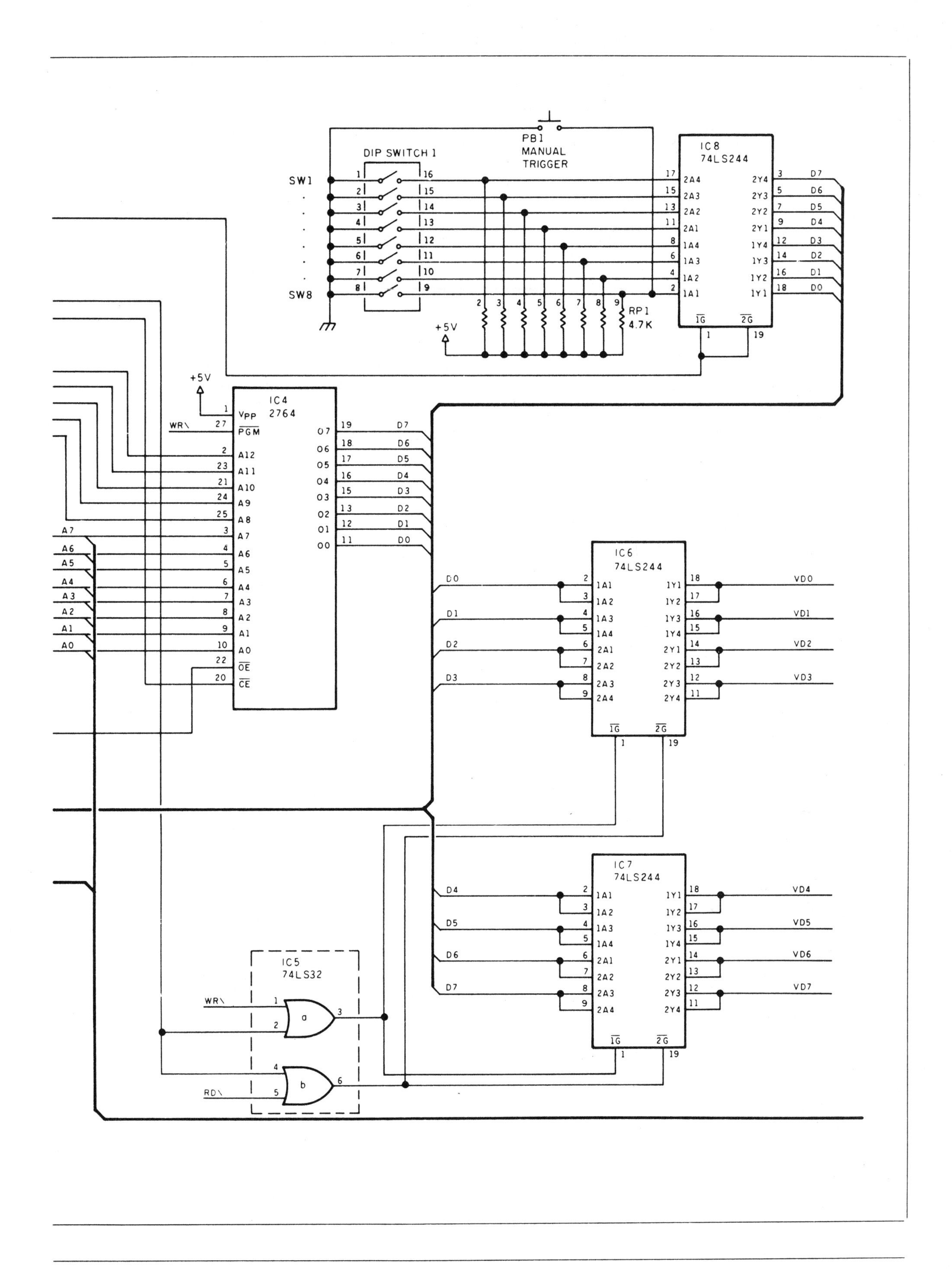
PB1
MANUAL
TRIGGER
DIP SWITCH 1
SW1
SW8
IC8
74LS244
RP1
4.7K
+5V
IC4
2764
IC6
74LS244
IC7
74LS244
IC5
74LS32
WR\
RD\

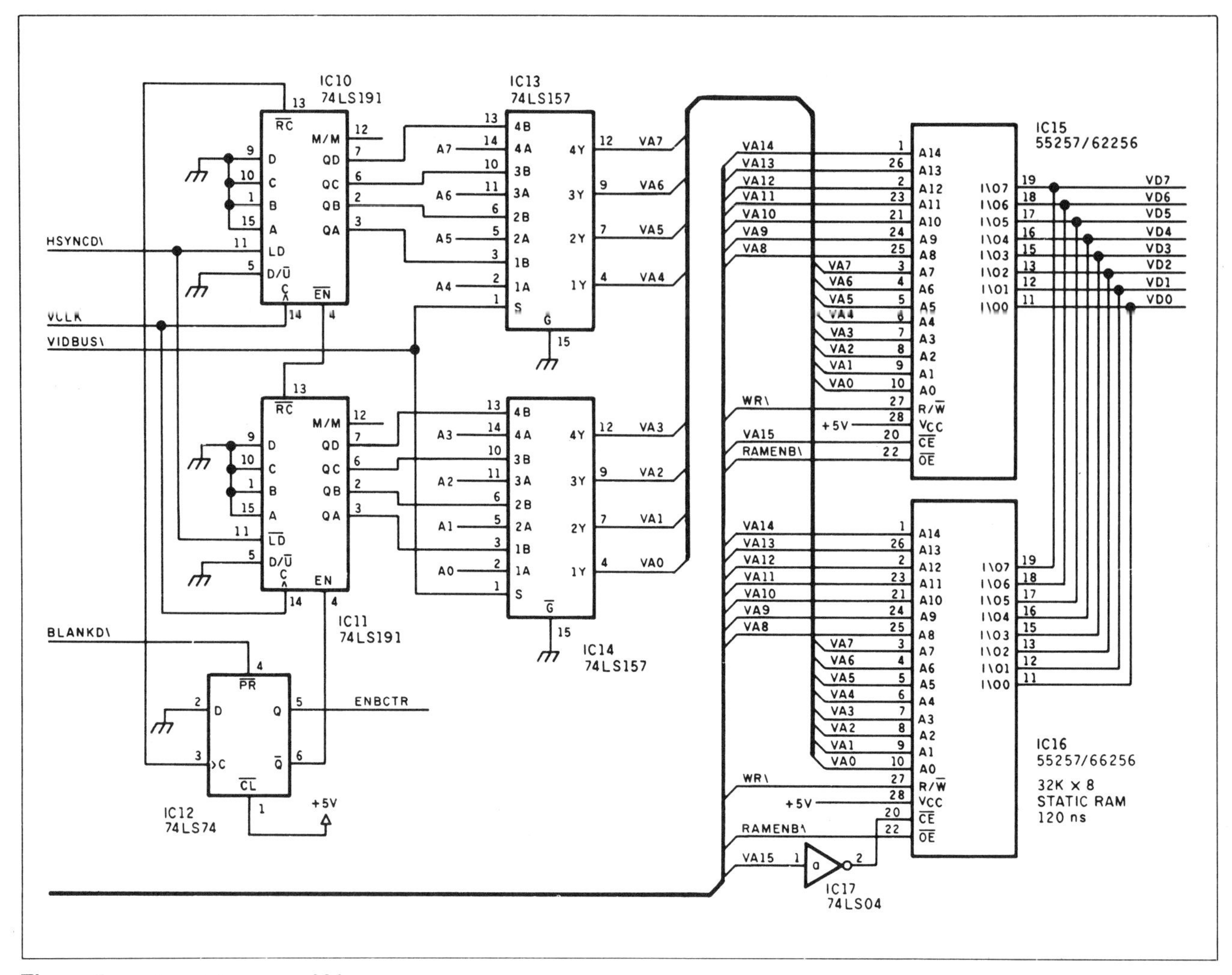

Figure 5: *continued from page 321*

except when the 8031 is reading or writing to the buffer. A third LS244 connects the DIP switches to the 8031's data bus. An 11.059-MHz crystal allows the 8031 to receive and generate standard RS-232 bit rates. The video data and sync timings are derived from a separate 10-MHz crystal oscillator circuit.

The divide-by-two counter that produces the video data clock is reset by the horizontal sync pulses from the 8254. This ensures that the pixel clock has the same phase in each line. Without the reset, the clock would alternate phases in successive lines because the length of each line is an odd multiple of the 500-ns clock. Worse, a given line would have a different phase in each frame because the frame length is also an odd multiple of the clock.

Serial Data

Each video field has 256 pixels on each line and 244 lines, for a total of 62,464 pixels (we round the half line up to a full

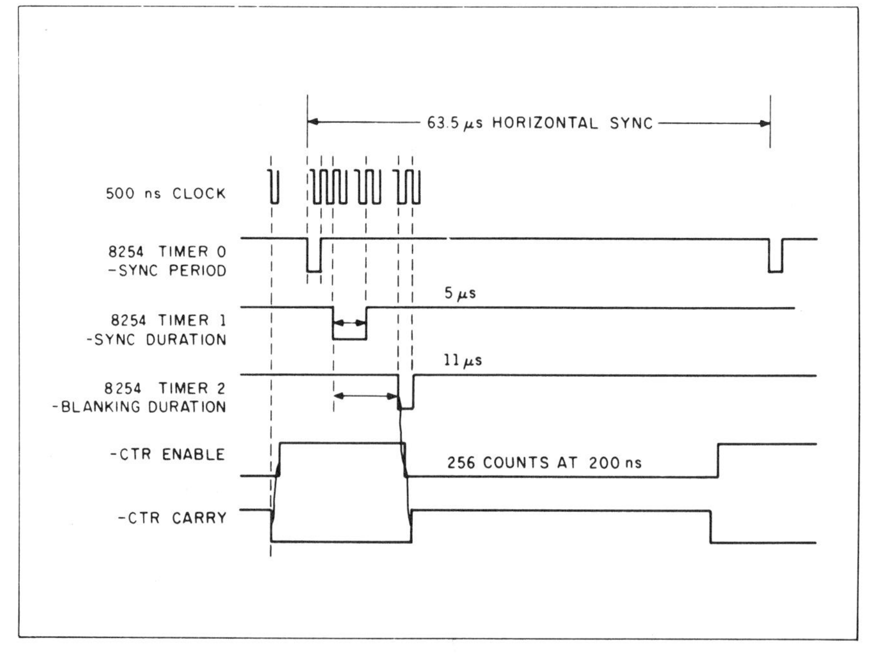

Figure 6: *Timing pulses generated by the display/receiver's 8254 PIT.*

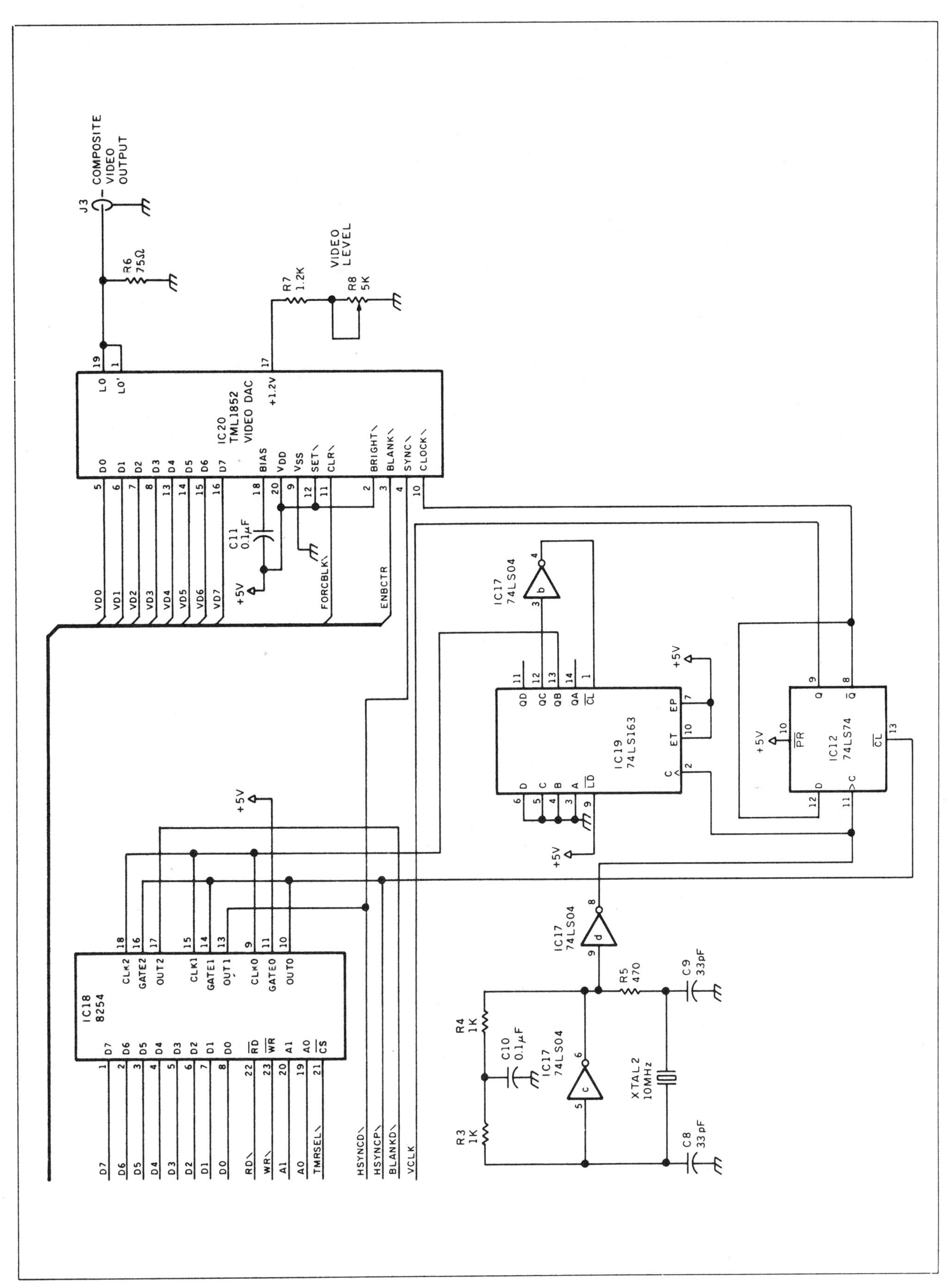

Figure 5: *continued from page 322*

line). Each pixel is contained in 1 byte, so there are 62,464 bytes in each field. If you use a serial rate of 3840 bytes per second (38.4k bps), a complete field will take about 16.2 seconds to transmit (it takes 10.8 seconds at 57.6k bps).

Fortunately, ImageWise takes advantage of the fact that most scenes have large areas of the same shade. The digitizer/transmitter (which I'll describe in detail in part 2 of this chapter) can compress each line of data by representing repeated bytes by a value and repetition count. In actual practice, the amount of data in a field can be reduced by a factor of two to four, with a corresponding reduction in transmission time.

The serial data format is 8 data bits, 1 start bit, no parity, and 1 stop bit. I did not build error checking into the system because it is intended for relatively short, robust connections. In any event, an error will generally be confined to a single line on the display. Because the video data itself has only 6 bits, 2 bits in each byte can be used for control information. Table 1 details the byte encoding used by ImageWise. The 8031 shifts the video data left so that it goes to the high-order 6 bits of the D/A converter.

The receiver puts the bytes into the field buffer as fast as it can, but at the faster rates it's possible for the transmitter to get ahead of the receiver. A circular buffer in the 8031's internal RAM holds up to 48 bytes until they can be processed. If this buffer begins to fill, the receiver sends an XOFF character to tell the transmitter to stop sending data. The receiver will continue transferring bytes from the circular buffer to the frame buffer until the former is nearly empty. Just before the circular buffer runs dry, the receiver sends an XON character to tell the transmitter to resume sending. If the circular buffer does empty completely, the receiver will simply wait for more bytes to show up.

As we'll see in part 2 of this chapter, the transmitter waits for an XON from the receiver before beginning to send data. The receiver will send the XON sometime after the circular buffer empties, even if it has sent one before. A DIP-switch setting determines the time between emptying the buffer and sending the XON. The choices are continuous pictures, every 4 seconds, every 8 seconds, or manually triggered (see table 2). A push button is used to trigger a new field from the transmitter in manual mode.

Table 1: *ImageWise serial data encoding. (a) This data flows from the digitizer/transmitter to the display/receiver. (b) This data flows from the display/receiver to the digitizer/transmitter or is sent by a computer connected to the digitizer/transmitter. All other characters are ignored.*

(a)

Bit number 7	6	5	4	3	2	1	0	Bit Definition
0	0	x	x	x	x	x	x	Video data byte
0	1	0	0	0	0	0	0	Start of video field
0	1	0	0	0	0	0	1	Start of video line
0	1	0	0	0	0	1	0	End of video field data
0	1	1	x	x	x	x	x	Reserved
1	0	0	0	x	x	x	x	Repeat previous byte *x* times (0 = 16 reps)
1	0	0	1	x	x	x	x	Repeat previous byte 16*x* times (0 = 256 reps)
1	1	x	x	x	x	x	x	Reserved

(b)

Bit number 7	6	5	4	3	2	1	0	Bit Definition
0	0	0	1	0	0	0	1	XON, starts or restarts transmission
0	0	0	1	0	0	1	1	XOFF, halts transmission
1	0	0	0	0	0	0	0	Use 256 by 244 resolution (full)
1	0	0	0	0	0	0	1	Use 128 by 122 resolution (half)
1	0	0	0	0	0	1	0	Use 64 by 61 resolution (quarter)

The Software

It's worthwhile to describe how the software pulls the receiver hardware together. The two main jobs are maintain-

Table 2: *Receiver DIP-switch settings. ON and OFF refer to switch positions. (a) SW1, SW2, and SW3 select the serial bit rate (must match transmitter rate). (b) SW4 and SW5 select the time-out interval. (c) SW6 and SW7 select the transmitter resolution. (Note: A manual push button is connected to the SW8 position, so SW8 must be OFF.)*

(a)

SW1	SW2	SW3	Serial transmission rate (bits/second)
OFF	OFF	OFF	300
OFF	OFF	ON	600
OFF	ON	OFF	1200
OFF	ON	ON	2400
ON	OFF	OFF	9600
ON	OFF	ON	19.2k
ON	ON	OFF	28.8k
ON	ON	ON	57.6k

(Note: 4800 bps intentionally omitted.)

(b)

SW4	SW5	Time-out interval
OFF	OFF	Continuous pictures, no delay
OFF	ON	4-second delay between pictures
ON	OFF	8-second delay between pictures
ON	ON	Send picture by manual push-button trigger

(c)

SW6	SW7	Transmitter resolution
OFF	OFF	Full: 256 by 244
OFF	ON	Half: 128 by 122
ON	OFF	Quarter: 64 by 61
ON	ON	Reserved

ing stable video sync and accepting bytes from the serial interface. The code is written in assembly language to maximize the performance of the 8031. Figure 7 is a flowchart of the software's major components.

The 8254 PIT generates the precise sync signals for each line, so the 8031 need only reprogram the PIT when a change is needed. Because changes to the 8254's settings take effect with the next 8254 sync output, the 8031 must make the changes one sync pulse before they're actually needed. All timings are determined by counting sync pulses, which are connected to the 8031's INTO interrupt request pin.

The INTO interrupt handler decrements a counter and checks to see if it's 0. If so, an 8254 change is required; otherwise, the handler simply returns to the mainline code. Each change to the 8254 involves writing a few bytes and reloading the counter to tell how many interrupts will pass before the next change.

Each 8031 instruction takes 1 or 2 μs. At most, only about 50 instructions can be executed per horizontal line. During the vertical retrace interval the sync pulses are only 31.5 μs apart, giving time for only 20 instructions per sync pulse. The interrupt routine must have enough

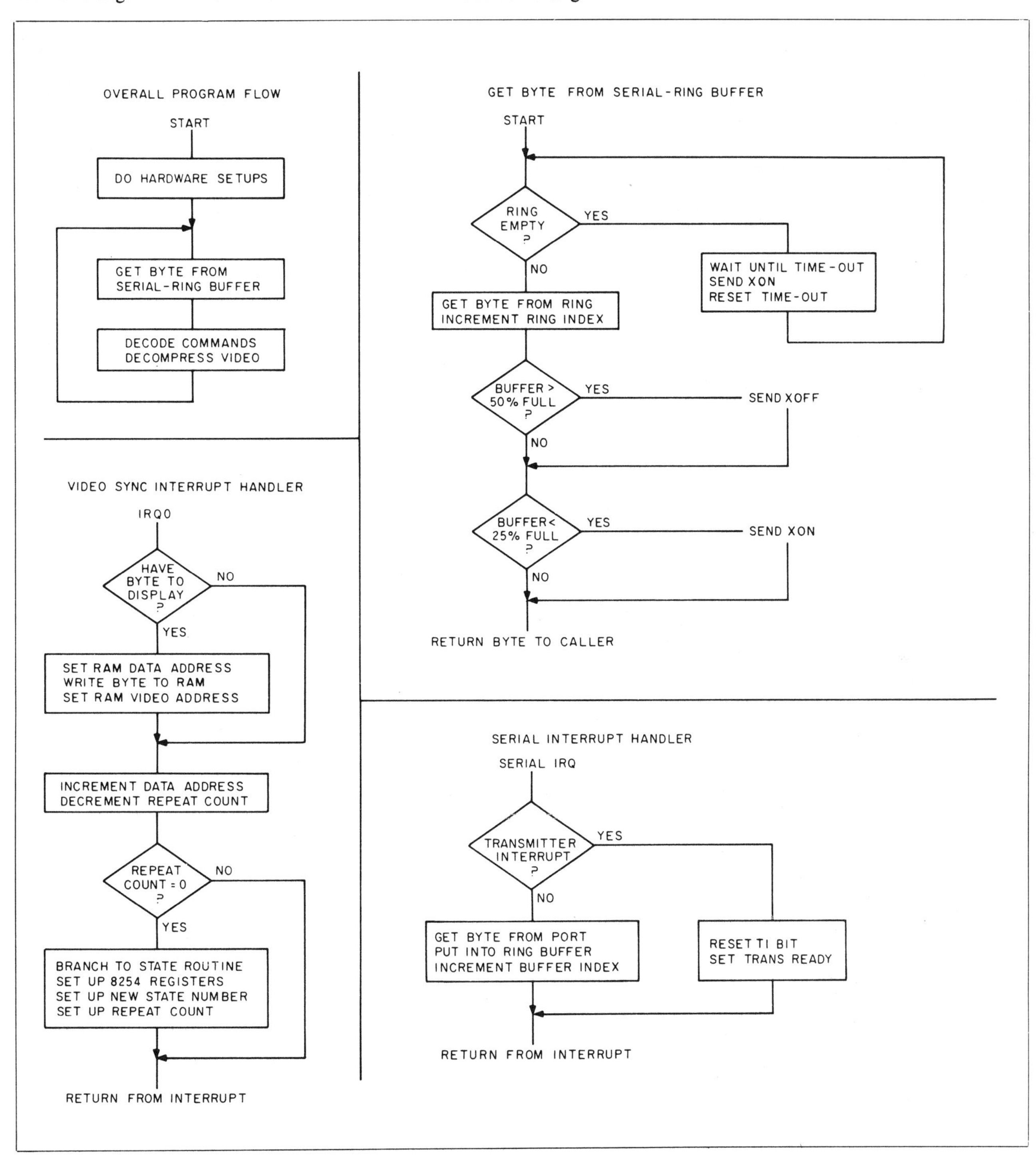

Figure 7: *Flowchart for the ImageWise display/receiver system's software.*

time to get ready for the 8254 loading during the short sync pulses in the vertical retrace interval, so control is passed to the routine two sync pulses before the change is needed. The interrupt routine then uses a polling loop to detect the last sync pulse.

Another interrupt is generated within the 8031 whenever a byte is received on the serial port. This interrupt awakens the serial interrupt handler routine, which reads the byte from the port and places it in the circular buffer in the 8031's internal RAM. The serial interrupt handler has a lower priority than the sync interrupt; consequently, the serial interrupt handler can be interrupted whenever a sync pulse occurs.

The sync and serial interrupt routines are linked by a background task that simply waits for bytes to show up in the circular buffer. Whenever a byte appears in the buffer, the background task takes it out and decides what to do with it. In most cases, the byte is either video data that should be put in the field buffer or a count that tells how many times the previous data byte should be repeated.

The ordinary way of putting a byte in the field buffer would be to have a subroutine that saves all the registers, sets up the buffer address, does the write, restores the registers, and returns to the caller. Unfortunately, this scheme doesn't work in our application because the writes to the frame buffer have to occur just after the video syncs to reduce sparkles in the display. Additionally, the sync interrupt routine must get control at the same time to reset the 8254. Something has to give!

The solution is to combine the two functions in the video sync interrupt handler. Whenever the background routine has a byte to be written in the buffer, it sets up the registers and turns on a flag bit. The sync routine checks the flag, does the write if it's on, then turns the flag off. The background routine sits in a loop until it sees that the flag has been reset, then continues on its way. Because the background routine has handled all the register setups, the interrupt routine can proceed at full speed and write the byte immediately without saving or restoring any registers.

The possibility arises that the serial interrupt handler will be interrupted by the video sync handler. Because the video sync handler assumes that the registers are set up for it, the serial interrupt handler has to take special precautions to make sure that the wrong byte doesn't get written at the wrong address.

The sync interrupt handler checks the switches once every frame (at the end of the second field) to see if anything's changed. If so, it drops what it's doing and runs through the power-on initialization routine again. If a picture is being received when you flip the switches, it will get garbled because the serial port will miss a few characters. The rule of thumb is to change switch settings only when nothing else is happening.

See page 447 for a listing of parts and products available from Circuit Cellar Inc.

29

BUILD A GRAY-SCALE VIDEO DIGITIZER

PART 2: DIGITIZER/TRANSMITTER

An imaging system with remarkable features for the price

In part 1 of this chapter I described the ImageWise video digitizer's display/receiver section. The display/receiver board accepts binary data from a serial RS-232 port and decodes that data to generate a gray-level display (with 64 levels) on a standard TV monitor. In part 2, I'll complete the project by describing the digitizer/transmitter board (see photo 1) and discussing some possible applications. (Note: Certain portions of this chapter depend heavily on material presented in part 1.)

As I mentioned in part 1, ImageWise is technically a "field" grabber rather than a "frame" grabber. The digitizer/transmitter board makes no distinction between the two fields in a frame: One is as good as the other. The digitizer/transmitter must decide when a new field is starting, wait until the first active line begins, then begin converting the analog video signal into digital bytes that are stored into the field buffer. Because the video can't be "slowed down," all this must happen when the video occurs rather than when the processor is ready.

You might think that you could locate the start of the first active video line by counting horizontal sync pulses after the conclusion of the vertical sync pulse, but it's not that easy. Some cameras do not produce "standard-width" vertical sync pulses, so counting pulses won't work. Instead, I used an internal timer on the 8031 to provide a fixed delay period (DIP-switch-selectable, either 16- or 20-line times). The first horizontal sync pulse after that delay becomes the first active line to be digitized.

Digitizer/Transmitter Hardware

The digitizer/transmitter has two main functions. First, it digitizes the analog video signal; second, it transmits the data to the display/receiver over a serial RS-232 link. Figure 1 shows the digitizer/transmitter circuitry.

The analog circuitry merits a detailed description. Many people who are familiar with the level of integration possible in digital circuitry are appalled at the number of components needed to accomplish even the simplest analog task. The whole mass of circuitry attached to the analog video input performs five functions: termination, clamping, filtering, buffering, and level comparison.

Standard composite video is communicated over coaxial cable with a character-

Photo 1: *The ImageWise digitizer/transmitter in prototype printed circuit form. The digitizer/transmitter board flash-digitizes the video output of a TV camera or other video source (connectors in upper right corner) and converts it to serial data that can be stored and manipulated by a computer or redisplayed on an ImageWise display/receiver board (see Chapter 29, part 1). The flash A/D converter is the 18-pin chip in the upper right corner; the two 28-pin chips on the left are 64K bytes of static RAM.*

istic impedance of 75 ohms. To prevent reflections from the end of the cable, it must be terminated in that impedance. The 75-ohm resistor at the video-input connector of the digitizer/transmitter (J3 and J4) accomplishes this goal. A jumper (JP1) disconnects the terminator if you have a terminated device (perhaps a monitor to watch "live" video) connected to the second parallel connector.

Video signals never seem to exhibit "textbook" profiles. Frequently, inexpensive cameras produce video signals with a DC offset that depends on the scene being viewed. The digitizer incorporates clamping circuitry that forces the tips of the sync pulses to 0 volts. This means that the brightest white will be in the area of 1 to 1.4 V (different cameras give different results).

Both color and black-and-white sig-

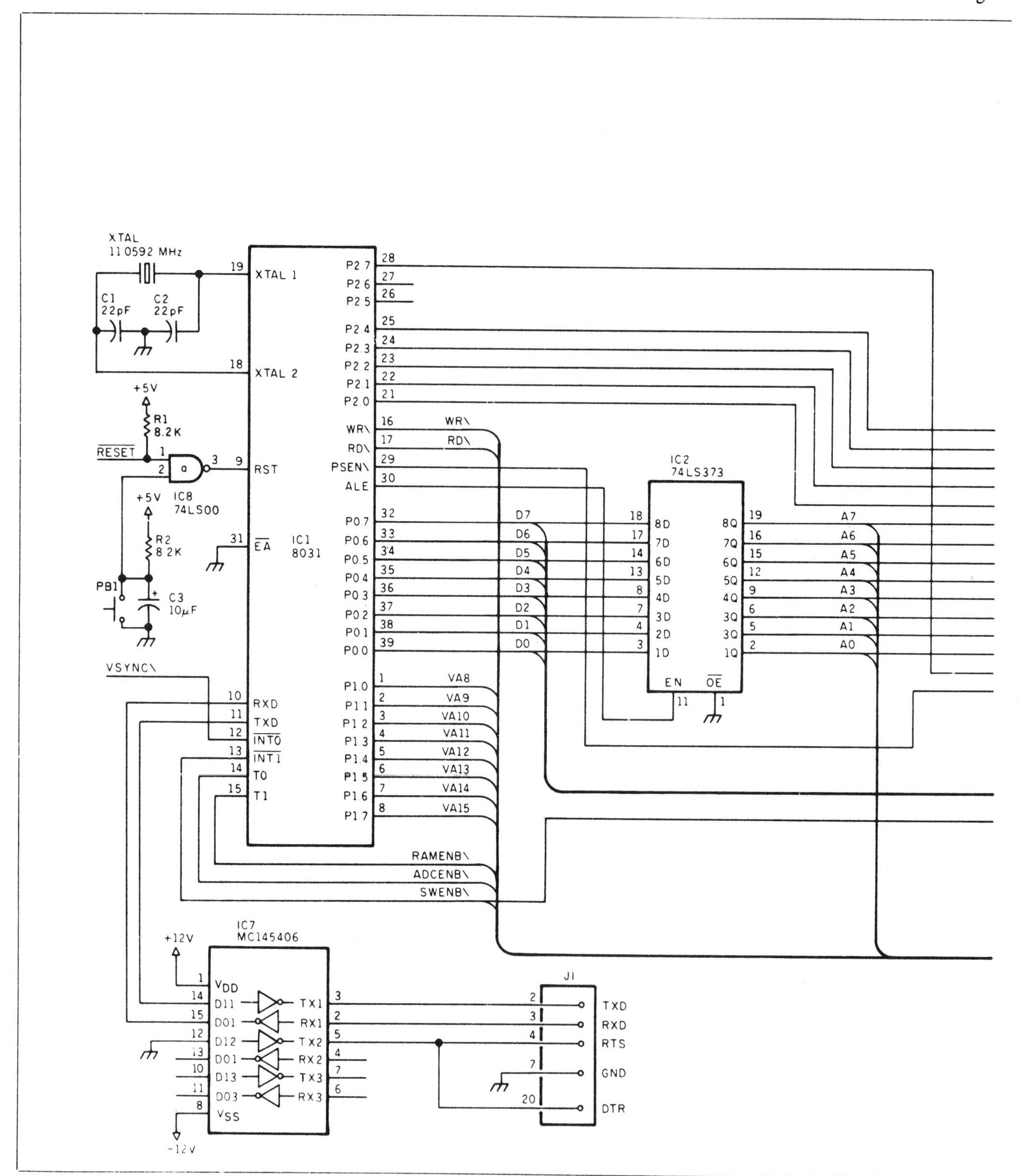

Figure 1: *Schematic for the ImageWise digitizer/transmitter board.*

nals will work with the ImageWise digitizer; however, a color signal contains more information than is necessary. This extra color information is detrimental because it can impart a herringbone pattern on the black-and-white digitized image. I solved this problem by using a filter that removes the frequencies used to encode the color. The remaining signal contains the important intensity voltage that the digitizer measures. You should disconnect this color filter (jumper-selectable JP2) when using a black-and-white camera because the filter doesn't have a sharp cutoff, and it can "soften" the picture by removing some of the finer details. Try it both ways and use the jumper setting that works best for your application.

The 2N4401 transistor serves as a power amplifier to ensure that the re-

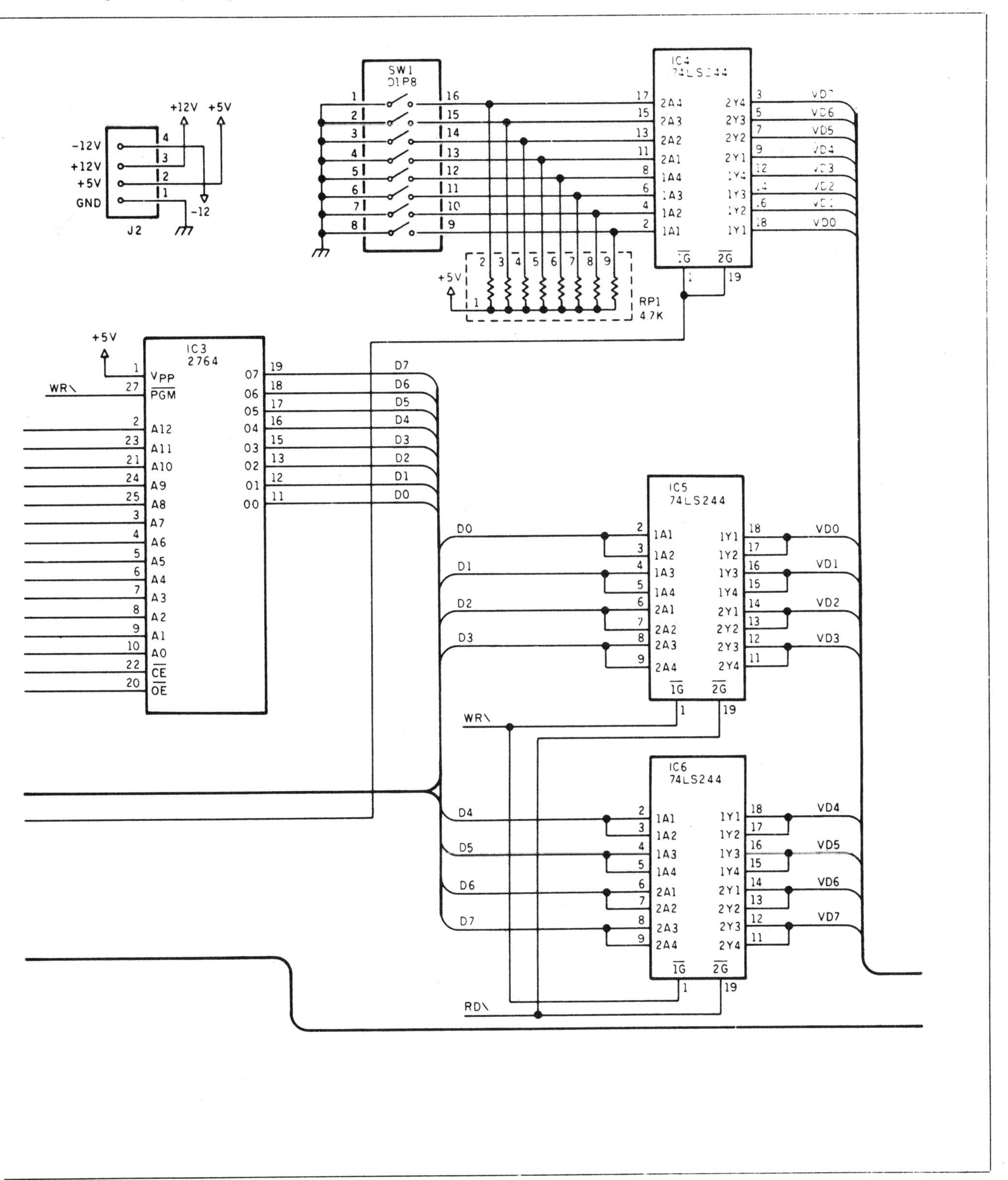

A flash A/D converter differs from slower converters in having more internal circuitry.

maining circuitry gets a clean signal without loading the input. Configured as an emitter follower, this transistor circuit supplies the relatively high input current required by the RCA CA3306 flash A/D converter.

A composite video signal contains both video and synchronization information. I used an LM311 comparator configured as a sync detector. Whenever the video drops below 200 millivolts (set by the resistors on pin 3), the LM311 output goes low. Because the clamping circuit forces the sync tips to ground, the LM311 output goes low only during horizontal or vertical sync pulses and never during video data.

Next, the buffered video signal is directed to the A/D converter. As we've already established, video data is extremely fast, and therefore the A/D conversion must be equally fast. I chose the CA3306 6-bit A/D converter for reasons of economy and the existence of readily available sources. The CA3306 is a special flash A/D converter.

A flash A/D converter is different from slower converters because it has more internal circuitry. Rather than use a D/A converter as an integral component in the conversion process, a 6-bit flash A/D converter contains 64 individual voltage comparators, each set to trigger at a specific level. Sophisticated decoding logic determines which comparator is triggered as a result of the applied signal and outputs the appropriate binary code. The ultimate speed of a flash A/D converter is the reaction time of the comparators and the decoding logic and is independent of any system clocks or other timing signals. In the case of the CA3306, its conversion time is 55 to 83 nanoseconds, or 12 million to 18 million samples per second. The ImageWise system requires a converter that samples at 5 megahertz.

The output of the CA3306 is 6 bits within a relative range defined between $+V^{ref}$ and $-V^{ref}$. Most often these limits are +5 V to +8 V and ground. Because we are digitizing only 64 gray levels, however, it is worthwhile to include only the active video range of +0.2 V to +1.5 V. This is easily accomplished with the white ($+V^{ref}$) and black ($-V^{ref}$) trim potentiometers that adjust the CA3306's conversion thresholds so that a bright white will be digitized as hexadecimal 3F and dark black will be hexadecimal 00. I'll describe the adjustment procedure later.

Finally, the delay trim potentiometer (R21) determines the blanking delay from the start of each horizontal sync pulse, which must be adjusted to match the camera. Video conversion begins when the LS221 one-shot times out. It ends exactly 256 pixels later. IC20 is a 20-MHz oscillator that is divided by four to produce the 200-ns clock that drives the pixel counters and supplies the RAM write signal.

The digital portion of the transmitter board is similar to that of the receiver board (see Chap. 29, part 1). Both boards must process the video data in the same way, albeit in opposite directions, so much of the circuitry can be the same.

Without the 8254 counter/timer used on the receiver board, fewer control lines were needed, so I was able to eliminate the 74LS138 decoder. Two 74LS244s isolate the processor data lines from the video data lines, and a third 74LS244 buffers the option switches. An 11.059-MHz crystal lets the 8031 receive and generate standard RS-232 bit rates.

As I mentioned earlier, the input video must be sampled at a rate of 5 MHz, which translates to around 200 ns between samples. A check of the 8031's execution speed shows there is no way that the processor can read the A/D converter, set up the proper address in memory, and store the byte in 200 ns. I chose instead to use the 8031 to select which line is currently being scanned and set up that

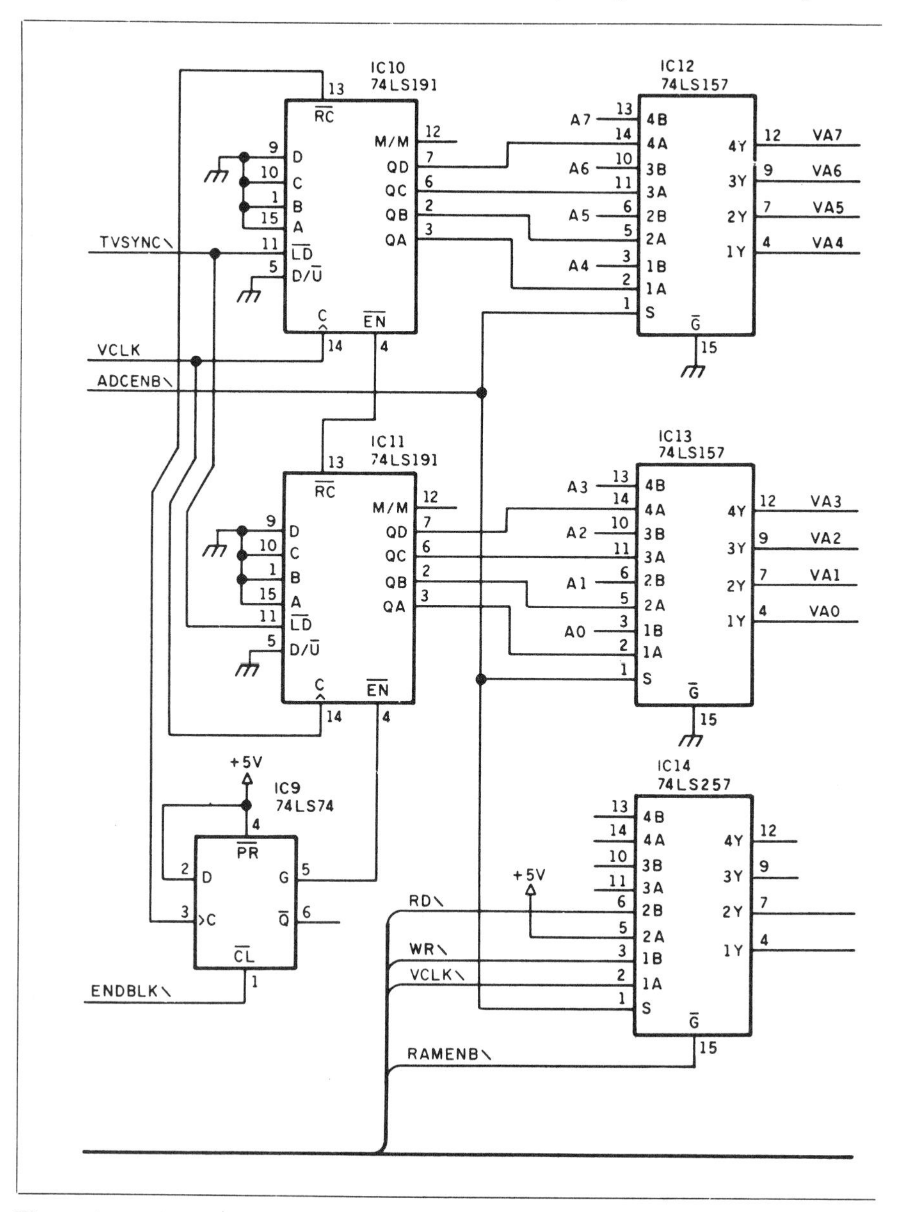

Figure 1: *continued from page 329*

address in the upper 8 bits of the memory address. Two 74LS191 4-bit counters provide the lower 8 bits automatically. At the start of each scan line, the counters are cleared, and the 8031 sets up the line number. After each pixel has been digitized and stored, the counters increment the address. Since there are about 66 microseconds between the start of each scan line, the processor has plenty of breathing room.

The 74LS257 ensures that we are always writing to the RAM during digitization. Normal processor reads and writes can be performed at any other time for image transmission or other processing.

The Software Connection

The digitizer/transmitter software is a simple loop that captures a video field in the RAM buffer, then compresses and transmits the data via serial RS-232. A DIP-switch setting determines whether the digitizer/transmitter will transmit continuously or wait for an XON from the display/receiver before starting each field (see table 1).

The software begins video data capture when it detects the first vertical sync pulse in a field, as described above. The program then waits for the vertical-blanking delay (determined by using the 8031's internal timer) before enabling the RAMs and the A/D converter. Next, the program counts sync pulses and increments the RAM line address after each pulse. When the buffer is full, it disables the RAMs and the A/D converter to prevent further buffer writes.

The flowchart shown in figure 2 describes the process used to compress the video data in each line. Notice that a unique sync byte designates the start of the field and the line within the field, as well as the end of the video data.

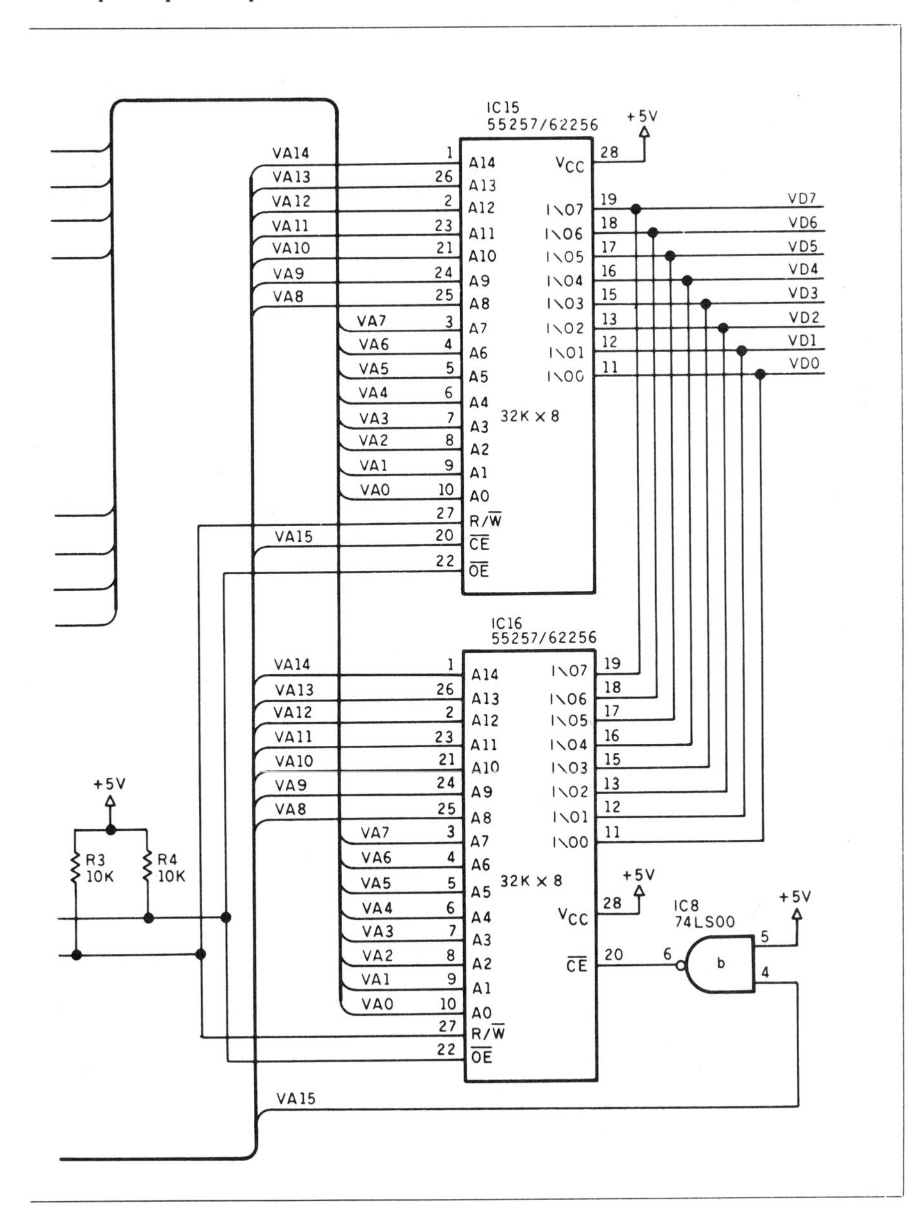

The software captures a video field, then it compresses and transmits the data.

Getting Started

Assuming that you have an ImageWise display/receiver in good working order from the project in part 1 of this chapter, use the procedure that follows to make your first digitized video connection.

First, connect the camera to the monitor (a coaxial cable without any fancy hardware between) and get a picture that's well-lighted and focused. The adage about "garbage in, garbage out" certainly applies to this operation! Make sure that you've got the monitor terminated in 75 ohms.

Connect the display/receiver to the monitor (remember to disconnect the camera first). Set the DIP switches to 28.8 kilobits per second and no time-out (continuous pictures; see Chap. 29, part 1 for a DIP-switch-setting guide for the display/receiver).

Turn the video level (R8) trim potentiometer to midrange and then plug in the power. Do not connect the digitizer/transmitter. The display/receiver will display a diagonal test pattern that includes a gray scale ranging from full white to full black. Adjust the video level trim potentiometer so that the monitor shows the complete range of shades. You may have to tinker with the monitor's hold, brightness, and contrast controls.

Connect the camera to the digitizer/transmitter and install the 75-ohm terminator jumper (JP1). If you're using a color camera, install the color filter jumper (JP2); otherwise, remove it. Set the DIP switches to

28.8k bps
16-line vertical delay
Compression enabled
Ignore +/−1 count changes
Paced mode disabled

Turn the delay (R21), black level (R18), and white level (R14) trim potentiometers to midrange; connect a serial cable between the digitizer/transmitter and display/receiver; and plug in the power. (Note: Set up the serial cable so that pins

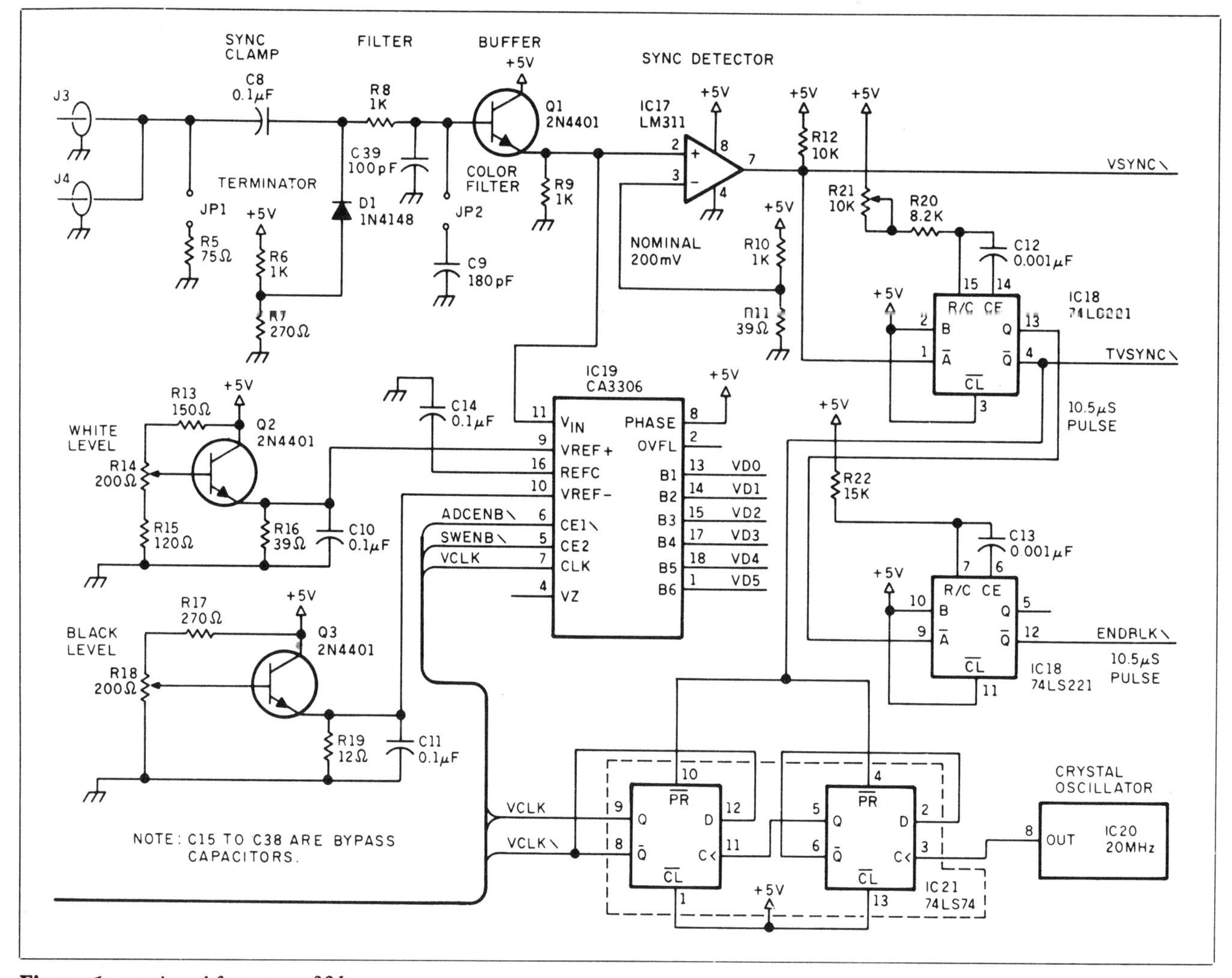

Figure 1: *continued from page 331*

Table 1: *ImageWise digitizer/transmitter DIP-switch settings. ON and OFF refer to switch positions.*

Switches 1, 2, and 3 select the serial bit rate (must match the receiver's rate).

1	2	3	Serial bit rate (must match receiver rate)
OFF	OFF	OFF	300 bps
OFF	OFF	ON	600 bps
OFF	ON	OFF	1200 bps
OFF	ON	ON	2400 bps
ON	OFF	OFF	9600 bps
ON	OFF	ON	19.2k bps
ON	ON	OFF	28.8k bps
ON	ON	ON	57.6k bps

(Note: 4800 bps was intentionally omitted.)

(Note: Switches 7 and 8 are not used and must be OFF.)

Switch 4 selects the vertical blanking delay from the first vertical sync pulse.

4	Vertical blanking delay
OFF	16 lines (normal)
ON	20 lines (extended)

Switch 5 enables or disables picture compression.

5	Run-length encoding
OFF	disabled (no compression)
ON	enabled (compression)

Switch 6 enables or disables the +/−1 count change compression.

6	Compress +/− 1 count changes
OFF	encode (less compression)
ON	ignore (more compression)

2 and 3 are exchanged and pins 5 and 7 are straight through.) You should see the digitized picture appearing on the monitor, painting from top to bottom. After the entire scene is done, another image will be sent. You'll be able to see a horizontal line marking the descending edge of the new picture overlaying the old one.

You should now adjust the black and white trim potentiometers to get the maximum amount of detail in the picture. If the black level is too high (clockwise), dark areas will have little detail, and the whole picture will be dark. If the white level is too low (counterclockwise), the bright areas will suffer, and the picture will be light. You should have a high-contrast scene in front of the camera to make sure you have the right levels. (On the other hand, don't make the black level too low or the white level too high, because that will reduce the number of digital levels in your scene. For example, if your camera's highest voltage is 1.4 V, it does no good to have the white setting at 2.0 V; that 0.6-V difference contains some digital codes that will never be used.)

Adjust the delay trim potentiometer so that the scene is horizontally centered on the monitor. If your monitor has a great deal of overscan, it won't matter too much what the delay setting is. You can also adjust the monitor's horizontal hold knob slightly.

If a line of trash (there's no better way to describe it) appears at the top of the monitor's display, try setting the digitizer/transmitter DIP switch for 20 lines of vertical delay instead of 16. Some cameras produce a few lines of trash at the beginning of the field, and the digitizer/transmitter faithfully digitizes it. If you don't see anything, or if the trash moves to the bottom, leave the DIP switch set for 16 lines.

If you have a monitor connected to the camera (a viewfinder doesn't count), you must make sure that either the monitor is terminated or the termination jumper is installed, but not both or neither. You should terminate the device at the end of the camera cable, not the one in the middle.

You may want to try the filter jumper to see what effect it has on the scene. If you have a color camera, the filter is probably going to be essential; if you have a monochrome camera, you may simply like a softer picture. Do not confuse the effect of the filter jumper with the output of an unfocused camera; make sure the scene is crisp to start with.

Try flipping the digitizer/transmitter switch that ignores +/−1 count changes

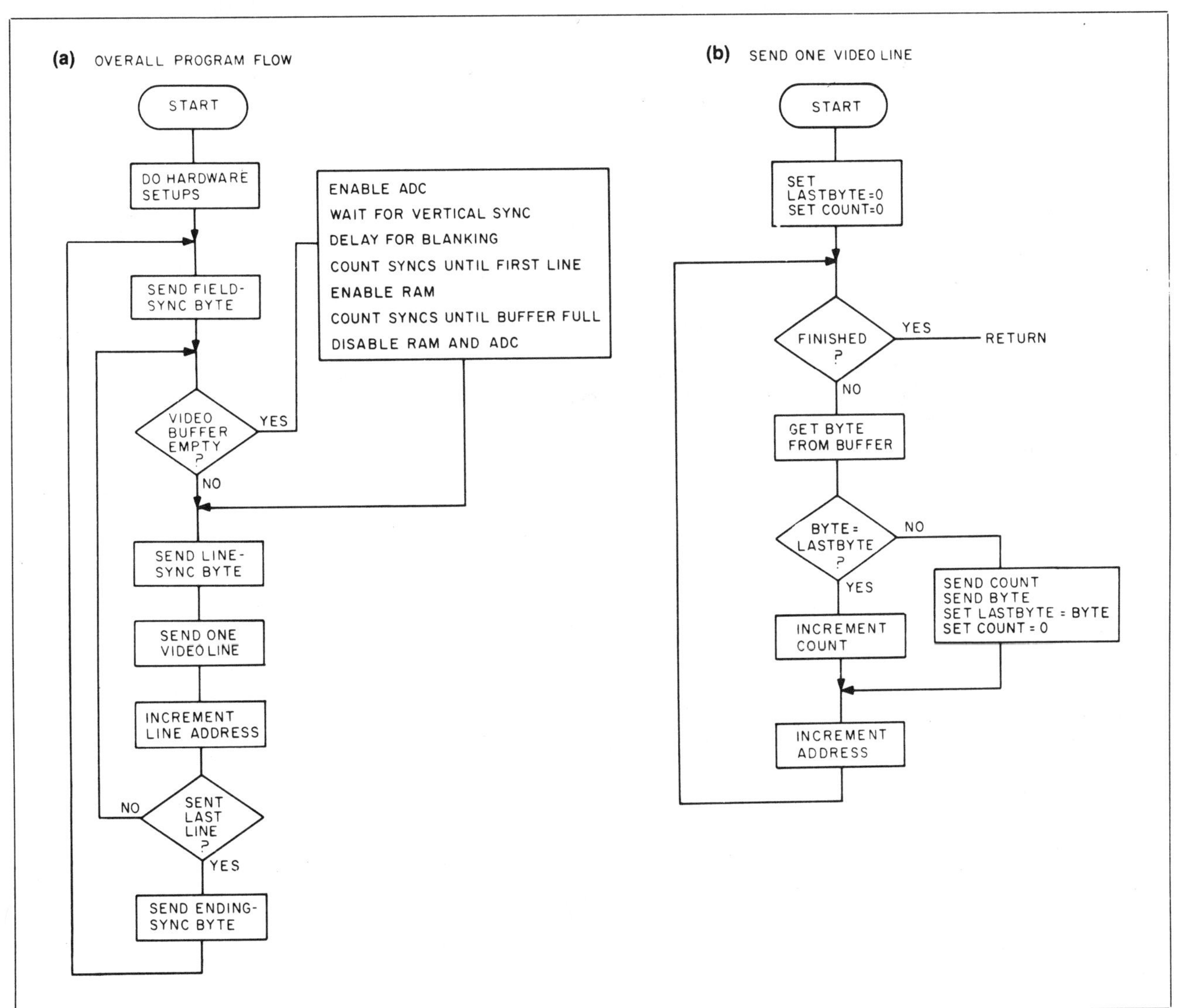

Figure 2: *Flowchart of the digitizer/transmitter software.* **(a)** *This outlines the overall flow of the software.* **(b)** *A flowchart of the routine to transmit a single video line.*

to see what happens to the flat areas in the picture. Disable compression and measure the increase in transmission time.

You can also try changing the bit rates to see which rates work for you. A direct connection can run at 28.8k bps, but for longer wires running through the house, you may need to use a slower rate. At 300 bps you can see the compression working. Remember that the display/receiver will get confused when you change the rate, so you may need to reset the system after changing any DIP-switch settings.

Cheap Buffer Option

I've described how the digitizer/transmitter works with a 64K-byte RAM field buffer made of two 32K-byte static RAM chips. It turns out that you can use an 8K-byte buffer in the digitizer/transmitter. Because the 32K-byte static RAM chips are still rather expensive, I thought it would be worthwhile to reduce the cost for applications that don't need the advantages of the full 64K-byte buffer (Note: The ImageWise kit contains 32K-byte static RAM chips.)

An 8K-byte RAM can hold 32 lines of video (8K bytes divided by 256 bytes per line). The digitizer/transmitter digitizes 32 lines, transmits them, captures another 32 lines, and so on. It keeps track of the last digitized line and starts with the next line for the next group. When it's done, it begins again with the first video line.

Using an 8K-byte RAM is simply a matter of plugging it into the first RAM socket. The 8031 program "feels around" after a reset to determine the RAM size during initialization and uses the proper addresses automatically; no DIP-switch settings are required.

The only difference between using 8K- and 64K-byte RAM buffers occurs when the scene contains moving objects. The 64K-byte buffer can hold a complete frame that is captured in 1/60 second, and it is the only configuration that can legitimately be called a frame grabber. The 8K-byte buffer holds 32 line groups that are digitized several seconds apart, so it's possible to get confused images. If you've ever seen those "panoramic" shots of a line of people with the same person at both ends of the line, you'll recognize the problem right away. But if your application doesn't involve rapid motion, you can save some money on buffer RAMs.

This trick doesn't work in the display/receiver because it must have the entire picture in RAM at all times. Replacing the 64K-byte buffer with an 8K-byte buffer would simply give you 32 lines' worth of picture.

Table 2: *Hayes Smartmodem DIP-switch settings for use with the ImageWise system.*

Switch	Setting	Description
1	DOWN	Smartmodem DTR input on pin 20 is forced active.
2	UP	Don't care, UP for English responses to commands.
3	UP	Suppress responses to commands.
4	DOWN	Do not echo characters.
5	UP	Smartmodem will auto-answer on first ring.
6	UP	Don't care, UP for CD when carrier detected.
7	UP	Single-line phone connection.
8	UP	Disable Smartmodem command recognition.

Photo 2: *The ImageWise digitizer captures a high-quality gray-scale image that can be used in areas like security and pattern recognition. In the surveillance application shown here, an empty room is entered, used, and then exited. To save space in this presentation, all four scenes are displayed on a GT180 (in 16-level gray scale).*

Using a Modem

Because both the digitizer/transmitter and the display/receiver use a two-wire serial interface and XON/XOFFs to control data flow, they can be connected through a pair of modems as well as by direct wiring. The only problem is the low data rate that modems can handle. This is an ideal application for 2400- and 9600-bps modems.

One application might be to have the ImageWise digitizer/transmitter wait for a phone call, auto-answer, and then transmit the picture it sees at that remote location. Voilà, video security system or videotelephone. The switch settings for a Hayes Smartmodem 1200 are shown in table 2. Note that you can use other modems as long as they have auto-answer capabilities.

Using ImageWise

The ImageWise system can be used as a digitizer and display board pair or as individual components of some higher-function system. Used as a pair, "teleimaging" becomes a reality. By adding the sense of sight to our ordinary audio communication, we add a new level of communication and understanding. No longer does the field engineer have to be frustrated trying to justify replacement rather than repair of an expensive electrical component. A quick digitized image flashed back to the head office verifies significant fire damage and gets the proper authorization for immediate action.

The key factor in this new level of communication is the old saying that "seeing is believing." Consider another example: You are a consultant at a customer's site, and some question arises as to the actual wording and the date of a revision note on an important schematic. Rather than waste a day with express delivery, you can call your office and have them transmit a digitized image of the portion of the schematic in question complete with the authorization signature and date as they appear.

ImageWise has an infinite number of stand-alone uses. It can be used to instantly communicate fingerprints and ID photos, monitor traffic at remote intersections, monitor remote security risk zones (see photo 2) via auto-answer modems, aid in conducting company-wide lectures (standard audio teleconference with pictures of the blackboard periodically sent to all locations; see photo 3), or send x-rays and CAT scan pictures to medical personnel for corroborating diagnosis.

Some of these uses might seem ambitious for ImageWise, but similar more expensive units are already being applied in these areas. My immediate application may seem mediocre by comparison. Recently, I've been spending time out of the Circuit Cellar at an office across town. Since I already had a TV camera in my driveway (no windows in the cellar, remember), I simply attached it to the digitizer/transmitter and a 2400-bps auto-answer modem. Now I can call the house and get the latest snapshot or simply leave it on all the time as a real-time display of all the activity around the house (see photo 4). (I have four telephone lines; I can call the Home Control System on its own line if I want to have some real fun with someone like a delivery man. These guys all think my house is haunted.)

You'll note that I have described ImageWise only in terms of a 256- by 256- by 6-bit digitizer. Because I intended to use it with a modem, I felt the need to increase the picture-transmission speed. One way is to reduce the resolution to 128 by 121 or 64 by 61 bits. Even though such resolutions produce grainy images, they are still quite recognizable, especially if they are of familiar faces or geography (the recently advertised Mitsubishi video telephone has a 94- by 94- by 4-bit resolution by comparison). The 64- by 61-bit image is transmitted eight times faster than a 256- by 256-bit image and is suitable for monitoring gross changes in a driveway scene when a car or a person approaches.

When something appears, I can immediately change the DIP switches on the display board for a higher resolution and trigger another picture while the form is still in view. (The frame is grabbed instantly and is independent of the transmission time.) The picture-repeat rate and resolution, remember, are commanded from the receiver and not fixed by the transmitter. The interaction is completely dynamic. My next activity is to connect the ImageWise digitizer to a computer and let it decide what's happening for itself and make all the decisions.

Fortunately, this is as easily said as done. Probably the most significant feature of ImageWise is that it is computer-nonspecific. It is a serial RS-232 I/O device that does not depend on any computer-specific bus. The ImageWise digitizer/transmitter's serial port can be connected to any personal computer. The computer can receive image data and store it on disk or send it to a similarly connected display/receiver board. So far we've written the software for my SB180 and the IBM PC. Others will follow.

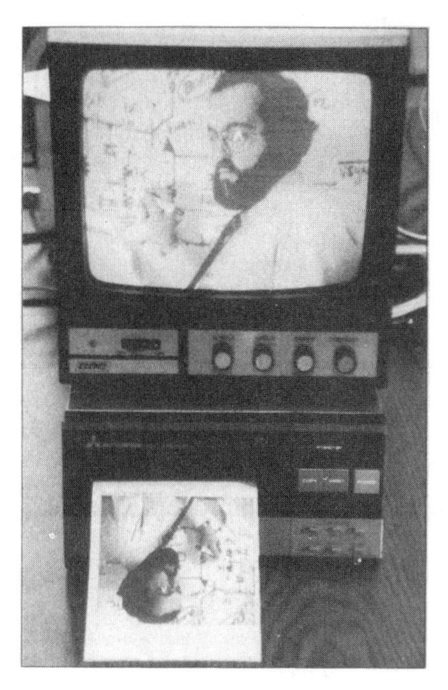

Photo 3: *This is how a picture of me standing in front of a blackboard in Connecticut would be received by an ImageWise display/receiver board in California. A video printer like the Mitsubishi unit shown here can save the current scene while another is being transmitted.*

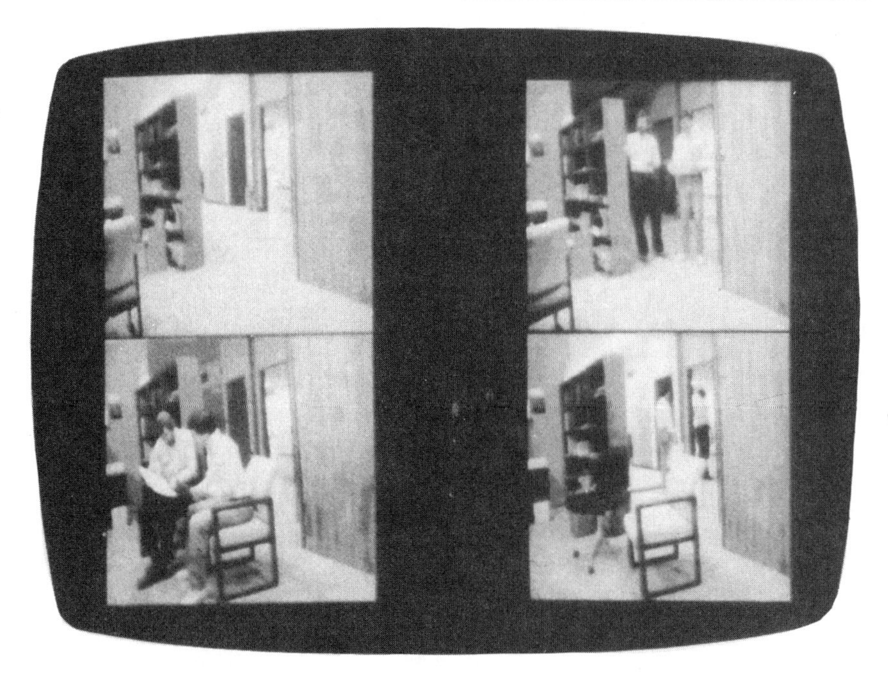

Photo 4: *The equipment counter in the Circuit Cellar where I took most of the photos with the setup you see. The interesting point to realize here is that this picture is completely digitized. A video camera (out of view) is pointed at the two stacked monochrome monitors and the film camera. The monitor on the bottom displays the picture as it is produced by the video camera and input to the digitizer/transmitter board. The display on top shows the output of the display/receiver board after it receives the data from the digitizer. It is the object of view by the film camera. This picture is the screen of the top monitor.*

In Conclusion

Once you've got a picture in digital format, you can write programs that perform magic tricks with it (hardware people like to think that way). By manipulating the binary data making up the picture, you can transform it into another picture that may be more meaningful. You can even combine two pictures to find differences—this is called image processing.

Now that we have the ImageWise digitizer, we have the means to perform some real experiments. I know many tutorial articles on image processing have been published, but the true Circuit Cellar creed is to build it yourself. Using ImageWise, in Chap. 30, I'll demonstrate how the basics of picture comparison, enhancement, and other image-processing fundamentals can be done for real.

Special thanks to Ed Nisley for his expert collaboration on this project.

See page 447 for a listing of parts and products available from Circuit Cellar Inc.

30

USING THE IMAGE WISE VIDEO DIGITIZER: IMAGE PROCESSING

This digitization and display process is easy to duplicate

While I was writing the second part of the ImageWise project, a BYTE editor sent me copies of the image processing theme articles used in Chap. 32. After I got over my first reaction to the common thread of the articles (it's almost all software—yech!), I realized that, while I was covering the hardware specifics of ImageWise more than adequately, to do real justice to the subject I should include more on using and processing the data created from the digitizer.

Getting from here to there constituted a problem, however. While many people can read and instantly visualize the image transformations described in these image processing tutorial articles, some people prefer an alternative approach to such presentations. Although Chapter 32 is devoted to image processing, I'd like to think there is a difference when I discuss a subject.

I describe ideas, but also I try to include a little hands-on experience. Unlike a tutorial that contains little mention of the hardware you might use to duplicate such feats, all of the picture data used in this chapter was digitized on the ImageWise digitizer/transmitter and displayed on its companion display/receiver board (except for the zoom shots, which are displayed on a GT180). You should be able to easily duplicate the process.

I have expanded the original two-part ImageWise hardware project to include two more articles with a little software. Admittedly, I am out of my element, and I ask you to bear with me if I drop a few bits now and then (think of it as poetic license). I couldn't pass up an opportunity to string together such interesting ideas as image processing and colorization—which I can actually demonstrate.

In this chapter, I will focus on image processing. As in the related tutorials, I'll take a digitized image and detect edges, enhance it, filter it, enlarge it, subtract it, and create other more useful images.

First, a quick hardware review of ImageWise will show you what the data is that I am processing (see Chap. 29, parts 1 and 2, for more details).

Picture Format

The ImageWise digitizer/transmitter digitizes a single field of the camera's video signal on-the-fly, converting it into 244 rows of 256 pixels each. The rows are numbered from 0 to 243, and the pixels are numbered from 0 to 255 in each row.

A pixel's brightness is represented by one of 64 gray levels, with a black pixel equal to 0 and a bright white pixel equal to 63. Each pixel requires 1 byte of storage, so there are 62,464 pixel bytes per image. Software adds some additional control-information codes to simplify the display/receiver's job, giving a total of 62,710 bytes in an image.

The digitizer/transmitter compresses the video data using run-length encoding to reduce the time needed to send it over the RS-232 serial link. When the digitizer/transmitter finds a gray-level value repeated more than twice in adjacent pixels (a "run") in the line, it replaces the repetitions with a count. Typical scenes are reduced by 50 percent to 75 percent, with a corresponding speedup in transmission.

The display/receiver accepts RS-232 data, decompresses it into a RAM display buffer, and generates the synchronization signals required to show the images on a standard composite-video TV monitor. The result is a TV picture that looks remarkably like the original scene.

The Personal Computer Connection

Because both the digitizer/transmitter and display/receiver communicate over a standard RS-232 line, you can connect either one to a serial port on a personal computer (the unit can connect to any computer with a serial port, but all my examples use an IBM PC). When the computer is connected to the digitizer/transmitter, it acts as a display/receiver, storing the image data on disk. When it's connected to the display/receiver, it acts as a digitizer/transmitter and sends the stored images out for display.

The computer can accentuate or suppress details in an image by performing simple arithmetic on the numeric values for the pixels. For example, a program can compare two scenes by subtraction, and a count of nonzero values in the result can tell you whether something has moved into (or out of) the picture.

This chapter demonstrates a tool kit of programs that you can use to develop a complete image processing application. The programs are written in Turbo Pascal for an IBM PC, but you can easily convert them for use on other computers. I used an 8-megahertz IBM PC AT with 640K bytes of RAM, a 10-MHz 80287 math coprocessor, and a 1.2-megabyte RAM disk to develop these programs. They will work on any computer that runs Turbo Pascal and has sufficient RAM (about 512K bytes) but might take somewhat longer to run. Because the images are displayed on a TV monitor connected to the display/receiver, you don't need a graphics display on the computer.

Serial Setup

You set the data rate on the serial link using DIP switches on the digitizer/transmitter and display/receiver boards. Although the maximum data rate is 57.6k bits per second, the PC simply can't keep up at that rate with the present software. While I could have used some computer assembly code to tweak the critical loops, I felt it was better to use a more easily understood technique. So the programs are limited to half the maximum rate: 28.8k bps. If your computer can't handle this rate, you must recompile the programs to use a lower rate.

Only two programs actually communicate with the ImageWise boards. The Grab program prompts the digitizer/transmitter to send an image and stores it on disk. The Show program reads the disk file and sends it to the display/receiver. Both use the COM1 serial port, so you'll have to swap cables when you use each program. (I used a serial-port switch box, but you can easily recompile the programs to grab images from COM1 and show them on COM2.)

Photo 1: *An image captured by the ImageWise digitizing system.*

Photo 2: *I created this image by multiplying the pixels in photo 1 by 2 using the Multiply program.*

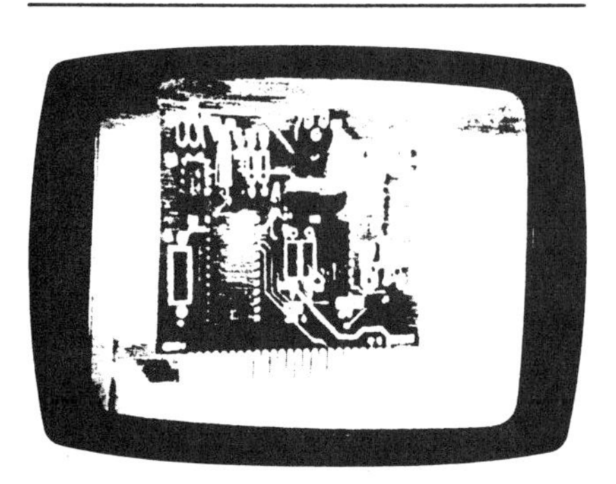

Photo 3a: *The digitized image of a circuit board.*

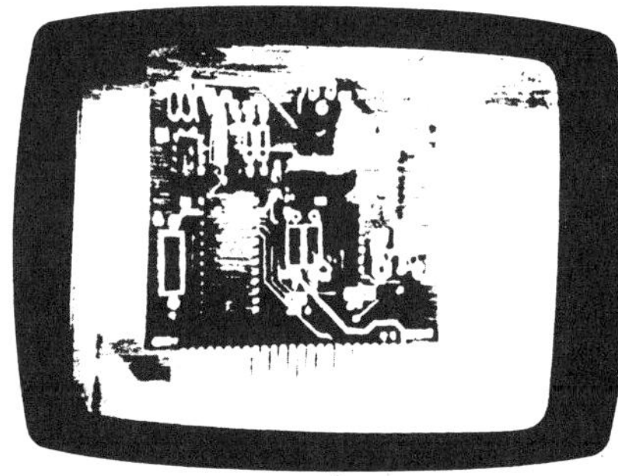

Photo 3b: *I have added something new.*

Photo 3c: *You can use the Subtract program to discover what has changed.*

Photo 4: *You can run the Thresh program on the image in photo 3c to remove background clutter.*

One of the first tasks I have is undoing one of ImageWise's features. Although the compressed data format reduces the transmission time, it's not well-adapted to image processing. The programs must examine every image pixel, something that's not easily done with run-length-encoded data. So Grab decompresses the images before it stores them on disk, creating a 62,720-byte file for each picture. There are 62,710 image and control bytes, with 10 padding bytes added to fill out the file's last 128-byte block.

The Show program and the display/receiver can handle either run-length-encoded or expanded files, so there's no problem sending them to the display/receiver, except for the increased transmission time. The Compress and Expand programs convert between the two formats.

Taking Pictures

In addition to the ImageWise digitizer/transmitter and display/receiver boards, you'll need a TV camera and monitor, a tripod for the camera, and some RS-232 and video cables. A color TV camera will work fine, even though the digitizer/transmitter is designed for monochrome. If you see herringbone patterns on the display/receiver, install the Filter jumper on the digitizer/transmitter to remove the color information from the camera signal.

A zoom lens is a great help because you can adjust the focal length to fill the screen with the scene. If you are taking pictures of small objects, you might also need a macro lens or attachment. Most consumer TV cameras come with a macro-focusing zoom lens, so you're probably in good shape if you have one.

I captured the scenes in this article using a monochrome camera equipped with a 15- to 75-millimeter zoom lens. I used a 75-watt desk lamp for illumination. The camera lens was usually opened wide to f/2.1.

The first rule of photography is to get enough light on the subject. While you can use light meters and judgment, checking the actual results is better. The Histo program analyzes an image and reports on the number of pixels having each of the 64 possible brightness levels. Figure 1 shows the output of Histo for the image in photo 1.

The large peak is created by the desktop and background areas that are all more or less the same shade. There are relatively few black areas (near 0) and relatively few white areas (near 63). The peaks on every other pixel count indicate a little bit of noise in the A/D circuits.

Notice the small number of pixels brighter than about 30. Although it's better to increase the amount of light on the scene, you can achieve a similar result by multiplying each pixel by a constant. Photo 2 is the same as photo 1, with each pixel multiplied by 2 using the Multiply program. This is nearly equivalent to increasing the exposure by one f-stop, thus doubling the brightness.

Figure 2 is Histo's output for the image in photo 2; notice that the pixel values are all even (multiples of two) except for the pixels that "stuck" at 63. The brightest areas of photo 2 look flat because they are all the same value. Increasing the illumination would have filled in the odd-numbered pixels.

When you're setting up a new picture-taking session, always use Histo to make sure you're getting enough light on the scene. It's all too easy to twist the brightness knob on the monitor, which doesn't do anything for the digitizer/transmitter.

What's New and Different?

One of the more interesting things you can do with two images is to find the differences between them. Photo 3a shows a small circuit board, and photo 3b has something new added. By using the Subtract program to produce the image in photo 3c, you can see exactly what changed.

Often there will be minor, inconsequential differences between the images. You can see some background clutter in photo 3c resulting from small differences in lighting and position. Regardless of how careful you are, these differences will occur. What you need is a program to get rid of the irrelevant details.

The Thresh program sets pixel values below a specified threshold level to 0 (black). Running Thresh to remove all pixels below 40 gives the image in photo 4. In addition to suppressing the clutter, Thresh removed the face inside the helmet. This should serve as a reminder that Thresh is concerned only with the brightness of each pixel: Because the face pixels are less than 40, they are set to 0 just like the background clutter.

Photo 4 contains only the parts of photo 3b that aren't in photo 3a, but there are some shadows and reflections in addition to the figurine. The pixel values represent the brightness of the scene and don't "know" whether they are part of an interesting object or the background. Your image-recognition software must distinguish between the actual objects and their shadows and reflections.

Inspection Applications

I know that many of you are interested in using video for inspection, so the next example shows what's needed to compare two pictures to find differences. Since I occasionally digress, I thought inspecting printed circuit boards for missing components was a suitable example.

A critical ingredient in any inspection task is a reference standard that "looks right." All other items are compared to that standard; anything different is regarded as an error. Of course, the differences have to be visible to be detected.

The image in photo 5a is the reference circuit board (anything less than a perfect image is due to lousy lighting). Photo 5b shows a test board with one IC missing. Notice that the ICs are darker than both the board and the silk-screen print below them, but that the capacitors are lighter than the board.

The Compare function is the same as Subtract, except that it returns the absolute value of the difference. Thus, any change will show up as a bright pixel. Photo 5c shows the results of using Compare to process the images of the reference and test boards. A lot of background clutter is due to minor variations in the boards, the lighting, and positions. A simple threshold won't remove the clutter because some of it is quite bright. The trick is to know where the important

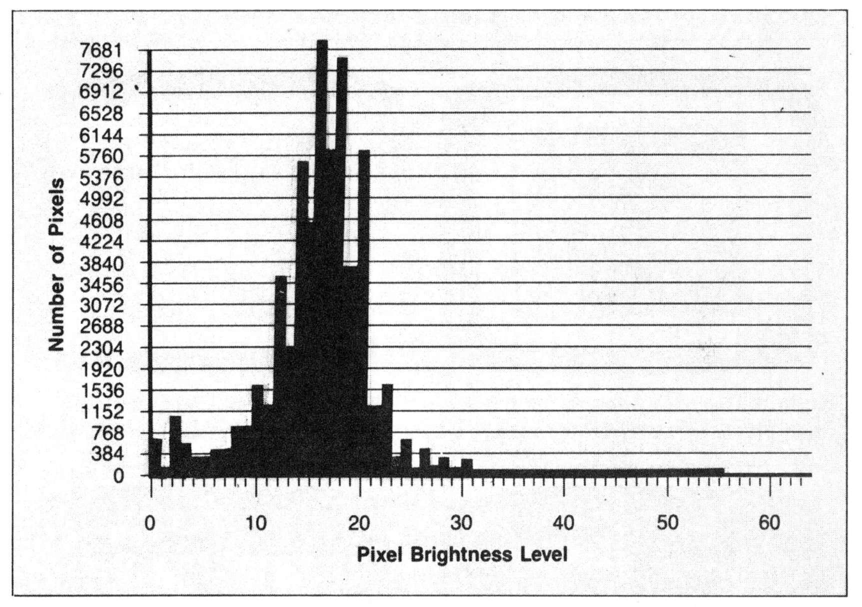

Figure 1: *Output of the Histo program: a pixel-intensity histogram for the image in photo 1.*

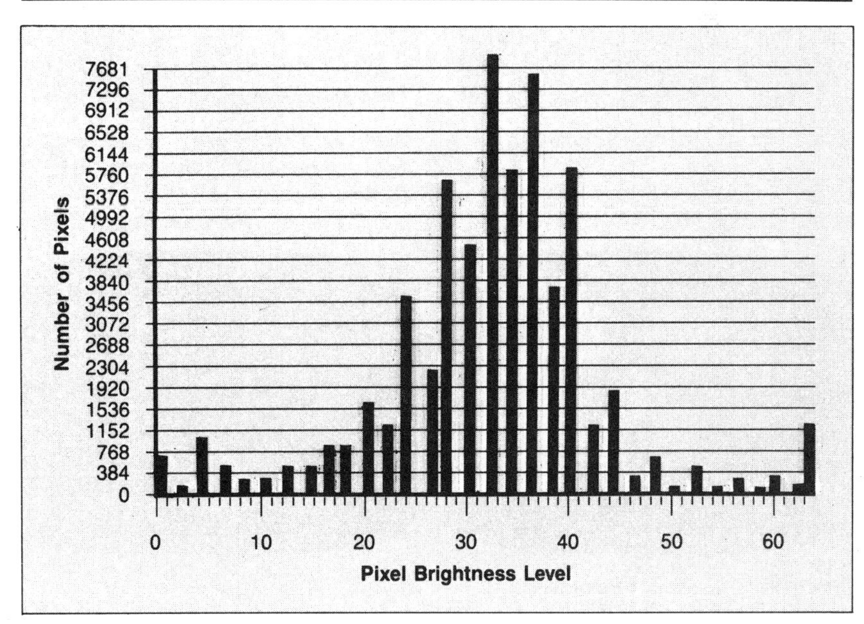

Figure 2: *If you process the image in photo 2 through the Histo program, this is what you get. Notice that all pixels are multiples of 2.*

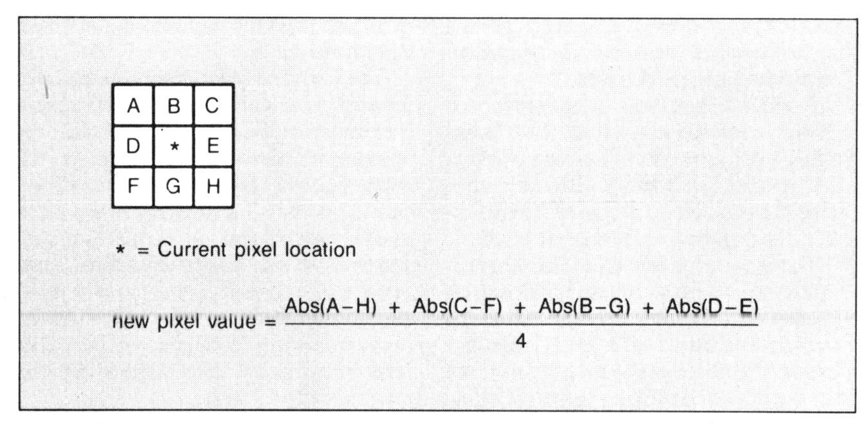

Figure 3: *The formula used by the Edge program.*

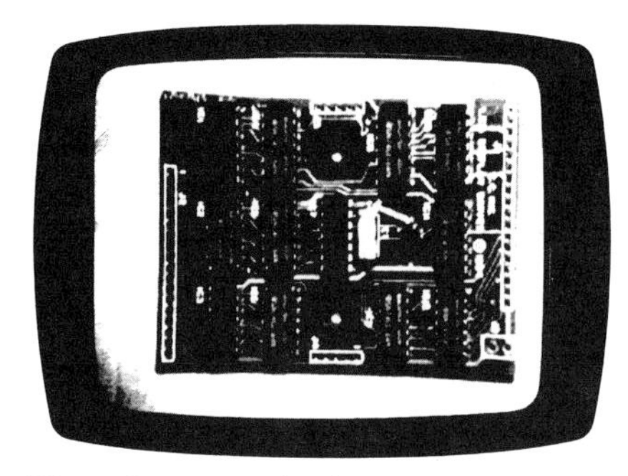

Photo 5a: *Using the Compare program, I have processed the reference image.*

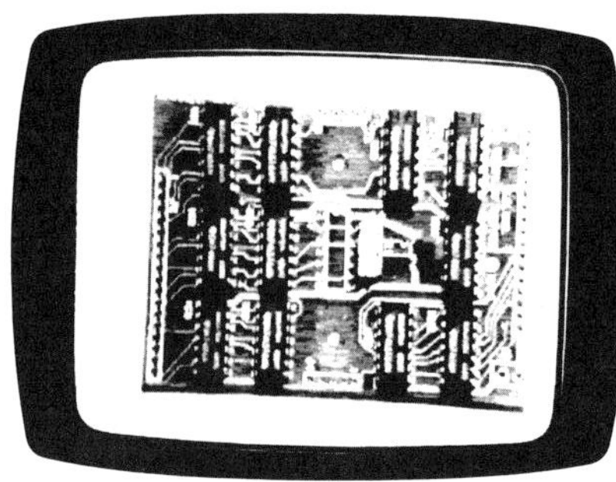

Photo 5b: *This was done with a test image from which I have removed an IC.*

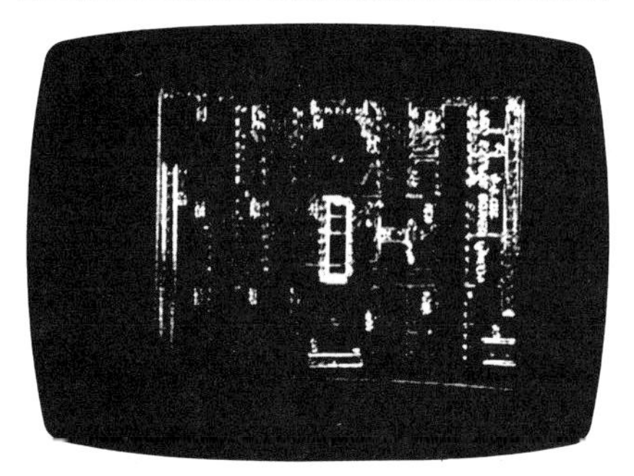

Photo 5c: *The missing IC stands out.*

Photo 6a: *I've prepared a mask image and used it to isolate the important elements in photo **5c**.*

Photo 6b: *The result of processing the image in photo **5c**.*

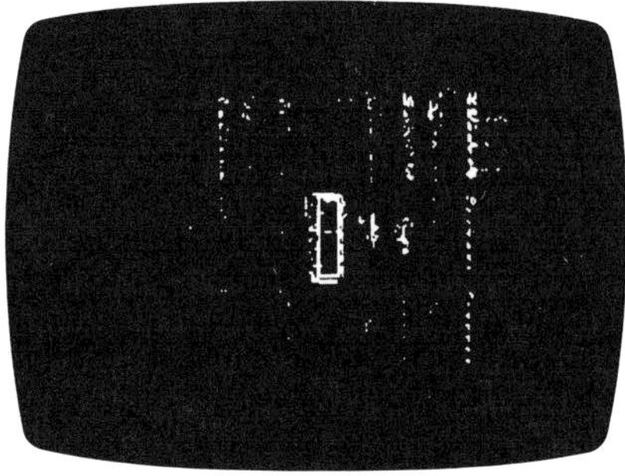

Photo 7: *Here, I use the Thresh program to remove background noise from the image in photo **6b**. The missing IC in **5b** now stands out clearly.*

areas of the picture are and ignore the rest.

Photo 6a shows a mask with bright areas surrounding each component location, prepared by putting white tape on a blank printed circuit board and processing the image to remove the board traces. The Mask program will suppress any pixels in one image lying outside the masked areas in a second image. The image in photo 6b is the result of using the Mask program with the image in photo 6a on the image in photo 5c.

The final step is to apply Thresh to reveal the missing IC in photo 7. What you're seeing is the white silk-screen image printed on the board. It is difficult to see the difference between a dark gray IC and a dark blue circuit board; anything you can do to increase the contrast will help. You will also need to ensure that the two images are accurately aligned and lighted to reduce background clutter. A fixture to hold the boards at an exact location relative to the camera and lights will be essential. You can use Histo, Compare, and Thresh to set up.

Edges and Filters

In some cases, you might be interested in the location of the edges of an object. For example, you might want to know that a pattern is correctly positioned without caring what color (or gray shade) it is. The Edge program produces an image that contains the difference between a pixel's neighbors, calculated as shown in figure 3. A sharp junction between a light and a dark area will result in a bright line, while a uniform area will be reduced to black. The actual shades are not important, only the differences between them.

The Edge routine is a bit more complex than Compare. It finds the absolute value of the differences between the eight pixels surrounding the current pixel in all four directions: vertical, horizontal, and the two diagonals. This is a simple example of a more complex operation called a convolution, which you can use to identify other features in an image.

I restricted Edge to a 3 by 3 set of pixels to reduce the amount of time required to get the answer: It works well enough for these examples. You might want to experiment with a 5 by 5 or larger array, which will let you identify edges more precisely, particularly diagonals at other than 45 degrees. You can also identify the direction of the edges by removing the absolute value function. An edge-detector algorithm that performs this operation (see figure 3) is

$$\text{edge} = \frac{(A-F) + (B-G) + (C-H)}{3}$$

This operation would return a positive value for the horizontal edge between an upper, bright object and a lower, dark object. Reverse the two objects, and the sign becomes negative. Because the results must be returned as pixel values, you will need to add a fixed offset before setting the final pixel value.

Usually, you will have to multiply the result of Edge by 2 or 3 to make all the edges visible on the monitor. Thresh can then suppress all the "soft" edges. Photo 8b is the result of running Edge on the image in photo 8a, multiplying by 3, then using Thresh to remove pixels below 30.

It's also possible to remove edges and textures. Filter averages four pixels using the algorithm shown in figure 4 to produce the output image. Compare the images in photos 9a and 9b to see how Filter reduces "crispness" and fine details. This can be useful if you have an object with fine detail that is not needed by the rest of the processing.

Intruder Alert!

One obvious application for image processing is in a security system that can compare two images, decide when something has changed enough to warrant human inspection, and sound an alert (or fire the laser, or whatever).

You've seen most of the pieces already:

1. Grab a reference image.
2. Grab a test image.
3. Compare the images.
4. Thresh the result to remove clutter.
5. Count the number of changed pixels.
6. If the count is high enough, take action.
7. Replace the reference image with the test image.
8. Go to step 2.

The key program is Count, which examines an image and counts the number of pixels that exceed a threshold level. The preceding image processing steps must create an image with high-intensity pixels identifying the intruder. When the count is high (corresponding to a new shape of a person on the screen), it's time to sound the alarm.

If the images don't differ by too much, you use the test image as the reference image for the next loop. This lets the system cope with small, slow changes in lighting and motion. Obviously, you could defeat this system by easing slowly into the picture, but in practice it's hard to fool.

WATCHDOG.BAT (see listing 1) combines the programs we've used so far to automate that process. The batch file can

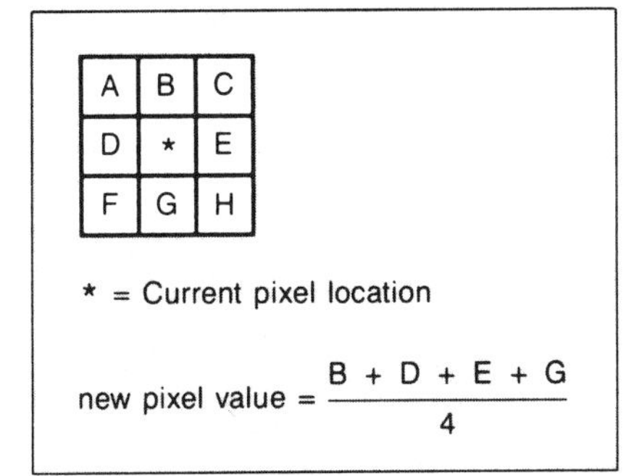

Figure 4: *The formula used by the Filter program.*

Photo 8: *Edge detection with ImageWise. The image in* **(a)** *is processed through Edge, Multiply, and Thresh to produce* **(b)**.

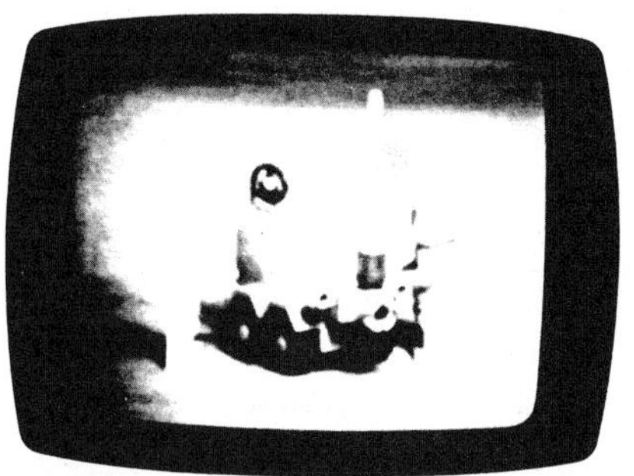

Photo 9: *Running the Filter program on* **(a)** *produces the image in* **(b)**.

Photo 10: *An image digitized on the ImageWise digitizer/transmitter is displayed on the GT180* **(a)**. *This image is magnified 2 times* **(b)**, *4 times* **(c)**, *and 8 times* **(d)**.

Image Processing Routines

The following is a list of the programs described in this chapter, plus some additional image processing software you might find interesting. File specs between angle brackets are optional. Results will be stored in the first file spec if the target file spec is omitted. An *n* indicates a numeric value. These programs are available from the Circuit Cellar BBS, BIX, and BYTEnet.

ADD.PAS pic1 pic2 <pic3>
Function: pic3 = pic1 + pic2.

COMPARE.PAS pic1 *n* <pic2>
Function: pic2 = pic1 if pixel >= *n*
= 0 otherwise.

COMPRESS.PAS pic1 <pic2>
Function: pic2 is the run-length-encoded version of pic1. Compressed files cannot be used by the other programs.

COUNT.PAS pic1 *n*
Function: DOS ERRORLEVEL variable = number of pixels >= *n*.

DUMPER.PAS pic1
Function: Produces formatted print dump of pic1 for hand analysis; use redirection to send output to a disk file.

EDGE.PAS pic1 <pic2>
Function: pic2 contains edge-intensity information from pic1.

EXPAND.PAS pic1 <pic2>
Function: pic2 is the non-RLE version of pic1. Expanded files are required by the other programs.

FASTDOG.PAS *n*1 *n*2
Function: Watches a scene, reports an intruder when *n*2 changed pixels exceed *n*1.

FILTER.PAS pic1 <pic2>
Function: pic2 is a low-pass filtered version of pic1.

GRAB.PAS pic1 /*n* /c
Function: Accepts picture from transmitter board and stores the expanded data in pic1. Switch /*n* prevents showing the picture on the receiver; switch /c stores the image without expanding it.

HISTO.PAS pic1
Function: Displays a pixel-intensity histogram for pic1.

INVERT.PAS pic1 <pic2>
Function: pic2 = 63 − pic1.

MASK.PAS pic1 pic2 <pic3>
Function: pic3 = pic1 if pic2 > 0
= 0 otherwise.

MULTIPLY.PAS pic1 *n* <pic2>
Function: pic2 = pic1 * *n*.

SHOW.PAS pic1
Function: Sends pic1 to the display board.

SUBTRACT.PAS pic1 pic2 <pic3>
Function: pic3 = pic1 − pic2.

THRESH.PAS pic1 *n* <pic2>
Function: pic2 = pic1 if pic1 >= *n*
= 0 otherwise.

Listing 1: *The WATCHDOG.BAT program.*

```
ECHO off
REM Syntax is:
REM   WATCHDOG brightness pixels
REM brightness is COUNT's threshold level
REM pixels is # of pixels >= brightness, in units of 100
REM   WATCHDOG 10 4
REM will alarm when 400 pixels or more are brighter than 10
REM runs best with image files on a RAM disk!
ECHO Make sure serial cable is connected to transmitter
PAUSE
:newref
GRAB ref /n
:newtest
GRAB test /n
COMPARE ref test deltas
COUNT deltas %1
IF errorlevel %2 goto gotcha
ECHO no intruder so far...
ERASE ref
RENAME test ref
GOTO newtest
:gotcha
ECHO --- Intruder alert!!! ---
ECHO Switch serial cable to receiver for display
PAUSE
SHOW test
ECHO Switch serial cable to transmitter
PAUSE
goto newref
```

examine one picture every 30 seconds or so, which might be adequate for most purposes. (If you need more speed, I have a faster program called FASTDOG.BAT in the downloadable software.)

The Count program returns the number of qualifying pixels (divided by 100) in the DOS ERRORLEVEL variable to let the IF statement decide whether an intruder is present. You should replace the ECHO statement with a program that does something useful, like turn on the lights, sound a loud alarm, or whatever you choose.

You'll need to do some experimentation to pick the best values for the threshold and count levels. Count can't tell the difference between one large change and several smaller ones, nor can it decide what the change "looks like." You'll have to mask areas of the picture or pick compromise values that don't generate too many false alarms but still never miss a real intruder. Put on your skulking suit and try to fool it.

Hardware Image Processing

While we generally think of image processing solely as software-dependent tasks, many of the newest graphics-display chips incorporate some of these functions in hardware. Most prominent among such features is the hardware zoom or image-expansion function. The GT180 color graphics board I presented in the November 1986 Circuit Cellar has a hardware zoom that can expand an image up to 16 times.

Photo 10a shows a standard-resolution 256- by 244-pixel picture (no, it's not me this time) digitized on the digitizer/transmitter board and displayed in 16-level gray scale on a GT180 high-resolution graphics-display board. Because the GT180 has a resolution of 640 by 480, the lower-resolution digitized picture fills only the top left corner, but it expands to fill and then overflow the screen as it is zoomed. Photo 10b is 2 times magnification, photo 10c is 4 times, and photo 10d is 8 times the original image.

Conclusions

As anyone who owns a TV camera can attest, video is fascinating. Until now, small computer users haven't been able to work with pictures of the real world because the video hardware was frightfully expensive. With the hardware and software I've provided, you can take digital pictures, enhance them to pick out interesting objects, and save them for later. I'm sure you'll find many more ways of tweaking the video.

The complete source code for all the programs described in the text box on page 28 is available from the Circuit Cellar BBS.

Special thanks to Ed Nisley for his expert collaboration on this project.

See page 447 for a listing of parts and products available from Circuit Cellar Inc.

31

BUILD AN INFRARED REMOTE CONTROLLER

A custom hand-held infrared transmitter and receiver

In the last few years, I have installed a lot of automatic features in my home and the Circuit Cellar. Besides the sophisticated alarm system, computer-controlled wood stove, and perimeter lighting system (see "Living in a Sensible Environment" in the July 1985 BYTE), the Home Run Control System (HCS) has significantly increased the convenience of living in my house (see "Build the Home Run Control System" in the April through June 1985 BYTEs). I have the usual remote-controlled stereos, TVs, FM radios, etc., but it is the small things like lights that automatically go on and off as I walk through the house and a voice-synthesized central monitoring system that I can call and hear a verbal status report when I'm away that have forever erased any pleasure derived from a manually oriented existence.

Just when I thought I had come to grips with electronic living, I ran into two new problems: automated living user intervention anarchy (ALUIA) and hand-held infrared controller overpopulation (HIROP). Generally speaking, these are high-tech diseases common among gadget-happy technocrats and overzealous inventor-authors who are insane enough to try to install all their designs into one house.

ALUIA is like gridlock. So many things are controlled within an environment that the only way to activate something is through the control system (if you can remember which system is controlling what device at a particular time). Unless you physically intervene, the control sequence will remain as set, and you must live with the consequences. Perhaps you have a timer controlling a series of outside lights, but one night you want them on at different times. Do you reset the automatic timer to a new cycle, override the automatic system manually at each setting, or disconnect the automatic system? The problem with many automatic systems is that they make little provision for unpredicted and unprogrammed user behavior.

In the case of the HCS, I tried to provide for such possibilities. Rather than just a timer that turns lights on and off, for example, the HCS can turn a light on in a room as long as it senses your presence via a motion detector connected to one of its hard-wired inputs. Such a control system adapts itself to your behavior—up to a certain point! It cannot control what it cannot sense, nor can it be expected to accommodate all your wishes through passive sensors. If you want the stereo to go on when you walk into the room, it could be coordinated with the automatic lighting, but would you want it to come on even when you merely pass through the room?

Some automatic control functions are more applicable to lighting than to stereo equipment. Is the only alternative manual control if the system is not fully automatic? Not if it is designed to allow independent user input in addition to automatic control.

This control gridlock comes about when you and the system have different ideas as to what should be happening. Sometimes the only alternative is to design a system with shared control—one that executes a preprogrammed sequence but accommodates itself to selective manual in-

Photo 1: *Enclosed IRCOMM transmitter and receiver units.*

tervention. By allowing you a means to override or direct the activities of the control system, the system retains the flexibility of independent control in the absence of user directives and lessens its potential for becoming a frustrating obstacle as your needs change.

Perhaps you want the stereo to automatically come on when you walk in the room any time between 5 and 6 p.m., but not at any other time unless you physically turn it on. You could easily program the HCS to use the motion sensor in the room to activate the stereo and lock out execution of the order except between 5 and 6 p.m. But how do you manually turn the stereo on yet still advise the control system of your action? The system needs to know that the stereo should be ON, even if it has been turned on manually. Any well-designed control system like the HCS

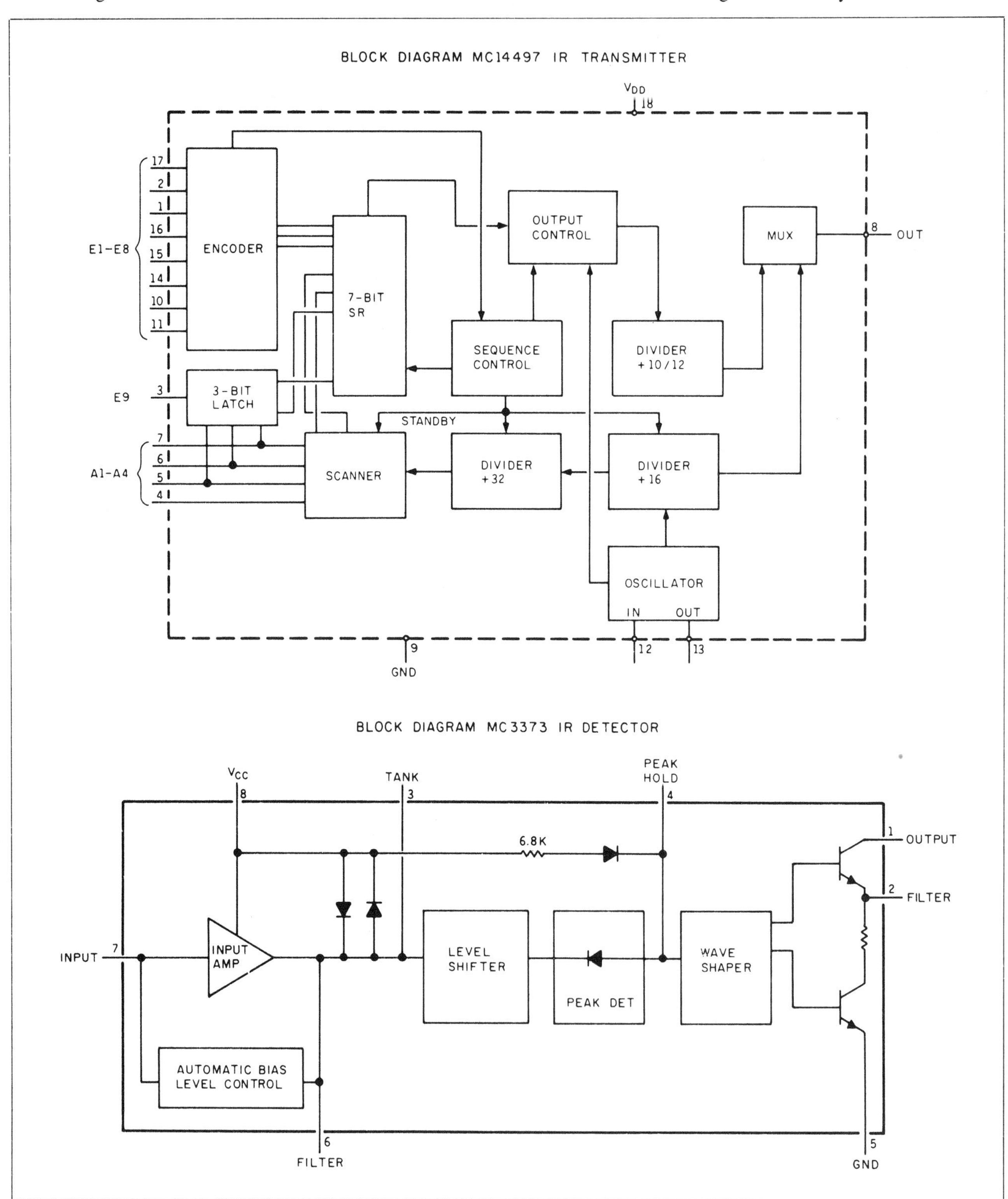

Figure 1: *Block diagrams of the chips used in the IRCOMM.*

periodically "refreshes" or retransmits the current status (ON or OFF) to all controlled devices, reducing concern about false activation/deactivation of the BSR X-10 remote-control modules by transients. The best way to coordinate such automatic/manual control is to do everything through the control system.

I don't expect you to run down to your Circuit Cellar and type a manual override every time you want to turn the stereo on. Since we are intending a direct and specific control action, "turn the stereo on," it only makes sense to use a direct input to the control system similar to a motion detector. A simple push button connected to one of the HCS inputs can turn on the stereo. When you press it, the control system is actually turning the stereo on, and it will coordinate this with its other functions.

While ultimately successful, if you follow this tack, you will soon find that it falls short as an easily implementable procedure. The interest in using a BSR X-10 wireless remote-control device with the HCS in the first place was to eliminate the need for wires. The more convenient alternative would be a wireless remote-control device that communicates specific commands from you to the HCS or other computerized control system. A typical example of such a device is the infrared remote control used with most TV sets.

This chapter's project is the design and construction of a custom hand-held infrared transmitter and receiver, called the IRCOMM (see photo 1). The transmitter circuit can be constructed as a small inexpensive hand-held controller or expanded to implement a 62-key wireless keyboard. The receiver is equally uncomplicated and intended to provide a convenient link between the user and the home control system.

An unfortunate side effect of creating the IRCOMM is that it adds one more IR remote-control unit to the pile you probably have and contributes to HIROP, as I stated earlier. Therefore, in Chap. 32. I will make amends for contributing to IR remotes on every table and chair with my own form of population control.

Generally Speaking

My primary consideration in the design of IRCOMM was to use it with the HCS. Therefore, hundreds of remote function keys and a 20-mile effective range were of little importance as design criteria. Much like remote controls for TVs or VCRs, the IRCOMM controller needed only to be short-range and command basic functions like "stereo system power ON," "surround sound system power ON," "projection TV ON," "mood lights ON," "entertainment system all power OFF," "room-to-room sound tracking GO," etc.

When we speak of the functions that a remote controller performs, we are actually describing what the device being controlled (a TV, for example) does as a result of your pressing a key on the remote. The remote control is nothing other than a wireless keyboard. When you press a key, a stream of data is transmitted either as an ultrasonic, radio-frequency, or infrared signal. Present-day consumer electronic devices primarily use infrared signaling because of its low cost and limited interference with other remote-controlled appliances (like accidentally turning on the TV in the next apartment).

Consumer electronic devices use infrared signaling because of its low cost and limited interference with other remote-controlled appliances.

Infrared controllers generally use pulse position modulation (PPM) or pulse code modulation (PCM). The actual technique used is significant only to a person designing a receiver/decoder.

In designing the IRCOMM, I tried to keep both my needs and the intelligence

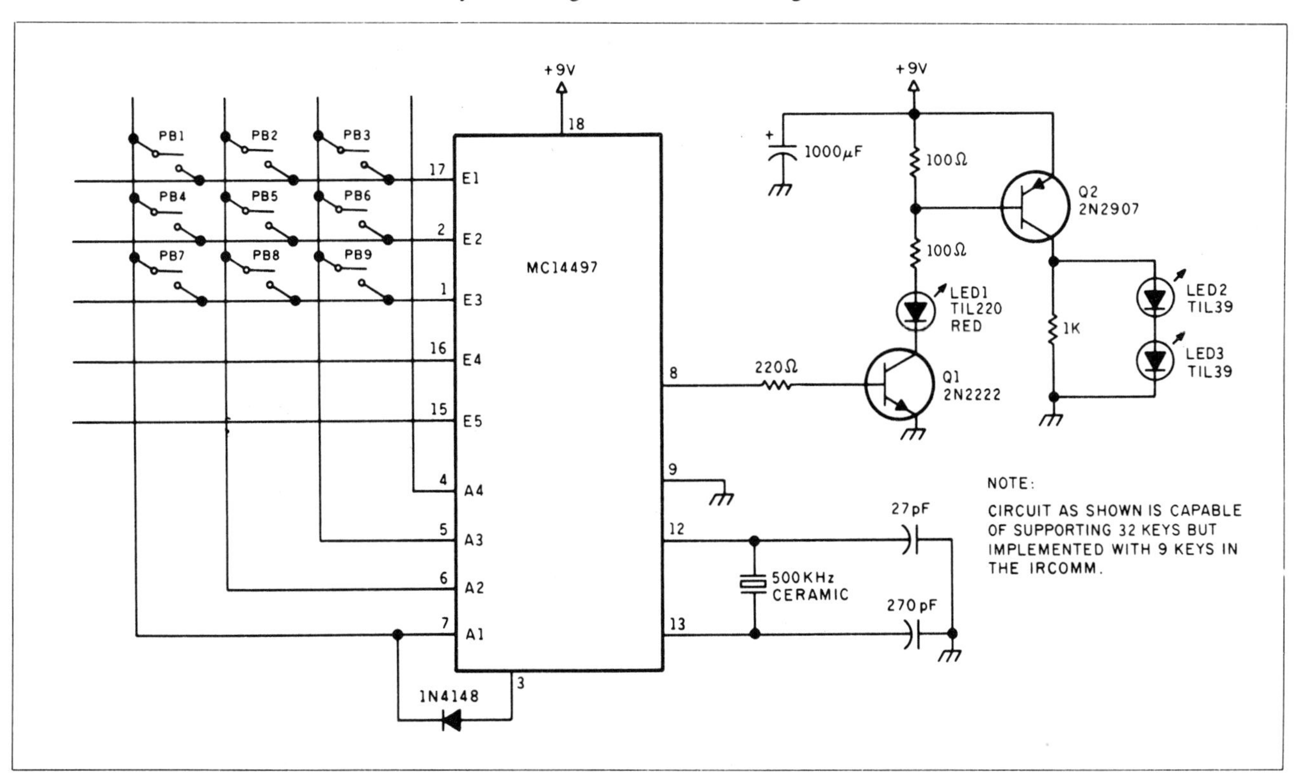

Figure 2: *Schematic of the IRCOMM hand-held transmitter.*

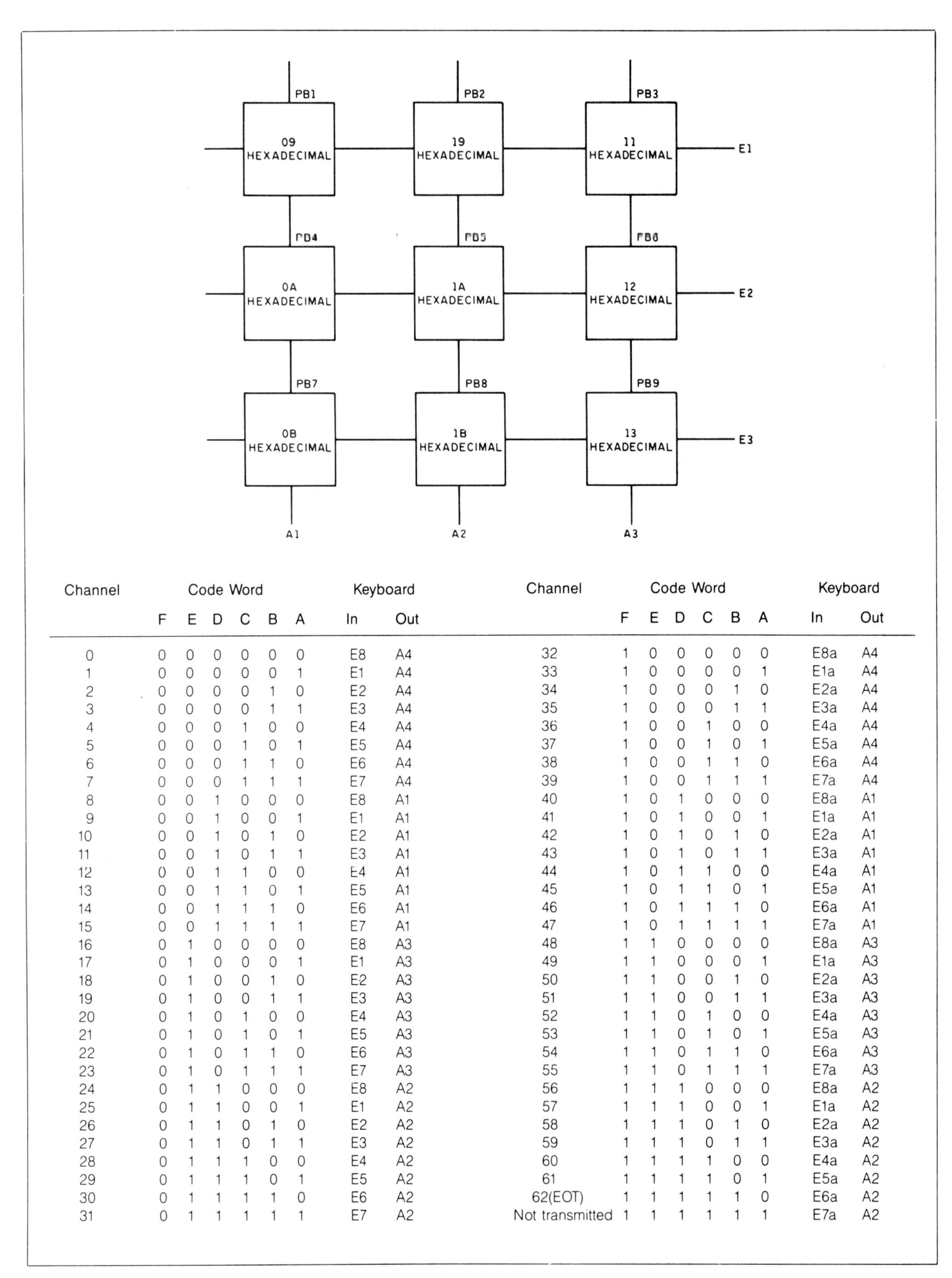

Channel	Code Word						Keyboard	
	F	E	D	C	B	A	In	Out
0	0	0	0	0	0	0	E8	A4
1	0	0	0	0	0	1	E1	A4
2	0	0	0	0	1	0	E2	A4
3	0	0	0	0	1	1	E3	A4
4	0	0	0	1	0	0	E4	A4
5	0	0	0	1	0	1	E5	A4
6	0	0	0	1	1	0	E6	A4
7	0	0	0	1	1	1	E7	A4
8	0	0	1	0	0	0	E8	A1
9	0	0	1	0	0	1	E1	A1
10	0	0	1	0	1	0	E2	A1
11	0	0	1	0	1	1	E3	A1
12	0	0	1	1	0	0	E4	A1
13	0	0	1	1	0	1	E5	A1
14	0	0	1	1	1	0	E6	A1
15	0	0	1	1	1	1	E7	A1
16	0	1	0	0	0	0	E8	A3
17	0	1	0	0	0	1	E1	A3
18	0	1	0	0	1	0	E2	A3
19	0	1	0	0	1	1	E3	A3
20	0	1	0	1	0	0	E4	A3
21	0	1	0	1	0	1	E5	A3
22	0	1	0	1	1	0	E6	A3
23	0	1	0	1	1	1	E7	A3
24	0	1	1	0	0	0	E8	A2
25	0	1	1	0	0	1	E1	A2
26	0	1	1	0	1	0	E2	A2
27	0	1	1	0	1	1	E3	A2
28	0	1	1	1	0	0	E4	A2
29	0	1	1	1	0	1	E5	A2
30	0	1	1	1	1	0	E6	A2
31	0	1	1	1	1	1	E7	A2
32	1	0	0	0	0	0	E8a	A4
33	1	0	0	0	0	1	E1a	A4
34	1	0	0	0	1	0	E2a	A4
35	1	0	0	0	1	1	E3a	A4
36	1	0	0	1	0	0	E4a	A4
37	1	0	0	1	0	1	E5a	A4
38	1	0	0	1	1	0	E6a	A4
39	1	0	0	1	1	1	E7a	A4
40	1	0	1	0	0	0	E8a	A1
41	1	0	1	0	0	1	E1a	A1
42	1	0	1	0	1	0	E2a	A1
43	1	0	1	0	1	1	E3a	A1
44	1	0	1	1	0	0	E4a	A1
45	1	0	1	1	0	1	E5a	A1
46	1	0	1	1	1	0	E6a	A1
47	1	0	1	1	1	1	E7a	A1
48	1	1	0	0	0	0	E8a	A3
49	1	1	0	0	0	1	E1a	A3
50	1	1	0	0	1	0	E2a	A3
51	1	1	0	0	1	1	E3a	A3
52	1	1	0	1	0	0	E4a	A3
53	1	1	0	1	0	1	E5a	A3
54	1	1	0	1	1	0	E6a	A3
55	1	1	0	1	1	1	E7a	A3
56	1	1	1	0	0	0	E8a	A2
57	1	1	1	0	0	1	E1a	A2
58	1	1	1	0	1	0	E2a	A2
59	1	1	1	0	1	1	E3a	A2
60	1	1	1	1	0	0	E4a	A2
61	1	1	1	1	0	1	E5a	A2
62(EOT)	1	1	1	1	1	0	E6a	A2
Not transmitted	1	1	1	1	1	1	E7a	A2

Figure 3: *Codes generated by the MC14497 and the IRCOMM keypad.*

of the BYTE readership in mind. What I am presenting should be considered as a model and a sample application of infrared remote control and not as the only way to implement it. The usual article approach to this subject is to buy off-the-shelf remote-control chip sets designed for the TV industry. Such an approach is valid, but it better serves the author than the user. TV remote-control chips are designed for a specific application, and their receivers are often bus or multiplexed output devices. Additional glue logic is frequently added to provide 1-of-24 signal lines or 4- to 16-bit decoded outputs. Whatever the decoding technique employed, the resulting receiver outputs must still be read through a parallel input port. Ever try to find a parallel input port on your IBM PC?

The Circuit Cellar IRCOMM

We aren't intending to use the IRCOMM to mimic a TV remote control, so why bother to spend the money or carry the overhead of decoding circuitry intended for TVs? A better alternative is to merely condition the incoming signal and allow the control computer to decode the signal. Motorola manufactures a pair of general-purpose CMOS IR remote-control chips, the MC14497 and the MC3373, that fit the bill exactly (see figure 1).

Figure 2 is the schematic of the hand-held IRCOMM transmitter. As I have it shown, the MC14497 is hard-wired for use with up to 32 keys, AM modulation, and a logic 1 start bit. The MC14497 is a CMOS biphase PCM remote-control transmitter chip in an 18-pin package. In standby mode it draws a mere 10 microamperes and operates anywhere within a range of 4 to 10 volts. Transmission and internal timing are controlled with a 500-kilohertz ceramic resonator.

The basic configuration of the MC14497 will support 32 keys (described as channel 0 through channel 31). With two additional diodes and switches between pins 3 to 6 and 3 to 5, the capacity can be increased to a maximum of 62 keys. Either option is far more than I needed. Perhaps more as a result of the plastic box I had on hand than any calculated requirement, I ended up with 9 keys (PB1–PB9) connected as shown. If you want more keys, simply add more push buttons at the cross points of the Ex and Ax lines.

The IRCOMM has both infrared and visible outputs when it transmits. A two-transistor driver circuit simultaneously pulses visible LED1 and infrared LEDs, LED2 and LED3. The MC14497 transmits either in FSK (frequency-shift keying) or AM mode. We are restricted to AM transmission because of the receiver I used, so that is all I will address.

Figure 3 shows the layout and hexadecimal codes for the 9-key IRCOMM. Since I arbitrarily selected the matrix lines they do not correspond to channels 0, 1, 2, etc. (Since the receiver cares only *which* nine channels it has to identify—not that they be sequentially ordered—keys can be placed anywhere in the matrix.)

Figure 4 illustrates what the PCM coding looks like as it is transmitted. Biphase PCM is relatively easy to read once you get the hang of it. The most important part in reading PCM is keeping track of the bit times and noticing at what point the 0-to-1 or 1-to-0 logic transition occurs during the bit time. If the pulse burst is sent/received during the first half bit time, the bit is a logic 1. Conversely, if the pulse occurs during the second half bit time, the bit is a logic 0.

When a key is pressed, the transmitter sends an AGC (automatic gain control) burst lasting a half bit time, a start bit (logic 1), and a 6-bit PCM data word. The purpose of the AGC pulse preceding the PCM data is to set up the AGC loop in the receiver in time for the start bit. The 6 data bits are designated as A (least significant bit) through F (most significant bit) and are shifted in LSB first. While each bit is represented as a logic 0 or 1 level pulse having a duration of 0.5 or 1 millisecond, the actual output of the transmitter is a 41.66-kHz pulse burst for the duration of any logic 1 level. Only after the data is conditioned by the receiver will it appear as discrete logic levels.

The IRCOMM repeatedly transmits the same code as long as the key is pressed. When the key is released, a channel 62 EOT (end of transmission) code is automatically sent. Channel 63 is not used.

The fact that the coding is simplified allows some license to be taken with the receiver circuitry.

The IRCOMM Receiver

As you can see, the IRCOMM chip transmits an easily recognizable and repeatable code. Many TV-style IR remote chips send the pulse burst once or send it once and repeatedly transmit the EOT signal. It is easy to read the code from the IRCOMM with an oscilloscope. Try reading your TV remote manually.

The fact that the coding is simplified allows some license to be taken with the receiver circuitry. Rather than the expensive LSI hardware often required in pulse position coding, only the bit timing is relevant. To acquire this data, we merely convert the 41.66-kHz pulse bursts to TTL logic levels through an envelope detector and apply this signal to a computer that monitors the bit timing. I could decode the signal all in the hardware, but if the end result is still to connect it to a computer,

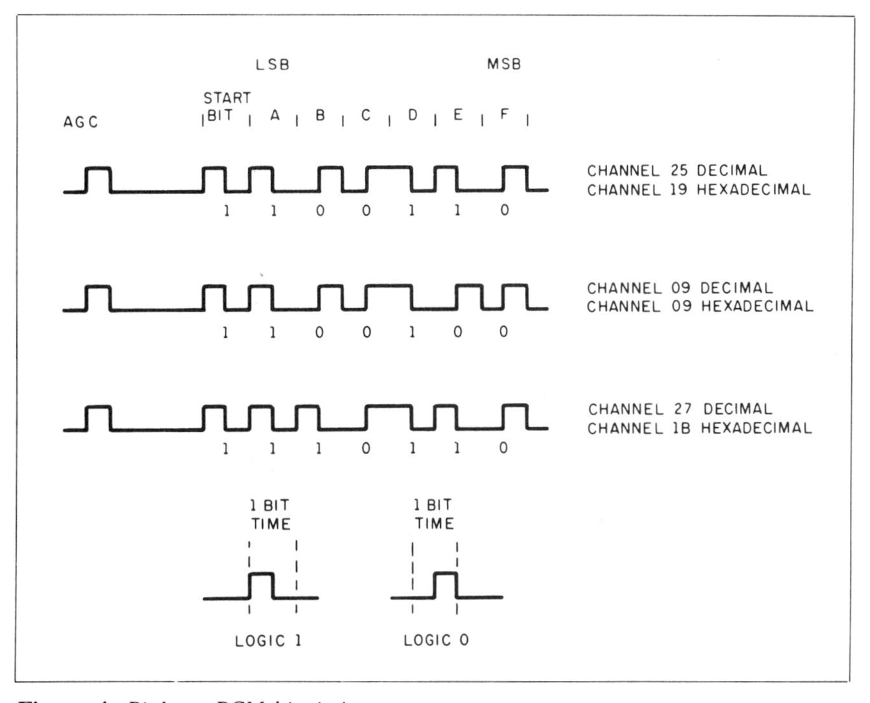

Figure 4: *Biphase PCM bit timing.*

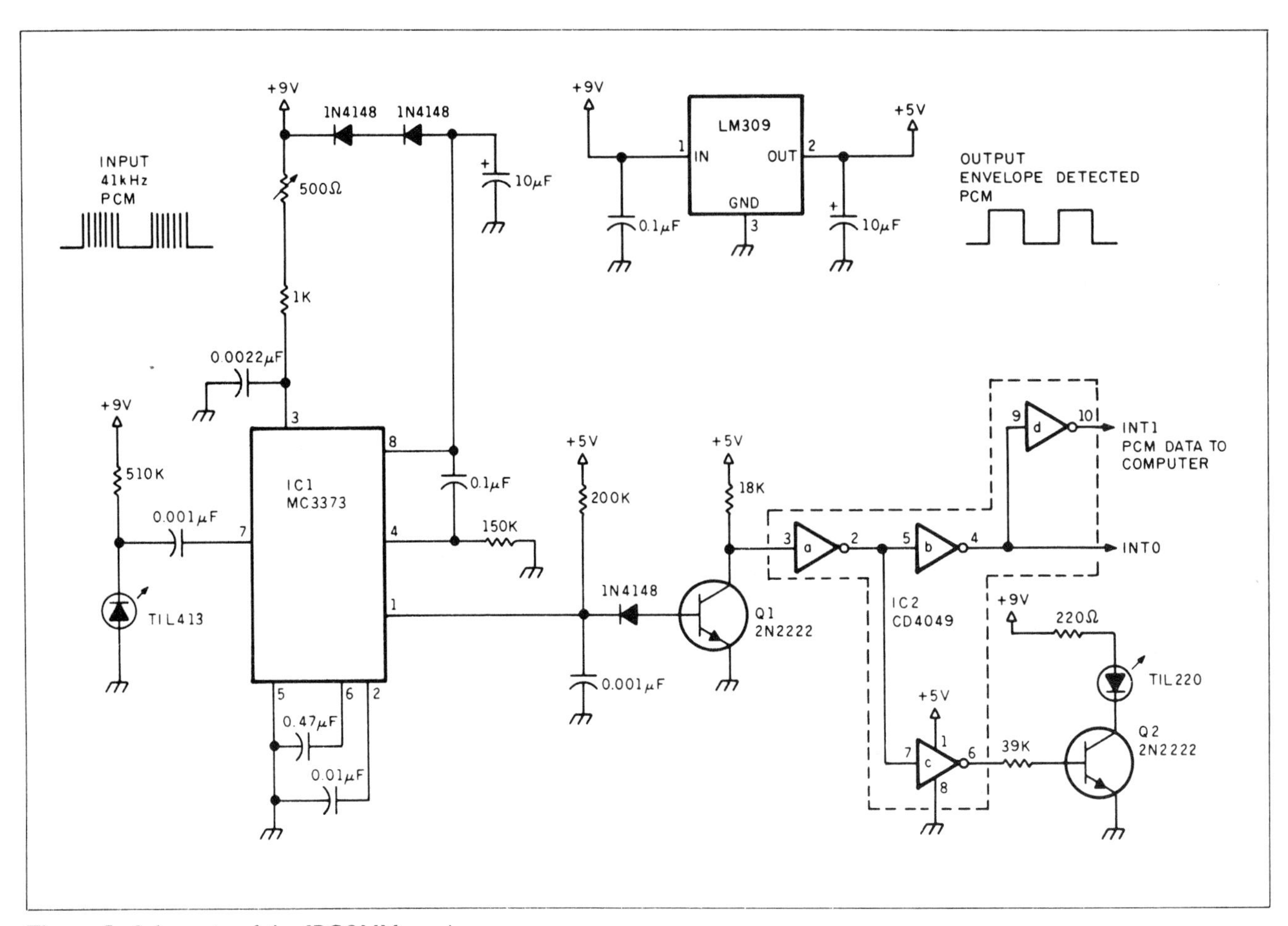

Figure 5: *Schematic of the IRCOMM receiver.*

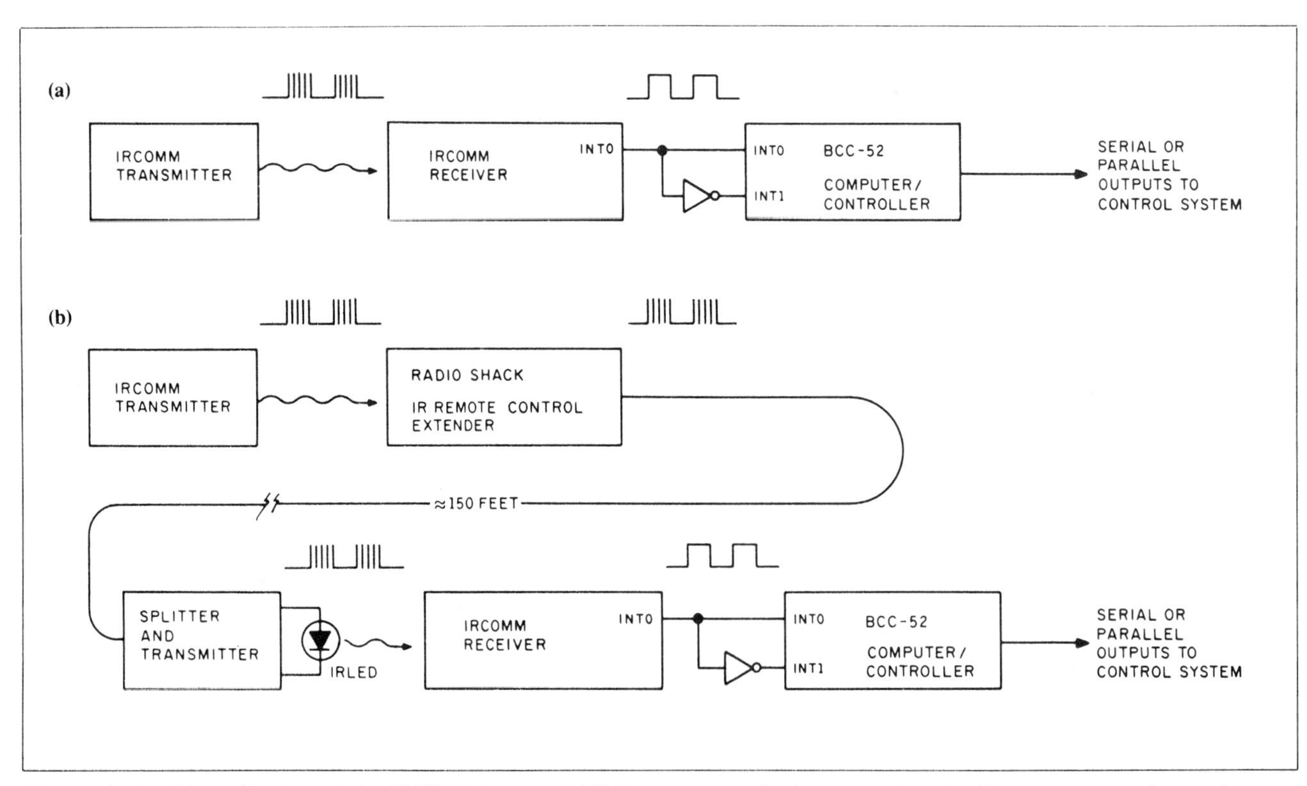

Figure 6: *(a) Direct hookup of the IRCOMM to the BCC-52 computer; (b) hookup using the IR remote-control extender.*

why not use the computer to decode the raw transmission in the first place? I never thought I'd be advocating a software solution, but I support reality.

The Motorola MC3373 wideband amplifier-detector chip is designed for use with infrared pulse-burst transmissions. The entire receiver circuit, shown in figure 5, is a two-chip envelope detector and TTL level shifter. An AC-coupled photodiode receives the infrared pulses from the IRCOMM transmitter and amplifies them. When an infrared signal of approximately 40 kHz is perceived, the output goes low. Q1 inverts this signal and applies it to a series of 4049 CMOS inverters that are capable of driving the LSTTL (low-power Schottky transistor-transistor logic) input load of the computer and lighting an additional visible LED so that you can see that data is being received.

Decoding the PCM in Software

As I mentioned earlier, my intended application of the IRCOMM was to add remote-control features to my home control system by using it to trigger direct inputs to the HCS. Of course, the HCS was not designed with facility for IR remote control, but it does have 16 parallel input lines that can be used to trigger events. To use the IRCOMM with the parallel input of the HCS, however, another computer must be interposed between them. This special-function computer translates the IRCOMM receiver's PCM output into 9 (or 32 if you used that many keys) parallel signal lines that are attached to the HCS. I chose the BCC-52 computer/controller for this task (see "Build the BASIC-52 Computer/Controller" in the August 1985 BYTE). Figure 6a illustrates a block diagram of the connection.

The BCC-52 is programmable in BASIC or 8051 assembly language. To achieve the processing speed necessary to analyze PCM bit times, we must use an assembly language routine. The flowchart of the conversion process is given in figure 7, and a complete source listing of the necessary routine is provided in listing 1.

I used the BCC-52 because it was convenient and cost-effective. Since it is bus-compatible with a variety of A/D, display, and power-control peripherals that have also been presented as Circuit Cellar projects, this suggests a far more powerful future application for the IRCOMM remote control. The IRCOMM and the computer that decodes the PCM signal are

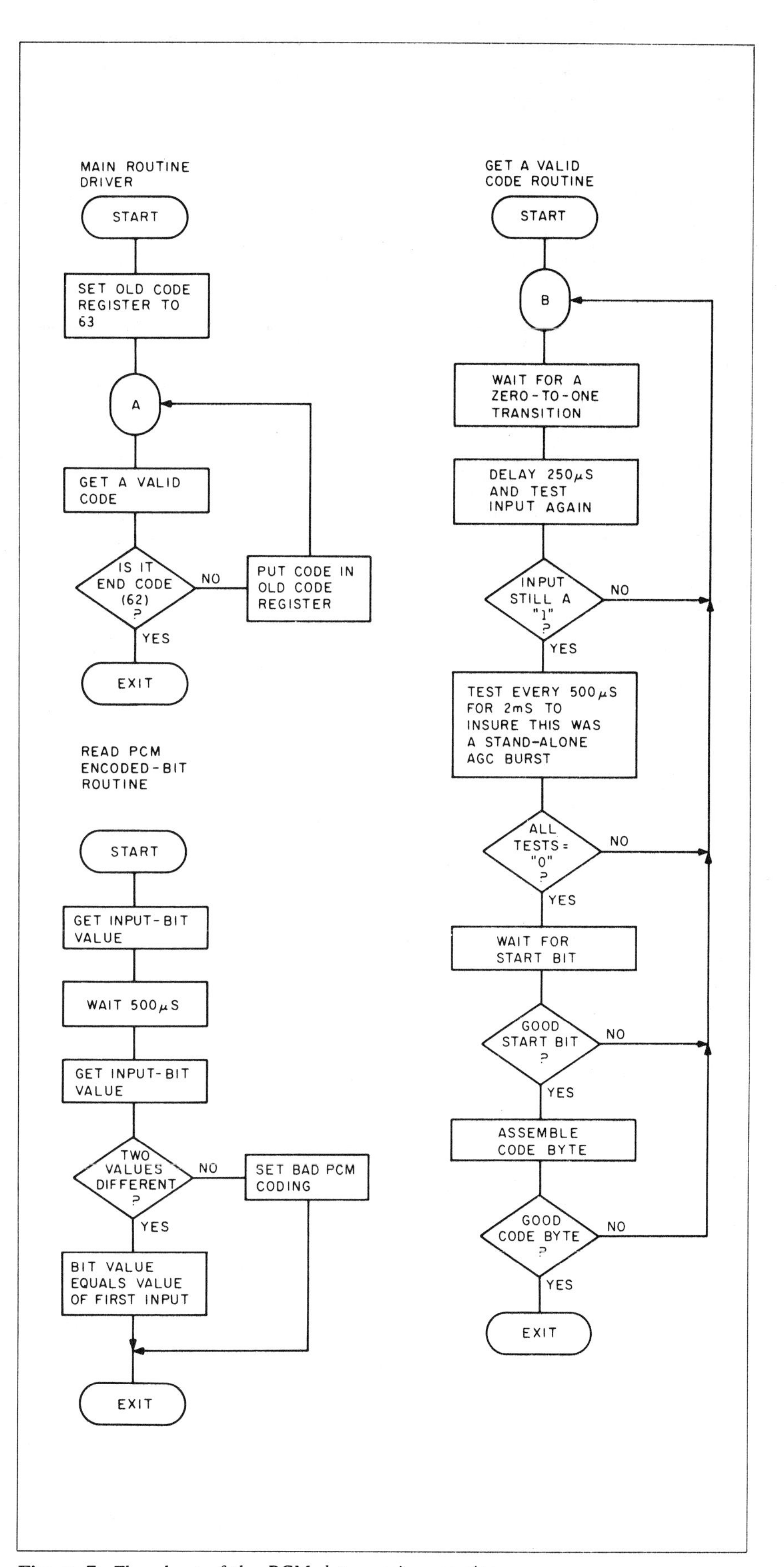

Figure 7: *Flowchart of the PCM data-receiver routine.*

separate, however. I chose to use the BCC-52 so that the complete flow diagram and source code could be presented as an example. It could just as easily be connected to an input bit on the parallel printer port of an IBM PC or other computer. If anyone implements the code for another computer, please upload the routine to my BBS at (203) 871-1988 so that I might share it with other readers.

The 100-byte PCM decoding routine is located at 4000 hexadecimal in an EPROM. The output of the IRCOMM receiver is attached to the INT0 input line of the BCC-52.

There are two entry points to the decoder routine: 4000 and 4011 hexadecimal. When you execute a CALL4000H, the software reads the incoming IR data but waits for a channel 62 EOT code before returning to BASIC. The code for the key that was pressed is in memory location 1F hexadecimal. If you execute a CALL4011H instead, the software returns immediately after it receives the first valid PCM signal. This code is stored in location 19 hexadecimal. A three-line BASIC program is all that is required to print the received code using either of these calls.

Print the hexadecimal key pressed code after the EOT is received:

```
10 CALL4000H
20 PH0. DBY(1FH)
30 GOTO 10
```

Print the hexadecimal key pressed code as it is received:

```
10 CALL4011H
20 PH0. DBY(19H)
30 GOTO 10
```

Of course, it would hardly be worth your time to consider dedicating a whole computer like the BCC-52 just to scan one input bit. A far superior way of connecting the IRCOMM is to use the PCM data to trigger an interrupt on the BCC-52. This is accomplished by inverting the PCM data and connecting this signal to the INT1 input in addition to the connection already made. When the IRCOMM key is pressed, the AGC pulse will trigger an interrupt and call the decoder routine. Using the interrupt goto (ONEX1) and the GET command, the keyboard can be scanned on the fly without waiting at an INPUT statement, and fast BASIC programs with input provided either through the keyboard or IRCOMM can be written. The following BASIC program demonstrates the combined use of the IRCOMM and the keyboard. It prints any key pressed, either on the keyboard or the terminal, as it's entered:

Listing 1: *Assembly language listing for the PCM data-decoder routine.*

```
;
; LEDPCM - PCM DATA DECODER ROUTINE FOR READING IR LED
;          CODES USING A BCC-52 COMPUTER/CONTROLLER
;          - WRITTEN BY WILLIAM D. CURLEW
;
; COPYRIGHT CIARCIA'S CIRCUIT CELLAR 1986
;
; READ THE SERIAL IR RECEIVER INPUT. REMEMBER THE LAST CODE
; UNTIL AN END CODE (DECIMAL 62) IS RECEIVED. RETURN THE
; REMEMBERED CODE IN REGISTER 1FH. A RETURNED CODE OF 63
; DECIMAL MEANS AN END CODE WAS THE FIRST THING DETECTED.
;
; ASSUMES AMPLITUDE MODULATION MODE IS BEING USED.
;
         ORG 4000H
LEDPCM   EQU $
         MOV OLDCODE,#INVALID   ; SET UP OLD CODE AS 63
CODELOOP EQU $
         CALL GETCODE           ; GET A CODE
         MOV A,CODE             ; PUT CODE IN A REG
         CJNE A,#ENDCODE,STORE  ; IF NOT END CODE, STORE & WAIT
         RET                    ; RETURN TO CALLER
STORE    EQU $
         MOV OLDCODE,CODE       ; SAVE CODE AS OLDCODE
         JMP CODELOOP           ; AND WAIT FOR THE NEXT CODE
;
; END OF CODE LOOP ROUTINE
;
;
; THIS ROUTINE RECEIVES THE NEXT VALID CODE.
; INVALID CODE GROUPS ARE IGNORED
;
GETCODE  EQU $
WAITAGC  EQU $                  ; WAIT FOR AGC BURST
         CALL SAMPLE            ; GET THE VALUE AT THE INPUT PORT
         CJNE A,#ZERO,WAITAGC   ; WAIT UNTIL IT IS A 0
ONEWAIT  EQU $
         CALL SAMPLE            ; GET ANOTHER VALUE
         CJNE A,#ONE,ONEWAIT    ; IF NOT ONE, WAIT FOR A 1
         MOV B,#1               ; SET DELAY TO 250 US
         CALL WAIT              ; AND WAIT FOR THAT TIME
         CALL SAMPLE            ; RE-TEST INPUT BIT
         CJNE A,#ONE,WAITAGC    ; IF NOT 1, FALSE AGC BURST.
         MOV B,#3               ; SET UP FOR 3 * 500 US
AFTERAGC EQU $
         PUSH B                 ; SAVE ON STACK
         MOV B,#2               ; SET DELAY TO 500 US
         CALL WAIT              ; AND WAIT FOR THAT TIME
         CALL SAMPLE            ; CHECK INPUT VALUE
         POP B                  ; RECOVER COUNT VALUE
         CJNE A,#ZERO,WAITAGC   ; IF 1, FALSE WAIT AFTER AGC
         DJNZ B,AFTERAGC        ; CHECK IF DONE WITH WAIT
;
; RECEIVE START BIT
;
STARTBIT EQU $
         CALL SAMPLE            ; GET INPUT BIT
         CJNE A,#ONE,STARTBIT   ; IF NOT 1, TRY AGAIN
         MOV B,#1               ; SET DELAY TO 250 US
         CALL WAIT              ; AND WAIT FOR THAT TIME
         CALL READBIT           ; DO PCM BIT INPUT
         JC WAITAGC             ; IF INVALID PCM, START OVER
         MOV A,IN1              ; GET BIT VALUE
         CJNE A,#ONE,WAITAGC    ; IF NOT 1, BAD START BIT
;
; READ CODE BITS
```

```
;
READCODE EQU $
         MOV CODE,#00           ; RESET CODE BYTE
         MOV B,#6               ; DO 6 BIT INPUTS
READLOOP EQU $
         PUSH B                 ; SAVE B VALUE
         MOV B,#2               ; WAIT 500 US
         CALL WAIT              ; AND WAIT FOR THAT TIME
         CALL READBIT           ; GET PCM BIT VALUE
         JC BADCODE             ; IF INVALID PCM, THROW AWAY CODE
         CLR C                  ; CLEAR CARRY FLAG
         MOV A,IN1              ; GET PCM BIT VALUE
         CJNE A,#ZERO,LOAD1     ; IF BIT=1 THEN DO 1
         JMP ROTATE             ; ELSE JUST ROTATE
LOAD1    EQU $
         SETB C                 ; SET CARRY FLAG
ROTATE   EQU $
         MOV A,CODE             ; GET CODE BYTE
         RRC A                  ; ROTATE XTER RIGHT THROUGH CARRY
         MOV CODE,A             ; STORE IN CODE REG
         POP B                  ; RECOVER BIT COUNT
         DJNZ B,READLOOP        ; IF NOT ALL BITS, DO AGAIN
         RR A                   ; ROTATE TWO MORE
         RR A                   ; BITS RIGHT
         MOV CODE,A             ; STORE IN CODE REG
         CLR C                  ; CLEAR CARRY FLAG
         JMP CODEEND            ; AND EXIT
BADCODE  EQU $
         POP B                  ; RECOVER B REG FROM STACK
         SETB C                 ; BAD CODE INDICATOR
CODEEND  EQU $
         RET                    ; RETURN TO CALLER
;
; END OF CODEEND ROUTINE
;
;
; THIS ROUTINE RECEIVES A VALID PCM BIT
; CARRY IS SET IF THE PCM ENCODING IS NOT VALID
;
READBIT  EQU $
         CALL SAMPLE            ; GET INPUT VALUE
         MOV IN1,A              ; STORE IN IN1 REG
         MOV B,#2               ; WAIT ANOTHER 500 US
         CALL WAIT              ; AND WAIT FOR THAT TIME
         CALL SAMPLE            ; GET ANOTHER SAMPLE
         CJNE A,IN1,GOODPCM     ; IF 2 INPUTS <>, GOOD PCM CODE
BADPCM   EQU $
         SETB C                 ; SET CARRY FLAG (BAD PCM)
         JMP READEND            ; AND EXIT
GOODPCM  EQU $
         CLR C                  ; CLEAR CARRY FLAG
READEND  EQU $
         RET                    ; RETURN TO CALLER
;
; END OF READBIT ROUTINE
;
;
; THIS ROUTINE SAMPLES THE INPUT BIT
;
SAMPLE   EQU $
         MOV A,INPORT           ; READ BYTE AT PORT
         ANL A,#ONE             ; MASK OFF OUR BIT
         RET                    ; AND RETURN TO CALLER
;
;  END OF SAMPLE ROUTINE
;
```

Continued on page 354

```
 10 ONEX1 40
 20 A=GET : IF A<>0 THEN
    PRINT A
 30 GOTO 10
 40 PRINT"IRCOMM INTERRUPT
    RECEIVED" : GOSUB 100
 50 PRINT"INTERRUPT
    PROCESSED"
 60 RETI
100 CALL4000H
110 PHO. DBY(1FH)
120 RETURN
```

Extending the Capabilities

With most infrared remotes, the range is limited to about 25 to 30 feet. Extending the range beyond that involves more powerful transmitters and more sensitive receivers. While I was considering doing just that, I came across a Radio Shack product called the Video Remote Control Extender that seems to adequately solve the problem.

The Extender (catalog number 15-1289) is an IR repeater. As shown in figure 6b, it consists of an infrared receiver, amplifier, and transmitter. One end, located in a room where you might also use the IRCOMM, is the receiver and the amplifier. At the other end is a splitter box with an attached infrared LED. The splitter is connected to the receiver with antenna wire. When a pulse burst is received, it is amplified and conveyed through the antenna wire to the splitter box where it is retransmitted via that IRLED. It is easy to visualize the HCS mounted in the Circuit Cellar with the IRCOMM and BCC-52 next to it. The Extender would be upstairs in the entertainment room.

More than one Extender can be used (I have three on the IRCOMM). Since their outputs are IRLEDs aimed at the IRCOMM receiver, Extenders from other rooms can also be used with their IRLEDs mounted next to each other all aimed at the IRCOMM.

In Conclusion

I'm becoming as dependent now on the IRCOMM as I am on my automatic lighting. With it, the entertainment room comes alive in a programmed and orderly manner. Without it and the HCS, I bump into cold, dark walls of a house bathed in utter silence.

Indeed, I've solved the problem of coordinating the control of the multitude of electronic boxes in the entertainment room, but I've created an overabundance of IR remotes. The result is that I've created a terminal condition of HIROP.

In Chapter 32 we'll throw away the IRCOMM and the rest of your remotes and replace them with a single Circuit Cellar IR Master. The IR Master is a trainable remote controller that has the capacity to retain the

```
; THIS ROUTINE DELAYS 250 MICROSECONDS FOR EACH
; COUNT IN THE B REGISTER
;
WAIT      EQU $
          PUSH B              ; SAVE B REG COUNT
DELAY250 EQU $
          MOV B,#DELAYCNT     ; LOAD WITH DELAY COUNT VALUE
LOOP250  EQU $
          DJNZ B,LOOP250      ;  BURN UP CYCLES
          POP B               ; RECOVER COUNTS
          DJNZ B,WAIT         ; IF NOT DONE, WAIT MORE
          RET                 ; RETURN TO CALLER
;
; END OF WAIT ROUTINE
;
; SYSTEM EQUATES
;
; REGISTER ALIASES
;
IN1       EQU 18H             ; FIRST HALF PCM INPUT
CODE      EQU 19H             ; CURRENT CODE
OLDCODE   EQU 1FH             ; LAST CODE READ
;
; CONSTANTS
;
ZERO      EQU  0              ; ZERO BIT VALUE
ONE       EQU  4              ; PORT 3, BIT 2 MASK VALUE
DELAYCNT EQU  111             ; 250 US CONSTANT
ENDCODE   EQU  62             ; END CODE FROM TRANSMITTER
INVALID   EQU  63             ; INVALID CODE SETTING
;
; HARDWARE PORTS
;

INPORT    EQU P3              ; MEMORY MAPPED INPUT PORT
;
; END OF GLOBAL EQUATES
;
; END OF LEDPCM PROGRAM
;
ZZZZ      EQU $
          END
```

command functions of up to 16 (yes, 16!) independent IR remotes. It uses a scrolling LCD to indicate command function and control device. A single execute ("DO IT") key is the only command button.

Special thanks to Bill Curlew for his software expertise.

See page 447 for a listing of parts and products available from Circuit Cellar Inc.

There is an on-line Circuit Cellar bulletin board system that supports past and present projects. You are invited to call and exchange ideas and comments with other Circuit Cellar supporters. The 300/1200/2400-bps BBS is on-line 24 hours a day at (203) 871-1988.

32

BUILD A TRAINABLE INFRARED MASTER CONTROLLER

This device can control all your home entertainment equipment

First of all, this is not the second part of a two-part chapter. As you will come to understand, this project is the solution to a problem I aggravated by building the project in Chapter 31 (an infrared remote control for my home control system). Confused? Let me explain.

While people residing in warm climates tend toward Jacuzzis and hot tubs, some of us who live in colder climates prefer not to tempt fate and brave the elements for about six months of the year. Of course, I could succumb to the winter sports thing. You know, skiing, skating, snow this, and snow that, but it would be much too great a chore at this stage to reorient my sedentary lifestyle to enjoy northeast winters. I hibernate like most indigenous mammals and wait for the color outside the window to metamorphose from white to green.

About a year ago, I decided that holing up in the cellar for six months a year was antisocial. While the isolation proved beneficial in coming up with great projects for the summer and fall issues, I did find that by the time March rolled around, I looked very much like a bear that was leaving his cave, and I communicated just about as well.

In an attempt to improve the quality of winter life and break the cycle of hibernation, this last year I decided to spend some of the time aboveground (upstairs) in an environment that allowed me to observe the realities of my existence (through the windows) and absorb the cumulative knowledge of our culture (watch TV).

In layman's terms, I built a media room. Not just a TV den, mind you, but a room where I could be immersed in a synthesized environment so far from the ice and snow that six months seemed like overnight. Of course, this audiovisual experience was tastefully produced by massive amounts of electronic equipment.

The beautiful scene of the tropical island was accurately reproduced on a Kloss 2000 projection TV. You'd think you were sitting next to that tinkling waterfall as the music moves above and around you in complete surround sound. And when the warm breeze of island spring (actually the heat wafting from the seven amplifiers) is tumultuously interrupted by a hurricane faithfully reproduced with 2400 watts of Nakamichi audio power through a pair of B&W 808s (180 pounds each), two Speakerlab subwoofers, and 11 Canton surround speakers, you feel like the walls are about to explode. Sometimes it is good not to have neighbors.

Enough of warm breezes. I now had a new problem. In addition to all the audiovisual stuff, there were a couple of VCRs, an FM tuner, and a CD player. All this equipment required the 14 remote handheld controls shown in photo 1. Media rooms are a great idea, but you can't expect people to glue a dozen remotes on a long board. There had to be a better way.

IR Master Controller to the Rescue

This chapter's project, an infrared Master Controller that takes charge of all your gadgets, can prevent "controller clutter." It "learns" the infrared signals for each function and plays them back on command. It uses a six-button keypad to select the device and functions, shown on a two-line LCD, and a single button, Do It, to execute what's selected.

I am not the first person to design a trainable remote control. More than a year ago, I bought a similar device made by General Electric, called Control Central. This device could be trained to simulate the functions of four remotes.

Control Central and similar commercial units have two major shortcomings. First, all the acquisition, data-reduction, processing, and memory circuitry is contained in the single hand-held unit. Given the finite physical size of today's integrated circuitry, there is a limit to the capacity of such a device that allows it to still be cost-effective. Second, it is designed for use by a mass audience assumed to have a finite set of electronic devices. The buttons have predesignated nomenclature, so it is not user-programmable.

You can still train your GE controller to simulate the remote control for your CD player. The Mute button on the GE unit could be trained to be the Auto Repeat on your CD player remote, for example. Unfortunately, every time you want to repeat a CD, you'll have to remember to press Mute since there is no Repeat button on the GE.

I am not criticizing the GE Control Central. I am merely making a case for designing something different for a very vertical, gadget-happy, affluent audience: BYTE readers. Why tie a design to the lowest common denominator. Instead, yell "let them eat cake" and demand the remote to end all remotes: the Circuit Cellar Master Controller!

The shortcomings of the GE and other trainable remotes are the strengths of the Master Controller. Rather than attempt to contain all the necessary intelligence and processing circuitry, the Master Controller temporarily utilizes an external computer

Infrared controllers are not compatible because each manufacturer speaks in a different language.

as a user-programmable interface. Also, rather than having buttons with fixed-function nomenclature, the Master Controller incorporates a scrolling LCD to identify unit designations (devices) and functions (commands). Device designations like "Bedroom VCR" or "Nakamichi preamp" and commands like "CD repeat/all" or "slow motion" are used instead of remembering what the Mute button was supposed to do.

The Master Controller uses an IBM PC for training. After that, it is battery-operated and completely independent. The IBM PC is connected to the Master Controller via an RS-232 interface and is used to set up menus of devices (receivers, CD players, tape decks) and functions for each device (turn on, play forward, etc.). After a menu is downloaded to the Master Controller, each function is "taught" and tested. Next, the completed menu and synthesis data are then uploaded to the IBM PC and stored on disk (in case you want to load it into another Master Controller or add another device later without retraining all of them).

The Master Controller's IBM PC program can also combine sets of infrared signals once they are trained for their respective devices. I can now use a single Master Controller button to turn on the audio system, route the output to the living room, select the CD player, move to the third selection, and repeat it forever. Compared to other commercial controllers, the Master Controller solves the IR remote plague hands down. Because it uses an external computer for functional modifications by the user, more room is available for its ultimate task. Instead of four remotes, the Master Controller can be trained to simulate the functions of 16 individual remote controls complete with descriptive command designations.

An Infrared Introduction

Most infrared remote controls are functionally similar. The microprocessor in the remote controller creates a stream of bits that is turned into on/off pulses of IR light from an IR LED. An IR-sensitive photodiode in the receiver turns the light pulses back into an electrical signal from which the original bits can be extracted.

The IR LED's fast on-and-off action creates a carrier signal. The carrier is then turned on and off to form the individual bits of the message. Each controller uses a different carrier frequency, sets different bit timings, and assigns different meanings to the bits in the message. The reason that controllers are not compatible is that there is no standard for the format of the bits in the message. Each manufacturer speaks in a different language.

The Master Controller sidesteps this problem by simply recording and playing back the infrared signals without attempting to decode the messages. It's just like a tape recorder. You can record English, Russian, and Spanish on the same tape because they all occupy the same frequency bands, and you don't have to understand the languages to play them back.

This scheme works because of a limited range of differences in the IR signals. The controllers I've tested had carrier frequencies ranging from 32 to 48 kilohertz. Each message bit has between 10 and 30 carrier cycles, and there are two different carrier modulation systems: pulse-width modulation and pulse-position modulation. In Chap. 31 I explained pulse-width modulation. Pulse-position modulation works by determining the time when a bit occurs relative to a fixed starting point.

Photo 1: *The six-button Master Controller can duplicate the functions of the 14 controllers shown in the background.*

The Heart of the Master

An Intel 8031 single-chip microprocessor running a program stored in a 2764 EPROM directs the operation of the rest of the circuitry. (See photos 2 and 3 and figure 1.) The menus and IR signals are stored in a single 32K-byte battery-backed static RAM. The user interface consists of a two-line LCD and a six-button keypad. The keypad is either a simple membrane matrix or individual keys arranged in a matrix that is scanned by the 8031. This eliminates the need for a keyboard encoder. Two keys each are used for device and function up/down scrolling on the LCD. A fifth button, Do It, executes the device/function command appearing on the display. A sixth button, Learn, is used for training.

The LCD has 20 characters on each of two lines. The interface to it requires only six wires: four data bits, one timing strobe, and an address line. The display's internal character generator converts ASCII data into character dots, so the 8031 can communicate directly in ASCII.

A TIL413 photodiode converts the IR signals from other remote controllers into a discernible logic signal. Because the Master Controller and the remote are placed close together during training, there is no need for the sophisticated signal processing that's required to detect weak IR signals across a room. The Master Controller photodiode circuitry was designed to accept strong IR signals only. You should position the remote-control unit within a few inches of the Master Controller's photodiode. If it's too far away, the Master Controller will "see" nothing. If it's too close, the Master Controller will receive a distorted signal. A little experimenting with each controller will locate the correct position.

An LM311 comparator converts the photodiode input signal to a TTL-level signal. A 74LS164 shift register samples the output from the LM311 at a 1-megahertz rate and converts the data into parallel format. The 8031 reads the shift register every 8 microseconds while it is learning a new IR signal. This data is stored in RAM for later analysis.

At the transmitting end, the process is reversed. Although the 8031 is a fast microprocessor, it cannot generate both the carrier and bit timing of the IR signals instantaneously. To lighten the processing overhead, an 8254 programmable interval timer controls the IR carrier frequency and duty cycle as well as the duration of each message bit. The 8031 sets up the 8254's registers for each bit of the IR message.

A pair of TIL39 infrared LEDs produce the IR signal. Because the human eye cannot see IR light, a visible LED is connected in parallel as an indicator. The LEDs are switched by a field-effect transistor driven by a standard logic gate. The FET is an efficient way to interface logic levels with real-world devices because it directly translates an input voltage into an output current.

Power

Power is an important consideration in any battery-operated device. The Master Controller was designed to use either 74LS or 74HC devices at 5 volts. The 5 V is derived from a 6-V battery (four AA cells) using a special low-dropout voltage regulator. While LS takes considerably more power than HC, the duty cycle is low. The Master Controller need only be powered up long enough to set the device and function and press Do It. It can be shut off afterward. Admittedly, I could have spent more time developing automatic power up/down circuitry, but it would have complicated the design and added more software. Feature-specific circuit tailoring will have to wait.

Turning the power on and off is not a problem. The 8031's system software is contained in a 2764 EPROM, and the LCD and IR data are contained in battery-backed RAM. The memory is a 32K by 8-bit static low-power CMOS RAM chip. The backup circuit consists of two 3-V lithium batteries and a Dallas Semiconductor DS1210 battery-backup controller chip. The DS1210 senses loss of the +5-V supply voltage and automatically write-protects the RAM as it switches power to the battery. The second battery is necessary only if the first one fails.

Signal Processing

As you can see from the schematic in figure 1, most of the Master Controller's functions are done in software. It's worthwhile to look more closely at the processing required for the learning and reproduction of the IR signals. The IR carrier frequency is about 40 kHz, giving a period of about 25 μs. The particular frequency used by a controller must be measured precisely because each microsecond of error changes the reproduced frequency by about 4 percent. While this doesn't sound like much, when the Master Controller reproduces the IR signal, the receiver could completely ignore it. The reason for this is that the IR receivers in consumer electronic gear must detect faint IR signals.

Generally, the receivers use a phase-locked loop, tuned to the remote unit's carrier frequency. The PLL can handle a 10 to 20 percent frequency error, but the design margins include errors due to temperature, voltage, and other effects.

Photo 2: *Component side of the Master Controller, showing the 2764 EPROM and 8031 CPU (center left and right). The backup batteries and Dallas Semiconductor DS1210 battery controller are in the bottom center and right.*

Photo 3: *The etch side of the Master Controller showing the 2-line by 20-character LCD and six control buttons.*

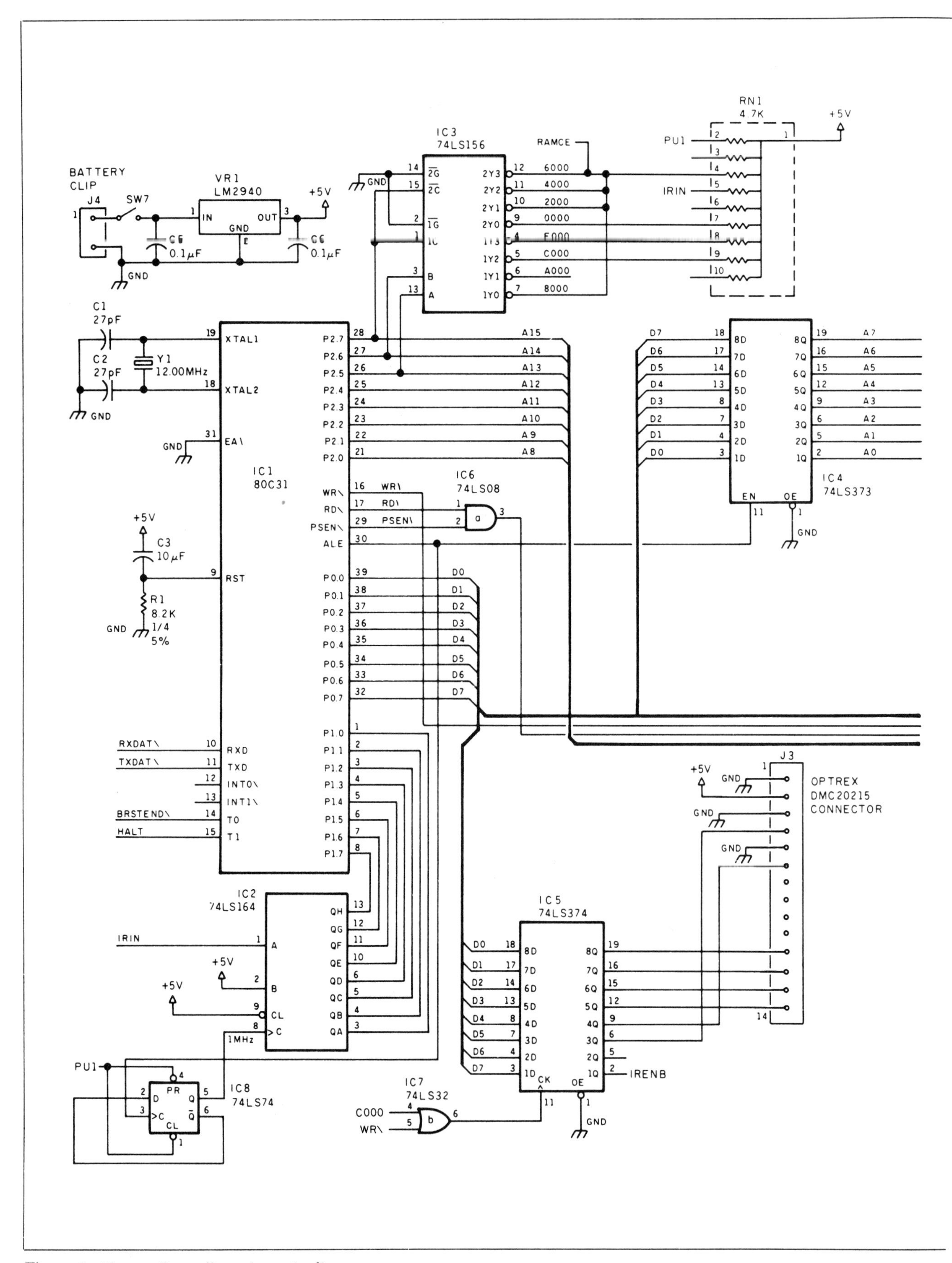

Figure 1: *Master Controller schematic diagram.*

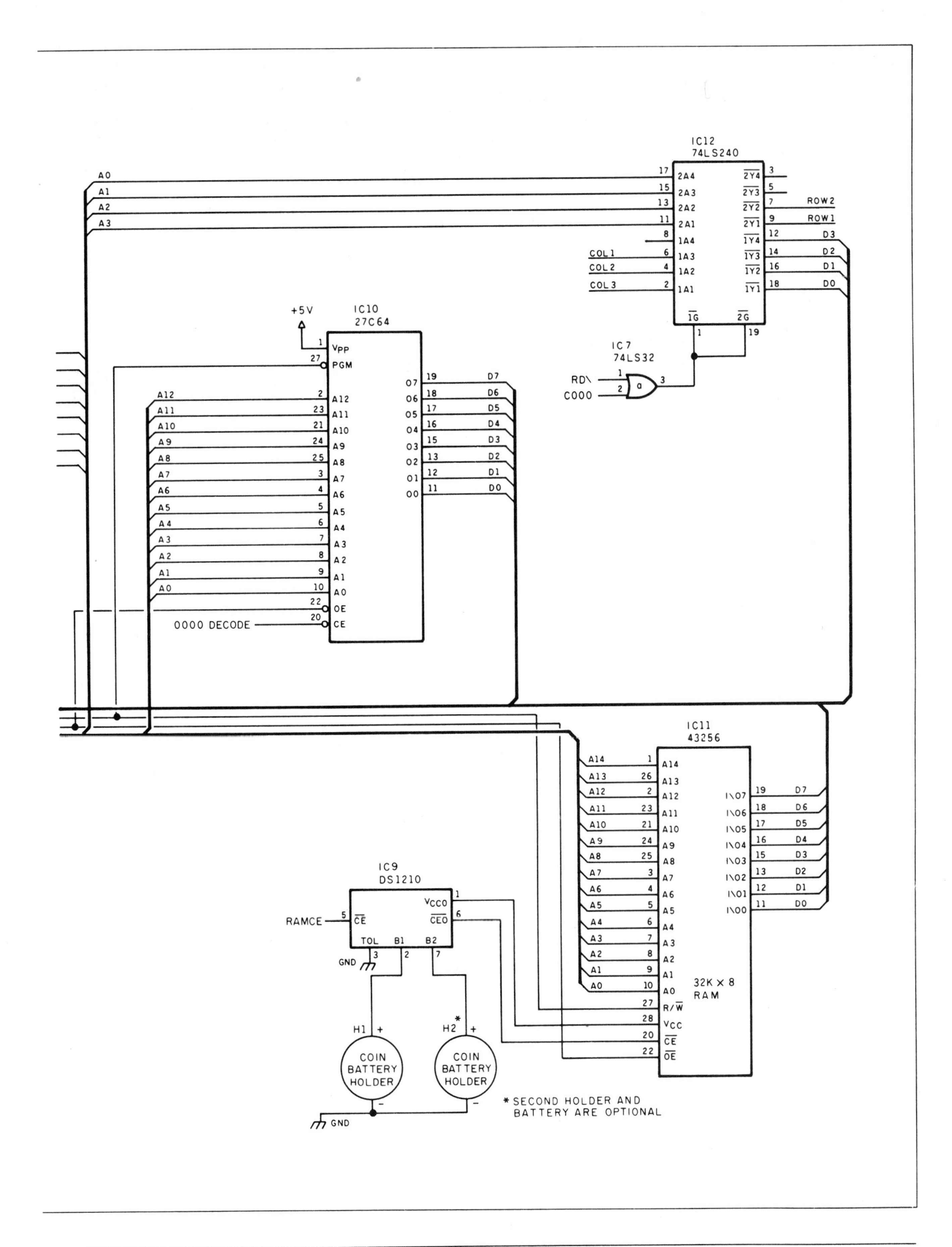
IC12
74LS240
ROW2
ROW1
COL1
COL2
COL3
IC10
27C64
+5V
IC7
74LS32
RD\
C000
0000 DECODE
IC11
43256
32K × 8
RAM
IC9
DS1210
RAMCE
GND
COIN
BATTERY
HOLDER
* SECOND HOLDER AND
BATTERY ARE OPTIONAL

The whole margin isn't available for the Master Controller. An 8031 with a 12-MHz crystal can execute most instructions in 1 or 2 μs. The shortest possible loop used to sample an input pin takes 2 μs. While this might seem very fast to you, reading and storing the value takes much longer and necessitates some form of input buffering.

The solution involves using a 74LS164 shift register to accumulate 8 bits of IR signal at a 1-MHz clock rate (see figure 2). The shift register in turn is sampled once every 8 μs, a requirement that's easily met. The 1-MHz clock for the shift register (and also the 8254, which I'll describe in a moment) is derived from the 8031's address latch enable output. The ALE signal occurs at a 2-MHz rate and is divided down to 1 MHz by half of the 74LS74 flip-flop. Exactly 32 samples of the shift register data are copied into internal RAM, a process that accumulates 256 μs of IR signal. The software then examines the data to pick out the start and stop of each carrier pulse. The 256-μs sample will include 8 to 12 complete carrier pulses, depending on the exact frequency (generally, 32 to 48 kHz). The software averages the length of the pulses to compute the carrier period and also determines the average duty cycle. Using several samples reduces the effect of noise on the final average.

Once the carrier frequency is determined, the next step is to measure the length of each bit in the message. The shortest bits we've measured contain at least 10 carrier pulses, and the average seems to be about 20 (although some contain more than 60 pulses). Given the variability in carrier frequency and pulse length, the main problem lies in determining when the bit ends. The software assumes that 32 μs without an IR signal marks the end of a bit.

The duration of each bit and the following pause are recorded in the external RAM of the 8031. Each IR message can contain up to 256 bits (and the following pauses) and can last up to half a second. Because most remote controllers repeat the message as long as the key is held down, it is very important to tap that key lightly.

Although the carrier frequency and message analysis could be done on one sample of the IR signal, the Master Controller requires two separate samples. The first is analyzed for carrier frequency, the second for message bits (see figure 3). This reduces the chances that a partial signal will be recorded in case the first bit is less than 256 μs long. Each IR signal is summarized by its carrier frequency and duty cycle and up to 256 pairs of 16-bit numbers that record the bit times. Therefore, each signal can occupy up to about 1K byte of RAM. Typical signals have a few dozen bits and require only about 100 bytes. This allows the Master Controller to easily accommodate 16 remote-control units with 16 commands each.

Just as the 8031 isn't quite fast enough to directly record the IR signals, it needs a little help creating them. An 8254 programmable interval timer provides the high-speed logic required to generate signals with microsecond timing resolution. The 8254 PIT contains three identical timers that can be set up in a bewildering variety of modes. The Master

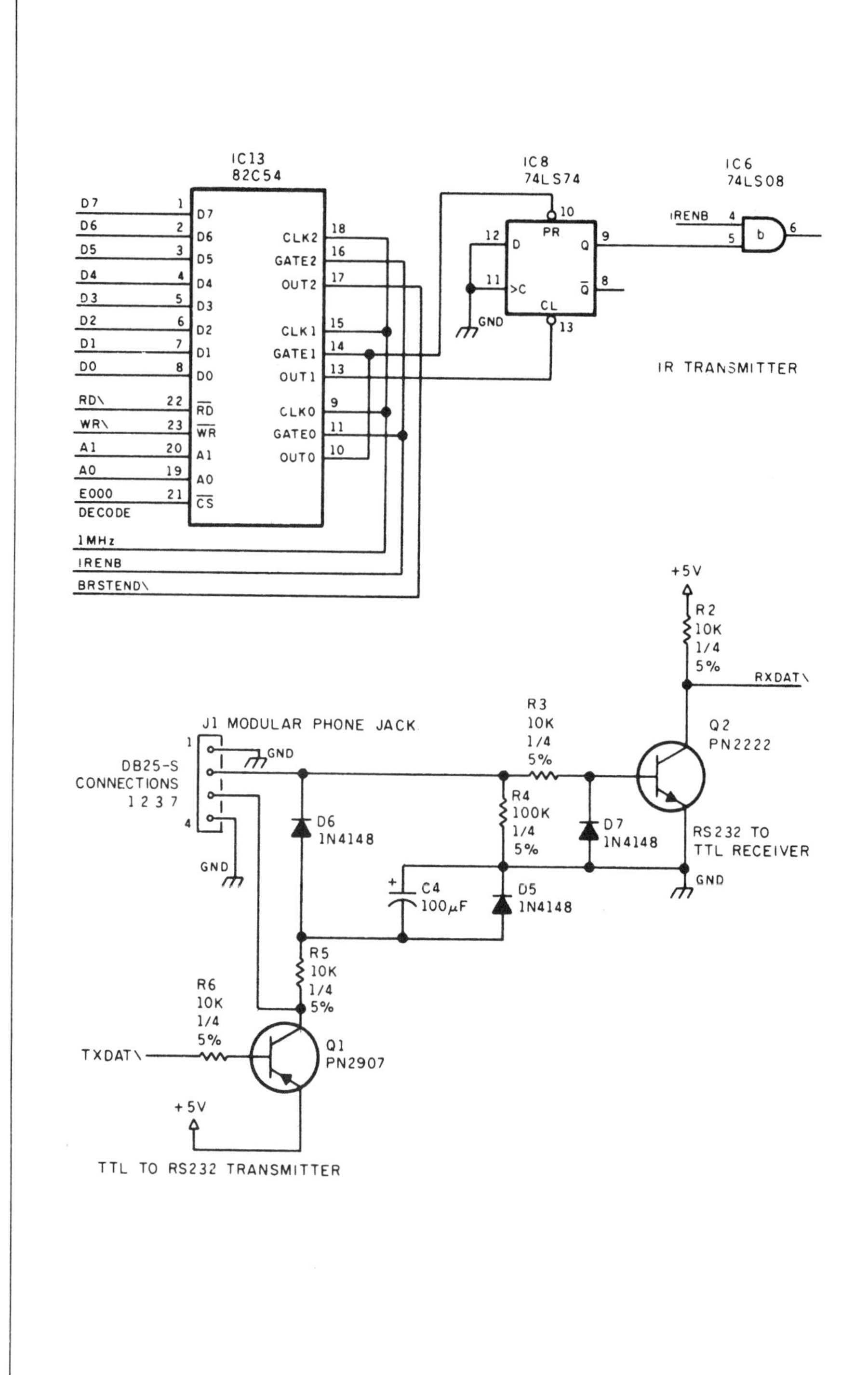

Figure 1: ***Continued from page*** 359

Controller software uses all three of them, as well as a timer inside the 8031, during the IR playback.

Timers 0 and 1 in the 8254 (IC13) set the IR carrier frequency and duty cycle, respectively. Timer 2 determines the duration of each message bit, and the 8031 timer controls the pause following each bit. The first two timers are set once at the beginning of the message, while the last two are set for each bit. The times are stored in external data RAM accessed only when the 8031 is running.

Because the 8254 produces a pulse only at the end of each timer's count, a 74LS74 (IC8) is used to create the actual IR pulses. The Timer 0 pulse (pin 10 of IC13) sets the LS74's output at the start of the carrier cycle, with Timer 1 (pin 13 of IC13) resetting the output at the end of the carrier pulse. The LS74's output is combined with a gating signal (IRENB) and sent to the IR LED drivers. As you can see, a great deal of code is required to handle the IR signal analysis. The code is written in 8031 assembly language.

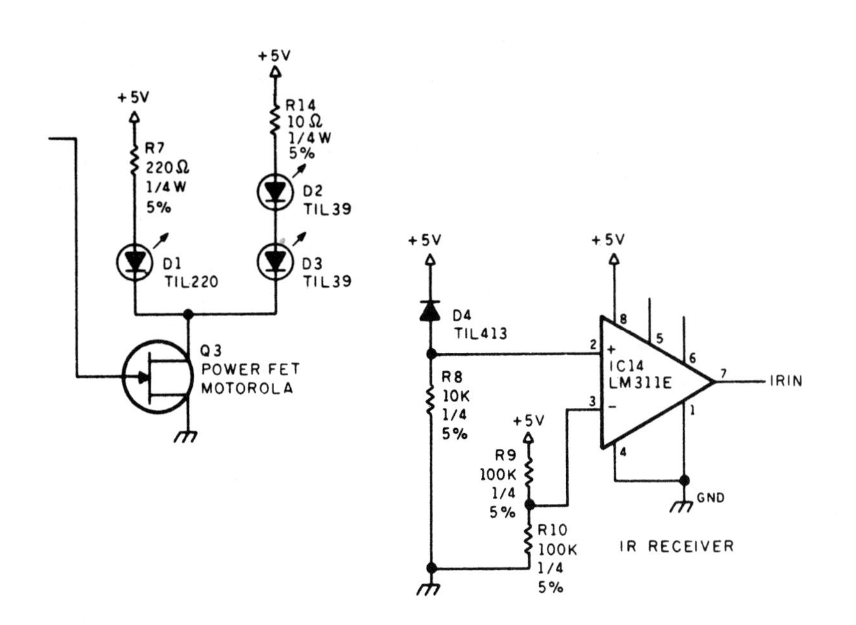

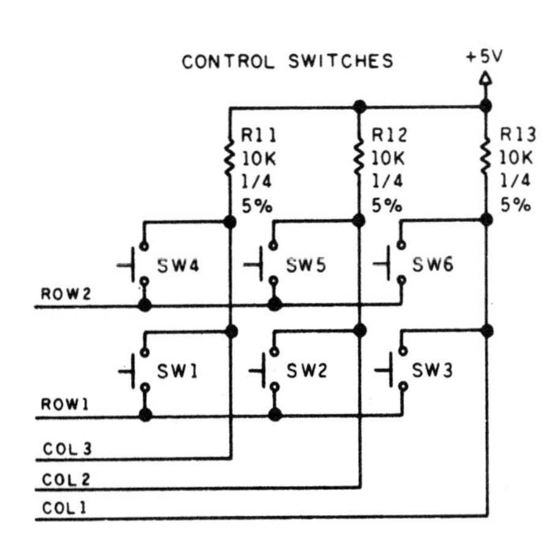

The PC Connection

Using an IBM PC to create menus for the Master Controller may seem like overkill, but it really simplified the logic. The PC has a full keyboard and display, disk storage, and, best of all, high-level programming languages. Writing a PC program is much easier than writing an 8031 program, so I decided to put as little code in the Master Controller as possible.

The PC program (called MASTER) provides three main operations: creating and editing menus of devices and functions, saving and loading these menus in disk files, and transferring them to and from the Master Controller. (See photos 4, 5, and 6.) The Master Controller is connected to the PC only when uploading and downloading menus. Under normal use, the Master Controller doesn't have any wires trailing out of it.

I'll have to admit to taking a little poetic license in the design of the RS-232 circuit. Rather than include a separate negative-voltage power supply for the RS-232 signal levels, I used a diode and capacitor to "borrow" the negative voltage from the PC's transmitted data line. A pair of transistors are simpler than the power-hungry level converters normally used to translate between logic levels and RS-232 levels. This is an important consideration in battery-operated devices. Because I know the communication will always be with an IBM PC, a worst-case, tolerant RS-232 circuit is not a necessity.

The serial connector is an RJ-11 telephone jack instead of the usual 25-pin DB-25 connector. Only three wires are required: data from the PC, data to the PC, and signal ground. Because the MASTER program and the Master Controller were designed together, they use an efficient method of passing data that doesn't require the normal RS-232 RTS/CTS and DSR/DTR status lines.

The 8031 serial interface includes a bit-rate generator. The exact bit rate depends on a number programmed into a register as well as the frequency of the 8031's clock crystal. I used a 12-MHz crystal to get the highest resolution possible for the IR signal-processing circuitry, but that's not the optimum crystal for the serial interface. As a result, the 8031 transmits data

The MASTER program, which is written in Turbo Pascal, can create menus for up to 16 devices.

to the PC at 10,417 bits per second. If you're familiar with normal RS-232 data rates, you'll recognize that 10,417 isn't one of the choices. Fortunately, the PC's serial port also determines the bit rate from a number in a register. The closest match is 10,473 bps, but everything works just fine.

Here I should point out that, although the link between the Master Controller and the PC is called RS-232, it is surely not standard. For example, using mismatched bit rates and voltage levels is acceptable only if you've carefully checked out the consequences and assured yourself that both ends of the connection are still compatible. If you elect to build a Master Controller, you shouldn't try to stretch the limits of the connection too much. A 50-foot cable probably won't work at all!

The Master in Action

Perhaps an example of how to use the Master Controller is in order. I'll show how to set up the first menu, then how to combine IR signals to produce customized effects.

The MASTER program, written in Turbo Pascal, can create menus for up to 16 devices: receivers, CD players, tape decks, and so on. Each device can have up to 16 functions (on/off, play, rewind, volume up, etc.). While up to 256 functions are possible, the ultimate limit to the number of devices and functions is the size of the Master Controller's RAM. MASTER and the Master Controller cooperate to make sure that you don't download a menu that's too big.

MASTER treats the devices (Bedroom VCR, Kitchen TV, etc.) and functions (volume up, power on, etc.) as a collection of lists. Function keys let you "cut" an item from one list and "paste" it elsewhere (see photo 4). You can delete an item permanently, and you can insert a new item and give it a name. Devices and functions are treated as different items, so you can't cut a function and then paste it into the device list.

Table 1 shows the complete list of MASTER function keys on the IBM PC. As an example, you might use the MASTER screen for the Sony RM-S750 controller menu. The Tape Deck line in the Devices list is highlighted, and the Functions list details all the tape deck's functions. The word "new" at the end of each function indicates that the IR signals have not been learned yet (see photo 6).

The Remote Keyboard item in the Devices list contains the general functions like power on/off and volume up/down as well as the digits from 0 to 9 to allow direct radio tuning. You can duplicate functions under more than one device to make the Master Controller easier to use. After all the devices and function names have been entered, you should save the menu on disk. A good choice for a filename is the manufacturer's model number, so RM-S750 is a good choice for this one. The MASTER program will automatically supply an .MC file extension.

The next step is to download the menu to the Master Controller. The MASTER program and the 8031 program first verify that each other exists, exchange some status information, and finally transmit the menu. (The PC cable can be disconnected after the download is complete.) The Master Controller learns one function at a time. Use the Select Device keys to scroll through the Devices list, then the Select Function keys to pick a function for the device. The Learn key will record an IR signal for the selected function. You can test the signal and relearn it until it's correct, but you can learn only one signal for each function.

As I described earlier, the Master Controller requires two samples of the IR signal to find the carrier frequency and message bits. You should tap the remote's keys quickly to avoid filling the Master Controller's RAM with repetitions of the same signal. Because most remote controllers will repeat their IR signal as long as the key is held down, you should tap the Remote key and release it immediately. You should see the remote's LED blink briefly to indicate that it sent a signal. There's no point recording repetitions because the Master Controller will repeat the signal as long as you hold down the Do It key.

After the second tap, you can test the Master Controller's stored signal by aiming its IR LEDs at your VCR or TV system and pressing the Do It key. If the function works correctly, the Master Controller has a valid IR signal in RAM. You can repeat the learning process by tapping the Learn key again. Once you have a good signal stored in RAM, tap any key other than Learn or Do It to return to the

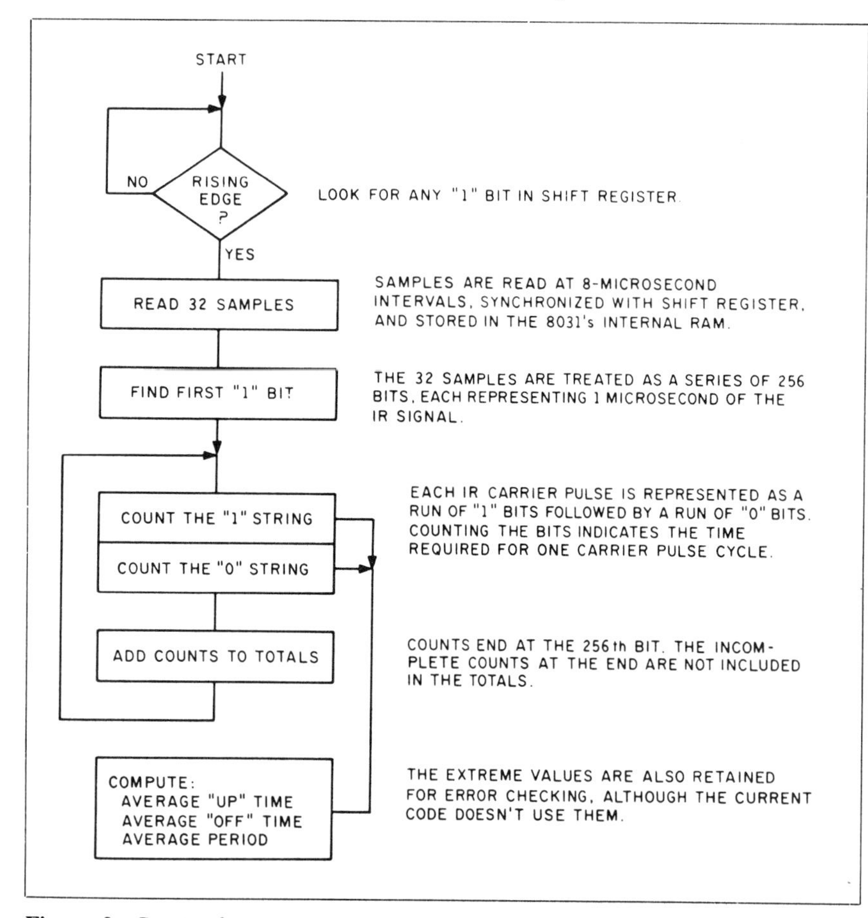

Figure 2: *Carrier-frequency measurement flowchart.*

normal display. Select the next function and repeat the learning process again. The cycle takes only a few seconds once you get the hang of it.

When all the functions are learned, reconnect the RS-232 connection to the PC and upload the menu using the MASTER program. The word "new" after each function is replaced with the length of the IR signal in bits. The Sony RM-S750 produces three repeats of 13 bits for each key press, so each function shows a length of 39. Other remotes will, of course, have different signal lengths.

Menu Modifications

At this point, you have a Master Controller menu that duplicates the functions of the RM-S750 as its first device. The MASTER program can combine the IR signals for two functions to produce the same effect as pressing two keys on the remote in sequence. This comes in handy for operations that you normally do in sequence, like turn on the power to the FM tuner and select your favorite station, for example.

To combine two IR signals, first "cut" one signal from the function menu by pressing F7. Then position the cursor over the other signal and press F10. Notice that the signal length is now the sum of the two old signals. You might want to use F9 to change the function name to reflect the new signal.

You can combine any number of function signals, with the only restriction being that the total length of the combined signals cannot exceed 1024 bits. The "cut" signal (or signals) is put at the end of the combined sequence. (Because the Master Controller uses a single carrier frequency for an entire IR signal, a combined signal may not work correctly if the signals came from different controllers.) MASTER will warn you if the carrier frequencies differ by more than about 10 percent but will allow you to shoot yourself in the foot. The Master Controller will use the first signal's carrier frequency for all the combined signals.

Some tape decks require pressing two keys (usually Record and Play) simultaneously to start recording. Generally, you can't get the same effect by pressing the Record key followed by the Play key. The reason is that the remote sends out a different message when the two keys are pressed simultaneously than it does for either of them separately. If you combine the Record and Play signals using Master, it won't work any better than the two separate keys will. You must "learn" the correct signal by pressing the two keys simultaneously. You've got to be quick on the keys to avoid filling the Master Con-

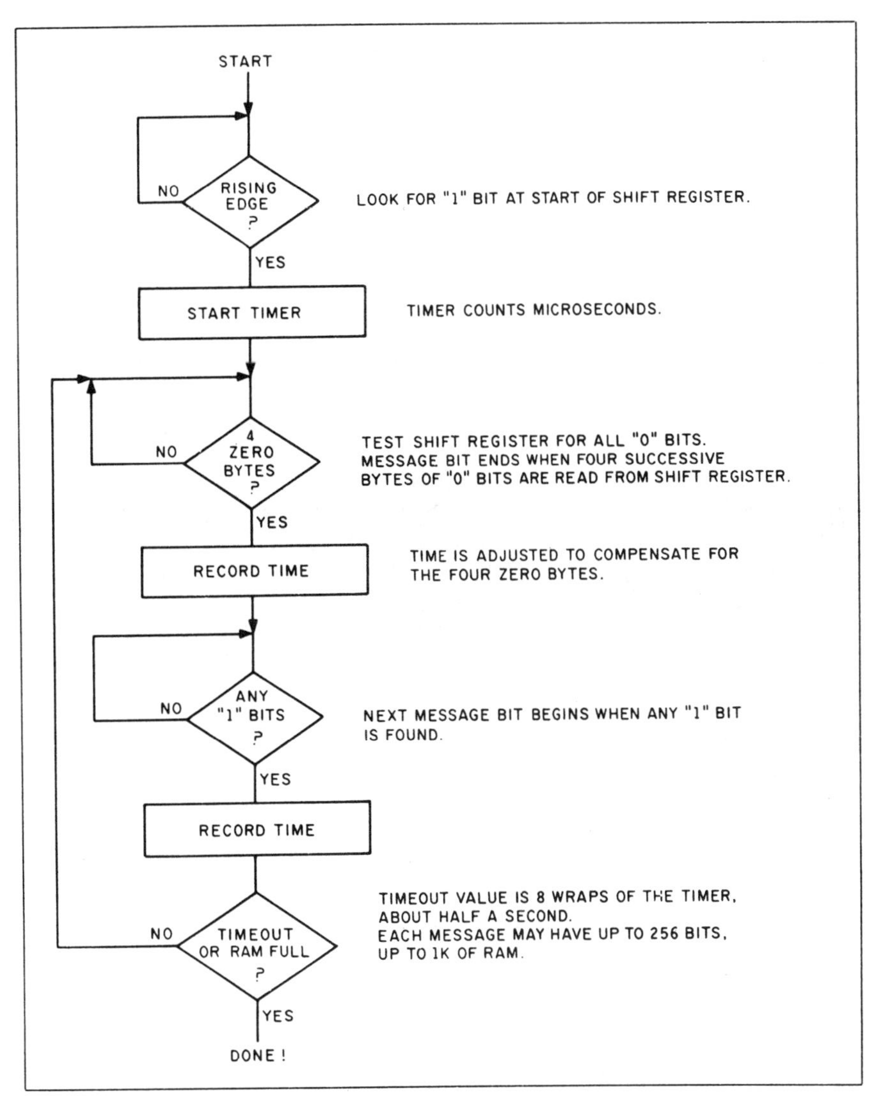

Figure 3: *Signal-capture flowchart.*

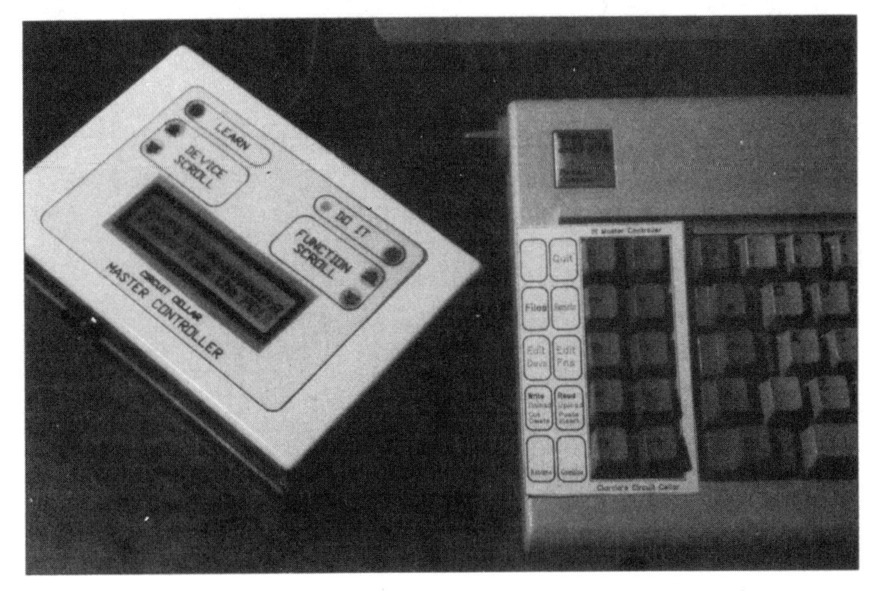

Photo 4: *The Master Controller and the IBM PC keyboard showing the MASTER program function-key template.*

troller's RAM with repetitions, though. As usual, practice makes perfect.

Conclusion

Once you've recorded the basic controller functions, you can use MASTER to combine them in wonderful ways. Although I've been calling the menu selections "devices" and "functions," you don't have to. You might wind up with a device called "Coming Home" with functions ranging from "Tired" to "Exhilarated" to turn on your system and select just the right lighting and music. Get the idea?

The Master Controller was designed and prototyped as a Circuit Cellar project. While it has some obvious and immediate consumer market potential, without a clear goal in mind it is hard to convince someone to go through the expense of manufacturing it (especially producing a custom enclosure). I have only a short time between projects, and I don't have the time to speculate on the eventual market niche or the specific configuration the Master Controller will take (Sharper Image, are you listening?). However, unless there is some way to evaluate the present device, another generation of the Master Controller will never be built.

To facilitate these evaluations, I've made a printed circuit board for the Master Controller. My intention is to populate a few more and circulate them in the proper consumer channels. While the Master Controller is not available as a kit per se, these PC boards are available if you want to build your own plague antidote.

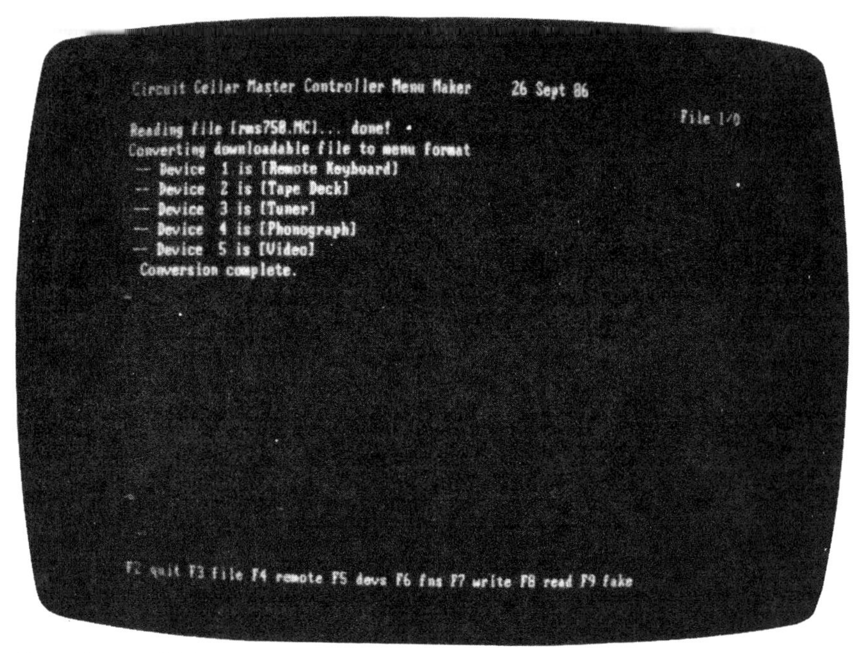

Photo 5: *MASTER program menu-maker screen with a five-device menu. Line 25 displays the function-key menu.*

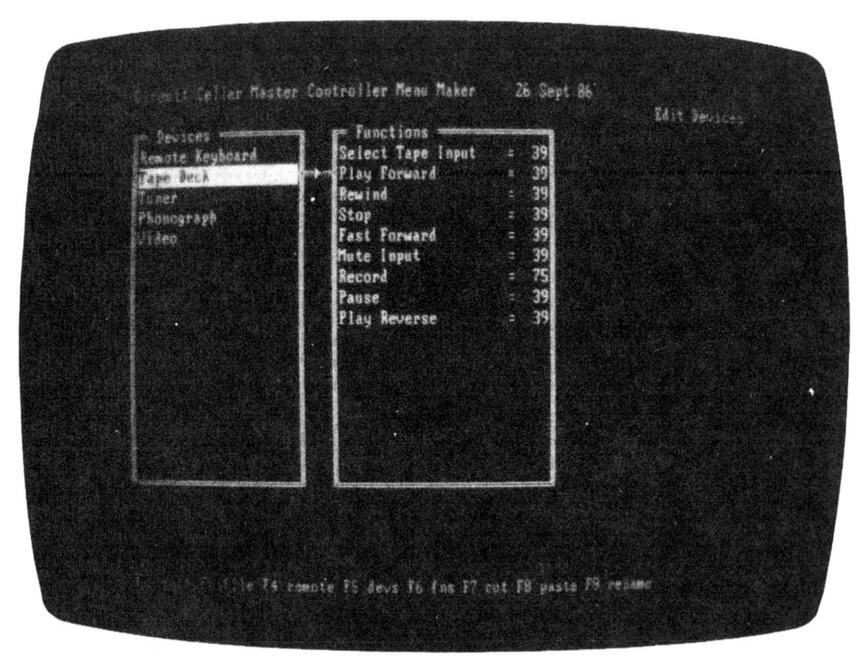

Photo 6: *MASTER program device-editor screen showing device and function menus. The Master Controller has just been trained to emulate the Sony RM-S750 controller.*

Table 1: *IBM PC function keys and their MASTER program functions.*

IBM PC function keys and their MASTER program functions

Rule of thumb: F2 through F6 select what you want to do
F7 through F10 actually do something
F1 does nothing
To get out of MASTER, press F2

F3 selects file I/O, then:

F7	writes the current menu to disk
F8	reads a menu file from disk
F9	creates a menu with simple names and random IR data (may be too big to fit into the remote RAM, though; this is of no use to anyone other than me)

F4 selects remote I/O, then:

F7	downloads the current menu to the remote
F8	uploads the remote's menu

F5 selects device editing, then:

F7	cuts a device and puts it on the junk list; if there was a device on the junk list, it's gone now
Ctrl-F7	deletes a device forever

F8 pastes a device from the junk list. If there's no device on the junk list, creates a new one:

F9	lets you rename a device
F6	selects function (and message) editing
F7	cuts a function and puts it on the junk list; if there was a function on the junk list, it's gone now
Ctrl-F7	deletes a function forever
Alt-F7	wipes out the IR message data for the current function
F8	pastes a function from the junk list; if there's no function on the junk list, creates a new one
Ctrl-F8	inserts a new function, leaving junk list alone
F9	lets you rename a function
F10	tacks the IR message from the junk list onto the end of the current function, leaving the junk list unchanged, so you can combine it with any number of functions

See page 447 for a listing of parts and products available from Circuit Cellar Inc.

33

COMPUTERS ON THE BRAIN

PART 1

Clever signal amplifiers, noise rejection, and A/D conversion are all part of the HAL EEG

A lot has been written recently about artificial intelligence (AI). Some writers declare that we are on the threshold of the most important advances in computing since Boole and Babbage began fooling around with two-valued logic and the difference engine. Others decry the hype and note that the majority of recent software releases are now touting some form of AI influence in their design or execution, weakening the meaning of the term in order to sell products.

Even after 100 years of study, not all psychologists are in complete agreement as to what constitutes intelligent behavior (look around you—how much have you seen lately?). Intelligence has generally been defined as the global ability to solve problems, to adapt to new situations, to form concepts, and to profit from experience.

However, it is obvious that there are many different types of behavior—many different ways of responding to the same problem—that can be called intelligent. Within the last 20 years, experts have paid much attention to the basic types of intelligence and how they are mediated by the biological substrate of the human brain.

Experts have long supposed that human beings use two major modes of thought: the way of reason and the way of emotion. A commonsense view is that these two ways of thought occasionally conflict. Some writers conceptualize the differences as analytic versus synthetic, successive versus simultaneous, or even digital versus analogical.

Paralleling the conceptualization of two modes of thought have been the results of research on the two hemispheres of the brain. Psychobiologist Roger Sperry of the California Institute of Technology won the 1981 Nobel Prize for Physiology and Medicine for his studies on the functions of the two hemispheres of the brain.

Essentially, Sperry and his colleagues studied individuals who had undergone a commissurotomy, an operation that severs the main bundle of nerve fibers that support the bulk of neural communication between the left and right hemispheres. They found that each hemisphere seems somewhat specialized for different tasks. For approximately 95 percent of the population (right-handed individuals and two-thirds of left-handed individuals), it appears that the left hemisphere of the brain is better organized for executing tasks characterized as:

- Verbal: language skills, speech, reading and writing, recalling names and dates, and spelling.
- Analytical: logical and rational evaluation of factual material.
- Literal: literal interpretation of words.
- Linear: sequential information processing.
- Mathematical: numeric and symbolic processing.
- Contralateral movement: controlling movement on the right side of the body.

The right hemisphere is better organized for tasks characterized as:

- Nonverbal: using imagery rather than words.
- Holistic: processing information simultaneously, in parallel.
- Visuospatial: functions involving perceptions of location and spatial relationships.
- Emotional: experiencing feelings.
- Dreaming: imaginative and metaphoric visual image-making.
- Contralateral movement: controlling movement on the left side of the body.

Hemispheric Activation Level Detector

This chapter's Circuit Cellar project is a brain-wave-monitoring biofeedback device that provides real-time information about predominant hemisphere activation. That is, this Hemispheric Activation Level Detector (HAL, for short) graphically displays the relative amounts of brain-wave activity in each brain hemisphere (see photo 1).

HAL can distinguish among grossly different conscious states, such as between concentrated mental activity and pleasant daydreaming. For example, if you are debugging a program, HAL should show a predominance of left-hemispheric activity. If you are listening to some light music and daydreaming, it should show a predominance of right-hemispheric activity.

HAL is a relatively sophisticated, low-cost, stand-alone, fully isolated four-channel electroencephalogram (EEG) brain-wave monitor. It gathers analog brain-wave voltages from four sets of scalp contacts, filters them, converts them to digital values, and transmits them via an RS-232C port (making HAL compatible with any computer) for recording or analysis.

HAL includes a two-channel fast Fourier transform (FFT) analysis-and-display routine for an IBM PC. (HAL's PC software is intended only as a graphics display demonstration—and there are limitations in processing power when using a straight 4.77-MHz PC—so it displays only two channels, even though HAL sends data on four channels.) If you have a more powerful machine, you should be able to expand the software to display more channels.

When running this special analysis-and-display package, the PC separates out various amplitudes and frequencies of alpha, beta, and theta waves, as well as phase differences between the hemispheres. The result is a graphical representation of what is going on in your brain in real time.

Warning

HAL is presented as an engineering example of the design techniques used in acquiring brain-wave signals. It is not a medically approved device, no medical claims are made for it, and it should not be used for any medical diagnostic purposes. Furthermore, the safe use of HAL requires that the electrical power and communications isolation described in its design not be circumvented. HAL is designed to be battery-operated only. Do *not* substitute plug-in power supplies.

Analyzing HAL's circuitry illustrates practical design techniques, including differential amplifiers for low-level signal detection in a high-background-noise-level environment, a low-frequency band-pass filtering-rectifying-integrating detector, optoisolation for safety, and A/D conversion.

I'm presenting the HAL project in a two-part chapter. In part 1, I'll look at the problems involved in picking up microvolt-level signals, amplifying and digitizing them, and sending them to your computer.

Science and the Brain

As I investigated this area, I found that a great deal of serious research has been going on regarding what we know about how our brain works. Much of this thinking is finding its way into computer science; even the Macintosh and the IBM PC now have neural-network hardware and software available for the experimenter.

In his book *Megabrain*, Michael Hutchison quotes National Institute of Mental Health neurochemist Candace Pert (discoverer of the opiate receptor in the brain and researcher on endorphins—the brain's own painkillers):

> There's a revolution going on. There used to be two systems of knowledge: hard science—chemistry, physics, biophysics—on one hand, and on the other, a system of knowledge that included ethology, psychology, and psychiatry. And now it's as if a lightning bolt had connected the two. It's all one system—neuroscience. . . . The present era in neuroscience is comparable to the time when Louis Pasteur first found out that germs cause disease.

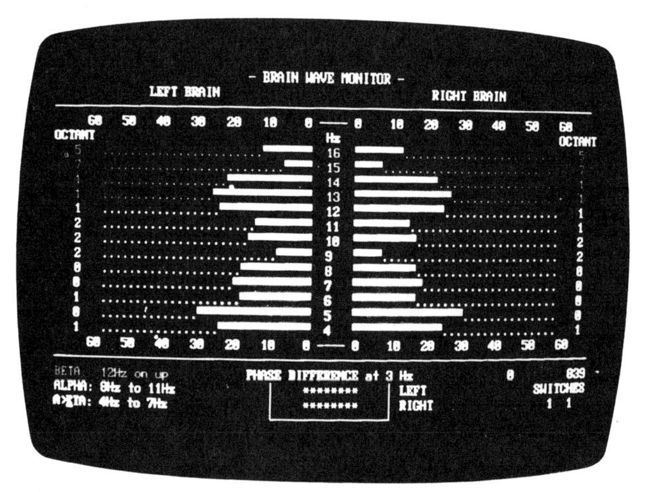

Photo 1: *HAL's output shows energy level by frequency of brain-wave signals for both sides of the brain.*

Hutchison further quotes neuroanatomist Floyd Bloom of the Scripps Clinic in La Jolla, California:

> A neuroscientist used to be like a man in a Goodyear blimp floating over a bowl game: He could hear the crowd roar, and that was about it. But now we are down in the stands. It's not too long before we'll be able to tell why one man gets a hot dog and one man gets a beer.

Much of the activity in this area has centered around the electrical characteristics of the brain. Advances in semiconductor technology have made available inexpensive ICs that let you design physiological monitoring equipment with laboratory quality at experimenter prices. When interest in alpha-wave biofeedback peaked about 15 years ago, a good-quality EEG feedback unit (which provided less information than HAL) cost $1000 for just one channel. Now, you can build four channels for under $200.

Digging into the Waves

The brain is a source of many electrical signals. An EEG is a recording (usually a strip chart) of the electrical potential differences between pairs of electrodes fastened to the scalp.

Silver-silver chloride electrodes pick up the signals. You must take some care to clean the area of the scalp with alcohol and perhaps use a mildly abrasive conducting cream to ensure good electrical contact. Ideally, there will be less than 10 kilohms impedance between any two electrodes, but anything under 25 kohms works (I'll describe placement of the electrodes in part 2 of this chapter).

It takes a trained eye to determine specific information about a person from an EEG. At present, we can only generalize as to what these recordings mean, and we are unable to correlate specific waveforms with intelligence. The observable electrical activity, however, does offer some clues.

According to medical and psychological research, by monitoring this activity, you could, in a gross way, investigate how the brain functions in a variety of circumstances. For example, if you monitor the two hemispheres while a person is solving problems, the type of problem could be indicated by the relative preponderance of one hemisphere's activity as compared to the other's. Sometimes you can even determine the activity (sleep versus reading; relaxed versus agitated).

The electrical signals we are currently able to monitor and identify from the brain are categorized as follows:

Alpha: Research has already indicated that in an awake person, the presence of alpha waves indicates a relaxed person with an absence of problem-oriented brain activity. (Alpha-wave activity describes electrical activity in the range of 8 to 12 Hz, a nearly sinusoidal signal at a voltage level of between 5 and 150 microvolts [μV]—typically 20 to 50 μV.)
Beta: When a person is thinking or attending to some stimulus, alpha-wave activity is replaced by beta-wave activity (14 to 25 Hz, activity of a lower amplitude).
Theta: Theta-wave activity (4 to 8 Hz, 20 μV and higher) usually appears during sleep, but it has been associated with deep reverie, mental imagery, creativity, dreaming, and enhanced learning ability.
Delta: Delta-wave activity (from 0.5 to 4 Hz) is seen in the deepest stages of sleep.

In addition, you must remember that I am describing an attempt to correlate cerebral electrical activity with subjectively observed events (types of cognitive tasks). While brain waves may be varying tens of times per second, our subjective experi-

To eliminate unwanted noise, HAL incorporates a band-pass filter that rejects frequencies under 4 Hz and over 20 Hz. While this compromises delta-wave acquisition, it does filter out most of the undesired signals.

ence varies more slowly. It may take a second or two to change concentration and to focus on a new task. Hence, you need to integrate the readings over a short period of time. Previous research in this area suggests that ¼ second to ½ second is reasonable.

A Noisy Environment

It's possible for HAL to "hear" more than we want. HAL is sensitive enough to detect artifactual signals: muscle activity from the forehead, eye and head movements, heart-rate activity, brain-wave "spikes" or irregular slow-wave activity, and—if you're not careful—60-cycle power-line hum. To eliminate this noise, HAL incorporates a band-pass filter that rejects frequencies under 4 Hz and over 20 Hz. While this compromises delta-wave acquisition, the benefit is that it filters out most of the undesired signals.

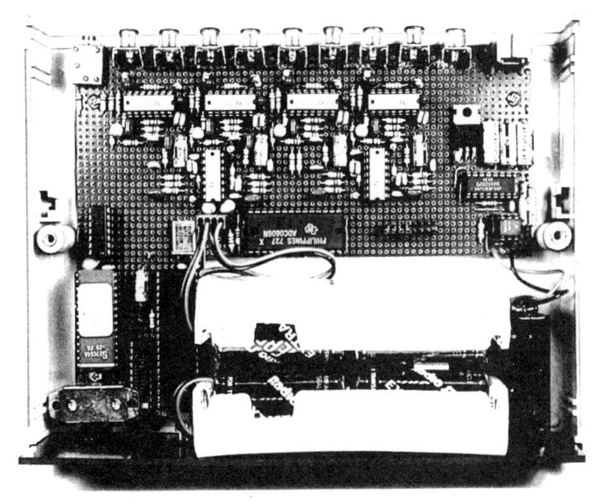

Photo 2: *HAL in the prototype stage. The input jacks are arranged along the top; HAL's battery is near the bottom.*

Photo 3: *HAL's optoisolated RS-232C circuitry.*

Detecting 4- to 20-Hz signals with a minimum amplitude of 5 μV from a source with approximately 10- to 20-kohm impedance is not an insignificant task. Ideally, the band-pass filter section should have a flat (±1-decibel [dB]) response across the passband; it should provide at least −18 dB per octave attenuation of signals outside the passband. The frequency response of the amplifier should be at least 50 to 60 dB down at 60 Hz. An equivalent input noise level of 0.5 μV or less would be good. Finally, input DC current should be less than 50 nanoamperes.

HAL's Circuitry

HAL's hardware circuitry is divided into two sections: preamplifier/filter and digitizer/control (see photo 2). The preamplifiers and filters acquire and boost the microvolt-level analog signals to useful levels. The digitizer section does the signal conditioning and A/D conversion and sends the data through an optocoupler to the host computer for analysis (see photo 3).

Several factors contributed to the evolution of the analog section of the circuitry. Initially, I planned to use narrow passband hardware filters to detect and measure only the alpha waves for each channel. Such an approach would discard a significant amount of information coming from the brain, essentially making the monitor capable of only simple "digital" discrimination—the presence or absence of alpha waves. This hardly seemed an achievement, since it merely duplicated the simple alpha biofeedback units available for the last 15 years.

Discussions with hardware and software experts eventually led to the conception of a more sophisticated system, one in which I considered the slowness of the EEG waveforms, the speed of the A/D conversion, and the analyzing power of an IBM PC. Ultimately, I decided that the HAL EEG monitor would function as a raw data accumulator and transmitter. The host computer would perform all signal analysis and display the results. (HAL's data output is RS-232C serial and can be analyzed and displayed on any computer. I chose to use an IBM PC here only for convenience.)

I expanded the bandwidth to allow the possibility of analyzing beta and theta waves. Even though these amplitudes are much lower than alpha waves, they are associated with some interesting phenomena.

To accomplish this task, I had to develop a special preamplifier/filter that would amplify only the specific EEG signals picked up from the scalp of the subject and amplify them to a level that is high enough for A/D conversion. Each HAL preamplifier/filter channel takes six operational amplifiers (op amps). Four of them provide amplification and impedance matching, and two others provide 60-Hz rejection filtering.

I designed the amplifiers and active filters in figure 1a around the TL-084 quad op amp and used as many common values as possible. The TL-084 provides junction-field-effect-transistor inputs with picoampere bias currents, low power consumption, and adequate input noise level. (If you are building this project, you should not substitute another type of op amp.) The bandwidth of the analog section is about 16 Hz (−3 dB at 20 Hz).

You can calculate the equivalent input noise by integrating the noise voltage as a function of frequency over the bandwidth. This 180-nanovolt equivalent noise, combined with the noise from the differential input stage multiplied by the system gain, yields a calculated output noise level of approximately 2.5 millivolts (mV).

Actual measurements of the noise output of the four-channel prototype were 3.5 mV root mean square, with a source impedance of 13-kohm impedance per input. I decided this was acceptable for the system with a 10-mV per bit A/D sensitivity.

I used three sections of IC1A to make a differential input instrumentation preamplifier. (Note that all six op amps associated with channel A are labeled IC1A and IC2A. Channel B's op amps are labeled IC1B and IC2B, respectively, and so forth.) An ideal difference amplifier will amplify only the voltage difference between the two inputs. Voltages that appear on both inputs when referenced to the ground lead are called common mode voltages.

For example, if the voltage on one input is +50 μV and the other input is +15 μV, the difference signal would be 35 μV and the common mode signal would be 15 μV. HAL measures the difference signal between the two electrode positions. The difference amplifier measures this difference by applying one signal to the inverting input of the op amp and the other signal to the noninverting input.

The ability of the op amp to amplify only the difference is specified as the common mode rejection ratio. In HAL, I measured this experimentally by shorting the inputs, applying an input signal between the shorted inputs and ground, and comparing the output with that obtained by applying the same signal across the two inputs. The common mode output was 43 dB down below the differential output. (You would correctly suspect that the major component of common mode voltages in HAL will be induced by the 60-Hz power line. I'll discuss how HAL rejects the 60-Hz signals later.)

I set the voltage gain of the preamplifier to 5800 and incorporated AC coupling between the stages to eliminate DC offset voltages and provide some low-frequency roll-off. Feedback capacitor C3 provides high-frequency roll-off, with the gain down 9 dB at 60 Hz. The third-order active filter stage has a −3-dB frequency of 22 Hz and is 30 dB down at 60 Hz. You'll find the same third-order filter at the input of each final amplifier to the A/D converter (ADC), thus providing another 30 dB, for a total of 69 dB attenuation at 60 Hz.

The interstage coupling capacitors set the low-frequency passband of the amplifiers. The low-frequency roll-off is 24 dB

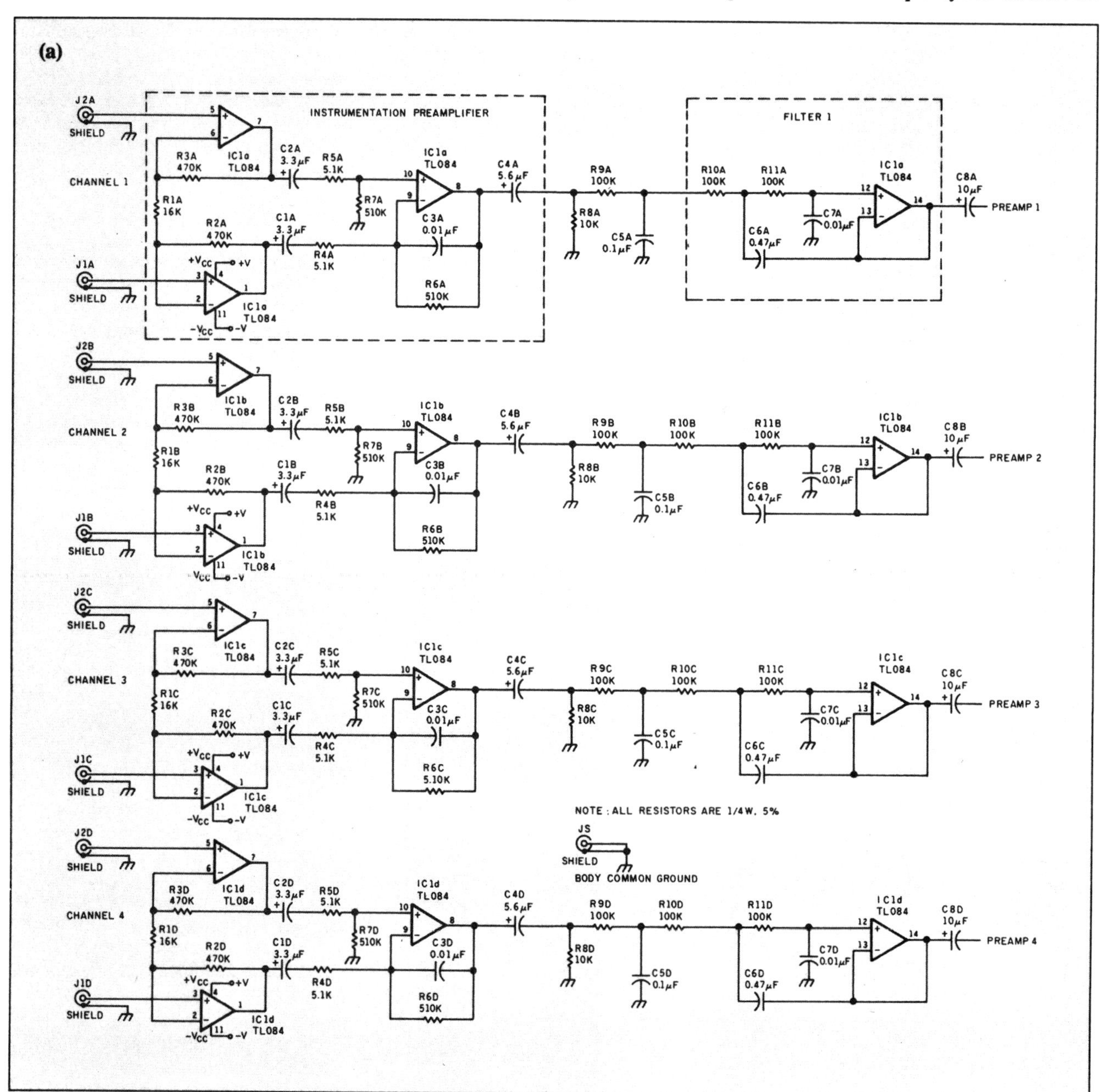

per octave, with the −3-dB point at 6 Hz. This is well above muscle activity and other noise.

A 2.5-volt reference diode sets the analog references to the ADC at 1.75 V and 3.25 V, or ±1.25 V of half the power supply. The last amplifier stage is DC offset to one half the power supply voltage, with the AC signals having a permissible peak value of 1.25 V. I set the overall gain of the amplifier stage to 12,500 so that a 100-μV signal would be the maximum input. This amounts to about 0.8 μV per bit sensitivity.

Since the ADC0808 is generally thought of as a DC converter and HAL measures AC signals, offsetting the reference to the ADC lets it measure signals that swing above and below some point designated as "zero" (offset binary converter). When you apply 0 V to the ADC, its output will be 80 hexadecimal.

A voltage gain of 12,500 corresponds to 82 dB (20×log Av). The 60-Hz rejection of 69 dB results in a 60-Hz gain of 13 dB (82 dB − 69 dB). The common mode rejection of 43 dB reduces the 60-Hz gain to a loss of −30 dB (13 dB − 43 dB). This all means that a 60-Hz common mode signal at the inputs is reduced by a factor of 0.03 in getting to the ADC.

To show up as a ±1-bit ripple on the data, the common mode input signal would have to have an amplitude of 300 mV peak to peak. This 300 mV would be reduced by a factor of 0.03 to become 10 mV at the ADC. When I connected a 1-inch unshielded lead to HAL's input, it picked up about 100 mV peak to peak of noise. This seems adequate, but all the same, don't use HAL while standing directly beneath a neon sign transformer!

The Digitizer and Control Section

The signals from the four preamplifier/filter channels go to four of the eight analog inputs of the ADC0808. An 80C31 CPU performs channel selection and transmission to the host CPU. (While it is possible to duplicate the preamplifier/filter section to ultimately produce an eight-channel version of HAL, the current level of software for the 80C31 is designed for only four channels.)

Figure 1c shows the microcontroller part of the headset cir-

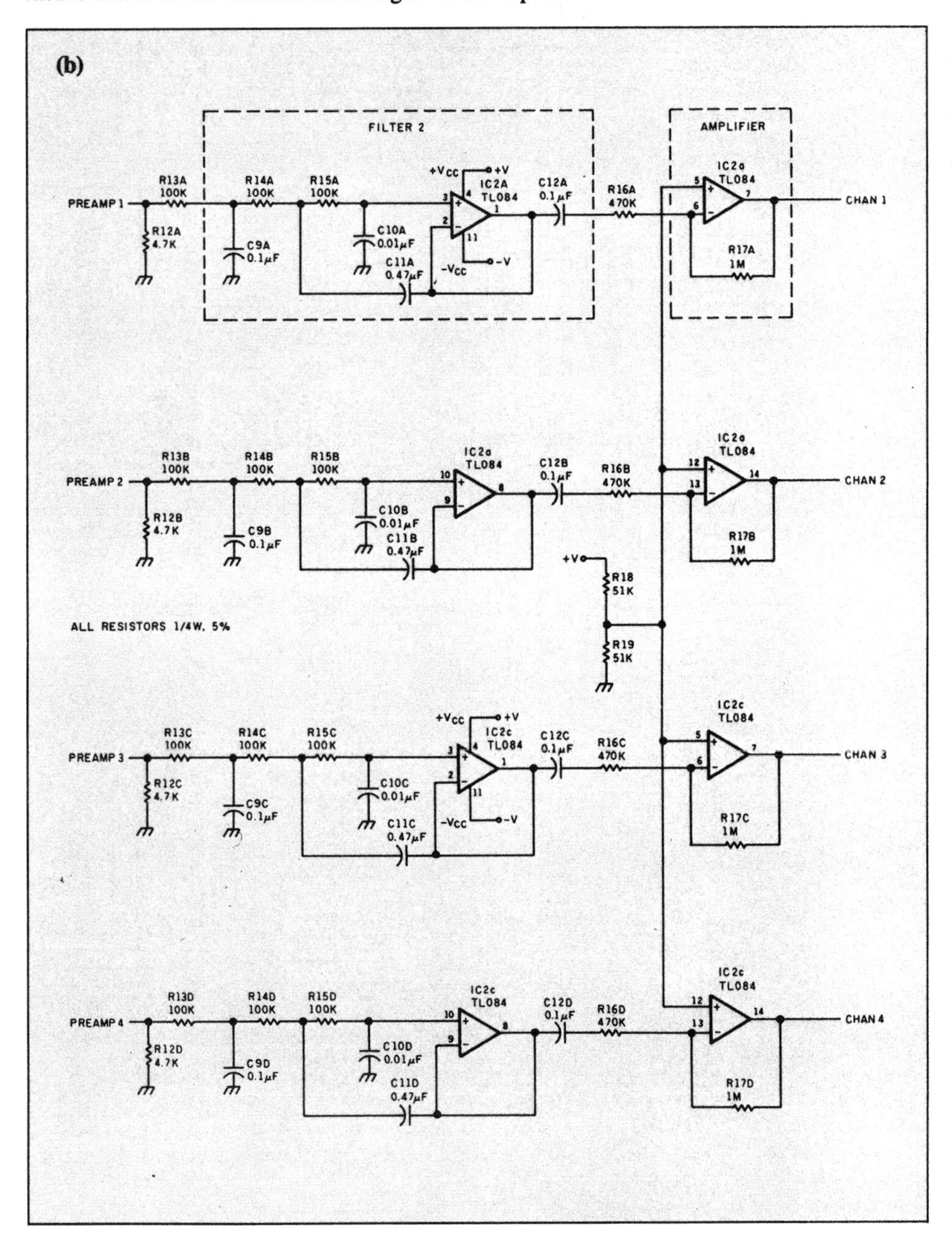

Figure 1: *The schematic for HAL.* **(a)** *Electrode pickup, preamplifier, and part of the filter stage.* **(b)** *More filtering and the final amplifier.*

cuitry. The 80C31's port 1 connects to the ADC0808's data outputs, with all the control and status bits handled by port 3. Bits from port 1 also drive the serial output line and the two event marker switch inputs. Because port 2 is dedicated to the upper half of the program address and port 0 is the EPROM data bus, no port bits are left for anything else.

The timing requirements are so simple that the code doesn't even need interrupts. It samples the two switches, reads the left and right hemisphere voltages from the ADC0808, and sends the results out serially. Each data sample consists of a 5-byte

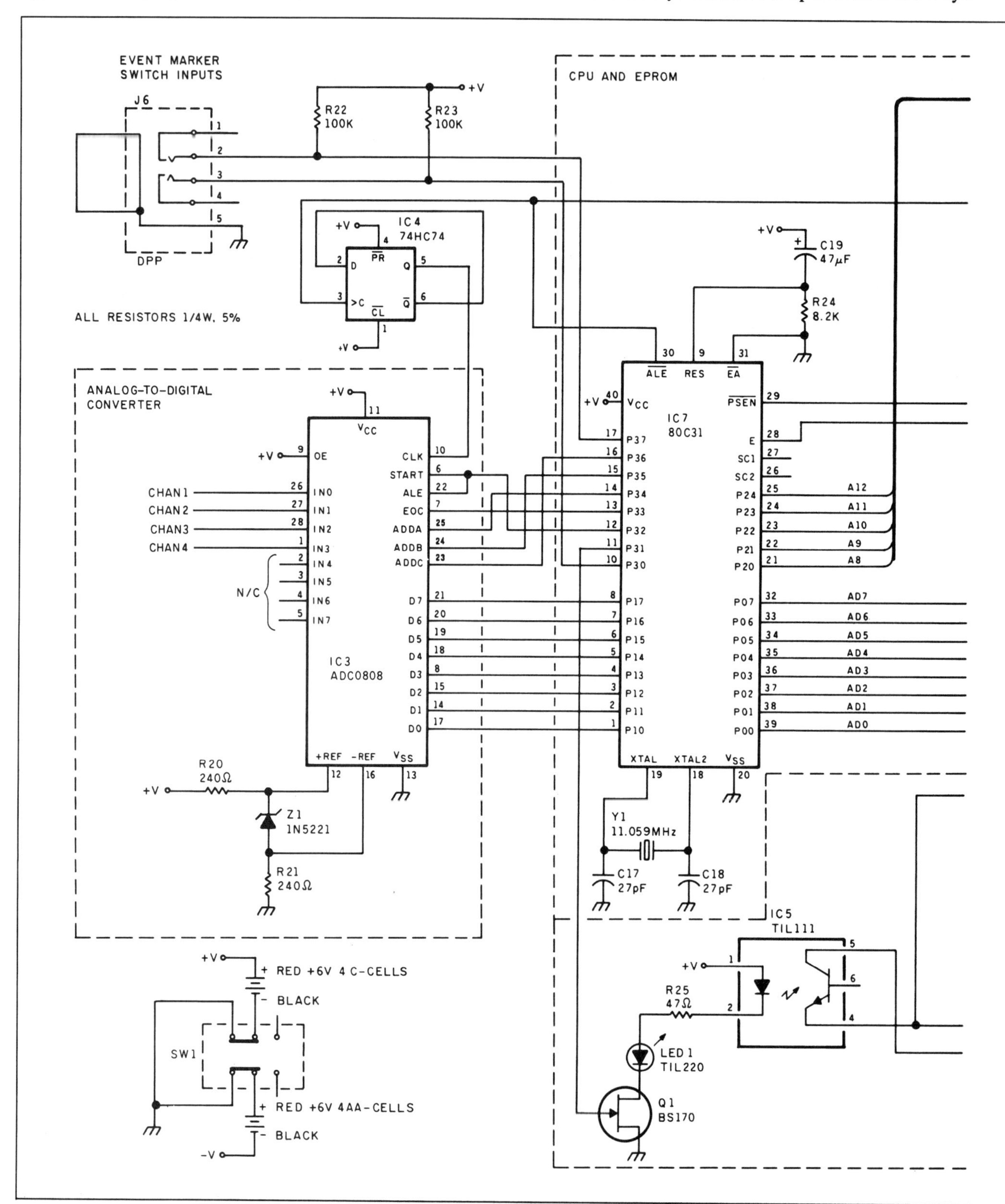

Figure 1c: *HAL's A/D converter, on-board CPU, and RS-232C port.*

transmission. The data sequence is first byte, two switch position codes with 6 bits of leading zeros, followed by 4 bytes of sequentially sampled A/D channels.

HAL is battery-powered. Four alkaline C cells provide +6 V, and 4 AA cells provide −6 V for the op amps. The CMOS digital circuitry runs from the 6-V supply. Current drain with all CMOS components is 225 milliamperes (mA) at 6 V and 50 mA at −6 V. (OK, I know that most chips like 5 V, but CMOS digital chips will work fine in this application at 6 V.) An addi-

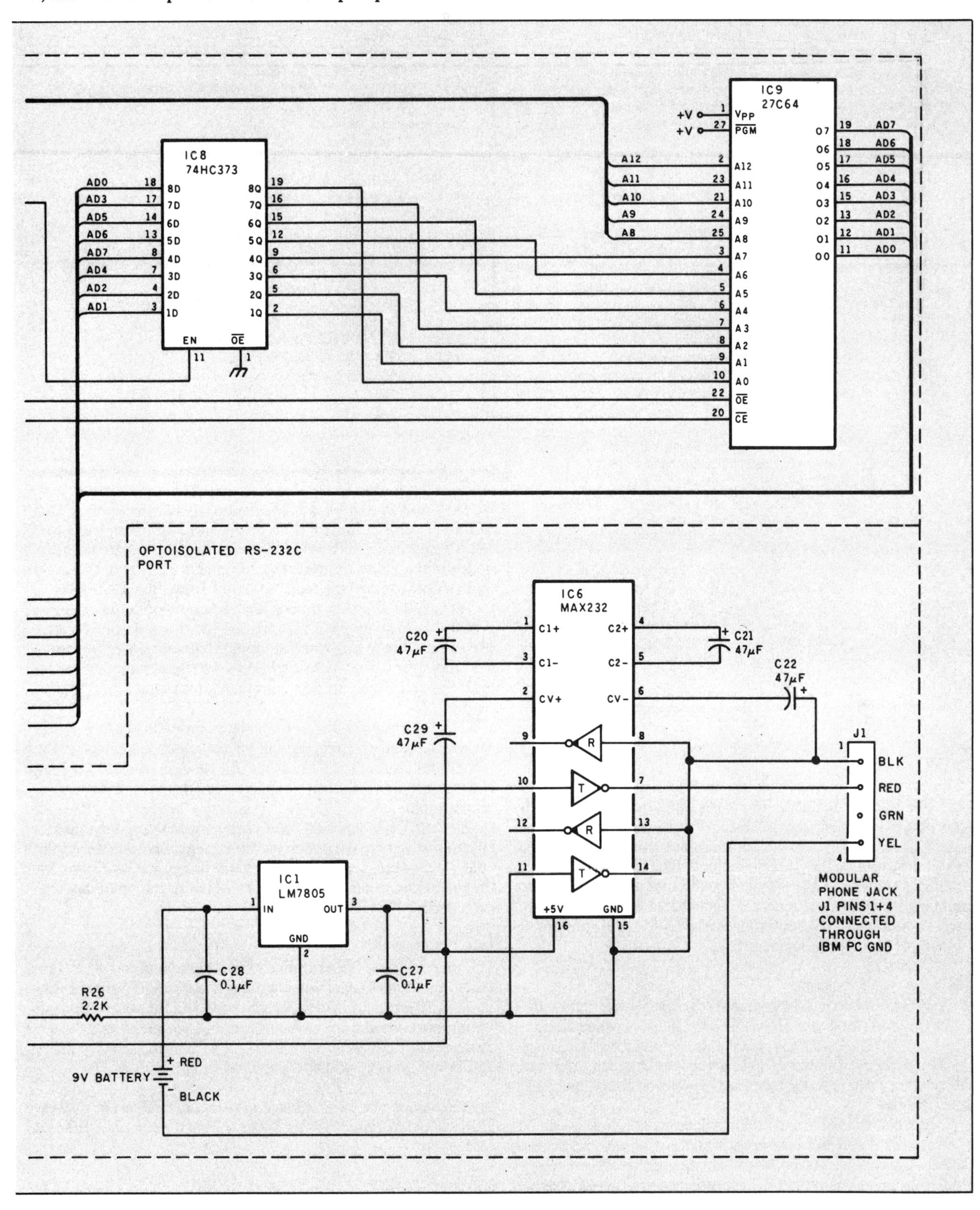

Listing 1: *Source code for HAL's main loop.*

```
; -- 80C31 main loop

ADCdata  EQU  P1    ; data inputs
ADCctls  EQU  P3    ; control I/O
start    EQU  2     ; + to start conversion
ready    EQU  3     ; + on end of conversion
switches EQU  81H   ; switch bit locations
ADC0     EQU  10H   ; low order address bit
ADCaddr  EQU  70H   ; address bit location
sync     EQU  P3.0  ; sync out of switch input!

;>> setup code omitted

;wait for Timer 0 to run out, then reload it
again EQU  $
      JNB  TF0,again      ; loop until timer tick
      CLR  sync           ; blip scope sync down
      NOP
      SETB sync           ; ... and back up again
      CLR  TR0
      MOV  TH0,#HIGH -T0period ; reload
      MOV  TL0,#LOW  -T0period
      CLR  TF0            ; clear end flag
      SETB TR0            ; restart counter
;--- read the channels in a great rush
      CALL getADC
;--- read the switches and send them out
      MOV  A,ADCctls      ; grab input bits
      ANL  A,#switches    ; strip switches
      RL   A              ; move to bits 1-0
Lsw   JNB  TI,Lsw         ; will set every time!
      MOV  SBUF,A         ; send byte
      CLR  TI             ; reset ready flag
;--- send analog data
L0    JNB  TI,L0          ; wait for trans ready
      MOV  SBUF,R0        ; drop in the value
      CLR  TI             ; reset trans ready
L1    JNB  TI,L1          ; repeat for channel 1
      MOV  SBUF,R0
      CLR  TI
      JMP  again
```

tional 9-V battery (10 mA) provides power for the serial communication.

The two-push-button switch inputs (J6) allow operator signaling to the host computer. Serial data output drives the TIL111 optocoupler by means of a BS-170 field effect transistor. A MAX232 (IC6) converts the optocoupler's output to RS-232C levels compatible with the serial input of the IBM PC host computer. The MAX232 is powered by the separate 9-V battery to maintain isolation between HAL and the PC when the serial port is connected. (Do not try to use the 6-V C cells that power the main HAL circuit to power IC6.)

The Control Program

HAL's firmware control program (the main loop is shown in listing 1), contained in a 2764 EPROM (IC9), is called BIO31. Nearly all BIO31's time is spent in line waiting for timer flag 0 (TF0) to become a logic 1. Whenever that happens, the code reloads timer 0 to produce the next 1/64-second delay and clears the flag again.

While the 5 bytes in each sample take only 6.25 milliseconds to transmit at 4800 bits per second and there's lots of idle time on the link (the PC code needs 64 samples per second; we don't send it faster because the PC analysis program would choke), the FFT software in the PC presumes that all the data points are sampled at the same instant in time. As a requirement, then,

Listing 2: *The code HAL uses to read its A/D converter.*

```
;----------------------------------
; Get channels from the ADC input
; Values are stashed in registers
getADC PROC
       MOV  A,ADCctls  ; reset address
       ANL  A,#NOT ADCaddr
       SETB ACC.ready  ; ensure this bit is a 1
       MOV  ADCctls,A
;--- grab channels
       SETB ADCctls.start ; blip start line
       CLR  ADCctls.start ;  with 1 us pulse
; Wait for EOC to go away
Lw0r   JB   ADCctls.ready,Lw0r
; Now wait for EOC active
Lw0e   JNB  ADCctls.ready,Lw0e
       MOV  R0,ADCdata ; save data in reg
       ADD  A,#ADC0    ; tick channel number
       MOV  ADCctls,A
       SETB ADCctls.start ; repeat for chan. 1
       CLR  ADCctls.start
Lw1r   JB   ADCctls.ready,Lw1r
Lw1e   JNB  ADCctls.ready,Lw1e
       MOV  R1,ADCdata
       RET
getADC ENDPROC
       END
```

BIO31 runs the ADC as fast as possible between samples. I used in-line code to eliminate the overhead of subroutine calls and returns, although I'll be the first to admit that the few microseconds probably don't make any difference at all. Listing 2 shows what's needed to grab channels 0 and 1 from the ADC0808.

Throughout the conversions, the accumulator holds a copy of port 3, so changing the ADC channel address is simply a matter of adding 1 to the proper accumulator bit and reloading port 3. The code sets the ADC ready bit to a 1 to make sure that the bit is always an input; writing a zero to that bit would turn it into an output.

Toggling the ADC's start bit using a pair of CLR/SETB instructions provides a 1-microsecond pulse on that output. One of the nice things about the 8031 is that you can tell exactly how long each instruction will take, so generating precise time intervals is quite simple.

The ADC0808 takes a few microseconds to drop the line that signals the end of conversion before starting the next one, so the code includes a loop to wait for that bit to go away before continuing. This is one of those cases where the computer can outrun the peripheral!

In Conclusion

In Chap. 33 part 2, I'll examine the software components of HAL including an 8088 machine language discrete FFT callable from BASIC. I'll provide BASIC source code so that you can design your own software and reconfigure HAL into a sophisticated brain-wave biofeedback monitor or a continuously recording EEG, or so you can add additional channels.

Special thanks for help provided on this article to Dr. Robert Stek, David Schulze, Rob Schenck, Jeff Bachiochi, and Ed Nisley.

BIBLIOGRAPHY

Hutchison, Michael. *Megabrain: New Tools and Techniques for Brain Growth and Mind Expansion*. New York: Morrow, 1986.

This book contains the results and speculations of some pioneers of brain-machine technology.
Meikson, Z. H., and Philip C. Thackray. *Electronic Design with Off-the-Shelf Integrated Circuits*, 2nd. ed. Englewood Cliffs, NJ: Prentice Hall, 1984.
1983-84 Opto-electronics Data Book. Texas Instruments.

See page 447 for a listing of parts and products available from Circuit Cellar Inc.

33

COMPUTERS ON THE BRAIN

PART 2

Making sense out of the data

In Chap. 33, part 1, I introduced you to the Circuit Cellar electroencephalogram (EEG) monitor called HAL (Hemispheric Activation Level Detector) and discussed its hardware design. In part 2, I'll look at the system software and provide some directions for HAL's use.

Software Overview

The overall operation of HAL is relatively straightforward. It samples four channels of analog brain-wave data 64 times per second and transmits this data serially to a host computer. The host analyzes two channels of the data to determine frequency components, and it then displays the results on a monitor as a series of continuously changing bar charts.

As I mentioned in part 1, the frequency range of interest in brain-wave detection is 4 Hz to 20 Hz, which includes alpha, beta, and theta waves. I could have designed HAL with many independent band-pass filters to separate each channel's complex waveform into discrete sinusoidal components, but that would have involved much more hardware. Instead, HAL's raw digitized analog data is analyzed on the host computer using a fast Fourier transform (FFT).

An FFT is like a mathematical prism. Just as a prism breaks down light into its component colors, the FFT breaks down a waveform into the sinusoidal components that make up the signal. In this way, I can extract amplitude and phase values for any given frequency in the target spectrum of 4 Hz to 16 Hz (I didn't display frequencies from 16 Hz to 20 Hz because of speed constraints on the IBM PC).

The system software coordinates four major tasks, which are graphically outlined in figure 1. The Intel 8031 microcontroller's task is precise sampling of hemispherical data and transmission via an optocoupled RS-232C link to the host IBM PC (the 8031 firmware is called BIO31). An IBM PC task called `COMMO` is an interrupt-driven communications routine that receives HAL's incoming data. Being interrupt-driven, `COMMO` operates independently of the activities of the FFT analysis-and-display routine (and any other tasks that happen to be running on the IBM PC).

As the data comes in, the host loads it into 1024-byte sample queues for left brain and right brain data. The FFT algorithm reads this incoming data, calculates amplitude and phase values per frequency, and updates integer arrays. Finally, the BIO program reads these arrays to continually update the bar graphs on the video display.

Simple—right? Just like using a hammer to make gravel out of a boulder. Simple in theory, at least.

While HAL is technically independent of the host computer (by virtue of its serial interface), I chose to present the demonstration display on an IBM PC for the sake of convenience. (I invite conversion to other computers and will make such conversions available through the Circuit Cellar bulletin board system.)

Because HAL is supposed to be both an educational and a functional presentation, my fundamental goal for the software was to keep it simple and fast. To that end, I used BASIC wherever possible. However, to maintain processing speed, I incorporated certain assembly language routines that are called from BASIC (i.e., the QuickBASIC compiler).

Refer again to figure 1, the overview block diagram of the system. The `COMMO` and FFT routines are written in assembly language. `COMMO` is really a group of subroutines; BASIC merely calls the initialization code.

I wrote the FFT block in assembly language for maximum speed in the FFT calculations. Using a BASIC program, a 64-point FFT took about 20 seconds to do. The existing FFT algorithm in assembly language will calculate 64-point transforms for both channels in about 240 milliseconds (ms). More about this later.

HAL's Brain

The 8031 is a remarkably capable processor that I have been using for many stand-alone projects lately. I described the inner workings of BIO31 last month with the hardware description, since BIO31 works so closely with the hardware. Basically, BIO31 is a simple read-transmit-wait loop (see figure 2). The read portion of the loop samples the switches and the A/D converter. It then places this information in the output buffer and transmits it out the serial port at 4800 bits per second (bps) to the host computer. The transmission is a 5-byte sequential string: switch position/sync, channel 1, channel 2, channel 3, and channel 4.

Once transmission is completed, the program waits for the 8031's internal timer to time out. Since proper sampling of a continuous waveform requires sampling at a fixed rate, this precise time delay is what provides the host program with data samples that the FFT can successfully analyze.

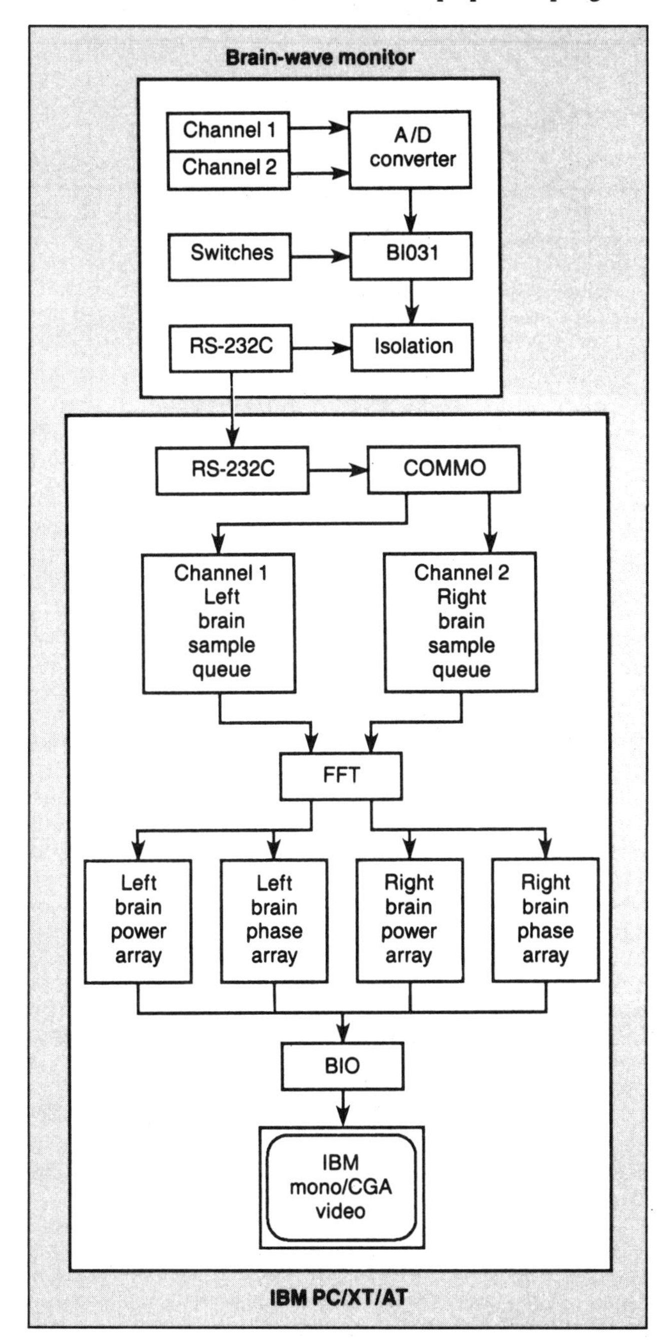

Figure 1: *A block diagram of the complete HAL system, showing major components of the brain-wave monitor and the host machine.*

The IBM PC Communications Interface

The IBM PC's COM1 port is configured as the headset interface port. Recall that the data coming from HAL arrives at 4800 bps in packets of 5 bytes, transmitted 64 times a second. The first byte indicates HAL's switch configuration, and the host computer uses this byte for synchronization purposes.

A switch position/sync byte value of 00000011 binary indicates that both the left and right switches are closed; a value of 00000000 indicates they are both open. The maximum value the switch position/sync byte can have is 3. BIO31, as one of its functions, tests all outgoing data and limits the lowest A/D channel data to a value of 4. Since any incoming data with a value of 3 or less must therefore be the switch position/sync byte, it is possible for COMMO to synchronize easily with the incoming data stream.

COMMO's service routine uses a state machine to synchronize itself with the incoming data (see figure 3). It reads the data byte from the Data Receive register and stores it according to which state the routine is in. If state equals 0, it saves the switch bits and advances to the next state. If state equals 1, it loads the sample data into the left brain sample queue. If state equals 2, it loads the sample data into the right brain sample queue. If a value of 3 or less is encountered, the state will automatically change to 0, and it will save the data as switch parameters.

The Software Prism

After the computer receives HAL's data, the FFT takes over, transforming the collected information into something useful. In this application, I wanted to see what frequencies are active in whatever brain is connected to the EEG and how powerful those frequencies are.

My first impression of the FFT was stated earlier: a mathematical prism. I've read articles that have used the FFT on everything from spectral analysis and digital filtering to biorhythms and random numbers. But what does the FFT do here?

Consider figure 4. The four sine waves on the bottom represent energy levels at four different frequencies plotted with respect to time. The vertical bars represent a 1-second sample window. (The 64 dots on each graph between the two vertical lines represent $\frac{1}{64}$-second intervals.) The top squiggle is an integrated waveform, the sum of the four waveforms below it at any given point in time.

This integrated waveform is similar to what we find in the real world. I used values generated from this integrated waveform to test HAL's software. The results of the FFT on this waveform appear on the screen shown in photo 1. The bars are read side to side, with their respective frequencies labeled in the middle. Energy levels are defined both above and below the bar charts. Notice that I applied the same test data to both the left brain and right brain sample queues. I will discuss the display in greater detail later.

There is a little more to structuring an FFT than just using an algorithm out of a book. Without getting into any mathematical formulas, I will explain the impact of modifying FFT parameters. The text box "Understanding the Fast Fourier Transform" on page 379 explains the FFT using a version written in BASIC. The initial lines of code are configuration data, and only the last 11 lines perform the actual transform on the data in the array RAWDATA.

You must address three major factors when building a simple FFT algorithm: the duration of the sample window, the sample

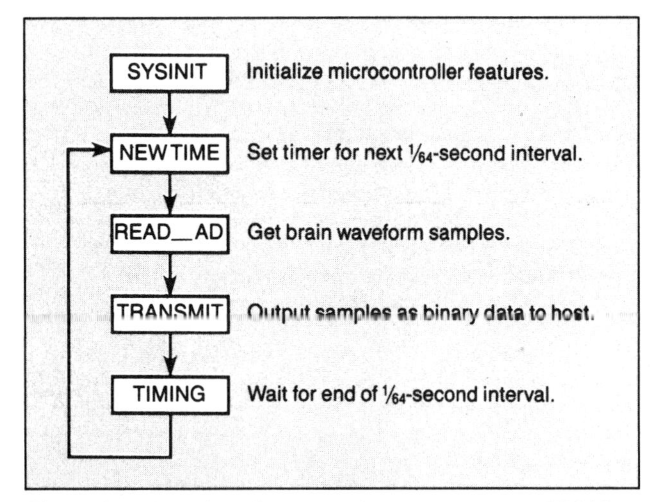

Figure 2: *A flowchart for HAL's firmware program, BIO31.*

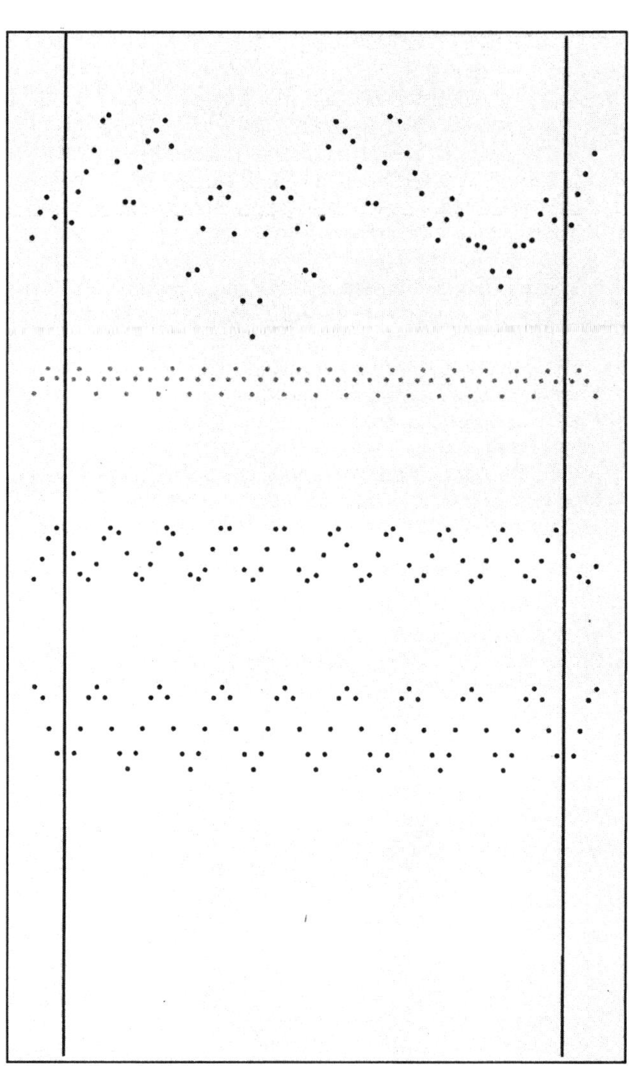

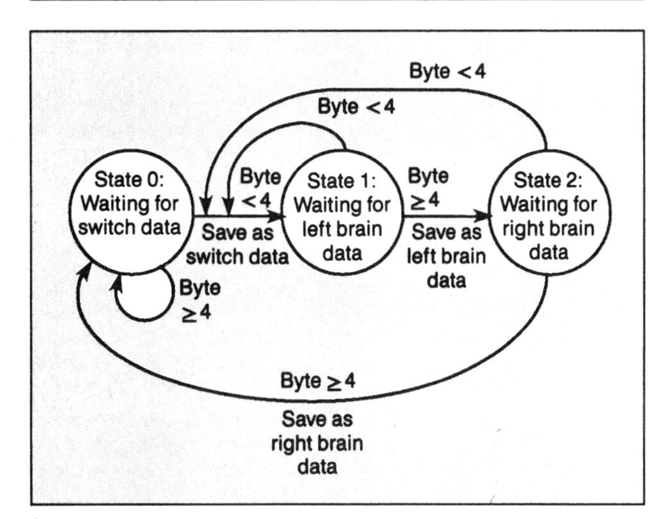

Figure 3: *The host computer's receiver routine executes a finite-state machine, shown here graphically. Each circular node of the graph represents a state, and transitions from state to state occur when the system receives a byte.*

Figure 4: *If you add together the lower four sine waves, you get the topmost, random-looking waveform. The FFT does just the reverse: It takes the composite waveform and extracts its components.*

rate, and the actual frequencies desired for output. These factors are identified as:

- `WDW`: The sample window in seconds.
- `PT`: The number of samples to be taken in the sample window.
- `MIN` and `MAX`: Lowest and highest frequencies to be evaluated.

The `FREQ` array represents the sinusoidal frequencies to be integrated into the waveform I talked about earlier, `A` represents the peak amplitude of these sine waves, and `P` serves to introduce a phase offset in a given channel. The variable `SM` specifies the starting offset of the points of data to be analyzed by the FFT. Graphically, `SM` locates the vertical bars on the display seen in figure 5.

Faster FFT

Speed was a key issue in this project. For a visual display to be of any value to a user, the software has to analyze and display the brain-wave data as quickly as possible. Ideally, you should

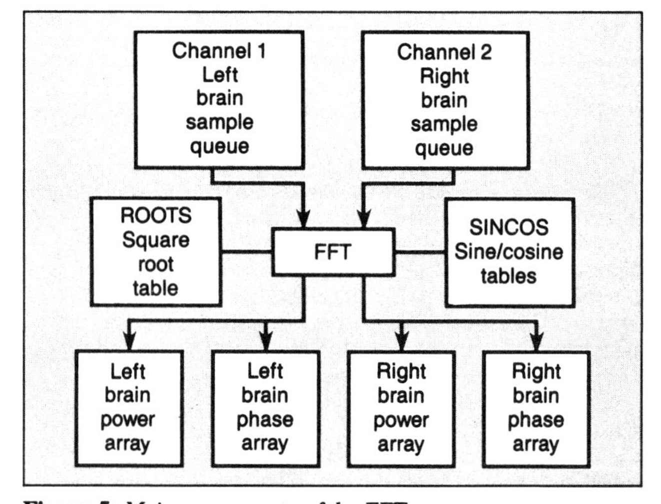

Figure 5: *Main components of the FFT.*

Understanding the Fast Fourier Transform

Listing A is a heavily commented code fragment in BASIC that performs the FFT on an array of data points (RAWDATA). The following are important program variables and their usage in the program.

WDW, the sample window: The time period over which the sample is taken will affect the resolution you can expect in your output. A period of 1 second will yield 1-Hz intervals of frequency. This is the way the program is set up now. If you reduce the sample window to half a second, the resolution is halved and the frequency output will be in 2-Hz increments. By increasing the sample window to 2 seconds, you get 0.5-Hz intervals.

PT, the sample rate: The number of points taken during the sampling period determines the range of frequency increments (dependent on window period) you can calculate. According to the Nyquist sampling theorem, you must sample at a frequency of at least twice as fast as the greatest frequency you expect to encounter. If your high end is 40 Hz, you must take at least 81 samples during that sample window. HAL takes 64 samples during a 1-second window, so it should be able to evaluate frequency output up to 30 Hz.

MIN and MAX, the output frequencies: The countervariable HZ need only be the frequencies desired. The FFT algorithm is really an accelerated version of what is called a discrete Fourier transform (DFT). The DFT calculates frequency output from 0 to the number of points taken. But this output includes something called negative frequencies. These mathematical monkeyshines can be ignored for the purpose of understanding this project. However, FFT.BAS can easily be tweaked to display this graphically, also.

Listing A: *This BASIC program performs the FFT on an array of data points.*

```
REM FREQUENCY(Hz)
FREQ(0)=16:  REM  Channel 1
FREQ(1)=9:   REM  Channel 2
FREQ(2)=8:   REM  Channel 3
FREQ(3)=2:   REM  Channel 4

REM AMPLITUDE
A(0)=10:  REM Channel 1
A(1)=20:  REM Channel 2
A(2)=30:  REM Channel 3
A(3)=40   REM Channel 4

REM PHASE  (Radians)
P(0)=0:   REM Channel 1
P(1)=0:   REM Channel 2
P(2)=0:   REM Channel 3
P(3)=0:   REM Channel 4

WDW=1.0: REM Sample period in seconds
PT=64:   REM Number of sample points
MIN=2:   REM Min freq to evaluate (1 - 25)
MAX=16:   REM Max freq to evaluate (MIN - 25)

SM=040:   REM Offset to start of sample period

FOR HZ = MIN TO MAX
REM ** Clear the real and imaginary
REM ** components.
  REAL = 0
  IMAG = 0
REM ** Begin the frequency analysis
REM ** loop.
  FOR X=0 TO PT-1
REM ** Calculate reference angle
    G = 2*PI*HZ*X/PT
REM ** Calculate real components
    REAL = REAL + RAWDATA(X)*COS(G)/PT
REM ** Calculate imaginary component
    IMAG = IMAG - RAWDATA(X)*SIN(G)/PT
  NEXT X
REM ** Power calculation
  FFT(HZ,0) = 2*SQR(IMAG*IMAG + REAL*REAL)
REM ** Phase calculation
  FFT(HZ,1) = ATN(IMAG/REAL)
NEXT HZ
```

be able to see changes in hemispherical activity every half second or sooner.

Clearly, a high-level language was not going to do the trick. (Running the FFT algorithm above in compiled BASIC takes about 5 minutes.) Ultimately, I created a machine language version of the FFT algorithm described in FFT.BAS. Figure 5 shows the flow of sample data from the left and right brain sample queues through the algorithm and into the power and phase arrays defined in the BASIC program. On an IBM PC XT, it takes about 2 seconds to complete the entire feedback cycle; an IBM PC AT processes a cycle in about half a second.

To get the necessary speed, I used a variety of techniques to optimize the FFT:

• *Number of data points*: I chose 64 data points as the sample size for a few reasons. First, you can perform integer division by 64 using shift operations instead of a divide instruction. When I used the 8088's integer-division instruction, it took about 340 ms to calculate a single 64-point FFT. In contrast, division using shift operations takes about 120 ms to accomplish the same thing.

Second, taking the division by shifting into account, 64 points per second will give the resolution and the range needed to evaluate frequencies from 4 Hz to 20 Hz. The Nyquist sampling theorem dictates that you must sample a waveform at a frequency of at least twice its highest frequency component. Consequently, 32 points would not be enough samples to analyze anything above 15 Hz. On the other end, 128 points would give you a greater frequency range for analysis, but it would take considerably longer to calculate.

• *Table lookup*: The FFT algorithm requires only a fixed number of sine and cosine values. By building these 64 values into a table rather than calculating them on the fly, you can save a lot

You can display HAL's output in either CGA or monochrome.

of time. Using a table of square roots is similarly much faster than calculating them.

• *Scaling*: When I built the tables, I took care to scale the integer values so power calculations can divide by 256. This magic number lets the software shift bytes instead of bits, giving considerable savings in time.

• *Phase octants*: Because the accuracy of this device is limited to integer calculations—and again because of speed—the program reports phase results in 45-degree increments. Instead of using the arctangent to calculate phase results, it is quicker in machine language to determine them using the sign and absolute amplitude of both real and imaginary components. The result is in phase octants.

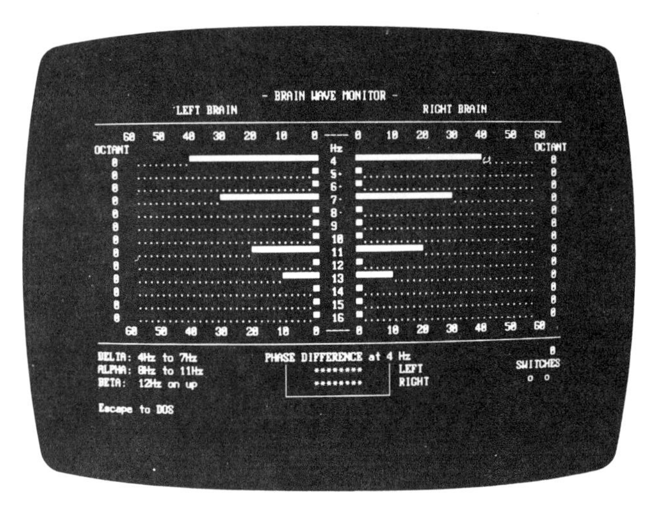

Photo 1: *HAL properly unravels the complex waveform shown in figure 4. Note the four long horizontal bars; each corresponds to one of the four component sinusoidal waves.*

Finally, the FFT is a much deeper subject than I can possibly cover in this chapter. I recommend a book entitled *The FFT: Fundamentals and Concepts* by Robert W. Ramirez (Prentice Hall, 1985). It provides an excellent look at the FFT and the theory behind it.

Driving the IBM Video Display

The last major block in the system diagram is called BIO. This module takes calculated output from the power and phase arrays just loaded by the FFT, converts this output into character strings, and writes them to the display. This part of the software displays the window opened by HAL into your brain.

Photo 2 shows a typical HAL display. You can display HAL's output in either CGA or monochrome. The horizontal bars represent the relative energy levels found in the frequencies evaluated. The bars that lead to the left represent the energy in the left hemisphere (channel 1); the bars to the right of center are the right hemisphere (channel 2).

Respective frequencies are labeled down the middle of the display under the "Hz" heading. The numbers 0 to 60 located in the rows above and below the energy bars represent relative energy levels only (pressing F3 changes the scale). They do not necessarily indicate absolute voltage levels that a much more expensive EEG machine might provide.

The phase angle of any given frequency is shown under the "OCTANT" heading for its respective hemisphere. An octant is nothing more than an eighth of a cycle. Octant 0 means the waveform is within the first 45 degrees of its cycle. A value of 1 puts it between 45 and 90 degrees, and so on.

Relative phase is indicated in a box at the bottom center of the screen under the "PHASE DIFFERENCE" heading. This feature selects the frequency from the left hemisphere with the greatest energy level and compares its phase angle with the phase angle of the corresponding frequency in the right hemisphere. The two rows of asterisks represent the position of the right hemisphere relative to the left hemisphere. If the right

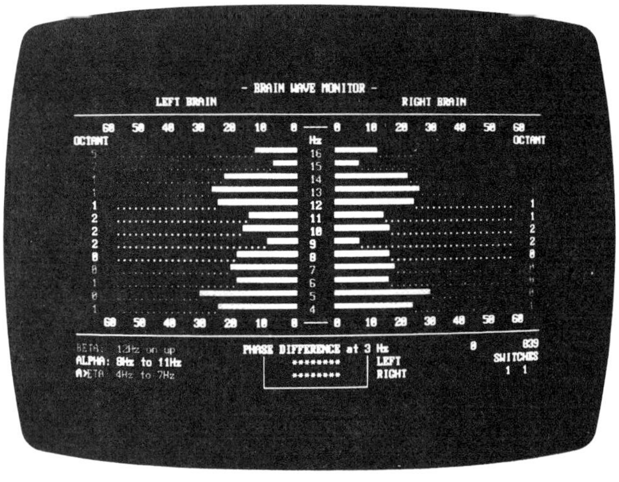

Photo 2: *HAL's output can be displayed on an IBM PC using the CGA.*

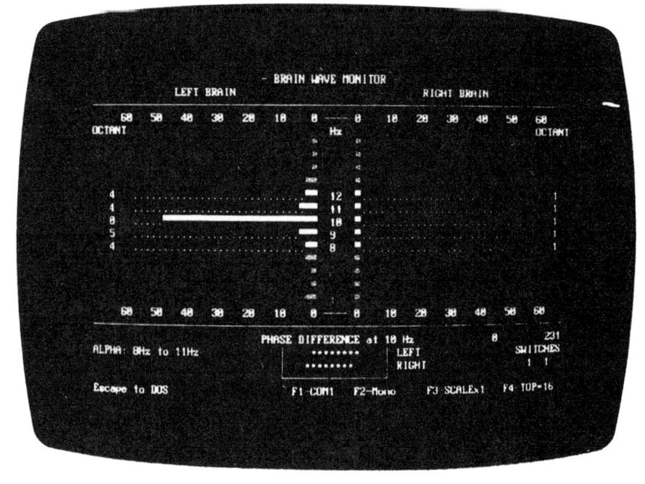

Photo 3: *While testing HAL, I injected a 10-Hz sine wave through the headset electronics. The resulting output, shown here, indicates that the system works properly.*

hemisphere is one octant ahead of the left hemisphere, the right hemisphere indicator will be advanced one column in front of the left hemisphere indicator.

Yes, But Does It Work?

How do we know that what we see on the display is really what is going on in our head? What makes data from the headset electronics different from a string of random data? Testing, of course.

I've already discussed the first test: Taking a waveform built of known sinusoidal components, passing it through the FFT algorithm, and verifying that the frequency and energy levels coming out are what I put in. The next level of confidence comes from using a sine-wave generator to force-feed the hardware with real data. Photo 3 is a sample of a 10-Hz sine wave being introduced into the headset electronics channel 1 input. Note that all other frequency bands are null.

Once we know how the FFT algorithm works—and we have the confidence that the hardware will deliver the data correctly—the rest is up to interpretation by the users. A sample session using the complete system seems to indicate that many frequencies are active. There also appears to be a difference between the right and left hemispheres in both amplitude and phase. But just what that means will have to be left to a person more involved with biofeedback.

Using HAL

HAL is one of the few Circuit Cellar projects in which the circuitry and software are relatively simple in comparison to the application. While the basic EEG apparatus has been in use for 60 years, we are just beginning to understand the "circuitry" and "software" operating within our own heads. Hundreds of volumes are devoted to clinical and research electroencephalography. I have included some possibilities for further reading at the end of the article; you may find them interesting as you explore this complex and fascinating field.

HAL provides an effective demonstration of how our gross behavior is in some way mirrored in the electrical activity of the brain. To view this behavior, however, you must learn how to connect HAL to your head.

You can purchase disposable EEG electrodes with adhesive pads and conductive gel from medical supply houses. Reusable silver/silver-chloride electrodes are also available. The reusable electrodes are more expensive initially, but you can use them almost indefinitely if you care for them properly.

Five wires are involved in a two-channel HAL connection: common reference, left-channel differential input pair (J1A and J2A), and right-channel differential input pair (J1B and J2B). You should construct the differential pair wires with shielded cable. I found that standard shielded microphone cable—with a male RCA connector at one end and a snap to mate with the electrode at the other end—is quite sufficient. The shield is, of course, grounded only at the RCA jack on HAL. Be careful not to accidentally short the shield to the differential input electrodes or HAL will produce erroneous results.

You can place the common reference electrode on the mastoid, the bony projection just behind the ear. You should first clean the area with soap and water to remove oils from the hair and scalp. Next, rub the area lightly with a piece of alcohol-dampened gauze. Finally, peel the adhesive from a disposable electrode, fill the well with conductive gel (but do not overfill), and place the electrode on the selected spot.

Follow a similar procedure to place the differential-pair electrodes on each hemisphere. One wire (J2A or J2B) goes over the frontal lobe—directly above the eye and just below where your hairline was before it started receding! The second electrode (J1A or J1B) is more difficult to place (unless you happen to be completely bald). You should put it over the occipital lobe, which in most people is covered with hair.

On the back of your head is a ridge, where your skull begins to bend inward toward your neck. Find a spot about a third of the way from the midline of your head to the common electrode placed on the mastoid and just on or below the ridge. Prepare the scalp as before, taking care to hold the hair carefully away from the site. You might try using a sweatband to hold the electrode in better contact with the scalp, making several small holes in the sweatband to allow easier access to the electrode connector.

Figure 6 shows the general location and nomenclature for commonly used electrode placements. If you build a multichannel model of HAL, you should refer to more detailed literature about electrode placement, monopolar and bipolar placement, and other esoteric subjects.

Assuming that you have successfully placed both channels of electrodes appropriately, it is time to fire up the software supplied with HAL. Remember, to retain its electrical isolation,

> **Warning**
>
> HAL is presented as an engineering example of the design techniques used in acquiring brain-wave signals. It is not a medically approved device, no medical claims are made for it, and it should not be used for any medical diagnostic purposes. Furthermore, the safe use of HAL requires that the electrical power and communications isolation described in its design not be circumvented. HAL is designed to be battery-operated only. Do *not* substitute plug-in power supplies.

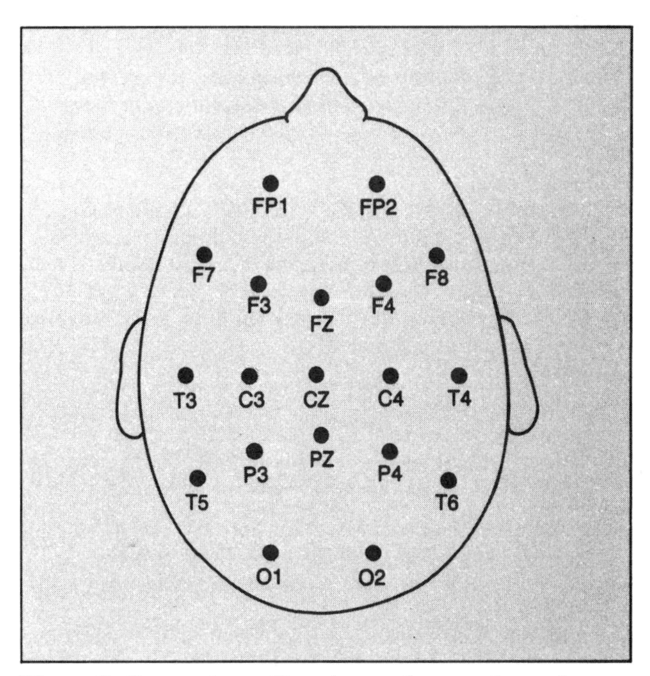

Figure 6: *Commonly used locations and nomenclature for electrode placement.*

you should operate HAL only on batteries, *exactly* as presented.

Attach the electrode wires to the electrode connectors, connect HAL to the computer's COM1 port, turn on HAL, load BIO.EXE into the computer, and watch your monitor. The sample display illustrated previously will let you see your own brain activity as you produce it. Actually, due to sampling and processing time, you will see it delayed by about half a second.

You may see little high-amplitude activity, and the dominant frequency may drift quite a bit (pressing F3 will change the amplitude range). But if you close your eyes and make your mind less blank than usual, your "lab partner" should see a dominant frequency appear somewhere in the alpha wave range. Of course, if you open your eyes to see it, it will immediately disappear. Somehow this is reminiscent of the old question about a tree falling in a forest making a sound or not. Since HAL's data is serial, you could store a session to disk and view it "off-line" later.

The display also shows the condition of the two optional switches that can be attached to HAL. The intent here was to provide a way for you to monitor either some external condition or internal state, so you can correlate it with your brain-wave activity. You could use two momentary contact switches, one for each hand perhaps, to see if you can correctly identify from internal cues some aspect of brain functioning (e.g., what band you are producing a dominant frequency in, whether the left or right hemisphere is showing greater power output for different mental tasks, and so forth).

Alternatively, some external source—the presence or absence of music, a strobe light, or other stimulus—could trigger the switches. Since you have two switches, you could test for four different conditions.

You can also modify the BIO code to turn HAL into a standard biofeedback device. In this case, you would want to add threshold controls. That is, the program could provide audio feedback only when you had achieved certain minimum values of relative amplitude within a certain frequency band. You could provide different feedback for each channel, or feedback could be tied into both channels meeting the same (or different) criteria.

A multichannel HAL would need a completely different display to provide meaningful feedback about the brain's activity. An EGA display could be mighty useful here, in some way showing activity in terms of color on a map of the brain.

A Final Thought

As I said earlier, we are only starting to understand the relationship between our brain-wave activity and our mental states. HAL is not intended to take the place of a $100,000 EEG analyzer in a modern neurologist's lab, but it certainly can provide the serious experimenter a valid vehicle for entry into the fascinating world of neuroscience.

BIBLIOGRAPHY

Brown, Barbara. *New Mind, New Body: Bio-Feedback; New Directions for the Mind.* New York: Harper & Row, 1974.

Kooi, Kenneth. *Fundamentals of Electroencephalography.* New York: Harper & Row, 1971.

Scientific American Editors. *Altered States of Awareness.* San Francisco: W. H. Freeman & Co., 1972.

Segalowitz, Sid J. *Two Sides of the Brain: Brain Lateralization Explored.* Englewood Cliffs, NJ: Prentice Hall, 1983.

Special thanks for help provided on this article to Dr. Robert Stek, David Schulze, Rob Schenck, Jeff Bachiochi, and Ed Nisley.

See page 447 for a listing of parts and products available from Circuit Cellar Inc.

34

BUILD THE CIRCUIT CELLAR IC TESTER

PART 1: HARDWARE

This versatile tester can save you hours of troubleshooting when building and debugging electronic systems

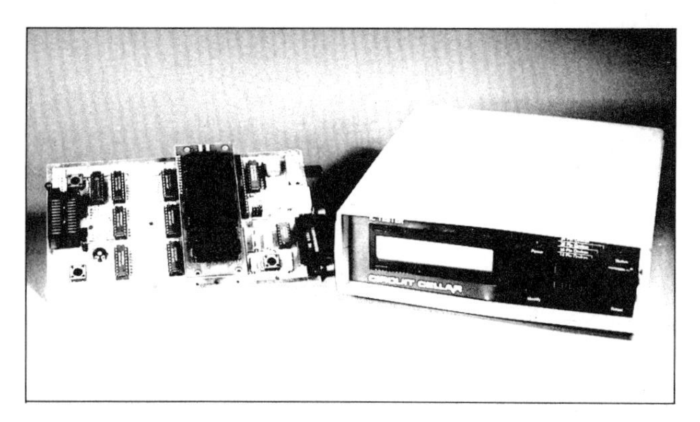

Having designed and debugged many electronic systems, I have seen more than my share of defective ICs. I have also wasted more time than I care to remember discovering that my latest creation was not deficient, but that one of the factory-fresh ICs I put in it was in fact defective. You'd think they'd test them, wouldn't you?

An IC tester can provide both time savings and increased confidence when building and debugging electronic systems. In fact, finding defective ICs before manufacturing an electronic product can also save a considerable amount of money by minimizing the labor and board damage costs involved with reworking electronic boards.

For the most part though, IC testers are used for repairing failed electronic circuits. My latest example was my home: While I was preparing this project, lightning struck my house and practically everything got blitzed. If it were not for my IC tester's help in finding the 29 blown chips in my home-control and automatic-lighting system, I'd still be sitting in a dark, dead house (I thought I had added every preventive measure I could, but I can see we'll need another project on transient protection). I was especially thankful that it could successfully test open-collector driver chips—a problem for most economical testers.

Having an IC tester saved my day, and it may be something you have always needed, too. In this part of Chapter 34, I will describe the design and construction of a digital IC tester with tutorial emphasis on the thinking I had to go through in the process of building it. In Chapter 34, part 2, I will conclude with a discussion of the specific operation of this tester and its advanced software.

Design Considerations

The first step in designing any project is to carefully consider and define what the device is to do. For the IC tester, I first looked at units already on the market and noted their features, prices, deficiencies, and benefits.

I found a price range that varies from less than $200 to several thousand dollars. They also vary considerably in their operation and capability. The low-cost units are generally bus-specific—plugged into a computer slot (Apple II or Commodore 64)—and include operating software. Up the scale from those are the stand-alone—but relatively "dumb"—IC identifiers. With these, if you put a good chip into the socket, a two-, three-, or four-digit number indicating its identification appears on the seven-segment LED display.

The low-end (less than $1000) testers I found have fixed device libraries and perform only simple digital tests (i.e., no AC-parametric tests and no logic-threshold tests). Most, however, indicate that they do provide "periodic" library updates as new standard parts become available.

The high-end testers, costing several thousand dollars, allow some AC-parametric testing, threshold testing, and testing of analog ICs. While they are probably incapable of verifying complete compliance to manufacturers' data sheets, they certainly come close. They can help identify chips with marginal timing specifications. The cost of these devices (including the cost of maintenance, special adapters, and new device support) makes them prohibitive to ordinary users; such devices typically find their home in large corporations with special testing requirements (often those involved with military or aerospace applications).

Flexibility at an Economical Price

My goal in developing the IC tester was to provide as much capability and flexibility as possible in an affordable device that can be used by small businesses and electronic experimenters.

Certainly, economics played its part in requiring compromises in the design. I decided that AC-parametric testing and threshold-level testing would put the device into a higher-price category than I was targeting, so these features were the first to go. Then, I needed to determine what the user interface should be like.

One possibility was to design a card that plugged into an IBM PC slot, with an external test box connected by a cable. This approach would let me develop and include PC software permitting users to develop tests for their own devices. This would include standard devices not yet in the master library and custom devices, like programmable array logics. Unfortunately, this limited the use of the tester to owners of PCs or compatibles (with a free backplane slot and a long extension cord), and the tester would hardly be portable.

Another possibility was to configure the tester to connect to a dumb terminal, or to any computer with terminal-emulation capability, via RS-232C. While this would broaden the number of potential users of the tester, and would give the tester a little more flexibility, it would also take away the flexibility of user-generated device tests unless that extra (and I might add, very intensive) software capability was provided within the tester.

Finally, I could choose the pure stand-alone approach. Such a configuration would be a self-contained portable tester with its own display and some form of entry panel. Even though it's an easier concept, a stand-alone unit would be more expensive to build and would potentially have the same limitations as terminal-based testers unless it also contained the "smarts" of a larger computer.

Three Units in One

After considering the various circuit possibilities, I concluded that my IC tester should support all three modes of operation. With only a slight increase in hardware complexity, I could present a single design that operates in different ways depending upon which peripheral components and software you install (see photo 1). The operating configurations are called PC-host mode, terminal mode, and stand-alone LCD mode.

The PC-host and terminal modes simply require a serial port for operation. In terminal mode, the tester presents all statements regarding test functions and results on the video terminal's display. The PC-host mode is similar, with the exception that it has the added flexibility of letting you directly modify and extend the device library.

In the stand-alone LCD mode, the tester shows device parameters and data on a 2-line by 20-character LCD. (It should be noted that the LCD is optional; you can operate the tester in the other two modes without it.)

In essence, the stand-alone LCD mode provides a portable (i.e., battery-operated) IC tester suitable for testing any chips that are precoded within its extensive EPROM-resident device library. (The Revision 1.0 library currently contains about 600 74xx00-series and CMOS 4000-series chips.)

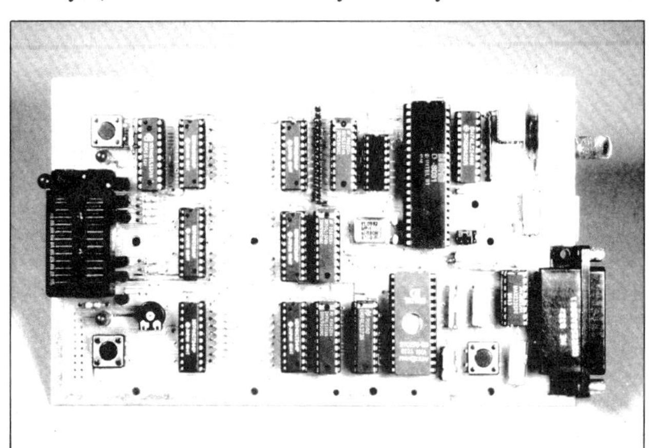

Photo 1: *The prototype IC tester printed circuit board configured for terminal operation. The IC under test is inserted into a special zero-insertion-force socket on the left side, and test information and menu selections are displayed on a terminal connected through the DB-25 connector on the right side.*

The terminal mode provides a menu format intended to maximize the information displayed, while the PC-host mode converts this otherwise stand-alone piece of hardware into an interactive and configurable diagnostic tool with the intelligence of a full computer.

Testing Logic ICs

How do you go about testing ICs? Certainly, I had to answer this question before I could design the tester.

Testing 7400-series logic devices appears relatively straightforward (I didn't consider AC and voltage threshold checking for reasons of economics). To test a two-input NAND gate, for example, you merely set specific logic levels on the gate inputs and check that the outputs are what they are supposed to be.

The process involves a series of test vectors. A test vector is a pattern of bits (0s and 1s) applied to the inputs of the device under test (DUT), to which the DUT responds with a response vector (a pattern of bits on the DUT's outputs). You then compare the response vector from the DUT to the expected response vector, with bit differences indicating pin failures.

You can specify any number of test vectors for a device, allowing you to test the chip as completely as you desire. For each test vector specified for a device, you must also specify a corresponding expected response vector. Since there are cases when some outputs of a device may be in an unknown state, you must also provide a "don't care" mask for each expected response vector, indicating which bit comparisons the tester should ignore.

One significant difference between my IC tester and others in the same price range is that mine does a full-function logic test using as many vectors as necessary to exercise all logic possibilities on the test device. Most inexpensive testers don't do this.

So Many Logic Families

Unfortunately, real-world electronics doesn't quite follow theory. Specifying test vectors is only part of the job. Dealing with all the electrical parameters of the various IC logic families is the real problem.

Since its initial development and introduction by Texas Instruments, the 7400 series of ICs has become an industry standard—at least in terms of device functions and pin-outs. These chips are composed of a large variety of SSI-, MSI-, and LSI-logic building blocks, which designers put together to produce the desired functions.

The original 7400-series family consisted primarily of simple functions, like gates and flip-flops. These were adequate for many applications, but designers kept demanding devices with increasingly greater complexity and functionality.

IC technology did not stand still as designers needed more devices with higher speed and lower power. These requirements led to the introduction of the 74H00-series (high-speed) and 74L00-series (low-power) devices. For the most part, these new series maintained the device pin-outs established by the standard-TTL predecessors (the 7400 series). However, the 74H00-series devices consumed substantially more power than, and the 74L00-series devices were slower than, the standard 7400-series devices.

As the technology improved, even more families appeared. A faster family using Schottky technology was established, the 74S00 series, along with a popular low-power Schottky family, the 74LS00 series.

Eventually, the very-low-power CMOS devices that had been manufactured with 4000-series numbering shifted over to the more popular 74xx00-series pin-out and numbering scheme with the introduction of the 74C00-series family of devices. These devices were slow and had low-current-drive outputs, but they filled a niche in designs requiring extremely low power consumption.

Other families include 74ALS (advanced low-power Schottky), 74AS (advanced Schottky), 74HC (high-speed CMOS), 74HCT (high-speed CMOS, TTL-compatible), 74AC (advanced CMOS), 74ACT/74AHCT (advanced CMOS, TTL-compatible), and 74F (Fairchild advanced Schottky).

Simple Concept, Tough Trade-offs

Digitally speaking, the logical parameters of a 74xx00 are the same regardless of its family, and you could easily be misled into thinking that we are designing a digital tester. However, each of these families has analog characteristics that differ from the other families. The IC tester is actually more an exercise in analog design. Let me explain.

Typical differences between logic families are power consumption, speed, output current drive, input current loading, input transition thresholds, and output voltage swings. Comparisons of some of these parameters for a 74x00 quad NAND gate from several families are shown in table 1. (While the parameters specified in table 1 for the 74x00 devices do not apply to all devices within the respective families, they are representative of the majority of the devices).

In effect, table 1 shows the wide variations of input and output parameters that the ideal IC tester must support. Low-level input currents range from 1 microampere to 2 milliamperes (and much higher on some device inputs), and low-level output currents range from 360 μA to 20 mA.

The tester's ability to identify a device presents an important consideration. If the tester is designed for 74ALS or 7400 "straight" TTL, you might smoke a 74C chip if you inserted it into the tester operated at the current levels of those devices.

Any truly general purpose (read usable) tester must accommodate the wide ranging voltage and current parameters of all the families. Since the tester may not know at the outset what device is installed in the ZIF (zero insertion force) socket (remember, one of the modes is to identify unmarked chips), it cannot make any assumptions as to which pins are inputs and which are outputs.

The tester requires a certain amount of trial and error to identify an unknown device, and it must employ current-limiting resistors between the DUT (in the ZIF socket) and the IC tester's vector-generation circuitry (for when a DUT and tester output are connected together).

Also, while most devices have totem-pole outputs, some have tristate, open-collector, or open-drain outputs. The tester must be able to pull tristate outputs high and low when they are in the high-impedance state to verify the state, and it must also be able to pull open-collector and open-drain device outputs high and low to verify proper operation.

The catch-22 is to determine a resistor value that will support the input and output current specifications for all the device families to be tested, yet not overstress the DUT. If you go strictly by the book, no single current-limiting resistor value works for both inputs and outputs in all families.

The device specifications provided in table 1 are the manufacturer's recommended operating conditions (ROCs). Looking further into the data sheets, however, we find more information regarding what the chips can do if they have to, such as limited-duration short-circuit output current.

In effect, if we take advantage of our regulated testing environment, we can stretch the ROC a little to choose a resistor that presents the best compromise for handling all the logic families. Think of it as the electronic equivalent of poetic license.

All things considered, I found that the resistor value should be in the 390- to 421-ohm range. Since 390 ohms is the nearest standard resistor value (5 percent tolerance), I chose it for the tester. (After I built the tester, I substituted all standard resistor values between 300 and 430 ohms, inclusive, and verified that the 390-ohm choice provides the best overall performance.)

How It Works

After determining the above, I had one more hurdle. The tester needed to be able to apply virtually any number of test vectors to the DUT without losing the device's state from the previous vector—and without causing undo stress on the DUT (i.e., without keeping any of the DUT outputs in a high-current output mode for an extended period of time). I solved this with what I like to refer to as a combinatorial-latch circuit.

Each ZIF-socket pin typically has three circuit connections to the IC tester (see figure 1). One connection (connection A) is to

Table 1: *Comparison of specifications for various 74xx00 devices. (Subscript identifiers are IL—input low, IH—input high, OL—output low, and OH output high.)*

Device name	I_{IL} max	I_{IH} max (μA)	V_{IL} max (V)	V_{IH} min (V)	V_{OL} max (V)	V_{OH} min (V)	I_{OL} max (mA)	I_{OH} max (mA)
74LS00	-0.4 mA	20	0.8	2.0	0.5	2.7	8.0	-0.4
74H00	-2.0 mA	50	0.8	2.0	0.4	2.4	20	-0.5
74L00	-0.18 mA	10	0.7	2.0	0.4	2.4	3.6	-0.2
74S00	-2.0 mA	50	0.8	2.0	0.5	2.7	20	-1.0
74AS00	-0.5 mA	20	0.8	2.0	0.5	2.5	20	-2.0
74ALS00	-0.1 mA	20	0.8	2.0	0.5	2.5	8.0	-0.4
74HC00	-1.0 μA	1.0	1.2	3.15	0.33	3.84	4.0	-4.0
74HCT00	-1.0 μA	1.0	0.8	2.0	0.33	3.84	4.0	-4.0
74F00	-0.6 mA	20	0.8	2.0	0.5	2.7	20	-0.36
74C00	-1.0 μA	1.0	1.5	3.5	0.4	2.4	0.36	-0.36
7400	-1.6 mA	40	0.8	2.0	0.4	2.4	16	-0.4

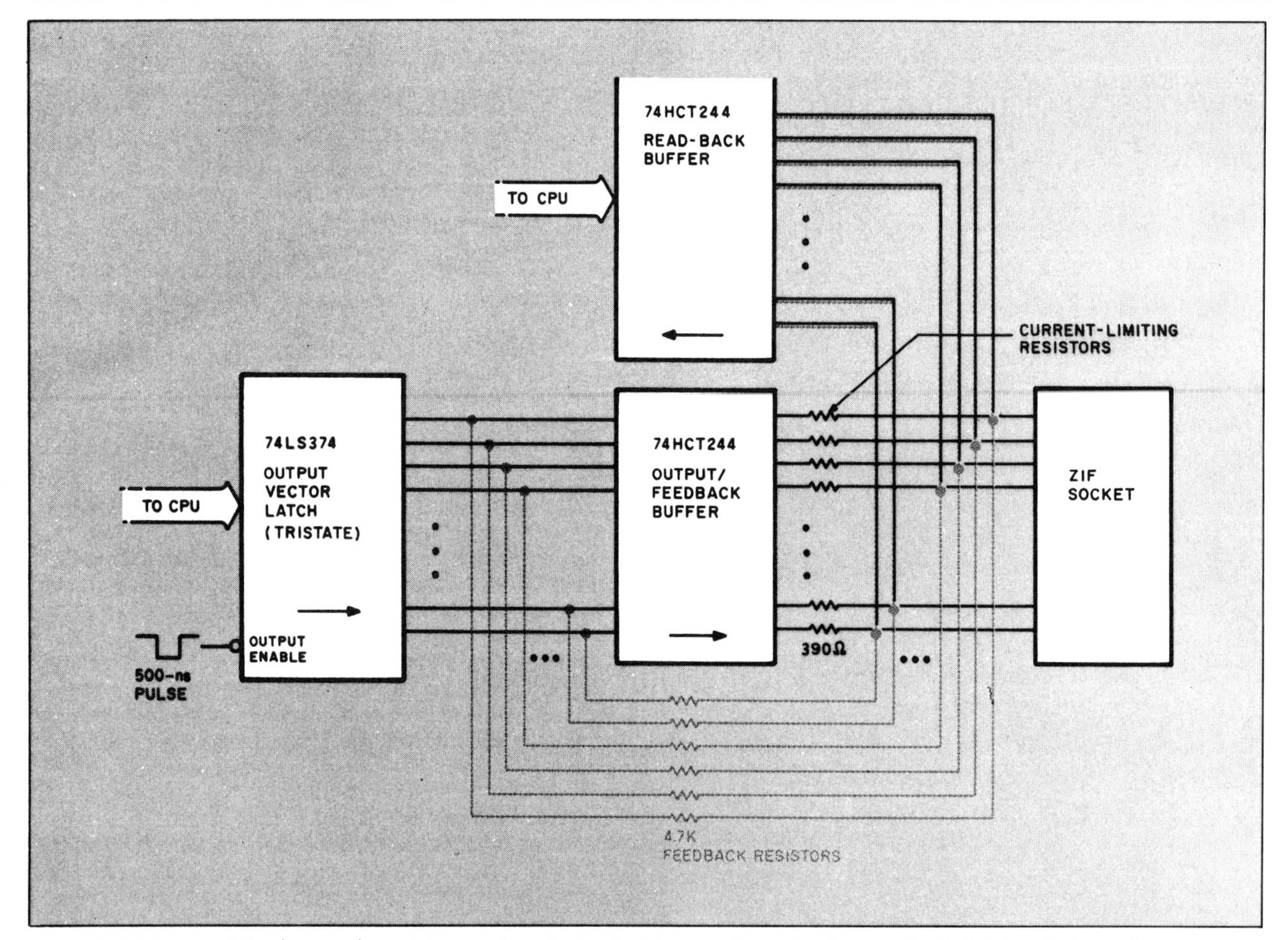

Figure 1: *Diagram of the IC tester's combinatorial-latch circuit. The zero-insertion-force socket holds the device under test.*

The DB-25S connector provides the RS-232C interface connection to an IBM PC or any dumb terminal.

an output of a 74HCT244 buffer—the feedback buffer—through a series 390-ohm current-limiting/load resistor. Another connection (connection B) is to the corresponding input of the same 74HCT244, through a 4.7-kilohm series feedback resistor. The 74HCT244 input is also connected to an output of a 74LS374 tristate latch.

The final ZIF-socket-pin connection (connection C) is directly to an input of another 74HCT244 tristate buffer—the read-back buffer. By reading the 74HCT244 read-back buffer, the processor can determine the logic levels of the DUT pins (the ZIF-socket pins).

The IC tester sends a test vector to the DUT by writing the desired bit pattern into the 74LS374 latch, while the latch's outputs remain in the high-impedance state. The system then enables the outputs of the 74LS374 (i.e., they are allowed to go active) for a period of 500 nanoseconds, applying the test-vector bit pattern to the inputs of the feedback 74HCT244 buffer.

During the 500-ns 74LS374-enable period, the relatively high value of the feedback resistors (4.7 kilohms) ensure that the 74HCT244 inputs will see the test-vector logic levels from the 74LS374, regardless of the logic levels present at the DUT pins.

Within a few nanoseconds (i.e., propagation time) of the time the feedback 74HCT244 first sees the new logic levels from the 74LS374, the same logic levels will appear on the outputs of the 74HCT244; these logic levels will remain on the 74HCT244 outputs for the duration of the 500-ns pulse.

If a DUT output in the ZIF socket is in the opposite logic state as the corresponding 74HCT244 output, the resistor between the 74HCT244 output and the DUT pin will present a load to the DUT output, possibly causing it to go into its "overdrive" mode in an attempt to retain its desired output logic level. The overdrive operation will continue until the end of the 500-ns pulse, when the 74LS374 outputs are finally disabled, returning to their high-impedance state.

When the 74LS374 outputs are disabled, the only inputs to the feedback 74HCT244 will be from the DUT feedback resistors. Since the feedback buffer is a 74HCT-series device, it presents negligible input current loading (about 1 μA), so the voltage levels reaching the 74HCT244 inputs through the feedback resistors will be nearly the same as those at the corresponding DUT pins.

If the voltage coming through a feedback resistor to the 74HCT244 is the same logic level as that presented previously by the enabled 74LS374 output (the case when the DUT pin is an output of the same logic level or when the DUT pin is an input), the 74HCT244 output will remain unchanged. Thus, the logic level is combinatorially latched by the 74HCT244.

If the voltage appearing at the 74HCT244 input from the feedback resistor is the opposite logic level of that presented previously by the 74LS374 (which is the case when the DUT pin is an output of the opposite logic level), the 74HCT244 will see the new logic level at its input and change its output to match. When this occurs, the 74HCT244 output then matches the output of the DUT pin, eliminating the loading that was present. Again, the new logic value will be combinatorially latched by the 74HCT244 using the feedback loop.

You can see that the loading duration on a DUT output will essentially be the duration of the enable pulse—only 500 ns. This keeps potential chip stress to a minimum, while verifying the ability of device outputs to operate properly under load conditions.

The IC Tester Hardware

The schematic for the IC tester is shown in figure 2. The 8031 single-chip microcontroller (IC1) is the brains of the tester. The firmware to run the tester is provided in an EPROM at IC6. The current standard device library (version 1.0) is supplied on a 27256, but IC6 can accommodate several EPROM types, including 2764, 27128, and 27512 devices. The type you would use is determined by the JP1's jumper configuration.

The ZIF socket (IC17) is an Aries universal socket. This specific socket supports devices up to 24 pins, having either 0.3- or 0.6-inch DIP-package widths. When you insert devices into the ZIF socket, you bottom-justify them.

Unfortunately, one problem with using a single ZIF socket on a tester is configuring the power pins for the DUT. Most ICs conform to the standard diagonally opposite corner-pin power/ground configuration: pins 24/12, 16/8, and 14/7. However, a number of devices have oddball power and ground pin-outs. These include 14-pin ICs with ground on pin 11 and power on pin 4, 16-pin ICs with ground on pin 12 and power on pin 5, and 16-pin ICs with ground on pin 13 and power on pin 5, among others (there are also devices with two power pins to support voltage-level conversion).

After reviewing the devices in each oddball pin-out category, I chose to support the two categories with the most devices: 14-pin devices with ground on pin 11 and power on pin 4 and 16-pin chips with ground on pin 12 and power on pin 5. This is, of course, in addition to supporting devices having corner power and ground pins. (In the stand-alone identify-unmarked-chip operating mode, the tester will successfully identify only corner-pin-powered chips.)

The DB-25S connector provides the RS-232C interface connection to an IBM PC or any dumb terminal. The connector is configured as a DCE (data communication equipment) device, allowing you to use a straight-through cable. You need only three pins on the connector (pins 2, 3, and 7—receive, transmit, and signal ground, respectively), but I've hard-wired the DTR (pin 6) handshaking line to a logic high for terminals that need it.

The IC tester has push buttons and some switch-selectable options. A four-position DIP switch (SW1) is used for several purposes, including data-transfer-rate selection, PC-host/terminal mode selection, and 74Cx mode selection (to be described next month). Push buttons PB1 and PB2 are for supporting stand-alone mode operation. PB1 is the identify button, and PB2 is the retest button.

J3 is the connector for the optional LCD, which uses the 8031's P1 connector as its data bus. I chose the P1 bus as the LCD's driver to meet the LCD's (relatively slow) timing requirements. The 74LS139 (IC7) is the address-decoding circuit for accessing several devices on the tester. It decodes the ZIF tristate latches (IC8 through IC10) and read-back buffers (IC14 through IC16), as well as the power/ground transistor latch (IC19).

The 74LS139 also provides a special signal that enables the outputs of the 74LS374 tristate latches for approximately 500 ns (the 8031 WR\ strobe duration), transferring the latched 74LS374 bits to the combinatorial latches formed by the 74HCT244s (IC11 through IC13) and their associated feedback resistors.

For the tester's buffers (IC11 through IC13), I chose 74HCT devices instead of 74LS (or other family) devices. Members of this family drive their outputs close to the power and ground rails, can source a lot of current, and provide negligible load on the resistor-feedback circuit. Similarly, the read-back buffers (IC14 through IC16) are 74HCT devices to keep loading to an

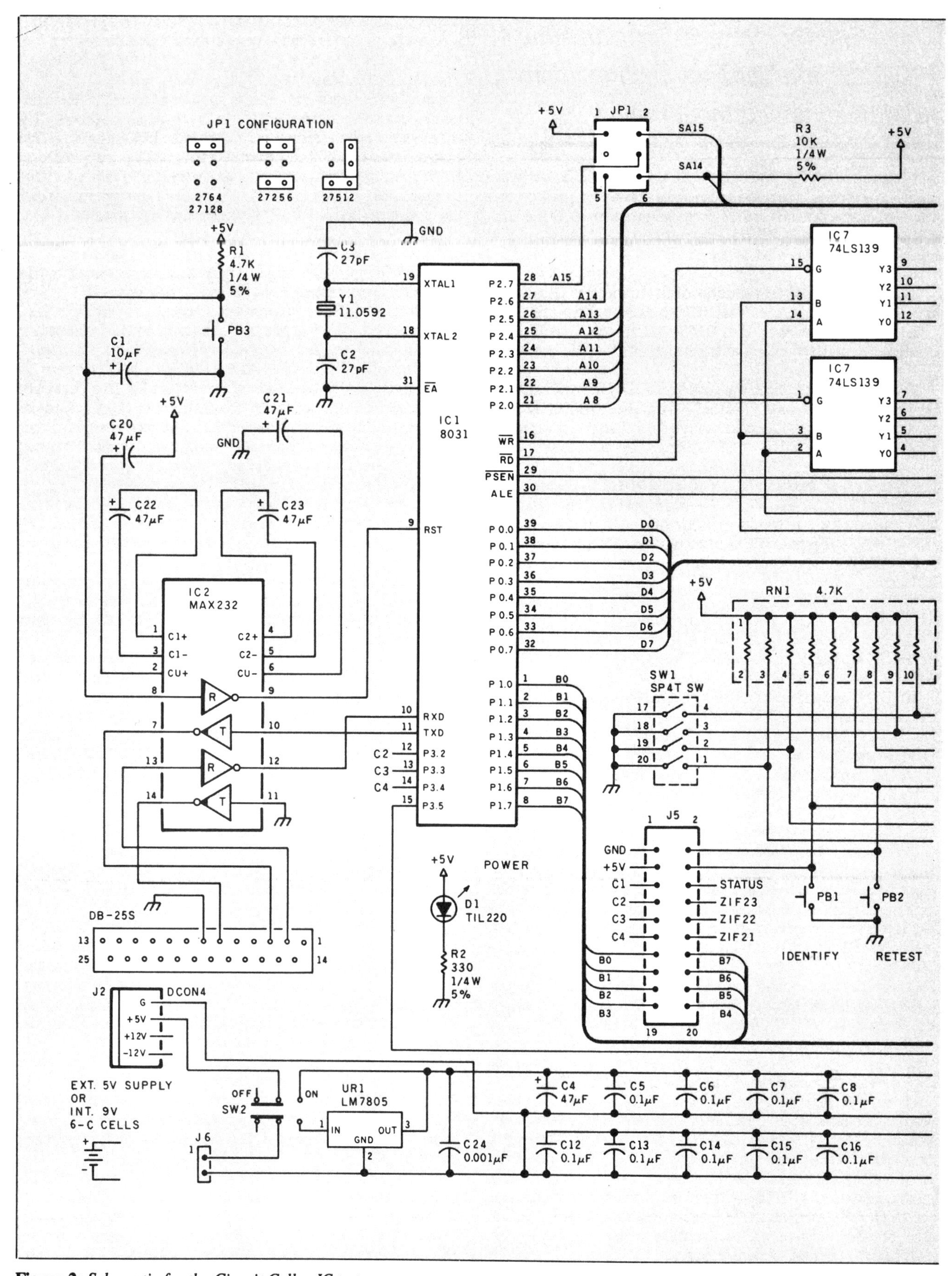

Figure 2: *Schematic for the Circuit Cellar IC tester.*

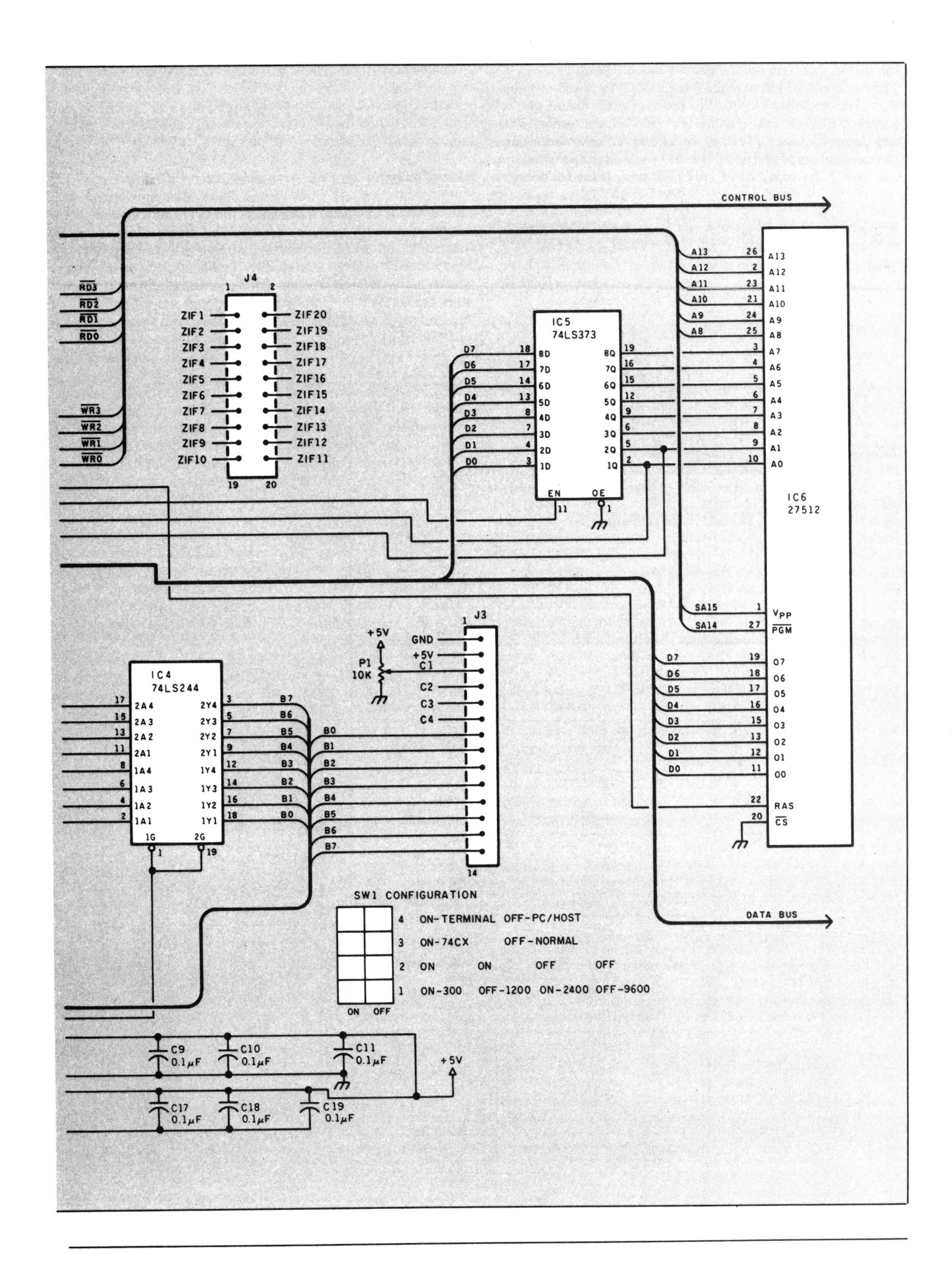

CONTROL BUS
J4
ZIF1
ZIF2
ZIF3
ZIF4
ZIF5
ZIF6
ZIF7
ZIF8
ZIF9
ZIF10
ZIF20
ZIF19
ZIF18
ZIF17
ZIF16
ZIF15
ZIF14
ZIF13
ZIF12
ZIF11
RD3
RD2
RD1
RD0
WR3
WR2
WR1
WR0
IC5
74LS373
IC6
27512
IC4
74LS244
J3
GND
+5V
C1
C2
C3
C4
P1
10K
SW1 CONFIGURATION
4 ON-TERMINAL OFF-PC/HOST
3 ON-74CX OFF-NORMAL
2 ON ON OFF OFF
1 ON-300 OFF-1200 ON-2400 OFF-9600
ON OFF
DATA BUS
C9 0.1μF
C10 0.1μF
C11 0.1μF
C17 0.1μF
C18 0.1μF
C19 0.1μF
+5V

absolute minimum (do *not* substitute 74LS devices).

The discrete transistors (Q1 through Q6) provide the power and ground switching for the ZIF socket (IC17). Pin 24 of the ZIF socket is connected directly to +5 volts, eliminating the need for an additional transistor. The PN2907s (Q3 through Q6) are for turning on power (+5 V) to various ZIF-socket pins (9, 19, 20, and 22), while the PN2222s (Q1 and Q2) are for turning on ground to two of the ZIF-socket pins (12 and 16).

The 74HCT374 latch (IC19) controls the transistors. As mentioned earlier, 74HCT devices can source and sink current equally well. This fact made the 74HCT374 a good choice for driving the transistors, since it can handle the transistor base currents equally well for the ground switches (high 74LS374 outputs) and the +5-V switches (low 74LS374 outputs).

The tester has two LEDs. D1 is merely a power-on indicator that lights whenever power is applied. D2 is a software-controlled status LED used to indicate when the device is operating in an RS-232C mode (PC-host or terminal, LED on) or a stand-alone mode (LED off).

Experimenters

While you can order printed circuit boards and kits for the Circuit Cellar IC tester, I encourage you to build your own. If you don't mind doing a little work, I will again support your efforts. A hexadecimal file of the executable code for the 8031 Revision 1.0 system EPROM code, suitable for stand-alone or terminal operation, is available for downloading from my bulletin board at (203) 871-1988.

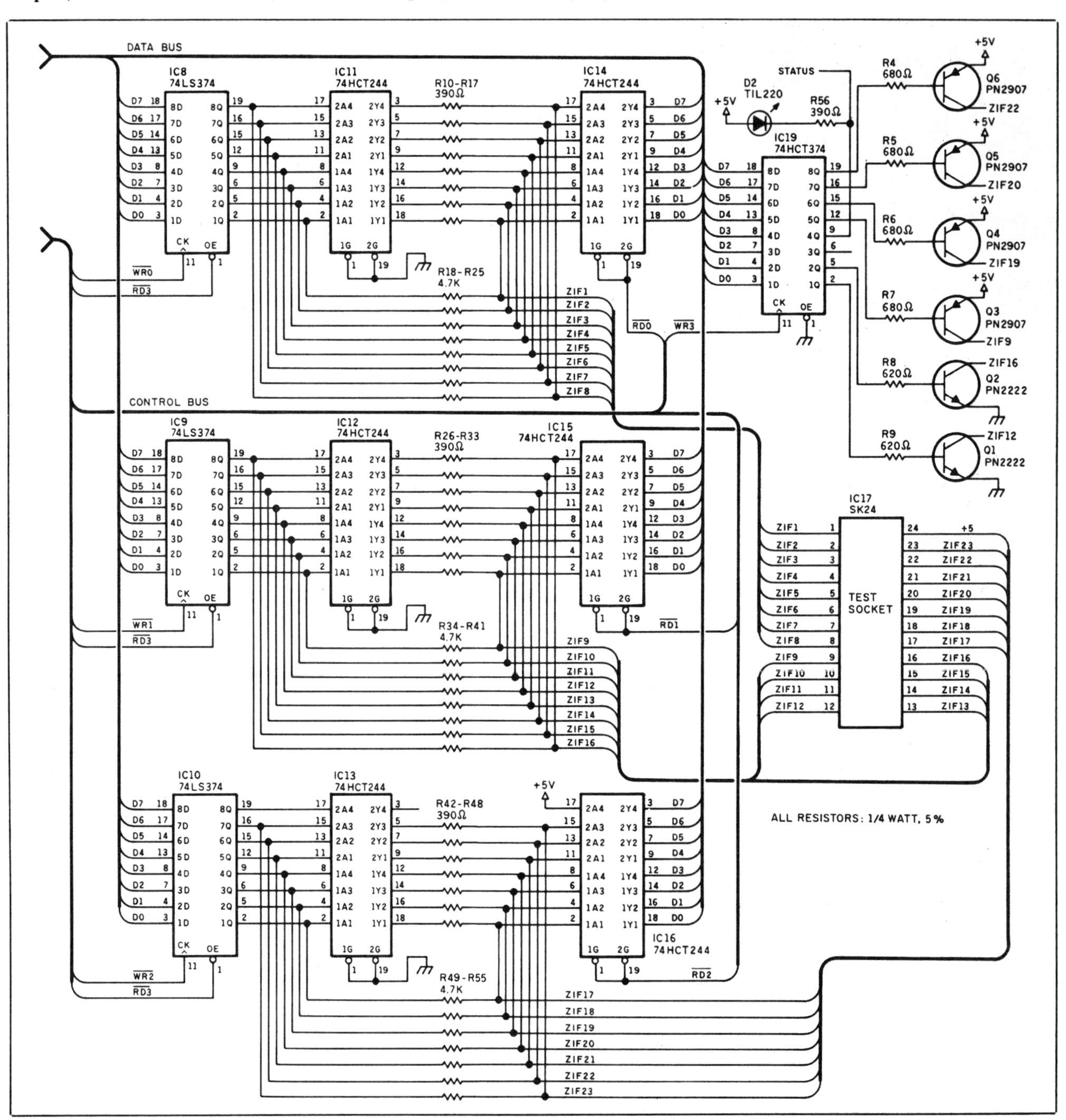

Figure 2: *Continued.*

34

BUILD THE CIRCUIT CELLAR IC TESTER

PART 2: SOFTWARE AND OPERATION

Steve guides us on a tour of the software that makes his inexpensive IC tester possible

In part 1 of this chapter, I talked about the design of my IC tester. In this part, I'll talk about its software and operation.

Three in One

To refresh your memory, the IC tester supports three modes of operation: PC-host mode, terminal mode, and stand-alone LCD mode.

PC-host mode requires that you connect the tester to a serial port on an IBM PC or compatible. In this mode, the PC handles all test-vector transfers and comparisons and provides the highest level of flexibility and power.

To operate the tester in terminal mode, you connect it to a dumb terminal or any microcomputer that emulates a terminal (see photo 1). The options are essentially the same as those offered in PC-host mode, although you can use only a fixed, ROM-resident device library.

The stand-alone mode of operation lets you operate the tester with only two push-button switches and a 2-line by 20-character LCD. As in terminal mode, this mode operates only with a fixed, ROM-resident device library. It lacks some features of the other two modes, but it permits device identification (using the Identify push button) and specified-device testing (using the Retest push button). The latter lets you determine specific pin failures on a bad IC and display this information on the LCD.

Much of the flexibility of the IC tester comes from its modifiable and expandable device library. While an IBM PC (or clone) is essential for PC-host mode operation, it is required if you're going to make any system software changes, like adding new chips to the library.

With the exception of a single assembly language serial-port driver, all the software was written in Turbo Pascal on an IBM PC. (While the programs do take advantage of some PC-specific features of Turbo Pascal, you shouldn't have much trouble converting them to other Pascal compilers.)

The Definition of a Test Vector

In order to define test vectors, it is important to develop a straightforward means of describing the vector information. What information do we need to define a device and its test vectors?

The device definition consists of the device name (e.g., 7400), the specific package size (e.g., 14 pins), the locations of the power and ground pins (e.g., 14 and 7), and which pins are inputs, outputs, or tri-state.

A test vector merely specifies the high (1) and low (0) logic levels to be written to the pins of the device under test (DUT). A test vector written to the DUT pins is referred to as an output vector.

To determine if the DUT responded properly to the output vector (i.e., to make sure outputs switched as expected and to verify that no inputs are shorted), the tester must read a corresponding read-back vector from the DUT and compare this to an expected read-back vector. Each complete test vector consists of an output vector and an expected read-back vector.

The format for specifying the vector-definition modules is shown in table 1. The order of the different line types is important, though you may freely intersperse comment lines. (Like many assemblers, all characters on a line following an asterisk are ignored by the test-vector compiler.)

The best way to understand the vector-definition module format is by example. Table 2 shows the vector-definition module for a 7400 quad two-input NAND gate. As its name implies, this device contains four two-input NAND gates (the pin-out is shown in figure 1).

Photo 1: *The Circuit Cellar IC tester shown here is operating in terminal mode, connected to a Tandy DT-100 terminal via the RS-232C port on the top of the tester.*

In table 2, the first line is the device name. The name can appear anywhere on the line after the pound sign (preceding and following spaces are ignored). As a general rule, you should keep device names as generic as possible. Instead of using the name "74LS00" use "7400," and so on. Since the tester will logically identify both a 74LS00 and 74HC00 as the same chip, it is better to display "7400," or perhaps "74xx00." There are, of course, cases where you can make exceptions.

The second noncomment line of the vector-definition module is the setup line, which has an *S* in the first column. Three numbers with delimiting spaces must follow the *S*; the first number indicates the number of pins the device has, the second indicates the ground pin number, and the third indicates the power pin number. These numbers tell the compiler (and the tester) what the chip's device is. As I described in Chapter 34, part 1, the tester supports six device types (see table 3).

Following the next comment line is the pin-function line, which has an *F* in the first column. This line specifies a pin-function identifier for each pin, with the identifiers being separated by one or more spaces. Valid identifiers are *I* for input pins, *O* for output pins, and *T* for tri-state pins.

The pin-function line also determines the columnization for the remainder of the device vector definition. All 1s and 0s in the test vectors must be aligned under these columns, and the pin numbers in the pin-number line (the next line in the definition) must also be aligned under these columns.

The next line in the vector-definition module is the pin-number line. It has the letter *P* in the first column. This line specifies the device pin numbers used in testing. The numbers must correspond to the pin-function identifiers specified in the pin-function line and must fall in the columns defined by the function identifiers. If the pin number for a column has two digits (e.g., pin 14), either of the two digits can fall in the column.

The next several lines in the vector-definition module are the actual test vectors. The lines beginning with *I* are initial vectors (output vectors), and the lines beginning with *R* are the expected read-back vectors.

For *I* vectors, the acceptable identifiers are 1 and 0, corresponding to high and low digital values, respectively. For *R* vectors, acceptable identifiers are 1, 0, and *X*, with *X* indicating "don't care." (*X* indicates that the tester should ignore the specified pin when comparing the actual read-back vector to the expected read-back vector. If the 1 or 0 bit value of a column does not change from one line to the next, leaving the column blank in the subsequent line[s] implies that the value should be the same as the last value explicitly stated for that column.)

The last line in the vector-definition module is the end line,

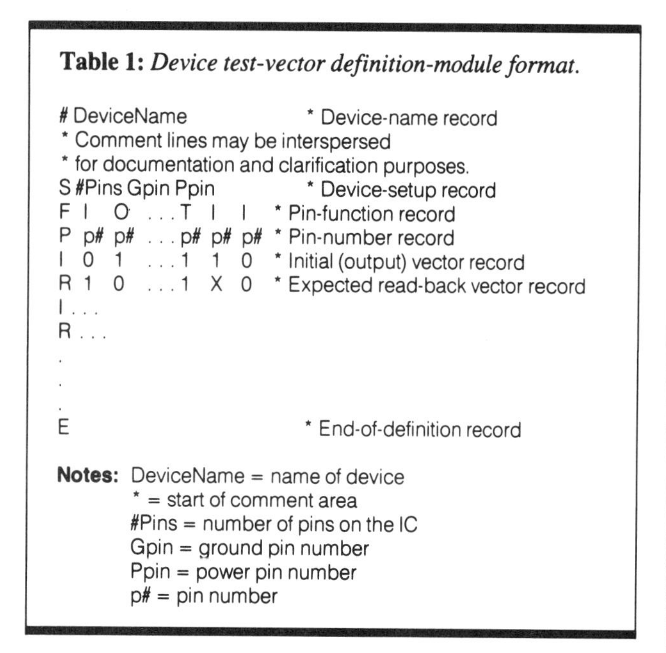

Table 1: *Device test-vector definition-module format.*

```
# DeviceName                 * Device-name record
* Comment lines may be interspersed
* for documentation and clarification purposes.
S #Pins Gpin Ppin            * Device-setup record
F  I   O  ...T   I   I    * Pin-function record
P  p#  p#  ...p# p#  p#   * Pin-number record
I  0   1   ...1  1   0    * Initial (output) vector record
R  1   0   ...1  X   0    * Expected read-back vector record
I...
R...
.
.
.
E                            * End-of-definition record
```

Notes: DeviceName = name of device
* = start of comment area
#Pins = number of pins on the IC
Gpin = ground pin number
Ppin = power pin number
p# = pin number

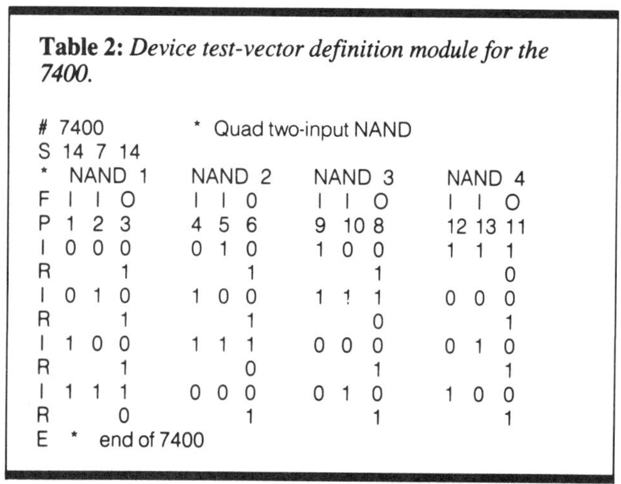

Table 2: *Device test-vector definition module for the 7400.*

```
# 7400            * Quad two-input NAND
S 14 7 14
*  NAND 1     NAND 2    NAND 3    NAND 4
F  I  I  O    I  I  O   I  I  O   I  I  O
P  1  2  3    4  5  6   9  10 8   12 13 11
I  0  0  0    0  1  0   1  0  0   1  1  1
R        1          1         1         0
I  0  1  0    1  0  0   1  1  1   0  0  0
R        1          1         0         1
I  1  0  0    1  1  1   0  0  0   0  1  0
R        1          0         1         1
I  1  1  1    0  0  0   0  1  0   1  0  0
R        0          1         1         1
E  *  end of 7400
```

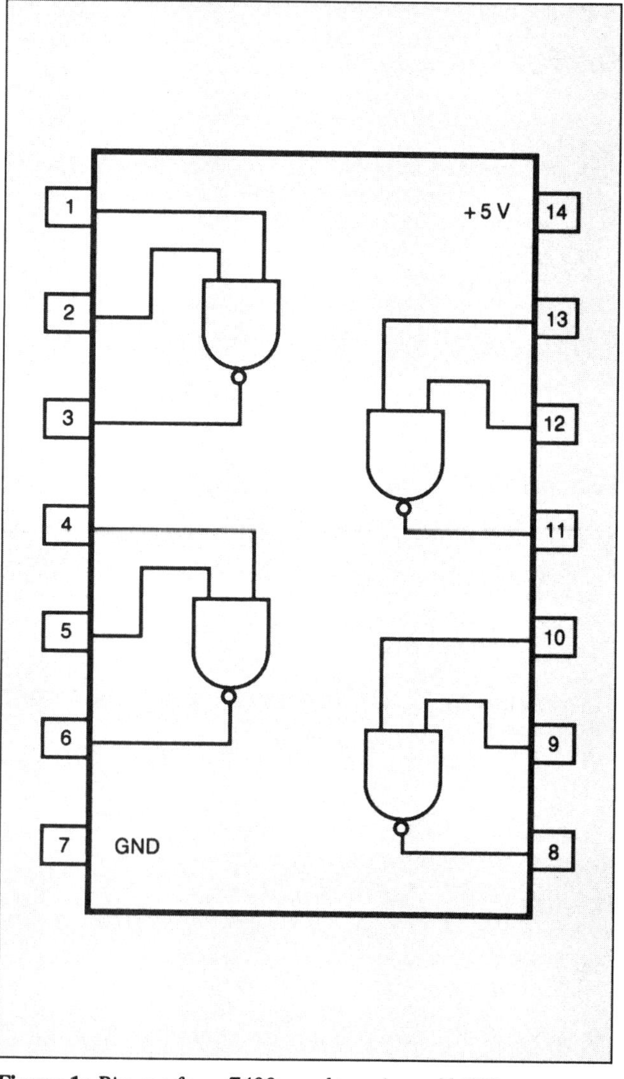

Figure 1: *Pin-out for a 7400 quad two-input NAND gate.*

which begins with an *E*. This is the only letter required to specify the end of the vector-definition module.

Finally, there is the issue of logically identical devices that have different part numbers (like 74LS04 and 74LS14). Differences typically lie in some of the special operational parameters, like Schmitt-trigger inputs or improved current drive capability, which cannot be detected by this IC tester.

Functionally identical devices (i.e., the same test vectors would pass on both devices) are declared to be clones of a specific device. An example of this is shown in table 4. The device-name and end lines are the same as the standard vector-definition module, but only the *C* line is found between them, indicating which device it is supposed to be cloned from.

Compiling Test Vectors for Use

The test-vector compaction compiler, VECCPT.PAS, is a Turbo Pascal program that accepts files conforming to the device test-vector definition format described above. It converts the device and vector information into a single compact module that the computer and tester use to test the devices.

VECCPT.PAS uses seven primary arrays to store the compacted vector information. The primary array, `VectorTable`, holds the actual test-vector information, including the device pin-function information (i.e., which pins are inputs, which are outputs, and which are tri-state), the output vector bytes, the input vector bytes, and the "don't care" mask bytes.

Because the ZIF (zero insertion force) socket has 24 pins, tester software uses 3 bytes for pin and vector information for every device, regardless of size. Consequently, the pin-function and test-vector information is stored as if a 24-pin device were being tested.

The "don't care" mask generated by VECCPT.PAS automatically masks the read-back vector pins not associated with the device being tested. For example, if a 20-pin device is being tested, the bits of the 3-byte read-back vector associated with ZIF-socket pins 1, 2, 23, and 24 will be masked by the "don't care" mask (power and ground pins are automatically masked).

Each of the six device types supported by the IC tester has its own associated array for storing device names and pointers into the `VectorTable` array. These arrays are called `DeviceType` arrays.

While the `VectorTable` array uses variable-length records, with each record being the information to support one device, the `DeviceType` arrays use fixed-length records, with each record containing a 9-byte field for the device name (8 bytes for the name and 1 byte for the string size) and an integer (2-byte) field for the `VectorTable` pointer.

Figure 2 illustrates the information stored in the various arrays and how the arrays interact. As shown, device names are stored in the appropriate `DeviceType` array, and the device pin-function and test-vector information is stored in the `VectorTable` array. A pointer in the `DeviceType` array indicates the start of the corresponding vector-information record in `VectorTable`.

The `VectorTable` device record begins with a 2-byte field indicating the number of bytes in the record. The next 3 bytes specify which pins are inputs and which are outputs (set bits are inputs, and cleared bits are outputs).

The following 3 bytes indicate which pins are tri-state (set bits are tri-state). If a pin is indicated as being tri-state, the 1/0 value in the corresponding bit position of the previous I/O definition bytes is irrelevant. By default, VECCPT.PAS specifies unused ZIF (zero insertion force)-socket pins as being tri-state.

Following the 2 record-size bytes and 6 device pin-function definition bytes, the actual test-vector information begins. Each complete test vector consists of 9 bytes in the record. The first 3 specify the output vector, the next 3 specify the expected read-back vector, and the last 3 specify the "don't care" mask.

As VECCPT.PAS executes, it stores device-name, pin-function, and test-vector information into the appropriate arrays. Notice that the program does not need to store device-type information, since a device's type is determined by which `DeviceType` array it is placed in.

Device clones are handled somewhat differently. When a device is specified as a clone of another device (the "original" device), the name of the clone is placed into the next available record of the appropriate `DeviceType` array. The record number of the original device (in the same array) is then determined, and the value 32,767 is subtracted from the record number; this value (always negative) is then stored in the pointer field of the clone record.

Thus, when the operating software finds a negative integer value in the pointer field of a device record, it will know the device is a clone of another device. It then adds 32,767 to the pointer value to get the record number of the original device.

I should point out that when VECCPT.PAS processes a clone, it looks through its arrays to find the named original device. If the specified original device is not found in any of the six `DeviceType` arrays, the software generates an error, and the clone device will not be stored in any array (the compiler would not even know which array should get the clone record). Thus, it is essential that you specify clone devices only after the corresponding original device.

When compaction of the test-vector files is complete, the compacted information is stored in a binary file. (The format of the data stored in the compacted file is shown in figure 3.)

Operating Software

Once the device test vectors have been developed and compiled into a compacted file, we are ready to use the tester for testing and identifying devices. This involves the cooperation of several programs.

First, there's a ROM-resident program on the IC tester. This program is written in 8031 assembly language and handles the three operating modes from the tester's vantage point. Then there's a Turbo Pascal program that executes on the IBM PC (or XT or AT) for operating the tester in PC-host mode.

Finally, another Turbo Pascal program converts the information in the output file produced by VECCPT.PAS into Intel hexadecimal ASCII format. This permits you to download to an EPROM burner. This lets you put new device vector information into the IC tester's ROM for operation in the terminal and stand-alone LCD modes.

Table 3: *The six device types supported by the tester.*

Device type	Number of pins	Gnd pin	+5-V pin
1	14	7	14
2	14	11	4
3	16	8	16
4	16	12	5
5	20	10	20
6	24	12	24

Table 4: *Definition module for the 7437, a "clone" of the 7400.*

```
#7437     * Quad two-input NAND buffers
C 7400    * Clone of 7400 (Dev. type=1)
E         * End of 7437 definition
```

Explaining all the software for the IC tester would involve considerably more space than I have available here (see the Circuit Cellar Ink applications publication for additional support materials). While my description here is tailored to the application and use of the IC tester, the user's manual and distribution software contain much source code and go into significant detail describing the process for creating a new device library and testing custom devices.

PC-Host Mode

The PC-host mode of operation is the most powerful of the three modes. I'll start with its description, because the basic testing technique is the same for all three modes.

The PC-host mode provides flexibility in letting you download and use different device libraries and offers test-vector debugging features not available in the other two modes. Functions like `Identify` and `Test Specified Device` differ only in the information displayed and are the same in all modes.

Once you give the PC-host mode operating program the name of the compacted test-vector file and the serial-port number (1 or 2), the software attempts to establish a communication link with the IC tester. If the tester does not respond, the PC will perform two retries (three tries total) before printing an error message and sounding a beep.

Once communication is established, the PC reads the specified compacted test-vector file, downloads it to the IC tester, and displays the version number and a formatted operation menu on the screen. The typical menu offers four device-testing options and two mode-selection options.

The display also shows three status/information lines. The first line, `Device:`, indicates the name of the current or most recent device being tested, or the name of an identified device. The second line, `Message:`, displays messages like `Device Passed` and `Device Not Found`. The third line, `Pin Failures:`,

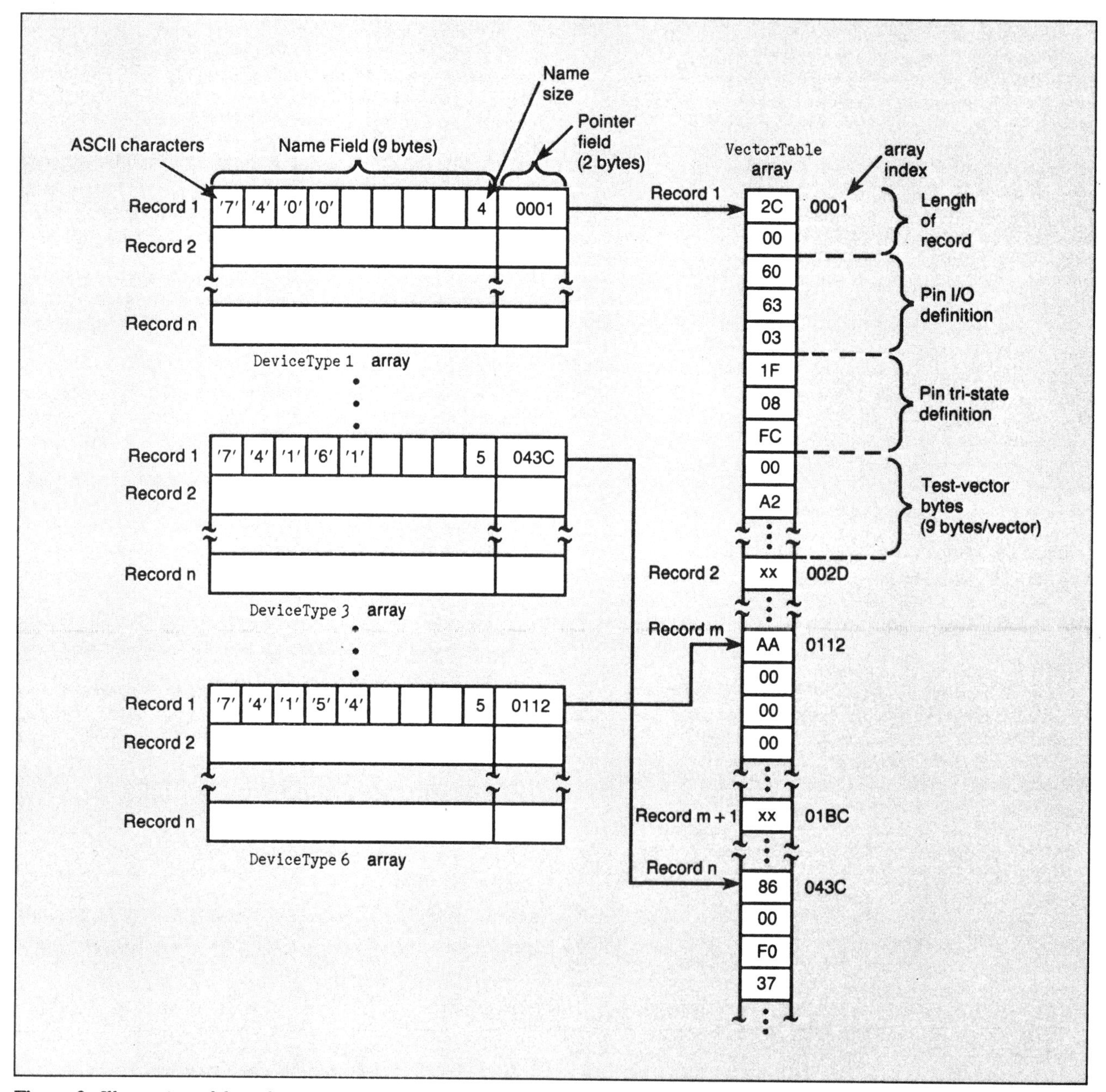

Figure 2: *Illustration of the information storage in VECCPT.PAS's primary arrays.*

displays pin numbers that failed vector tests when testing a specified device or an EPROM.

The first menu item, `Identify Device`, tells the tester to attempt to identify the device in the ZIF socket (the device-identification algorithm supports only devices having the corner power and ground pins). To identify a device, the system powers the ZIF socket for a 24-pin device and then applies the first 24-pin device test vector (if any) in the device library to the DUT.

If the read-back vector compares favorably to the expected read-back vector (along with the "don't care" mask), the next vector for the same device is applied, and so on. This continues until the DUT passes all the test vectors—indicating proper device identity—or until a vector failure occurs. If a vector failure occurs, a check is made to see which bits in the read-back vector, if any, are different from those sent out in the output vector. These bits represent pins that must be either output or tri-state pins, and the pin values are noted in an accuracy array.

If the DUT passes all the test vectors, the tester has identified the device; its name is displayed, and control returns to the menu. If the DUT fails a vector, the next device in the 24-pin library is checked.

Testing continues until the DUT is identified or no more 24-pin devices are left to test. If the program runs out of 24-pin devices, it clears the accuracy test array and repeats the same procedure with the 20-pin, then 16-pin, and finally 14-pin devices. Inability to finally identify the part is only the result of the device not being in the library, or because it is defective.

The second menu item, `Test Specified Device`, moves the cursor to the `Device` line. If any devices have already been tested or identified, the name of the last device tested is automatically displayed on the line. If you desire to retest the same part type, press Return (or Enter). If you wish to choose a different device, enter the new device name and press Return to test the DUT.

By telling the IC tester what type of chip is in the ZIF socket, all the test vectors for that device will be applied to the device and checked, regardless of whether they pass or fail. If vector failures do occur, you'll see the pin numbers on the `Pin Failures:` line.

The first two menu items represent the operations you will probably want to do 99 percent of the time and can be done in all three operating modes. Sometimes, however, you may have 2716 or 2732 EPROMs that you would like to verify are blank. Menu items 3 and 4 provide this capability.

In addition to performing a blank check on the EPROM, the EPROM tests also check for shorts on the EPROM input pins. If shorted pins are detected, an error message is displayed and the failed pins are displayed on the `Pin Failures:` line. Since the ZIF socket is only 24 pins, the tester cannot accommodate larger EPROMs.

The third menu selection deals with CMOS logic devices only. As I discussed in part 1 of this chapter, all the standard 74xx00-series logic families except the 74C00 series (and some specific devices within other families) are capable of sourcing and sinking enough current on their outputs for proper operation of the tester.

The 74C00-series devices (and the similar 4000-series CMOS devices) have a problem sinking enough current to switch logic states when an output is pulled up to +5 volts. Most of the tests for the 74xx00-series families attempt to load the device outputs in the direction opposite the expected state (if an output is expected to go low, it is loaded with a pull-up resistor), causing particular problems when testing the 74C00-series family devices when reading outputs that are expected low, but are being pulled up.

The remedy for the 4000-series devices is simple: Write all test vectors for these devices always using a pull-down load on all outputs. In order to keep the 74xx00-series tests the same for all families, however, I had to use a different approach. Menu item 5 lets you `Set 74Cx Mode`.

In this mode, regardless of the original output vector-bit levels, all output vector bits that correspond to device output (non-tri-state) pins are changed to low (pull-down). This allows the 74C00-series devices to pass the generic 74xx00-series tests. You can also select this mode for identifying 74C00-series devices.

The final menu option is `Set Diagnostic Mode`. This option is available only when operating the tester in PC-host mode. It adds an extra line to the bottom of the display, `Vector Failures:`, to indicate which test vectors failed when testing a device.

When testing a specified device (not when identifying a device) in diagnostic mode, the `Device:` line indicates the number of pins the device has, as well as the ground pin number and the power pin number. If the device is a clone of another device, this is also indicated, along with the device name of the original device.

If the device being tested fails, the `Message:` line indicates how many vectors failed (along with the normal failure message), and the `Vector Failures:` line indicates the vector numbers of the first 10 failed vectors (or all failed vector numbers, if fewer than 10 failed). The extra information can prove helpful when debugging new test vectors.

ROM-Resident Control Program

The IC tester's 8031 assembly language control program provides local support for all three modes. A software-readable, four-position DIP switch selects mode and data transfer rate, while a status LED indicates the tester's current operating disposition (a second LED acts as a power-on indicator).

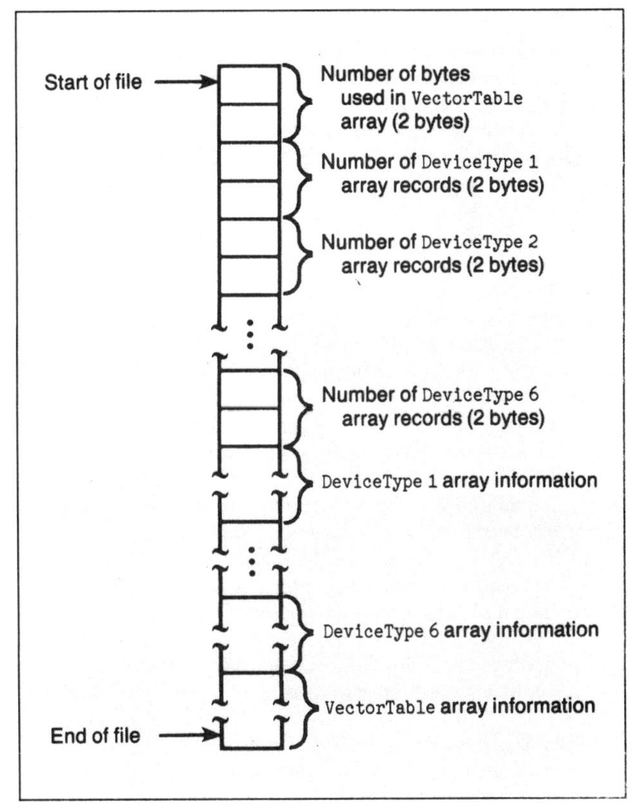

Figure 3: *Format of the vector-compaction file output by VECCPT.PAS.*

Upon power-up or reset (using the on-board reset button), the program initializes the 8031's on-chip ports to turn off all power and ground transistors to the ZIF socket and to place the LCD interface lines in their appropriate default states.

It then generates a brief delay (nominally, 1 second) to provide time for power to stabilize for all devices on the board. Software then checks two of the DIP switches to determine the desired data transfer rate and configures the 8031's on-chip UART to handle serial communications at the specified data transfer rate.

Once initialization is completed, the program checks another DIP switch to see if the user has selected PC-host mode or terminal mode. For terminal mode, the system turns on the status LED (to indicate that a serial operating mode, as opposed to stand-alone mode, is currently enabled) and sends a sign-on message and menu out the serial port to the attached terminal. For PC-host mode, no sign-on message is sent.

In either case, the tester also displays a sign-on message on the optional LCD, if present. In order to select the stand-alone mode, you merely press the "Identify" push button—which is constantly polled during both serial operating modes—and the system will turn off the status LED to indicate stand-alone mode operation. The only way to return to serial mode operation is by pressing Reset.

When operating in PC-host mode, the IC tester's ROM program merely responds to commands from the host. Various commands allow "reset" (power and ground transistors turned off), software version request, power and ground switch setup, and DUT output vector application and read-back vector reading. Terminal mode operation is similar to PC-host mode operation, with the exception that you are restricted to the device library stored in ROM, and the diagnostic mode described earlier is unavailable.

Stand-alone operation requires no connection to the serial port, but it does require that you have the LCD installed (see photo 2). All interaction is via the on-board "Identify" and "Retest" push-button switches and the LCD. A DIP switch enables or disables "74Cx" mode.

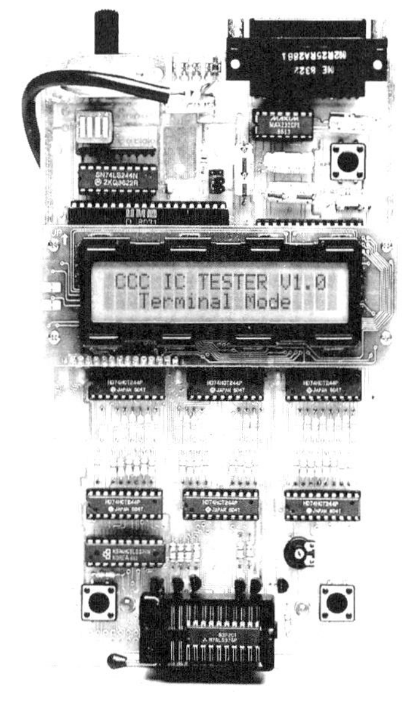

Pressing the "Identify" push button causes the tester to attempt to identify the device in the ZIF socket. If the identification is successful, the device name is displayed on the LCD; otherwise, an identification failure message is displayed.

Once a device has been identified, you can test other devices of the same type using the "Retest" push button. The test vectors for the identified device are then applied to the DUT, and detected pin failures, if any, are displayed on the LCD.

Photo 2: *You can use the IC tester in stand-alone mode, provided you have attached the tester's optional LCD. The push-button switches in the upper right, lower left, and lower right control the tester's operation.*

Flexibility

While the Circuit Cellar IC tester represents hundreds of hours of hardware and software development, the end result is something that was designed to be simple to operate. It clearly offers a great deal of flexibility for testing common devices, but it is also useful for developing tests for custom or proprietary devices like programmable array logic.

In order to test a PAL, you must develop a series of test vectors that apply bit patterns to the device inputs and watch for expected output values just like those from any standard 74xx logic device. The PAL test vectors are based on the logic-transfer functions (the logic equations) of the device.

You compile and name the test vectors and then add them to the device library. To test PALs, you run the IC tester in the normal way: Just insert the PAL to be tested in the ZIF socket (bottom-justified) and specify either the `Identify Device` option (the easier choice) or the `Test Specified Device` option, giving the device's name, "PAL1," for example.

In Conclusion

The powerful, yet easy-to-use, Circuit Cellar IC tester can provide testing and identification for innumerable standard and custom IC devices, in packages ranging from 14 to 24 pins. It's a tool that can save you time and money by catching potential problems during production, helping debug problem boards, and by identifying and/or verifying unknown devices or devices with uncertain operation. The flexibility and capability offered by this tester were previously available only to those willing to spend thousands of dollars.

In all honesty, I have to admit that the hardware for this project was trivial compared to the enormous software task involved in creating the operating system and device library. The initial Revision 1.0 ROM-resident library contains more than 200 generic entries. Considering that a generic entry of "7400" can cover 10 clone entries, the library physically covers about 800 chips. I owe a special debt of gratitude to those who helped put this project together and saved me from having to deal with all this software.

I would like to personally thank Roger Alford and Bill Potter for their collaborative efforts on this project. Bill Potter's tireless dedication creating the test-vector library and Roger Alford's clever programming expertise served to make the Circuit Cellar IC tester a true performer.

See page 447 for a listing of parts and products available from Circuit Cellar Inc.

35

WHY MICROCONTROLLERS?

PART 1

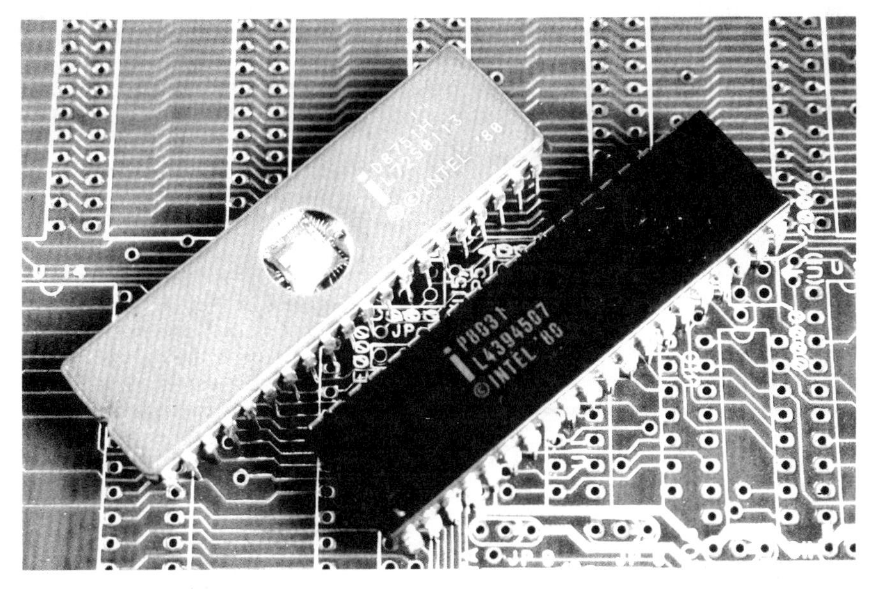

Photo 1: *Two versions of the Intel MCS-51 family: the 8031 ROMless microcontroller and the 8751 programmable (EPROM on-chip through the quartz window) microcontroller.*

Microcontrollers are used in keyboard and disk interfaces and in numerous other devices. Here's a tutorial on the 8031/8051 microcontroller family.

Gone are the days when a complex project required a suitcase full of TTL ICs and a wire-wrap gun. As with most of the recent Circuit Cellar designs, under the hood there are a few carefully chosen discrete ICs controlled by a microprocessor. The trade-off is simple: Hardware is expensive and software is cheap—once you get it right.

Many of you have noticed that I have been using two popular microcontrollers from Intel—the 8031 and the 8051—as the control elements in Circuit Cellar designs. I've received many requests for general information about these microprocessors and for guidance in applying them elsewhere. With that as an incentive, I decided to present this combination tutorial/development system project.

In this first part of a two-part chapter, I'll introduce the members of the 8031/8051 chip family and describe using them as embedded microcontrollers. The second part will present all the elements of a useful development system intended to simplify the process of designing systems using the 8031/8051 family.

The Difference in a Name

It's important to be clear on the distinction between microcomputers, microprocessors, and microcontrollers. That way, you'll understand where the 8031/8051 family fits in the big picture.

A microprocessor is just the CPU part of a computer, without the memory, I/O, and peripherals needed for a complete system. For example, 8088 and 80286 chips are microprocessors (the "micro" prefix designates that this CPU element is at the chip level). All other chips in an IBM PC are there to add features not found within the microprocessor chip itself. The hardware designer can choose different chips to implement those features in different ways, although a designer has little room for choice if the end result is supposed to be an IBM PC clone.

When a microprocessor is combined with I/O and memory peripheral functions, the combination is called a microcomputer. Of course, vendors anxious to designate that their computer is more powerful than others often shed the "micro" prefix, but it's still a microcomputer given today's definition. Ultimately, good economic sense suggests that all computers, including minicomputers and mainframes, will utilize the same basic elements, just differentiated by quantity.

The fact that combining a CPU with memory and I/O produces a microcomputer also holds true at the chip level. Many companies add these peripheral functions onto the same substrate with the CPU to make a complete microcomputer. These devices are called single-chip microcomputers to differentiate them from their big-cousin desktop microcomputers.

Generally speaking, microcomputer chips are designed for very small computer-based devices that don't need all the functions of a full computer system. In cost-sensitive control applications, even the few chips needed to support a CPU like an 8088 or Z80 are too many. Instead, designers often employ a single-chip microcomputer (or a slightly expanded circuit using one) to handle control-specific activities. When single-chip microcomputers are designed or used in industrial control systems, they are often called single-chip microcontrollers. Basically, there is no difference between microcomputers and microcontrollers; the name depends on how we use them.

Frequently, microcontrollers are used to replace circuit functions that ordinarily require many low-level chips or need the main CPU's attention each time the circuit is active. The IBM PC keyboard-interface circuit is a prime example of the use of a microcontroller chip. In the PC, a half dozen chips (excluding I/O addressing and decoding) are necessary to receive and decode the serial clock and data bit stream from the keyboard. In the IBM PC AT (and the CCAT I present in S/O 1987), this low-level circuitry is replaced with an 8742 microcontroller that completely simulates the old circuit and incorporates additional features in one chip.

The Intel 8051 Family

The Intel 8051 is a classic microcontroller (a generation more advanced than the 8742) and a true single-chip microcomputer containing parallel I/O, counter/timers, serial I/O, RAM, and EPROM or ROM (depending on the part type). The 8051 family contains several members (Intel refers to it as the MCS-51 family), each adapted for a specific type of system.

The different versions are outlined in table 1, and a block diagram of the 8051 is shown in figure 1. The 8051 has two close relatives, the 8751 and the 8031 (see photo 1), and a cousin, the 8052. All versions contain the same CPU, RAM, counter/timers, parallel ports, and serial I/O. The 8051 contains 4K bytes of ROM, which must be custom-masked when the chip is manufactured. In the 8751, the ROM is replaced with EPROM that you can program (the schematic for an 8751 programming adapter for the Circuit Cellar serial EPROM programmer presented in October 1986 is available by writing to me).

The 8031 is meant for expanded applications and uses external memory. The 8031 uses three of the four on-chip parallel ports to make a conventional address and data bus with appropriate control lines.

You might wonder why you'd choose a single-chip microcomputer in the first place if you end up converting it back to function as a CPU with other peripheral chips. Basically, it depends on the degree of expansion required. Since the 8031 still contains RAM, a parallel port, and a serial I/O port—even when functioning as the CPU core of an expanded circuit—the eventual number of chips necessary to expand the I/O or memory is still considerably less for the same ultimate capability than with a straight

Table 1: *Members of the 8051 microcontroller family tree.*

Device name	ROMless version	EPROM version	ROM bytes	RAM bytes	16-bit timers
8051	8031	(8751)	4K	128	2
8051AH	8031AH	8751H	4K	128	2
8052AH	8032AH	8752BH	8K	256	3

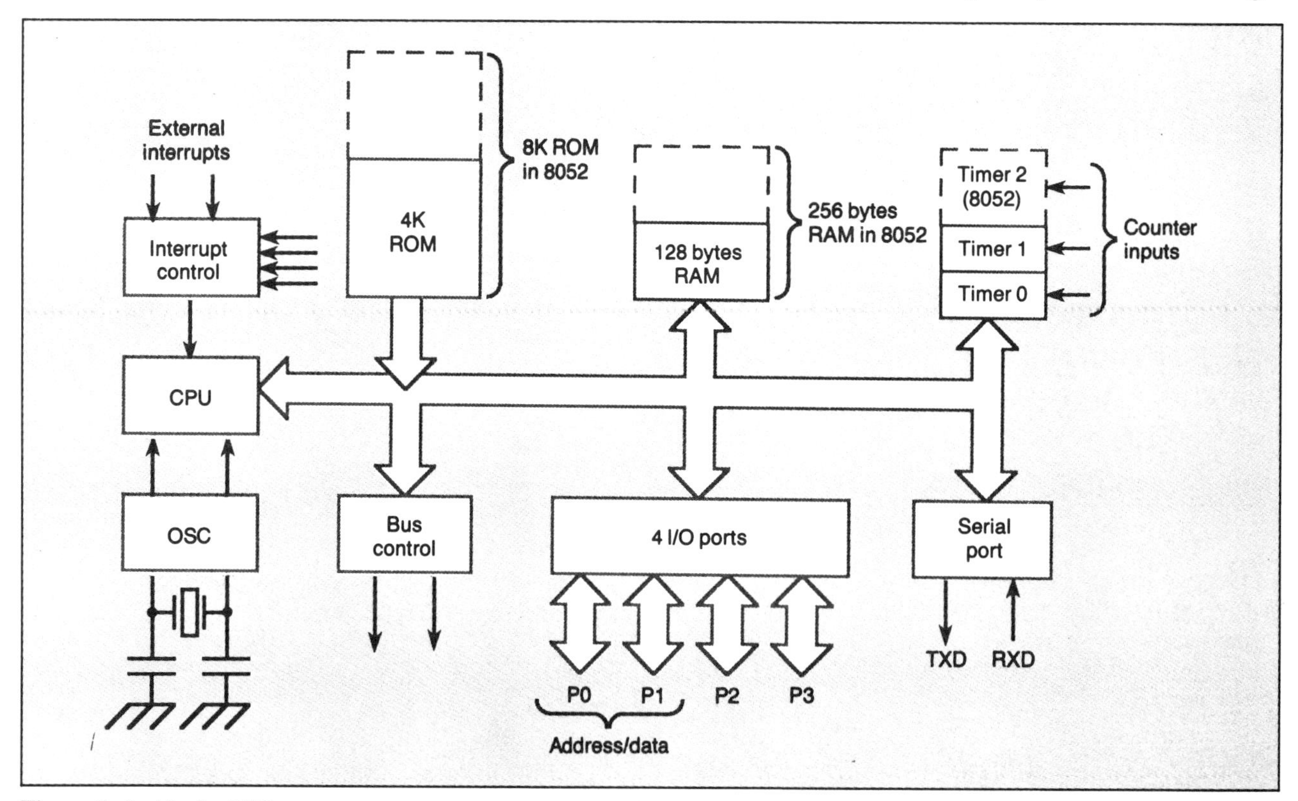

Figure 1: *Inside the 8051.*

microprocessor and peripheral chips. I'll go into the possible expansion techniques later.

Intel and other companies sell variations of the 8051 family with more internal memory, more I/O, lower power, and so forth. An 80C31 is a CMOS low-power version of the 8031, for example. The 8052, which I have also used in projects, is the same as the 8051, except that it has another counter/timer and additional RAM and ROM.

Using an 8051 is as simple as hooking up the power supply and clock crystal. Unlike the 8031, however, you have to supply Intel with a program for the 8051 so it can create a chip mask defining the internal ROM. (If your program is not quite right, it's time for another mask.)

As you might expect, both mask making and chip building take time and money. (Since I did this for a recent project, I thought you might be interested in some manufacturing information. If you plan on making an 8051 microcontroller and need to mask your own chip, the one-time masking charge is about $3000, with a minimum order of 1000 chips.) Mask programming makes sense for an application that uses thousands of identical 8051s a year, but it is not practical for low-volume systems and prototypes.

The 8751 has all the features of the 8051, except that an on-chip EPROM replaces the ROM program storage. Making a program change is as simple as erasing the EPROM with ultraviolet light and burning another program. Many developers use 8751s until the code works, then commit to a large 8051 order with the program in ROM. If the product volume is low enough, it's often worthwhile to use 8751s in the final product. An 8051 costs about $4 to $5 apiece in thousands; an 8751 is about $25 to $40.

The 8031 has no on-chip program storage at all. The system must include an external EPROM and an address latch. Considering the falling prices of EPROMs and the heavy costs of using either 8051s or 8751s in low volumes, the 8031 is a viable alternative despite the additional chips. For many small systems, the 8031/EPROM combination is far more cost-effective than an 8051. (This is the type of system I'll describe in my examples.)

As I mentioned before, all members of the 8051 family have the same core hardware and therefore use the same core instruction set. While some members have one or two additional instructions for features unique to the particular chip, I'll use the term "8051" to describe the "8051 chip family," unless I'm talking about a specific version with unique requirements.

Making It Real

With all that in mind, let's look at configuring a usable "computer/controller" using an 8031 microcontroller chip as part of the system. Remember that in a single-chip microcontroller, internal hardware replaces all the digital logic you'd normally add for control, timing, and so forth. You need add only the keyboard, display, relays, switches, and user-specific I/O that actually makes up the final product.

Figure 2a shows the bare-bones 8031 microcontroller system: the 8031, a 2764 EPROM to hold 8K bytes of program, and a 74LS373 latch to demultiplex the address/data bus. The system has 128 bytes of RAM on the 8031, a bidirectional parallel I/O port, a bidirectional serial port, two counter/timers, and two external interrupt inputs. With a 12-MHz crystal (most often, we select 11.0592 MHz for communications rate compatibility), it executes most instructions in one machine cycle—a peak rate of 1 million instructions per second. Not bad for three chips, is it?

A single I/O port can scan a 16-key matrix. With an additional output bit, it can drive a 2-line by 20-character smart liquid crystal display at the same time. The remaining I/O bits can handle triacs or power field-effect transistors for AC or DC control. Burn a program into the EPROM, and you have a real-time power controller. Run the serial port through a MAX232 RS-232C level converter, and you have a standard serial port for remote control or status monitoring at your master computer.

If you don't need the serial port, counter/timers, and external interrupts, the 8031 can use those special bits as a second parallel I/O port, so the minimum system can have up to 16 I/O bits. Each bit can be tested, set, and cleared individually under program control.

If one or two parallel I/O ports aren't enough for your application, figure 2b (an expansion of figure 2a) shows what's needed to get three more: Add a single 8255 programmable peripheral interface. The 8255's port C can be set up for automatic handshaking, so now you have the basis for a serial-to-parallel (and back) format converter or 24 more I/O bits for a bigger controller. Notice that no "glue" chips are needed between the 8031 and the 8255.

Because the 8255 uses the RD\ and WR\ bits, the second I/O port isn't completely free. The 6 remaining bits can still handle either general I/O or their individual special functions, though.

If your application requires more than 128 bytes of RAM, figure 2c (figure 2a expanded to include 2b and 2c) shows how to get 8K bytes of RAM by adding a 6264 static RAM chip. Now you can build a fancy buffering format converter, a data logger, or a serial-programmable power controller. A 62256 RAM would give 32K bytes with no more effort, still with no glue chips!

Finally, for those of you who need lots of RAM and I/O, figure 2d (figure 2a expanded to include the circuitry of 2b, 2c, and 2d) shows how to connect multiple I/O chips. The 74LS138 decoder generates chip select signals from the 8031's output addresses, with each select covering an 8K-byte range. The system shown has 16K bytes of RAM and seven bidirectional I/O ports. Pretty nice for seven chips.

The point of all this is that the "computer" part of your control system need not require elaborate hardware. For a unit of any reasonable size, you'll spend most of your hardware design time on the I/O devices rather than on the 8031 circuits, which is exactly as it should be.

Perhaps now you understand why I have been using the 8031 frequently. The main benefit of a microcontroller is the ease of adding new features to your system, just by changing the program, not changing the circuit-board connections. A new EPROM can give the hardware a completely new personality. Try doing that by rewiring a board of TTL control logic!

The Software Swamp

Every microprocessor has an instruction set exhibiting the conflict between all the instructions that could possibly be useful and the few that fit on the chip. The 8051 has many bit-manipulation instructions and few general instructions, reflecting its design as a controller rather than as a computer.

Most 8051 instructions are 1 or 2 bytes long, with the remainder requiring 3 bytes. All instructions except `MUL` and `DIV` execute in one or two instruction cycles. An instruction cycle is 1 microsecond (μs) at a 12-MHz clock rate. `MUL` and `DIV` lag along at 4 μs.

If you've written assembly language programs for any other microprocessor, you'll find some of the same instructions in the 8051's code. To understand the 8051's instructions, you must be familiar with the three main address spaces defined on the chip: 64K bytes of pro-

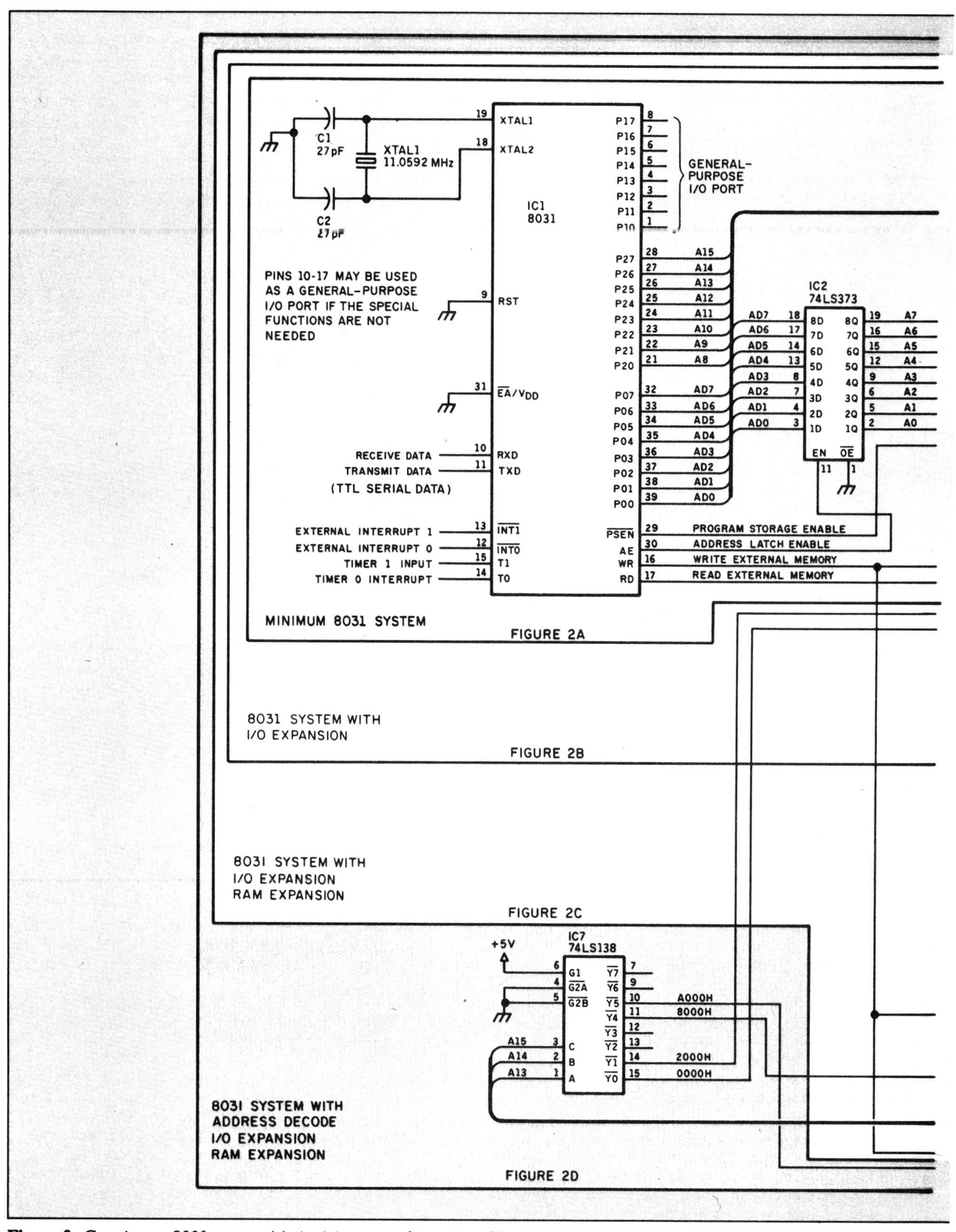

Figure 2: *Growing an 8031 system.* **(a)** *A minimum configuration,* **(b)** *minimum system with I/O expansion,* **(c)** *system with I/O and RAM expansion, and* **(d)** *system with address decode logic and even more RAM and I/O.*

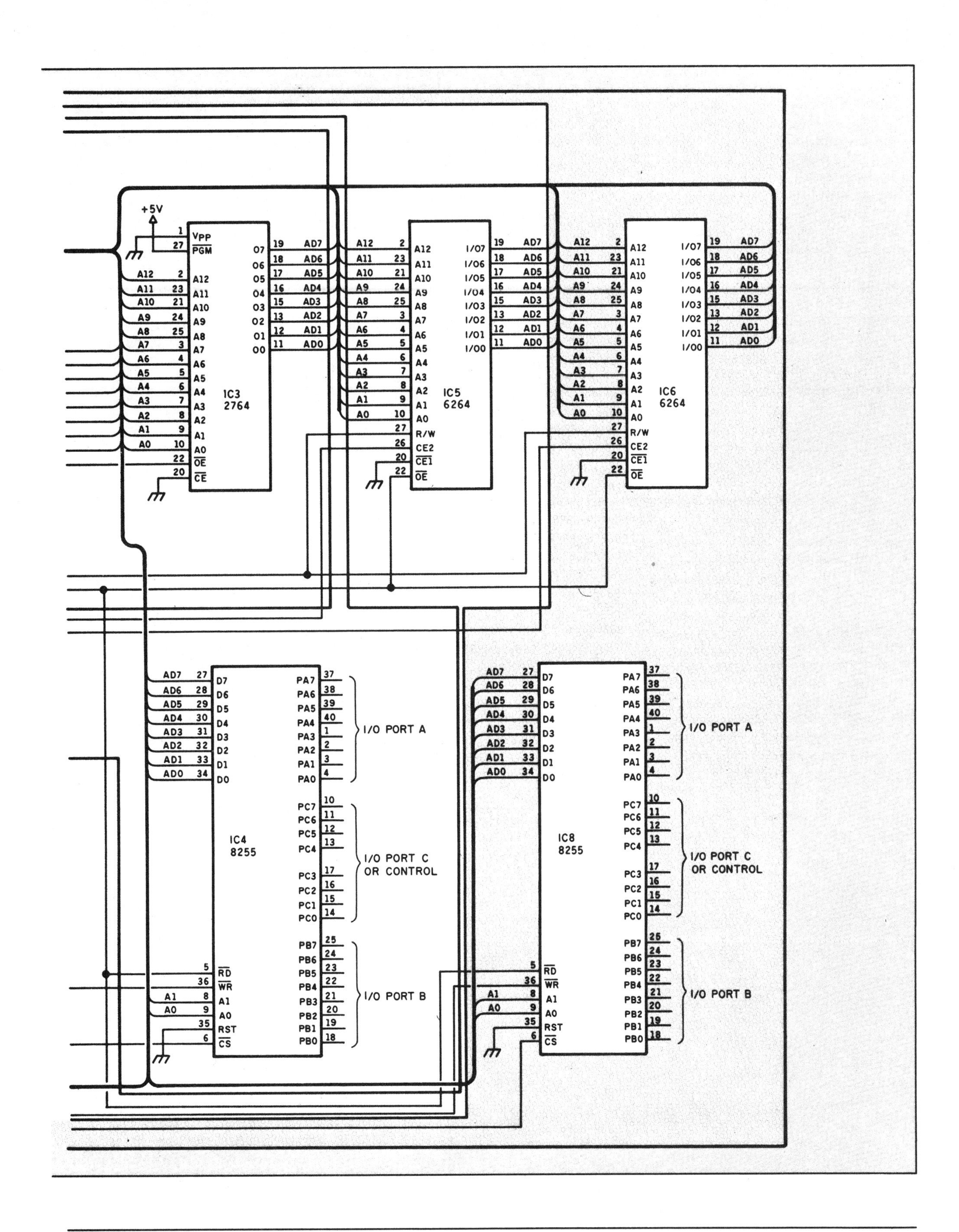
+5V
IC3
2764
IC5
6264
IC6
6264
IC4
8255
IC8
8255
I/O PORT A
I/O PORT C
OR CONTROL
I/O PORT B

gram memory addressed by the program counter (PC), 64K bytes of external data memory addressed by the data pointer (DPTR), and the on-chip internal data memory addressed in several different ways. Each instruction implies a particular address space, so you have to know where your data resides to select the right instruction.

There is a sharp distinction between internal and external data addresses. Internal addresses refer to locations on the 8051 chip, which can be accessed in a variety of ways. External addresses are located off the chip, in the 64K bytes of external data memory, and can be accessed only with MOVX instructions.

The four main internal data memory-addressing modes are direct, immediate, register, and register indirect. Direct mode embeds an internal RAM address in the instruction. Immediate mode uses the data value itself. Register and register indirect modes use a register number, with indirect addressing taking the contents of that register as a direct address to access the data.

The MOVX instructions transfer a single byte between the accumulator and external data memory. The DPTR register contains a 16-bit external data memory address, which can be either loaded by a single MOV or incremented. Unfortunately, there aren't any other 16-bit instructions.

The 8051 has a single accumulator, called ACC or A depending on the instruction. Nearly all instructions use the accumulator in one way or another. An auxiliary accumulator (called B, of course) is used by MUL and DIV. Many data-manipulation instructions can move data to or from one of the active banks of eight "working registers" in internal RAM. Four register banks are available.

Because most controller applications require handling at least a few I/O bits, the 8051 has a rich selection of bit-manipulation instructions that are completely separate from the standard byte instructions. A single instruction can set, clear, complement, or copy any bit in internal data memory. The on-chip I/O ports show up in that address space, so there's no need for the "read, mask, set, combine, write" instructions found in most other microprocessors.

Unlike the Intel 8088 or Z80 microprocessor families, the 8051 has no explicit I/O instructions. The on-chip I/O ports are mapped into the internal data memory-address space and accessed with the same MOV instructions used for other transfers. You have to map off-chip I/O into the external data memory-address space and access it with MOVX instructions.

Rather than belaboring the various instructions in detail, I'll introduce them in part 2 of this chapter in short chunks of code that do useful tasks, as we build some hardware. With those examples as a base, you should have little trouble designing your own system.

Ugly Reality

The trade-off for not wiring up a board of TTL gates is writing a program for the EPROM. That program tests the inputs, computes the outputs, and handles all the timing to make the system work correctly. Unless you are much better than average, your program won't do the right thing the first time you try it out.

The ugly reality of microcontroller systems is getting the software to work. It's made considerably more ugly by microcontrollers buried inside specialized systems—those never intended to look or act even vaguely like a computer.

For example, which system would you rather debug: an IBM PC AT with a full keyboard, EGA display, hard disk drive, and state-of-the-art editors and debuggers, or a microcontroller in a 3- by 5- by 4-inch box with four push buttons and two LEDs, cabled to a heater in a vat of photographic solution?

The traditional way to debug microcontroller programs is called "burn and crash." You burn the program into EPROM, plug it in, turn it on. . . and then try to figure out why it crashed. Doing a Sherlock Holmes on the listing is the only way to find bugs in the program, although a logic analyzer and an oscilloscope help a lot.

The major problem with burn-and-crash debugging is the damage caused by a crash if you are trying to debug a program when the controller is attached to actual machine hardware. Imagine what happens when your new 10-story hammerhead crane controller goes "full speed counterclockwise" and refuses to reset.

Obviously, burn and crash has its limitations. An 8051 simulator program running on a host computer development system removes most problems and simplifies finding program bugs. The simulator reads the EPROM's data and interprets the 8051 program one instruction at a time. Because all the 8051's registers, I/O ports, and memory are provided by simulator variables, you can display and modify memory contents at will. Even better, because the simulator's software replaces the 8051's hardware, there's no way that an errant program can damage anything.

With the simulator, you can use program breakpoints to stop execution at specific 8051 instructions or when a given condition occurs. Also, since the simulator records how the 8051 program got to a particular instruction, you can undo each step back to the source of the problem at the press of a key.

Unfortunately, while a simulator is a great step up from burning and crashing, it is not a true real-time test. Because the execution of each 8051 instruction requires the execution of many program instructions in the development system, the

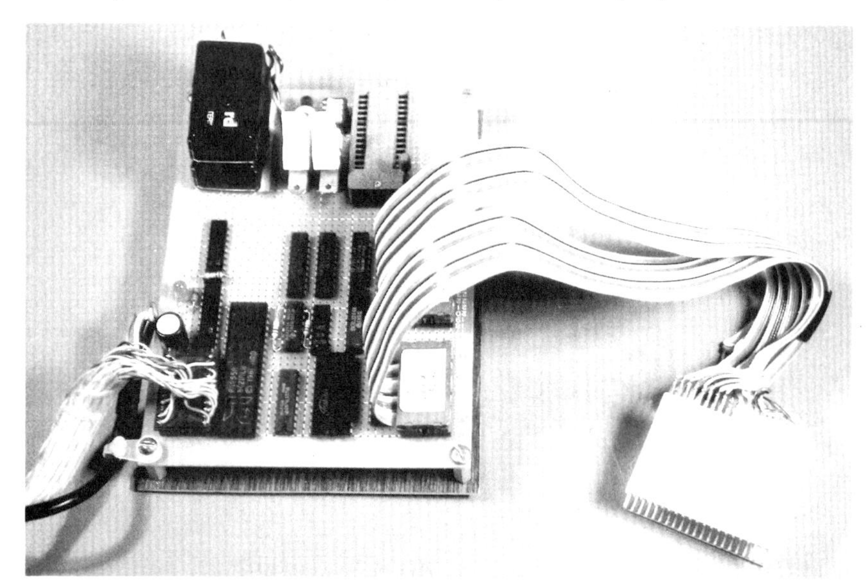

Photo 2: *A prototype of the DDT-51 development system. Note the DIP clip at the end of the cable for attaching onto the target system's processor.*

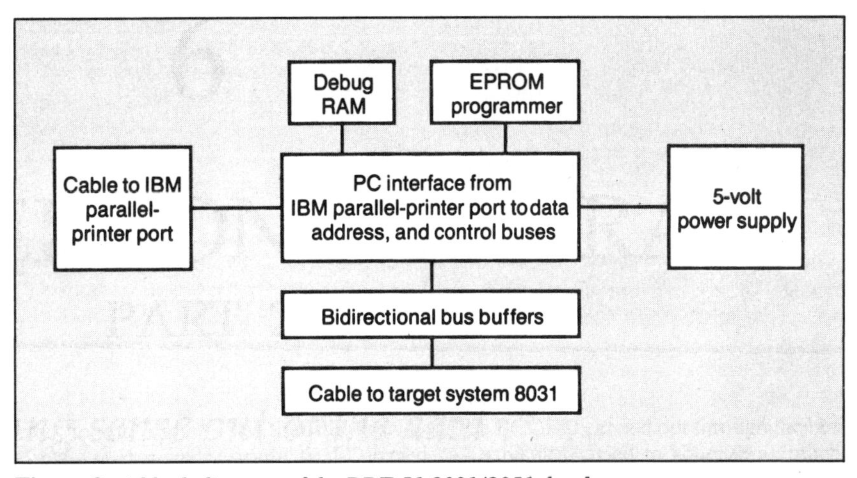

Figure 3: *A block diagram of the DDT-51 8031/8051 development system.*

simulated run time is much slower than the real-time rate on the actual hardware. A further complication is that interrupt timing is not easily duplicated on a simulator.

The ultimate solution is an in-circuit emulator (ICE), which is a special development system peripheral plugged into the 8051's socket in the actual hardware (called the target system). From the target's point of view, the ICE is an 8051 running at full speed. From the user's perspective, the ICE provides many of the features of a simulator, along with the ability to run programs at full speed using the real hardware.

Ideally, an ICE will have no effect on the target system, because all the 8051's features are provided in high-speed hardware. The 8051's internal registers and I/O ports are visible because the ICE uses discrete logic rather than a single chip. The development system directly monitors what's going on, logic comparators control the breakpoints, and there's no interference until the ICE stops at the selected instruction.

All this hardware makes ICE systems rather expensive. If you are developing many 8051 designs, an ICE is the only way to go. As a practical matter, however, an ICE is far beyond the typical user's budget and is generally reserved for the corporate lab.

A More Modest System

There is a middle ground between personal computer-based software simulators and hardware ICE systems. It's often enough to stop at a breakpoint and single-step through instructions while watching the target system's LEDs blink and relays click. By trading off some speed and hardware for time and money, a much simpler program development system can provide many features of an ICE at a fraction of the cost.

The DDT-51 system is an IBM PC-based 8031/8051 development and dynamic debugging tool (see photo 2). It uses a modified parallel printer port and a small circuit board to give the IBM PC complete control over the target system's hardware. The DDT-51 downloads the 8051 program into 8K-byte static RAMs, thus eliminating the need to burn an EPROM for each program change. An on-board 2K-byte static RAM holds the small amount of 8051 code required to support single-stepping and breakpoints. A disassembler shows the current 8051 instruction on the IBM PC's display while single-stepping, as well as the current 8051 registers and internal data memory values.

This system connects between the IBM PC and the target system (see figure 3). It has only about a dozen chips, including the world's simplest (and slowest) 2764 EPROM programmer.

The DDT-51 won't handle all possible 8051 target systems, but it will give you a good start. With that in hand, we can continue on with other interesting Circuit Cellar project designs.

Experimenters

As is the custom with Circuit Cellar projects, the software for the DDT-51 development system is available for downloading from my multiline bulletin board free of charge. The number is (203) 871-1988. Of course, this downloaded software is limited to noncommercial personal use unless licensed otherwise.

In Conclusion

The hardware and software specifics of DDT-51 will be presented in Chapter 35, part 2.

Special thanks for the technical contributions provided on this article from Jeff Bachiochi and Ed Nisley.

Diagrams specific to the Intel 8031/8051 architecture are reprinted by permission of Intel Corp.

See page 447 for a listing of parts and products available from Circuit Cellar Inc.

35

WHY MICROCONTROLLERS?

PART 2

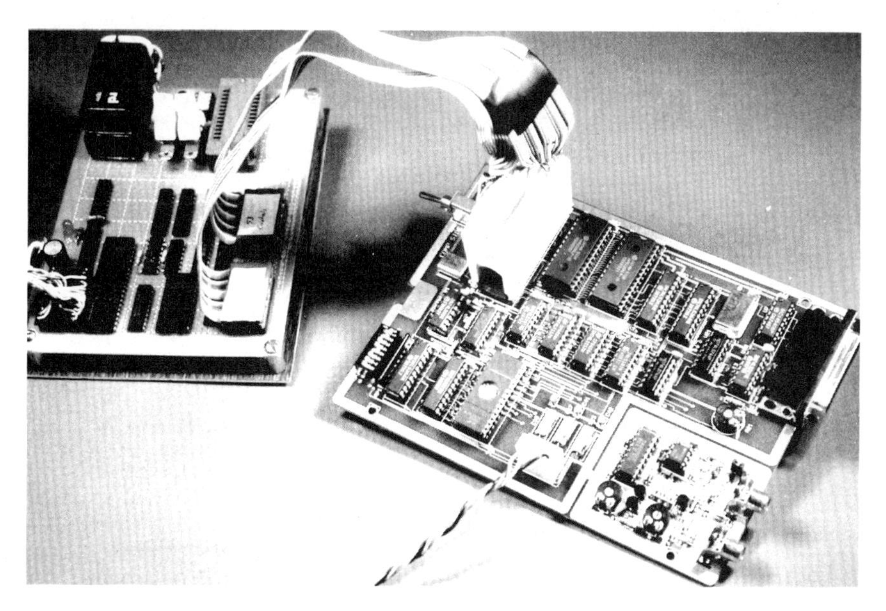

The DDT-51 is a low-cost IBM PC-based Intel 8031/8051 development system

Photo 1: *Prototype of the DDT-51 development system.*

Having explained in part 1 of Chapter 35 why we want to design projects around a single-chip microcontroller, it's time to talk about getting the software into the microcontroller. As I previously suggested, your options are either burn and crash (burn the program into an EPROM and figure out why it crashed) or use a development system.

The DDT–51 8031 development system I'll describe in part 2 of this chapter offers many of the features of more expensive software development systems, at a fraction of the price. Unlike those systems, you can build this one yourself, or modify the features to suit your tastes, and wind up with a system tailor-made for your application.

I'll start with the DDT-51 system hardware, then describe the software interface that lets an IBM PC control the target 8031 system.

The DDT-51 Development Tool

Nothing ever works the first time. When you are debugging microcontroller-based hardware, you need to be able to load and run a program, stop at breakpoints, single-step through critical parts of the code, disassemble instructions, examine and change registers, and continue where you left off. An ideal development system helps you do all that without using any 8031 system resources: no pins, no ICs, no program RAM—nothing at all. In-circuit emulators (ICEs) come closest to that ideal, because they replace the microcontroller with a hard-wired equivalent of the microcontroller chip—an equivalent that lets you directly observe the chip's (normally hidden) inner functions.

The DDT-51 system isn't quite that ideal, but it comes close. It needs one interrupt input, one output pin, and a 2K-byte RAM chip to hold debugging code and data. A 40-pin DIP clip attaches directly to the 8031 IC, and there are no other connectors to the target board. Photo 1 shows the prototype DDT-51 development system board, and figure 1 shows a block diagram of the hardware.

The software that runs on the host IBM PC is called DEBUG31. DEBUG31 shows all the 8031's internal RAM and the main registers when the microcontroller is stopped, as illustrated in figure 2. You can load new values into the registers, set a breakpoint to trap the program, and single-step through the code. You use function keys on the host to control all these operations.

Like an ICE, the DDT-51 supplies the data, address, and control lines needed to simulate an 8031's operations, although at a much slower rate. When the connections are disabled, the 8031 will run as though the DIP clip wasn't there.

Under the Hood

As described in part 1 of this chapter, an 8031 doesn't have an internal program ROM, so it needs an external EPROM (usually a 2764). If you replace that EPROM with a pin-compatible 6264 8K-byte RAM, the DDT–51/PC link can load and change the 8031's program. Once you've got the program loaded, the

8031 can execute it normally, just as though it were in EPROM.

The debug RAM is a key component of the DDT-51 system, but some 8031 systems may not need any RAM. Because the DIP clip connects to the 8031 bus control lines, it's easy enough to put the debug RAM on the DDT-51 board. I used a 2K-byte RAM simply because it is the smallest one that's readily available. An 8K-byte RAM would work just as well.

Unfortunately, the 8031 wasn't designed to share its bus control lines with other hardware. All the bus lines float when the RESET input is high (they use tristate drivers), but the program halts until RESET goes low again. The program restarts from address 0000 hexadecimal after a RESET, rather than continuing where it left off, which is exactly what a RESET should do. What's needed is a dual-ported RAM that can be accessed by either system independently, with separate control, data, and address lines. The two systems can then store and examine data in the common RAM without interaction.

Although the PC and the 8031 can both access the debug RAM, they can't do so simultaneously. To keep the hardware simple, there are no interlocks to prevent collisions (both systems attempting to access the RAM at the same time). If a collision does occur, one processor will always get bad data because the proper buffers are not active. The PC and 8031 programs must adhere to a software gentleman's agreement to prevent RAM conflicts. It turns out that this is not at all difficult to accomplish.

How do they do that? Once the 8031 program starts running, the DDT-51 can pull INT1 low to interrupt normal execution. A few debugging instructions added to the 8031 code handle the interrupt and examine a location in the debug RAM to determine what function was requested. The 8031 copies the results into the debug RAM and returns to normal operation, giving DEBUG31 (on the PC) free access to the RAM.

The gentleman's agreement is controlled by two wires: the IRQ line, triggering an 8031 interrupt, and the 8031 HALTED line, which is set while the interrupt handler is in control. Remember that the 8031 isn't halted after the interrupt, but is in the middle of the interrupt handler.

Figure 3a describes the physical connection between the PC and the target system; figure 3b shows the timing diagram that manages access to the debug RAM. DEBUG31 on the PC writes control information into the debug RAM and pulls the IRQ line low. The 8031 responds to the interrupt, examines the debug RAM, and executes whatever functions are needed. When it's done, it sets the HALTED line active and waits for DEBUG31 to restore the IRQ line to a high level. The 8031 then sets HALTED low, reloads the registers, and returns to normal operation in the interrupted program.

Figures 4a and 4b show the DDT-51 system circuit diagram. The DDT-51 system divides neatly into four sections: the PC parallel-port interface, bus buffers to control the various lines, the debug RAM, and an EPROM programmer.

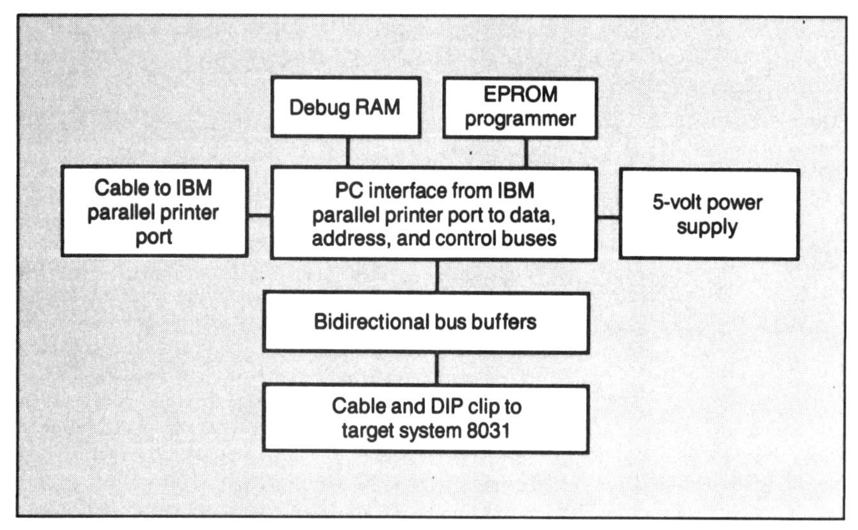

Figure 1: *A block diagram of the DDT-51 development system.*

The Port Swings Both Ways

The DDT-51 system trades hardware for software. The DDT-51's hardware pro-

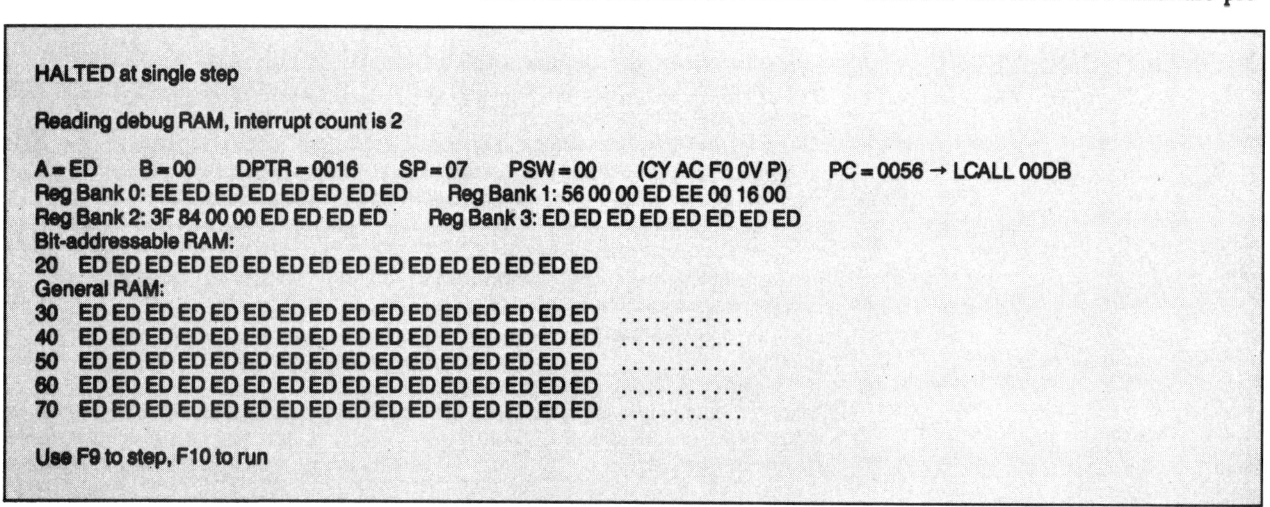

```
HALTED at single step

Reading debug RAM, interrupt count is 2

A=ED    B=00    DPTR=0016    SP=07    PSW=00    (CY AC F0 OV P)    PC=0056 → LCALL 00DB
Reg Bank 0: EE ED ED ED ED ED ED ED      Reg Bank 1: 56 00 00 ED EE 00 16 00
Reg Bank 2: 3F 84 00 00 ED ED ED ED      Reg Bank 3: ED ED ED ED ED ED ED ED
Bit-addressable RAM:
20  ED ED ED ED ED ED ED ED ED ED ED ED ED ED ED ED
General RAM:
30  ED ED ED ED ED ED ED ED ED ED ED ED ED ED ED ED  ...........
40  ED ED ED ED ED ED ED ED ED ED ED ED ED ED ED ED  ...........
50  ED ED ED ED ED ED ED ED ED ED ED ED ED ED ED ED  ...........
60  ED ED ED ED ED ED ED ED ED ED ED ED ED ED ED ED  ...........
70  ED ED ED ED ED ED ED ED ED ED ED ED ED ED ED ED  ...........

Use F9 to step, F10 to run
```

Figure 2: *A sample display from the DEBUG31 program, shown here in single-step mode.*

vides only essential functions; the software handles most of the logic. DEBUG31 merely sets and reads bits in the DDT-51 hardware to find out what's going on in the target 8031. Sounds simple enough.

Because speed of interaction between the host PC and the DDT-51 is important, the DDT-51 connects to the PC via a bidirectional parallel port (using the serial port would greatly increase the hardware complexity of the DDT-51). Unfortunately, an irritating problem with the design of the PC is that its standard parallel printer port is output only.

All the hardware needed to read or write 8 bits of data is already in place, but it lacks a connection to enable that function. While you can buy a custom parallel I/O board, I think changing a single trace on the existing board might be worth it to some readers. Figure 5 shows the single cut and addition to convert a standard parallel printer port to bidirectional operation.

Of course, all the clone boards (even the IBM boards) seem to use different IC numbers and assign the bits to different pins on the ICs, so this modification can be a real mystery. If you are unwilling to chop up your printer port card, buy a $50 clone printer port card. They are usually bidirectional. Some clone cards also omit the input connection between the data bus and pin 14 of the LS174. I suppose they figure that because the output is unused, the input is irrelevant. It's easy to find the two unused pins, however (a modified port will still work correctly with all your other software, simply because the code doesn't know about the change and won't take advantage of it).

DEBUG31 examines each of the three possible printer ports, working down from LPT3, to find which one has the modification installed. This provides a convenient way to check your work. DEBUG31 will tell you which card it's using.

Creating Five Ports from One

The DDT-51 system needs more than one I/O port to control all its hardware. Rather than burden the PC with more ports, I used an 8255 parallel peripheral interface (PPI) chip and a pair of LS374s to get five more ports. DEBUG31 controls these ports using the four standard printer control lines.

IC9, the control register, holds bits that must be active all the time, like the 8031 RESET and IRQ lines. It also supplies the 8255's register addresses and several bits for the DDT-51's bus logic.

IC12, the system register, holds the 8031 system bus control bits. There are 4 unused bits in this register that can be used for additional functions. The outputs are disabled whenever the 8031 drives the DDT-51's address and data buses, so that there is no conflict.

IC8, the 8255 bus interface, drives the high byte of the external address bus through port B and writes and reads the external data bus using port A. Port C samples the 8031 HALTED line. Other port C bits monitor the outputs of the DDT-51 logic circuits when running diagnostics from the PC.

Every 8031 system has a latch that demultiplexes the low-order byte of the address from the combined address/data bus. You might think that, because the outputs of that latch don't connect to the 8031, the DDT-51 system would need another DIP clip to get access to that part of the address bus. Instead, it is simpler to duplicate the latch, using IC10.

The DDT-51's debug RAM presents a similar problem, since the RAM requires a chip select line and the target 8031 system may not have an address decoder. I added IC11 to provide chip selects on 8K-byte boundaries throughout the 64K-byte address space. The EPROM programming socket, which is empty in normal use, also gets a chip select line from this decoder.

Bus Buffering

Both the 8031 and the DDT-51 system have control lines to read and write the program RAM (normally on the 8031 board) and the debug RAM (on the DDT-51 board). There are more combinations of data, address, and control line directions than might seem possible at first, which is why the logic isn't as simple as you'd expect.

IC9 supplies two control lines, called

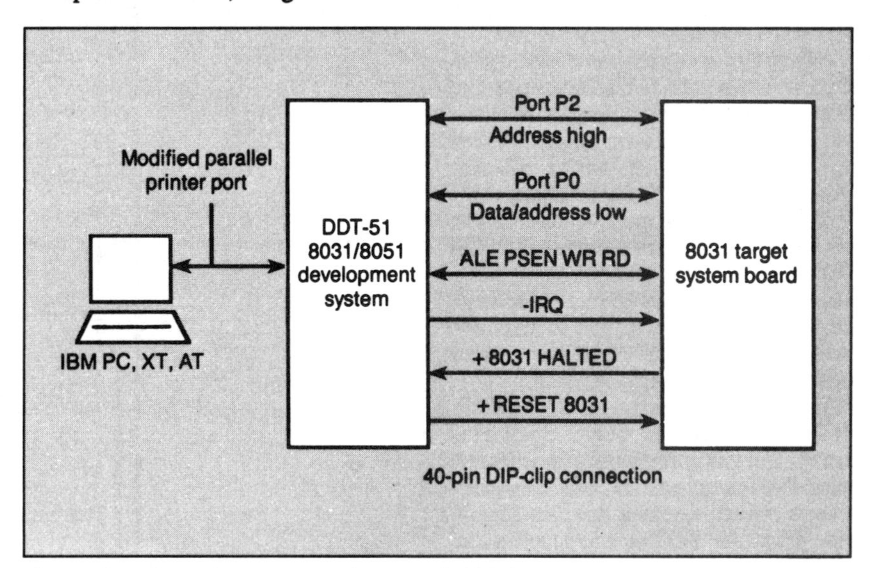

Figure 3a: *Physical connections between the DDT-51 and its host and target systems.*

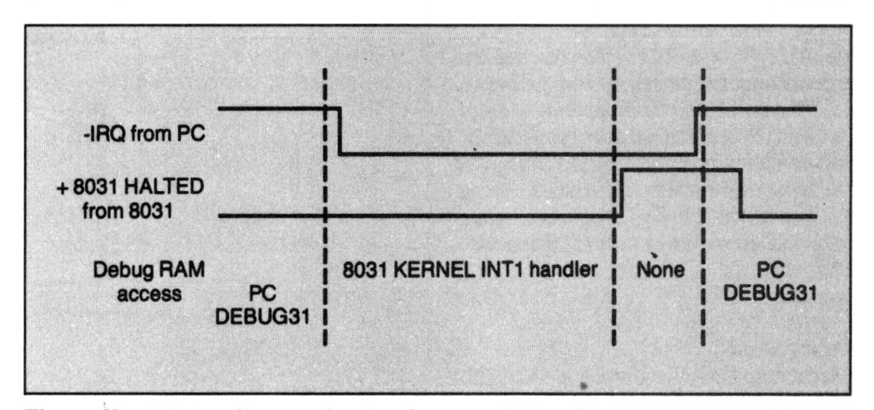

Figure 3b: *A timing diagram showing the system's "gentleman's agreement" for controlling access to the debug RAM.*

CTLS TO 8031 and CTLS FROM 8031, to select one of three states. The 8031 controls the DDT-51 logic when CTLS FROM 8031 is active, DDT-51 logic controls the 8031 system when CTLS TO 8031 is active, and the two systems are isolated from each other when both lines are inactive.

Setting the direction of the data and address lines isn't a simple task. For example, when the target 8031 system writes to the debug RAM, both the address and data come from the 8031. When it's reading from the debug RAM, the address comes and the data goes. A similar situation, in reverse, occurs when the DDT-51 reads the 8031's program RAM.

IC14 sets the DATA TO 8031 and DATA FROM 8031 lines by deciding which way the data must flow based on the current address, read or write activity, and signal source. This must be handled by hardware because it varies cycle by cycle as the 8031 runs its program.

When the two systems are isolated, the PC can access the debug RAM and the 8031 can access the program RAM without interacting. If either system tries to access the other while they're isolated, the data returned will be indefinite.

You've probably noticed that there's no hardware preventing both the CTLS TO 8031 and the CTLS FROM 8031 lines from being active at once. Further, as the schematic shows, this is definitely an illegal condition; it should never occur with properly operating PC code.

The reason for using an LS374 (IC10) to latch the address should now make sense. The bus buffers can't be activated until the control and address lines indicate in which direction the data should go. When the 8031 system supplies the ALE signal, the DDT-51 hardware accepts the 8031 data bus and tries to latch it in IC10. An LS373 is a *transparent* latch, meaning that the outputs track the inputs whenever the clock is high. When the 8031 lowers ALE, the bus buffers turn off and the data goes away before an LS373 can reliably latch it.

The LS374 is an *edge-triggered* latch, which means that the outputs change at the upward transition of the clock. Using ALE to clock the latches gives a sharp edge just before the data buffers turn off. This way, the LS374 records the inputs correctly.

Debug RAM

The debug RAM is simply a 6116 2K-byte static RAM chip, driven by the buffered address and data lines. Either sys-

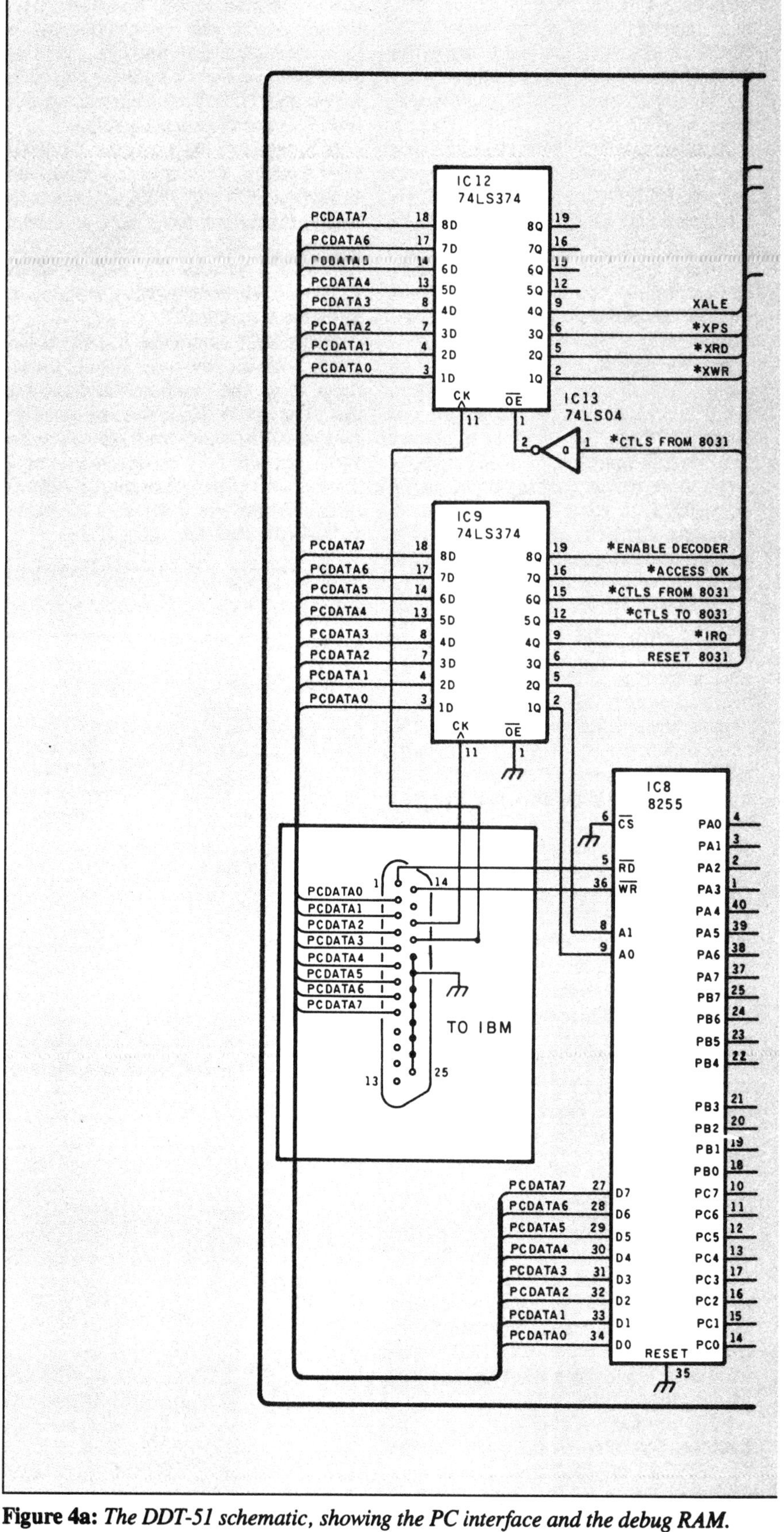

Figure 4a: *The DDT-51 schematic, showing the PC interface and the debug RAM.*

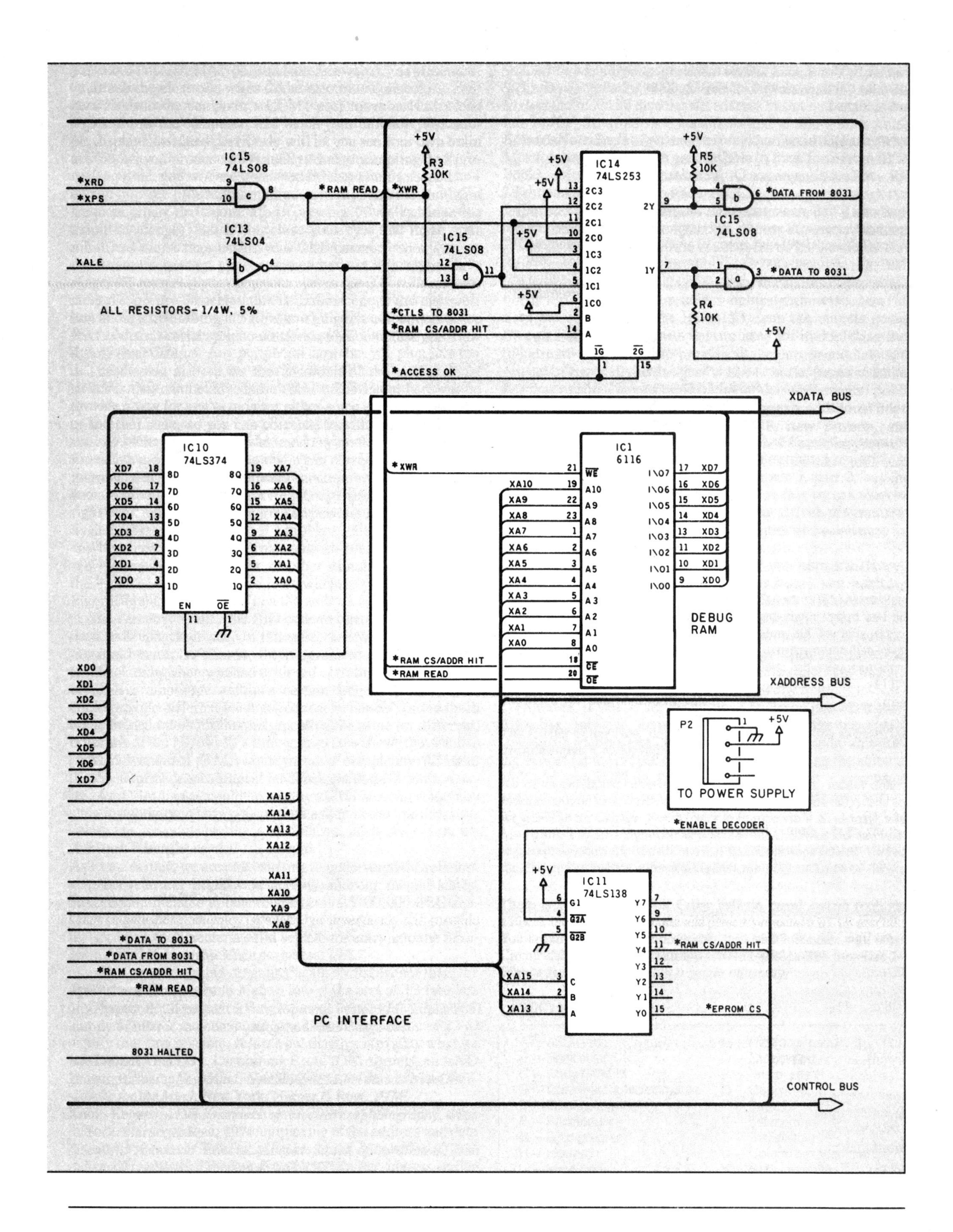
IC15
74LS08
*XRD
*XPS
*RAM READ
*XWR
+5V
R3
10K
IC13
74LS04
XALE
IC15
74LS08
ALL RESISTORS- 1/4W, 5%
IC14
74LS253
*CTLS TO 8031
*RAM CS/ADDR HIT
*ACCESS OK
R5
10K
R4
10K
*DATA FROM 8031
*DATA TO 8031
XDATA BUS
IC10
74LS374
IC1
6116
DEBUG
RAM
XADDRESS BUS
P2
TO POWER SUPPLY
*ENABLE DECODER
IC11
74LS138
*RAM CS/ADDR HIT
*EPROM CS
*DATA TO 8031
*DATA FROM 8031
*RAM CS/ADDR HIT
*RAM READ
PC INTERFACE
8031 HALTED
CONTROL BUS

tem can access it. DEBUG31 assumes that the debug RAM is located at 8000 hexadecimal.

EPROM Burning

Although the DDT-51 system is used to develop programs using RAM instead of EPROM, there (presumably) comes a time when you've finished the program and you're ready to ship it. At that point, you need an EPROM programmer. DDT-51 incorporates the world's least expensive 2764 EPROM programmer.

The programmer section of figure 4b shows all the hardware required to program a 2764 EPROM. The programming supply voltage comes from two 9-volt batteries hard-wired to add +18 V to the +5-V power supply. A simple emitter follower reduces this 23-V level to 21 V and provides enough current for the programming pulse. The 0.1-microfarad capacitors filter out noise glitches from the logic. You'll have no problem with battery life unless you start using this programmer for production quantities.

It turns out that programming a 2764 EPROM is much like writing to a rather slow RAM. First, you set up the address and data lines, select the chip, and pulse pin 27 low for 50 milliseconds. If pin 1 has +21.0 V on it while pin 27 is low, you've just programmed 1 byte of the EPROM. That's all there is to it.

DEBUG31 can already write data to any 8031 address, so programming an EPROM simply requires an "EPROM burn" software flag to lengthen the write pulse to 50 ms. A manual switch applies 21.0 V to the zero-insertion-force (ZIF) socket.

This system handles code that fits into a single 2764 EPROM. A DIP switch on the DDT-51 system board sets the chip-select address range so you can burn EPROMs at different addresses. And 8K bytes of EPROM holds a lot of 8031 code, particularly for smaller projects. Unlike the Circuit Cellar serial EPROM programmer (see the October 1986 Circuit Cellar), however, there are no circuits to prevent damage occurring when you insert or remove an EPROM from a "live" socket. Make sure you turn the DDT-51 system's power off and disconnect the DIP clip from the 8031 before inserting or removing the EPROM.

Of course, 2764s allow faster "intelligent" programming methods, but I haven't implemented them (I thought an EPROM programmer that required zero additional chips was pretty good). As a result, the DDT-51 system is both the world's least expensive and slowest

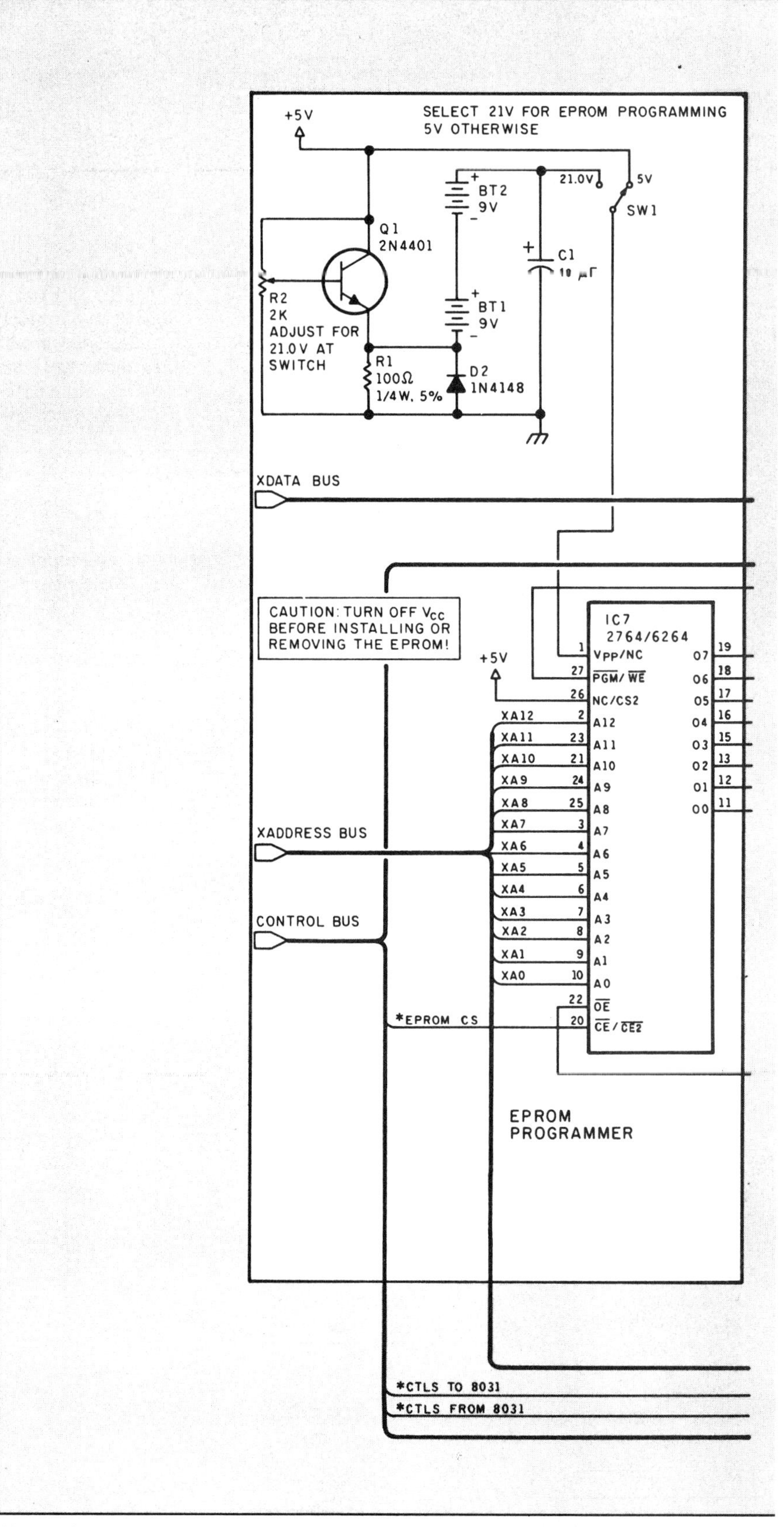

Figure 4b: *The EPROM programmer subsystem and bidirectional bus buffers.*

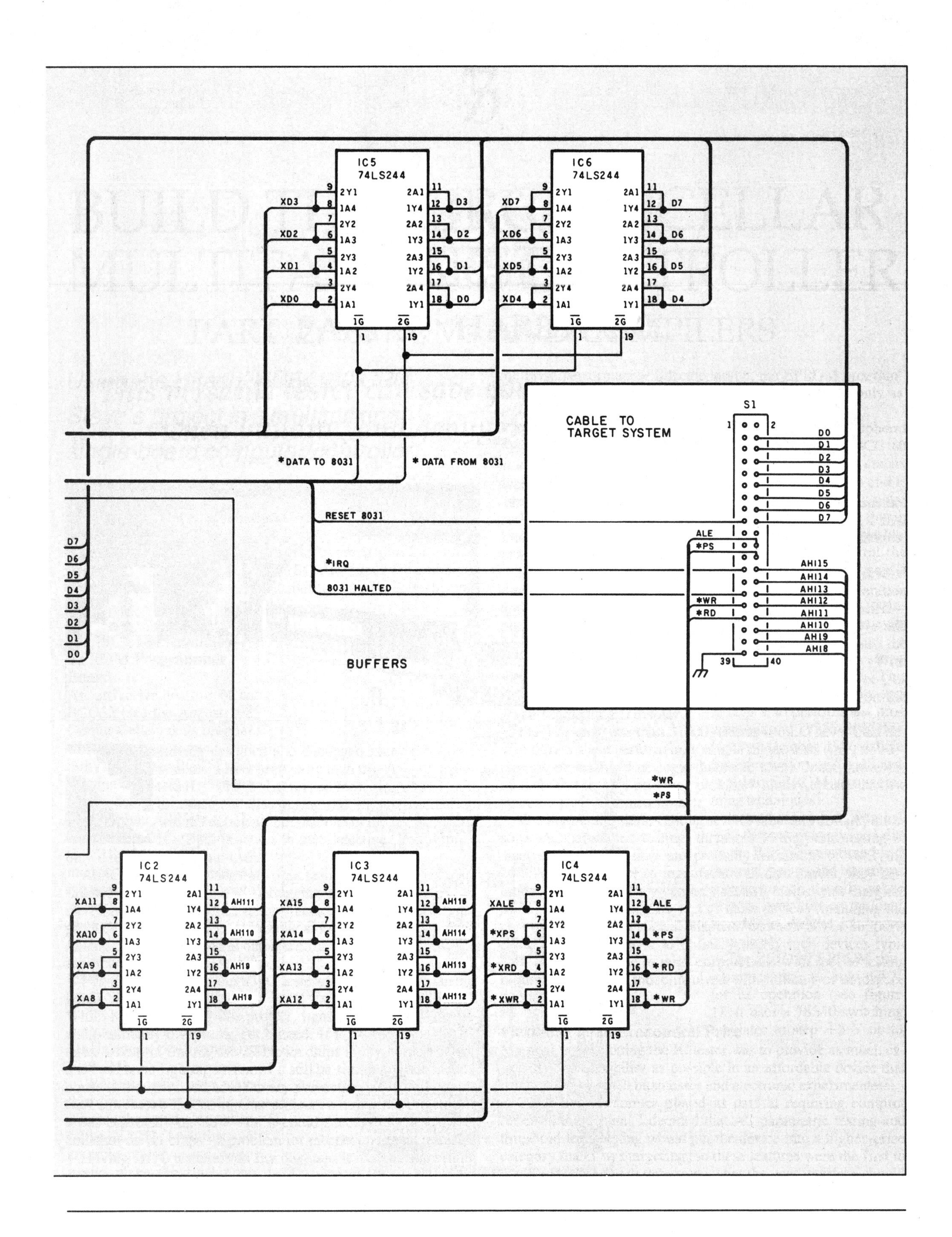
IC5
74LS244
IC6
74LS244
XD3
XD2
XD1
XD0
D3
D2
D1
D0
XD7
XD6
XD5
XD4
D7
D6
D5
D4
CABLE TO
TARGET SYSTEM
S1
*DATA TO 8031
*DATA FROM 8031
RESET 8031
*IRQ
8031 HALTED
ALE
*PS
*WR
*RD
AH115
AH114
AH113
AH112
AH111
AH110
AH19
AH18
BUFFERS
IC2
74LS244
IC3
74LS244
IC4
74LS244
XA11
XA10
XA9
XA8
XA15
XA14
XA13
XA12
XALE
*XPS
*XRD
*XWR

EPROM programmer. If you start writing long programs, you may want to step up to my serial EPROM programmer.

DEBUG31 Essentials

You might think that DEBUG31 on the PC is written in assembly language to handle all the bit manipulation needed to control the hardware. It turns out that it's all done in Turbo Pascal, which is a whole lot easier to read and understand. All the tricky stuff is handled by a few low-level routines.

At the highest level, DEBUG31.PAS is a simple loop that waits for function keys to be pressed. Each function key triggers a separate procedure to handle whatever is requested. Listing 1 shows the entire main loop, which is easy enough to understand.

Listing 2 shows a procedure called by the main loop (and elsewhere, too). SysReset activates the RESET 8031 line from IC9 and sets up DDT-51 hardware so that the PC has access to the 8031 through the bus buffers. When the RESET line is active, the 8031 has stopped operating and floated all its bus and control lines, so the PC can read or write any hardware on the shared data bus.

SysReset calls SetCR (set the control register IC9), SetSR (set the system register IC12), and Load8255 (interface to the 8255), which are some of the lowest level routines in the system. Each controls a single IC connected to the printer-port data lines. The routine handles all the operations required by other code and provides a convenient set of mnemonics that are easier to remember than the actual bit locations and values.

The 8031 Kernel

The DDT-51 system depends on a small kernel of 8031 code to handle interrupts and perform functions on behalf of the PC program. This code runs only during INT1 interrupts and must adhere closely to the gentleman's agreement interface to avoid hardware collisions.

Unlike DEBUG31.PAS, KERNEL.ASM must be written in 8031 assembly language and incorporated into any 8031 program that uses the debugging features of DEBUG31, but most of it is located in the 2K-byte debug RAM located on the DDT 51 board. The normal program memory must hold only about two dozen bytes of KERNEL's code, and you can move most of that chunk to any convenient location.

Because the 8031 registers and internal RAM are not directly accessible through the DDT-51 interface, KERNEL must copy those values into the debug RAM before DEBUG31 can see them. Once they've been copied, the PC can read, display, and change them as

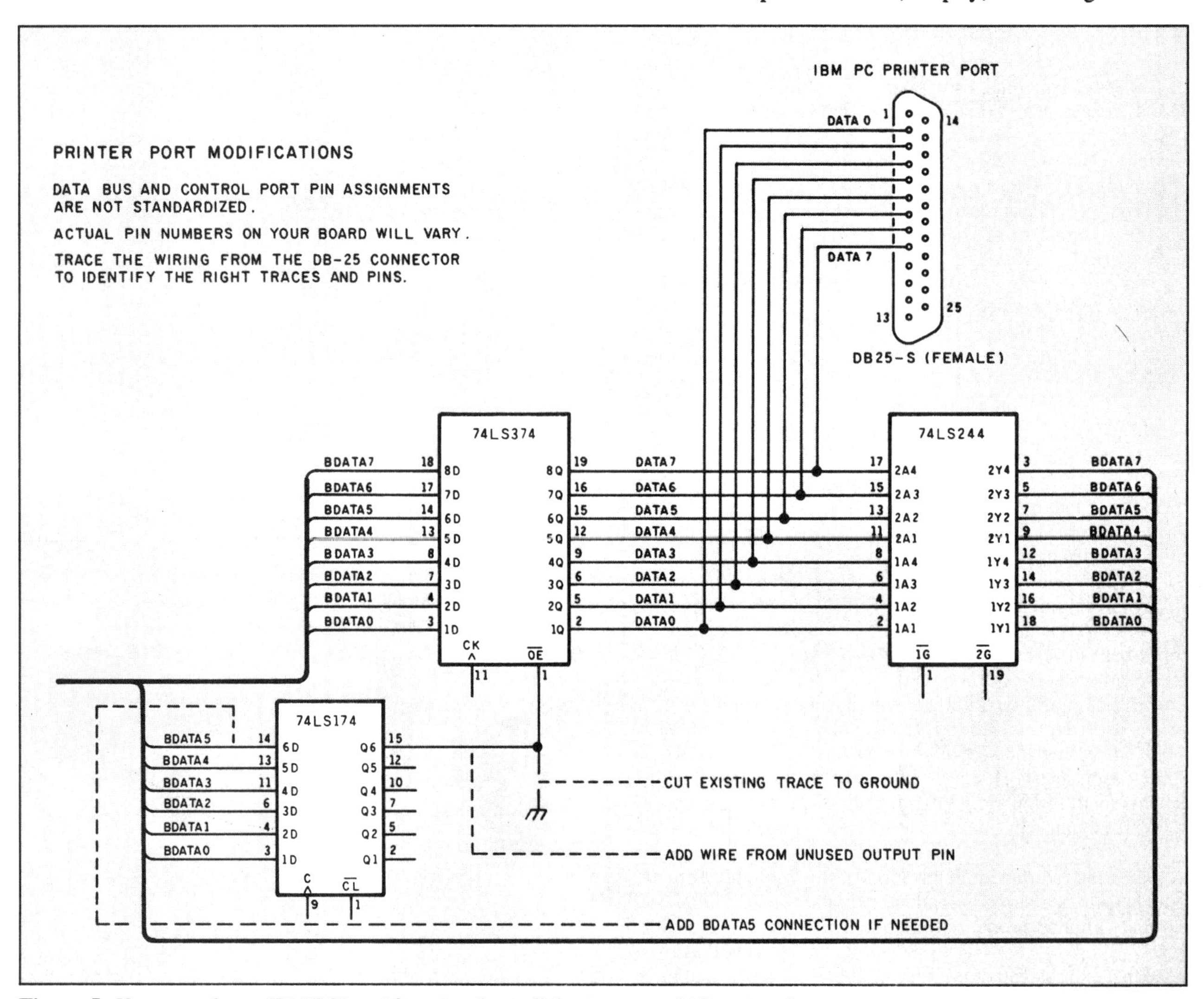

Figure 5: *How to make an IBM PC's unidirectional parallel printer port bidirectional.*

needed. KERNEL then copies the new values back into the internal locations and returns to the interrupted code.

The software handles breakpoints in a similar manner. DEBUG31 stores the breakpoint address in debug RAM and triggers an interrupt. KERNEL decodes the operation and writes an LJMP instruction at the specified address. The KERNEL breakpoint handler gets control when the 8031 executes the LJMP and copies the internal values into debug RAM.

At the heart of the interrupt-handler routine, the first order of business, as with any interrupt code, is to save all the CPU registers. Later on, these saved values will be copied into debug RAM so that DEBUG31 can reconstruct the actual state of the machine when the interrupt occurred.

If a breakpoint was active, the UndoBP routine removes the LJMP and restores the program RAM so that the next instruction is ready to go. Because an LJMP requires 3 bytes, it may have overwritten one or two instructions following the breakpoint. Regrettably, the 8031 does not have a single-byte software interrupt instruction like the 8088's special INT 3 breakpoint.

The DumpRAM routine copies the internal RAM values into debug RAM as described before, as well as extracting the registers saved on the stack. Remember that the internal RAM reflects the current registers rather than the ones used by the main program. Some internal RAM locations can't be read without introducing side effects: Reading the serial buffer will reset a pending interrupt, which would cause a malfunction in the main code.

Both KERNEL and DEBUG31 increment separate IRQ counters; KERNEL's counter is stored in debug RAM. If DEBUG31's counter differs from KERNEL's version, at least one IRQ got lost, which probably means that the 8031 program is out of control.

Because most of KERNEL is in the debug RAM, DEBUG31 can't read that RAM while the 8031 executes from it. The StepSpin routine is located in 8031 program RAM, which frees up the debug RAM for the PC. The StepSpin code sets 8031 HALTED to indicate that the debug RAM is available, then waits for the PC to raise INT1 when it's done changing the RAM. At that point, it's safe to begin executing from debug RAM again, so the code returns to SpinRet to continue.

The code following SpinRet reads the

Listing 1: *The* DEBUG31.PAS *main loop.*

```
(*** setup code omitted here ***)

 REPEAT
  { Get next key }
  GetKey(key1,key2);
  { Is it an extended key? }
  IF key1 = Chr(ESC)
   THEN BEGIN
    { Which function key? }
    CASE Ord(key2) OF
     59 : ShowHelp;
     60 : SysReset;
     61 : LoadKernel;
     62 : Download;
     63 : SetBurn;
     64 : Writeln('F6 not used');
     65 : SetRegs;
     66 : SetBP;
     67 : SysStep;
     68 : BEGIN
           { Normal, no IRQ }
           SysRun(running,FALSE);
           Writeln('Use F9 to stop/step');
          END;
     ELSE;
    END;
   END;

 UNTIL (key1 = Chr(ESC)) AND (key2 = Chr(0));

(*** some cleanup code omitted ***)
```

Listing 2: *The system reset procedure.*

```
{ Force reset and take control of bus }

PROCEDURE SysReset;

BEGIN

 { Ensure a reset }
 SetCR(Reset8031,ON);

 { Get control of bus }
 SetCR(IRQ,OFF);
 SetCR(CtlTo8031,ON);
 SetCR(CtlFrom8031,OFF);
 SetCR(AccessOK,ON);
 SetCR(EnableDec,ON);

 { Set up controls }
 SetSR(XRD,OFF);
 SetSR(XPSEN,OFF);
 SetSR(XALE,OFF);
 SetSR(XWR,OFF);

 { Our data and address }
 Load8255(I55Ctls,I55AoBoCi);

 { Reset IRQ counter }
 IRQctr := 0;
 PutRAMbyte(IRQctrB,$00);

 Writeln('8031 hardware is reset');

 { Indicate reset }
 state := reset;

END;
```

new run-mode byte from debug RAM and handles the request. This code also installs the new breakpoint LJMP instruction.

Finally, the StuffRAM routine copies the (possibly changed) values from debug RAM back into internal RAM. It inserts the new values for the registers into the right stack locations, so that the main program's registers can be altered by the PC. It does not restore registers with side effects (e.g., the serial transmitter buffer) for the same reason that DumpRAM didn't read from them.

After the registers are restored, the 8031 executes a RETI (return from interrupt) and the interrupted main code resumes execution, perhaps with new register values. (The complete listings for DEBUG31.PAS and KERNEL.ASM are available for downloading from the Circuit Cellar bulletin board system.)

A Middle Ground

Does working on the DDT-51 development system convert me into a software jockey? Certainly not. To maintain a high level of performance in Circuit Cellar designs at a time when the rest of the industry is moving to application-specific ICs (ASICs) and other development-intensive reduced-component custom designs, I have chosen a middle ground of using programmable microcontroller chips.

In the same way that a large board of IC logic chips is reduced functionally to an ASIC, I have lumped the controlling elements of the TTL circuit I usually use into a configurable application-specific controller that can simulate these functions. Using simple three- or four-chip microcontrollers as standard computer engines in my projects, I am now better able to concentrate on presenting applications of technology rather than providing redundant explanations of low-level logic.

The DDT-51 development tool will hopefully help other BYTE readers consider the benefits of this middle ground.

The DDT-51 development system won't handle all possible 8051 target systems, but it will give you a good start. With that in hand, we can continue with other interesting Circuit Cellar project designs.

Experimenters

As is the custom with all Circuit Cellar projects, the software for the DDT-51 development system is available for downloading from my multiline bulletin board free of charge. Call (203) 871-1988. Of course, this downloaded software is limited to noncommercial personal use unless licensed otherwise.

Special thanks for the technical contributions provided for this article by Jeff Bachiochi and Ed Nisley.

See page 447 for a listing of parts and products available from Circuit Cellar Inc.

36

THE SMARTSPOOLER

PART 1: THE SPOOLER HARDWARE

Steve clears your printer's throughput bottleneck and comes up with more than a buffer

The discrepancy in speed between fast CPUs and slow printers has existed since the early days of computing, and the magnitude of the speed difference is becoming ever larger. When you upgrade and add performance to a computer system, you generally add processing power rather than printing power. The net result is that many 20-MHz 80386 "million-instructions-per-second-class" machines out there are waiting for 100-character-per-second printers.

Admittedly, many computer users have light printing requirements, and they don't mind waiting 5 to 10 minutes for an occasional printout. Some, I'm sure, welcome this downtime as a mandatory coffee break. However, those who rely on their computers as a tool for making a living know that time is money.

Spooling to the Rescue

This problem is not new, and one solution has always been spooling. *Spool* stands for "simultaneous peripheral operations on-line." Pioneered on mainframes and minicomputers, spooling solves the CPU/printer speed gap by temporarily buffering printer output on intermediate storage (typically a disk drive) at high speed. Then, you move the stored data from storage to the printer at the latter's low speed. In the meantime, the CPU also runs foreground applications.

Traditional spooling has been tried on personal computers with limited success. One problem is that spooling requires an operating system with either multitasking capabilities or special software patches that effect the appearance of simultaneous operation. Even with a multitasking operating system or a patch, however, spooling often taxes the computer's processing power and affects system performance.

Spooling also creates problems in disk priority and data conflicts. If not properly coordinated through the operating system, a computer's disk can get "trashed" if it has to handle print data and application data at the same time. Given the declining cost of processing power and memory, it finally became obvious that a separate computer should handle the printing task, which led to the dedicated printer buffer.

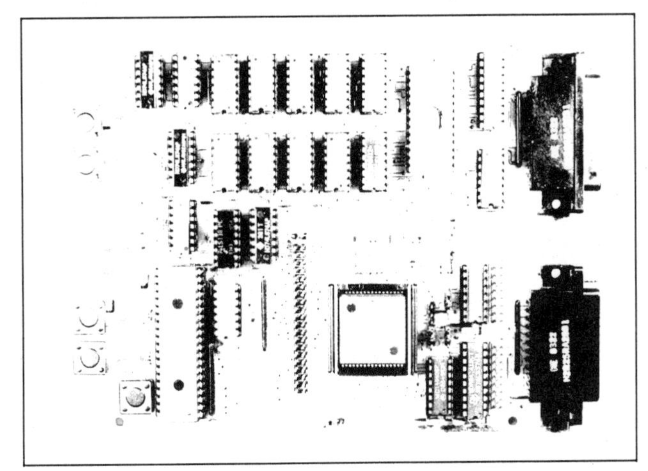

I have received many letters over the years asking when I would present a printer buffer as a project. Generally, I would avoid answering the question or jokingly say "never." Of course, it really wasn't a joking matter. It troubled me greatly that I couldn't do justice to such an important subject and had to live with such an embarrassing answer until I could invest the time and effort to present a Circuit Cellar–quality design. Let me explain.

Most printer buffer projects are cumbersome and simplistic. The ones I remember seeing offer only rudimentary functions and use 40 or 50 chips for a 64K-byte buffer.

Commercial units, on the other hand, have narrowly defined capabilities enhanced by software-implemented system functions and tricky hardware designs intended to reduce manufacturing cost and complexity.

I recall seeing one rather unique printer buffer design that used an 8051 single-chip microcomputer connected to dynamic RAMs (DRAMs) that operated *serially*. The 8051 would constantly cycle through the DRAM addresses at a rate that precluded any need for DRAM refresh. The system stored text serially in the individual 64K- by 1-bit DRAMs rather than in a parallel bank of them, as you'd usually expect. An 8K-byte text buffer therefore consisted of a single 64K-bit DRAM chip, while a 32K-byte buffer used four, and so on.

While I am not totally opposed to off-the-wall hardware designs, fancy technique has its price. Most computer-based technical designs are a trade-off between hardware and software. Anytime a company expects to be manufacturing something in a high-volume quantity, it tries to minimize recurring hardware costs by doing as much as possible in the one-time nonrecurring software development phase.

In a magazine article, however, there aren't any recurring expenses (beyond a few prototypes), therefore all costs associated with the one-time presentation are essentially nonrecurring. Since most magazine authors don't have infinite software resources (and deep pockets), published print spooler designs have usually ended up as a far cry from commercial units, both in technique and capability.

I could hardly criticize mediocre designs if I presented a piece of hardware that merely

While I've presented SmartSpooler primarily as a printer buffer, it can function as a complete remote data-processing computer.

met the objective with little regard for the hardware costs and complexity. At the same time, it would be insane to custom-mask an 8051 and present a three-chip printer buffer project that demonstrates nothing to readers who want to learn something about the architecture of printer buffers.

Fast Computers, Slow Printers

With the Circuit Cellar SmartSpooler, I have achieved my goal. This is an efficient integration of commercial and educational objectives. I've complemented its hardware design with interrupt-driven software that does not compromise performance to maintain cost-effectiveness. Designated as a spooler rather than as a simple printer buffer, SmartSpooler vastly improves computing throughput with slow peripherals.

SmartSpooler has 256K bytes of memory, serial and parallel I/O ports, and features that improve versatility and ease of use. These include "switchbox" capability for routing serial or parallel computer input to serial or parallel printer output, the ability to print multiple copies, single-sheet feeding mode, and buffer capacity indicators. Also, you can daisy-chain multiple SmartSpoolers to control a whole network of peripherals.

More important, SmartSpooler is intelligent. While I've presented it primarily as a printer buffer, a host computer can completely control its operation. You can even download executable code to SmartSpooler. It can function as a complete remote data-processing computer that analyzes and interprets the data flowing through it.

After a summary of SmartSpooler's features, I'll cover some printer buffer basics and then describe SmartSpooler's hardware and software specifics.

SmartSpooler has the following features:

- 256K-byte buffer capacity = 100 pages (typically)
- Switchable serial (RS-232C) or parallel (Centronics) inputs and outputs
- Serial port data rates: 300, 600, 1200, 2400, 4800, 9600, 19,200, and 38,400 bits per second (bps)
- Serial port handshaking: hardware ($\overline{RTS}$, $\overline{CTS}$, and so on) and software (XON-XOFF)
- Multiple copy capability: 0 to 4 copies plus original
- Pause mode for single-sheet feeding
- Real-time clock (power required)
- Front-panel buffer capacity gauge
- Host programmed configuration and operation mode
- Built-in interactive diagnostic monitor program
- Low power (5 volts [V] at 0.5 ampere [A], ±12 V at 0.1 A typical)

Generic printer buffers come with a variety of functions. SmartSpooler combines the best features of these other printer buffers and adds a few of its own. SmartSpooler sits between a computer printer port and the printer. Since it provides both parallel and serial I/O ports, it adapts between computers and printers of each type.

When the computer prints, SmartSpooler accepts the data at high speed, buffers it, and outputs it to the low-speed printer. From the computer's point of view, printing a document takes only seconds. During the printout, SmartSpooler can make copies, temporarily suspend printing (pause button), and handle embedded formfeeds for single-sheet printing.

Hardware—Keep It Simple

The hardware requirements for SmartSpooler are simple: a CPU, buffer memory, and host/printer ports (serial and parallel). We also need some parallel I/O ports to handle switches and LEDs for SmartSpooler's front-panel operation.

I used the complementary-metal-oxide-semiconductor HD64180 CPU because it is ideally suited to the task and significantly reduces the hardware complexity. First, it includes two RS-232C serial ports on-chip, eliminating an external dual-UART (universal asynchronous receiver/transmitter) chip.

Second, the need for a large, low-cost buffer demands dynamic memory chips. The HD64180, with its built-in refresh, easily accommodates DRAMs and offers a performance level that can meet today's high-speed communications requirements. (Though the task seems simple, many computers choke when running terminal emulator programs at 19,200 or 38,400 bps.)

The SmartSpooler block diagram in figure 1 shows the three basic components: CPU (i.e., HD64180), memory (i.e., DRAM and SmartSpooler EPROM), and I/O (i.e., host, printer, switches, and LEDs) interface. The SmartSpooler also provides an SB180 XBUS I/O expansion connector for additional I/O (more on that later). See figure 2 for the complete SmartSpooler schematic (less the section covering the DIP-switch settings and LEDs, which I will cover in Chapter 36, part 2).

The EPROM socket (IC7) accepts 8K-byte (2764) to 64K-byte (27512) EPROMs, though the SmartSpooler code presently requires only 16K bytes (27128). Because of the on-chip DRAM refresh controller, connecting 256K bytes of DRAM is relatively easy. All the HD64180 needs are three 74LS157 row/column address multiplexers (IC8 through IC10) and a CAS\

(text continued on page 420)

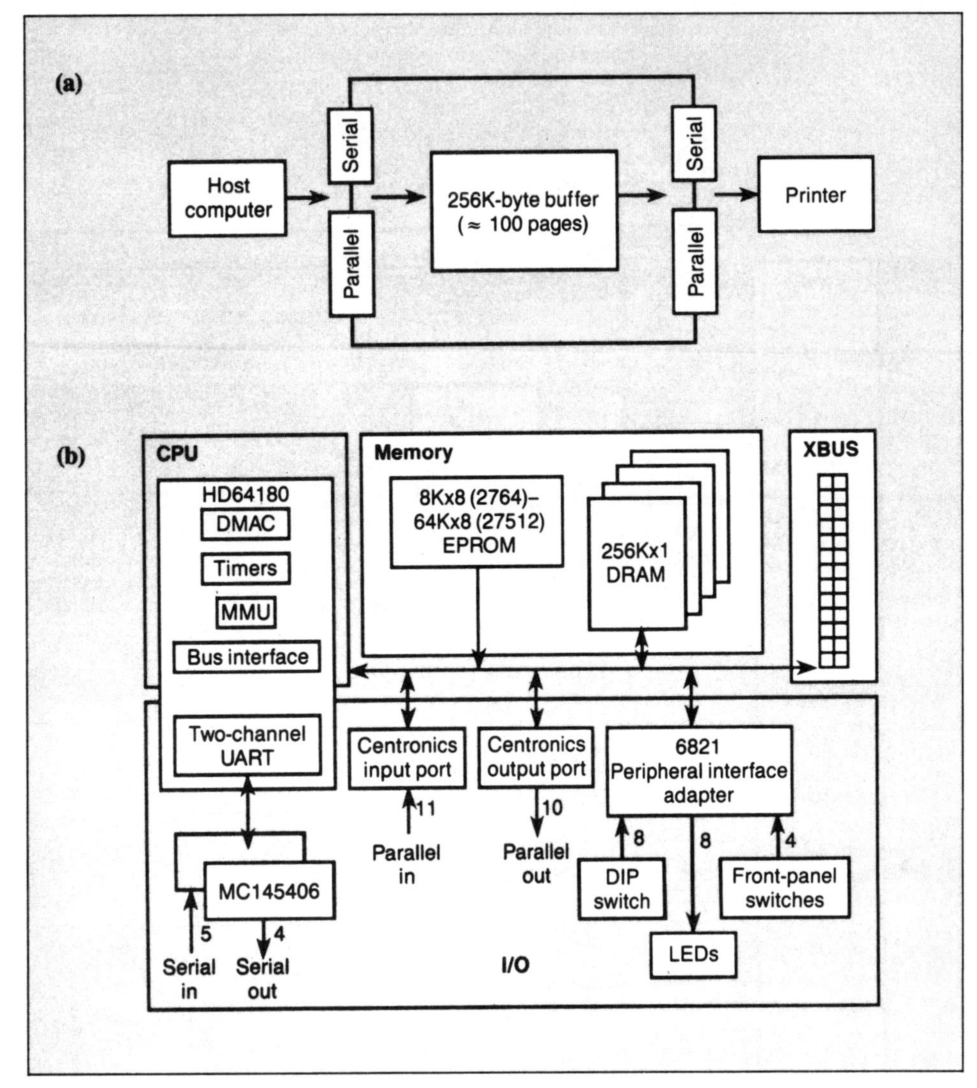

Figure 1: (a) *The SmartSpooler combines the function of a 256K-byte printer buffer and a serial/parallel converter switch.* **(b)** *Its HD64180 CPU—with built-in serial ports, extensive interrupt handling, address range beyond 64K bytes using direct memory access, and easy DRAM interface—is ideal for this application.*

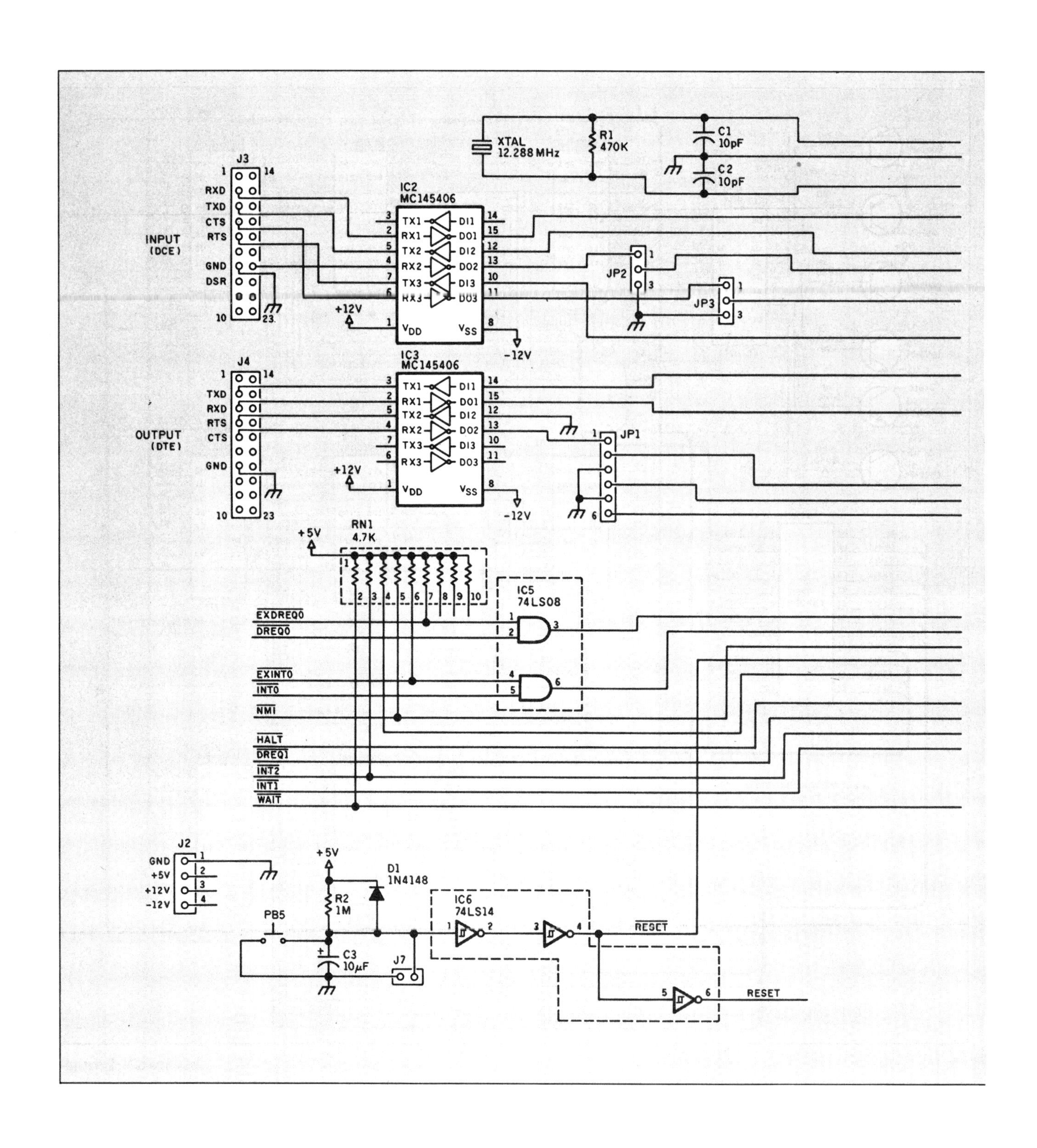

XTAL
12.288 MHz
R1
470K
C1
10pF
C2
10pF
J3
INPUT
(DCE)
RXD
TXD
CTS
RTS
GND
DSR
IC2
MC145406
TX1
RX1
TX2
RX2
TX3
RX3
DI1
DO1
DI2
DO2
DI3
DO3
VDD
VSS
+12V
-12V
JP2
JP3
J4
OUTPUT
(DTE)
IC3
MC145406
JP1
+5V
RN1
4.7K
IC5
74LS08
EXDREQ0
DREQ0
EXINT0
INT0
NMI
HALT
DREQ1
INT2
INT1
WAIT
J2
GND
+5V
+12V
-12V
D1
1N4148
R2
1M
PB5
C3
10μF
J7
IC6
74LS14
RESET
RESET

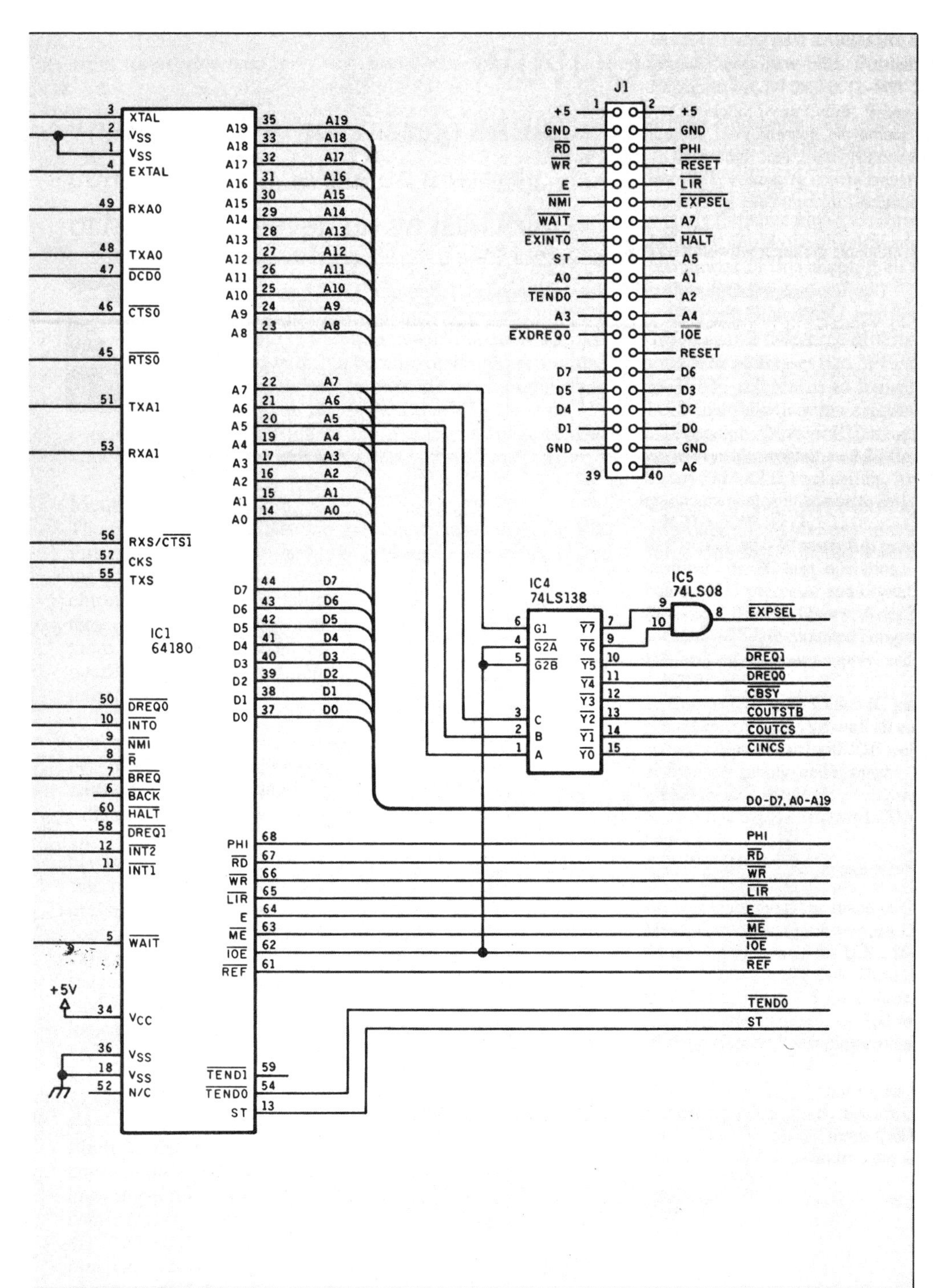

Figure 2a: *Schematic for the SmartSpooler's CPU and serial I/O section. (Not shown here are the 6821 peripheral interface adapter and other circuitry associated with the switches, push buttons, and LEDs; that portion will appear in* Chapter 36, part 2).

generator consisting of a flip-flop and a couple of gates (IC5, IC6, and IC19). The plastic-leaded-chip-carrier version of the HD64180 can address 1 megabyte, but I feel that 256K bytes is more than adequate for typical printing applications, and it keeps power consumption low.

If you examine the SmartSpooler's architecture closely, you will note some similarities with the SB180 single-board computer (see the September 1985 Circuit Cellar). Considering the performance and popularity of the SB180, any similarities are purely intentional. The SmartSpooler's core architecture lets it be more or less software-compatible with the BCC180 multitasking controller (see Jan.–Mar. 198[illegible]) and the SB180. Remember that you can download executable code to the SmartSpooler using its host programmed mode. This mode could be a compiled BASIC program generated on a BCC180 just as easily as any HD64180 assembly code produced on an IBM PC cross assembler or the SB180.

SmartSpooler's I/O

SmartSpooler has two serial ports and two parallel printer ports. Since the HD64180 has two on-chip serial ports, the only extra hardware needed for the RS-232C ports are a pair of MC145406 RS-232C level shifters (IC2 and IC3). The rest of the I/O hardware consists of the Centronics I/O ports and the front-panel switch/LED interface.

A 74LS138 (IC4) I/O address decoder generates eight chip selects. Two outputs are reserved for XBUS addressing: I/O addresses E0 to FF hexadecimal. Two chip selects are connected to the HD64180 direct-memory-access request inputs, and the remaining four are associated with the Centronics I/O ports.

The Centronics output port consists of a 74LS374 (IC22), which latches the output data, plus both halves of a 74LS74 (IC23) for controlling the handshaking. To send a character to the device connected to the port, the control software first

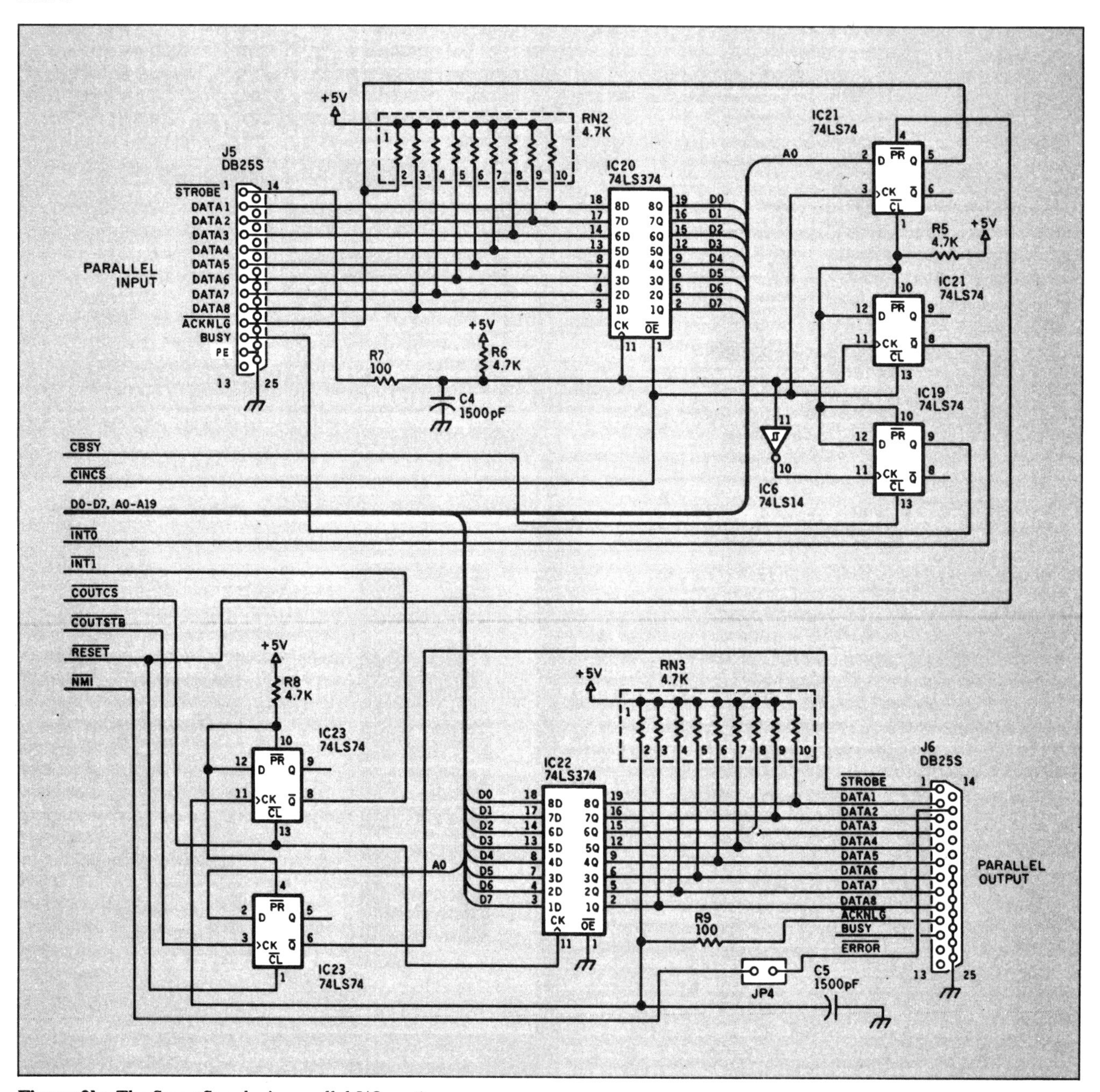

Figure 2b: *The SmartSpooler's parallel I/O section.*

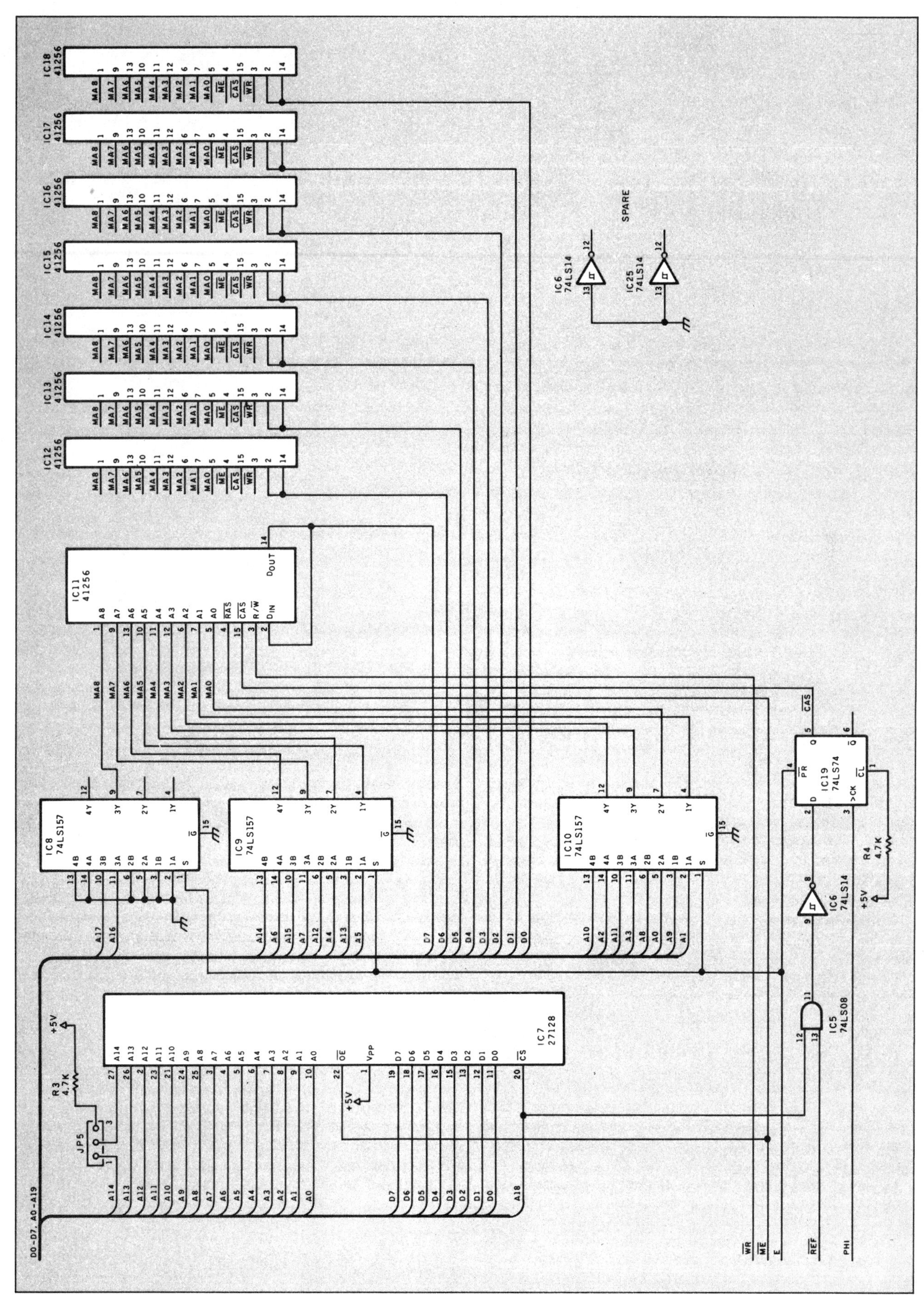

Figure 2c: *The EPROM and DRAM circuitry for the SmartSpooler.*

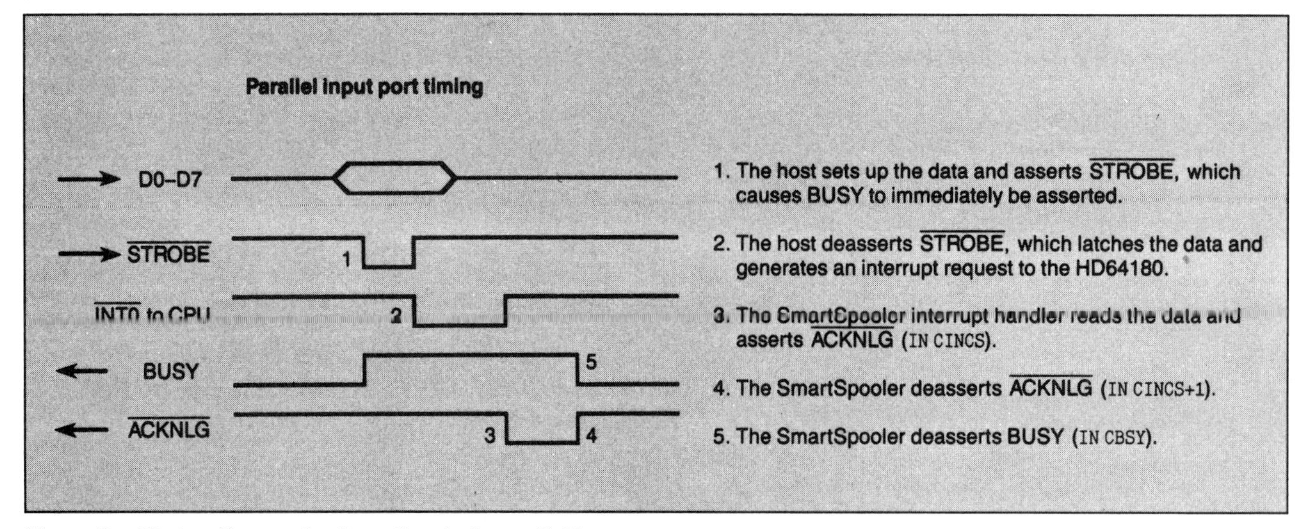

Figure 3a: *Timing diagram for SmartSpooler's parallel input port.*

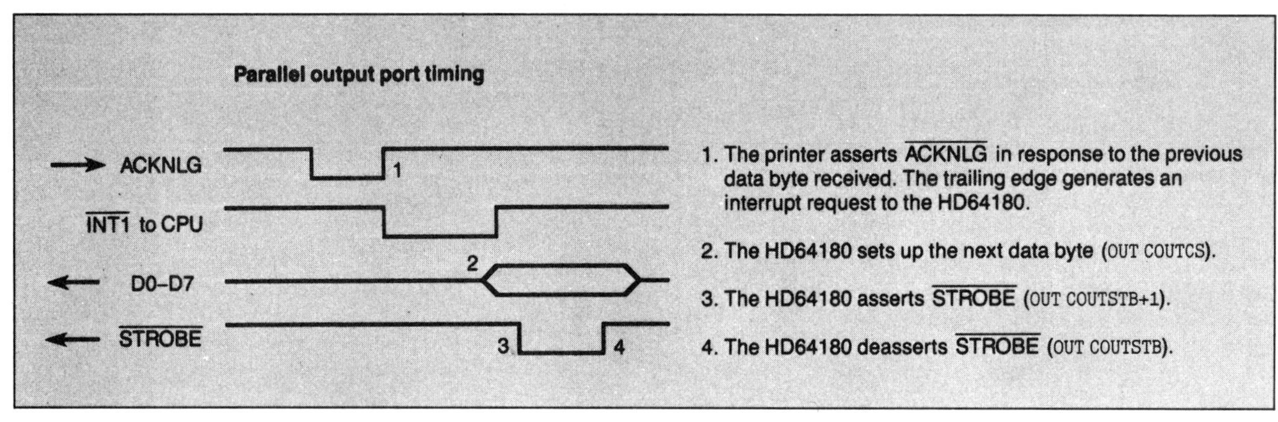

Figure 3b: *Timing diagram for the parallel output port.*

writes the byte out to port 90h. This latches the byte into the 74LS374 and also clears INT1\ high. (INT1\ is used during the acknowledge sequence that I'll describe next.)

Next, the program accesses (either reads or writes) port A1h to assert STROBE\ on J6 low. Finally, the software accesses port A0h to cause STROBE\ to go high, signaling the target device that a byte is ready.

Once the target printer device has retrieved the byte and wants to tell the spooler that it is ready for another byte, the device asserts the ACKNLG\ line on J6 low, then sets it high again. The rising edge clocks the 74LS74 and forces INT1\ low, generating an interrupt. Writing the next byte out to the Centronics port (usually part of the interrupt service routine) clears the interrupt.

The Centronics input port is similar to the output port, but it works in reverse. When the device connected to J5 sends the spooler a character, the device simultaneously generates a low-to-high signal on the STROBE\ input. This latches the data into a 74LS374 (IC20) and sets INT0\ low, generating an interrupt to the HD64180 processor. It also clocks another flip-flop, sending the BUSY line to the input device into a high state.

As part of the interrupt service routine, the processor first reads port 80h to get the character being sent to it by the input device. The port access also clears the interrupt and causes the ACKNLG\ output to go low as the first part of the acknowledge sequence. Next, an access to port 81h sets ACKNLG\ high again, signaling the input device that the character has been read and the spooler is ready for another. Finally, it accesses I/O port B0h, causing the BUSY output to go low. This signals those devices that rely on the BUSY signal that SmartSpooler is ready for another character. Figure 3 shows the timing details of the parallel I/O ports.

Purists may argue that SmartSpooler does not support the full Centronics interface, which, technically, includes additional signals like Paper Error, Error, and Init. However, these signals (and other vendor-unique signals) are inconsistent for different computers, printers, and software. Just in case, SmartSpooler's output port does provide an option for handling a Printer Error input signal. You can jumper it to NMI\.

Experimenters

While printed circuit board kits for the SmartSpooler are available, I encourage you to build your own. If you don't mind doing a little work, I will support your efforts as usual. You can download a hexadecimal file of the executable code for SmartSpooler's system EPROM (27128) from my bulletin board at (203) 871-1988. Alternatively, you can send me a preformatted IBM PC disk with return postage, and I'll put all the files on it for you (add $6 for the SmartSpooler User's Manual). Of course, this free software is for noncommercial personal use.

In Conclusion

In part 2 of this chapter, I'll finish the hardware by explaining the switch and LED configuration and describe SmartSpooler's software.

36

THE SMART SPOOLER

PART 2: SOFTWARE AND OPERATION

SmartSpooler can function as a complete remote data-processing computer to analyze data

In part 1 of Chapter 36, I introduced my version of the ultimate printer buffer: SmartSpooler. While SmartSpooler has 256K bytes of memory and supports both serial and parallel printers, it also has features that improve its versatility and ease of use. These features include a "switchbox" capability (i.e., for routing serial or parallel computer input to a serial or parallel printer output), multiple-copy printing, a single-sheet-feeding mode, and buffer capacity indicators. Also, you can daisy chain multiple SmartSpoolers to control a whole network of peripherals.

Most important, SmartSpooler is intelligent. While I've presented it primarily as a printer buffer, SmartSpooler lets a host computer completely control its operation. You can even download executable code to SmartSpooler. Rather than being merely a simple printer buffer, SmartSpooler can function as a complete remote data-processing computer that analyzes and interprets the data flowing through it.

In part 2 of this chapter, I'll finish the hardware discussion by explaining the user interface and then briefly describe SmartSpooler's operation.

Push Buttons and LEDs

A 6821 peripheral interface adapter (PIA) neatly connects four front-panel push buttons, eight front-panel LEDs, and an eight-position DIP switch. The PIA, which is IC24 in figure 1, provides two 8-bit ports (port A and port B) and four multipurpose handshaking lines (CA1, CA2, CB1, and CB2). The DIP switches connect to port A, which is programmed as an 8-bit input port. The switches specify options like data transfer rates and operating modes. Figure 2 contains switch settings and functions.

The four front-panel push buttons set configuration, pause the output, enter copy requests, and clear present settings. The combination of a simple resistor/capacitor circuit and a Schmitt-trigger inverter debounces each push-button input. The conditioned inputs are connected to the handshaking lines, which are programmed so that any switch closure will generate a CPU interrupt.

The eight LEDs with current-limiting resistors connect to port B. SmartSpooler sets this port up as an 8-bit output port. (Be aware that port B has the extra 10-milliampere current needed to drive the LEDs.) Four of the LEDs signify SmartSpooler's operating mode: Config, Copy, Pause, and Clear.

The remaining four LEDs indicate which ports are enabled: Parin (parallel in), Serin (serial in), Parout (parallel out), and Serout (serial out). During initial setup (Clear), these LEDs display the I/O port configuration (serial/parallel). When SmartSpooler is making copies (Copy), these LEDs display the number of copies requested (1 to 4) and then the number of copies remaining to be printed. During normal operation, the LEDs indicate how full the SmartSpooler buffer is: 0 percent, 25 percent, 50 percent, 75 percent, or 100 percent.

The two DB-25S IBM PC-compatible parallel printer input and output connectors are mounted on the printed circuit board. I've mounted two 20-pin headers behind these connectors for the serial ports. These headers accommodate a pair of optional ribbon cables with DB-25P serial connectors on the end. You can operate SmartSpooler with either or both pairs of connectors installed. If you need only parallel-to-parallel operation, you use only the DB-25S connectors.

Buffer-Manager Software

The basic algorithm at the core of SmartSpooler is a FIFO buffer manager. A FIFO, whether a single chip or a box like SmartSpooler, consists of an input port and an output port, connected by a buffer memory (perhaps 256 bytes for a FIFO chip and 256K bytes for SmartSpooler). The buffer memory decouples the input and output data rates: fast dump from the computer, slow dump to the printer.

A good analogy for a FIFO is a water tank with fill (input) and drain (output) pipes, each pipe having a pump (see figure 3). The input can pump faster than the output, so the rate difference is absorbed as the tank fills. As is true with the water tank, the FIFO has to handle two special cases: full and empty. When the tank is full or empty, the respective pump (input or output) should be turned off.

We can immediately dismiss the intuitive software algorithm for implementing a FIFO (i.e., actually moving the data). While suitable for very small FIFOs, such an algorithm would choke on a full 256K-byte buffer. Instead, we use a scheme called a *ring buffer*, which manipulates input and output pointers, instead of actually moving the data (see figure 4).

Interrupts and Direct Memory Access

You can divide the implementation of the FIFO into three components: determining when an I/O port is ready for transfer, performing the transfer, and updating the pointers.

You could use a pure software approach in which you poll the I/O ports for readiness, transfer data with `IN` and `OUT` instructions, and have the program update buffer pointers. However, this scheme has some problems.

First, polling is extremely inefficient. Consider the typical case of simultaneous high-speed input and low-speed output. For each input character, you have to check whether the output is ready, even though it normally won't be. Actually, it's much worse, since you also have to check for handshaking, front-panel switch closure, and a number of other mundane events. The overhead adds up quickly, limiting performance.

Second, software to access the buffer and maintain the buffer pointers is difficult to write for buffer sizes larger than 64K bytes. Eight-bit CPUs like the HD64180—and even 16-bit CPUs like the 8086/80286—must monitor each access and calculate memory-management-unit (HD64180) or segment-register (80286) reload values to manage a large buffer.

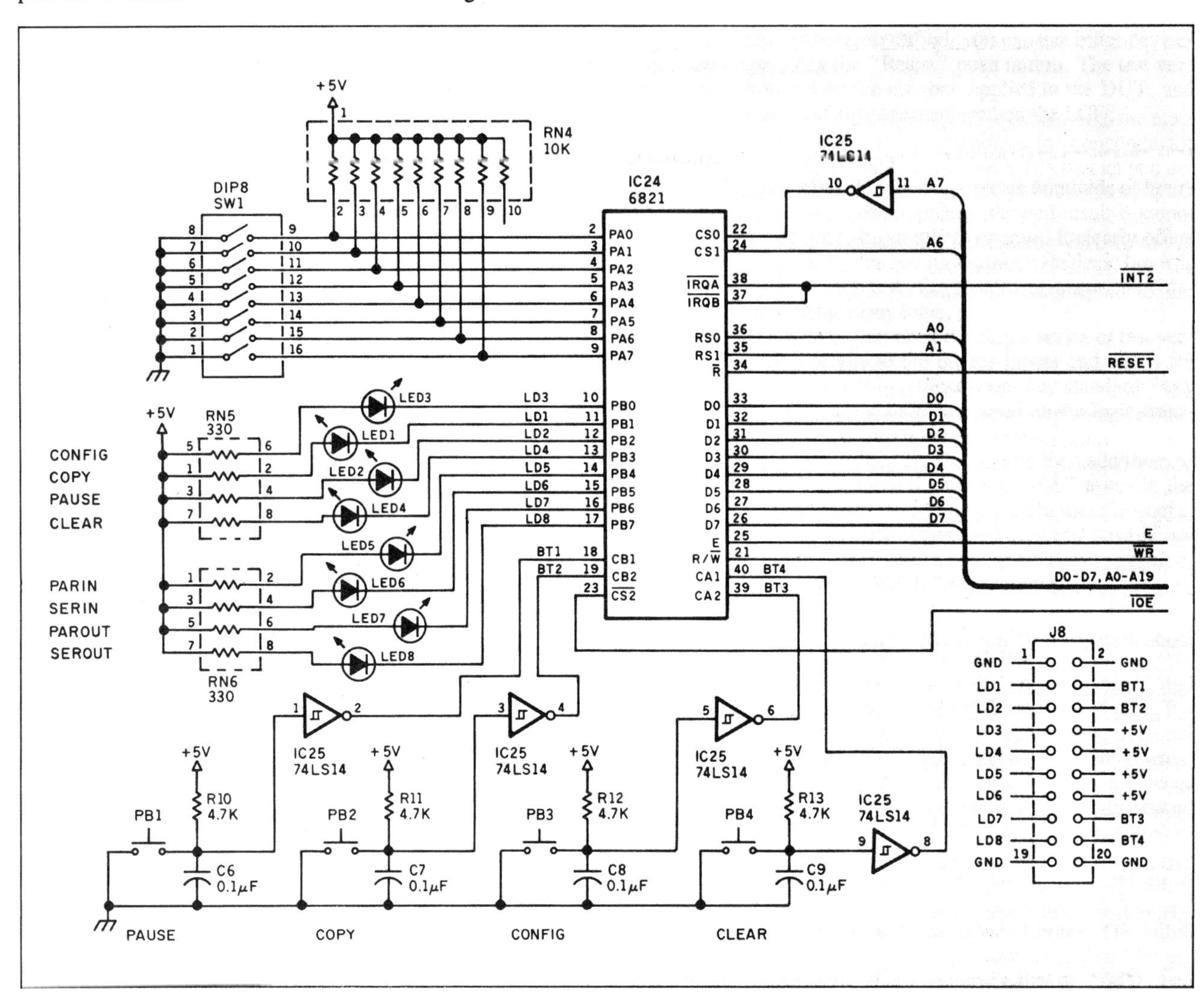

Figure 1: *Schematic for SmartSpooler's DIP-switch and LED I/O circuitry. A single 6821 peripheral-interface-adapter chip handles all the switches and lights.*

Switch	Off	On	Local (Mode)	Host (Mode)	Test (Mode)
(MSB) 1			IBR2	On	On
2			IBR1	On	On
3			IBR0	On	On
4			OBR2	Note	On
5			OBR1	Note	On
6			OBR0	Note	On
7			IPORT	Note	On
(LSB) 8			OPORT	Off	On

(Shipping position, all off: Parallel → parallel mode)

Local mode

	38,400	19,200	9600	4800	2400	1200	300	N/A
I/O BR2	Off	Off	Off	Off	On	On	On	On
I/O BR1	Off	Off	On	On	Off	Off	On	On
I/O BR0	Off	On	Off	On	Off	On	Off	On

I/O port Off = parallel, On = serial

Host mode

Note: Positions 4 to 7 off = last node in chain

Test mode

Push any button → memory test

Terminal on serial output port (<cr> . . <cr>) → monitor

Figure 2: *DIP-switch settings and their effects on SmartSpooler's various operating modes.*

Figure 3: *A FIFO buffer is like a water tank, absorbing the difference between input and output rates. In a typical printer buffer task, the input is faster than the output. A basic function of the FIFO/tank is to control the pumps for the special cases of tank full and tank empty.*

Figure 4: (a) *SmartSpooler's FIFO buffer operates using a ring-buffer algorithm that manipulates input and output data pointers rather than actually moving the data. The flowcharts shown here illustrate two key functions:* **(b)** *handling full and empty conditions and* **(c)** *checking for buffer wraparound.*

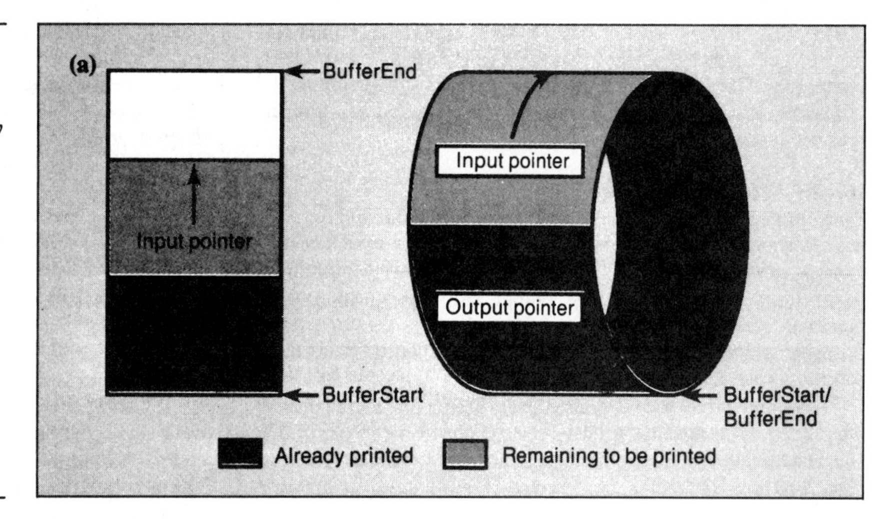

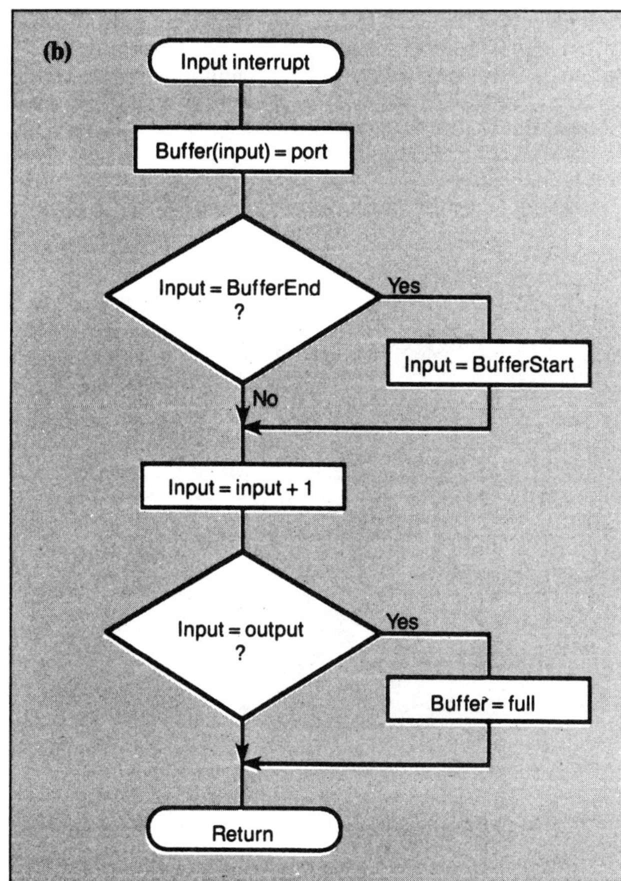

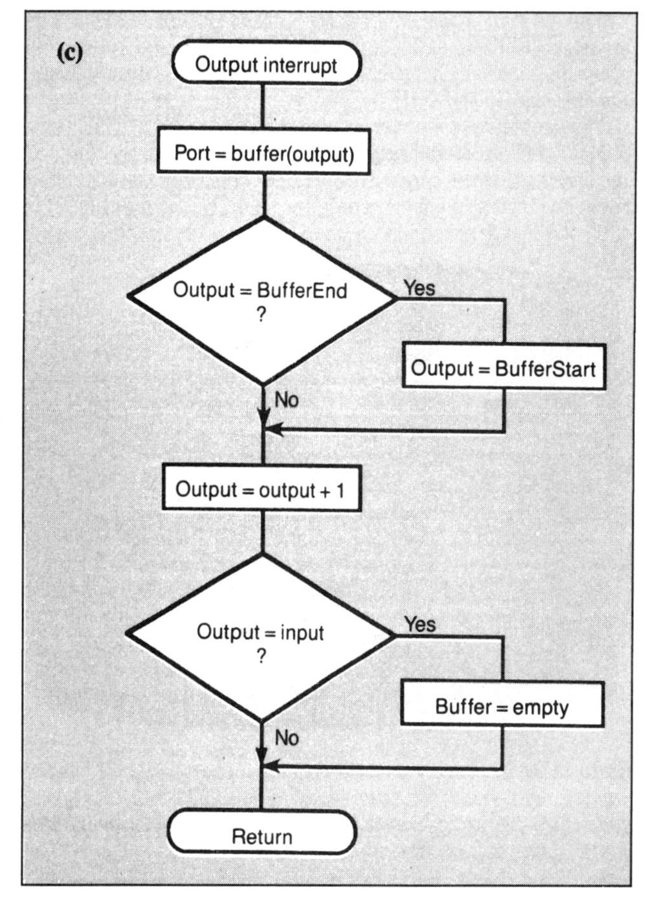

The solution to the problem of inefficiency is to exploit the HD64180's excellent interrupt capabilities. SmartSpooler's I/O is totally interrupt-driven, including the serial and parallel ports as well as the front-panel switches. Table 1 lists these interrupt assignments.

To solve the problem of large buffer maintenance, SmartSpooler uses the HD64180's direct-memory-access controller (DMAC), which has direct access to the entire physical address space. Besides performing the actual `IN` and `OUT` operations, the DMAC maintains the buffer pointers (using built-in DMA address registers with auto-increment). Channel 0 is the input channel, configured to perform I/O-to-memory DMA. Channel 1, as output, is configured for memory-to-I/O DMA.

Usually, I/O DMA occurs by request of the I/O device itself. Unfortunately, this doesn't easily handle special cases like buffer full, buffer empty, Pause button, copies, and handshaking. To get more flexibility, SmartSpooler uses a "soft" DMA technique. The HD64180 $\overline{\text{DREQ}}$ inputs are connected to CPU I/O ports instead of directly to the I/O peripheral ports. This lets the software initiate DMA.

Table 1: *SmartSpooler takes full advantage of the HD64180's interrupt capabilities. Note that the NMI and INT0 signals are also gated with XBUS inputs, allowing I/O expansion boards to use them.*

$\overline{\text{NMI}}$: Not used (parallel output port $\overline{\text{ERROR}}$ input is optional)
$\overline{\text{INT0}}$: Parallel input port
$\overline{\text{INT1}}$: Parallel output port
INT2: Front-panel switches
PRT(Timer)0: Software delay timer
PRT(Timer)1: Real-time timer
DMA0: Not used
DMA1: Not used
CSI/O: Not used
ASCI(UART)0: Serial input port
ASCI(UART)1: Serial output port

Hands Across the Buffer

Both input and output ports need to provide handshaking. On the input side (host to SmartSpooler), the host must be signaled to stop when the buffer fills, to prevent overflow (remember the water tank example). On the output side (SmartSpooler to printer), SmartSpooler needs to pause when the printer is busy printing or goes off-line. In the formal world of data communications, this is known as flow control.

Flow control requires handshaking, which is a way of conveying start/stop information between the various devices. Hardware handshaking uses extra signal lines dedicated to flow control. Software handshaking conveys flow-control information over the data channel itself.

The Centronics parallel interface uses hardware handshaking signals: $\overline{\text{STROBE}}$, $\overline{\text{ACK}}$, and BUSY. The RS-232C ports provide both hardware handshaking ($\overline{\text{RTS}}$ and $\overline{\text{CTS}}$) and software handshaking (XON/XOFF).

The problem with serial handshaking ($\overline{\text{RTS}}$, $\overline{\text{CTS}}$, and XON/XOFF) protocols occurs when the receiver can't shut off the sender in time to prevent overflow. Those of you who have spent time trying to get terminals or computers to run at 19,200 or 38,400 bits per second know what I mean (the beeping terminal syndrome). Also, some sending devices check for handshaking only at the end of each line, rather than for each character. To avoid overflow, SmartSpooler's serial port drivers incorporate a 256-byte "pad," allowing plenty of time for handshaking delays.

Local, Test, and Host Modes

SmartSpooler's operating mode is determined at power-on by DIP-switch settings.

Local mode is the normal mode of operation. In this case, SmartSpooler enters the default port configuration (parallel to parallel) and is ready to spool incoming data. Using the front-panel switches and LEDs, you can enter commands to change the port configuration, pause the output, and request copies.

Test mode works with a standard RS-232C terminal connected to one of the serial ports (see photos 1a and 1b). Instead of entering the spooler routines, SmartSpooler executes a built-in monitor program, which contains routines to test the ports, switches, and LEDs, as well as a complement of traditional monitor commands (`display`, `enter memory`, and so on). Test mode is useful for diagnosing hardware, cable, and host driver software problems.

Host mode lets the host computer download commands to

(a)

```
?
A = ASCII Table        >A
B = Bank Select        >Bbank# (bank#=BBR)
D = Display Memory     >D[startaddr],[endaddr]
E = Emulate Terminal   >E[baudrate]
F = Fill Memory        >Fstartaddr,endaddr,data8
G = Goto Program       >G[goaddr]  >GB brkaddr,[goaddr]
H = Hexmath            >Hdata16,data16
I = Input Port         >Iportaddr
M = Move Memory        >Mstartaddr,endaddr,destaddr
O = Output Port        >Oportaddr,data8
P = Printer Select     >P
Q = Query Memory       >Qdata8,[data8],[data8],[data8]
S = Set Memory         >Saddr
T = Test System        >Tdev (dev=A)sync,B)Button/LED,C)en,)mem)
U = Upload Hex File    >U[C]
V = Verify Memory      >Vstartaddr,endaddr,destaddr
X = Examine Registers >X
Y = Yank I/O Registers>Y

3C>
```

(b)

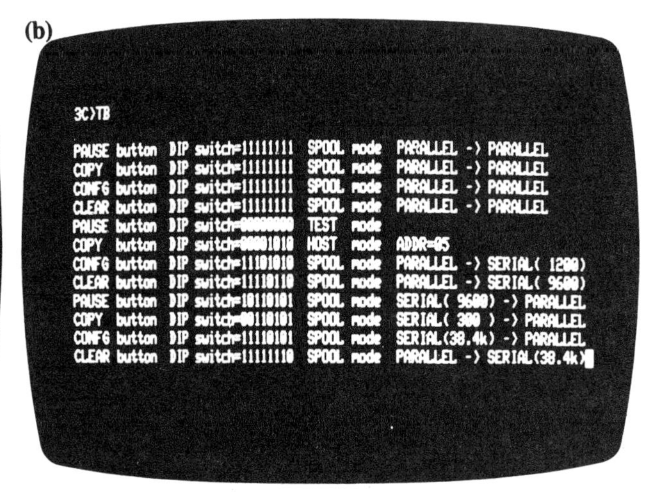

Photo 1: **(a)** *This screen shows the commands available while SmartSpooler is in test mode.* **(b)** *From this screen, you can individually test all four buttons, all eight LEDs, and all eight DIP switches. The right-hand column shows the SmartSpooler modes for several different DIP-switch settings. Whatever setting is last selected determines the default mode SmartSpooler enters when you reset it or power it up.*

Operation Summary for SmartSpooler

Power-on:
- Clear LED goes on.
- Config LED goes on, and the default port configuration is shown.
- Clear LED goes off, and the LEDs switch to show buffer capacity.
- SmartSpooler is ready for operation.

Changing the port configuration, changing the pause mode, and aborting printout:
- Push the Clear button.
- Push the Config button to change the port configuration and the Pause button to toggle the pause mode on/off.
- Push the Clear button (Clear LED off) to start SmartSpooler operation.
- The LEDs switch from showing the port configuration to monitoring the buffer capacity.

Suspending/resuming printing:
- Push the Pause button to suspend printing.
- Push the Pause button again to resume printing.

Making copies:
- Press Clear prior to sending document to copy.
- Send the document to SmartSpooler.
- Push the Copy button.
- Push the Config button to select the number of copies (0 to 4) desired.
- Push the Copy button again (Config LED off) to finish the copy request.
- The Copy LED will remain on until all copies are printed.

Single-sheet printing:
- Make sure you've selected the pause mode during power-on or Clear setup.
- Make sure your computer transmits a formfeed character to SmartSpooler prior to each new page (including the first).
- When the Pause LED goes on, insert a new page into the printer.
- Push the Pause button to print the next page.

Check buffer capacity, port configuration, and the number of copies remaining:
- During normal operation (Clear LED off), the parallel/serial In/Out LEDs show percent full (0, 25, 50, 75, or 100) the SmartSpooler buffer is, and the Pause LED controls print suspend/resume.

SmartSpooler. One benefit of host mode is that it lets you use software—instead of switches and LEDs—to set the port configuration and serial port format (data transfer rate, start/stop bits, parity, and so on). In fact, you can remove SmartSpooler's switches, LEDs, and corresponding circuits if you never use SmartSpooler for local mode operation.

For the ultimate in versatility, the host can even download a new control program to totally replace SmartSpooler's control program. SmartSpooler's ROM vectors all HD64180 interrupts through a RAM-based vector table, letting the new control program take over interrupt response. Combining SmartSpooler's hardware with optional XBUS expansion boards (e.g., the Circuit Cellar GT180 color graphics display or the COMM180 modem/small-computer-system-interface [SCSI] board) and your own control program opens the door for lots of interesting applications.

Using SmartSpooler

SmartSpooler is easy to use. The following is a summary of specific button functions.

Pushing the Clear button stops any operation (I/O) in progress, initializes SmartSpooler, and lights the Clear LED. Any data in the buffer is lost upon Clear.

You use the Clear button in the following instances:

- to change the port configuration
- to change the pause mode
- before receiving a document that will be copied
- to cancel a printout
- to finish the Clear request (Clear, Config, and Pause LEDs off)

The Pause button has two functions, one after the Clear button is pushed and another during normal operation. After you've pushed the Clear button (Clear LED on), pushing Pause toggles the pause mode on or off. When pause mode is on (Pause LED on), SmartSpooler will suspend output after transmission of a formfeed character to the printer. Use this mode for single-sheet feeding; position the next sheet and push the Pause button (Pause LED off) to resume printing. When pause mode is off (Pause LED off), SmartSpooler will not check formfeed characters. Use this mode when printing continuous (i.e., platen or tractor-fed) forms.

Push the Copy button (Copy LED on) to make copies of everything SmartSpooler has received since the last Clear. Then, increment the number of copies desired by pushing the Config button (Config LED blinks).

The LEDs show how many copies are selected: One LED on means one copy (plus original); four LEDs on means four copies (plus original). After entering the number of copies desired, push the Copy button again (Config LED off) to complete the copy request.

The Config button toggles the I/O (serial or parallel) port configuration when the Clear LED is on. The configuration is reflected on the parallel/serial In/Out LEDs.

Two Functions

SmartSpooler is actually two projects. One of these is a high-performance printer buffer; the second is a configurable intelligent peripheral controller. Most people will assemble it as a printer buffer, but others will find applications ideally suited for the host programmed mode. While SmartSpooler is not a trivial project, you can extend it to perform tasks that separate it from a mere buffer. One possibility is to use SmartSpooler between a host computer and a modem to filter incoming data, initiate and time calls, or encrypt and decrypt data.

As a printer buffer, 256K bytes is more than adequate. However, as a specific-application peripheral controller, SmartSpooler might need additional capability.

As I mentioned previously, it is not inconceivable to add 8 or 16 additional I/O ports, a 20-megabyte SCSI hard disk drive, or the GT180 color graphics display to SmartSpooler through its XBUS expansion connector. The necessary hardware for such peripherals already exists for SmartSpooler from previous Circuit Cellar articles.

I'd like to personally thank Tom Cantrell for his extensive work on this project. Without his software expertise, I'd be hopelessly mired in an ocean of bits forever.

See page 447 for a listing of parts and products available from Circuit Cellar Inc.

37

BUILD AN INTELLIGENT SERIAL EPROM PROGRAMMER

Steve's new and improved device includes on-board CPU and intelligent firmware

I don't like admitting that I made a mistake, but apparently I did. Well, not actually. You see, I was dragged into . . . Let me start from the beginning.

My February 1985 Circuit Cellar article was a project on how to build a serial EPROM programmer, about which I said: "The latest Circuit Cellar EPROM programmer is a serial-port programmer that has the speed of a turtle, the intelligence of the mightiest computer (that is, it has absolutely no smarts of its own), and is as functional as a doorstop between uses. On the positive side, it's fully documented, universally applicable, and easily expandable to accommodate future EPROM types."

What a mess after it was published! Everybody must have built this programmer. BYTEnet almost shut down the Peterborough phone company as people downloaded the BASIC listings, and my staff developed "postage tongue" replying to the correspondence. Needless to say, the project was well received.

In truth, it was an experimenter's project intended to satisfy a certain core of supporters yet enlighten the larger audience of readers about EPROM programming in general. Because I could not gauge its potential reception, and also because I didn't see it as having any greater performance than low-cost bus-compatible programmer boards, I didn't arrange to have it made into a printed circuit board as are most of my projects. I'm embarrassed to say that even after all these years I underestimated the number of experimenters who wanted to build a serial EPROM programmer.

It's too late to go back now, but I have to make up for past indiscretion and find some way to save face. I know that there are warmed soldering irons all across the country waiting for me to apologize appropriately. I trust you'll accept this improved rendition on an old theme as proper recompense.

As the title indicates, this programmer is still intended for serial-port operation. Thus, it retains computer and bus independence. The primary difference between then and now, however, is the addition of a microprocessor that greatly enhances its functions. The new Circuit Cellar intelligent serial EPROM programmer (CCSP for short) programs more types of EPROMs faster and more reliably. It also functions as a stand-alone programmer for copying or verifying EPROMs. (See photo 1.) The following is a list of CCSP features:

- RS-232–compatible (no handshaking necessary)
- internal V_{pp} power generation
- menu-selectable EPROM types (no programming configuration jumpers)
- default power-up selectable data rates
- automatic power-down of EPROM for installation/removal
- stand-alone or computer system/terminal–connected operation
- menu-driven operation
- single-byte or full-buffer write modes
- 32K-byte on-board memory buffer
- read, copy, or verify EPROM
- Intel hexadecimal file upload/download
- verify after write
- verify EPROM erasure
- screen dump by page or byte
- BASIC driver that can be modified by the user
- program EPROMs in standard 50-millisecond and 1-ms fast algorithm modes
- support V_{pp} settings of 25, 21, and 12.5 volts
- program all 27xxx 5-V single-supply EPROMs, including 2716, 2732, 2732A, 2764, 2764A, 27C64, 27128, 27128A, 27C128, 27256, 27512, and any functional equivalents

Obviously, a list this impressive would take a great deal of effort to put together as a single chapter's project. The potential software development nightmares of assembly language serial drivers, menu displays, and table manipulations hardly made it worth adding a microprocessor to my original BASIC-language-manipulated unit. Besides, how could it be done in one month?

BASIC allowed a significant level of interactive menus and help displays while requiring little software overhead. Unfortunately, using a high-level-language interpreter to simplify software development is of little value when the primary goal of producing a better programmer requires fast data manipulations that are best accomplished in assembly language.

Rather than be thwarted by this apparent dilemma, I decided to design a hybrid system that used both BASIC and assembly language. The obvious choice was the BASIC-52 computer/controller I presented in the August 1985 project. With the help of software guru and friend Bill Curlew, the CCSP was designed, built, and tested in two weeks flat.

The CCSP uses an Intel 8052AH-BASIC microprocessor that contains an 8K-byte ROM-based BASIC interpreter. Besides manipulating strings, tables, and menus, the BASIC contains serial communication drivers and easily links to assembly language routines. It seemed the perfect engine for a quickly designed user-modifiable project.

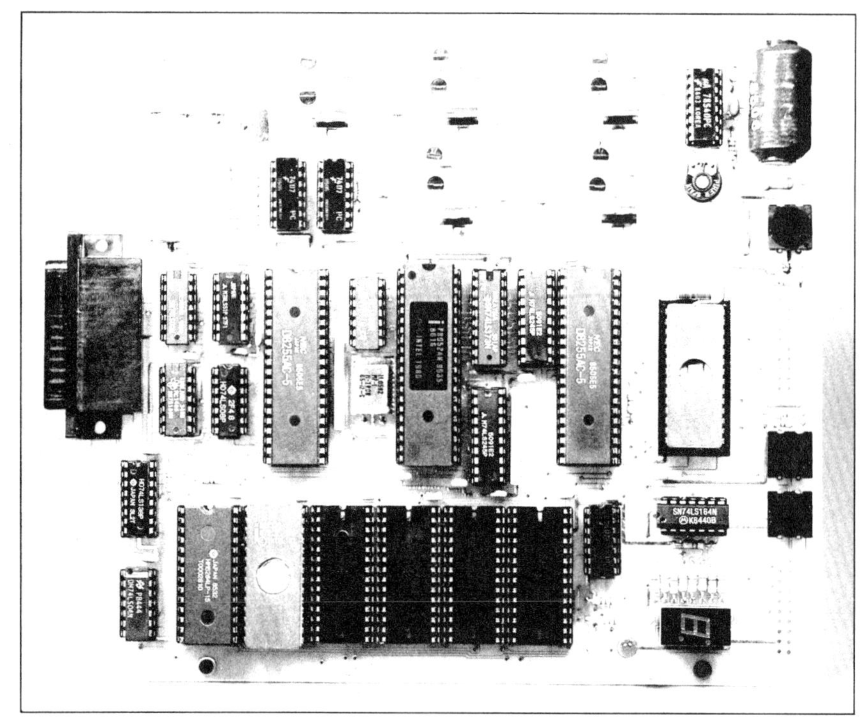

Photo 1: *Finished printed circuit prototype of the serial EPROM programmer. The digital section and memory buffer are at the bottom center. The analog-voltage-level switching section is at the top center.*

A Hybrid Approach

The CCSP is a stand-alone microcomputer with an application-specific I/O configuration. It supports 40K bytes of operating system and buffer RAM and 16K bytes of program ROM. It uses six parallel I/O ports to drive the programming-pin level-shifter voltage-control circuitry, EPROM address and data lines, and user-interactive buttons and display. The CCSP can be used by itself to copy EPROMs or, when connected to a terminal or computer, as a full-function programmer/verifier. It requires no programming jumpers or personality modules and is completely automatic. It programs/examines/verifies all 5-V EPROMs from 2716s through 27512s in both standard and fast modes (on applicable devices).

In the sections that follow, I'll describe the configuration of the microcomputer and its unique I/O structure. Once you have the hardware in hand, I'll describe the system software and how the different modes operate. First, I'll go over some EPROM basics.

A Review

A personal computer, even in its minimum configuration, always contains some user-programmable memory, or RAM, usually in the form of semiconductor-memory integrated circuits. This memory can contain both programs and data. Any machine-word-level storage element within the memory can be individually read or modified (written) as needed.

Any of several kinds of electronic components can function as bit-storage elements in this kind of memory. TTL-type 7474 flip-flops, bistable relays, or tiny ferrite toroids (memory cores) are suitable, but they all cost too much, are hard to use, and have other disadvantages.

In personal computer and other microprocessor-based applications, the most cost-effective memory is made from MOS integrated circuits. Unfortunately, data stored in these semiconductor RAMs is volatile. When the power is turned off, the data is lost. Many ways of dealing with this problem have been devised, with essential programs and data usually stored in some nonvolatile medium.

In most computer systems, some data or programs are stored in ROM. A semiconductor ROM can be randomly accessed for reading in the same manner as the volatile memory, but the data in the ROM is permanent. The data in a mask-programmed ROM is determined during the manufacturing process. Whenever power is supplied to the ROM, this permanent data (or program) is available. In small computer systems, ROM is chiefly used to contain operating systems and/or BASIC interpreters—programs that don't need to be changed.

Another type of ROM is the PROM, which is delivered from the factory containing no data. The user decides what data to put in it and permanently programs it with a special device. Once programmed, PROMs exhibit the characteristics of mask-programmed ROMs. You might label such PROMs "write-once" memories.

The ultraviolet-light erasable EPROM is a compromise between the "write-once" kind of PROM and volatile memory. You can think of the EPROM as a "read-mostly" memory, used in read-only mode most of the time but occasionally erased and reprogrammed as necessary. The EPROM is erased by exposing the silicon chip to ultraviolet light at a wavelength of 2537 angstroms. Conveniently, most EPROM chips are packaged in an enclosure with a transparent quartz window.

How an EPROM Works

EPROMs from several manufacturers store data bits in cells formed from stored-charge FAMOS (floating-gate avalanche-injection metal-oxide semiconductor) transistors. Such transistors are similar to positive-channel silicon-gate field-effect transistors, but with two gates. The lower or floating gate is completely surrounded by an insulating layer of silicon dioxide, and the upper control or select gate is connected to external circuitry.

The amount of electric charge stored on the floating gate determines whether the bit cell contains a 1 or a 0. Charged cells are read as 0s; uncharged cells are read as 1s. When the EPROM chip comes from the factory, all bit locations are cleared of charge and are read as logic 1s; each byte contains hexadecimal FF.

When a given bit cell is to be burned from a 1 to a 0, a current is passed through the transistor's channel from the source to the gate. (The electrons, of course, move the opposite way.) At the same time, a relatively high-voltage potential is placed on the transistor's upper select gate, creating a strong electric field within the layers of semiconductor material. (This is the function of the +12.5-V, +21-V, or +25-V V_{pp} charging potential applied to the EPROM.) In the presence of this strong electric field, some of the electrons passing through the source-drain channel gain enough energy to tunnel through the insulating layer that normally isolates the floating gate. As the tunneling electrons accumulate on the floating gate, it takes on a negative charge, which makes the cell contain a 0.

When data is to be erased from the chip, it is exposed to ultraviolet light, which contains photons of relatively high energy. The incident photons excite the electrons on the floating gate to sufficiently high-energy states that they can tunnel back through the insulating layer, removing the charge from the gate and returning the cell to a state of 1.

When data is to be erased from the chip, it is exposed to ultraviolet light.

The 27xxx EPROMs contain bit-storage cells configured as individually addressable bytes. This organization is often called "2K by 8" for a 2716 or "8K by 8" for a 2764. The completely static operation of the device requires no clock signals. The primary operating modes include read, standby, and program (program-inhibit and program-verify modes are important primarily in high-volume applications).

Control inputs are used to select the chip and configure it for one of these operating modes. In the program mode, particular bit cells are induced to contain 0 values. Both 1s and 0s are in the data word presented on the data lines, but only a 0 causes action to take place. For example, the 27128 is in the programming mode when V_{pp} input is at 21 V and CE and PGM are both at TTL low. The data to be programmed is applied 8 bits in parallel to the data output pins. For regular programming, CE should be kept TTL low at all times while V_{pp} is kept at 21 V. When the address and data are stable, a 50-ms (55 ms maximum) active-low TTL program pulse is applied to the $\overline{PGM}$ input. A program pulse must be applied at each address location to be programmed.

Standard vs. Fast

In the old days, all we had to contend with were 50-ms timing pulses (neglecting obsolete 1702 and 2708 EPROMs). Today, the newest EPROMs can use a fast closed-loop programming algorithm that lessens programming time (realize that a 27512 takes about 1 hour to program in 50-ms increments). The CCSP supports fast programming.

The fast algorithm uses closed-loop margin checking. To ensure reliable program margin, the fast algorithm utilizes two different pulse types: initial and overprogram. The duration of the initial PGM pulse(s) is 1 ms, which will then be followed by a longer overprogram pulse of length 4*x* ms; some chip types use 3*x* (*x* is the number of initial 1-ms pulses applied to a particular location before a correct verify occurs). Once it is verified, four times that number of pulses are applied to the same location to permanently burn the data. If 15 (some chip types require 25 pulses) 1-ms pulses are applied to any single-byte location without reaching the margin, the overprogram pulse is applied automatically.

The entire sequence of program pulses and byte verifications is performed at V_{cc} = 6.0 V and V_{pp} = 21.0 V (V_{pp} may be 12.5 V on some EPROMs). When the fast programming cycle has been completed, all bytes should be compared to the original data with V_{cc} = V_{pp} = 5.0 V.

The fast algorithm may be the preferred programming method since it allows certain EPROMs to be programmed in significantly less time than the standard 50-ms-per-byte programming routine. Typical programming times for 27128s, for example, are on the order of 2 minutes, a sixfold reduction in programming time from the standard method.

CONFIGURATION MAZE

The first problem encountered in any EPROM programmer design is to compare the pins of the various EPROMs (see figure 1b). Among the 28 defined pins (four are unused on 24-pin devices), 21 are used for the same functions (address and data). Evidently, semiconductor manufacturers never thought very far ahead or talked to each other, because the remaining seven pins are a complicated switching maze. Among the different EPROMs, the same pin location can

(a)

PIN #	2716	2732	2764	27128	27256	27512
1	N/C	N/C	VPP 21V 12.5V 5V	VPP 21V 12.5V 5V	VPP 12.5V 5V	A15 5V 0V
28	N/C	N/C	VCC* 6V 5V	VCC* 6V 5V	VCC* 6V 5V	VCC* 6V 5V
27	N/C	N/C	$\overline{PGM}$ 5V 0V	$\overline{PGM}$ 5V 0V	A14 5V 0V	A14 5V 0V
26/24	VCC 5V	VCC 5V	N/C	A13 5V 0V	A13 5V 0V	A13 5V 0V
23/21	VPP 25V 5V	A11 5V 0V	A11 5V 0V	A11 5V 0V	A11 5V 0V	A11 5V 0V
22/20	$\overline{OE}$ 5V 0V	$\overline{OE}$/VPP 25V 21V 0V	$\overline{OE}$ 5V 0V	$\overline{OE}$ 5V 0V	$\overline{OE}$ 5V 0V	$\overline{OE}$/VPP 12.5V 0V
20/18	$\overline{CE}$ 5V 0V	$\overline{CE}$ 5V 0V	$\overline{CE}$ 5V 0V	$\overline{CE}$ 5V 0V	$\overline{CE}$ 5V 0V	$\overline{CE}$ 5V 0V

*6.0 VOLTS ONLY ON EPROMS THAT ALLOW FAST PROGRAMMING

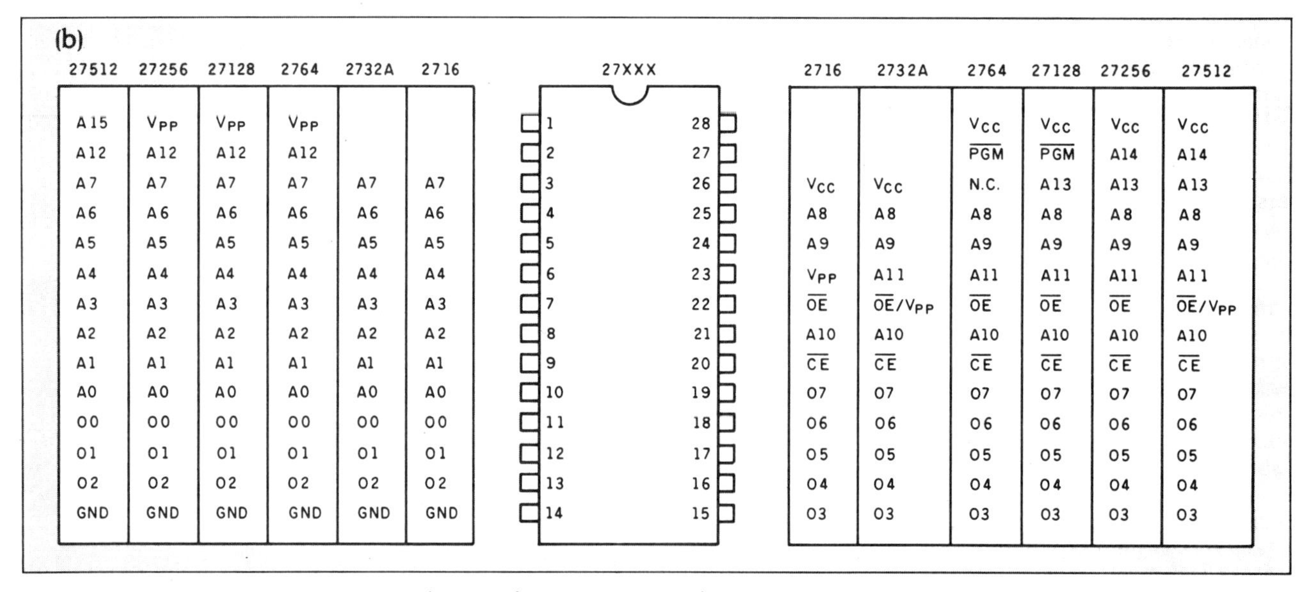

(b)

27512	27256	27128	2764	2732A	2716	27XXX pin
A15	VPP	VPP	VPP			1
A12	A12	A12	A12			2
A7	A7	A7	A7	A7	A7	3
A6	A6	A6	A6	A6	A6	4
A5	A5	A5	A5	A5	A5	5
A4	A4	A4	A4	A4	A4	6
A3	A3	A3	A3	A3	A3	7
A2	A2	A2	A2	A2	A2	8
A1	A1	A1	A1	A1	A1	9
A0	A0	A0	A0	A0	A0	10
O0	O0	O0	O0	O0	O0	11
O1	O1	O1	O1	O1	O1	12
O2	O2	O2	O2	O2	O2	13
GND	GND	GND	GND	GND	GND	14

27XXX pin	2716	2732A	2764	27128	27256	27512
28			VCC	VCC	VCC	VCC
27			$\overline{PGM}$	$\overline{PGM}$	A14	A14
26	VCC	VCC	N.C.	A13	A13	A13
25	A8	A8	A8	A8	A8	A8
24	A9	A9	A9	A9	A9	A9
23	VPP	A11	A11	A11	A11	A11
22	$\overline{OE}$	$\overline{OE}$/VPP	$\overline{OE}$	$\overline{OE}$	$\overline{OE}$	$\overline{OE}$/VPP
21	A10	A10	A10	A10	A10	A10
20	$\overline{CE}$	$\overline{CE}$	$\overline{CE}$	$\overline{CE}$	$\overline{CE}$	$\overline{CE}$
19	O7	O7	O7	O7	O7	O7
18	O6	O6	O6	O6	O6	O6
17	O5	O5	O5	O5	O5	O5
16	O4	O4	O4	O4	O4	O4
15	O3	O3	O3	O3	O3	O3

Figure 1: *(a) EPROM programming-pin functions by EPROM type. (b) The great EPROM pin-out maze, illustrating the configuration of those EPROMs the CCSP is designed to handle.*

supply power, address, or programming pulses. Figure 1a illustrates the differences in detail.

In inexpensive programmers, configuration jumpers are frequently used to select the specific wiring configuration for different EPROM types. Wire jumpers rather than semiconductor switches are used because of the high currents involved. Take pin 26 (pin 24 on 24-pin EPROMs) with either a 2732 or 27128 installed, for example. In both cases, the voltage level is 5 V. On a 27128 it is a TTL A13 address line; on a 2732 it is a 150-milliampere V_{cc} power line. Similarly, pin 22 (all pin numbers are referenced to a 28-pin layout) has to be set at 0 V, 5 V, 12.5 V, 21 V, or 25 V at currents ranging from 400 microamperes to 50 mA, depending upon the EPROM.

Fortunately, only five of the seven configuration pins require elaborate voltage and current control. Rather than use mechanical jumpers, I designed a voltage-control circuit that could be preset to the voltage limits of the desired EPROM type and easily pass high current when required. Figure 2 illustrates this basic circuit that is duplicated for each of the five pins (pins 28, 26, 1, 22, and 23).

The level shifter uses an LM317 voltage regulator as a programmable voltage controller. The basic LM317 output voltage is set by two resistors: R1 between the adjustment pin and ground and R2 between the adjustment pin and the output. As the formula shows, with R1=665 ohms and R2=221 ohms, the output is 5.0 V.

In this configuration, various R1 resistors can be connected from the adjustment pin to ground through open-collector 7407 drivers. These were used since they operate at up to 30 V (don't substitute a 7417). The four drivers from top to bottom set 5 V, 12.5 V, 21 V, and 25 V, respectively (not all sections are required for each EPROM pin). Their inputs are fed by a parallel output port.

Normally, the regulated output of an LM317 is 1.2 V to 32 V. An additional two-transistor control circuit is added to allow the output to go to 0 V on command. Rather than providing a resistance path to ground, however, this is accomplished by applying a negative 1.2 V to the adjustment pin. Because there is no way to know how many of the control circuits will be set to 0 V at any one time or if the 7407 drivers are enabled concurrently, the −1.4-V bias source is itself a regulated supply.

The CCSP level-shifter circuit can simulate a variety of programmable conditions. For example, by setting the 7407 driver that limits the output to 5 V and pulsing the 0-V enable line, we have a TTL-level $\overline{PGM}$, $\overline{OE}$, or $\overline{CE}$ control line. (In the tests I conducted, the circuit easily responded to control input changes of 20 kilohertz with little overshoot on the output. At those speeds, however, the output filter capacitor should be small.) Since the circuit is also capable of supplying 500 mA at 5 V, it is also appropriate to use this same circuit to supply and control V_{cc}.

The heart of the CCSP is found in the analog switching system and the management of the seven control lines in figure 1a. While I haven't explained yet how these level shifters are individually controlled, it still seems appropriate to show how they are ultimately configured. Figure 3a demonstrates how they are connected to the ZIF socket (zero insertion force programming socket), and figure 3b outlines their power source connections.

I designed a circuit that could be preset to the voltage limits of the desired EPROM.

8-Bit Microcomputer Intelligence

As I mentioned earlier, the CCSP's intelligence is provided by an Intel 8052AH microcomputer. BASIC-52 is particularly suited for this application.

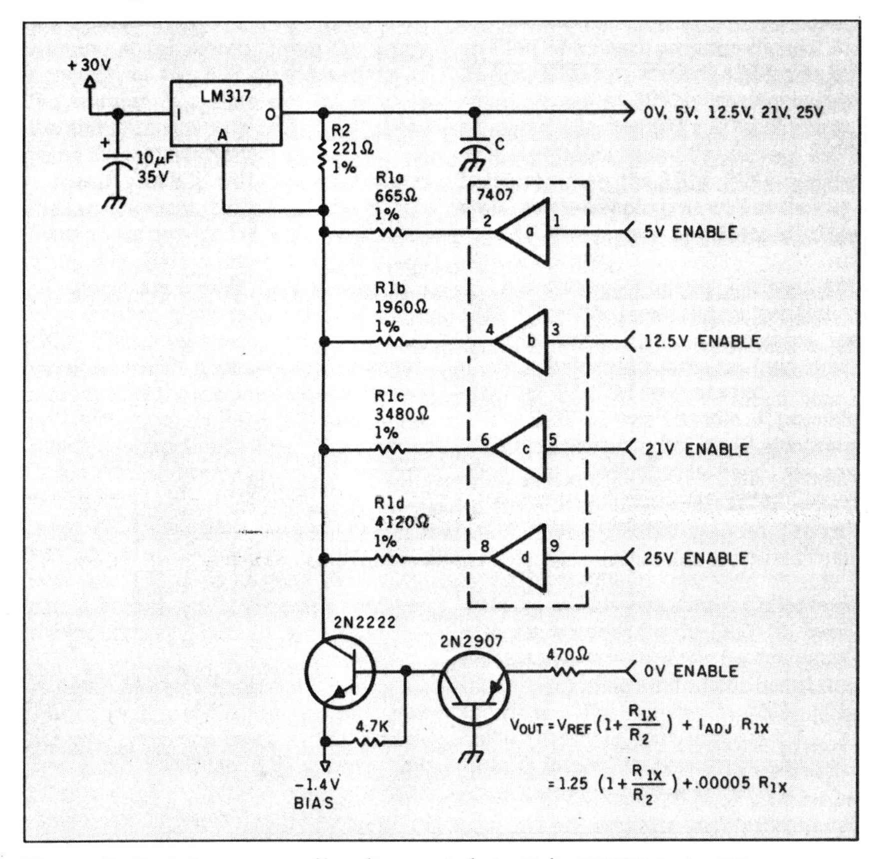

Figure 2: *Typical programmable voltage-control circuit for EPROM pins 28, 26, 1, 22, and 23.*

Three control lines—$\overline{\text{RD}}$, $\overline{\text{WR}}$, and $\overline{\text{PSEN}}$—are gated to allow 64K bytes of combined program and data memory.

providing IF. . .THEN, FOR. . .NEXT, DO. . .WHILE/UNTIL, ONTIME, and CALL statements as well as a broad repertoire of 8051 assembly language instructions. Calculations can be handled in integer or floating-point math.

The 8052AH contains an 8K-byte BASIC interpreter in ROM, 256 bytes of RAM, three 16-bit counter/timers, six interrupts, and 32 I/O lines that are redefined as a 16-bit address and an 8-bit data bus. A minimum of 1K byte of RAM is required for BASIC-52 to function, and any RAM must be located starting at 0000 hexadecimal. (I won't go into great detail on this computer since it closely resembles the BCC-52 presented in August 1985.) The microcomputer section of the CCSP is outlined in figure 4.

Three control lines—$\overline{\text{RD}}$ (pin 17), $\overline{\text{WR}}$ (pin 16), and $\overline{\text{PSEN}}$ (pin 29)—are gated to allow 64K bytes of combined program and data memory. The three most significant address lines (A13–A15) are connected to a 74LS138 decoder chip, IC4, which separates the addressable range into eight 8K-byte memory segments, each with its own chip select (Y0–Y7). The four most significant chip selects are connected to 8K-byte 6264 static RAMs, ICs 7–10. This area is the RAM buffer for reading or writing EPROMs. IC6, addressed at 0000 hexadecimal, must be another 6264 RAM for BASIC-52 to function. IC11 (2000–3FFF hexadecimal) contains the programmer software and is intended for either a 2764 or 27128.

All together, 56K bytes of memory are defined on the CCSP if you use five 6264 RAMs (as ICs 6–10) and a 27128 EPROM in IC11. To use the programmer, you need only the one RAM chip installed in IC6 (such a limited buffer area will require many passes to write or copy any large

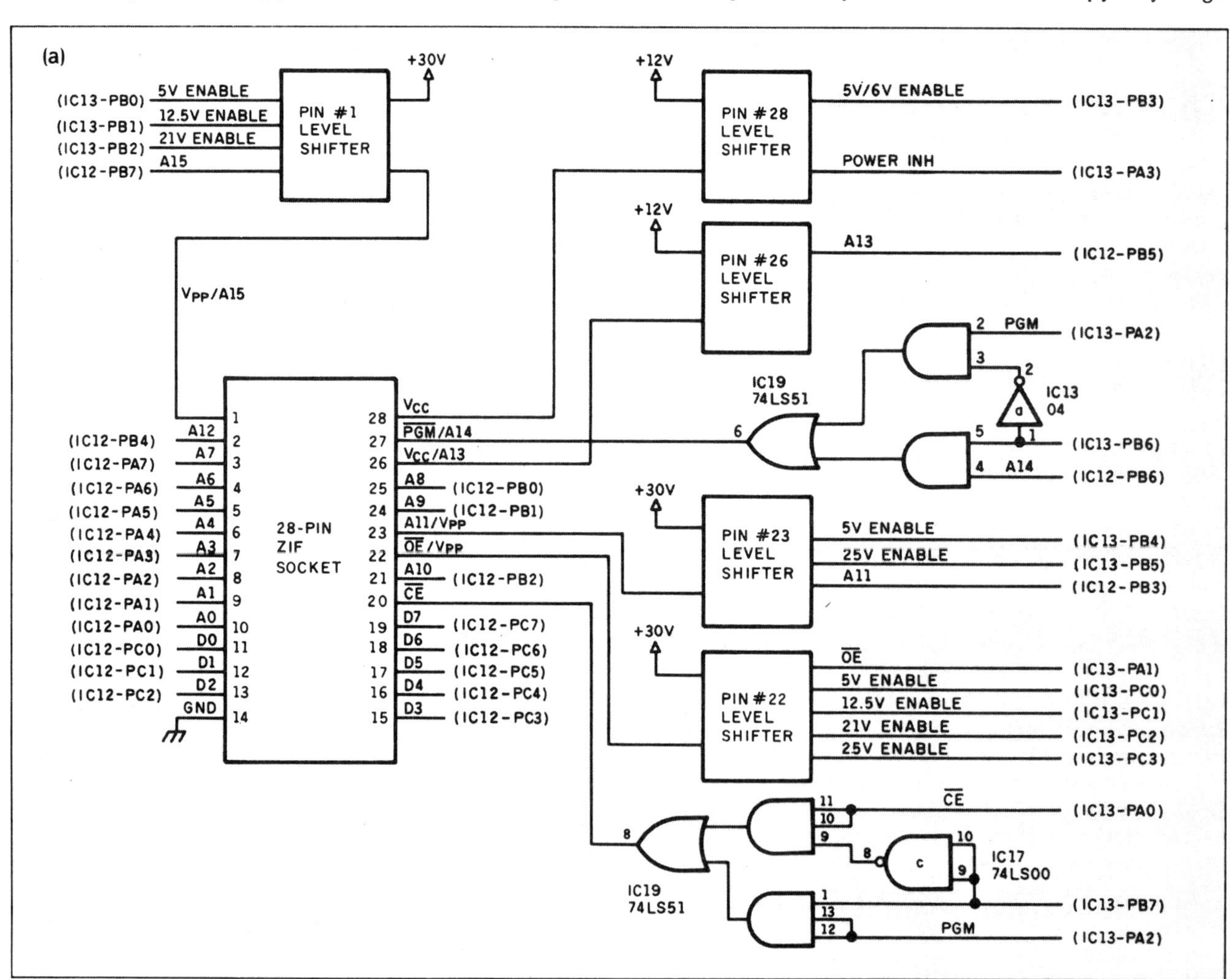

Figure 3: *(a) Block diagram showing the connections to the ZIF socket. Note the level-shifter circuitry connections for those pins that require programming voltages or that differ across EPROM types.*

EPROM). The memory cannot be expanded since the rest of the address space is decoded as I/O.

The address range of 6000-7FFF hexadecimal is divided into two I/O strobes at 6000 and 7000 through IC17. [*Editor's note: For the remainder of the chapter, all addresses will be in hexadecimal.*] Two 8255A-5 peripheral interface adapters providing three 8-bit I/O parallel ports each are controlled by a strobe line. The three I/O ports—labeled A, B, and C—and a write-only mode-configuration port on each 8255 occupy four consecutive addresses at 6000–6003 (IC12) and 7000–7003 (IC13), respectively. The ZIF socket and level-shifting circuitry outlined in figure 2 are connected to 41 of these parallel I/O bits. The lines attached to IC12 (the control PIA) are used primarily for presetting the level shifters and providing the programming pulses. IC13 (the address and data PIA) supplies the address and data bus lines to the EPROM. Figure 5 details the configuration and connection of the level shifters and power distribution.

The CCSP communicates with a terminal or host computer through an on-board serial port. The port's data rate is hard-coded in the program ROM and is preset at 1200 bits per second, but you can reprogram it to any standard value between 300 and 19,200 bps. (The 8052AH chip has the capability for automatic data-rate selection on the console port. Because the CCSP has both a local and a remote operating capability triggered by the GET command, the automatic data rate cannot be used.) MC1488 and MC1489 drivers/receivers (ICs 14 and 15) convert the 8052's serial I/O line TTL logic levels to RS-232.

Power supplies with the required output are readily available, so I excluded an on-board supply to keep costs lower.

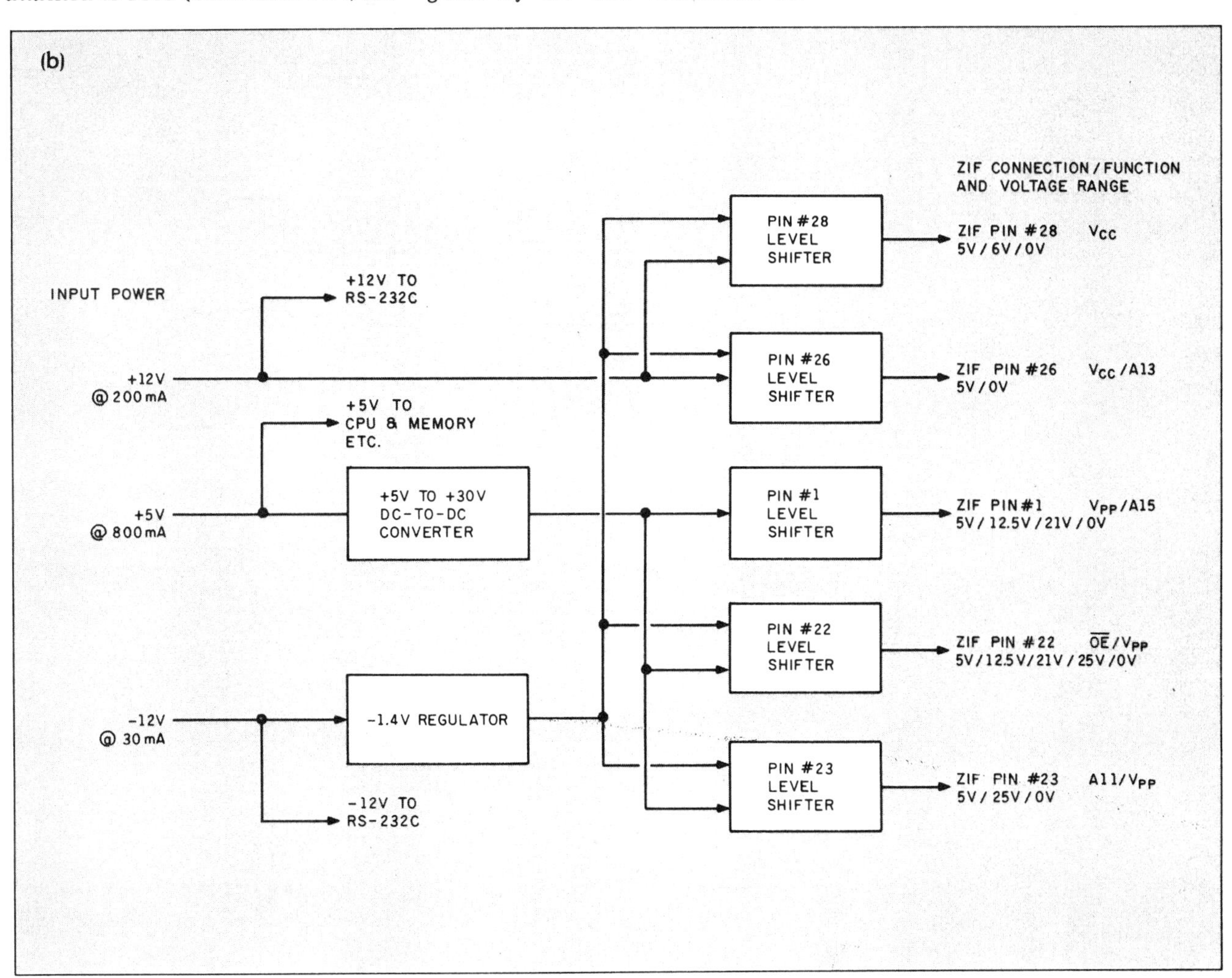

(*b*) CCSP *power-distribution block diagram.*

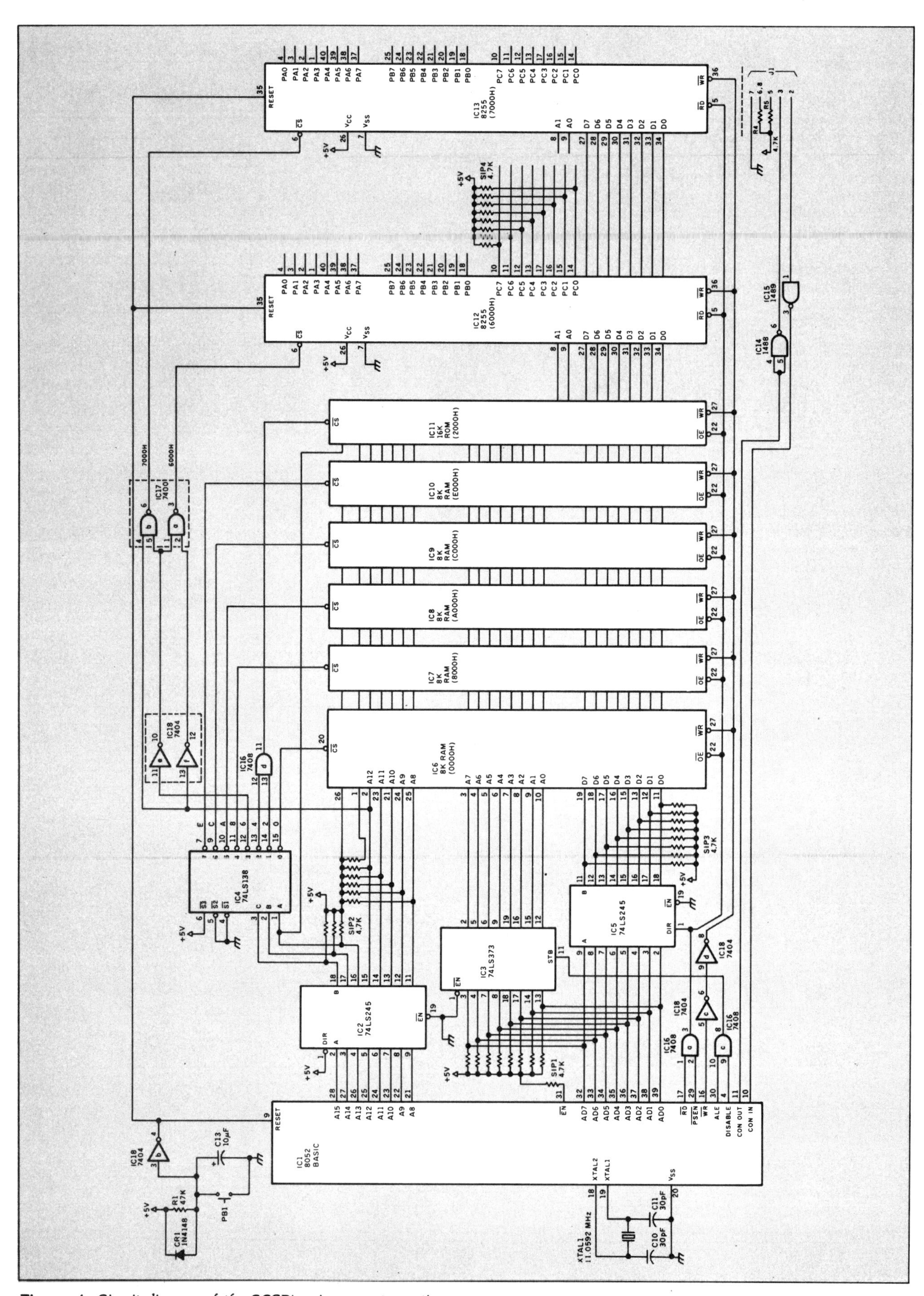

Figure 4: *Circuit diagram of the CCSP's microcomputer section.*

Power for the CCSP is provided by an external supply that must deliver +12 V at 200 mA, +5 V at 800 mA, and −5 V to −12 V at 30 mA. Power supplies with these outputs are readily available on the surplus market, so I excluded an on-board supply to keep costs lower. In fact, a perfect unit is the Coleco computer power supply available from Radio Shack for $4.95 (part #277-1022).

Three V_{pp} voltages must be contended with: 12.5 V, 21 V, and 25 V. All are derived from the +30-V output of the DC-to-DC converter circuit shown in figure 5. IC24 is a 78S40 switching regulator configured as a voltage multiplier. This circuit is capable of producing 30 V at 50 mA from a 5-V input. (For more information on this regulator and this specific circuit, see my November 1981 article, "Switching Power Supplies: An Introduction.")

The user entry/display interface is shown in figure 6. It consists of a two-button entry panel through which you operate the programmer in local mode, a local/remote LED indicator, EPROM power-on indicator, and a seven-segment display through which the computer displays EPROM type and errors. To save I/O bits, I used a somewhat unorthodox display driver rather than the usual parallel port and seven-segment decoder configuration. The seven-segment LED is attached to an 8-bit shift register that has each output connected to drive an individual segment and the decimal point. To display a character, the seven-segment information is extracted from a memory-resident table and quickly shifted into the shift register. Ordinarily, I wouldn't use such a software-intensive approach, but I didn't have to write the software.

Programmer Software

The CCSP is controlled by a program that is a combination of BASIC and 8051 assembly language. The BASIC-52 program provides all initialization and control functions, including local mode support and menu processing in the remote mode. The assembly language routines are used only where speed is critical, as in reading, comparing, verifying erasure, and programming EPROMs. In addition, the Intel hexadecimal file upload and download routines are written in

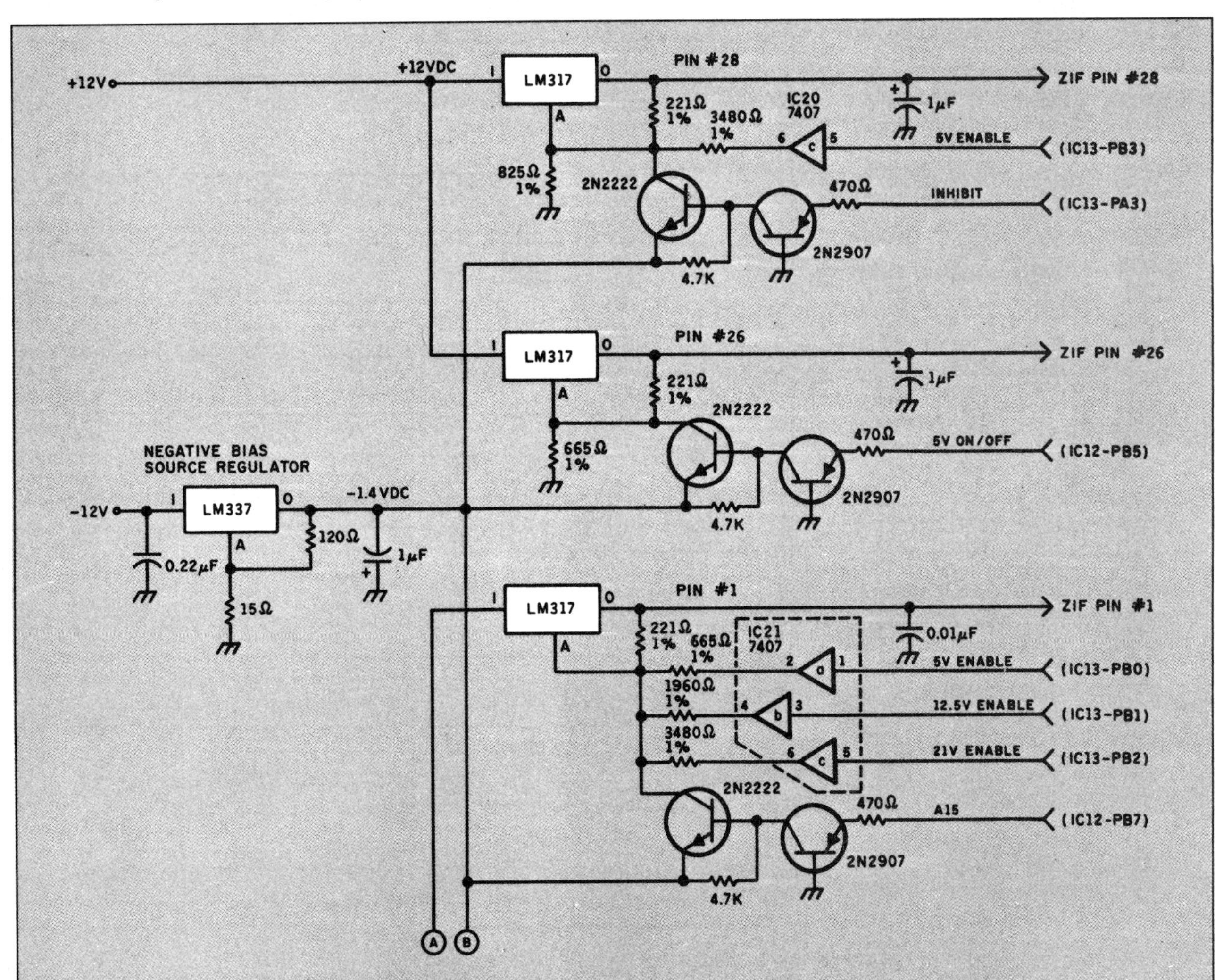

Figure 5: *Detailed schematic of the programmer's level-shifting circuitry.*

When working in local mode, you can copy any 27xxx EPROM by swapping the original and copy EPROMs multiple times.

assembly language to keep up with the attached ASCII terminal device. Figure 7 outlines the CCSP software logic flow.

The software that drives the CCSP is capable of running in two basic modes: local, where the CCSP acts as an EPROM copier controlled by buttons, and remote, where the CCSP acts as a full-featured programming workstation serially connected to the user's terminal. When working in local mode, you can copy any 27xxx EPROM regardless of its size by swapping the original and copy EPROMs multiple times. The larger the RAM buffer is, the fewer times you will have to change the EPROMs.

Power-up and Reset

When the CCSP is first powered up or reset, its software configures itself for a 2716 EPROM, the default type. After setting up the hardware, the software outputs a 0 in the seven-segment LED display to indicate the EPROM type, turns on the local mode LED, and sizes the RAM buffer.

If no memory is located at 8000 (the buffer area), the CCSP allocates 4K bytes of system RAM in IC6 as the buffer area. If it is unable to accomplish this, it will stop and display an alternating error code, E and 0, in the seven-segment LED display. Pressing a button or sending a character to the serial port will force the CCSP to reattempt sizing memory (memory sizing is destructive). If you have RAM chips plugged into locations IC7 through IC10, this will provide additional buffer memory. After memory is sized, the CCSP enters a loop to determine what mode you want the programmer to operate in.

During the mode-setting loop, the CCSP will decide if it is going to run in local or remote mode. The mode selected is determined by which event occurs first: If one of the buttons is pressed first, the CCSP establishes local mode; if a character is detected at the serial port first (via the BASIC-52 GET command), the pro-

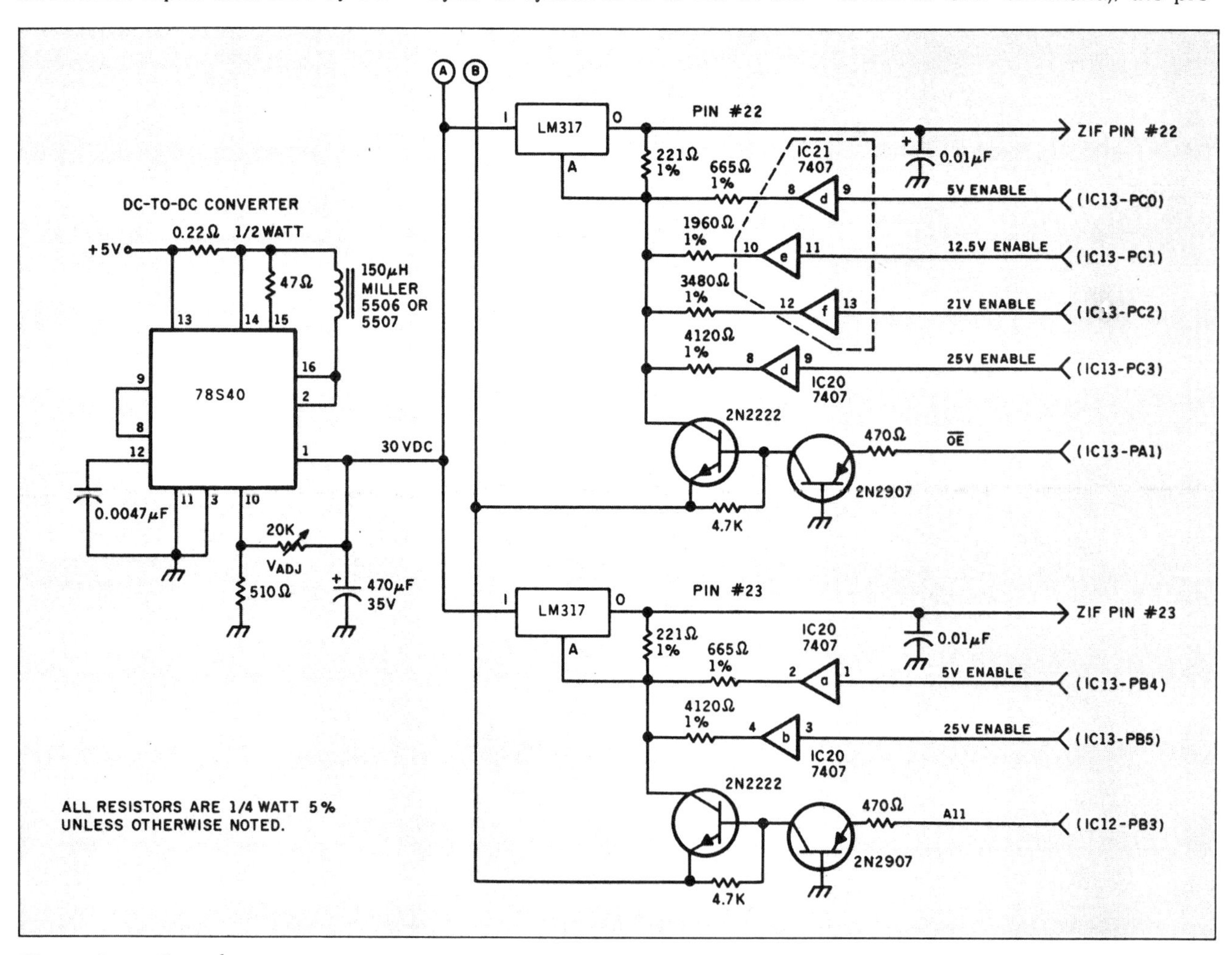

Figure 5 continued.

grammer enters remote mode. Once a mode has been selected, the CCSP must be reset or powered off/on to change modes.

Stand-Alone Local Mode

In local mode, the CCSP is controlled by two buttons called Type and Start/Next. Displays to the user are made via the seven-segment LED display. When local mode is initially entered, or at any point between completed programming cycles, you can change the designated EPROM type by pressing Type. Each press of the button steps the CCSP to the next EPROM type, and the seven-segment LED display is updated with the number that indicates the currently selected type. The designations are shown in table 1.

After setting the type of EPROM to work with, you begin the copy cycle by pressing Start/Next. At this point the seven-segment LED will display an alternating L and O, indicating that you should insert the original EPROM into the ZIF socket. You then load the original EPROM and press Start/Next again to begin the next step: reading the EPROM.

When the CCSP has read as much of the EPROM data as the memory buffer will allow, it signals you to remove the original EPROM and insert the copy EPROM by displaying an alternating L and C on the seven-segment display. After doing this, you again press Start/Next.

The CCSP will now attempt to program the contents of the RAM buffer into the copy EPROM. After verifying that the target area of the copy EPROM is erased, the letter "P" is displayed on the seven-segment display to indicate that programming is in progress (LED2 will be red, indicating that power is on to the EPROM and it should not be removed). When programming is complete, the contents of the EPROM are compared to the memory buffer contents. During this time, the letter "C" is displayed on the seven-segment display (LED2 will be green, indicating power off).

If the target EPROM is not erased or the programming was not successful (bad compare), the seven-segment LED will display an alternating E and a numeral, either a 1 for an unerased target EPROM or a 2 for a failed comparison. If an error does occur, you will be returned to the "between copies" state at the next press of the button.

Assuming all went well, the CCSP checks to see if the entire EPROM has been copied. If it has, the CCSP returns to the "between copies" state and displays the current EPROM type selected on the seven-segment display.

If the entire EPROM has not yet been copied, the effective starting address of the RAM buffer will be incremented by the size of the RAM buffer, and the CCSP will prompt you to insert the original EPROM again. This time, the programmer reads the EPROM starting at the new address. The amount of data read will be either the RAM buffer size or the remaining bytes to be copied from the EPROM, whichever is less. After reading the original, the CCSP calls for the copy EPROM, and programming continues as described above.

These steps will continue until the entire contents of the original EPROM have been transferred into the copy EPROM. Using this approach allows any size EPROM to be copied, regardless of the amount of memory in the RAM buffer.

Remote Mode Operation

When used in remote mode, the CCSP turns into a menu-driven programming workstation, controlled by an

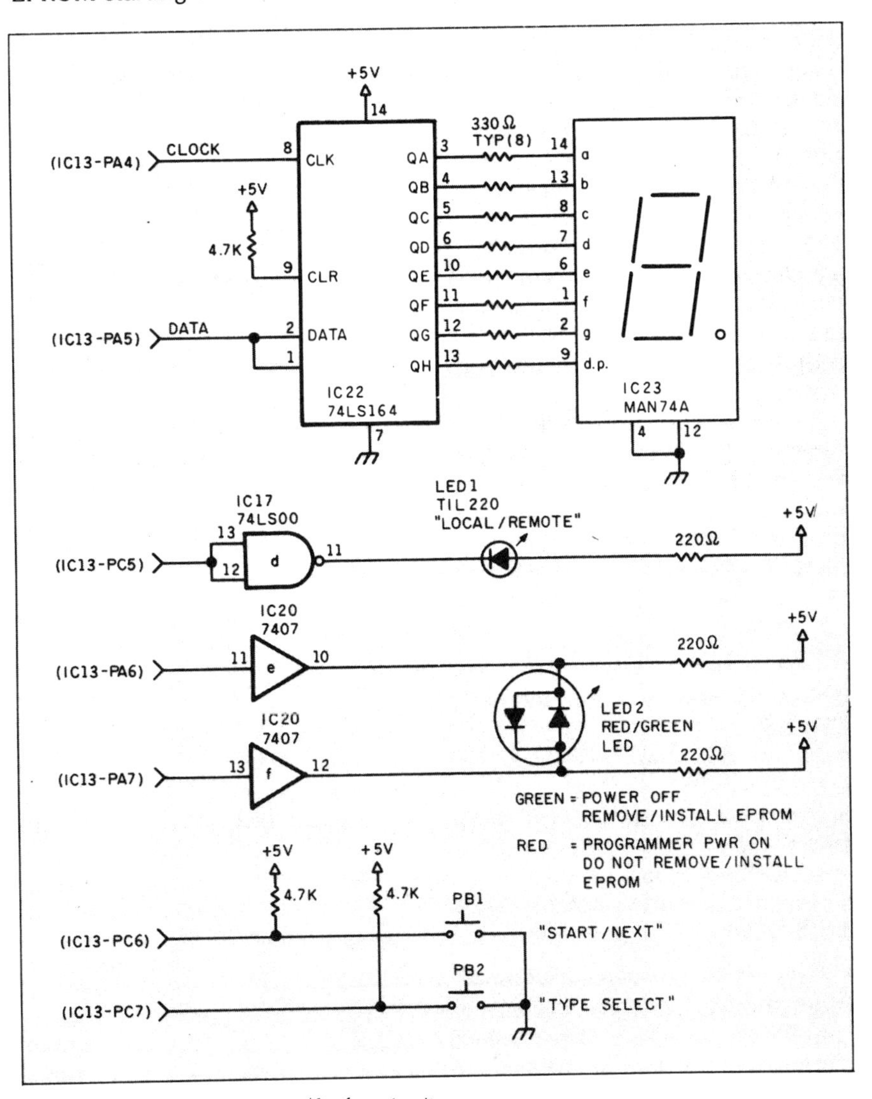

Figure 6: *The CCSP's entry/display circuit.*

All menus displayed on the terminal are generated by the CCSP. A terminal-emulation program is the only software necessary to use this programmer.

ASCII terminal. (See photo 2.) The data rate of the terminal must be hard-coded because the 8052 cannot automatically start the BASIC-52 program unless the data rate is defined. Using the automatic data-rate feature causes the 8052 to wait for a "space" character from the serial port before executing any program stored in it; this would eliminate the local mode of the CCSP. The data rate is set at 1200 bps, but you can change it to any standard value by reprogramming the system ROM with the default data-rate byte changed (details on this procedure are included with the software).

All the menus displayed on the screen of the terminal or computer are generated by the CCSP. No software other than a terminal-emulation program (if connected to a computer rather than a real terminal) is necessary to use this programmer. The remote mode menu provides the following options:

- read, compare, program, and verify EPROM
- display and change RAM buffer contents
- download and upload Intel hexadecimal files
- set EPROM type
- set effective starting address of the RAM buffer

The menu screens contain enough information to guide you through the use of most of these functions. Other pertinent information on the various options is given below.

Read, compare, program, and verify EPROM all depend on the RAM buffer, and they usually use the effective starting address and the length of the RAM buffer to determine the area of the EPROM that is being worked with. Think of the RAM buffer as a window into the contents of the EPROM. If the RAM buffer is not large enough to show you the whole EPROM, you can move it around by changing its effective starting address.

Let's use an example. The EPROM type is a 27512, which is 64K bytes, and the RAM buffer is 16K bytes. It should be pretty obvious that you can't get the whole 27512 into the RAM buffer at the same time. In this case, you would set the starting address of the RAM buffer to 0000 to work with the first quarter of the EPROM and then set it to 4000 to work with the second quarter, 8000 for the third, and C000 for the last. The READ, COMPARE, and PRO-

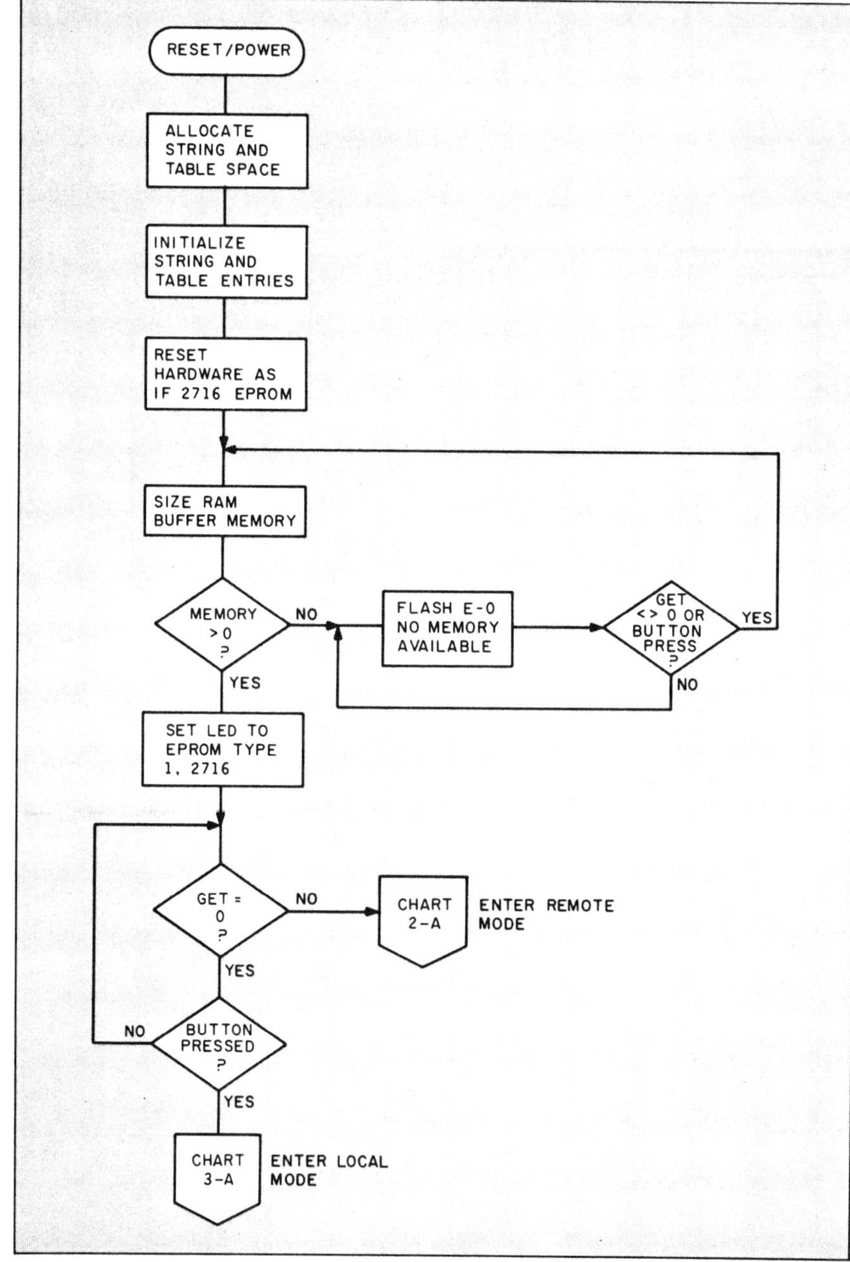

Figure 7a: *Flowchart 1 of the CCSP's overall logic flow, showing the power-up and reset routines.*

GRAM commands would use the starting address of the RAM buffer to see where to read data from or write data to the EPROM. The greatest length of the transferred data would be the size of the RAM buffer or the remaining number of bytes in the EPROM, whichever was smaller.

Even though the VERIFY command does not care about the size of the buffer, its default start and end addresses are controlled as described above. This is because VERIFY generally precedes a programming cycle (you use VERIFY to confirm that the EPROM is properly erased), and the RAM buffer addressing controls programming default start and end addresses.

The following functions—display and change RAM buffer contents and download/upload Intel hexadecimal files—are also tied into the RAM buffer. Since the RAM buffer is supposed to mirror the equivalent area of the EPROM, displays, changes, and uploads/downloads must be addressed to the RAM buffer, just as they would be in the real EPROM. This means that the software will reject addresses outside the range of the current RAM buffer area, which is especially important when doing uploads and downloads. These loads *must* be broken up to fit into the current RAM buffer area

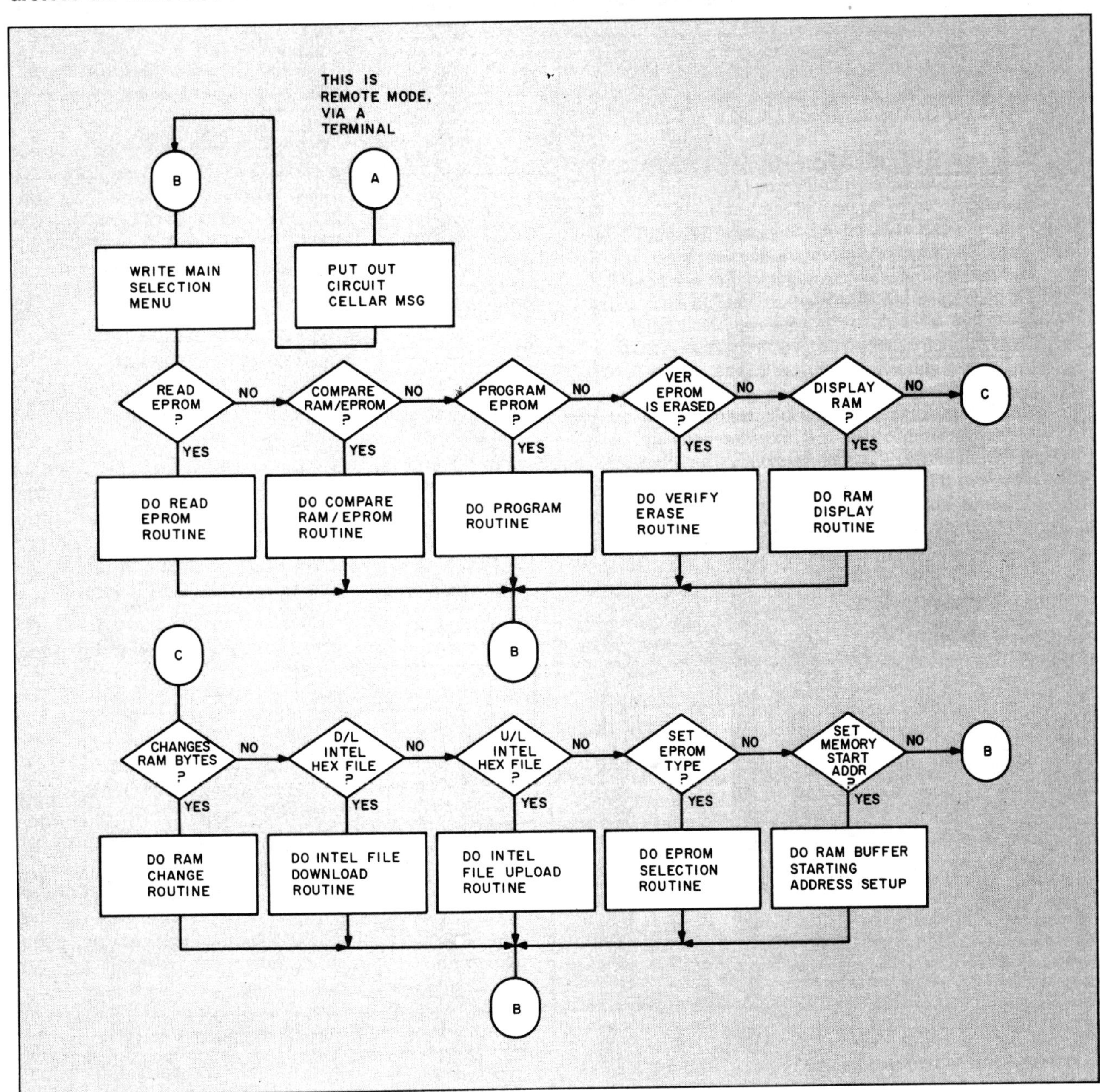

Figure 7b: *Flowchart 2, logic flow for the remote mode.*

Table 1: *The CCSP's EPROM selection number and corresponding EPROM types. Note that this list is of generic EPROM types, and other manufacturer designations should be cross-referenced to it. Also, since CMOS programming cycles are equivalent to those in standard EPROMs, separate 27Cxxx designations are not included.*

Number	EPROM	Type	Number	EPROM	Type
0	2716	25 V	5	27128	21 V
1	2732	25 V	6	27128A	12.5 V
2	2732A	21 V	7	27256	12.5 V
3	2764	21 V	8	27512	12.5 V
4	2764A	12.5 V			

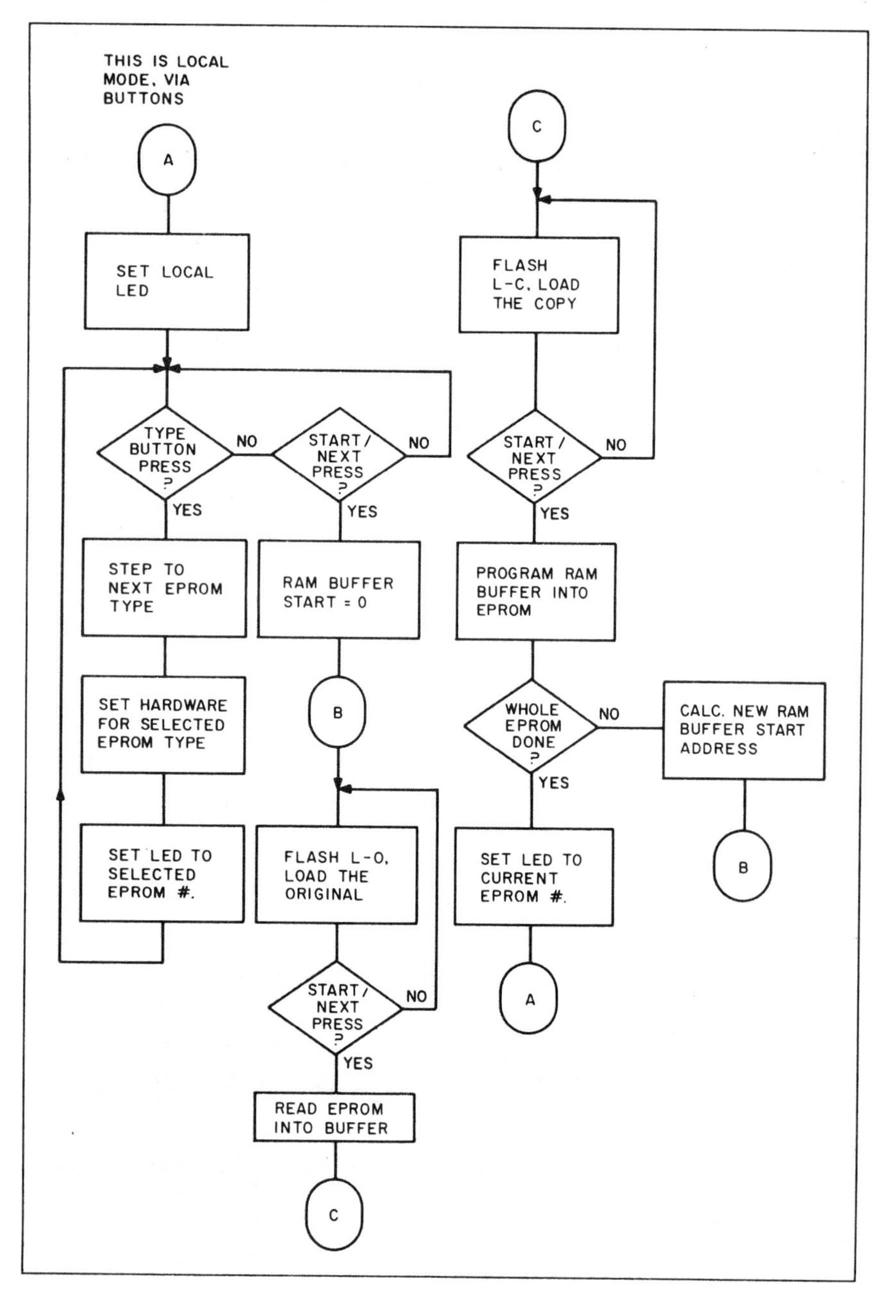

Figure 7c: *Flowchart 3, logic flow for the local mode.*

address range. Trying to go outside the range will abort the display/change/load processes.

The remaining two functions—set EPROM type and set RAM buffer starting address—let you deal with various EPROM types and manipulate the starting address of the RAM buffer. Setting the RAM buffer address lets you change the location of the window into the EPROM. This should be necessary only if the size of the EPROM exceeds the size of the RAM buffer. Otherwise, there is no reason to change the starting address from its default value of 0000.

UNDER THE COVERS

In order to handle the various combinations of sizes, programming voltages, and control lines used with different EPROM types, the software incorporates control tables. Four such tables are used in the CCSP:

- system global table G(*x*)
- LED character table L(*x*)
- EPROM string table $(*x*)
- EPROM data table E(*x*)

SYSTEM GLOBAL TABLE

The system global table contains information about current values for critical system information. The table's entries are set up as is shown in table 2.

The values for these items change based on the type of EPROM you are using, how much contiguous memory is at address 8000 (or the 4K bytes stolen from system RAM), and the last bytes written to the 8255s.

LED CHARACTER TABLE

The seven-segment LED display is controlled by a serial-to-parallel shift register. In order to create a character in the LED, the bits for the various segments must be shifted out in the correct order. This table contains the LED code byte needed to create the characters that can be displayed (see table 3).

EPROM STRING TABLE

BASIC-52 does not allow the mixing of text and numeric data in the same table, so the $(*x*) string table function is used to store this information. This table contains the EPROM designator and the programming voltage used with that type of EPROM. Actually, the programming voltage indicated in the table is only a reminder. You have to set the correct bits in the EPROM data table to ensure that the programmer uses the proper voltage.

EPROM Data Table

The EPROM data table contains all the information the system requires to work with the different EPROM types. The items in each record of the EPROM data table are shown in table 4.

Listing 1 illustrates how this is handled in BASIC. This data is maintained for the use of both the BASIC and assembly language routines. BASIC passes data from the EPROM data table to the assembly language routines via the free registers of the 8052 device.

The Assembly Language Routines

The CCSP software is a hybrid of BASIC and assembly language. Besides reading and verifying, the EPROM programming pulses are accurately timed in assembly language routines. The derivation of the timing's accuracy is given in table 5.

In Conclusion

At first look, the CCSP appears to be considerably more complicated than my programmer of 18 months ago. I think at this point I can change my new description to more accurately state that "this programmer is a serial-port programmer that has the speed of lightning, the intelligence of the mightiest computer (on-board), and is far too functional to be used as a doorstop between uses."

In actuality, only the explanation is more involved. With microcomputer intelligence, the CCSP achieves performance levels approaching kilobuck commercial units yet is flexible enough to be adapted to the next V_{pp} change when it happens.

I'm quite satisfied with my two-week miracle, but I still have to contend with a potential horde of builders. To make amends for my past indiscretion,

Table 2: *Contents of the system global table.*

Index	Use
0	Type number of the current EPROM.
1	Amount of RAM buffer available in 256-byte increments. increments.
2	Current starting address of RAM buffer.
3	Number of items in EPROM table entry.
4	Reserved.
5,6,7	Value of the last data byte written to IC12 address/data PIA (3 bytes, one for each port of the 8255).
8,9,10	Value of the last data byte written to IC13 control PIA.
11	Number of EPROM types in the EPROM string and data tables.

Table 3: *Contents of the LED character table.*

Index	Use
0–9	Characters 0–9 (no decimal point)
10–15	Characters A–F (no decimal point)
16–25	Characters 0–9 (decimal point)
26–31	Characters A–F (decimal point)
32	Blanks LED
33	Character H
34	Character L
35	Character P
36	Character U

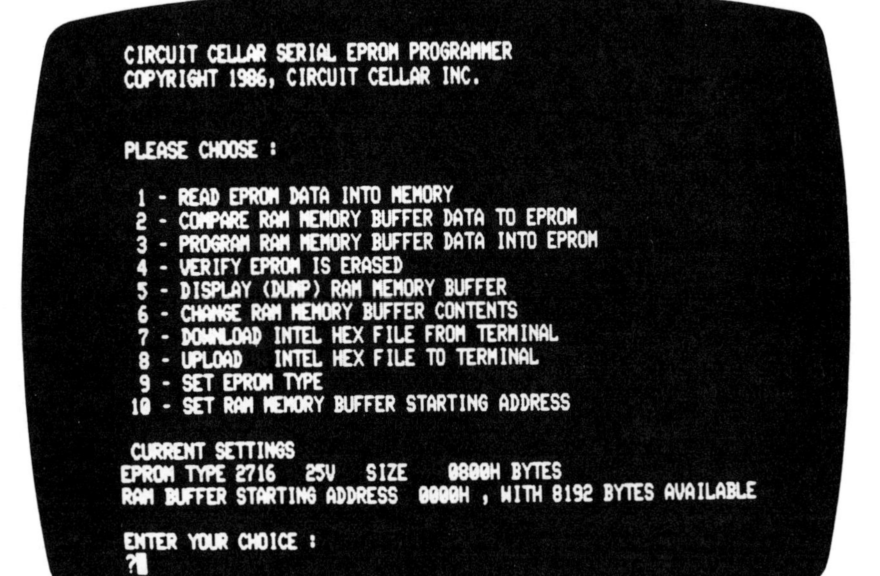

Photo 2: *Typical menu display presented when operating the serial* EPROM *programmer through a serial terminal or computer running in terminal-emulation mode.*

Table 4: *Contents of the EPROM data table.*

Index	Use
0	Number of the EPROM.
1	Size of the EPROM in 256-byte increments.
2	The pin V_{cc} is applied to, referenced to a 28-pin socket.
3,4,5	The initialization values for IC13 control PIA's ports A, B, and C.
6,7	The programming mode values for IC13's ports A and B.
8	Logic true value of the programming pin (CE and PGM).
9	Normal mode programming pulse width in milliseconds.
10	Fast mode programming pulse in milliseconds (0 if no fast programming allowed).
11	Maximum number of fast programming pulses before forced overprogram pulse occurs.
12	Overprogram pulse multiplication factor.

Table 5: *Derivation of the timing for the CCSP's 1-ms timing routine.*

Label	Instruction	Frequency of Execution	Clock Periods Used
PPLOOP	EQU $	A	0
	PUSH B	C	24
	CALL ONEMS	C	24
	POP B	C	24
	DJNZ B,PPLOOP	C	24
	RET	O	24
ONEMS	EQU $	A	0
	MOV B,#MSDLAY	C	24
MSLOOP	EQU $	A	0
	NOP	L	12
	NOP	L	12
	DJNZ B,MSLOOP	L	24
	RET	C	24
MSDLAY	EQU 227	A	0
	END	A	0

Frequency codes:
A—Assembler only, not executed
C—Executed once per 1-ms count
O—Executed only once per entry to subroutine
L—Executed in each loop of the 1-ms routine

Timing Calculations:

1MS=11059.2 clock periods (clock is 11059200 Hz)

$11059.2 = 24 + (24+24+24+24+24+24) + N(12+12+24)$

$11059.2 = 168 + (N \times 48)$

$(11059.2 - 168)/48 = N$

$N = 226.9$ (round to 227)

Error for 1-ms pulse is $0.1 \times 48 \times (1/11059200) = 0.000443$ ms

Cumulative error for 100-ms pulse is:

$$\frac{1105920-24-14400}{4800} = 227.395 \text{ (use 227)}$$

$0.395 \times 48 \times (1/11059200) = 0.00171$ ms

Listing 1: *The* BASIC-52 *code to initialize the* EPROM *data table.*

```
14170 REM
14180 REM INITIALIZE EPROM DATA TABLE
14190 REM
14200 FOR X=0 TO A:READ E(X):NEXT X
14210 REM  TYPE SIZE PWR 7XPA 7XPB 7XPC 7PCP 7PBP PLV NPL FP FXP FACTR
14220 DATA 2716,008H,026,0BBH,061H,001H,008H,000H,001,050,000,000,000
14230 DATA 2732,010H,026,0BFH,051H,001H,008H,000H,000,050,000,000,000
14240 DATA 2732,010H,026,0BFH,051H,001H,004H,000H,000,050,000,000,000
14250 DATA 2764,020H,028,0BFH,014H,001H,000H,000H,000,050,001,025,003
14255 DATA 2764,020H,028,0BBH,012H,001H,000H,000H,000,050,001,025,003
14260 DATA 27128,040H,028,0BBH,014H,001H,000H,000H,000,050,001,015,004
14265 DATA 27128,040H,028,0BBH,012H,001H,000H,000H,000,050,001,015,004
14270 DATA 27256,080H,028,0BBH,052H,001H,000H,000H,000,050,001,025,003
14280 DATA 27512,0100H,028,0BBH,051H,002H,000H,000H,000,050,001,025,003
14281 REM SET UP EPROM NAME TABLE
14283 $(1)="2716   25V  "
14284 $(2)="2732   25V  "
14285 $(3)="2732A  21V  "
14286 $(4)="2764   21V  "
14287 $(5)="2764A  12.5V"
14288 $(6)="27128  21V  "
14289 $(7)="27128A 12.5V"
14290 $(8)="27256  12.5V"
14291 $(9)="27512  12.5V"
14295 RETURN
```

there is indeed a printed circuit board and kit for this programmer.

See page 447 for a listing of parts and products available from Circuit Cellar Inc.

Products Available

The following items are available from:

Circuit Cellar Inc.
4 Park Street
Vernon, CT 06066
Tel: (203) 875-2751
Fax: (203) 872-2204

Ultrasonic Ranging System—TI01
BASIC-52 Computer/Controller—BCC52
SB180 Single-Board Computer
Analog-to-Digital Converter—BCC30
Gray-Scale Video Digitizer Transmitter and Receiver
 DT-01 and DR-01, Full Kit and Experimenter's Kit
Master Controller Experimenter's Kit
Computers on the Brain—HAL-4 Kit
Circuit Cellar IC Tester
 Full Kit and Experimenter's Kit
The SmartSpooler
 Full Kit and Experimenter's Kit
Build an Intelligent Serial EPROM Programmer
 Full Kit and Experimenter's Kit
DDT-51 Full Kit

PLEASE CALL FOR MORE INFORMATION

Additional projects from the Circuit Cellar are presented bimonthly in the *Computer Applications Journal*. For subscription information, please call (203) 875-2199.

About the Author

Steve Ciarcia (pronounced *see-are-see-ah*) is an electrical engineer who has built an international reputation as a consultant, engineer, and author in a career spanning nearly 20 years. Millions of people around the world have followed Steve as he built advanced computers, designed impressive intelligence into the control of his home, explored speech synthesis and communications, and explained leading-edge technology through the use of practical working projects.

Since 1988, Steve has been publishing *Circuit Cellar INK, The Computer Applications Journal*, which presents many of the same state-of-the-art articles for which he is so well known in the industry.

In his spare time, besides tinkering in the Circuit Cellar, Steve enjoys reading and occasionally conjuring up a gourmet delight in the kitchen.

Index

A/D (analog-to-digital) converter, 280–291
AC power lines:
carrier-current modem and, 197–198
power-line carrier-current modem and, 195–201
AC signal measurement, 24
AC to DC converter, 11–15, 22–23
ADC1205 A/D converter chips (National Semiconductor), 284–286
Address decoding in BCC-52, 245–246
Alarm system, stepper motor light scanner and, 43–44
Alford, Roger, 396
Aliases in ZCPR3, 274–276
Alpha waves, 368
Analog-to-digital (A/D) converter, 280–291
ADC1205, 284–286
BCC-30 board, 286, 289–291
D/A conversion, 281–283
Anemometer calibration, 145–146
Application program, 271
Application specific IC (ASIC), 414
Applications:
gray-scale video digitizer, 334–335, 337–343
HAL EEG, 381–382
image processing, 339–342
infrared master controller, 362–363
noncontact touch scanner, 37–40
RTC-4 controller, 192
ASIC (application specific IC), 414
Assembly language:
routines for CCSP, 443
SmartWatch firmware utility, 301–303
Automatic control, 184–194
Automobile warning device, 88–89

Bachiochi, Jeff, 194, 304, 374, 382, 404, 414
BASIC (language), 249
used in DVM, 15–17
used in Micromux control, 179, 181
used in noncontact touch scanner, 38
process-control, 110–112
used in SmartWatch, 301
used in Z8-BASIC microcomputer, 100–119, 183
BASIC/debug monitor, 107
BASIC-52 computer/controller (BCC-52), 232–251
address decoding, 245–246
analog-to-digital converter for, 280–291
board, 241, 244–245
EPROM programming, 247–249
expressions, 249–250
parallel I/O, 246–247
real-time operation, 250
serial I/O, 247
single-chip real-time clock for, 295–296
start-up, 249
variables, 249–250
Basic input/output system (BIOS) for SB180 Single-Board Computer, 276–277
BCC-30 A/D converter board, 286, 289–291
BCC-52 (BASIC-52 computer/controller), 232–251

Bell System Model 103 modem, 196
Beta waves, 368
BIOS (basic input/output system) for SB180 Single-Board Computer, 276–277
Biphase decoding, 67
Biphase encoding, 67
Biphase frequency-shift-keyed communication, 67–72
Blackouts, 203
Bloom, Floyd, 368
Brain-wave monitoring biofeedback device, 367–382
Breakout box, 166–175
Bridge rectifier, 233–235
Brown, Peter, 109, 119
Brownouts, 203
BSR X-10 Home Control System, 192–193
Buffers:
for gray-scale video digitizer, 334
printer buffer, 415–428
ring buffer, 423
for SmartSpooler, 423, 426
Buses:
DDT-51, 407–408
for SB180 Single-Board Computer, 265
Busy Box, 66
Busy signal of telephone, 212–213

Calibration:
of anemometer, 145–146
of noncontact touch scanner, 35–36
of wind vane, 145
(*See also* Testing)
Cantrell, Tom, 267, 279, 428
Carrier-current modem, 196–197
AC power lines and, 197–198
circuit, 197
power-line, 195–201
CCSP, 429–495
assembly language routines, 443
design criteria, 430
EPROM data table, 442, 443
EPROM string table, 442–443
features, 430
LED character table, 442
local mode, 439
memory, 433–435, 437
power-up, 438–439
remote mode, 439–442
reset, 438–439
software for, 437–438
system global table, 442
Centronics printer interface for SB180 Single-Board Computer, 259
Circuit Cellar Controller, 100–119
Cladding, 53
Clear pulse, 62
Clocks, real-time, 292–304
Closed-loop controller system, 66
CMOS LSI remote-control integrated circuits, 72–73
Code-activated switch, 176–183, 178
Coherent light, 55
Communication (*see* Data communication)
Communications interface for HAL EEG, 377

Communications multiplexer, 176
Communications port, code-activated switch for, 176–183
Computers:
BASIC-52 computer/controller, 232–251
data processing, 415–428
microcomputer, 397–398
personal computer, 431–432
power line disturbances and, 202–210
RTC-4 controller, 184–194
SB180 Single-Board, 252–279
single-chip, 101
Z8-BASIC microcomputer, 100–119, 183
Configurable intelligent peripheral controller, 427–428
Control Central, 355
Controllers:
BASIC-52 computer/controller, 232–251
Circuit Cellar, 100–119
configurable intelligent peripheral controller, 427–428
infrared master controller, 353–365
infrared remote, 345–354
RTC-4, 184–194
Conversions:
A/D converter, 280–291
D/A, 281–283
DC to DC converter, 1–6
dual voltage converter, 3–4
Core (fiber optics), 53
CP/M operating system, Z-System compatibility with, 254–256
Crafts, Bob, 88–89
Critical acceptance angle, 53
Crowbar transient suppressors, 208–209
Curlew, Bill, 155, 176, 304, 354

D/A (digital-to-analog) conversion, 281–283
Dallas Semiconductor, 297
Data acquisition, 115–116, 118
Data communication:
laser, 57–59
light beam, 49–59
long distance, 59
power-line carrier-current modem, 195–201
Data input devices, 30–31
Data processing computer, 415–428
DC to DC converter, 1–6
inverting supplies, 3
DDT-51, 403, 405–414
application specific IC, 414
bus buffering, 407–408
debug RAM, 408, 410
development tool, 405
8031 kernel, 412–414
EPROM burning, 410, 412
hardware for, 406–407
ports, 407
system, 405–406
Debug RAM in DDT-51, 408, 410
Delta waves, 368
Demodulator, Exar XR-2211, 198–201

Diagnostic tools:
IC tester, 383–396
RS-232C breakout box, 166–175
(*See also* Calibration; Testing)
Dialing of telephone, 212
Digital-microprocessor laboratory, 98–99
Digital thermometer, 146
Digital-to-analog (D/A) conversion, 281–283
Digital voltmeter (DVM), 7–17
Digital Wind Computer, 143
Digitalker, 120, 121
Digitizer, gray-scale video, 315–336
Direct memory access in SmartSpooler, 423–424, 426
DOS architecture in Z-System, 271
Driver/user area in SB180 Single-Board Computer, 277–278
DS1213 SmartSocket (Dallas Semiconductor), 297–299
DS1216 SmartWatch (Oki), 297, 299–303
Dual voltage converters, 3–4
DVM (digital voltmeter), 7–17
AC to DC converter, 11–15, 22–23
BASIC interpreter, 15–17
computer compatibility, 9
data format, 10
enhancement, 18–27
hardware for, 9–10, 18, 20
199.9 m V range, 23
software driver, 11–15, 23–24
Dynamic RAM in SB180 Single-Board Computer, 259

E product, 238
Echelon, Inc. Z-System, 269–279
Edges, 340–341
EEG (electroencephalogram), 367–382
8031/8051 microcontroller, 397–414
8031 kernel in DDT-51, 412–414
Electrical disturbances, 202–210
Electrical noise, 207
Electroencephalogram (EEG), 367–382
Electronic switches, 177–178
Electronic tape measure, 224–225, 227
Electronic wheelchair, 89
EPROM (erasable programmable read-only memory), 431
in BCC-52 programmer for, 247–249
burning in DDT-51, 410, 412
data table, 442, 443
intelligent serial EPROM programmer, 429–495
for microcontroller, 402–403
pins, 432–433
programming time, 431–432
string table, 442–443
in Z8-BASIC microcomputer, 114–115
Exar XR-2206 modulator, 198
Exar XR-2211 demodulator, 198–201
Expansion bus for SB180 Single-Board Computer, 265
Expressions in BCC-52, 249–250

Fast Fourier Transform (FFT), 376–380
FCP (flow control package) in Z-System, 272–273
FFT (Fast Fourier Transform), 376–380
FFT, The: Fundamentals and Concepts (Ramirez), 380
Fiber optics, 53–54
Field, 317, 327
Fields, R. H., 59
File type, 271
Filter, 207–208, 231, 340–341
Filter capacitor, 233
Firmware utility for SmartWatch, 301–303
Fixed-time programs, 190
Floppy disk drives attached to SB180 Single-Board Computer, 259, 264–265
Flow control package (FCP), 272–273
Formant speech synthesis, 120–121
Frame, 317, 327
Frequency shift keying, 196
Full-wave bridge filter, 231
Full-wave center-tap filter, 231

Gaude, Frank, 279
General Electric Control Central, 355
Gray-scale video digitizer, 315–336
applications, 334–335, 337–343
buffer, 334
design criteria, 318–319
digitizer/transmitter, 315, 327–336
display/receiver, 315–326
hardware for, 319, 322, 327–331
modem, 334
serial data, 322, 324
software for, 324–326, 331
video connection, 331, 333–334

HAL (Hemispheric Activation Level Detector) EEG, 367–382
applications, 381–382
circuitry, 369–371
communications interface, 377
control program, 374
control selection, 371–374
digitizer, 371–374
Fast Fourier Transform, 376–380
microprocessor, 376–377
noise filter, 369
safety precautions, 368, 381
software for, 376–382
video display for, 380–381
Half-wave filter, 231
Handheld infrared remote controller, 345–354
Handheld remote control for home management, 66–77
biphase frequency-shift-keyed communication, 67–72
CMOS LSI integrated circuits, 72–73
single-channel walkie-talkie interface, 73–75
Hardenbrook, Dave, 17
Hardware:
for DDT-51, 406–407
for DVM, 9–10, 18, 20
for gray-scale video digitizer, 319, 322, 327–331
for IC tester, 387, 390
for image processing, 342–343
for noncontact touch scanner, 37
for RTC-4 controller, 188–190
for SB180 Single-Board Computer, 252–267
for SmartSpooler, 415–418
HD64180 microprocessor for SB180 Single-Board Computer, 269–270
Heatsinking, 236–238
Hemispheric Activation Level Detector (HAL) EEG, 367–382
He-Ne laser, 56, 57
High resolution analog-to-digital joystick interface, 62, 65
High resolution static interference joystick interface, 62
Hitachi HD64180 microprocessor, 256–258
Home Control System (Sears), 66
Home-management/control systems, 305–313
handheld remote control, 66–77
infrared master controller, 353–365
infrared remote controller, 345–354
Host mode, 426–427
Hutchinson, Michael, 368

I/O:
parallel, 246–247
serial, 247
SmartSpooler, 420, 422
IC, 385 386
application specific IC, 414
IC tester, 383–396
design criteria, 383
hardware for, 383–390, 387, 390
operation, 386–387
PC-host mode, 384, 391, 394–395
ROM-resident control program, 395–396
software for, 391–396
stand-alone LCD mode, 384, 391
support for, 390
terminal mode, 384, 391
test vector, 391–393
testing of, 384, 396
Image processing, 337–343
applications, 339–343
computer connection, 337–338
edges, 340–341
filters, 340–341
hardware for, 342–343
picture format, 337
picture-taking, 338–339
routines, 342
security system, 341–342
serial setup, 338
Thresh program, 339
ImageWise, 315–343
Imaging system, 315–343
Indices of refraction, 53
Inductor coil, 138–140
Infrared, 356
Infrared master controller, 353–365
applications, 362–363
computer connection, 361–362
design criteria, 356–357
menu modifications, 363–364
power, 357
signal processing, 357, 360–361
Infrared remote controller, 345–354
capabilities extender, 353
PCM decoding in software, 351–353
receiver, 349, 351
support for, 354
Input/Output Package (IOP), 273
Intel 8031/8051 microcontroller, 397–414
Intelligent communication, 118–119
Intelligent peripheral controller, 427–428
Intelligent serial EPROM programmer, 429–495
Interfaces:
HAL EEG, 377
joystick, 60–65
logic analyzer, 93, 96
microcontroller and, 397–414
not required for real-time clock, 296–297
power supply, 1–6
SB180 Single-Board Computer, 259, 264–265
speech synthesis and, 129–131
Interrupts in SmartSpooler, 423–424, 426
IOP (Input/Output Package), 273
IR Master, 353–365
IRCOMM, 345–354

Jaffe, David L., 89
Joystick interfaces, 60–65
  high resolution analog-to-digital, 62, 65
  high resolution static interference, 62
  low resolution static interference, 61–62
  software driven, 62

Kernel, 8031, for DDT-51, 412–414
Keyboard, simulated, 40
Kildall, Gary, 254

Laser (light amplification by stimulated emission of radiation), 54–57
  data communication, 57–59
  safety precautions, 50
Lathers, Merrill, 267, 279
LCD mode, 384, 391
LED character table, 442
Light beam data communication, 49–59
Light pen, 30
Light scanner, 41–48, 86–88
Lightning, 204–205
Lightning rod, 205
Linear power supplies, 229–238
  bridge rectifier, 233–235
  filter, 231
  filter capacitor, 233
  heatsinking, 236–238
  layout, 235–236
  rectifier, 231
  regulator, 231
  transformer, 231, 232–233
  voltage of, 232
  voltage regulators, 235
Linear-predictive coding speech synthesis, 120, 121
Load pulse, 62
Local mode, 426, 439
Logic analyzer, 90–99
  interface, 93, 96
  software simulation, 98
  time-domain display, 96–97
  vector-display capability, 97
Long, Ray, 120
Long distance communication, 59
Low resolution static interference joystick interface, 61–62

McCord, Dave, 279
Maiman, Theodore, 55
Megabrain: New Tools and Techniques for Brain Growth and Mind Expansion (Hutchinson), 368
Memory:
  in CCSP, 433–435, 437
  quasi-static, 103, 106
  on SB180 Single-Board Computer, 259
  in Z8-BASIC microcomputer, 103, 106, 112
Metal-oxide varistor (MOV), 209
Microcomputers (*see* Computers)
Microcontroller, 397–414
  DDT-51, 403, 405–414
  design criteria, 399
  EPROM program, 402–403
  software for, 399, 402
  support for, 403, 414
Micromux, 178–179
  control in BASIC, 179, 181
  programming, 179
Microprocessors, 397
  for HAL EEG, 376–377
  for SB180 Single-Board Computer, 269–270
Microvox speech synthesizer, 166–167

Modem (modulator-demodulator), 196
  carrier-current, 196–197
  connected to gray-scale video digitizer, 334
  power-line carrier-current, 195–201
Modulator, Exar XR-2206, 198
Monochrome television signals, 316–318
Motion detection, 44–45
MOV (metal-oxide varistor), 209
Musical telephone bell, 211–217

Named Directory (NDR) in Z-System, 272
Nisley, Ed, 336, 343, 374, 382, 404, 414
Noncontact touch scanner, 28–40
  applications, 37–40
  BASIC programming, 38
  calibration of, 35–36
  hardware for, 37
  simulated keyboard, 40
  testing of, 35–36, 38
Nonvolatile RAM real-time clocks, 297–299

Open-loop controller system, 66
Oscilloscope, 99

Parallel I/O in BCC-52, 246–247
PC-host mode, 384, 391, 394–395
PCM decoding in software for infrared remote controller, 351–353
Peripheral controller, 427–428
Peripheral interface adapter (PIA) chips, 423
Peripherals, code-activated switch for, 176–183
Personal computer, 431–432
Pert, Candace, 368
Peterson, Greg, 120
Phoneme speech synthesis, 120–121
Phonemes, 129
PIA (peripheral interface adapter) chips, 423
Pioneer, CTF-750, 194
Polarity-inverting regulator, 134
Polaroid Ultrasonic Ranging System Demonstrator Kit, 78–89
Porter, Bill, 396
Ports in DDT-51, 407
Power-line carrier-current modem, 195–201
  Exar XR-2206 modulator, 198
  Exar XR-2211 demodulator, 198–201
  testing of, 201
Power-line disturbances, 202–210
  blackouts, 203
  brownouts, 203
  electrical noise, 207
  lightning, 204–205
  protection against, 207–209
    filters, 207–208
    metal-oxide varistors, 209
    transient suppressors, 201, 208–209
  voltage transients, 203, 205–207
Power supplies:
  linear, 229–238
  for SB180 Single-Board Computer, 265
  switching, 132–141
Printer buffer, 415–428
Printer interface for SB180 Single-Board Computer, 259
Process-control BASIC, 110–112
Processors (*see* Microprocessors)
Programming (*see* Software)
Prototype construction techniques, 192

Quasi-static memory, 103, 106

Radio Shack Video Remote Control Extender, 353
RAM (*see* Random-access memory)
RAM-disk initialization in SB180 Single-Board Computer, 277
Ramirez, Robert W., 380
Random-access memory (RAM):
  debug RAM in DDT-51, 408, 410
  nonvolatile real-time clocks, 297–299
  in SB180 Single-Board Computer, 259, 277
Ranging system, ultrasonic, 78–89, 218–228
RCP (Resident Command Package) in Z-System, 272
Read-only memory (ROM):
  in IC tester, 395–396
  in SB180 Single-Board Computer, 265–266
Real-time clocks, 292–304
  DS1216 SmartWatch, 297, 299–303
  interface not required for, 296–297
  nonvolatile RAM in, 297–299
  single-chip, 295–296
Receiver for infrared remote controller, 349, 351
Recording:
  tape recording, 193–194
  of voiceprints, 161, 164
Rectifier, 231
  bridge, 233–235
Regulator:
  in linear power supplies, 231
  polarity-inverting, 134
  series dissipative, 132
  series switching, 132–141
  78S40 switching regulator, 137–138
  step-down, 133
  step-up, 133–134
Remote-activated switch, 178
Remote control:
  for home management, 66–77
  infrared remote controller, 345–354
Remote mode (CCSP), 439–442
Reset in CCSP, 438–439
Resident Command Package (RCP) in Z-System, 272
Ring buffer, 423
Ringing detection in telephones, 213–214, 217
  ring-detector chips, 214, 216–217
Ringing signal in telephones, 212–213
ROM (*see* Read-only memory)
RS-232C breakout box, 166–175
  decade voltage-level indicator, 170–172
  terminal simulator, 172–175
RS-232C code-activated switch, 176–183
  machine-language control, 181, 183
  Micromux, 178–179, 181
  switches, 176–178
RS-232C ports on SB180 Single-Board Computer, 259
RTC-4 controller, 184–194
  applications, 192
  direct switch control, 190–191
  fixed-time programs, 191
  hardware for, 188–190
  internal programs, 191
  keyboard programming, 190
  program display and errors, 191–192
  prototype construction techniques, 192
Ruby laser, 55–56

SB180 Single-Board Computer, 252–279
  BIOS, 276–277
  design criteria, 258–259
  driver/user area, 277–278

SB180 Single-Board Computer (*Cont.*):
  dynamic RAM, 259
  expansion bus, 265
  floppy-disk interface, 259, 264–265
  hardware for, 252–267
  HD64180 microprocessor, 269–270
  memory interface, 259
  packaging, 266–267
  power supplies, 265
  printer interface, 259
  RAM-disk initialization, 277
  ROM monitor, 265–266
  RS-232C ports, 259
  security for, 278
  software for, 269–279
  start-up, 277
  support for, 278–279
  user customization, 278
  ZRDOS, 276–277
Scanners:
  noncontact touch scanner, 28–40
  stepper motor light scanner, 41–48, 86–88
  ultrasonic ranging system, 78–89, 218–228
Schenk, Rob, 374, 382
Schmuldt, Mel K., 99
Schulze, David, 374, 382
Security:
  image processing system, 341–342
  for SB180 Single-Board Computer, 278
  (*See also* Home-management/control systems)
Serial ports:
  on BASIC-52 computer/controller (BCC-52), 247
  code-activated switch for, 176–183
  EPROM programmer connected to, 429–495
  on gray-scale video digitizer, 322, 324
  RS-232C breakout box and, 166–175
Series dissipative regulator, 132
Series switching regulator, 132–141
7400 IC, 384
78S40 switching regulator, 137–138
Shells in Z-System, 274–276
Signal processing for infrared master controller, 357, 360–361
Simulated keyboard, 40
Simultaneous peripheral operations on-line (spool), 415–416
Single-Board Computer (*see* SB180 Single-Board Computer)
Single-channel walkie-talkie remote control interface, 73–75
Single-chip microcomputers, 101
Single-chip real-time clock, 295–296
SmarTime, 301–303
SmartSpooler, 415–428
  buffer, 426
  buffer management, 423
  design criteria, 416
  direct memory access, 423–424, 426
  hardware for, 415–418
  host mode, 426–427
  I/O, 420, 422
  interrupts, 423–424, 426
  LED, 423
  local mode, 426
  operation, 427
  push buttons, 423
  software for, 423–428
  support for, 422
  test mode, 426
SmartWatch, 297, 299–301
  assembly-language firmware utility, 301–303
SmartWatch (*Cont.*):
  BASIC programming, 301
SN28827 (sonar-ranging module), 218–228
Software:
  for CCSP, 437–438
  for gray-scale video digitizer, 324–326, 331
  for HAL EEG, 376–382
  for IC tester, 391–396
  for microcontroller, 399, 402
  PCM decoding for infrared remote controller, 351–353
  for SB180 Single-Board Computer, 269–279
  for SmartSpooler, 423–428
  software driver in DVM, 11–15, 23–24
Software driven joystick interface, 62
Software simulated logic analyzer, 98
Sonar-ranging module (SN28827), 218–228
Sonar sensor for seeing-impaired, 227–228
SonarTape, 218–228
Spectogram, 156–165
Speech analysis, 156–165
Speech synthesis, 120–122
Speech synthesizer, 120–131
  interfaces and, 129–131
  phonemes, 129
  weather station and, 147–148
Sperry, Roger, 367
Spool (simultaneous peripheral operations on-line), 415–416
Stand-alone LCD mode, 384, 391
Start-up:
  of BASIC-52 computer/controller (BCC-52), 249
  of CCSP, 438–439
  of SB180 Single-Board Computer, 277
Stek, Robert, 267, 374, 382
Step-down regulator, 133
Step-up regulator, 133–134
Stepper motor light scanner, 41–48, 86–88
  motion detection, 44–45
  scanner system, 47–48
Support:
  for IC tester, 390
  for infrared remote controller, 354
  for microcontroller, 403, 414
  for SB180 Single-Board Computer, 278–279
  for SmartSpooler, 422
Sweet Talker speech synthesizer, 120–131, 147–148
Switches:
  code-activated, 178
  electronic, 177–178
  remote-activated, 178
  RS-232C code-activated, 176–183
Switching power supplies, 132–141
  inductor coil, 138–140
  78S40 switching regulator, 137–138
System global table, 442

Take My Computer...Please! (Ciarcia), 313
Tape measure, electronic, 224–225, 227
Tape recording, 193–194
TCAP (Terminal Capabilities) on Z-System, 273–274
Teleimaging, 334–335
Telephone, 211–212
  busy signal, 212–213
  dialing, 212
  musical telephone bell, 211–217
  ringing detection, 213–214, 216–217
  ringing signal, 212–213
Television signals, 316–318
Terminal Capabilities (TCAP) on Z-System, 273–274
Terminal mode, 384, 391
Test equipment (*see* Diagnostic tools)
Test mode, 426
Test vector, 391–393
Testing:
  of IC tester, 384, 396
  of noncontact touch scanner, 35–36, 38
  of power-line carrier-current modem, 201
  (*See also* Calibration)
Texas Instruments:
  7400 IC, 384
  sonar-ranging module (SN28827), 218–228
  TMS1000, 184
  TMS1121C UTC, 184, 187
Thermometer, digital, 146
Theta waves, 368
Thresh program, 339
Time constant, 62
Timed automatic control, 184–194
TMS1000, 184
TMS1121C UTC, 184, 187
Touch scanner, 28–40
TPA (transient program area), 271
Trainable infrared master controller, 353–365
Transformers, 231, 232–233
Transient program area (TPA), 271
Transient suppressors, 201, 208–209
Transitions (biphase encoding), 67
Troubleshooting equipment (*see* Diagnostic tools)
TTL compatibility, 1–2

Ultrasonic ranging system, 78–89, 218–228
  computer connection, 222–224
  demonstration board, 80, 85
  EDB output decoding, 85–86
  electronic tape measure, 224–225, 227
  module, 218, 219, 221–222
  scanner, 86–88
Uninterruptible power supply (UPS), 207
User area, 271
User customization of SB180 Single-Board Computer, 278
Utilities:
  for SmartWatch, 301–303
  in Z-System, 274, 275

Variables in BASIC-52, 249–250
Video digitizer, 315–336
Video display:
  for HAL EEG, 380–381
  noncontact touch scanner, 28–40
Video Remote Control Extender, 353
Video signals, 316–318
Voice synthesis (*see* Speech synthesizer)
Voiceprints, 156–165
  experimental results, 164
  recording of, 161, 164
  voiceprint display, 158–159
Voltage-clamping transient suppressors, 209
Voltage of linear power supplies, 232
Voltage regulators, 235
Voltage transients, 203, 205–207
Voltmeter, digital (DVM), 7–27
Votrax SC-01, 120–131

Walkie-talkie remote control interface, 73–75
Walters, Steve, 109, 119

Walton, Phil, 120
Ward, Terry A., 98–99
WATCHDOG.BAT, 341–342
Waveform digitization speech synthesis, 120, 121
Weather instrumentation, 142–144
Weather station, 142–155
  anemometer calibration, 145–146
  digital thermometer, 146
  enhancement, 150, 153–154
  roof serial link, 146–147
  speech synthesis and, 147–148
  system configuration, 148, 150
  weather instrumentation, 142–144
  wind sensors, 144–145
  wind vane calibration, 145
Wheel of Z-System, 274
Wheelchair, electronic, 89
Whimsi-Bell, 217
Wind sensors, 144–145
Wind vane calibration, 145
Word template, 156
Wright, Joe, 279

XR-2206 modulator, 198
XR-2211 demodulator, 198–201

Z-Nodes, 279
Z-System, 269–279
  aliases, 274–276
  DOS architecture, 271
  environment, 271–272
  flow control package, 272–273
  input/output package, 273
  named directory, 272
  path, 274
  resident command package, 272
  shells, 274–276
Z-System (*Cont.*):
  terminal capabilities, 273–274
  utilities, 274, 275
  wheel, 274
  ZCPR3, 271
ZCPR3, 271
Z8-BASIC microcomputer, 100–119, 183
  BASIC/debug monitor, 107
  data acquisition, 115–116, 118
  dedicated-controller, 113–114
  EPROM programming, 114–115
  intelligent communication, 118–119
  memory allocation, 112
  process-control BASIC, 110–112
  program storage, 112–113
  quasi-static memory, 103, 106
  single-chip microcomputer, 101
  Z8671, 100–102
ZRDOS, 276–277